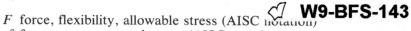
$F$ force, flexibility, allowable stress (AISC notation)
$f$ frequency, computed stress (AISC notation)
$G$ modulus of elasticity in shear
$g$ acceleration of gravity
$h$ height, depth of beam
$I$ moment of inertia of cross-sectional area
$J$ polar moment of inertia of circular cross-sectional area
$K$ stress concentration factor, effective length factor for columns
$k$ spring constant, constant
$L$ length
$M$ moment, bending moment, mass
$M_p$ plastic moment
$m$ mass, moment caused by virtual unit force
$N$ number of revolutions per minute
$P$ force, concentrated load
$p$ pressure intensity, axial force due to unit force
$Q$ first or statical moment of area $A_{fghj}$ around neutral axis
$q$ distributed load intensity, shear flow
$R$ reaction, radius
$S$ elastic section-modulus ($S = I/c$)
$S$ S-shape (standard) steel beam
$s$ second(s)
$r$ radius, radius of gyration
$T$ torque, temperature
$t$ thicknesss, width, tangential deviation
$U$ strain energy
$u$ internal force caused by virtual unit load, axial or radial displacement
$V$ shear force (often vertical), volume
$v$ deflection of beam, velocity
$W$ total weight, work
$W$ W-shape (wide flange) steel beam
$w$ weight or load per unit of length
$Z$ plastic section modulus

## GREEK LETTER SYMBOLS

$\alpha$     (alpha) coefficient of thermal expansion, general angle
$\gamma$     (gamma) shear strain, weight per unit volume
$\Delta$     (delta) total deformation or deflection, change of any designated function
$\epsilon$     (epsilon) normal strain
$\theta$     (theta) slope angle for elastic curve, angle of inclination of line on body
$\kappa$     (kappa) curvature
$\lambda$     (lambda) eigenvalue in column buckling problems
$\nu$     (nu) Poisson's ratio
$\rho$     (rho) radius, radius of curvature
$\sigma$     (sigma) tensile or compressive stress (i.e., normal stress)
$\tau$     (tau) shear stress
$\phi$     (phi) total angle of twist, general angle
$\omega$     (omega) angular velocity

# Engineering Mechanics of Solids

**PRENTICE-HALL INTERNATIONAL SERIES
IN CIVIL ENGINEERING AND ENGINEERING MECHANICS**

**William J. Hall,** *Editor*

# Engineering Mechanics of Solids

**EGOR P. POPOV**

*University of California, Berkeley*

PRENTICE HALL, Englewood Cliffs, New Jersey 07632

Popov, E. P. (Egor Paul)
   Engineering mechanics of solids / Egor P. Popov.
      p.    cm. — (Prentice-Hall international series in civil
engineering and engineering mechanics)
   Bibliography: p.
   Includes index.
   ISBN 0-13-279258-3
   1. Strength of materials.   I. Title.   II. Series.
TA405.P677   1990
620.1'12—dc20                                        89-8860
                                                        CIP

Editorial/production supervision: Sophie Papanikolaou
Interior design: Jules Perlmutter; Off-Broadway Graphics
Cover design: Bruce Kenselaar
Manufacturing buyer: Mary Noonan
Cover Illustration: Artist's Conception of stress transformation. See figure 8–16

© 1990 by Prentice-Hall, Inc.
A Division of Simon & Shuster
Englewood Clifs, New Jersey 07632

Printed in the United States of America
10  9  8  7  6  5  4  3  2  1

ISBN   0-13-279258-3   NBZI

Prentice-Hall International (UK) Limited, *London*
Prentice-Hall of Australia Pty. Limited, *Sydney*
Prentice-Hall Canada Inc., *Toronto*
Prentice-Hall Hispanoamericana, S.A., *Mexico*
Prentice-Hall of India Private Limited, *New Delhi*
Prentice-Hall of Japan, Inc., *Tokyo*
Simon & Schuster Asia Pte. Ltd., *Singapore*

# Contents

# 2 Axial Strains and Deformations in Bars                60

# 3 Generalized Hooke's Law, Pressure Vessels, and Thick-Walled Cylinders               139

# 5 Axial Force, Shear, and Bending Moment 224

# Preface

This book is an update of two of the author's earlier texts, *Mechanics of Materials* (Prentice-Hall, Inc., 2nd Ed., 1976) and *Introduction to Mechanics of Solids* (Prentice-Hall, Inc., 1968). It was felt important to supplement the traditional topics with some exposure to newly emerging disciplines. Among these, some treatment of the probabilistic basis for structural analysis, modest exposure to the matrix methods, and illustrations using the method of finite elements are discussed. Further, to conform with the more mathematical trend in teaching this subject, more rigorous treatment is selectively provided. A few more advanced topics have also been introduced. As a result, the book is larger than its predecessors. This has an advantage in that the user of this text has a larger choice for study, according to needs. Moreover, experience shows that the serious student retains the text for use as a reference in professional life.

This book is larger than what can easily be covered in a one quarter or one-semester course. Therefore, it should prove useful for a follow-up course on the subject at an intermediate level. As an aid in selecting text material for a basic course that is consecutive, with no gaps in the logical development of the subject, numerous sections, examples, and problems marked with a ** can be omitted, To a lesser extent, this also applies to material marked with a *. These guides to possibilities for deletion are provided throughout the text. In a few instances, suggestions for an alternative sequence in studying the subject are also given. The text is carefully integrated by means of cross-referencing.

It is the belief of the author that the serious student, because of the wealth of available material in the text, even in an abbreviated course, should become more knowledgeable. Several illustrations can be mentioned in this regard. For example, while the student is studying the allowable stress design of axially loaded members in Chapter 1, a mere glance at Fig. 1-26, showing histograms for two materials, should reveal the limitations of such a design. The same is true for the student studying thin-walled pressure vessels; even a superficial examination of Fig. 3-24 suggests why limitations are place by the ASME on the use of elementary formulas for thin-walled pressure vessels. Modest exposure to some matrix solutions and illustrations obtained using finite-element methods should arouse interest. Some exposure to the plastic-limit-state methods given in the last section of the last chapter warrants attention. In the hands of an instructor, these side issues can be discussed in a minimum of time and brought in wherever desired. Next, some remarks on the philosophy of the subject and issues of possible controversy are raised.

Chapter 2 forms the cornerstone of the subject and has to be studied carefully. The introduced concepts are repeatedly used in the remainder of the text. Further, the sequence of study for this chapter can be varied, depending on preference. For example, by studying Section 2-19 immediately following Section 2-7, the distinction between statically determinate and indeterminate systems becomes less important. This approach can be useful in introducing the displacement method of analysis. The text as written, however, follows the traditional approach. The suggested variation in the sequence would probably require assistance from an instructor.

The more controversial issue encountered in developing this text deals with the adopted shear sign convention for beams. The one used is thoroughly entrenched in U.S. practice; however, it is in conflict with the right-hand sign convention for axes. If needed, it can easily be modified for use with a computer. The engineering sign convention for shear used, in addition to its virtually universal use in design, requires no sign changes in consecutive integrations. Experience has shown that fewer mistakes are made in using it in hand calculations.

The introduction of Mohr's circles of stress and strain presented a problem. Whereas the basic algebra and comprehensive meaning of the construction of the circles is the same, two alternative methods are in general use, and there are strong advocates for each method. Therefore, both approaches are developed; the choice of procedure is left to the reader, with the alternative one remaining as a reference.

In the preparation of this book, over 30 people at more than a dozen universities contributed to its development. Among these, W. Bickford (ASU)[†], M. E. Criswell (CSU), J. Dempsey (CU), H. D. Eberhart (UCB and UCSB), J. J. Tuma (ASU), and G. A. Wempner (GIT), reviewed the entire manuscript and offered numerous valuable suggestions; F. Filippou (UCB), J. L. Lubliner (UCB), and A. C. Scordelis (UCB) provided much encouragement and made useful suggestions for clarifying the text; A.

---

[†] Letters in parentheses identify the following universities: ASU, Arizona State University; CSU, Colorado State University; CU, Clemson University; GIT, Georgia Institute of Technology; LSU, Louisiana State University; NTU, National Taiwan University; UCB, University of California, Berkeley; UCD, University of California at Davis; UCLA, University of California at Los Angeles; UCSB, University of California at Santa Barbara; USC, University of Southern California; UTA, University of Texas, Austin; UQ, University of Queensland; and UW, University of Washington.

der Kiureghian (UCB) provided valuable assistance for the section on probabilistic methods in Chapter 1; M. D. Engelhardt (UTA), L. R. Herrmann (UCD), and J. M. Ricles (UCSD) gave useful suggestions for Chapter 2; E. L. Wilson (UCB) offered useful comments on Chapter 4; S. B. Dong (UCLA) encouraged more rigorous development for treatment of composite beams resulting in significant improvements; Y. F. Dafalias (UCD) suggested useful refinements for Chapter 8; J. L. Meek (UQ) encouraged presentation of the matrix method in Chapter 12; and C. W. Roeder (UW) carefully reviewed Chapter 13 and provided useful suggestions.

In addition to these, the following also greatly contributed to the development of the text: M. S. Agbabian (USC), H. Astaneh (UCB), D. O. Brush (UCD), A. K. Chopra (UCB), F. Hauser (UCB), J. M. Kelly (UCB), P. Monteiro (UCB), F. Moffitt (UCB), J. L. Sackman (UCB), R. Stephen (UCB), R. L. Taylor (UCB), and G. Voyiadjis (LSU). Dr. K. C. Tsai (NTU) provided valuable assistance in supervising the assembly of problem solutions for the first nine chapters, the remainder was compiled by J-H. Shen (UCB). Among the proceeding, M. D. Engelhardt, R. L. Taylor, J. M. Ricles also assisted with the preparation of finite element results for figures 2–31, 7–13, 7–14, 9–7 and 9–8.

The author sincerely thanks all and feels a debt of gratitude to each for many suggested improvements. The author also thanks his collaborators on one of the earlier books, Drs. S. Nagarajan and Z. A. Lu, who indirectly contributed to this text also.

In producing this book, Douglas Humphrey and Sophie Papanikolaou of Prentice-Hall spared no effort in preparing an excellent publication. Lastly, as in all previous books, the author again is deeply indebted to his wife, Irene, for unstinting support and continual help with the manuscript.

EGOR P. POPOV
*Berkeley, California*

# Engineering Mechanics of Solids

_chapter_ **1**

# Stress, Axial Loads, and Safety Concepts

## 1-1. Introduction

In all engineering construction, the component parts of a structure or a machine must be assigned definite physical sizes. Such parts must be properly proportioned to resist the actual or probable forces that may be imposed upon them. Thus, the walls of a pressure vessel must be of adequate strength to withstand the internal pressure; the floors of a building must be sufficiently strong for their intended purpose; the shaft of a machine must be of adequate size to carry the required torque; a wing of an airplane must safely withstand the aerodynamic loads that may come upon it in takeoff, flight, and landing. Likewise, the parts of a composite structure must be rigid enough so as not to deflect or "sag" excessively when in operation under the imposed loads. A floor of a building may be strong enough but yet may deflect excessively, which in some instances may cause misalignment of manufacturing equipment, or in other cases result in the cracking of a plaster ceiling attached underneath. Also a member may be so thin or slender that, upon being subjected to compressive loading, it will collapse through buckling, i.e., the initial configuration of a member may become unstable. The ability to determine the maximum load that a slender column can carry before buckling occurs or the safe level of vacuum that can be maintained by a vessel is of great practical importance.

In engineering practice, such requirements must be met with the minimum expenditure of a given material. Aside from cost, at times—as in the design of satellites—the feasibility and success of the whole mission may depend on the weight of a package. The subject of _mechanics of_

**1**

*materials,* or the *strength of materials,* as it has been traditionally called in the past, involves analytical methods for determining the **strength,** **stiffness** (deformation characteristics), and **stability** of the various load-carrying members. Alternately, the subject may be called the *mechanics of solid deformable bodies,* or simply *mechanics of solids.*

Mechanics of solids is a fairly old subject, generally dated from the work of Galileo in the early part of the seventeenth century. Prior to his investigations into the behavior of solid bodies under loads, constructors followed precedents and empirical rules. Galileo was the first to attempt to explain the behavior of some of the members under load on a rational basis. He studied members in tension and compression, and notably beams used in the construction of hulls of ships for the Italian navy. Of course, much progress has been made since that time, but it must be noted in passing that much is owed in the development of this subject to the French investigators, among whom a group of outstanding men such as Coulomb, Poisson, Navier, St. Venant, and Cauchy, who worked at the break of the nineteenth century, has left an indelible impression on this subject.

The subject of mechanics of solids cuts broadly across all branches of the engineering profession with remarkably many applications. Its methods are needed by designers of offshore structures; by civil engineers in the design of bridges and buildings; by mining engineers and architectural engineers, each of whom is interested in structures; by nuclear engineers in the design of reactor components; by mechanical and chemical engineers, who rely upon the methods of this subject for the design of machinery and pressure vessels; by metallurgists, who need the fundamental concepts of this subject in order to understand how to improve existing materials further; finally, by electrical engineers, who need the methods of this subject because of the importance of the mechanical engineering phases of many portions of electrical equipment. Engineering mechanics of solids, contrasted with the mathematical theory of continuum mechanics, has characteristic methods all its own, although the two approaches overlap. It is a definite discipline and one of the most fundamental subjects of an engineering curriculum, standing alongside such other basic subjects as fluid mechanics, thermodynamics, as well as electrical theory.

The behavior of a member subjected to forces depends not only on the fundamental laws of Newtonian mechanics that govern the equilibrium of the forces, but also on the mechanical *characteristics* of the materials of which the member is fabricated. The necessary information regarding the latter comes from the laboratory, where materials are subjected to the action of accurately known forces and the behavior of test specimens is observed with particular regard to such phenomena as the occurrence of breaks, deformations, etc. Determination of such phenomena is a vital part of the subject, but this branch is left to other books.[1] Here the end results of such investigations are of interest, and this book is concerned with the analytical or mathematical part of the subject in contradistinction to experimentation. For these reasons, it is seen that mechanics of solids is a blended science of experiment and Newtonian postulates of analytical mechanics. It is presumed that the reader has some familiarity in both of these areas. In the development of this subject, statics plays a particularly dominant role.

This text will be limited to the simpler topics of the subject. In spite of the relative simplicity of the methods employed here, the resulting techniques are unusually useful as they apply to a vast number of technically important problems.

The subject matter can be mastered best by solving numerous problems. The number of basic formulas necessary for the analysis and design of structural and machine members by the methods of engineering mechanics of solids is relatively small; however, throughout this study, the reader must develop an ability to *visualize* a problem and the nature of the quantities being computed. *Complete, carefully drawn diagrammatic sketches of problems to be solved will pay large dividends in a quicker and more complete mastery of this subject.*

There are three major parts in this chapter. The general concepts of stress are treated first. This is followed with a particular case of stress distribution in axially loaded members. Strength design criteria based on stress are discussed in the last part of the chapter.

## Part A    GENERAL CONCEPTS: STRESS

## 1-2. Method of Sections

One of the main problems of engineering mechanics of solids is the investigation of the internal resistance of a body, that is, *the nature of forces set up within a body to balance the effect of the externally applied forces.* For this purpose, a uniform method of approach is employed. A complete diagrammatic sketch of the member to be investigated is prepared, on which *all* of the external forces acting on a body are shown at their respective points of application. Such a sketch is called a *free-body* diagram. All forces acting on a body, including the reactive forces caused by the

---

[1] W. D. Callister, *Materials Science and Engineering* (New York: Wiley, 1985). J. F. Shackelford, *Introduction to Materials Science for Engineers* (New York: Macmillan, 1985). L. H. Van Vlack, *Materials Science for Engineers,* 5th ed., Reading, MA: Addison-Wesley, 1985).

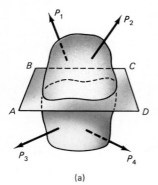

(a)

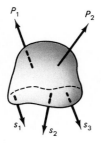

(b)

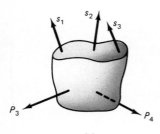

(c)

**Fig. 1-1** Sectioning of a body.

supports and the weight[2] of the body itself, are considered external forces. Moreover, since a stable body at rest is in equilibrium, the forces acting on it satisfy the equations of static equilibrium. Thus, if the forces acting on a body such as shown in Fig. 1-1(a) satisfy the equations of static equilibrium and are all shown acting on it, the sketch represents a free-body diagram. Next, since a determination of the internal forces caused by the external ones is one of the principal concerns of this subject, an arbitrary section is passed through the body, completely separating it into two parts. The result of such a process can be seen in Figs. 1-1(b) and (c), where an arbitrary plane *ABCD* separates the original solid body of Fig. 1-1(a) into two *distinct* parts. This process will be referred to as the *method of sections*. Then, if the body as a whole is in equilibrium, *any part* of it must also be in equilibrium. For such parts of a body, however, some of the forces necessary to maintain equilibrium must act at the cut section. These considerations lead to the following fundamental conclusion: *the externally applied forces to one side of an arbitrary cut must be balanced by the internal forces developed at the cut,* or, briefly, the external forces are balanced by the internal forces. Later it will be seen that the cutting planes will be oriented in particular directions to fit special requirements. However, the method of sections will be relied upon as a first step in solving *all* problems where internal forces are being investigated.

In discussing the method of sections, it is significant to note that some moving bodies, although not in static equilibrium, are in dynamic equilibrium. These problems can be reduced to problems of static equilibrium. First, the acceleration *a* of the part in question is computed; then it is multiplied by the mass *m* of the body, giving a force $F = ma$. If the force so computed is applied to the body at its mass center in a direction opposite to the acceleration, the dynamic problem is reduced to one of statics. This is the so-called *d'Alembert principle*. With this point of view, all bodies can be thought of as being instantaneously in a state of static equilibrium. Hence, for any body, whether in static or dynamic equilibrium, a free-body diagram can be prepared on which the necessary forces to maintain the body as a whole in equilibrium can be shown. From then on, the problem is the same as discussed before.

## 1-3. Definition of Stress

In general, the internal forces acting on infinitesimal areas of a cut are of varying magnitudes and directions, as was shown earlier in Figs. 1-1(b) and (c), and as is again shown in Fig. 1-2(a). These forces are vectorial

---

[2] Strictly speaking, the weight of the body, or, more generally, the inertial forces due to acceleration, etc., are "body forces," and act throughout the body in a manner associated with the units of volume of the body. However, in most instances, these body forces can be considered as external loads acting through the body's center of mass.

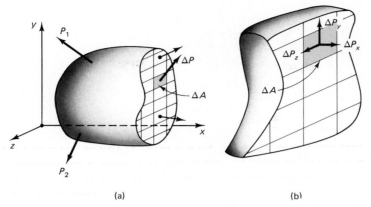

**Fig. 1-2** Sectioned body: (a) free body with some internal forces, (b) enlarged view with components of $\Delta P$.

in nature and they maintain the externally applied forces in equilibrium. In mechanics of solids it is particularly significant to determine the intensity of these forces on the various portions of a section as resistance to deformation and to forces depends on these intensities. In general, they vary from point to point and are inclined with respect to the plane of the section. It is advantageous to resolve these intensities perpendicular and parallel to the section investigated. As an example, the components of a force vector $\Delta P$ acting on an area $\Delta A$ are shown in Fig. 1-2(b). In this particular diagram, the section through the body is perpendicular to the $x$ axis, and the directions of $\Delta P_x$ and of the normal to $\Delta A$ coincide. The component parallel to the section is further resolved into components along the $y$ and $z$ axes.

Since the components of the intensity of force per unit area—i.e., of *stress*—hold true only at a point, the mathematical definition[3] of stress is

$$\tau_{xx} = \lim_{\Delta A \to 0} \frac{\Delta P_x}{\Delta A} \qquad \tau_{xy} = \lim_{\Delta A \to 0} \frac{\Delta P_y}{\Delta A} \qquad \text{and} \qquad \tau_{xz} = \lim_{\Delta A \to 0} \frac{\Delta P_z}{\Delta A}$$

where, in all three cases, the first subscript of $\tau$ (tau) indicates that the plane perpendicular to the $x$ axis is considered, and the second designates the direction of the stress component. In the next section, all possible combinations of subscripts for stress will be considered.

The intensity of the force perpendicular to or normal to the section is called the *normal stress* at a point. It is customary to refer to normal stresses that cause traction or tension on the surface of a section as *tensile stresses*. On the other hand, those that are pushing against it are *compressive stresses*. In this book, normal stresses will usually be designated by the letter $\sigma$ (sigma) instead of by a double subscript on $\tau$. A single

---

[3] As $\Delta A \to 0$, some question from the atomic point of view exists in defining stress in this manner. However, a homogeneous (uniform) model for nonhomogeneous matter appears to have worked well.

subscript then suffices to designate the direction of the axis. The other components of the intensity of force act parallel to the plane of the elementary area. These components are called *shear* or *shearing stresses*. Shear stresses will be always designated by $\tau$.

The reader should form a clear mental picture of the stresses called normal and those called shearing. To repeat, normal stresses result from force components perpendicular to the plane of the cut, and shear stresses result from components tangential to the plane of the cut.

It is seen from the definitions that since they represent the intensity of force on an area, stresses are measured in units of force divided by units of area. In the U.S. customary system, units for stress are pounds per square inch, abbreviated *psi*. In many cases, it will be found convenient to use as a unit of force the coined word *kip*, meaning kilopound, or 1000 lb. The stress in kips per square inch is abbreviated *ksi*. It should be noted that the unit pound referred to here implies a pound-force, not a pound-mass. Such ambiguities are avoided in the modernized version of the metric system referred to as the International System of Units or SI units.[4] SI units are being increasingly adopted and will be used in this text along with the U.S. customary system of units in order to facilitate a smooth transition. The base units in SI are *meter*[5] (m) for length, *kilogram* (kg) for mass, and *second* (s) for time. The derived unit for area is a *square meter* (m²), and for acceleration, a *meter per second squared* (m/s²). The unit of force is defined as a unit mass subjected to a unit acceleration, i.e., *kilogram-meter per second squared* (kg·m/s²), and is designated a *newton* (N). The unit of stress is the *newton per square meter* (N/m²), also designated a *pascal* (Pa). Multiple and submultiple prefixes representing steps of 1000 are recommended. For example, force can be shown in *millinewtons* (1 mN = 0.001 N), *newtons,* or *kilonewtons* (1 kN = 1000 N), length in *millimeters* (1 mm = 0.001 m), *meters,* or *kilometers* (1 km = 1000 m), and stress in *kilopascals* (1 kPa = $10^3$ Pa), *megapascals* (1 MPa = $10^6$ Pa), or *gigapascals* (1 GPa = $10^9$ Pa), etc.[6]

The stress expressed numerically in units of N/m² may appear to be unusually small to those familiar with the U.S. customary system of units. This is because the force of 1 newton is small in relation to a pound-force, and 1 square meter is associated with a much larger area than 1 square inch. Therefore, it is often more convenient in most applications to think in terms of a force of 1 newton acting on 1 square millimeter. The units for such a quantity are N/mm², or, in preferred notation, *megapascals* (MPa).

[4] From the French, Systéme International d'Unités.

[5] Also spelled *metre*.

[6] A detailed discussion of SI units, including conversion factors, rules for SI style, and usage can be found in a comprehensive guide published by the American Society for Testing and Materials as ASTM *Standard for Metric Practice* E-380-86. For convenience, a short table of conversion factors is included on the inside back cover.

Some conversion factors from U.S. customary to SI units are given on the inside of the back cover. It may be useful to note that approximately 1 in ≈ 25 mm, 1 pound-force ≈ 4.4 newtons, and 1 psi ≈ 7000 Pa.

It should be emphasized that *stresses multiplied by the respective areas on which they act give forces*. At an *imaginary* section, a vector sum of these forces, called *stress resultants, keeps a body in equilibrium*. In engineering mechanics of solid, the stress resultants at a selected section are generally determined first, and then, using established formulas, stresses are determined.

## 1-4. Stress Tensor

If, in addition to the section implied in the free body of Fig. 1-2, another plane an infinitesimal distance away and parallel to the first were passed through the body, an elementary slice would be isolated. Then, if an additional two pairs of planes were passed normal to the first pair, a cube of infinitesimal dimensions would be isolated from the body. Such a cube is shown in Fig. 1-3(a). All stresses acting on this cube are identified on the diagram. As noted earlier, the first subscripts on the $\tau$'s associate the stress with a plane perpendicular to a given axis; the second subscripts designate the direction of the stress. On the *near faces* of the cube, i.e., on the faces away from the origin, the directions of stress are positive if they coincide with the positive directions of the axes. On the faces of the cube toward the origin, from the action-reaction equilibrium concept, positive stresses act in the direction opposite to the positive directions of the axes. (Note that for normal stresses, by changing the symbol for stress from $\tau$ to $\sigma$, a single subscript on $\sigma$ suffices to define this quantity without ambiguity.) The designations for stresses shown in Fig. 1-3(a) are widely used in the mathematical theories of elasticity and plasticity.

If at a point in question a different set of axes are chosen, the corre-

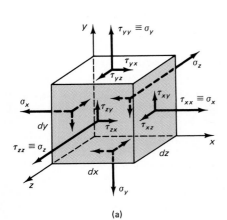

(a)

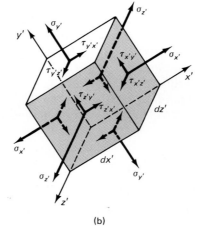

(b)

**Fig. 1-3** (a) General state of stress acting on an infinitesimal element in the initial coordinate system. (b) General state of stress acting on an infinitesimal element defined in a rotated system of coordinate axes. All stresses have positive sense.

sponding stresses are as shown in Fig. 1-3(b). These stresses are related, but are not generally equal, to those shown in Fig. 1-3(a). The process of changing stresses from one set of coordinate axes to another is termed *stress transformation*. The state of stress at a point which can be defined by three components on each of the three mutually perpendicular (orthogonal) axes in mathematical terminology is called a *tensor*. Precise mathematical processes apply for transforming tensors, including stresses, from one set of axes to another. A simple case of stress transformation will be encountered in the next section, and a more complete discussion is given in Chapter 8.

An examination of the stress symbols in Fig. 1-3(a) shows that there are three normal stresses: $\tau_{xx} \equiv \sigma_x$, $\tau_{yy} \equiv \sigma_y$, $\tau_{zz} \equiv \sigma_z$; and six shearing stresses: $\tau_{xy}$, $\tau_{yx}$, $\tau_{yz}$, $\tau_{zy}$, $\tau_{zx}$, $\tau_{xz}$. By contrast, a force vector $P$ has only three components: $P_x$, $P_y$, and $P_z$. These can be written in an orderly manner as a column vector:

$$\begin{pmatrix} P_x \\ P_y \\ P_z \end{pmatrix} \tag{1-1a}$$

Analogously, the stress components can be assembled as follows:

$$\begin{pmatrix} \tau_{xx} & \tau_{xy} & \tau_{xz} \\ \tau_{yx} & \tau_{yy} & \tau_{yz} \\ \tau_{zx} & \tau_{zy} & \tau_{zz} \end{pmatrix} \equiv \begin{pmatrix} \sigma_x & \tau_{xy} & \tau_{xz} \\ \tau_{yx} & \sigma_y & \tau_{yz} \\ \tau_{zx} & \tau_{zy} & \sigma_z \end{pmatrix} \tag{1-1b}$$

This is a matrix representation of the *stress tensor*. It is a second-rank tensor requiring two indices to identify its elements or components. A vector is a first-rank tensor, and a scalar is a zero-rank tensor. Sometimes, for brevity, a stress tensor is written in indicial notation as $\tau_{ij}$, where it is understood that $i$ and $j$ can assume designations $x$, $y$, and $z$ as noted in Eq. (1-1b).

Next, it will be shown that the stress tensor is symmetric, i.e., $\tau_{ij} = \tau_{ji}$. This follows directly from the equilibrium requirements for an element. For this purpose, let the dimensions of the infinitesimal element be $dx$, $dy$, and $dz$, and sum the moments of forces about an axis such as the $z$ axis in Fig. 1-4. Only the stresses entering the problem are shown in the figure. By neglecting the infinitesimals of higher order,[7] this process is equivalent to taking the moment about the $z$ axis in Fig. 1-4(a) or, about point $C$ in its two-dimensional representation in Fig. 1-4(b). Thus,

---

[7] The possibility of an infinitesimal change in stress from one face of the cube to another and the possibility of the presence of body (inertial) forces exist. By first considering an element $\Delta x \, \Delta y \, \Delta z$ and proceeding to the limit, it can be shown rigorously that these quantities are of higher order and therefore negligible.

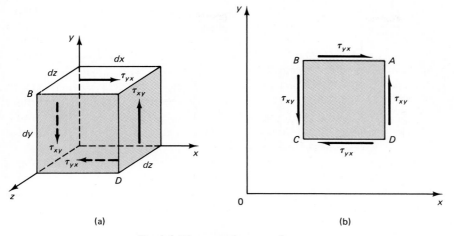

(a)                                                    (b)

**Fig. 1-4** Elements in pure shear.

$$M_C = 0 \circlearrowright + \quad + (\tau_{yx})(dx\ dz)(dy) - (\tau_{xy})(dy\ dz)(dx) = 0$$

where the expressions in parentheses correspond respectively to stress, area, and moment arm. Simplifying,

$$\boxed{\tau_{yx} = \tau_{xy}} \qquad\qquad (1\text{-}2)$$

Similarly, it can be shown that $\tau_{xz} = \tau_{zx}$ and $\tau_{yz} = \tau_{zy}$. Hence, the subscripts for the shear stresses are commutative, i.e., their order may be interchanged, and the stress tensor is symmetric.

The implication of Eq. 1-2 is very important. The fact that subscripts are commutative signifies that shear stresses on mutually perpendicular planes of an infinitesimal element are numerically equal, and $\sum M_z = 0$ is not satisfied by a single pair of shear stresses. On diagrams, as in Fig. 1-4(b), the arrowheads of the shear stresses must meet at diametrically opposite corners of an element to satisfy equilibrium conditions.

In most subsequent situations considered in this text, more than two pairs of shear stresses will seldom act on an element simultaneously. Hence, the subscripts used before to identify the planes and the directions of the shear stresses become superfluous. In such cases, shear stresses will be designated by $\tau$ without any subscripts. However, one must remember that shear stresses always occur in two pairs.

This notation simplification can be used to advantage for the state of stress shown in Fig. 1-5. The two-dimensional stress shown in the figure is referred to as *plane stress*. In matrix representation such a stress can be written as

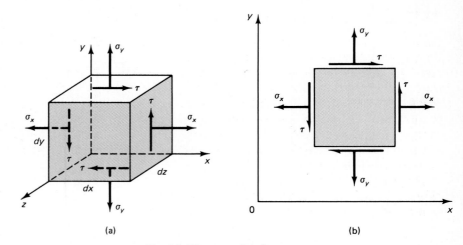

**Fig. 1-5** Elements in plane stress.

$$\begin{pmatrix} \sigma_x & \tau & 0 \\ \tau & \sigma_y & 0 \\ 0 & 0 & 0 \end{pmatrix} \qquad (1\text{-}3)$$

It should be noted that the initially selected system of axes may not yield the most significant information about the stress at a point. Therefore, by using the procedures of stress transformation, the stresses are examined on other planes. Using such procedures, it will be shown later that a particular set of coordinates exists which diagonalize the stress tensor to read

$$\begin{pmatrix} \sigma_1 & 0 & 0 \\ 0 & \sigma_2 & 0 \\ 0 & 0 & \sigma_3 \end{pmatrix} \qquad (1\text{-}4)$$

Note the absence of shear stresses. For the three-dimensional case, the stresses are said to be *triaxial*, since three stresses are necessary to describe the state of stress completely.

For plane stress $\sigma_3 = 0$ and the state of stress is *biaxial*. Such stresses occur, for example, in thin sheets stressed in two mutually perpendicular directions. For axially loaded members, discussed in the next section, only one element of the stress tensor survives; such a state of stress is referred to as *uniaxial*. In Chapter 8, an inverse problem[8] will be discussed: how this one term can be resolved to yield four or more elements of a stress tensor.

[8] Some readers may prefer at this time to study the first several sections in Chapter 8.

## **\*\***[9]**1-5. Differential Equations of Equilibrium**

An infinitesimal element of a body must be in equilibrium. For the two-dimensional case, the system of stresses acting on an infinitesimal element $(dx)(dy)(1)$ is shown in Fig. 1-6. In this derivation, the element is of unit thickness in the direction perpendicular to the plane of the paper. Note that the possibility of an increment in stresses from one face of the element to another is accounted for. For example, since the rate of change of $\sigma_x$ in the $x$ direction is $\partial\sigma_x/\partial x$ and a step of $dx$ is made, the increment is $(\partial\sigma_x/\partial x)\,dx$. The partial derivative notation has to be used to differentiate between the directions.

The inertial or body forces, such as those caused by the weight or the magnetic effect, are designated $X$ and $Y$ and are associated with the unit volume of the material. With these notations,

$$\sum F_x = 0 \rightarrow +, \qquad \left(\sigma_x + \frac{\partial\sigma_x}{\partial x}\,dx\right)(dy \times 1) - \sigma_x(dy \times 1)$$

$$+ \left(\tau_{yx} + \frac{\partial\tau_{yx}}{\partial y}\,dy\right)(dx \times 1) - \tau_{yx}(dx \times 1) + X(dx\,dy \times 1) = 0$$

Simplifying and recalling that $\tau_{xy} = \tau_{yx}$ holds true, one obtains the basic equilibrium equation for the $x$ direction. This equation, together with an analogous one for the $y$ direction, reads

$$\frac{\partial\sigma_x}{\partial x} + \frac{\partial\tau_{yx}}{\partial y} + X = 0 \tag{1-5}$$

$$\frac{\partial\tau_{xy}}{\partial x} + \frac{\partial\sigma_y}{\partial y} + Y = 0$$

[9] Sections identified with ** can be omitted without loss of continuity in the text.

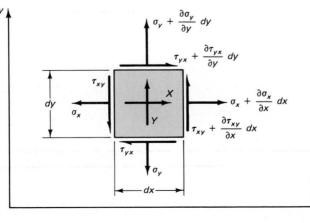

Fig. 1-6 Infinitesimal element with stresses and body forces.

The moment equilibrium of the element requiring $\sum M_z = 0$ is assured by having $\tau_{xy} = \tau_{yx}$.

It can be shown that for the three-dimensional case, a typical equation from a set of three is

$$\frac{\partial \sigma_x}{\partial x} + \frac{\partial \tau_{yx}}{\partial y} + \frac{\partial \tau_{zx}}{\partial z} + X = 0$$

Note that in deriving the previous equations, mechanical properties of the material have not been used. This means that these equations are applicable whether a material is elastic, plastic, or viscoelastic. Also it is very important to note that there are not enough equations of equilibrium to solve for the unknown stresses. In the two-dimensional case, given by Eq. 1-5, there are three unknown stresses, $\sigma_x$, $\sigma_y$, and $\tau_{xy}$, and only two equations. For the three-dimensional case, there are six stresses, but only three equations. Thus, all problems in stress analysis are internally statically intractable or *indeterminate*. A simple example as to how *a static equilibrium* equation is supplemented by *kinematic* requirements and *mechanical properties of a material* for the solution of a problem is given in Section 3-14. In engineering mechanics of solids, such as that presented in this text, this indeterminacy is eliminated by introducing appropriate assumptions, which is equivalent to having additional equations.

A *numerical procedure* that involves discretizing a body into a large number of *small finite elements,* instead of the infinitesimal ones as above, is now often used in complex problems. Such finite element analyses rely on high-speed electronic computers for solving large systems of simultaneous equations. In the finite element method, just as in the mathematical approach, the equations of statics are supplemented by the kinematic relations and mechanical properties of a material. A few examples given later in this book show comparisons among the "exact" solutions of the mathematical theory of elasticity, and those found using the finite element technique and/or conventional solutions based on the methods of engineering mechanics of solids.

## Part B    STRESS ANALYSIS OF AXIALLY LOADED BARS

## 1-6. Stresses on Inclined Sections in Axially Loaded Bars

The traditional approach of engineering mechanics of solids will be used for determining the internal stresses on arbitrarily inclined sections in axially loaded bars. The first steps in this procedure are illustrated in Fig. 1-7. Here, since an axial force $P$ is applied on the right end of a prismatic

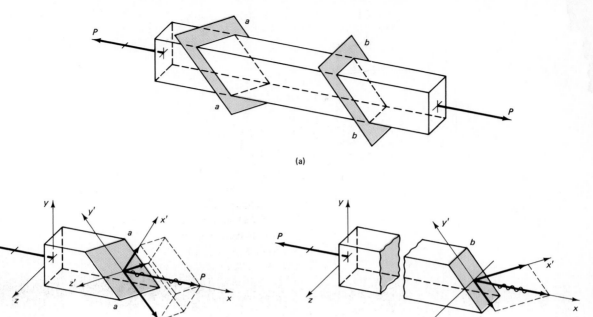

Fig. 1-7 Sectioning of a prismatic bar on arbitrary planes.

bar, for equilibrium, an equal but opposite force $P$ must act on the left end. To distinguish between the applied force and the reaction, a slash is drawn across the reaction force vector $P$. This form of identification of reactions will be used frequently in this text. Finding the reactions is usually the first essential step in solving a problem.

In the problem at hand, after the reactive force $P$ is determined, free-body diagrams for the bar segments, isolated by sections such as $a$–$a$ or $b$–$b$, are prepared. In both cases, the force $P$ required for equilibrium is shown at the sections. However, in order to obtain the conventional stresses, which are the most convenient ones in stress analysis, the force $P$ is replaced by its components along the selected axes. A wavy line through the vectors $P$ indicates their replacement by components. For illustrative purposes, little is gained by considering the case shown in Fig. 1-7(b) requiring three force components. The analysis simply becomes more cumbersome. Instead, the case shown in Fig. 1-7(c), having only two components of $P$ in the plane of symmetry of the bar cross section, is considered in detail. One of these components is normal to the section; the other is in the plane of the section.

As an example of a detailed analysis of stresses in a bar on inclined planes, consider two sections 90 degrees apart perpendicular to the bar sides, as shown in Fig. 1-8(a). The section $a$–$a$ is at an angle $\theta$ with the

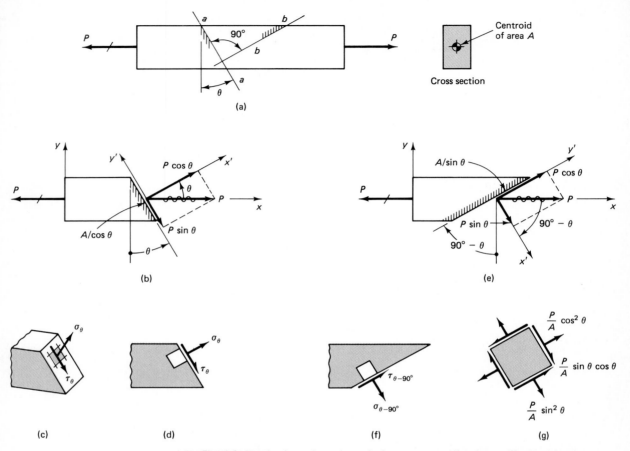

**Fig. 1-8** Sectioning of a prismatic bar on mutually perpendicular planes.

vertical. An isolated part of the bar to the left of this section is shown in Fig. 1-8(b). Note that the normal to the section coinciding with the $x$ axis also forms an angle $\theta$ with the $x$ axis. *The applied force, the reaction, as well as the equilibrating force $P$ at the section all act through the centroid of the bar section.* As shown in Fig. 1-8(b), the equilibrating force $P$ is resolved into two components: the normal force component, $P \cos \theta$, and the shear component, $P \sin \theta$. The area of the inclined cross section is $A/\cos \theta$. Therefore, the normal stress $\sigma_\theta$ and the shear stress $\tau_\theta$ are given by the following two equations:

$$\sigma_\theta = \frac{\text{force}}{\text{area}} = \frac{P \cos \theta}{A/\cos \theta} = \frac{P}{A} \cos^2 \theta \qquad (1\text{-}6)$$

and

$$\tau_\theta = -\frac{P \sin \theta}{A/\cos \theta} = \frac{P}{A} \sin \theta \cos \theta \qquad (1\text{-}7)$$

The negative sign in Eq. 1-7 is used to conform to the sign convention for shear stresses introduced earlier. See, for example, Fig. 1-5. The need for a negative sign is evident by noting that the shear force $P \sin \theta$ acts in the direction opposite to that of the $y$ axis.

It is important to note that the basic procedure of engineering mechanics of solids used here gives the *average* or *mean* stress at a section. These stresses are determined from the axial forces necessary for equilibrium at a section. Therefore they *always* satisfy *statics*. However based on the additional requirements of *kinematics* (geometric deformations) and *mechanical properties of a material,* large *local* stresses are known to arise in the proximity of concentrated forces. This also occurs at abrupt changes in cross-sectional areas. The average stresses at a section are accurate at a distance about equal to the depth of the member from the concentrated forces or abrupt changes in cross-sectional area. The use of this simplified procedure will be rationalized in Section 2-10 as *Saint Venant's principle.*

Equations 1-6 and 1-7 show that the normal and shear stresses vary with the angle $\theta$. The sense of these stresses is shown in Figs. 1-8(c) and (d). The normal stress $\sigma_\theta$ reaches its maximum value for $\theta = 0°$, i.e., when the section is perpendicular to the axis of the rod. The shear stress then correspondingly would be zero. This leads to the conclusion that the maximum normal stress $\sigma_{max}$ in an axially loaded bar can be simply determined from the following equation:

$$\sigma_{max} = \sigma_x = \frac{P}{A} \qquad (1\text{-}8)$$

where $P$ is the applied force, and $A$ is the cross-sectional area of the bar.

Equations 1-6 and 1-7 also show that for $\theta = \pm 90°$, both the normal and the shear stresses vanish. This is as it should be, since no stresses act along the top and bottom free boundaries (surfaces) of the bar.

To find the maximum shear stress acting in a bar, one must differentiate Eq. 1-7 with respect to $\theta$, and set the derivative equal to zero. On carrying out this operation and simplifying the results, one obtains

$$\tan \theta = \pm 1 \qquad (1\text{-}9)$$

leading to the conclusion that $\tau_{max}$ occurs on planes of either $+45°$ or $-45°$ with the axis of the bar. Since the sense in which a shear stress acts is usually immaterial, on substituting either one of the above values of $\theta$ into Eq. 1-7, one finds

$$\tau_{max} = \frac{P}{2A} = \frac{\sigma_x}{2} \qquad (1\text{-}10)$$

Therefore, the maximum shear stress in an axially loaded bar is only half as large as the maximum normal stress. The variation of $\tau_\theta$ with $\theta$ can be studied using Eq. 1-7.

Following the same procedure, the normal and shear stresses can be found on the section $b\text{–}b$. On noting that the angle locating this plane from the vertical is best measured clockwise, instead of counterclockwise as in the former case, this angle should be treated as a negative quantity in Eq. 1-7. Hence, the subscript $-(90° - \theta) = \theta - 90°$ will be used in designating the stresses. From Fig. 1-8(e), one obtains

$$\sigma_{\theta - 90°} = \frac{P \sin \theta}{A/\sin \theta} = \frac{P}{A} \sin^2 \theta \qquad (1\text{-}11)$$

and

$$\tau_{\theta - 90°} = \frac{P \cos \theta}{A/\sin \theta} = \frac{P}{A} \sin \theta \cos \theta \qquad (1\text{-}12)$$

Note that in this case, since the direction of the shear force and the $y$ axis have the same sense, the expression in Eq. 1-12 is positive. Equation 1-12 can be obtained from Eq. 1-7 by substituting the angle $\theta - 90°$. The sense of $\sigma_{\theta - 90°}$ and $\tau_{\theta - 90°}$ is shown in Fig. 1-8(f).

The combined results of the analysis for sections $a\text{–}a$ and $b\text{–}b$ are shown on an infinitesimal element in Fig. 1-8(g). Note that the normal stresses on the adjoining element faces are not equal, whereas the shear stresses are. The latter finding is in complete agreement with the earlier general conclusion reached in Section 1-4, showing that shear stresses on mutually perpendicular planes must be equal.

## 1-7. Maximum Normal Stress in Axially Loaded Bars

In most practical situations with axially loaded bars, it is expedient to directly determine the maximum normal stress. As has been demonstrated in the previous section, these stresses develop *on sections perpendicular to the bar axis*. For such sections, the cross-sectional area of a bar is a minimum and the force component is a maximum, resulting in a maximum normal stress. The procedure for determining this stress directly is shown in Fig. 1-9. Similar to the steps discussed earlier for the general case, a free-body diagram is prepared either for the left or the right part of the bar, as illustrated in Fig. 1-9(b). All force vectors $P$ pass through the bar's *centroid*. As shown in Fig. 1-9(c), the reaction on the left end is equilibrated at section $a\text{–}a$ by a *uniformly distributed normal stress* $\sigma$. The sum of these stresses multiplied by their respective areas generate a stress resultant that is *statically equivalent* to the force $P$. A thin slice of the bar with equal uniformly distributed normal stresses of opposite sense on the two parallel sections is shown in Fig. 1-9(d). This *uniaxial* state of stress may be represented on an infinitesimal cube, as shown in Fig.

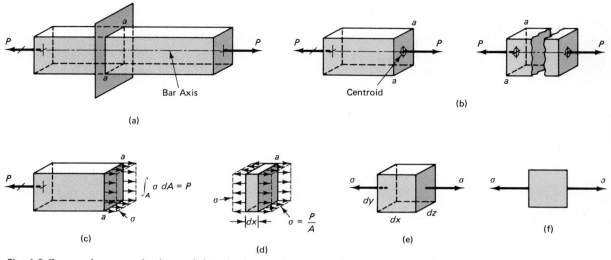

(a)

(b)

(c)

(d)

(e)

(f)

**Fig. 1-9** Successive steps in determining the largest normal stress in an axially loaded bar.

1-9(e). However, a simplified diagram such as shown in Fig. 1-9(f) is commonly used.

For future reference, the relevant Eq. 1-8 for determining directly the maximum normal stress in an axially loaded bar is restated in customary form without any subscript on $\sigma$. Subscripts, however, are frequently added to indicate the direction of the bar axis. This equation gives the largest normal stress at a section taken perpendicular to the axis of a member. Thus,

$$\sigma = \frac{\text{force}}{\text{area}} = \frac{P}{A} \qquad \left[\frac{\text{N}}{\text{m}^2}\right] \text{ or } \left[\frac{\text{lb}}{\text{in}^2}\right] \qquad (1\text{-}13)$$

where, as before, $P$ is the applied axial force, and $A$ is the cross-sectional area of the member. In calculations, it is often convenient to use N/mm$^2$ = MPa in the SI system of units and *ksi* in the U.S. customary system.

It is instructive to note that the normal stress $\sigma$ given by Eq. 1-13, and schematically represented in Fig. 1-9(e), is a complete description of the state of stress in an axially loaded bar. Therefore, only one diagonal term remains in the matrix representation of the stress tensor given by Eq. 1-1b. This remaining term is associated with the direction of the bar axis. If different axes are chosen for isolating an element, as in Fig. 1-8(g), the stress tensor would resemble Eq. 1-3. A detailed study of this topic will be pursued in Chapter 8.

Equation 1-13 strictly applies only to prismatic bars, i.e., to bars having a constant cross-sectional area. However, the equation is reasonably ac-

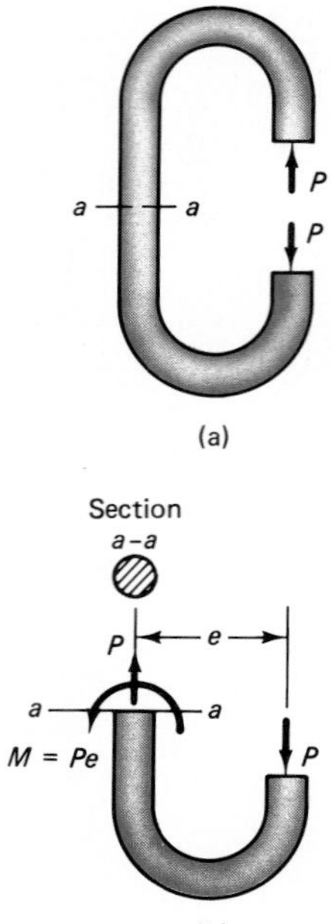

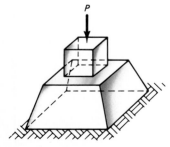

(a)

Section
$a$–$a$

(b)

Fig. 1-10 A member with a nonuniform stress distribution at Section $a$-$a$.

curate for slightly tapered members.[10] For a discussion of situations where an abrupt change in the cross-sectional area occurs, causing severe perturbation in stress, see Section 2-10.

As noted before, the stress resultant for a uniformly distributed stress acts through the centroid of a cross-sectional area and assures the equilibrium of an axially loaded member. If the loading is more complex, such as that, for example, for the machine part shown in Fig. 1-10, the stress distribution is nonuniform. Here, at section $a$–$a$, in addition to the axial force $P$, a bending couple, or moment, $M$ must also be developed. Such problems will be treated in Chapter 6.

Similar reasoning applies to axially loaded compression members and Eq. 1-13 can be used. However, one must exercise additional care when compression members are investigated. These may be so slender that they may not behave in the fashion considered. For example, an ordinary fishing rod under a rather small axial compression force has a tendency to buckle sideways and could collapse. The consideration of such *instability* of compression members is deferred until Chapter 11. *Equation 1-13 is applicable only for axially loaded compression members that are rather chunky,* i.e., to short blocks. As will be shown in Chapter 11, a block whose *least* dimension is approximately one-tenth of its length may usually be considered a short block. For example, a 2 by 4 in wooden piece may be 20 in long and still be considered a short block.

Sometimes compressive stresses arise where one body is supported by another. If the resultant of the applied forces coincides with the centroid of the contact area between the two bodies, the intensity of force, or stress, between the two bodies can again be determined from Eq. 1-13. It is customary to refer to this normal stress as a *bearing stress*. Figure 1-11, where a short block bears on a concrete pier and the latter bears on the soil, illustrates such a stress. Numerous similar situations arise in mechanical problems under washers used for distributing concentrated forces. These bearing stresses can be approximated by dividing the applied force $P$ by the corresponding contact area giving a useful *nominal bearing stress*.

In accepting Eq. 1-13, it must be kept in mind that the material's behavior is *idealized*. Each and every particle of a body is assumed to contribute equally to the resistance of the force. A perfect *homogeneity* of the material is implied by such an assumption. Real materials, such as metals, consist of a great many grains, whereas wood is fibrous. In real materials, some particles will contribute more to the resistance of a force than others. Ideal stress distributions such as shown in Figs. 1-9(d) and (e) actually do not exist if the scale chosen is sufficiently small. The true stress distribution varies in each particular case and is a highly irregular, jagged affair somewhat, as shown in Fig. 1-12(a). However, on the av-

Fig. 1-11 Bearing stresses occur between the block and pier as well as between the pier and soil.

[10] For accurate solutions for tapered bars, see S. P. Timoshenko, and J. N. Goodier, *Theory of Elasticity,* 3rd ed. (New York: McGraw-Hill, 1970) 109.

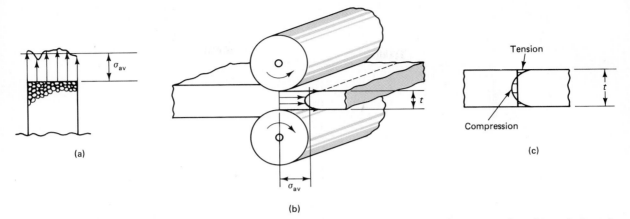

**Fig. 1-12** (a) Schematic illustration of stress irregularity in material due to lack of homogeneity, (b) variation of tensile stress across a plate during a rolling operation, and (c) residual stress in a rolled plate.

erage, statistically speaking, computations based on Eq. 1-13 are correct, and, hence, the computed average stress represents a highly significant quantity.

It is also important to note that the basic equations for determining stresses, such as given by Eq. 1-13, *assume initially stress-free material*. However, in reality, as materials are being manufactured, they are often rolled, extruded, forged, welded, peened, and hammered. In castings, materials cool unevenly. These processes can set up high internal stresses called *residual stresses*. For example, hot steel plates during a rolling operation are pulled between rollers, as shown schematically in Fig. 1-12(b). This process causes the development of larger normal stresses near the outer surfaces than in the middle of a plate. These stresses are equivalent to an average normal stress $\sigma_{av}$ that may be considered to generate a force that propels a plate through the rolls. On leaving the rolls, the plate shown in Fig. 1-12(c) is relieved of this force, and as per Eq. 1-13, the $\sigma_{av}$ is subtracted from the stresses that existed during rolling. The stress pattern of the residual normal stresses is shown in Fig. 1-12(c). These residual stresses are self-equilibrating, i.e., they are in equilibrium without any externally applied forces. In real problems, such residual stresses may be large and should be carefully investigated and then added to the calculated stresses for the initially stress-free material.

## 1-8. Shear Stresses

Some engineering materials, for example, low-carbon steel, are weaker in shear than in tension, and, at large loads, slip develops along the planes of maximum shear stress. According to Eqs. 1-9 and 1-10, these glide or slip planes in a tensile specimen form 45° angles with the axis of a bar, where the maximum shear stress $\tau_{max} = P/2A$ occurs. On the polished

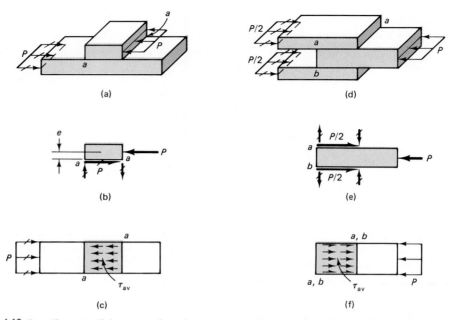

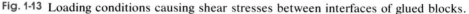

**Fig. 1-13** Loading conditions causing shear stresses between interfaces of glued blocks.

surface of a specimen, these lines can be readily observed and are called *Lüders lines*.[11] This kind of material behavior exhibits a *ductile failure*.

In many routine engineering applications, large shear stresses may develop at critical locations. To determine such stresses precisely is often difficult. However, by *assuming* that in the plane of a section, a *uniformly distributed shear stress* develops, a solution can readily be found. By using this approach, the average shear stress $\tau_{av}$ is determined by dividing the shear force $V$ in the plane of the section by the corresponding area $A$.

$$\tau_{av} = \frac{\text{force}}{\text{area}} = \frac{V}{A} \qquad \left[\frac{N}{m^2}\right] \text{ or } \left[\frac{lb}{in^2}\right] \qquad (1\text{-}14)$$

Some examples as to where Eq. 1-14 can be used to advantage are shown in Figs. 1-13 to 1-15. In Fig. 1-13(a), a small block is shown glued to a larger one. By separating the upper block from the lower one by an imaginary section, the equilibrium diagram shown in Fig. 1-13(b) is obtained. The small applied couple $Pe$, causing small normal stresses acting perpendicular to the section $a$–$a$, is commonly neglected. On this basis

---

[11] Also known as *Piobert lines*. Named in honor, respectively, of German and French nineteenth-century investigators.

$\tau_{av}$, shown in Fig. 1-13(c), can be found using Eq. 1-14 by dividing $P$ by the area $A$ of the section $a$–$a$. A similar procedure is used for determining $\tau_{av}$ for the problem shown in Fig. 1-13(d). However in this case, *two* glued surfaces are available for transferring the applied force $P$. The same approach, employing imaginary sections, is applicable to solid members.

Examples of two bolted connections are shown in Figs. 1-14(a) and (e). These connections can be analyzed in two different ways. In one approach, it is assumed that a tightened bolt develops a sufficiently large clamping force, so that the friction developed between the faying (contacting) surfaces prevents a joint from slipping. For such designs, high-strength bolts are commonly employed. This approach is discussed in Section 1-13. An alternative widely used approach assumes enough slippage occurs, such that the applied force is transferred first to a bolt and then from the bolt to the connecting plate, as illustrated in Figs. 1-14(b) and (f). To determine $\tau_{av}$ in these bolts, a similar procedure as discussed before is applicable. One simply uses the cross-sectional area $A$ of a bolt instead of the area of the joint contact surface to compute the average shear stress. The bolt shown in Fig. 1-14(a) is said to be in *single* shear, whereas the one in Fig. 1-14(e) is in *double* shear.

In bolted connections, another aspect of the problem requires consideration. In cases such as those in Figs. 1-14(a) and (e), as the force $P$ is applied, a highly irregular pressure develops between a bolt and the plates. The *average* nominal intensity of this pressure is obtained by dividing the force transmitted by the projected area of the bolt onto the plate. This is referred to as the *bearing stress*. The bearing stress in Fig. 1-14(a) is $\sigma_b = P/td$, where $t$ is the thickness of the plate, and $d$ is the diameter of the bolt. For the case in Fig. 1-14(e), the bearing stresses for the middle plate and the outer plates are $\sigma_1 = P/t_1 d$ and $\sigma_2 = P/2t_2 d$, respectively.

The same procedure is also applicable for riveted assemblies.

Another manner of joining members together is welding. An example of a connection with fillet welds is shown in Fig. 1-15. The maximum shear stress occurs in the planes $a$–$a$ and $b$–$b$, as shown in Fig. 1-15(b).

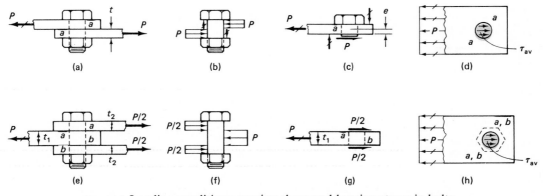

**Fig. 1-14** Loading conditions causing shear and bearing stress in bolts.

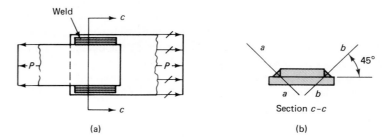

**Fig. 1-15** Loading condition causing critical shear in two planes of fillet welds.

(a)          (b)

The capacity of such welds is usually given in units of force per unit length of weld. Additional discussion on welded connections is given in Section 1-14.

## 1-9. Analysis for Normal and Shear Stresses

Once the axial force $P$ or the shear force $V$, as well as the area $A$, are determined in a given problem, Eqs. 1-13 and 1-14 for normal and shear stresses can be readily applied. These equations giving, respectively, the maximum magnitudes of normal and shear stress are particularly important as they appraise the greatest imposition on the strength of a material. These greatest stresses occur at a section of _minimum_ cross-sectional area and/or the greatest axial force. Such sections are called _critical sections_. The critical section for the particular arrangement being analyzed can usually be found by inspection. However, to determine the force $P$ or $V$ that acts through a member is usually a more difficult task. In the majority of problems treated in this text, the latter information is obtained from statics.

For the equilibrium of a body in space, the equations of statics require the fulfillment of the following conditions:

$$
\begin{array}{ll}
\sum F_x = 0 & \sum M_x = 0 \\
\sum F_y = 0 & \sum M_y = 0 \\
\sum F_z = 0 & \sum M_z = 0
\end{array}
\tag{1-15}
$$

The first column of Eq. 1-15 states that the sum of _all_ forces acting on a body in any $(x, y, z)$ direction must be zero. The second column notes that the summation of moments of _all_ forces around _any_ axis parallel to any $(x, y, z)$ direction must also be zero for equilibrium. In a _planar_ problem, i.e., all members and forces lie in a single plane, such as the $x$-$y$ plane, relations $\sum F_z = 0$, $\sum M_x = 0$, and $\sum M_y = 0$, while still valid, are trivial.

These equations of statics are directly applicable to deformable solid bodies. The deformations tolerated in engineering structures are usually negligible in comparison with the overall dimensions of structures. Therefore, *for the purposes of obtaining the forces in members, the initial undeformed dimensions of members are used in computations.*

If the equations of statics suffice for determining the external reactions as well as the internal stress resultants, a structural system is *statically determinate.* An example is shown in Fig. 1-16(a). However, if for the same beam and loading conditions, additional supports are provided, as in Figs. 1-16(b) and (c), the number of *independent* equations of statics is insufficient to solve for the reactions. In Fig. 1-16(b), any one of the vertical reactions can be removed and the structural system remains stable and tractable. Similarly, any two reactions can be dispensed with for the beam in Fig. 1-16(c). Both of these beams are statically *indeterminate.* The reactions that can be removed leaving a stable system statically determinate are superfluous or *redundant.* Such redundancies can also arise within the internal system of forces. Depending on the number of the redundant internal forces or reactions, the system is said to be indeterminate to the *first degree,* as in Fig. 1-16(b), to the *second degree,* as in Fig. 1-16(c), etc. Multiple degrees of statical indeterminacy frequently arise in practice, and one of the important objectives of this subject is to provide an introduction to the methods of solution for such problems. Procedures for solving such problems will be introduced gradually beginning with the next chapter. Problems with multiple degrees of indeterminacy are considered in Chapters 10, 12, and 13.

Equations 1-15 should already be familiar to the reader. However, several examples where they are applied will now be given, emphasizing solution techniques generally used in engineering mechanics of solids. These statically determinate examples will serve as an informal review of some of the principles of statics and will show applications of Eqs. 1-13 and 1-14.

Additional examples for determining shear stresses in bolts and welds are given in Sections 1-13 and 1-14.

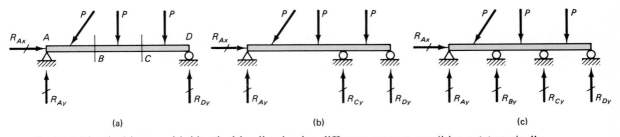

**Fig. 1-16** Identical beam with identical loading having different support conditions: (a) statically determinate, (b) statically indeterminate to the first degree, (c) statically indeterminate to the second degree.

### EXAMPLE 1-1

The beam *BE* in Fig. 1-17(a) is used for hoisting machinery. It is anchored by two bolts at *B*, and at *C*, it rests on a parapet wall. The essential details are given in the figure. Note that the bolts are threaded, as shown in Fig. 1-17(d), with *d* = 16 mm at the root of the threads. If this hoist can be subjected to a force of 10 kN, determine the stress in bolts *BD* and the bearing stress at *C*. Assume that the weight of the beam is negligible in comparison with the loads handled.

### Solution

To solve this problem, the actual situation is idealized and a free-body diagram is made on which all known and unknown forces are indicated. This is shown in Fig. 1-17(b). The vertical reactions of *B* and *C* are unknown. They are indicated, respectively, as $R_{By}$ and $R_{Cy}$, where the first subscript identifies the location, and the second the line of action of the unknown force. As the long bolts *BD* are not effective in resisting the horizontal force, only an unknown horizontal reaction at *C* is assumed and marked as $R_{Cx}$. The applied known force *P* is shown in its proper location. After a free-body diagram is prepared, the equations of statics are applied and solved for the unknown forces.

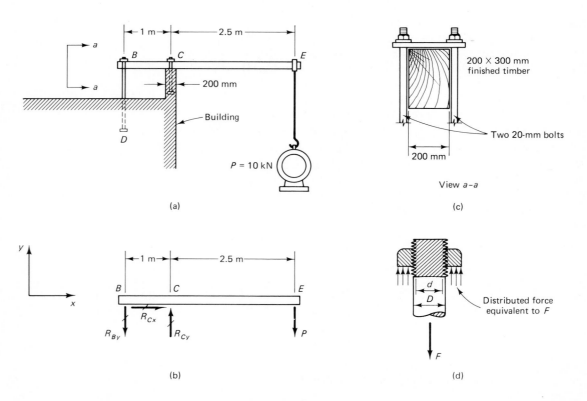

(a)

(c)

(b)

(d)

View *a–a*

**Fig. 1-17**

$$\sum F_x = 0 \qquad\qquad\qquad\qquad\qquad\qquad R_{Cx} = 0$$
$$\sum M_B = 0 \; \circlearrowleft + \qquad 10(2.5 + 1) - R_{Cy} \times 1 = 0 \qquad R_{Cy} = 35 \text{ kN} \uparrow$$
$$\sum M_C = 0 \; \circlearrowleft + \qquad\quad 10 \times 2.5 - R_{By} \times 1 = 0 \qquad R_{By} = 25 \text{ kN} \downarrow$$
$$Check: \; \sum F_y = 0 \uparrow + \qquad -25 + 35 - 10 = 0$$

These steps complete and check the work of determining the forces. The various areas of the material that resist these forces are determined next, and Eq. 1-13 is applied.

Cross-sectional area of one 20-mm bolt: $A = \pi 10^2 = 314 \text{ mm}^2$. This is not the minimum area of a bolt; threads reduce it.

The cross-sectional area of one 20-mm bolt at the root of the threads is

$$A_{net} = \pi \, 8^2 = 201 \text{ mm}^2$$

Maximum normal tensile stress† in each of the two bolts *BD*:

$$\sigma_{max} = \frac{R_{By}}{2A} = \frac{25 \times 10^3}{2 \times 201} = 62 \text{ N/mm}^2 = 62 \text{ MPa}$$

Tensile stress in the shank of the bolts *BD*:

$$\sigma = \frac{25 \times 10^3}{2 \times 314} = 39.8 \text{ N/mm}^2 = 39.8 \text{ MPa}$$

Contact area at *C*:

$$A = 200 \times 200 = 40 \times 10^3 \text{ mm}^2$$

Bearing stress at *C*:

$$\sigma_b = \frac{R_{Cy}}{A} = \frac{35 \times 10^3}{40 \times 10^3} = 0.875 \text{ N/mm}^2 = 0.875 \text{ MPa}$$

The calculated stress for the bolt shank can be represented in the manner of Eq. 1-1b as

$$\begin{pmatrix} 0 & 0 & 0 \\ 0 & +39.8 & 0 \\ 0 & 0 & 0 \end{pmatrix} \text{ MPa}$$

where the *y* axis is taken in the direction of the applied load. In ordinary problems, the complete result is implied but is seldom written down in such detail.

† See also discussion on stress concentrations, Section 2-10.

### EXAMPLE 1-2

The concrete pier shown in Fig. 1-18(a) is loaded at the top with a uniformly distributed load of 20 kN/m². Investigate the state of stress at a level 1 m above the base. Concrete weighs approximately 25 kN/m³.

### Solution

In this problem, the weight of the structure itself is appreciable and must be included in the calculations.
Weight of the whole pier:

$$W = [(0.5 + 1.5)/2] \times 0.5 \times 2 \times 25 = 25 \text{ kN}$$

Total applied force:

$$P = 20 \times 0.5 \times 0.5 = 5 \text{ kN}$$

From $\sum F_y = 0$, reaction at the base:

$$R = W + P = 30 \text{ kN}$$

These forces are shown schematically in the diagrams as concentrated forces acting through their respective centroids. Then, to determine the stress at the desired level, the body is cut into two separate parts. A free-body diagram for

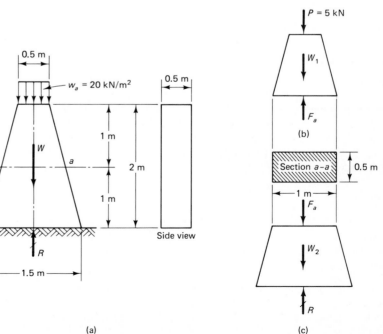

Side view

(a)

(b)

(c)

Fig. 1-18

either part is sufficient to solve the problem. For comparison, the problem is solved both ways.

Using the upper part of the pier as a free body, Fig. 1-18(b), the weight of the pier above the section:

$$W_1 = (0.5 + 1) \times 0.5 \times 1 \times 25/2 = 9.4 \text{ kN}$$

From $\sum F_y = 0$, the force at the section:

$$F_a = P + W_1 = 14.4 \text{ kN}$$

Hence, using Eq. 1-13, the normal stress at the level $a$–$a$ is

$$\sigma_a = \frac{F_a}{A} = \frac{14.4}{0.5 \times 1} = 28.8 \text{ kN/m}^2$$

This stress is compressive as $F_a$ acts on the section.

Using the lower part of the pier as a free body, Fig. 1-18(c), the weight of the pier below the section:

$$W_2 = (1 + 1.5) \times 0.5 \times 1 \times 25/2 = 15.6 \text{ kN}$$

From $\sum F_y = 0$, the force at the section:

$$F_a = R - W_2 = 14.4 \text{ kN}$$

The remainder of the problem is the same as before. The pier considered here has a vertical axis of symmetry, making the application of Eq. 1-13 possible.[12]

## EXAMPLE 1-3

A bracket of negligible weight shown in Fig. 1-19(a) is loaded with a vertical force $P$ of 3 kips. For interconnection purposes, the bar ends are clevised (forked). Pertinent dimensions are shown in the figure. Find the axial stresses in members $AB$ and $BC$ and the bearing and shear stresses for pin $C$. All pins are 0.375 in in diameter.

## Solution

First, an idealized free-body diagram consisting of the two bars pinned at the ends is prepared, see Fig. 1-19(b). As there are no intermediate forces acting on the bars and the applied force acts through the joint at $B$, the forces in the bars are directed along the lines $AB$ and $BC$, and the bars $AB$ and $BC$ are loaded axially.

[12] Strictly speaking, the solution obtained is not exact, as the sides of the pier are sloping. If the included angle between these sides is large, this solution is altogether inadequate. For further details, see S. Timoshenko and J. N. Goodier, *Theory of Elasticity,* 3rd ed. (New York: McGraw-Hill, 1970) 139.

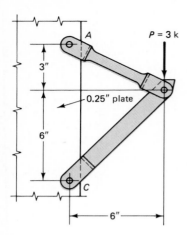

(a)

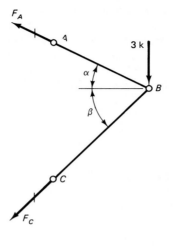

(b)

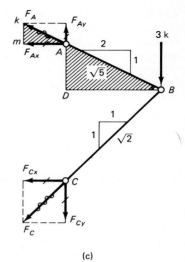

(c)

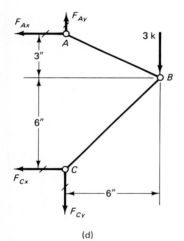

(d)

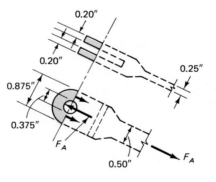

(e)

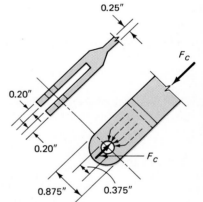

(f)

**Fig. 1-19**

The magnitudes of the forces are unknown and are labeled $F_A$ and $F_C$ in the diagram.[13] These forces can be determined graphically by completing a triangle of forces $F_A$, $F_C$, and $P$. These forces may also be found analytically from two simultaneous equations $\sum F_y = 0$ and $\sum F_x = 0$, written in terms of the unknowns $F_A$ and $F_C$, a known force $P$, and two known angles $\alpha$ and $\beta$. Both these procedures are possible. However, in this book, it will usually be found advantageous to proceed in a different way. Instead of treating forces $F_A$ and $F_C$ directly, their components are used; and instead of $\sum F = 0$, $\sum M = 0$ becomes the main tool.

Any force can be resolved into components. For example, $F_A$ can be resolved into $F_{Ax}$ and $F_{Ay}$, as in Fig. 1-19(c). Conversely, if any one of the components of a directed force is known, the force itself can be determined. This follows from similarity of dimensions and force triangles. In Fig. 1-19(c), the triangles $Akm$ and $BAD$ are similar triangles (both are shaded in the diagram). Hence, if $F_{Ax}$ is known,

$$F_A = (AB/DB)F_{Ax}$$

Similarly, $F_{Ay} = (AD/DB)F_{Ax}$. Note further that $AB/DB$ or $AD/DB$ are ratios; hence, relative dimensions of members can be used. Such relative dimensions are shown by a little triangle on member $AB$ and again on $BC$. In the problem at hand,

$$F_A = (\sqrt{5}/2)F_{Ax} \quad \text{and} \quad F_{Ay} = F_{Ax}/2$$

Adopting the procedure of resolving forces, a revised free-body diagram, Fig. 1-19(d), is prepared. Two components of force are necessary at the pin joints. After the forces are determined by statics, Eq. 1-13 is applied several times, thinking in terms of a free body of an individual member:

$$\sum M_C = 0 \circlearrowright + \quad + F_{Ax}(3 + 6) - 3(6) = 0 \qquad F_{Ax} = +2 \text{ k}$$
$$F_{Ay} = F_{Ax}/2 = 2/2 = +1 \text{ k}$$
$$F_A = 2(\sqrt{5}/2) = +2.23 \text{ k}$$

$$\sum M_A = 0 \circlearrowleft + \quad + 3(6) + F_{Cx}(9) = 0, \qquad F_{Cx} = -2 \text{ k}$$
$$F_{Cy} = F_{Cx} = -2 \text{ k}$$
$$F_C = \sqrt{2}(-2) = -2.83 \text{ k}$$

*Check:* 
$$\sum F_x = 0 \qquad F_{Ax} + F_{Cx} = 2 - 2 = 0$$
$$\sum F_y = 0 \qquad F_{Ay} - F_{Cy} - P = 1 - (-2) - 3 = 0$$

Tensile stress in main bar $AB$:

$$\sigma_{AB} = \frac{F_A}{A} = \frac{2.23}{0.25 \times 0.50} = 17.8 \text{ ksi}$$

[13] In frameworks it is convenient to assume all unknown forces are tensile. A negative answer in the solution then indicates that the bar is in compression.

Tensile stress in clevis of bar $AB$, Fig. 1-19(e):

$$(\sigma_{AB})_{\text{clevis}} = \frac{F_A}{A_{\text{net}}} = \frac{2.23}{2 \times 0.20 \times (0.875 - 0.375)} = 11.2 \text{ ksi}$$

Compressive stress in main bar $BC$:

$$\sigma_{BC} = \frac{F_C}{A} = \frac{2.83}{0.875 \times 0.25} = 12.9 \text{ ksi}$$

In the compression member, the net section at the clevis need not be investigated; see Fig. 1-19(f) for the transfer of forces. The bearing stress at the pin is more critical. Bearing between pin $C$ and the clevis:

$$\sigma_b = \frac{F_C}{A_{\text{bearing}}} = \frac{2.83}{0.375 \times 0.20 \times 2} = 18.8 \text{ ksi}$$

Bearing between the pin $C$ and the main plate:

$$\sigma_b = \frac{F_C}{A} = \frac{2.83}{0.375 \times 0.25} = 30.2 \text{ ksi}$$

Double shear in pin $C$:

$$\tau = \frac{F_C}{A} = \frac{2.83}{2\pi(0.375/2)^2} = 12.9 \text{ ksi}$$

For a complete analysis of this bracket, other pins should be investigated. However, it can be seen by inspection that the other pins in this case are stressed either the same amount as computed or less.

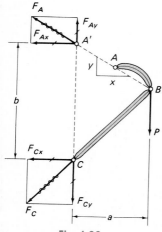

Fig. 1-20

The advantages of the method used in the last example for finding forces in members should now be apparent. It can also be applied with success in a problem such as the one shown in Fig. 1-20. The force $F_A$ transmitted by the curved member $AB$ acts through points $A$ and $B$, since the forces applied at $A$ and $B$ must be collinear. By resolving this force at $A'$, the same procedure can be followed. Wavy lines through $F_A$ and $F_C$ indicate that these forces are replaced by the two components shown. Alternatively, the force $F_A$ can be resolved at $A$, and since $F_{Ay} = (y/x)F_{Ax}$, the application of $\sum M_C = 0$ yields $F_{Ax}$.

In frames, where the applied forces do not act through a joint, proceed as before as far as possible. Then isolate an individual member, and using its free-body diagram, complete the determination of forces. If inclined forces are acting on the structure, resolve them into convenient components.

## Part C DETERMINISTIC AND PROBABILISTIC DESIGN BASES

### 1-10. Member Strength as a Design Criterion

The purpose for calculating stresses in members of a structural system is to compare them with the experimentally determined material strengths in order to assure desired performance. Physical testing of materials in a laboratory provides information regarding a material's resistance to stress. In a laboratory, specimens of known material, manufacturing process, and heat treatment are carefully prepared to desired dimensions. Then these specimens are subjected to successively increasing known forces. In the most widely used test, a round rod is subjected to tension and the specimen is loaded until it finally ruptures. The force necessary to cause rupture is called the *ultimate* load. By dividing this ultimate load by the *original* cross-sectional area of the specimen, the *ultimate strength* (stress) of a material is obtained. Figure 1-21 shows a testing machine used for this purpose. Figure 1-22 shows a tension-test specimen. The

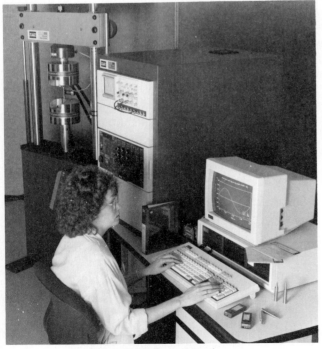

Fig. 1-21 Universal testing machine (Courtesy of MTS Systems Corporation).

Fig. 1-22 A typical tension test specimen of mild steel before and after fracture.

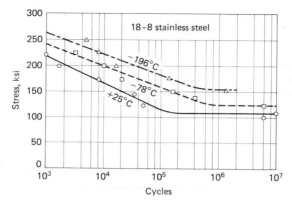

**Fig. 1-23** Fatigue strength of 18-8 stainless steel at various temperatures (reciprocating beam test).

tensile test is used most widely. However, compression, bending, torsion, and shearing tests are also employed.[14] Tables 1A and B of the Appendix gives ultimate strengths and other physical properties for a few materials.

For applications where a force comes on and off the structure a number of times, the materials cannot withstand the ultimate stress of a static test. In such cases, the "ultimate strength" depends on the number of times the force is applied as the material works at a particular stress level. Figure 1-23 shows the results of tests[15] on a number of the same kind of specimens at different stresses. Experimental points indicate the number of cycles required to break the specimen at a particular stress under the application of a fluctuating load. Such tests are called "fatigue tests," and the corresponding curves are termed *S-N* (stress-number) diagrams. As can be seen from Fig. 1-23, at smaller stresses, the material can withstand an ever-increasing number of cycles of load application. For some materials, notably steels, the *S-N* curve for low stresses becomes essentially horizontal. This means that at a low stress, an infinitely large number of reversals of stress can take place before the material fractures. The limiting stress at which this occurs is called the *endurance limit* of the material. This limit, being dependent on stress, is measured in *ksi* or MPa.

Some care must be exercised in interpreting *S-N* diagrams, particularly with regard to the range of the applied stress. In some tests, complete reversal (tension to compression) of stress is made; in others, the applied load is varied in a different manner, such as tension to no load and back to tension. The major part of fatigue testing done on specimens is bending. Stress-dependent deformations may also play a key role in selecting the permissible or allowable stress for a given material, since some materials deform an unpermissible amount prior to fracture. Some materials

---

[14] ASTM (American Society for Testing and Materials) issues an *Annual Book of ASTM Standards* now consisting of 66 volumes, divided into 16 sections, giving classification of materials, ASTM standard specifications, and detailed test methods. ASTM material designation such as A36 steel is frequently used in this book.

[15] J. L. Zambrow, and M. G. Fontana, "Mechanical Properties, Including Fatigue, of Aircraft Alloys at Very Low Temperatures," *Trans. ASM* 41 (1949): 498.

deform plastically under a sustained load, a phenomenon called *creep*. Experience with turbines, tightened bolts in mechanical equipment, wooden or reinforced concrete beams indicates some of the examples where creep may be a problem. In some instances, the rate of load application has a major effect, as some materials become considerably stronger at very rapidly applied loads. Likewise, the effect of temperature usually has a very important effect on the endurance limit. Some of these issues are discussed further in Sections 2-3 and 2-5. At the design level, most of these problems can be controlled by reducing design stresses.

The aforementioned facts, coupled with the impossibility of determining stresses accurately in complicated structures and machines, necessitate a substantial reduction of stress compared to the ultimate strength of a material in a static test. For example, ordinary steel will withstand an ultimate stress in tension of 60 *ksi* and more. However, it deforms rather suddenly and severely at the stress level of about 36 *ksi*, and it is customary in the United States to use an allowable stress of around 22 *ksi* for structural work. This allowable stress is even further reduced to about 12 *ksi* for parts that are subjected to alternating loads because of the fatigue characteristics of the material. *Fatigue properties of materials are of utmost importance in mechanical equipment.* Many failures in machine parts can be traced to disregard of this important consideration. (See also Section 2-10.) Low-cycle fatigue ($10^4$ cycles or less) also cannot be excluded from design considerations in seismically resistant structures.

As pointed out in Section 1-7, in some situations, it is also appropriate to consider residual stresses.

The decision process in choosing an appropriate allowable stress is further complicated since there is *great uncertainty in the magnitudes of the applied loads*. During the life of a machine or a structure, occasional overloads are almost a certainty, but their magnitudes can only be estimated at best.

These difficult problems are now resolved using two alternative approaches. In the traditional approach, in the spirit of classical mechanics, *unique magnitudes* are assigned to the applied forces as well as to the *allowable stresses*. In this manner, these two principal parameters are precisely known, i.e., determinate, in the design process. This *deterministic* approach is commonly used in current practice and will be largely adhered to in this text. However, as the complexity of engineering hardware systems increases, less reliance can be placed on past experience and a limited number of experiments. Instead, after identification of the main parameters in a given stress-analysis problem, their statistical variability is assessed, leading to the *probabilistic* method of estimating structural safety. This approach has found favor in the design of advanced aircraft, offshore structures, and is emerging in structural design of buildings and bridges. A brief discussion of the probabilistic approach to structural design is given in Section 1-12. The traditional deterministic approach is discussed next.

## 1-11. Deterministic Design of Members: Axially Loaded Bars

In the deterministic design of members, a stress resultant is determined at the highest stressed section using conventional mechanics. For axially loaded bars, it means determining the largest internal axial force $P$ at a minimum cross section. Then, for the selected material, an allowable stress $\sigma_{\text{allow}}$ must be chosen.

Professional engineering groups, large companies, as well as city, state, and federal authorities, prescribe or recommend[16] allowable stresses for different materials, depending on the application. Often such stresses are called the allowable *fiber*[17] stresses.

Since according to Eq. 1-13, stress times area is equal to a force, the allowable and ultimate stresses may be converted into the allowable and ultimate forces or "loads," respectively, that a member can resist. Also a significant ratio may be formed:

$$\frac{\text{ultimate load for a member}}{\text{allowable load for a member}}$$

This is the basic definition of the *factor of safety*, F.S. This ratio must always be greater than unity. Traditionally this factor is recast in terms of stresses as

$$\text{F.S.} = \frac{\text{maximum useful material strength (stress)}}{\text{allowable stress}}$$

and is widely used not only for axially loaded members, but also for any type of member and loading conditions. As will become apparent from subsequent reading, whereas this definition of F.S. in terms of elastic stresses is satisfactory for some cases, it can be misleading in others.

In the aircraft industry, the term factor of safety is replaced by another, defined as

$$\frac{\text{ultimate load}}{\text{design load}} - 1$$

[16] For example, see the American Institute of Steel Construction *Manual*, Building Construction Code of any large city, ANC-5 *Strength of Aircraft Elements* issued by the Army-Navy Civil Committee on Aircraft Design Criteria, etc.

[17] The adjective *fiber* in this sense is used for two reasons. Many original experiments were made on wood, which is fibrous in character. Also, in several derivations that follow, the concept of a continuous filament or fiber in a member is a convenient device for visualizing its action.

and is known as the *margin of safety*. In the past, this ratio was usually recast to read

$$\frac{\text{ultimate stress}}{\text{maximum stress caused by the design load}} - 1$$

The newer analytical methods, some of which will be pointed out in the text as they occur, can provide reasonable estimates of the ultimate loads for complex systems and should be used in the basic definition of F.S. as well as of margin of safety. For example, for static loadings, instead of designing members at working loads using allowable stress, an alternative approach consisting of selecting member sizes for their *ultimate* or *limit load* is becoming widely adopted. In such cases, the ultimate load is usually obtained by multiplying the working loads by a suitably chosen *load factors*. For bars in simple tension or compression, this leads to the same results. Significantly different results may be obtained in many other cases where inelastic behavior is more complex. In this text, however, the customary *allowable stress design* (ASD) approach will be largely followed.

The application of the ASD approach for axially loaded members is both simple and direct. From Eq. 1-13, it follows that the required net area $A$ of a member is

$$A = \frac{P}{\sigma_{\text{allow}}} \tag{1-16}$$

where $P$ is the applied axial force, and $\sigma_{\text{allow}}$ is the allowable stress. Equation 1-16 is generally applicable to tension members and short compression blocks. *For slender compression members, the question of their stability arises and the methods discussed in Chapter 11 must be used.*

The simplicity of Eq. 1-16 is unrelated to its importance. A large number of problems requiring its use occurs in practice. The following problems illustrate some application of Eq. 1-16 as well as provide additional review in statics.

## EXAMPLE 1-4

Reduce the size of bar *AB* in Example 1-3 by using a better material such as chrome-vanadium steel. The ultimate strength of this steel is approximately 120 *ksi*. Use a factor of safety of 2.5.

### Solution

$\sigma_{\text{allow}} = 120/2.5 = 48$ ksi. From Example 1-3, the force in the bar $AB$: $F_A = +2.23$ kips. Required area: $A_{\text{net}} = 2.23/48 = 0.0464$ in$^2$. Adopt: 0.20-in by 0.25-in bar. This provides an area of $(0.20)(0.25) = 0.050$ in$^2$, which is slightly in excess of the required area. Many other proportions of the bar are possible.

With the cross-sectional area selected, the actual or working stress is somewhat below the allowable stress: $\sigma_{\text{actual}} = 2.23/(0.050) = 44.6$ ksi. The actual factor of safety is $120/(44.6) = 2.69$, and the actual margin of safety is 1.69.

In a complete redesign, clevis and pins should also be reviewed and, if possible, decreased in dimensions.

### EXAMPLE 1-5

Select members $FC$ and $CB$ in the truss of Fig. 1-24(a) to carry an inclined force $P$ of 650 kN. Set the allowable tensile stress at 140 MPa.

### Solution

If all members of the truss were to be designed, forces in all members would have to be found. In practice, this is now done by employing computer programs developed on the basis of matrix structural analysis[18] or by directly analyzing the truss by the method of joints. However, if only a few members are to be designed or checked, the method of sections illustrated here is quicker.

It is generally understood that a planar truss, such as shown in the figure, is stable in the direction perpendicular to the plane of the paper. Practically, this is accomplished by introducing braces at right angles to the plane of the truss. In this example, the design of compression members is avoided, as this will be treated in the chapter on columns.

To determine the forces in the members to be designed, the reactions for the whole structure are computed first. This is done by completely disregarding the interior framing. Only reaction and force components definitely located at their points of application are indicated on a free-body diagram of the whole structure; see Fig. 1-24(b). After the reactions are determined, free-body diagrams of a part of the structure are used to determine the forces in the members considered; see Figs. 1-24(c) and (d).

Using the free-body diagram in Fig. 1-24(b):

$$\sum F_x = 0 \qquad R_{Dx} - 520 = 0 \qquad R_{Dx} = 520 \text{ kN}$$

$$\sum M_E = 0 \,\circlearrowleft + \qquad R_{Dy} \times 3 - 390 \times 0.5 - 520 \times 1.5 = 0$$

$$R_{Dy} = 325 \text{ kN}$$

$$\sum M_D = 0 \,\circlearrowright + \qquad R_E \times 3 + 520 \times 1.5 - 390 \times 2.5 = 0$$

$$R_E = 65 \text{kN}$$

*Check:* $\sum F_y = 0 \qquad 325 - 390 + 65 = 0$

[18] See, for example, O. C. Zienkiewicz, *The Finite Element Method,* 3rd ed. (London: McGraw-Hill, 1977).

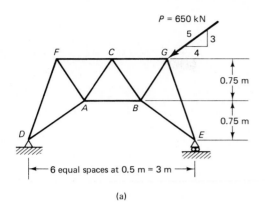

(a)

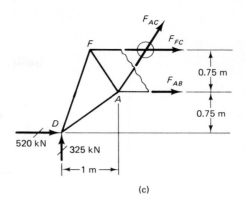

(c)

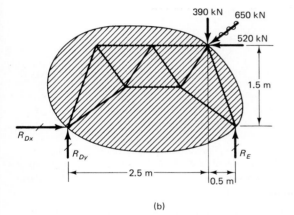

(b)

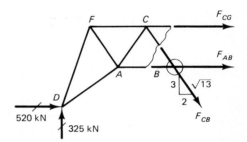

(d)

**Fig. 1-24**

Using the free-body diagram in Fig. 1-24(c):

$$\sum M_A = 0 \; \circlearrowleft + \qquad F_{FC} \times 0.75 + 325 \times 1 - 520 \times 0.75 = 0$$

$$F_{FC} = +86.7 \text{ kN}$$

$$A_{FC} = F_{FC}/\sigma_{\text{allow}} = 86.7 \times 10^3/140 = 620 \text{ mm}^2$$

(use 12.5 × 50-mm bar)

Using the free-body diagram in Fig. 1-24(d):

$$\sum F_y = 0 \qquad -(F_{CB})_y + 325 = 0 \qquad (F_{CB})_y = +325 \text{ kN}$$

$$F_{CB} = \sqrt{13}(F_{CB})_y/3 = +391 \text{ kN}$$

$$A_{CB} = F_{CB}/\sigma_{\text{allow}} = 391 \times 10^3/140 = 2790 \text{ mm}^2$$

(use two bars 30 × 50 mm)

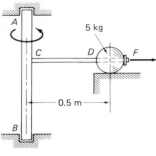

**Fig. 1-25**

## EXAMPLE 1-6

Consider the idealized system shown in Fig. 1-25, where a 5-kg mass is to be spun on a frictionless plane at 10 Hz.[19] If a light rod $CD$ is attached at $C$, and the allowable stress is 200 MPa, what is the required size of the rod? Neglect the weight of the rod and assume that the rod is enlarged at the ends to compensate for the threads.

### Solution

The rod angular velocity $\omega$ is $20\pi$ rad/s. The acceleration $a$ of the mass toward the center of rotation is $\omega^2 R$, where $R$ is the distance $CD$. By multiplying the mass $m$ by the acceleration, the force $F$ acting on the rod is obtained. As shown in the figure, according to the d'Alembert's principle, this force acts in the opposite direction to that of the acceleration. Therefore,

$$F = ma = m\omega^2 R = 5 \times (20\pi)^2 \times 0.500 = 9870 \text{ kg·m/s}^2 = 9870 \text{ N}$$

$$A_{net} = \frac{9870}{200} = 49.3 \text{ mm}^2$$

An 8-mm round rod having an area $A = 50.3 \text{ mm}^2$ would be satisfactory.

The additional pull at $C$ caused by the mass of the rod, which was not considered, is

$$F_1 = \int_0^R (m_1 \, dr)\omega^2 r$$

where $m_1$ is the mass of the rod per unit length, and ($m_1 \, dr$) is its infinitesimal mass at a variable distance $r$ from the vertical rod $AB$. The total pull at $C$ caused by the rod and the mass of 5-kg at the end is $F + F_1$.

---

## **[20]1-12. Probabilistic Basis for Structural Design

In the conventional (deterministic) design of members, the possibility of failure is reduced to acceptably small levels by factors of safety based on judgment derived from past successful and unsuccessful performances. By contrast, in the probabilistic approach, variability in material properties, fabrication-size tolerances, as well as uncertainties in loading and even design approximations, can be appraised on a statistical basis. As far as possible, the proposed criteria are calibrated against well-established cases, as disregard of past successful applications is out of the question. The probabilistic approach has the advantage of consistency in the factors of safety, not only for individual members, but also for complex

---

[19] Hz (abbreviation for hertz), or cycles per second, is the SI unit for frequency.
[20] The remainder of this chapter can be omitted.

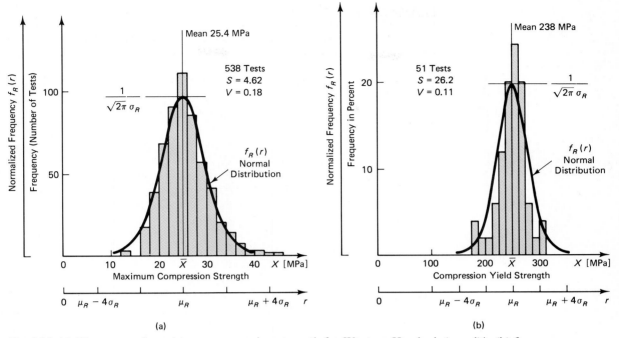

**Fig. 1-26** (a) Histogram of maximum compression strength for Western Hemlock (wood)*; (b) frequency diagram of compression yield strength of ASTM grades A7 and A36 steels.**

structural assemblies. Important risk analyses of complete engineering systems are based on the same premises.

### Experimental Evidence

As an example of the probabilistic approach based on statistics, consider the behavior of specimens for two sets of similar experiments. For one set, experimental results of several compression tests for identical short wooden blocks are plotted in Fig. 1-26(a).[21] Similar results are shown for steel stub columns in Fig. 1-26(b).[22] The bar widths in these histograms correspond to a narrow range of compression stress for which a given number of specimens were either crushed (wood) or have yielded[23] (steel). In these diagrams, the inner scales apply to direct experimental results. The meaning of the outer scales will be discussed later.

[21] J. M. Illston, J. M. Dinwoodie, and A. A. Smith, *Concrete, Timber, and Metals* (New York: Van Nostrand Reinhold, 1979), Fig. 14.3, p. 439, © Crown Copyright, Building Research Establishment, U.K.

[22] T. V. Galambos, and M. K. Ravindra, *Tentative Load and Resistance Design Criteria for Steel Buildings,* Research Report No. 18, Structural Division, Washington University, September 1973.

[23] Since yielding is accompanied by a large amount of deformation, this condition can in many applications be considered failure. For further discussion, see Section 2-3.

In statistical terminology, the test results are termed "population" samples. In the analysis of such data, several quantities of major importance are generally computed. One of these is *sample mean* (average), $\overline{X}$; another is *sample variance, $S^2$*. For $n$ samples (tests), these quantities are defined as

$$\overline{X} = \frac{1}{n} \sum_{i=1}^{n} X_i \tag{1-17}$$

and

$$S^2 = \frac{1}{n} \sum_{i=1}^{n} (X_i - \overline{X})^2 \tag{1-18}$$

where $X_i$ is an $i$th sample.[24]

A square root of the variance, i.e., $S$, is called the *standard deviation*. Dividing $S$ by $\overline{X}$, one obtains the *coefficient of variation*,[25] $V$, i.e.,

$$V = S/\overline{X} \tag{1-19}$$

$\overline{X}$, $S$ (or $S^2$), and $V$ play dominant roles in the theory of probability. The *expected sample value* is $\overline{X}$, the mean; $S$ is a *measure of dispersion* (scatter) of the data, and $V$ is its dimensionless measure.

### Theoretical Basis

In Fig. 1-26, in addition to the histograms, theoretical curves for the two cases are also shown. These bell-shaped curves of *probability density functions* (PDFs) are based on *normal or Gaussian*[26] *distribution*. These continuous PDFs for approximating the dispersion of observed data are the most widely used model in applied probability theory. In analytical form, the PDF of $Z$, i.e., $f_Z(z)$, is given as

$$f_Z(z) = \frac{1}{\sqrt{2\pi}\,\sigma_Z} \exp\left[ -\frac{1}{2}\left(\frac{z - \mu_Z}{\sigma_Z}\right)^2 \right] \tag{1-20}$$

where

$$\mu_Z = \int_{-\infty}^{+\infty} z f_Z(z)\, dz \tag{1-21}$$

and

$$\sigma_Z^2 = \int_{-\infty}^{+\infty} (z - \mu_Z)^2 f_Z(z)\, dz \tag{1-22}$$

---

[24] In order to remove bias in $\overline{X}$, instead of dividing by $n$, one uses $n - 1$. For large values of $n$, the difference in results is small.

[25] In this section, the notation differs from that used in the remainder of the text.

[26] So named in honor of the great German mathematician Karl Friedrich Gauss (1777–1855), who first introduced this function based on theoretical considerations.

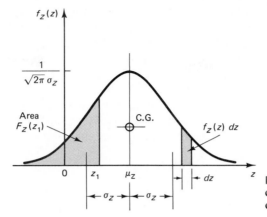

**Fig. 1-27** Normal probability density function (PDF) of Z.

The constant $1/\sqrt{2\pi}$ in Eq. 1-20 is selected so that the normalized frequency diagram encloses a unit area, i.e.,

$$\int_{-\infty}^{+\infty} f_Z(z) \, dz = 1 \qquad (1\text{-}23)$$

which means that the occurrence of $z$ within its entire range is a certainty.

In the previous equations $\mu_Z$ is the *mean* and $\sigma_Z$ is the *standard deviation*. A typical PDF of $Z$ with normal distribution is shown in Fig. 1-27. Illustrations of normal PDFs of resistances $R$ relating them to experimental results are shown in Fig. 1-26. In applications, the theoretical model is usually selected by setting $\mu_Z = \overline{X}$, and $\sigma_Z = S$. For the theoretical model, the *coefficient of variation* will be designated by $\delta_Z$ and is equal to the previously defined experimental $V$.

Some interesting properties of $f_Z(z)$ are illustrated in Fig. 1-28. Thus,

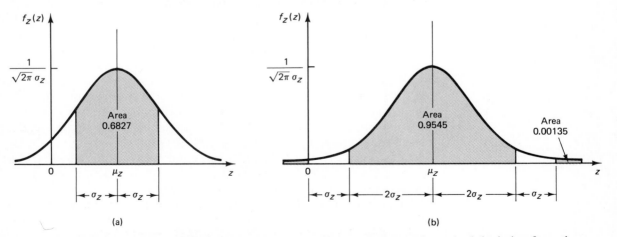

**Fig. 1-28** Examples of probabilities of outcomes at different amounts of standard deviation from the mean.

from Fig. 1-28(a), it can be seen that the probability of the occurrence of an outcome between one standard deviation on either side of the mean is 68.27%. Whereas, as shown in Fig. 1-28(b), between two standard deviations on either side of the mean, this value becomes 95.45%. The areas enclosed under the curve tails that are three standard deviations from the mean are only 0.135% of the total outcomes. As will become apparent later, the small number of outcomes likely to take place under $f_Z(z)$ several standard deviations away from the mean is of the utmost importance in appraising structural safety.

### Practical Formulations

For a probabilistic appraisal of the structural safety of a member or a structure, one must have a statistically determined resistance PDF $f_R(r)$, such as discussed before, and a corresponding load effect PDF. Again statistical studies show that since the loads are susceptible to variations, their effect on a member or a structure can be expressed in probabilistic form. Such load effects, resembling $f_R(r)$, will be designated as $f_Q(q)$. For a given member or a structure, these functions define the behavior of the same critical parameter such as a force, stress, or deflection. Two such functions probabilistically defining the load effect $f_Q(q)$ and the resistance $f_R(r)$ for a force acting on a member are shown in Fig. 1-29. For purposes of illustration, it is assumed that the load effect $f_Q(q)$ has a larger standard deviation, i.e., larger dispersion of the load, than that for the member resistance.

In conventional (deterministic) design, the load magnitudes are usually set above the observed mean. This condition is represented by $Q_n$ in Fig. 1-29. On the other hand, in order to avoid possible rejections, a supplier will typically provide a material with an average strength slightly greater than specified. For this reason, calculated nominal member resistance $R_n$ would be below the mean. On this basis, the conventional factor of safety is simply defined as $R_n/Q_n$.

In reality, both $Q$ and $R$ are uncertain quantities and there is no unique answer to the safety problem. To illustrate the interaction between the two main variables in Fig. 1-31, $f_R(r)$ is shown along the horizontal axis and $f_Q(q)$ is plotted along the vertical axis. For the ensemble of an infinite number of possible outcomes, a line at 45° corresponding to $R = Q$ divides the graph into two regions. For $R > Q$, no failure can occur. For example, for the range of small and large outcomes $Q_1$, $Q_2$, $Q_3$, the resistance outcomes, respectively, $R_1$, $R_2$, $R_3$ suffice to preserve the integrity of a member. However, for outcomes $Q_3$ and $R_1$ with a common point at $D$ and falling in the region where $R < Q$, a failure would take place.

While enlightening, the above process is difficult to apply in practice. Fortunately, however, it can be mathematically demonstrated that for *normal distribution* of $R$ and $Q$ their difference, i.e., $R - Q$, is also a normal distribution. In this manner, the information implied in Fig. 1-30

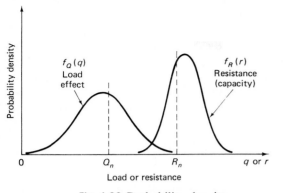

**Fig. 1-29** Probability density functions for the two main random variables (load and resistance).

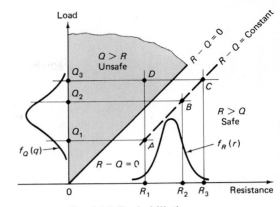

**Fig. 1-30** Probabilistic definition of safe and unsafe structural regions.

can be compressed into a single normal PDF such as that shown in Fig. 1-31(a). In this diagram the *probability of failure*, $p_f$, is given by the area under the tail of the curve to the left of the origin. A possible magnitude of a $p_f$ may be surmised from Fig. 1-28(b). A member would survive in all instances to the right of the origin.

As can be seen from Fig. 1-31(a), $\beta \sigma_{R-Q} = \mu_{R-Q}$, where $\beta$ is a constant and $\sigma_{R-Q}$ is standard deviation. For applications, this relation can be put into a more convenient form by noting that the variance of a linear function of two independent normal variables, $\sigma^2_{R-Q}$, is the sum of the variances

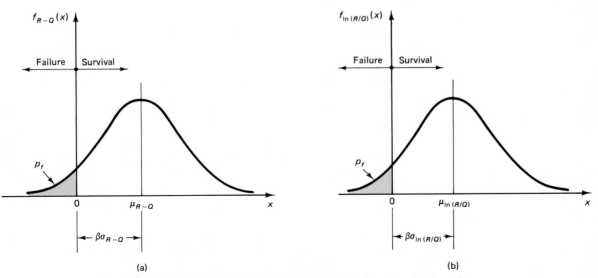

**Fig. 1-31** (a) Normal and (b) lognormal probability density functions.

of its parts.[27] Moreover, since variance is a square of standard deviation, one has the following expression for the *safety index* β.

$$\beta = \frac{\mu_{R-Q}}{\sigma_{R-Q}} = \frac{\mu_R - \mu_Q}{\sqrt{\sigma_R^2 + \sigma_Q^2}} \qquad (1\text{-}24)$$

where $\sigma_R$ and $\sigma_Q$ are, respectively, the standard deviations for the resistance $R$ and the load effect $Q$. A larger β results in fewer failures, and thus, a more conservative design.

An alternative appoach for establishing the formulation for the safety index β can be based on the more widely used concept of the *factor of safety* defined as the ratio $R/Q$. This approach is particularly useful when the distributions of $R$ and $Q$ are skewed and the lognormal distribution[28] rather than the normal is appropriate. In this formulation, for reasons of mathematical convenience, it is preferable to work with the logarithm of the ratio $R/Q$, Fig. 1-31(b). By carrying out this approach and making use of first order, mean-value approximations, the expression for the safety index, β, reads

$$\beta \approx \frac{\mu_{\ln(R/Q)}}{\sqrt{\delta_R^2 + \delta_Q^2}} = \frac{\ln \mu_R - \ln \mu_Q}{\sqrt{\delta_R^2 + \delta_Q^2}} \qquad (1\text{-}26)$$

where, as before, $\mu_R$ and $\mu_Q$ are the mean values for the respective functions, and $\delta_R$ and $\delta_Q$ are, respectively, the coefficients of variation for $R$ and $Q$.

It can be noted that Eqs. 1-24 and 1-26 resemble each other. A graphical interpretation for a solution based on the use of $\ln(R/Q)$ is shown in Fig. 1-31(b). Analogous to the first approach, the probability of failure $p_f$ is given by the area under the tail of the curve to the left of the origin. For routine applications, a β on the order of 3 is considered appropriate.

It must be recognized that the safety index, β, is only a relative measure of reliability and cannot be considered exact. Nevertheless, uncertainties

---

[27] A. H-S. Ang, and W. H. Tang, *Probability Concepts in Engineering Planning and Design*, Vol. 1 (New York: John Wiley and Sons, 1975).

[28] Lognormal distribution for a random variable $R$ is defined as

$$f_R(r) = \frac{1}{\sqrt{2\pi}\, \xi_R r} \exp\left[ -\frac{1}{2}\left( \frac{\ln r - \lambda_R}{\xi_R} \right)^2 \right] \qquad (1\text{-}25)$$

where $\lambda_R = \ln \mu_R / \sqrt{1 + \delta_R^2}$ and $\xi_R^2 = \ln(1 + \delta_R^2)$ are, respectively, the mean and standard deviation of $\ln R$. Similar expressions apply for $f_Q(q)$. However, for lognormal $R$ and $Q$, it can be shown that $Z = \ln(Q/R)$ has the normal distribution. See A. H-S. Ang and W. H. Tang, *Probability Concepts in Engineering Planning and Design*, Vol. 2—*Decision, Risk, and Reliability* (New York: Wiley, 1983).

in design variables can be explicitly included by using the coefficients of variation in the design parameters, resulting in more consistent reliability of structures and machines.

In addition to the failure limit states emphasized before, the probabilistic approach is suitable for other situations. Important among these are the serviceability limit states. Among these, control of maximum deflections or limitations on undesirable vibrations can also be treated in probabilistic terms.

### EXAMPLE 1-7

Consider two kinds of loading to be suspended by steel tension rods. In both cases, a nominal permanent, or *dead* load, $D_n$, is 5 kips. In one case, however, a nominal intermittent, or *live* load, $L_n$, is 1 kip, whereas in the other, $L_a$ is 15 kips. Assume that for the design of these rods, American Institute of Steel Construction (AISC) provisions for the design of buildings using ASTM Grade A36 steel apply.

(a) Determine the cross-sectional areas for the rods using the conventional allowable stress design (ASD) approach, for which $\sigma_{allow} = 22$ ksi.[29]

(b) Find the cross-sectional areas for the same rods using an approach deduced from the basics of probabilistic concepts. According to AISC/LRFD,[30] this requires the use of the following relation:

$$\phi R_n \geq \sum_{i=1}^{k} \gamma_i Q_i \qquad (1\text{-}27)$$

where $R_n$ is the nominal strength of the structure, and $\phi < 1$ is the resistance factor such that $\phi R_n$ is the design resistance of the member; the load factors $\gamma_i > 1$ account for possible overloads over the nominal load effect $Q_i$. Since in this case only two types of loading are considered, Eq. 1-27 reduces to

$$\phi R_n \geq 1.2 D_n + 1.6 L_n \qquad (1\text{-}28)$$

where, according to the code for this case, $\phi$ is 0.90, $\gamma_i$'s are 1.2 and 1.6, and the yield strength of the steel, $\sigma_{yp}$, is 36 ksi.

(c) For the four solutions found before, calculate the corresponding safety indices, $\beta$, using Eq. 1-26. This equation is based on lognormal distribution for the variables associated with the load and resistance per AISC/LRFD.

### Solution

(a) Since the total axial force $P$ is caused by the dead and live loads, $P = D_n + L_n$, and, on applying Eq. 1-16, the required areas are

---

[29] AISC, *Manual of Steel Construction*, 9th ed. (Chicago, 1989).
[30] AISC, *Manual of Steel Construction, Load and Resistance Factor Design (LRFD)*, 1st ed. (Chicago: 1986).

$$A_1 = \frac{D_n + L_n}{\sigma_{\text{allow}}} = \frac{5 + 1}{22} = 0.273 \text{ in}^2 \qquad \text{for } L_n = 1 \text{ kip}$$

$$A_2 = \frac{5 + 15}{22} = 0.909 \text{ in}^2 \qquad \text{for } L_n = 15 \text{ kips}$$

(b) Since $R_n = A\sigma_y$, again from Eq. 1-16:

$$A = \frac{1.2D_n + 1.6L_n}{\phi\sigma_y}$$

and

$$A_1^* = \frac{1.2 \times 5 + 1.6 \times 1}{0.90 \times 36} = 0.235 \text{ in}^2 \qquad \text{for } L_n = 1 \text{ kip}$$

Similarly, $A_2^* = 0.926 \text{ in}^2$ for $L_n = 15$ kips.

The coefficients 1.2 for $D_n$, 1.6 for $L_n$, and 0.90 for $\phi$ have been statistically determined to approximate probabilistic solutions to various problems. Such studies show that dead load is more predictable than live load, and, for that reason, has a smaller multiplier, 1.2, for obtaining the most probable maximum load. The coefficient $\phi$ varies from 0.60 to 1.00, depending on the statistically determined strength of the type of member. If in addition to dead and live loads, other loading conditions such as those caused by wind, snow, or earthquakes should be considered, additional $\gamma_i Q_i$ terms appear in Eq. 1-27.

(c) In order to solve this part of the problem, additional information is needed. The nominal values of $R_n$ and $Q_n$ should be transformed into the mean values $\mu_R$ and $\mu_Q$ for the probabilistic formulation. For this reason, based on statistical information,[31] $R_n$ is multiplied by a factor of 1.05 to obtain $\mu_R$, and $\mu_Q$ is set arbitrarily equal to $Q_n$. The coefficient of variation, $\delta_R$, for $\mu_R$ due to the variation in $\sigma_{yp}$ and the cross-sectional area is taken as 0.11, whereas the coefficient of variation $\delta_L$ for $\overline{L}_n$ is taken as 0.25, and $\delta_D$ for $D_n$ is 0.10. To combine $\delta_L$ and $\delta_D$ into a coefficient of variation $\delta_Q$ for both loads requires the use of the following relation employing the notation of this problem[32]:

$$\delta_Q = (\delta_D^2 \overline{D}_n^2 + \delta_L^2 \overline{L}_n^2)^{1/2}/(\overline{D}_n + \overline{L}_n) \qquad (1\text{-}29)$$

On substitution, for the light 1-kip live load, $\delta_{Q1} = 0.093$, and, for the 15-kip live load, $\delta_{Q2} = 0.189$.

Based on the information for the part (a) and recalling that $\sigma_{yp}$ is 36 ksi, $\mu_{R1} = 1.05 \times 0.273 \times 36 = 10.3$ kips; $\delta_R = 0.11$; $\mu_{Q1} = 6$ kips; and $\delta_{Q1}$

[31] B. R. Ellingwood et al., *Development of a Probability Based Load Criterion for American National Standard A58*, National Bureau of Standards, Special Publication No. 577, June 1980.
[32] J. R. Benjamin, and C. A. Cornell, *Probabilistic Statistics and Decisions for Civil Engineers* (New York: McGraw-Hill, 1970).

= 0.093. Alternatively, $\mu_{R2} = 1.05 \times 0.909 \times 36 = 34.4$ kips; $\delta_R = 0.11$; $\mu_{Q2} = 20$ kips; and $\delta_{Q2} = 0.189$. On substituting into Eq. 1-26, the safety indices, respectively, are

$$\beta_1 = \frac{\ln(10.3/6)}{\sqrt{0.11^2 + 0.093^2}} = 3.75$$

and

$$\beta_2 = \frac{\ln(34.4/20)}{\sqrt{0.11^2 + 0.189^2}} = 2.48$$

Similarly, for part (b), $\mu_{R1}^* = 1.05 \times 0.235 \times 36 = 8.88$ kips; $\delta_R = 0.11$; $\mu_{Q1}^* = 6$ kips; $\delta_{Q1} = 0.093$; and, alternatively, $\mu_{R2}^* = 1.05 \times 0.926 \times 36 = 35.0$ kips; $\mu_{Q2}^* = 20$ kips; and $\delta_{Q2} = 0.189$. Hence,

$$\beta_1^* = \frac{\ln(8.88/6)}{\sqrt{0.11^2 + 0.093^2}} = 2.72$$

and

$$\beta_2^* = \frac{\ln(35/20)}{\sqrt{0.11^2 + 0.189^2}} = 2.56$$

By comparing the safety indices for the two solutions, it can be seen that they are far apart using the conventional approach. On the other hand, the $\beta^*$'s are very near one another. Considering that many approximations are made to deduce $\gamma_i$'s and $\phi$ factors for code use, it is encouraging that a solution based on the probabilistic approach lead to such a good result.[33]

## **1-13. Bolted and Riveted Connections

In Section 1-8, some basic aspects in analyzing the behavior of bolted connections were given. Further details of such analyses are discussed here. The same procedures are applicable in the design of riveted connections. The usually assumed behavior of a bolted or riveted joint is summarized in Fig. 1-32.[34] A connection design approach based on preventing slippage between the faying surfaces is discussed later in this section.

The total force acting concentrically on a joint is assumed to be equally distributed between connectors (bolts or rivets) of equal size. In many cases, this cannot be justified by elastic analysis, however, ductile deformations and/or slip between the faying surfaces permits an equal redistribution of the applied force before the ultimate capacity of a con-

[33] H. Madsen, S. Krenk, and N. Lind, *Methods of Structural Safety* (New York: McGraw-Hill, 1986).

[34] From G. Dreyer, *Festigkeitslehre und Elastizitätslehre* (Leipzig: Jänecke, 1938) 34.

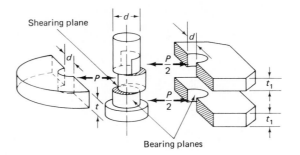

**Fig. 1-32** Assumed action for a bolted or a riveted connection.

nection is reached. This assumption has been justified on the basis of tests.[35]

In contrast to the illustration shown in Fig. 1-32, in simple lap joints, the connectors are in single shear, and the plates near the connector tend to bend to maintain the axial force concentric. However, bending in the connected plates is commonly neglected (see Fig. 1-33). Numerous bolted connections of this type are used in steel construction, and riveted ones are used for joining aluminum alloy sheets in aircraft.

When connectors are arranged as shown in Fig. 1-34(a), determining the net section in tension poses no difficulty. However, if the rows for bolt holes are closely spaced and staggered, as shown in Fig. 1-34(b), a *zig-zag* tear may be more likely to occur than a tear across the normal section *b–b*. Methods for treating such cases are available.[36] It is also necessary to have a sufficient *edge distance e* to prevent a shear failure across the *c–c* planes shown in Fig. 1-34(c).

An illustration of a failure in bearing is given in Fig. 1-35. Although the actual stress distribution is very complex, as noted in Section 1-8, in practice, it is approximated on the basis of an *average bearing stress acting over the projected area of the connector's shank onto a plate,* i.e.,

**Fig. 1-33** Bending of plates commonly neglected in lap joints.

[35] A conclusive experimental verification of this assumption may be found in the paper by R. E. Davis, G. B. Woodruff, and H. E. Davis, "Tension Tests of Large Riveted Joints," *Trans. ASCE* 105 (1940): 1193.

[36] For details, for example, see AISC, *Manual of Steel Construction,* 9th ed. (Chicago, 1989).

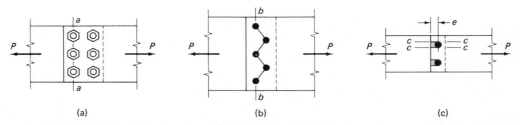

**Fig. 1-34** Possible modes of failure in bolted joints (connections): (a) net section, (b) zig-zag tear, and (c) tear out due to insufficient edge (end) distance along lines *c-c*.

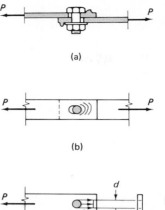

Fig. 1-35 (a),(b) Illustration of a bearing failure, and (c) assumed stress distribution.

on area $td$. It is difficult to justify this procedure theoretically. However, the allowable bearing stress is determined from experiments and is interpreted on the basis of this *average stress* acting on the projected area of a rivet. Therefore, the inverse process used in design is satisfactory.

In the previous design approach, the frictional resistance between the faying surfaces at the connectors has been neglected. However, if the clamping force developed by a connector is both sufficiently large and reliable, the capacity of a joint can be determined on the basis of the friction force between the faying surfaces. This condition is illustrated in Fig. 1-36. With the use of high-strength bolts with yield strength on the order of 100 ksi (700 MPa), this is an acceptable method in structural steel design. The required tightening of such bolts is usually specified to be about 70 percent of their tensile strength. For the purposes of simplified analysis, an allowable shear stress based on the nominal area of a bolt is specified. These stresses are based on experiments. This enables the design of connections using high-strength bolts to be carried out in the same manner as that for ordinary bolts or rivets.

The procedure for analyzing bearing-type bolted and riveted joints by the AISC/LRFD probabilistically based approach remains essentially the

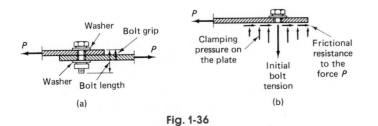

Fig. 1-36

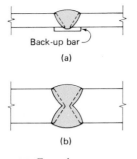

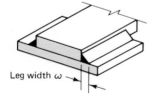

Fig. 1-37 Complete penetration butt welds. (a) Single V-groove weld, (b) double V-groove weld.

same. However, the applied forces are increased using load factors (see Section 1-12 and Eq. 1-27), and stresses are multiplied by appropriate resistance factors. For slip-critical investigation of bolted joints at working loads, neither the net section nor bearing are checked. These details are beyond the scope of this text.[37]

## **1-14. Welded Joints

Steel and aluminum alloy connections by means of welding are very widely used. Butt welds, such as shown in Fig. 1-37, and fillet welds, illustrated in Fig. 1-38, are particularly common. The strength of butt welds is simply found by multiplying the cross-sectional area of the thinner plate being connected by the allowable stress for welds. The allowable stresses are usually expressed as a certain percentage of the strength of the original solid plate of the parent material. This percentage factor varies greatly, depending on the workmanship. For ordinary work, a 20-percent reduction in the allowable stress for the weld compared to the solid plate may be used. For this factor, the efficiency of the joint is said to be 80 percent. On high-grade work, some of the specifications allow 100-percent efficiency for the welded joint. Most pressure vessels are manufactured using such welds. Similar joints are used in some structural frames. In such work, the AISC specifications, based on the recommendations of the American Welding Society (AWS), allow the same tensile stress in the weld as in the base metal in the case of butt welds subjected to static loads.

Fillet welds are designed on a semiempirical basis. These welds are designated by the size of the legs, Fig. 1-38(b), which are usually made of equal width ω. The smallest dimension across a weld is called its *throat*. For example, a standard $\frac{1}{2}$-in weld has both legs $\frac{1}{2}$ in wide and a throat equal to $0.5 \sin \theta = 0.5 \sin 45° = 0.707 \times 0.5$ in. The strength of a fillet weld, *regardless of the direction of the applied force*,[38] is based on the cross-sectional area at the *throat* multiplied by the allowable *shear stress* for the weld metal. The AWS allowable shear stress is 0.3 times the electrode tensile strength. For example, E70 electrodes (i.e., tensile strength of 70 ksi) used as weld metal has an allowable shear stress of $0.3 \times 70 = 21$ ksi. The allowable force $q$ per inch of the weld is then given as

$$q = 21 \times 0.707\omega = 14.85\omega \qquad [\text{k/in}] \qquad (1\text{-}30)$$

where ω is the width of the legs. For a $\frac{1}{4}$-in fillet weld, this reduces to 3.71 kips per in; for a $\frac{3}{8}$-in fillet weld, 5.56 kips per in, etc.

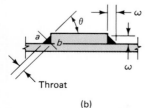

Leg width ω

(a)

θ

ω

a

b

ω

Throat

(b)

Fig. 1-38 An example of a fillet weld.

---

[37] AISC/LRFD *Manual of Steel Construction,* 1st ed. (Chicago, 1986).
[38] This is a considerable simplification of the real problem.

## EXAMPLE 1-8

Determine the required lengths of welds for the connection of a 3 in by 2 in by $\frac{7}{16}$ in steel angle to a steel plate, as shown in Fig. 1-39. The connection is to develop the full strength in the angle uniformly stressed to 20 ksi. Use $\frac{3}{8}$-in fillet welds, whose allowable strength per AWS specification is 5.56 kips per linear inch.

### Solution

Many arrangements of welds are possible. If two welds of length $L_1$ and $L_2$ are to be used, their strength must be such as to maintain the applied force $P$ in equilibrium without any tendency to twist the connection. This requires the resultant of the forces $R_1$ and $R_2$ developed by the welds to be equal and opposite to $P$. For the optimum performance of the angle, force $P$ must act through the centroid of the cross-sectional area (see Table 7 of the Appendix). For the purposes of computation, the welds are assumed to have only linear dimensions.

$$A_{\text{angle}} = 2.00 \text{ in}^2 \qquad \sigma_{\text{allow}} = 20 \text{ ksi}$$
$$P = A\sigma_{\text{allow}} = 2 \times 20 = 40 \text{ k}$$

$\sum M_d = 0 \circlearrowleft + \qquad R_1 \times 3 - 40 \times 1.06 = 0 \qquad R_1 = 14.1 \text{ k}$

$\sum M_a = 0 \circlearrowright + \qquad R_2 \times 3 - 40 \times (3 - 1.06) = 0 \qquad R_2 = 25.9 \text{ k}$

*Check:* $R_1 + R_2 = 14.1 + 25.9 = 40 \text{ k} = P$

Hence, by using the specified value for the strength of the $\frac{3}{8}$-in weld, note that $L_1 = 14.1/5.56 = 2.54$ in and $L_2 = 25.9/5.56 = 4.66$ in. The actual length of welds is usually increased a small amount over the lengths computed to account for craters at the beginning and end of the welds. The eccentricity of the force $P$ with respect to the plane of the welds is neglected.

To reduce the length of the connection, end fillet welds are sometimes used. Thus, in this example, a weld along the line *ad* could be added. The centroid of the resistance for this weld is midway between *a* and *d*. For this arrangement, lengths $L_1$ and $L_2$ are so reduced that the resultant force for all three welds coincides with the resultant of $R_1$ and $R_2$ of the former case. To accomplish the same purpose, slots and notches in the attached member are also occasionally used.

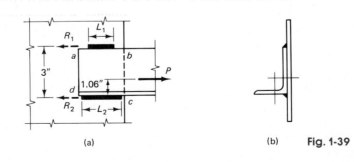

(a)          (b)    **Fig. 1-39**

# Problems

## Section 1-5

**1-1.** Verify equilibrium Eq. 1-5a for the $x$ direction with the aid of a sketch, similar to Fig. 1-3(a), where the stress increments for three-dimensional stresses are shown.

**1-2.** Show that the differential equations of equilibrium for a two-dimensional plane stress problem in polar coordinates are

$$\frac{\partial \sigma_r}{\partial r} + \frac{1}{r}\frac{\partial \tau_{r\theta}}{\partial \theta} + \frac{\sigma_r - \sigma_\theta}{r} = 0$$

$$\frac{1}{r}\frac{\partial \sigma_\theta}{\partial \theta} + \frac{2\,\tau_{r\theta}}{r} = 0$$

The symbols are defined in the figure. Body forces are neglected in this formulation.

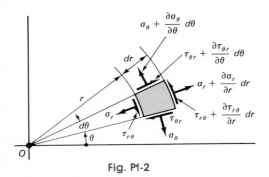

**Fig. P1-2**

## Section 1-6

**1-3.** On the same graph, plot the normal stress $\sigma_\theta$ and the shear stress $\tau_\theta$ as functions of the angle $\theta$ defined in Fig. 1-8. Angle $\theta$ should range from 0° to 360° on the abscissa. Identify the maxima and minima for these functions.

**1-4.** In Fig. 1-8(a), determine the angles $\theta$ where the magnitudes of $\sigma_\theta$ and $\tau_\theta$ are equal.

**1-5.** Using polar coordinate axes, on the same graph, plot $\sigma_\theta$ and $\tau_\theta$ as functions of angle $\theta$ defined in Fig. 1-8. Identify the maxima and minima for these functions.

**1-6.** A 10-mm square bar is subjected to a tensile force $P = 20$ kN, as shown in Fig. 1-8(a). (a) Using statics, determine the normal and shear stress acting on sections $a$–$a$ and $b$–$b$ for $\theta = 30°$. (b) Verify the results

using Eqs. 1-6 and 1-7. (c) Show the results as in Fig. 1-8(g).

**1-7.** Repeat Problem 1-6 for a $\frac{1}{2}$-in square bar if $P = 5$ kips and $\theta = 20°$.

**1-8.** A glued lap splice is to be made in a $10 \times 20$ mm rectangular member at $\alpha = 20°$, as shown in the figure. Assuming that the shear strength of the glued joint controls the design, what axial force $P$ can be applied to the member? Assume the shear strength of the glued joint to be 10 MPa.

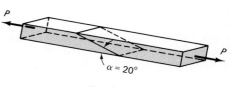

**Fig. P1-8**

## Section 1-7

**1-9.** If an axial tensile force of 110 kips is applied to a member made of a W $8 \times 31$ section, what will the tensile stress be? What will the stress be if the member is a C $12 \times 20.7$ section? For designation and cross-sectional areas of these members, see Tables 4 and 5 in the Appendix.

**1-10 and 1-11.** Short steel members have the cross-sectional dimensions shown in the figures. If they are subjected to axial compressive forces of 100 kN each, find the points of application for these forces to cause no bending, and determine the normal stresses. All dimensions are in mm.

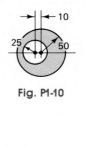

**Fig. P1-10**

**Fig. P1-11**

**1-12.** A bar of variable cross section, held on the left, is subjected to two concentrated forces, $P_1$ and $P_2$, as shown in the figure. (a) Find the maximum axial stress if $P_1 = 10$ kips, $P_2 = 8$ kips, $A_1 = 2$ in$^2$, and $A_2 = 1$ in$^2$. (b) On two separate diagrams, plot the axial force and the axial stress along the length of the bar.

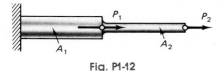

**Fig. P1-12**

**1-13.** A bar of variable cross section, held on the left, is subjected to three forces, $P_1 = 4$ kN, $P_2 = -2$ kN, and $P_3 = 3$ kN, as shown in the figure. On two separate diagrams, plot the axial force and the axial stress along the length of the bar. Let $A_1 = 200$ mm$^2$, $A_2 = 100$ mm$^2$, and $A_3 = 150$ mm$^2$.

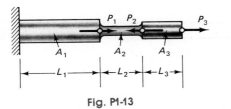

**Fig. P1-13**

**1-14.** Rework Problem 1-13 by reversing the direction of the force $P_2$.

**1-15.** A 2-mm thick hollow circular tube of 40 mm outside diameter is subjected on the outside surface to a constant shear of 10 Pa in the axial direction, as shown in the figure. If the tube is 400 mm long, what is the maximum axial stress? Plot the variation of the axial stress along the tube.

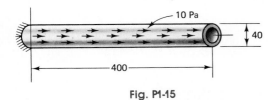

**Fig. P1-15**

**1-16.** A short compression member is made up of two standard steel pipes, as shown in the figure. If the allowable stress in compression is 15 ksi, (a) what is the allowable axial load $P_1$ if the axial load $P_2 = 50$ kips;

(b) what is the allowable load $P_1$ if load $P_2 = 15$ kips? See Table 8 in the Appendix for cross-sectional areas of U.S. standard pipes.

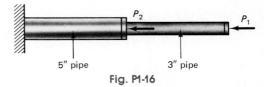

**Fig. P1-16**

**1-17.** Determine the bearing stresses caused by the applied force at $A$, $B$, and $C$ for the wooden structure shown in the figure. All member sizes shown are nominal. See Table 10 in the Appendix for U.S. standard sizes of lumber.

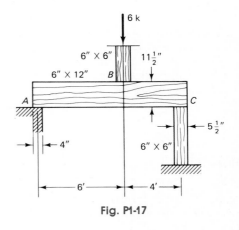

**Fig. P1-17**

## Section 1-8

**1-18.** A 40 × 80 mm wooden plank is glued to two 20 × 80 mm planks, as shown in Fig. 1-13(d). If each of the two glued surfaces is 40 × 80 mm and the applied force $P = 20$ kN, what is the average shear stress in the joints?

**1-19.** Two 10-mm thick steel plates are fastened together, as shown in the figure, by means of two 20-mm bolts that fit tightly into the holes. If the joint transmits a tensile force of 45 kN, determine (a) the

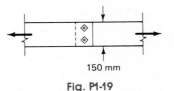

**Fig. P1-19**

average normal stress in the plates at a section where no holes occur; (b) the average normal stress at the critical section; (c) the average shearing stress in the bolts; and (d) the average bearing stress between the bolts and the plates.

**1-20.** A gear transmitting a torque of 4000 in-lb to a $2\frac{3}{16}$-in shaft is keyed to it, as shown in the figure. The $\frac{1}{2}$-in square key is 2 in long. Determine the shear stress in the key.

**Fig. P1-20**

**1-21.** A $\frac{1}{2} \times 6$ in steel plate is to be attached to the main body of a machine, as shown in Fig. 1-15. If the applied force $P = 72$ kips and the welds to be used are good for 5.56 kips/in, see Eq. 1-30, how long should the welds be? Due to symmetry, each weld line resists the same force.

## Section 1-9

**1-22.** What is the shear stress in bolt $A$ caused by the applied load shown in the figure? The bolt is 6 mm in diameter, and it acts in double shear. All dimensions are in mm.

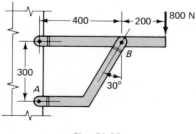

**Fig. P1-22**

**1-23.** Calculate the shear stress in pin $A$ of the bulldozer if the total forces acting on the blade are as shown in the figure. Note that there is a $1\frac{1}{2}$-in-diameter pin on each side of the bulldozer. Each pin is in single shear.

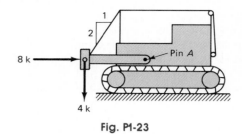

**Fig. P1-23**

**1-24.** A control pedal for actuating a spring mechanism is shown in the figure. Calculate the shear stress in pins $A$ and $B$ due to force $P$ when it causes a stress of 10,000 psi in rod $AB$. Both pins are in double shear.

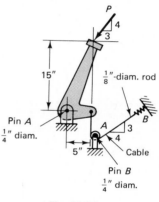

**Fig. P1-24**

**1-25.** A 6-ft-diameter cylindrical tank is to be supported at each end of a hanger arranged as shown in the figure. The total weight supported by the two hangers is 15 k. Determine the shear stresses in the 1-in-diameter pins at points $A$ and $B$ due to the weight of the tank. Neglect the weight of the hangers and assume that contact between the tank and the hangers is frictionless.

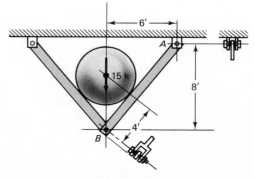

**Fig. P1-25**

**1-26.** For the planar frame loaded as shown in the figure, determine the axial stress in member *BC*. The cross section of member *BC* is 400 mm². The dimensions are given in mm.

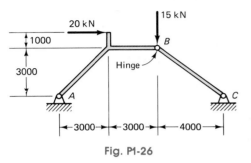

**Fig. P1-26**

**1-27.** Two steel wires with well-designed attachments and a joint are subjected to an external force of 700 N, as shown in the figure. The diameter of wire *AB* is 2.68 mm and that of wire *BC* is 2.52 mm. (a) Determine the stresses in the wires caused by the applied vertical force. (b) Are the wire sizes well-chosen?

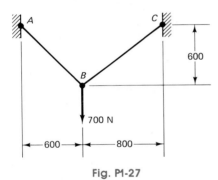

**Fig. P1-27**

✓**1-28.** Find the stress in the mast of the derrick shown in the figure. All members are in the same vertical

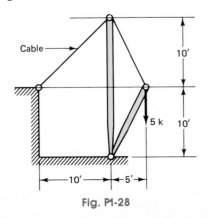

**Fig. P1-28**

plane and are joined by pins. The mast is made from an 8-in standard steel pipe weighing 28.55 lb/ft. (See Table 8 in the Appendix.) Neglect the weight of the members.

**1-29.** A signboard 15 by 20 ft in area is supported by two wooden frames, as shown in the figure. All wooden members are 3 by 8 in. (See Table 10 in the Appendix for actual lumber sizes.) Calculate the stress in each member due to a horizontal wind load of 20 lb/ft² on the sign. Assume all truss joints are pinned and that two-sixths of the total wind force acts at *B* and one-sixth at *C*. Neglect the possibility of buckling of the compression members. Neglect the weight of the structure.

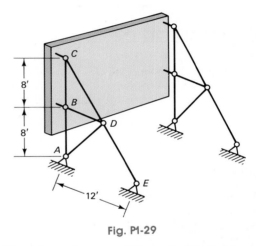

**Fig. P1-29**

\***1-30.** A braced structural frame is designed to resist the lateral forces shown in the figure. Neglecting the

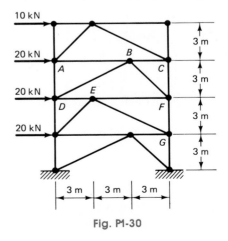

**Fig. P1-30**

frame weight, determine the axial stresses in members *BD*, *FG*, and *DE*; the respective areas for these members are 160, 400, and 130 mm².

**\*1-31.** A planar system consists of a rectangular beam *AC* suported by steel members *AE* and *BE* and a pin at *C*, as shown in the figure. Member *AE* is made up of two $\frac{1}{4}$ by 1 in parallel flat bars, and pin *C*, acting in double shear, is $\frac{3}{4}$ in in diameter. Determine the axial stress in bars *AE* and the shear in pin *C*.

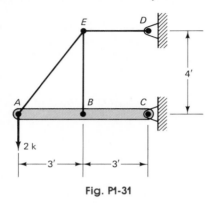

**Fig. P1-31**

**\*1-32.** By means of numerous vertical hangers, the cable shown in the figure is designed to support a continuously distributed load. This load, together with the cable and hangers, can be approximated as a uniformly distributed load of 2 kN/m. Determine the cross section required for the cable if the yield strength of the material is 1000 MPa and the required factor of safety is 2. (*Hint:* The cable assumes the shape of a parabola and develops only a horizontal force *H* at its lowest point. The larger resultant at a support is equal to the largest force in the cable.)

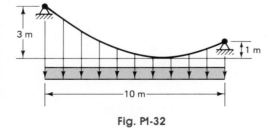

**Fig. P1-32**

## Section 1-11

**1-33.** A 150 mm square wooden post delivers a force of 50 kN to a concrete footing, as shown in Fig. 1-11.

(a) Find the bearing stress of the wood on the concrete.
(b) If the allowable pressure on the soil is 100 kN/m², determine in plan view the required dimensions of a square footing. Neglect the weight of the footing.

**1-34.** For the structure shown in the figure, calculate the size of the bolt and area of the bearing plates required if the allowable stresses are 18,000 psi in tension and 500 psi in bearing. Neglect the weight of the beams.

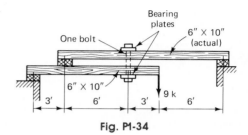

**Fig. P1-34**

**1-35.** What minimum distances *a* and *b* are required beyond the notches in the horizontal member of the truss shown? All members are nominally 8 by 8 in in cross section. (See Table 10 in the Appendix for the actual size.) Assume the ultimate strength of wood in shear parallel to the grain to be 500 psi. Use a factor of safety of 5. (This detail is not recommended.)

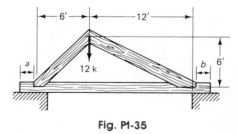

**Fig. P1-35**

**1-36.** A steel bar of 1 in diameter is loaded in double shear until failure; the ultimate load is found to be 100,000 lb. If the allowable stress is to be based on a safety factor of 3, what must be the diameter of a pin designed for an allowable load of 6000 lb in single shear?

**1-37.** What is the required diameter of pin *B* for the bell crank mechanism shown in the figure if an applied force of 60 kN at *A* is resisted by a force *P* at C? The allowable shear stress is 100 MPa.

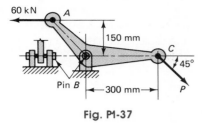

**Fig. P1-37**

**1-38.** A joint for transmitting a tensile force is to be made by means of a pin, as shown in the figure. If the diameter of the rods being connected is $D$, what should be the diameter $d$ of the pin? Assume that the allowable shear stress in the pin is one-half the maximum tensile stress in the rods. (In Section 8-16, it will be shown that this ratio for the allowable stresses is an excellent assumption for many materials.)

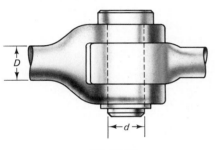

**Fig. P1-38**

**1-39.** Two high-strength steel rods of different diameters are attached at $A$ and $C$ and support a mass $M$ at $B$, as shown in the figure. What mass $M$ can be

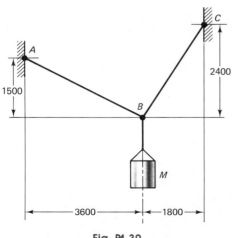

**Fig. P1-39**

supported? The ultimate strength of the rods is 800 MPa, and the factor of safety is to be 2. Rod $AB$ has $A = 200$ mm$^2$; rod $BC$ has $A = 400$ mm$^2$. (The ends of the wires in such applications require special attachments.)

**1-40.** Find the required cross-sectional areas for all tension members in Example 1-5. The allowable stress is 140 MPa.

**1-41.** A tower used for a highline is shown in the figure. If it is subjected to a horizontal force of 540 kN and the allowable stresses are 100 MPa in compression and 140 MPa in tension, what is the required cross-sectional area of each member? All members are pin-connected.

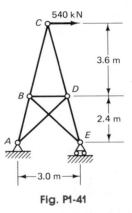

**Fig. P1-41**

**1-42.** For the frame shown for Problem 1-30, find the required cross-sectional areas for members $AB$, $AD$, and $BF$. The allowable stress in tension is 120 MPa and that in compression is 75 MPa.

***1-43.** A planar truss system has the dimensions shown in the figure. Member $AE$ is continuous and can

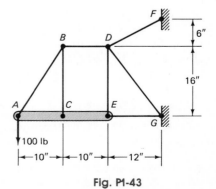

**Fig. P1-43**

resist bending. All joints are pinned. Determine the diameter required for tension member *AB* to carry the applied force at *A*. The allowable stress is 20 ksi.

**\*1-44.** A planar frame has the dimensions shown in the figure. Members *AC* and *DF* are continuous and can resist bending. All joints are pinned. Determine the diameter required of a high-strength steel rod for member *CD*. Assume that the ultimate strength for the rod is 1250 MPa and that the efficiency of the end attachments is 80%. The safety factor for the rod is 2.

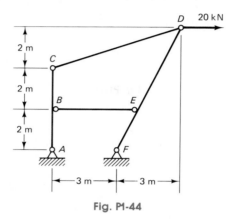

**Fig. P1-44**

**1-45.** To support a load $P = 180$ kN, determine the necessary diameter for rods *AB* and *AC* for the tripod shown in the figure. Neglect the weight of the structure and assume that the joints are pin-connected. No allowance has to be made for threads. The allowable tensile stress is 125 MPa. All dimensions are in meters.

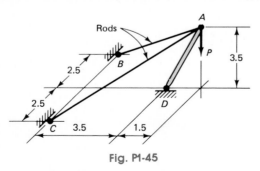

**Fig. P1-45**

**\*1-46.** A pin-connected frame for supporting a force *P* is shown in the figure. Stress $\sigma$ in both members *AB* and *BC* is to be the same. Determine the angle $\alpha$ necessary to achieve the minimum weight of construction. Members *AB* and *BC* have a constant cross section.

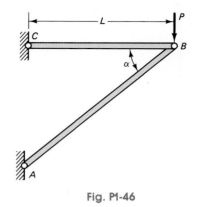

**Fig. P1-46**

**1-47.** Three equal 0.5-kg masses are attached to a 10-mm-diameter wire, as shown in the figure, and are rotated around a vertical axis, as shown in Fig. 1-25, on a frictionless plane at 4 Hz. Determine the axial stresses in the three segments of the wire and plot the results on a diagram as a function of *r*. Consider the masses to be concentrated as points.

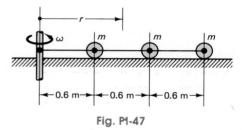

**Fig. P1-47**

**\*1-48.** A bar of constant cross-sectional area *A* is rotated around one of its ends in a horizontal plane with a constant angular velocity $\omega$. The unit weight of the material is $\gamma$. Determine the variation of the stress $\sigma$ along the bar and plot the result on a diagram as a function of *r*.

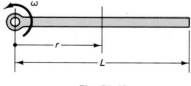

**Fig. P1-48**

## Section 1-12

**1-49.** Rework Example 1-7 for $D_n = L_n = 10$ kips. With the help of this additional solution, what conclusions can be drawn regarding the ASD and LRFD design approaches?

## Section 1-13

**1-50.** Find the capacity of tension member *AB* of the Fink truss shown in the figure if it is made from two 3 by 2 by $\frac{5}{16}$ in angles (see Table 7 in the Appendix) attached to a $\frac{3}{8}$-in-thick gusset plate by four $\frac{3}{4}$-in high-strength bolts in $\frac{13}{16}$-in diameter holes. The allowable stresses are 22 ksi in tension, 15 ksi in shear, and 87 ksi in bearing on the angles as well as the gusset.

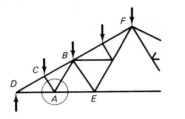

(a)

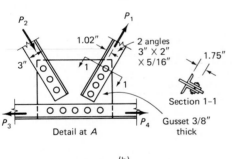

(b)

**Fig. P1-50**

✓ **1-51.** Find the capacity of a standard connection for a W 12 × 36 beam shown in the figure. The connection

consists of two 4 × 3$\frac{1}{2}$ × $\frac{3}{8}$ in angles, each 8$\frac{1}{2}$ in long; $\frac{7}{8}$ in high-strength bolts spaced 3 in apart are used in $\frac{15}{16}$-in holes. Use the allowable stresses given in Problem 1-50.

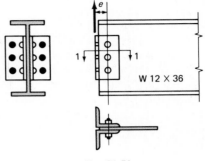

**Fig. P1-51**

✓ **1-52.** A structural multiple-riveted lap joint, such as is shown in the figure, is designed for a 42-kip load. The plates are $\frac{3}{8}$ in thick by 10 in wide. The rivets in $\frac{7}{8}$-in holes are $\frac{3}{4}$ in. (a) What is the shear stress in the middle rivet? (b) What are the tensile stresses in the *upper* plate in rows 1-1 and 2-2?

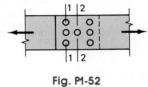

**Fig. P1-52**

## Section 1-14

**1-53.** Rework Example 1-8 for an 8 × 6 × $\frac{3}{4}$ in angle using $\frac{1}{2}$-in fillet welds.

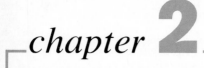

## chapter 2

# Axial Strains and Deformations in Bars

## 2-1. Introduction

This chapter is subdivided into two parts. In Part A, extensional strain for axially loaded members is defined and some typical experimental stress-strain relationships are illustrated for selected materials. Analytical idealizations for stress-strain behavior follow. These provide the basis for calculating deflections in axially loaded members. Statically determinate cases are considered first. Statically indeterminate situations encountered in axially loaded members are discussed in Part B.

## Part A    STRAINS AND DEFORMATIONS IN AXIALLY LOADED BARS

## 2-2. Normal Strain

A solid body subjected to a change of temperature or to an external load deforms. For example, while a specimen is being subjected to an increasing force $P$ as shown in Fig. 2-1, a change in length of the specimen occurs between any two points, such as $A$ and $B$. Initially, two such points can be selected an arbitrary distance apart. Thus, depending on the test, either 1-, 2-, 4-, or 8-in lengths are commonly used. This initial distance between the two points is called a *gage length*. In an experiment, the change in the length of this distance is measured. Mechanical dial gages, such as shown in Fig. 2-1, have been largely replaced by electronic extensometers for measuring these deformations. An example of a small clip-on extensometer is shown in Fig. 2-2.

During an experiment, the change in gage length is noted as a function of the applied force. With the same load and a longer gage length, a larger deformation is observed, than when the gage length is small. Therefore, it is more fundamental to refer to the observed deformation per unit of length of the gage, i.e., to the intensity of deformation.

If $L_o$ is the initial gage length and $L$ is the observed length under a given load, the gage elongation $\Delta L = L - L_o$. The elongation $\varepsilon$ per unit of initial gage length is then given as

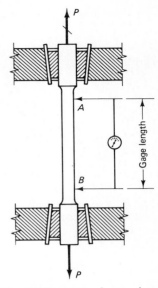

Fig. 2-1 Diagram of a tension specimen in a testing machine.

$$\varepsilon = \frac{L - L_o}{L_o} = \frac{\Delta L}{L_o} \tag{2-1}$$

This expression defines the _extensional strain_. Since this strain is associated with the normal stress, it is usually called the _normal strain_. It is a dimensionless quantity, but it is customary to refer to it as having the dimensions of in/in, m/m, or $\mu$m/m (microstrain). Sometimes it is given as a percentage. The quantity $\bar{\varepsilon}$ generally is very small. In most engineering applications of the type considered in this text, it is of the order of magnitude of 0.1 percent.

It is of interest to note that in some engineering applications, as, for example, in metal forming, the strains may be large. For such purposes, one defines the so-called _natural_ or _true strain_ $\bar{\varepsilon}$. The strain increment $d\varepsilon$ for this strain is defined as $dL/L$, where $L$ is the instantaneous length of the specimen, and $dL$ is the incremental change in length $L$. Analytically,

$$\bar{\varepsilon} = \int_{L_o}^{L} dL/L = \ln L/L_o = \ln(1 + \varepsilon) \tag{2-2}$$

For small strains, this definition essentially coincides with the conventional strain $\varepsilon$. If under the integral, the length $L$ is set equal to $L_o$, the strain definition given by Eq. 2-1 is obtained.

Natural strains are useful in theories of viscosity and viscoplasticity for expressing an instantaneous rate of deformation. Natural strains are not discussed elsewhere in this text.[1]

Fig. 2-2 Small clip-on extensometer (courtesy of MTS Systems Corporation).

Since the strains generally encountered are very small, it is possible to employ a highly versatile means for measuring them, using expendable electric strain gages. These are made of very fine wire or foil that is glued to the member being investigated. As the forces are applied to the member, elongation or contraction of the wires or foil takes place concurrently with similar changes in the material. These changes in length alter the electrical resistance of the gage, which can be measured and calibrated

[1] Natural strains were introduced by P. Ludwik in 1909. See A. Nadai, _Theory of Flow and Fracture of Solids_, Vol. 1, 2nd ed. (New York: McGraw-Hill, 1950), and L. E. Malven, _Introduction to the Mechanics of a Continuous Medium_ (Englewood Cliffs, NJ: Prentice-Hall, 1969).

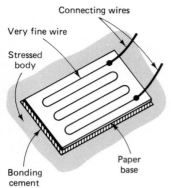

Fig. 2-3 Wire strain gage (protective top cover not shown).

to indicate the strain taking place. Such gages, suitable for different environmental conditions, are available in a range of lengths, varying from 4 to 150 mm (0.15 to 6 in). A schematic diagram of a wire gage is shown in Fig. 2-3, and a photograph of a typical small foil gage is shown in Fig. 2-4.[2]

## 2.3. Stress-Strain Relationships

In solid mechanics, the mechanical behavior of real materials under load is of primary importance. Experiments, mainly tension or compression tests, provide basic information on this behavior. In these experiments, macroscopic (overall) response of specimens to the applied loads is observed in order to determine empirical force-deformation relationships. Researchers in material science[3] attempt to provide reasons for the observed behavior.

It should be apparent from the previous discussion that for general purposes, it is more fundamental to report the strain of a member in tension or compression than to report the elongation of its gage. Similarly, stress is a more significant parameter than force since the effect on a material of an applied force $P$ depends primarily on the cross-sectional area of the member. As a consequence, in the experimental study of the mechanical properties of materials, it is customary to plot diagrams of the relationship between stress and strain in a particular test. Such diagrams, for most practical purposes, are assumed to be independent of the size of the specimen and of its gage length. In these diagrams, it is customary to use the ordinate scale for stress and the abscissa for strain.

Experimentally determined stress-strain diagrams differ widely for different materials. Even for the same material they differ depending on the temperature at which the test was conducted, the speed of the test, and a number of other variables. Conventional stress-strain diagrams for a few representative materials are illustrated in Figs. 2-5 and 2-6. These are shown to larger scale in Fig. 2-6, particularly for strain. Since for most

[2] See Society for Experimental Mechanics (SEM), A. S. Kobayashi (ed.), *Handbook on Experimental Mechanics* (Englewood Cliffs, NJ: Prentice-Hall, 1987).

[3] See, for example, references given on page 3.

Fig. 2-4 Typical single-element metal-foil electrical-resistance strain gage (courtesy of Micro-Measurements Division, Measurements Group, Inc., Raleigh, North Carolina, USA).

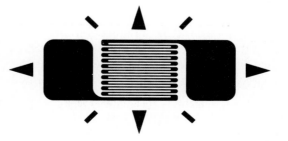

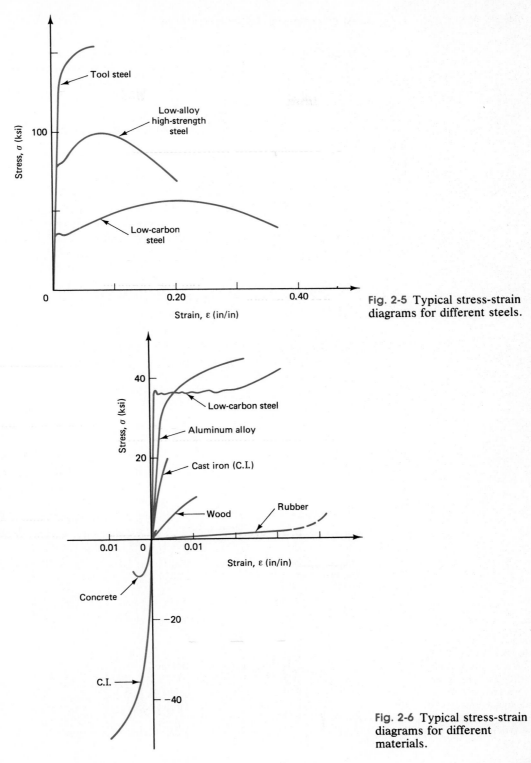

Fig. 2-5 Typical stress-strain
diagrams for different steels.

Fig. 2-6 Typical stress-strain
diagrams for different
materials.

**63**

engineering applications, deformations must be limited, the lower range of strains is particularly important. The large deformations of materials in the analysis of such operations as forging, forming, and drawing are not pursued.

An illustration of fractured tension specimens after static tension tests, i.e., where the loads were gradually applied, is shown in Fig. 2-7. Steel and aluminum alloy specimens exhibit ductile behavior, and a fracture occurs only after a considerable amount of deformation. This behavior is clearly exemplified in their respective stress-strain diagrams; see Fig. 2-6. These failures occur primarily due to slip in shear along the planes forming approximately 45° angles with the axis of the rod (see Fig. 1-8). A typical "cup and cone" fracture may be detected in the photographs of steel and aluminum alloy specimens. By contrast, the failure of a cast-iron specimen typically occurs very suddenly, exhibiting a square fracture across the cross section. Such cleavage or separation fractures are typical of brittle materials.

Several types of stress-strain diagrams may be identified from static tests at constant temperature. The curve shown in Fig. 2-8(a) is characteristic of mild steel, whereas the curves shown in Fig. 2-8(b) cover a wide range of diverse materials. The upper curve is representative of some brittle tool steels or concrete in tension, the middle one of aluminum alloys or plastics, and the lower curve of Fig. 2-8(b) is representative of rubber. However, the extreme values of strain that these materials can withstand

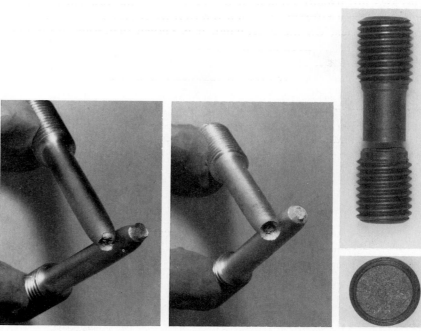

**Fig. 2-7** Ductile fractures for (a) A572 steel and (b) 6061-T6 aluminum alloy. Brittle fracture for (c) cast iron. (Numbers refer to ASTM designations for steel and that of Aluminum Association for aluminum alloy).

            (a)                           (b)                         (c)

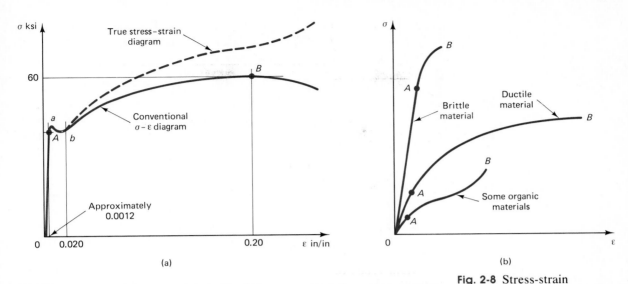

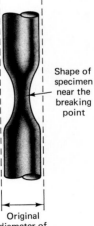

**Fig. 2-8** Stress-strain diagrams. (a) Mild steel. (b) Typical materials.

differ drastically. The "steepness" of these curves also varies greatly. Numerically, each material has its own characteristic curve. The terminal point on a stress-strain diagram represents the complete failure (rupture) of a specimen. Materials capable of withstanding large strains without a significant increase in stress are referred to as *ductile materials*. The converse applies to *brittle materials*.

Stresses are usually computed on the basis of the original area of a specimen[4]; such stresses are often referred to as *conventional* or *engineering* stresses. On the other hand, it is known that some transverse contraction or expansion of a material always takes place. For mild steel or aluminum, especially near the breaking point, this effect, referred to as *necking*, is particularly pronounced; see Fig. 2-9. Brittle materials do not exhibit it at usual temperatures, although they too contract transversely a little in a tension test and expand in a compression test. Dividing the applied force, at a given point in the test, by the corresponding actual area of a specimen at the same instant gives the so-called *true stress*. A plot of true stress vs. strain is called a *true stress-strain diagram*; see Fig. 2-8(a).

## 2.4. Hooke's Law

For a limited range from the origin, the experimental values of stress vs. strain lie essentially on a straight line. This holds true almost without reservations for the entire range for glass at room temperature. It is true

[4] These are referred to as Cauchy stresses, named in honor of the great French mathematician (1789–1857). Definition of stress recognizing the change in cross-sectional area during straining is associated with the names of Piola (1833), the Italian elastician, and Kirchhoff (1852), the renowned German physicist.

**Fig. 2-9** Typical contraction of a specimen of mild steel in tension near the breaking point.

for mild steel up to some point, as $A$ in Fig. 2-8(a). It holds nearly true up to very close to the failure point for many high-grade alloy steels. On the other hand, the straight part of the curve hardly exists in concrete, soil, annealed copper, aluminum, or cast iron. Nevertheless, for all practical purposes, up to some such point as $A$, also in Fig. 2-8(b), *the relationship between stress and strain may be said to be linear for all materials*. This sweeping idealization and generalization applicable to all materials is known as *Hooke's law*.[5] is Symbolically, this law can be expressed by the equation

$$\sigma = E\varepsilon \qquad (2\text{-}3)$$

which simply means that stress is directly proportional to strain, where the constant of proportionality is $E$. This constant $E$ is called the *elastic modulus,* modulus of elasticity, or Young's modulus.[6] As $\varepsilon$ is dimensionless, $E$ has the units of stress in this relation. In the U.S. customary system of units, it is usually measured in pounds per square inch, and in the SI units, it is measured in newtons per square meter (or pascals).

Graphically, $E$ is interpreted as the slope of a straight line from the origin to the rather vague point $A$ on a uniaxial stress-strain diagram. The stress corresponding to the latter point is termed the *proportional or elastic limit* of the material. Physically, the elastic modulus represents the stiffness of the material to an imposed load. *The value of the elastic modulus is a definite property of a material.* From experiments, it is known that $\varepsilon$ is *always a very small quantity*; hence, $E$ must be large. Its approximate values are tabulated for a few materials in Tables 1A and B of the Appendix. For all steels, $E$ at room temperature is between 29 and $30 \times 10^6$ psi, or 200 and 207 GPa.

It follows from the foregoing discussion that *Hooke's law applies only up to the proportional limit of the material*. This is highly significant as in most of the subsequent treatment, the derived formulas are based on this law. Clearly, then, such formulas are limited to the material's behavior in the lower range of stresses.

Some materials, notably single crystals and wood, possess different elastic moduli in different directions. Such materials, having different physical properties in different directions, are called *anisotropic*. A consideration of such materials is *excluded* from this text. The vast majority of engineering materials consist of a large number of *randomly* oriented

[5] Actually, Robert Hooke, an English scientist, worked with springs and not with rods. In 1676, he announced an anagram "c e i i i n o s s s t t u v," which in Latin is *Ut Tensio sic Vis* (the force varies as the stretch).

[6] Young's modulus is so called in honor of Thomas Young, the English scientist. His *Lectures on Natural Philosophy,* published in 1807, contain a definition of the modulus of elasticity.

crystals. Because of this random orientation, properties of materials become essentially alike in any direction.[7] Such materials are called *isotropic*. With some exceptions, such as wood, in this text, *complete homogeneity* (sameness from point to point) *and isotropy of materials is generally assumed.*

## 2.5. Further Remarks on Stress-Strain Relationships

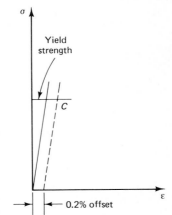

Fig. 2-10 Offset method of determining the yield strength of a material.

In addition to the proportional limit defined in Section 2-4, several other interesting points can be observed on the stress-strain diagrams. For instance, the highest points (*B* in Fig. 2-8) correspond to the *ultimate* strength of a material. *Stress* associated with the long plateau *ab* in Fig. 2-8(a) is called the *yield strength* of a material. As will be brought out later, this remarkable property of mild steel, in common with other *ductile* materials, is significant in stress analysis. For the present, note that at an essentially constant stress, strains 15 to 20 times those that take place up to the proportional limit occur during yielding. At the yield stress, a large amount of deformation takes place at a constant stress. The yielding phenomenon is absent in most materials.

A study of stress-strain diagrams shows that the yield strength (stress) is so near the proportional limit that, for most purposes, the two may be taken to be the same. However, it is much easier to locate the former. For materials that do not possess a well-defined yield strength, one is sometimes "invented" by the use of the so-called "offset method." This is illustrated in Fig. 2-10, where a line offset an *arbitrary* amount of 0.2 percent of strain is drawn parallel to the straight-line portion of the initial stress-strain diagram. Point *C* is then taken as the *yield strength* of the material at 0.2-percent offset.

That a material is elastic usually implies that stress is directly proportional to strain, as in Hooke's law. Such materials are *linearly elastic* or Hookean. A material responding in a nonlinear manner and yet, when unloaded, returning back along the loading path to its initial stress-free state of deformation is also an elastic material. Such materials are called *nonlinearly elastic*. The difference between the two types of elastic materials is highlighted in Figs. 2-11(a) and (b). If in stressing a material its elastic limit is exceeded, on unloading it usually responds approximately in a linearly elastic manner, as shown in Fig. 2-11(c), and a permanent deformation, or set, develops at no external load. As will become apparent after the study of Section 2-11, the area enclosed by the loop corresponds to dissipated energy released through heat. Ideal elastic materials are considered not to dissipate any energy under monotonic or cyclic loading.

For ductile materials, stress-strain diagrams obtained for short compressions blocks are reasonably close to those found in tension. Brittle

---

[7] Rolling operations produce preferential orientation of crystalline grains in some materials.

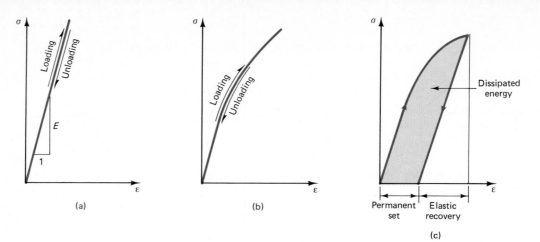

Fig. 2-11 Stress-strain diagrams: (a) linear elastic material, (b) nonlinear elastic material, and (c) inelastic or plastic material.

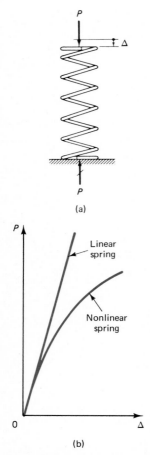

Fig. 2-12 Linear (Hookean) and nonlinear spring response.

materials, such as cast iron and concrete are very weak in tension but not in compression. For these materials, the diagrams differ considerably, depending on the sense of the applied force.

It is well to note that in some of the subsequent analyses, it will be advantageous to refer to elastic bodies and systems as springs. Sketches such as shown in Fig. 2-12 are frequently used in practice for interpreting the physical behavior of mechanical systems.

## 2.6. Other Idealizations of Constitutive Relations

In an increasingly larger number of technical problems, stress analyses based on the assumption of linearly elastic behavior are insufficient. For this reason, several additional stress-strain relations are now in general use. Such relations are frequently referred to as *constitutive relations* or *laws*. The three idealized stress-strain relations shown in Fig. 2-13 are encountered particularly often. The two shown in Figs. 2-13(a) and (b) will be used in this text; the one in Fig. 2-13(c) is often more realistic, however, its use is considerably more complicated and generally will be avoided because of the introductory nature of this book.

The idealized $\sigma$–$\varepsilon$ relationship shown in Fig. 2-13(a) is applicable to problems in which the elastic strains can be neglected in relation to the plastic ones. This occurs if plastic (inelastic) strains are dominant. Perfectly (ideally) plastic behavior means that a large amount of unbounded deformation can take place at a constant stress. The idealization shown in Fig. 2-13(b) is particularly useful if both the elastic and plastic strains have to be included. This situation frequently arises in analysis. Both of the previous idealizations are patterned after the behavior of low-carbon steel (see Figs. 2-6 and 2-8), where at the yield stress $\sigma_{yp}$, a substantial plateau in the stress-strain diagram is generally observed. In both instances, it is assumed that the mechanical properties of the material are the same in tension and in compression, and $\sigma_{yp} = | -\sigma_{yp} |$. It is also assumed that during unloading, the material behaves elastically. In such

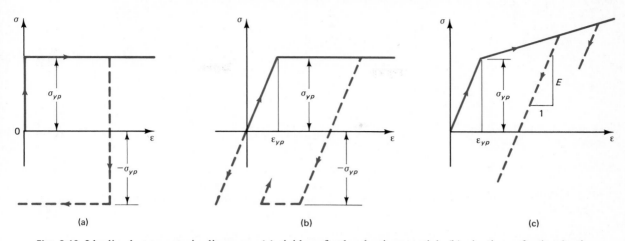

**Fig. 2-13** Idealized stress-strain diagrams: (a) rigid-perfectly plastic material; (b) elastic-perfectly plastic material; and (c) elastic-linearly hardening material.

a case, a stress can range and terminate anywhere between $+\sigma_{yp}$ and $-\sigma_{yp}$. For moderate amounts of plastic straining, this assumption is in good agreement with experimental observations.

The $\sigma$–$\varepsilon$ idealization shown in Fig. 2-13(c) provides a reasonable approximation for many materials and is more accurate than the previous models over a wider range of strain. Beyond the elastic range, on an increase in strain, many materials resist additional stress, a phenomenon referred to as *strain hardening*.

In some refined analyses, the stress-strain idealization shown in Fig. 2-13 may not be sufficiently accurate. Fortunately, with the use of computers, much better modeling of constitutive relations for real material is possible. For completeness, one such well-known algebraic formulation follows. In as much as implementation of such formulations requires a considerable amount of computer programming, this approach is not intended for general use in this text.

An equation capable of representing a wide range of stress-strain curves has been developed by Ramberg and Osgood.[8] This equation[9] is

$$\frac{\varepsilon}{\varepsilon_o} = \frac{\sigma}{\sigma_o} + \frac{3}{7}\left(\frac{\sigma}{\sigma_o}\right)^n \tag{2-4}$$

where $\varepsilon_o$, $\sigma_o$, and $n$ are characteristic constants for a material. The constants $\varepsilon_o$ and $\sigma_o$ correspond to the yield point, which, for all cases other

[8] W. Ramberg and W. R. Osgood, *Description of Stress-Strain Curves by Three Parameters,* National Advisory Committee on Aeronautics, TN 902, 1943.
[9] The coefficient 3/7 is chosen somewhat arbitrarily; different values have been used in some investigations. In this formulation, a discontinuity in the function arises when $n = \infty$.

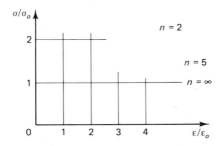

**Fig. 2-14** Ramberg-Osgood stress-strain diagrams.

than that of ideal plasticity, is found by the offset method (see Fig. 2-10). The exponent $n$ determines the shape of the curve, Fig. 2-14. Note that Eq. 2-4 is written in dimensionless form, a convenient scheme in general analysis. One of the important advantages of Eq. 2-4 is that it is a continuous mathematical function. For example, an *instantaneous* or *tangent modulus $E_t$* defined as

$$E_t = \frac{d\sigma}{d\varepsilon} \qquad (2\text{-}5)$$

can be uniquely determined.

In most applications, it is advantageous to work with the inverse of Eq. 2-4, i.e., to express stress as a function of strain. With the aid of such an equation, developed by Menegotto and Pinto,[10] remarkably accurate simulations of cyclic stress-strain diagrams can be obtained. An example is shown in Fig. 2-15.[11] In this diagram, a series of characteristic loops, referred to as hysteretic loops since they represent the dissipation of energy (see Section 2-11), are clearly evident.

Regardless of the idealization used for a stress-strain diagram, it must be recognized that it is strongly dependent on ambient temperature. An example of such an effect is illustrated in Fig. 2-16.[12] It is also important to be aware of the fact that no time-dependent phenomena in the behavior of materials is considered in this text. For example, with time-dependent

[10] See M. Manegotto, and P. E. Pinto, "Method of Analysis for Cyclically Loaded Reinforced Concrete Plane Frames Including Changes in Geometry and Nonelastic Behavior of Elements under Combined Normal Force and Bending" in IABSE (International Association for Bridge and Structural Engineering) Symposium on Resistance and Ultimate Deformability of Structures Acted on by Well-Defined Repeated Loads, Lisbon, 1973.

[11] F. C. Filippou, E. P. Popov, and V. V. Bertero, "Effects of Bond Deterioration on Hysteric Behavior of Reinforced Concrete Joint." Report No. UCB/EERC-83/19, August, 1983, p. 119.

[12] K. G. Hoge, "Influence of Strain Rate on Mechanical Properties of 6061-T6 Aluminum Under Uniaxial and Biaxial States of Stress, " *Experimental Mechanics*, **6**, no. 10 (April 1966), p. 204.

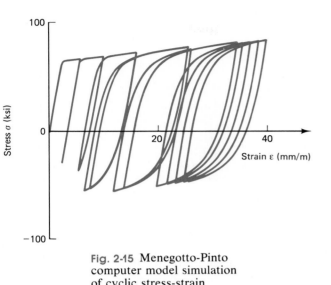

**Fig. 2-15** Menegotto-Pinto computer model simulation of cyclic stress-strain diagrams for steel.

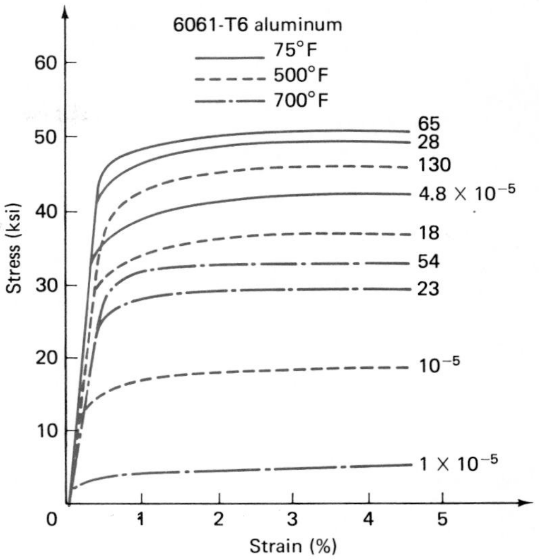

**Fig. 2-16** Effect of strain rate and temperature on stress-strain curves for 6061-T6 aluminum alloy.

behavior and a member subjected to a constant stress, the elongations or deflections continue to increase with time; see Fig. 2-17. This phenomenon is referred to as _creep_. Creep is observed in reinforced concrete floors and in turbine discs, for example. Likewise, the prestress in bolts of mechanical assemblies operating at high temperatures, as well as prestress in steel tendons in reinforced concrete, tend to decrease gradually with time. This phenomenon is referred to as _relaxation_; see Fig. 2-18.

## 2-7. Deformation of Axially Loaded Bars

When the deflection of an axially loaded member is a design parameter, it is necessary to determine the deformations. Axial deformations are also required in the analysis of statically indeterminate bars. The deflection

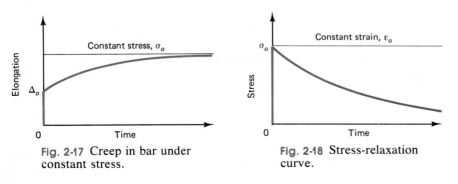

**Fig. 2-17** Creep in bar under constant stress.

**Fig. 2-18** Stress-relaxation curve.

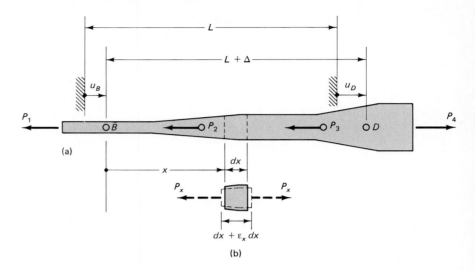

**Fig. 2-19** An axially loaded bar

characteristics of bars also provide necessary information for determining the stiffness of systems in mechanical vibration analysis.

Consider the axially loaded bar shown in Fig. 2-19(a) for deriving a relation for axial bar deformation. The applied forces $P_1$, $P_2$, and $P_3$ are held in equilibrium by the force $P_4$. The cross-sectional area $A$ of the bar is permitted to gradually change. The change in length that takes place in the bar between points $B$ and $D$ due to the applied forces is to be determined.

In order to formulate the relation, Eq. 2-1 for the normal strain is recast for a differential element $dx$. Thus the normal strain $\varepsilon_x$ in the $x$ direction is

$$\varepsilon_x = \frac{du}{dx} \tag{2-6}$$

where, due to the applied forces, $u$ is the absolute displacement of a point on a bar from an initial fixed location in space, and $du$ is the axial deformation of the infinitesimal element. This is the governing differential equation for axially loaded bars.

It is to be noted that the deformations considered in this text are generally *very small* (infinitesimal). This should become apparent from numerical examples throughout this text. Therefore in calculations the *initial* (undeformed) dimensions of members can be used for calculating deformations. In the following derivation this permits the use of the initial length $L$, between points such as $B$ and $D$ in Fig. 2-19, rather than its deformed length.

Rearranging Eq. 2-6 as $du = \varepsilon_x \, dx$, assuming the origin of $x$ at $B$, and integrating,

$$\int_0^L du = u(L) - u(0) = \int_0^L \varepsilon_x \, dx$$

where $u(L) = u_D$ and $u(0) = u_B$ are the absolute or global displacements of points $D$ and $B$, respectively. As can be seen from the figure, $u(0)$ is a *rigid body axial translation* of the bar. The difference between these displacements is the change in length $\Delta$ between points $D$ and $B$. Hence

$$\Delta = \int_0^L \varepsilon_x \, dx \qquad (2\text{-}7)$$

Any appropriate constitutive relations can be used to define $\varepsilon_x$.

For linearly elastic materials, according to Hooke's law, $\varepsilon_x = \sigma_x/E$, Eq. 2-3, where $\sigma_x = P_x/A_x$, Eq. 1-13. By substituting these relations into Eq. 2-7 and simplifying,

$$\Delta = \int_0^L \frac{P_x \, dx}{A_x E_x} \qquad (2\text{-}8)$$

where $\Delta$ is the change in length of an *elastic* bar of length $L$, and the force $P_x = P(x)$, the cross-sectional area $A_x = A(x)$, and the elastic modulus $E_x = E(x)$ can vary along the length of a bar.

**Procedure Summary**

It should be emphasized that the central theme in engineering mechanics of solids consists of repeatedly applying *three basic concepts*. In developing the theory for axially loaded bars these basic concepts can be summarized as follows:

1. *Equilibrium conditions* are used for determining the internal resisting forces at a section, first introduced in Chapter 1. As shown later in this chapter, this may require solution of a statically indeterminate problem.
2. *Geometry of deformation* is used in deriving the change in length of a bar due to axial forces by assuming that sections initially perpendicular to the axis of a bar remain perpendicular after straining, see Fig. 2-19(b).
3. *Material properties* (constitutive relations) are used in relating axial normal stresses to axial normal strain and permit calculation of axial deformations between sections.

Solutions based on this theory give correct *average* stresses at a section, see Section 1-6. However, at concentrated forces and abrupt changes in cross section irregular *local* stresses (and strains) arise. Only at distances about equal to the depth of the member from such disturbances are the stresses and strains in agreement with the developed theory. Therefore solutions based on the concepts of engineering mechanics of solids are best suited for relatively slender members. The use of this simplified procedure is rationalized in Section 2-10 as *Saint-Venant's principle*.

Several examples showing application of Eq. 2-8 follow.

### EXAMPLE 2-1

Consider bar *BC* of constant cross-sectional area *A* and of length *L* shown in Fig. 2-20(a). Determine the deflection of the free end, caused by the application of a concentrated force *P*. The elastic modulus of the material is *E*.

### Solution

The deformed bar is shown in Fig. 2-20(b). Conceptually, it is often convenient to think of such elastic systems as springs; see Fig. 2-20(e).

A free-body diagram for an isolated part of the loaded bar to the left of an arbitrary section *a–a* is shown in Fig. 2-20(c). From this diagram, it can be concluded that the axial force $P_x$ is the same everywhere and is equal to *P*. It is given that $A_x = A$, a constant. By applying Eq. 2-8,

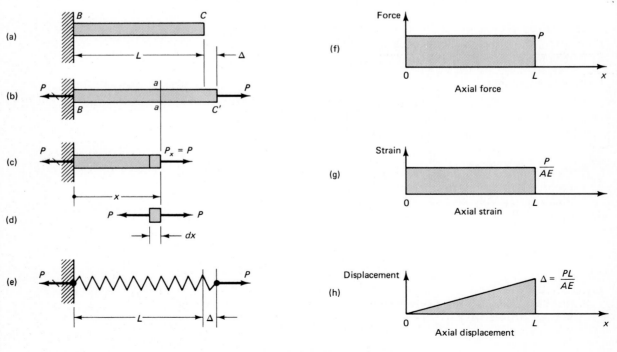

Fig. 2-20

$$\Delta = \int_A^B \frac{P_x \, dx}{A_x E} = \frac{P}{AE} \int_0^L dx = \frac{P}{AE} \left. x \right|_0^L = \frac{PL}{AE}$$

Hence,
$$\Delta = \frac{PL}{AE} \qquad\qquad (2\text{-}9)$$

A graphic interpretation of the solution is shown in Figs. 2-20(f)–(h). The constant axial bar strain follows by dividing the constant axial force $P$ by $AE$. Since the axial strain is constant, the displacements of the points on the bar increase directly with the distance from the origin of $x$ at a constant rate. No displacement is possible at the left end.

It is seen from Eq. 2-9 that the deflection of the rod is directly proportional to the applied force and the length and is inversely proportional to $A$ and $E$.

Since Eq. 2-9 frequently occurs in practice, it is meaningful to recast it into the following form:

$$P = (AE/L)\Delta \qquad\qquad (2\text{-}10)$$

This equation is related to the familiar definition for the *spring constant* or *stiffness k* reading

$$k = P/\Delta \qquad [\text{lb/in}] \text{ or } [\text{N/m}] \qquad\qquad (2\text{-}11)$$

This constant represents the force required to produce a unit deflection, i.e., $\Delta = 1$. Therefore, for an axially loaded $i$th bar or bar segment of length $L_i$ and constant cross section,

$$k_i = \frac{A_i E_i}{L_i} \qquad\qquad (2\text{-}12)$$

and the analogy between such a bar and a spring shown in Fig. 2-20(e) is evident. The reciprocal of $k$ defines the *flexibility f*, i.e.,

$$f = 1/k = \Delta/P \qquad [\text{in/lb}] \text{ or } [\text{N/m}] \qquad\qquad (2\text{-}13)$$

The constant $f$ represents the deflection resulting from the application of a unit force, i.e., $P = 1$.

For the particular case of an axially loaded $i$th bar of constant cross section,

$$f_i = \frac{L_i}{A_i E_i} \qquad\qquad (2\text{-}14)$$

The concepts of structural stiffness and flexibility are widely used in structural analysis, including mechanical-vibration problems. For more complex structural systems, the expressions for $k$ and $f$ become more involved.

### EXAMPLE 2-2

Determine the relative displacement of point $D$ from $O$ for the elastic steel bar of variable cross section shown in Fig. 2-21(a) caused by the application of concentrated forces $P_1 = 100$ kN and $P_3 = 200$ kN acting to the left, and $P_2 = 250$ kN and $P_4 = 50$ kN acting to the right. The respective areas for bar segments $OB$, $BC$, and $CD$ are 1000, 2000, and 1000 mm$^2$. Let $E = 200$ GPa.

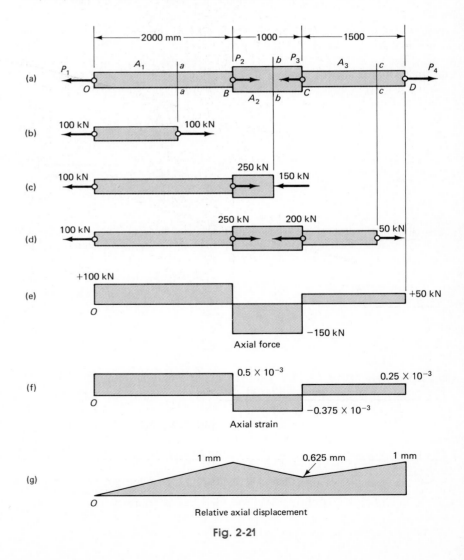

Fig. 2-21

## Solution

By inspection, it can be seen that the bar is in equilibrium. *Such a check must always be made before starting a problem.* The variation in $P_x$ along the length of the bar is determined by taking three sections, *a–a*, *b–b*, and *c–c* in Fig. 2-21(a) and determining the necessary forces for equilibrium in the free-body diagrams in Figs. 2-21(b)–(d). This leads to the conclusion that *within each bar segment,* the forces are *constant,* resulting in the axial force diagram shown in Fig. 2-21(e). Therefore, the solution of the deformation problem consists of adding algebraically the individual deformations for the three segments. Equation 2-9 is applicable for *each* segment. Hence, the total axial deformation for the bar can be written as

$$\Delta = \sum_i \frac{P_i L_i}{A_i E} = \frac{P_{OB} L_{OB}}{A_{OB} E} + \frac{P_{BC} L_{BC}}{A_{BC} E} + \frac{P_{CD} L_{CD}}{A_{CD} EK}$$

where the subscripts identify the segments.

Using this relation, the relative displacement between $O$ and $D$ is

$$\Delta = + \frac{100 \times 10^3 \times 2000}{1000 \times 200 \times 10^3} - \frac{150 \times 10^3 \times 1000}{2000 \times 200 \times 10^3} + \frac{50 \times 10^3 \times 1500}{1000 \times 200 \times 10^3}$$
$$= +1.000 - 0.375 + 0.375 = +1.000 \text{ mm}$$

Note that in spite of large stresses in the bar, the elongation is very small.

A graphic interpretation of the solution is shown in Figs. 2-21(f) and (g). By dividing the axial forces in the bar segments by the corresponding $AE$, the axial strains along the bar are obtained. These strains are constant within each bar segment. The area of the strain diagram for each segment of the bar gives the change in length for that segment. These values correspond to those displayed numerically before.

## EXAMPLE 2-3

Determine the deflection of free end $B$ of elastic bar $OB$ caused by its own weight $w$ lb/in; see Fig. 2-22. The constant cross-sectional area is $A$. Assume that $E$ is given.

## Solution

The free-body diagrams of the bar and its truncated segment are shown, respectively, in Figs. 2-22(a) and (b). These two steps are essential in the solution of such problems. The graph for the axial force $P_x = w(L - x)$ is in Fig. 2-22(c). By applying Eq. 2-8, the change in bar length $\Delta(x)$ at a generic point $x$,

$$\Delta(x) = \int_0^x \frac{P_x \, dx}{A_x E} = \frac{1}{AE} \int w(L - x) \, dx = \frac{w}{AE}\left(Lx - \frac{x^2}{2}\right)$$

A plot of this function is shown in Fig. 2-22(d), with its maximum as $B$.

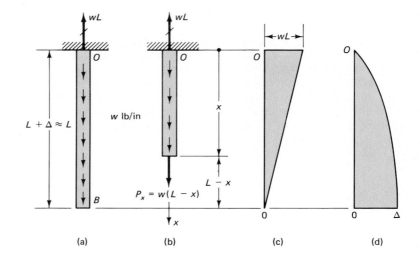

**Fig. 2-22**          (a)          (b)          (c)          (d)

The deflection of $B$ is

$$\Delta = \Delta(L) = \frac{w}{AE}\left(L^2 - \frac{L^2}{2}\right) = \frac{wL^2}{2AE} = \frac{WL}{2AE}$$

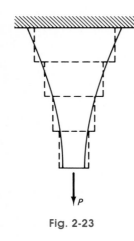

where $W = wL$ is the *total* weight of the bar.

If a concentrated force $P$, in *addition* to the bar's own weight, were acting on bar $OB$ at end $B$, the total deflection due to the *two causes* would be obtained by *superposition* as

$$\Delta = \frac{PL}{AE} + \frac{WL}{2AE} = \frac{[P + (W/2)]L}{AE}$$

In problems where the area of a rod is variable, a proper *function* for it must be substituted into Eq. 2-8 to determine deflections. In practice, it is sometimes sufficiently accurate to analyze such problems by approximating the shape of a rod by a *finite number* of elements, as shown in Fig. 2-23. The deflections for each one of these elements are added to obtain the total deflection. Because of the rapid variation in the cross section shown, the solution would be approximate.

**Fig. 2-23**

---

## EXAMPLE 2-4

For the bracket analyzed for stresses in Example 1-3, determine the deflection of point $B$ caused by the applied vertical force $P = 3$ kips. Also determine the vertical stiffness of the bracket at $B$. Assume that the members are made of 2024-T4 aluminum alloy and that they have constant cross-sectional areas, i.e., neglect the enlargements at the connections. See idealization in Fig. 2-24(a).

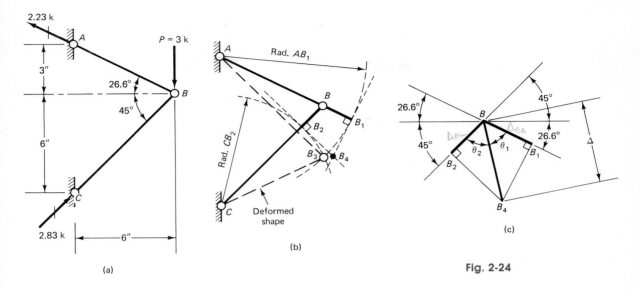

Fig. 2-24

## Solution

As found in Example 1-3, the axial forces in the bars of the bracket are $\sigma_{AB} = $ 17.8 ksi and $\sigma_{BC} = 12.9$ ksi. The length of member $AB$ is 6.71 in and that of $BC$ is 8.49 in. Per Table 1A in the Appendix, for the specified material, $E = 10.6 \times 10^3$ ksi. Therefore, according to Eq. 2-9, the individual member length changes are

$$\Delta_{AB} = \left[\frac{PL}{AE}\right]_{AB} = \left[\sigma \frac{L}{E}\right]_{AB} = \frac{17.8 \times 6.71}{10.6 \times 10^3} = 11.3 \times 10^{-3} \text{ in}$$

$$\text{(elongation)}$$

$$\Delta_{BC} = -\frac{12.9 \times 8.29}{10.6 \times 10^3} = -10.3 \times 10^{-3} \text{ in} \quad \text{(contraction)}$$

These length changes, as $BB_1$ and $BB_2$, are shown to a greatly exaggerated scale in relation to the bar lengths in Fig. 2-24(b). The indicated locations of points $B_1$ and $B_2$ are incompatible with the physical requirements of the problem. Therefore, elongated bar $AB_1$ and shortened bar $CB_2$ must be rotated around their respective support points $A$ and $C$ such that points $B_1$ and $B_2$ meet at common point $B_3$. This is shown schematically in Fig. 2-24(b). However, since in classical solid mechanics, one deals with small (infinitesimal) deformations, an approximation can be introduced. In such analyses, it is customary to assume that short arcs of large circles can be approximated by normals to the members along which the bar ends move to achieve compatibility at the joints. This construction[13] is indicated in Fig. 2-24(b), locating point $B_4$. An enlarged detail of the changes in bar lengths and this approach for locating point $B_4$ is shown in Fig. 2-24(c). The

[13] First introduced by M. Williot in 1877.

required numerical results can be obtained either graphically or by using trigo-nometry. Here the latter procedure is followed.

If $\Delta$ is the deflection or displacement of point $B$ to position $B_4$, Fig. 2-24(c), and changes in bar lengths $\Delta_{BC} = BB_2$ and $\Delta_{AB} = BB_1$,

$$\Delta_{BC} = \Delta \cos \theta_2 \quad \text{and} \quad \Delta_{AB} = \Delta \cos \theta_1$$

On forming equal ratios for both sides of these equations, substituting the numerical values for $\Delta_{BC}$ and $\Delta_{AB}$ found earlier, and simplifying, one obtains

$$\frac{\cos \theta_2}{\cos \theta_1} = \frac{\Delta_{BC}}{\Delta_{AB}} = \frac{10.3 \times 10^{-3}}{11.3 \times 10^{-3}} = 0.912$$

However, since

$$\theta_2 = 180° - 45° - 26.6° - \theta_1 = 108.4° - \theta_1$$

it follows that

$$\cos \theta_2 = \cos 108.4° \cos \theta_1 + \sin 108.4° \sin \theta_1$$

and

$$\frac{\cos \theta_2}{\cos \theta_1} = \cos 108.4° + \sin 108.4° \tan \theta_1 = 0.912$$

Therefore,

$$\tan \theta_1 = 1.29 \quad \text{and} \quad \theta_1 = 52.2°$$

Based on this result,

$$\Delta = \Delta_{AB}/\cos \theta_1 = 18.4 \times 10^{-3} \text{ in}$$

forming an angle of 11.2° with the vertical.

Since $\Delta_{\text{vert}} = \Delta \cos 11.2° = 18.0 \times 10^{-3}$ in, the vertical stiffness of the bracket is given by the spring constant

$$k = \frac{P}{\Delta_{\text{vert}}} = \frac{3}{18.0 \times 10^{-3}} = 167 \text{ kips/in}$$

This problem contains geometric nonlinearity in displacement, which has been neglected; therefore, the solution is accurate only for small deformations, a common practice for many engineering problems.

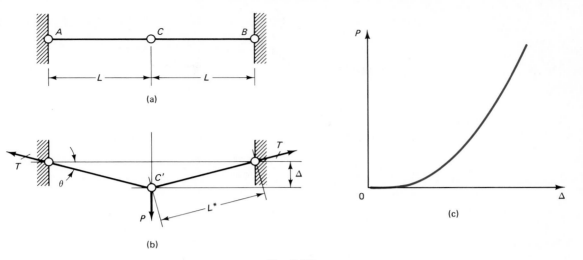

**Fig. 2-25**

## **EXAMPLE 2-5

Two hinge-ended elastic bars of equal lengths and cross-sectional areas attached to immovable supports are joined in the middle by a pin, as shown in Fig. 2-25(a). Initially, points $A$, $B$, and $C$ are on a straight line. Determine the vertical deflection $\Delta$ of point $C$ as a function of applied force $P$. Consider small deflections only.

### Solution

The given structural system is incapable of supporting any vertical force in its initial configuration. Therefore, equilibrium of the system in a slightly deflected condition must be examined, Fig. 2-25(b), where initial bar lengths $L$ become $L^*$. For this position of the bars, one can write an equation of equilibrium for joint $C'$ and express elongations of the bars via two different paths. One such relation for elongation of each bar follows from Eq. 2-9 and the other from purely geometric considerations. On these bases, from equilibrium,

$$P = 2T \sin \theta$$

and

$$\frac{TL^*}{AE} = L^* - L = L^* - L^* \cos \theta$$

Hence,

$$T = AE(1 - \cos \theta)$$

On substituting this expression for $T$ into the first equation,

$$P = 2AE(1 - \cos \theta) \sin \theta$$

Further, by expanding cos θ and sin θ into Taylor's series,

$$P = 2AE \left( \frac{\theta^2}{2!} - \frac{\theta^4}{4!} + \cdots \right) \left( \theta - \frac{\theta^3}{3!} + \cdots \right)$$

On retaining only one term in each series,

$$P \approx AE\theta^3$$

However, since the analysis is being made for small deflections, angle $\theta \approx \Delta/L$. Therefore,

$$P \approx \frac{AE}{L^3} \Delta^3 \qquad \text{or} \qquad \Delta = L \sqrt[3]{\frac{P}{AE}} \qquad (2\text{-}15)$$

This result, shown qualitatively in Fig. 2-25(c), clearly exhibits the highly nonlinear relationship between $P$ and $\Delta$. By contrast, most of the problems that will be encountered in this text will lead to linear relationships between loads and displacements. The more accurate solutions of this problem show that the approximate solution just obtained gives good results for $\Delta/L$, on the order 0.3.

In this problem, the effect of geometry change on equilibrium was considered, whereas in Example 2-4, it was neglected because the displacement was very small.

## 2-8. Poisson's Ratio

In addition to the deformation of materials in the direction of the applied force, another remarkable property can be observed in all solid materials, namely, that at right angles to the applied uniaxial force, a certain amount of lateral (transverse) expansion or contraction takes place. This phenomenon is illustrated in Fig. 2-26, where the deformations are *greatly exaggerated*. For clarity, this physical fact may be restated thus: if a solid body is subjected to an axial tension, it contracts laterally; on the other hand, if it is compressed, the material "squashes out" sideways. With this in mind, directions of lateral deformations are easily determined, depending on the sense of the applied force.

For a general theory, it is preferable to refer to these lateral deformations on the basis of deformations per *unit* of length of the transverse dimension. Thus, the lateral deformations on a *relative* basis can be expressed in in/in or m/m. These relative unit lateral deformations are termed *lateral strains*. Moreover, it is known from experiments that lateral strains bear a *constant* relationship to the longitudinal or axial strains caused by an axial force, provided a material remains *elastic* and is homogeneous and isotropic. This constant is a definite property of a material, just like the elastic modulus $E$, and is called *Poisson's ratio*.[14] It will be denoted by $\nu$ (nu) and is defined as follows:

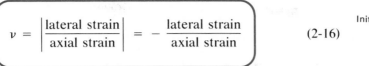

$$\nu = \left| \frac{\text{lateral strain}}{\text{axial strain}} \right| = - \frac{\text{lateral strain}}{\text{axial strain}} \qquad (2\text{-}16)$$

*where the axial strains are caused by uniaxial stress only,* i.e., by simple tension or compression. The second, alternative form of Eq. 2-16 is true because the lateral and axial strains are always of opposite sign for uniaxial stress.

The value of $\nu$ fluctuates for different materials over a relatively narrow range. Generally, it is on the order of 0.25 to 0.35. In extreme cases, values as low as 0.1 (some concretes) and as high as 0.5 (rubber) occur. The latter value is the *largest possible*. It is normally attained by materials during plastic flow and signifies constancy of volume.[15] In this text, Poisson's ratio will be used only when materials behave elastically.

In conclusion, note that the Poisson effect exhibited by materials causes *no additional stresses* other than those considered earlier *unless the transverse deformation is inhibited or prevented*.

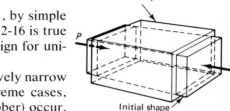

(a)

(b)

**Fig. 2-26** (a) Lateral contraction and (b) lateral expansion of solid bodies subjected to axial forces (Poisson's effect).

## EXAMPLE 2-6

Consider a carefully conducted experiment where an aluminum bar of 50-mm diameter is stressed in a testing machine, as shown in Fig. 2-27. At a certain instant the applied force $P$ is 100 kN, while the measured elongation of the rod is 0.219 mm in a 300-mm gage length, and the diameter's dimension is decreased by 0.01215 mm. Calculate the two physical constants $\nu$ and $E$ of the material.

### Solution

*Transverse or lateral strain:*

$$\varepsilon_t = \frac{\Delta_t}{D} = - \frac{0.01215}{50} = -0.000243 \text{ mm/mm}$$

In this case, the lateral strain $\varepsilon_t$ is negative, since the diameter of the bar *decreases* by $\Delta_t$.

[14] Named after S. D. Poisson, the French scientist who formulated this concept in 1828.
[15] A. Nadai, *Theory of Flow and Fracture of Solids*, Vol. 1 (New York: McGraw-Hill, 1950).

$L = 300$ mm

$D = 50$ mm

**Fig. 2-27**

*Axial strain:*

$$\varepsilon_a = \frac{\Delta}{L} = + \frac{0.219}{300} = 0.00073 \text{ mm/mm}$$

*Poisson's ratio:*

$$\nu = -\frac{\varepsilon_t}{\varepsilon_a} = -\frac{(-0.000243)}{0.00073} = 0.333$$

Next, since the area of the rod $A = \frac{1}{4}\pi \times 50^2 = 1960 \text{ mm}^2$, from Eq. 2-9,

$$E = \frac{PL}{A\,\Delta} = \frac{100 \times 10^3 \times 300}{1960 \times 0.219} = 70 \times 10^3 \text{ N/mm}^2 = 70 \text{ GPa}$$

In practice, when a study of physical quantities, such as $E$ and $\nu$, is being made, it is best to work with the corresponding stress-strain diagram to be assured that the quantities determined are associated with the elastic range of the material behavior. Also note that it makes no difference whether the initial or the final lengths are used in computing strains, since the deformations are very small.

## 2-9. Thermal Strain and Deformation

With changes of temperature, solid bodies expand on increase of temperature and contract on its decrease. The thermal strain $\varepsilon_T$ caused by a change in temperature from $T_o$ to $T$ measured in degrees Celsius or Fahrenheit, can be expressed as

$$\boxed{\varepsilon_T = \alpha(T - T_o)} \tag{2-17}$$

where $\alpha$ is an experimentally determined coefficient of linear thermal expansion. For moderately narrow ranges in temperature, $\alpha$ remains reasonably constant.

Equal thermal strains develop in every direction for unconstrained homogeneous isotropic materials. For a body of length $L$ subjected to a uniform temperature, the extensional deformation $\Delta_T$ due to a change in temperature of $\delta T = T - T_o$ is

$$\boxed{\Delta_T = \alpha(\delta T)L} \tag{2-18}$$

For a decrease in temperature, $\delta T$ assumes negative values.

An illustration of the thermal effect on deformation of bars due to an increase in temperature is shown in Fig. 2-28.

## EXAMPLE 2-7

Determine the displacement of point $B$ in Example 2-4 caused by an increase in temperature of 100°F. See Fig. 2-29(a).

### Solution

Determining the deflection at point $B$ due to an increase in temperature is similar to the solution of Example 2-4 for finding the deflection of the same point caused by stress. Per Table 1A in the Appendix, the coefficient of thermal expansion for 2024-T4 aluminum alloy is $12.9 \times 10^{-6}$ per °F. Hence, from Eq. 2-18,

$$\Delta_{AB} = 12.9 \times 10^{-6} \times 100 \times 6.71 = 8.656 \times 10^{-3} \text{ in}$$
$$\Delta_{BC} = 12.9 \times 10^{-6} \times 100 \times 8.49 = 10.95 \times 10^{-3} \text{ in}$$

Here the displacement $\Delta_T$ of point $B$ to position $B_4$, Fig. 2-29(b), caused by a change in temperature, is related to the bar elongations in the following manner:

$$\Delta_T \cos \theta_2 = \Delta_{AB} \quad \text{and} \quad \Delta_T \cos \theta_1 = \Delta_{BC}$$

Forming equal ratios for both sides of these equations, substituting numerical values for $\Delta_{AB}$ and $\Delta_{BC}$, and simplifying leads to the following result:

$$\frac{\cos \theta_2}{\cos \theta_1} = \frac{\Delta_{AB}}{\Delta_{BC}} = \frac{8.656 \times 10^{-3}}{10.95 \times 10^{-3}} = 0.7905$$

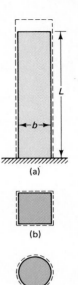

Fig. 2-28 Thermal expansions of bars resting on frictionless surface. Dashed lines represent final shape for an increase in temperature.

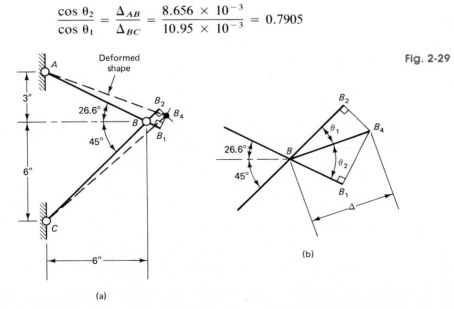

Fig. 2-29

Here, however, $\theta_2 = 45° + 26.6° - \theta_1 = 71.6° - \theta_1$; therefore,

$$\cos \theta_2 = \cos 71.6° \cos \theta_1 + \sin 71.6° \sin \theta_1$$

and $\quad \dfrac{\cos \theta_2}{\cos \theta_1} = \cos 71.6° + \sin 71.6° \tan \theta_1 = 0.7905$

Hence, $\qquad \tan \theta_1 = 0.500 \quad$ and $\quad \theta_1 = 26.6°$

Based on this result,

$$\Delta_T = \Delta_{BC}/\cos \theta_1 = 12.2 \times 10^{-3} \text{ in}$$

forming an angle of $45° - \theta_1 = 18.4°$ with the horizontal.

It is interesting to note that the small displacement $\Delta_T$ is of comparable order of magnitude to that found due to the applied vertical force $P$ in Example 2-4.

## 2-10. Saint-Venant's Principle and Stress Concentrations

The analysis of axially loaded bars based on engineering mechanics of solids is very accurate for bars of constant cross section when transmitting uniformly distributed end forces. For such ideal conditions stresses and strains are uniform everywhere. In reality, however, applied forces often approximate concentrated forces, and the cross sections of members can change abruptly. This causes stress and strain disturbances in the proximity of such forces and changes in cross sections. In the past these situations were studied analytically using the *mathematical theory of elasticity*. In such an approach, the behavior of two or three dimensional *infinitesimal elements* is formulated and the conditions of equilibrium, deformation and mechanical properties of material[16] are satisfied subject to the prescribed boundary conditions. More recently a powerful numerical procedure has been developed, where a body is subdivided into a *discrete number* of finite elements, such as squares or cubes, and the analysis is carried out with a computer. This is called the *finite element method* of analysis. The end results of analyses by either one of these two methods can be very effectively used to supplement solutions in engineering mechanics of solids. An example showing the more accurate solutions by these two advanced methods for the nature of stress distribution at concentrated force follows. These solutions provide comparison with those found by applying the method of engineering mechanics of solids.

A short block is shown in Fig. 2-30(a) acted upon by concentrated forces at its ends. Analyzing this block for stresses as a two dimensional problem

---

[16] These are the same basic concepts as used in engineering mechanics of solids.

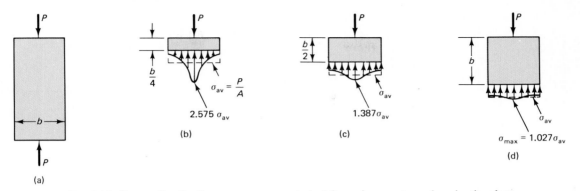

**Fig. 2-30** Stress distribution near a concentrated force in a rectangular elastic plate.

using the methods of the theory of elasticity gives the results shown in Figs. 2-30(b), (c), and (d).[17] The *average* stress $\sigma_{av}$ as given by Eq. 1-13 is also shown on these diagrams. From these it can be noted that at a section a distance $b/4$ from an end, Fig. 2-30(b), the maximum normal stress greatly exceeds the average. For a purely elastic material the maximum stress theoretically becomes infinite right under the concentrated force, since a finite force acts on a zero area. In real situations, however, a truly concentrated force is not possible and virtually all materials exhibit some plastic behavior; therefore the attainment of an infinite stress is impossible.

It is important to note two basic aspects from this solution. First, the *average* stress for all cases, being based on conditions of equilibrium, is always correct. Second, the normal stresses at a distance equal to the width of the member are essentially uniform.

The second observation illustrates the famed *Saint-Venant's principle.* It was enunciated by the great French elastician in 1855. In common engineering terms it simply means that the manner of force application on stresses is important only in the vicinity of the region where the force is applied. This also holds true for the disturbances caused by changes in cross section. Consciously or unconciously this principle is nearly always applied in idealizing load carrying systems.

Using the finite element method,[18] the results of a solution for the same problem are shown in Fig. 2-31. The initial undeformed mesh into which the planar block is arbitrarily subdivided, and the *greatly exaggerated deformed mesh* caused by the applied force are shown in Fig. 2-31(a). By placing the mesh on rollers as shown, only the upper half of the block

[17] S. Timoshenko, and J. N. Goodier, *Theory of Elasticity,* 3rd. ed., New York: McGraw-Hill, 1970, p. 60. Fig. 2-30 is adapted from this source.

[18] For this subject see for example, O. C. Zienkiewicz, *The Finite Element Method,* 3rd ed. (London: McGraw-Hill Ltd., 1977). K. J. Bathe and E. L. Wilson, *Numerical Methods in Finite Element Analysis* (Englewood Cliffs, N.J.: Prentice-Hall, 1976). R. H. Gallagher, *Finite Element Fundamentals* (Englewood Cliffs, N.J.: Prentice-Hall, 1975).

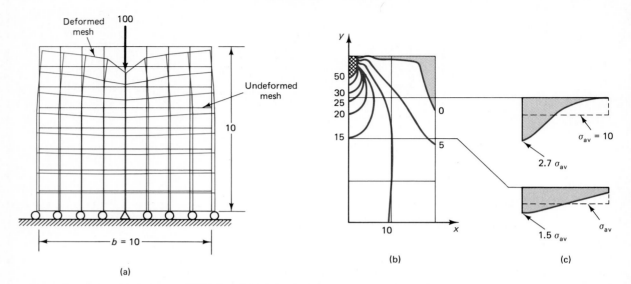

**Fig. 2-31** (a) Undeformed and deformed mesh of an elastic plate. (b) $\sigma_y$ contours, (c) Normal stress distributions at $b/4$ and $b/2$ below top.

needed to be analyzed because of symmetry around the mid-section. The calculated stress contours in Fig. 2-31(b) clearly show the development of large stresses in the vicinity of the concentrated force. Unlike the solution based on mathematical elasticity, in the finite element model the stresses at the applied force are very large, but finite, because of finite mesh size. As to be expected, the corners carry no stress. The stress distribution at b/4 and b/2 below the top, shown in Fig. 2-31(c), are in reasonable agreement with the more accurate results given in Figs. 2-30(b) and (c). Better agreement can be achieved by using a finer mesh. This versatile method can be applied to bodies of any shape and for any load distribution. Its use in accurate stress analysis problems is gaining an ever wider use. However, because of the simplicity of the procedures discussed in this text, at least for preliminary design, they remain indispensable.

The example cited above is extreme, since theoretically infinite stresses appear to be possible at the concentrated force. There are numerous situations, however, such as at bolt holes or changes in cross section, where the *maximum* normal stresses are finite. These maximum stresses, in relation to the *average* stress as given by Eq. 1-13 for linearly elastic materials, depend *only* on the *geometrical proportions* of a member. The ratio of the maximum to the average stress is called the *stress-concentration factor*, designated in this text as *K*. Many such factors are available in technical literature[19] as functions of the geometrical parameters of

[19] R. J. Roark and W. C. Young, *Formulas for Stress and Strain,* 5th ed. (New York: McGraw-Hill, 1975).

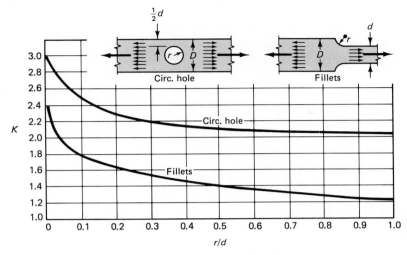

**Fig. 2-32** Stress-concentration factors for flat bars in tension.

members. For the example given before, at a depth below the top equal to one-quarter width, $K = 2.575$. Hence $\sigma_{max} = 2.575\,\sigma_{av}$. Generalizing this scheme, the maximum normal stress at a section is

$$\sigma_{max} = K\sigma_{av} = K\,\frac{P}{A} \qquad (2\text{-}19)$$

where $K$ is an appropriate stress-concentration factor, and $P/A$ is the average stress per Eq. 1-13.

Two particularly significant stress-concentration factors for *flat* axially loaded bars are shown in Fig. 2-32.[20] The $K$s that may be read from the graphs give the *ratio* of the maximum normal stress to the average stress on the net section as shown in Fig. 2-33. A considerable stress concentration also occurs at the root of threads. This depends to a large degree upon the sharpness of the cut. For ordinary threads, the stress-concen-

[20] This figure is adapted from M. M. Frocht, "Factors of Stress Concentration Photoelastically Determined." *Trans., ASME*, 1935, vol. 57, p. A-67.

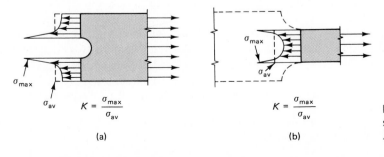

$$K = \frac{\sigma_{max}}{\sigma_{av}}$$

(a)

$$K = \frac{\sigma_{max}}{\sigma_{av}}$$

(b)

**Fig. 2-33** Measing of the stress-concentration factor $K$.

tration factor is on the order of 2 to 3. The application of Eq. 2-19 presents no difficulties, provided proper graphs or tables of $K$ are available. In the past many such factors have been determined using the methods of photoelasticity (see Sec. 9-4).

An example of low-cycle fatigue fracture in tension of a high-strength bolt with a minimum specified strength of 120 ksi (830 MPa) is shown in Fig. 2-34. Note that the fracture occurred at the root of the threads.

### EXAMPLE 2-8

Find the maximum stress in member $AB$ in the forked end $A$ in Example 1-3.

### Solution

*Geometrical proportions:*

$$\frac{\text{radius of the hole}}{\text{net width}} = \frac{3/16}{1/2} = 0.375$$

**Fig. 2-34** Low-cycle tensile fatigue fracture of 7/8 in A325 steel bolt.

*From Fig. 2-32:*[21] $K \approx 2.15$ for $r/d = 0.375$.
*Average stress from Example 1-3:* $\sigma_{av} = P/A_{net} = 11.2$ ksi.
*Maximum stress, Eq. 2-19:* $\sigma_{max} = K\sigma_{av} = 2.15 \times 11.2 = 24.1$ ksi.

This answer indicates that actually a large local increase in stress occurs at this hole, a fact that may be highly significant.

---

In considering stress-concentration factors in design, it must be remembered that their theoretical or photoelastic determination is based on the use of Hooke's law. If members are *gradually* stressed beyond the proportional limit of a *ductile* material, these factors lose their significance. For example, consider a flat bar of *mild steel,* of the proportions shown in Fig. 2-35, that is subjected to a gradually increasing force $P$. The stress distribution will be geometrically similar to that shown in Fig. 2-33 until $\sigma_{max}$ reaches the yield point of the material. This is illustrated in the top diagram in Fig. 2-35. However, with a further increase in the applied force, $\sigma_{max}$ remains the same, as a great deal of deformation can take place while the material yields. Therefore, the stress at $A$ remains virtually frozen at the same value. Nevertheless, for equilibrium, stresses acting over the net area must be high enough to resist the increased $P$. This condition is shown in the middle diagram of Fig. 2-35. Finally, for ideally plastic material, stress becomes uniform across the entire net section. Hence, for ductile materials prior to rupture, the local stress con-

---

[21] Actually, the stress concentration depends on the condition of the hole, whether it is empty or filled with a bolt or pin.

centration is practically eliminated, and a nearly uniform distribution of stress across the net section occurs prior to necking.

The previous argument is not quite as true for materials less ductile than mild steel. Nevertheless, the tendency is in that direction unless the material is unusually brittle, like glass. The argument presented applies to situations where the force is gradually applied or is static in character. *It is not applicable for fluctuating loads, as found in some machine parts.* For fatigue loadings, the working stress level that is actually reached *locally* determines the fatigue behavior of the member. The maximum permissible stress is set from an *S-N* diagram (Section 1-9). *Failure of most machine parts can be traced to progressive cracking that originates at points of high stress.* In machine design, then, stress concentrations are of paramount importance, although some machine designers feel that the theoretical stress concentration factors are somewhat high. Apparently, some tendency is present to smooth out the stress peaks, even in members subjected to cyclic loads.

From the previous discussion and accompanying charts, it should be apparent why a competent machine designer tries to "streamline" the junctures and transitions of elements that make up a structure.

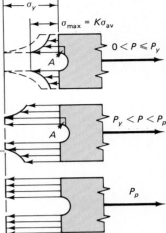

**Fig. 2-35** Stress distribution at a hole in an elastic-ideally plastic flat bar with progressively increasing applied force *P*.

## *[22]2-11. Elastic Strain Energy for Uniaxial Stress

In mechanics, energy is defined as the capacity to do work, and work is the product of a force times the distance in the direction that the force moves. In solid deformable bodies, stresses multiplied by their respective areas are forces, and deformations are distances. The product of these two quantities is the *internal work* done in a body by externally applied forces. This internal work is stored in an elastic body as the *internal elastic energy of deformation,* or the *elastic strain energy.* A procedure for computing the internal energy in axially loaded bars is discussed next.

Consider an infinitesimal element, such as shown in Fig. 2-36(a), subjected to a normal stress $\sigma_x$. The force acting on the right or the left face of this element is $\sigma_x \, dy \, dz$, where $dy \, dz$ is an infinitesimal area of the element. Because of this force, the element elongates an amount $\varepsilon_x \, dx$, where $\varepsilon_x$ is normal strain in the $x$ direction. If the element is made of a linearly elastic material, stress is proportional to strain; Fig. 2-36(b). Therefore, if the element is initially free of stress, the force that finally acts on the element increases linearly from zero until it attains its full value. The average force acting on the element while deformation is taking place is $\frac{1}{2}\sigma_x \, dy \, dz$. This average force multiplied by the distance through which it acts is the work done on the element. For a perfectly elastic body, no energy is dissipated and the work done on the element is stored as recoverable internal strain energy. Thus, the internal elastic strain energy $U$ for an infinitesimal element subjected to uniaxial stress is

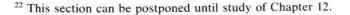

[22] This section can be postponed until study of Chapter 12.

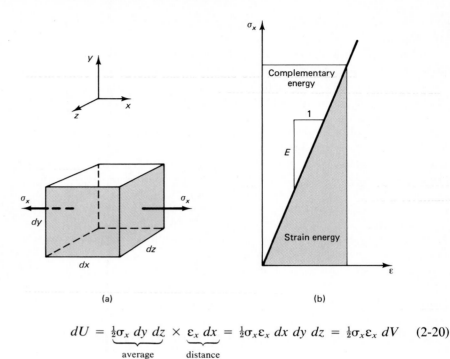

**Fig. 2-36** (a) An element in uniaxial tension and (b) a Hookean stress-strain diagram.

(a)                                    (b)

$$dU = \underbrace{\tfrac{1}{2}\sigma_x \, dy \, dz}_{\substack{\text{average} \\ \text{force}}} \times \underbrace{\varepsilon_x \, dx}_{\text{distance}} = \tfrac{1}{2}\sigma_x \varepsilon_x \, dx \, dy \, dz = \tfrac{1}{2}\sigma_x \varepsilon_x \, dV \quad (2\text{-}20)$$

$$\underbrace{\hspace{6cm}}_{\text{work}}$$

where $dV$ is the volume of the element.

By recasting Eq. 2-20, one obtains the strain energy stored in an elastic body per unit volume of the material, or its *strain-energy density* $U_o$. Thus,

$$U_o = \frac{dU}{dV} = \frac{\sigma_x \varepsilon_x}{2} \quad (2\text{-}21)$$

This expression may be graphically interpreted as an area under the inclined line on the stress-strain diagram; Fig. 2-36(b). The corresponding area enclosed by the inclined line and the vertical axis is called the *complementary energy,* a concept to be used in Chapter 12. For linearly elastic materials, the two areas are equal. Expressions analogous to Eq. 2-21 apply to the normal stresses $\sigma_y$ and $\sigma_z$ and to the corresponding normal strains $\varepsilon_y$ and $\varepsilon_z$.

Since in the elastic range, Hooke's law applies, $\sigma_x = E\varepsilon_x$, Eq. 2-21 may be written as

$$U_o = \frac{dU}{dV} = \frac{E\varepsilon_x^2}{2} = \frac{\sigma_x^2}{2E} \quad (2\text{-}22)$$

or
$$U = \int_{\text{vol}} \frac{\sigma_x^2}{2E} \, dV \qquad (2\text{-}23)$$

These forms of the equation for the elastic strain energy are convenient in applications, although they mask the dependence of the energy expression on force and distance.

For a particular material, substitution into Eq. 2-22 of the value of the stress at the proportional limit gives an index of the material's ability to store or absorb energy without permanent deformation. The quantity so found is called the *modulus of resilience* and is used to differentiate materials for applications where energy must be absorbed by members. For example, a steel with a proportional limit of 30,000 psi and an $E$ of $30 \times 10^6$ psi has a modulus of resilience of $\sigma^2/2E = (30,000)^2/2(30)10^6 = 15$ in-lb/in$^3$, whereas a good grade of Douglas fir, having a proportional limit of 6450 psi and an $E$ of 1,920,000 psi has a modulus of resilience of $(6,450)^2/(1,920,000) = 10.8$ in-lb/in$^3$.

By reasoning analogous to that before, the area under a complete stress-strain diagram, Fig. 2-37, gives a measure of a material's ability to absorb energy up to fracture and is called its *toughness*. The larger the total area under the stress-strain diagram, the tougher the material. In the inelastic range, only a small part of the energy absorbed by a material is recoverable. Most of the energy is *dissipated* in permanently deforming the material and is lost in heat. The energy that may be recovered when a specimen has been stressed to some such point as $A$ in Fig. 2-37(b) is represented by the triangle $ABC$. Line $AB$ of this triangle is parallel to line $OD$, since all materials essentially behave elastically upon the release of stress.

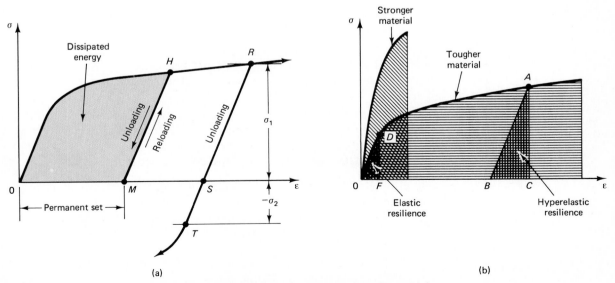

(a)

(b)

**Fig. 2-37** Some typical properties of materials.

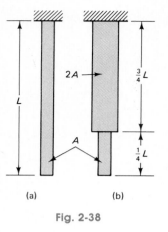

(a)                (b)

Fig. 2-38

### EXAMPLE 2-9

Two elastic bars, whose proportions are shown in Fig. 2-38, are to absorb the same amount of energy delivered by axial forces. Neglecting stress concentrations, compare the stresses in the two bars. The cross-sectional area of the left bar is $A$, and that of the right bar is $A$ and $2A$ as shown.

### Solution

The bar shown in Fig. 2-38(a) is of uniform cross-sectional area, therefore, the normal stress $\sigma_1$ is constant throughout. Using Eq. 2-23 and integrating over the volume $V$ of the bar, one can write the total energy for the bar as

$$U_1 = \int_V \frac{\sigma_1^2}{2E}\, dV = \frac{\sigma_1^2}{2E} \int_V dV = \frac{\sigma_1^2}{2E}(AL)$$

where $A$ is the cross-sectional area of the bar, and $L$ is its length.

The bar shown in Fig. 2-38(b) is of variable cross section. Therefore, if the stress $\sigma_2$ acts in the lower part of the bar, the stress in the upper part is $\frac{1}{2}\sigma_2$. Again, by using Eq. 2-23 and integrating over the volume of the bar, it is found that the total energy that this bar will absorb in terms of the stress $\sigma_2$ is

$$U_2 = \int_V \frac{\sigma^2}{2E}\, dV = \frac{\sigma_2^2}{2E} \int_{\text{lower part}} dV + \frac{(\sigma_2/2)^2}{2E} \int_{\text{upper part}} dV$$

$$= \frac{\sigma_2^2}{2E}\left(\frac{AL}{4}\right) + \frac{(\sigma_2/2)^2}{2E}\left(2A\,\frac{3L}{4}\right) = \frac{\sigma_2^2}{2E}\left(\frac{5}{8}\,AL\right)$$

If both bars are to absorb the same amount of energy, $U_1 = U_2$ and

$$\frac{\sigma_1^2}{2E}(AL) = \frac{\sigma_2^2}{2E}\left(\frac{5}{8}\,AL\right) \qquad \text{or} \qquad \sigma_2 = 1.265\sigma_1$$

Hence, for the same energy load, the stress in the "reinforced" bar is 26.5 percent higher than in the plain bar. The enlargement of the cross-sectional area over a part of the bar is actually detrimental. This situation is not found in the design of members for static loads.

---

## *2-12. Deflections by the Energy Method

The principle of conservation of energy may be very effectively used for finding deflections of elastic members due to applied forces. General methods for accomplishing this will be discussed in Chapter 12. Here a more limited objective, determining the deflection caused by the application of a single axial force, is considered. For such a purpose, the internal strain energy $U$ for a member is simply equated to the external work $W_e$ due to the applied force, i.e.,

$$U = W_e \tag{2-24}$$

In this treatment, it is assumed that the external force is gradually applied. This means that, as it is being applied, its full effect on a member is reached in a manner similar to that shown in Fig. 2-36(b) for stress. Therefore, the external work $W_e$ is equal to one-half of the total force multiplied by the deflection in the direction of the force action. In the next section, this approach will be generalized for dynamic loads.

## EXAMPLE 2-10

Find the deflection of the free end of an elastic rod of constant cross-sectional area $A$ and length $L$ due to axial force $P$ applied at the free end.

## Solution

If force $P$ is gradually applied to the rod, external work, $W_e = \frac{1}{2} P \Delta$, where $\Delta$ is the deflection of the end of the rod. The expression for the internal strain energy $U$ of the rod was found in Example 2-9, and since $\sigma_1 = P/A$, it is

$$U = \frac{\sigma_1^2}{2E} AL = \frac{P^2 L}{2AE}$$

Then, from $W_e = U$,

$$\frac{P\Delta}{2} = \frac{P^2 L}{2AE} \quad \text{and} \quad \Delta = \frac{PL}{AE}$$

which is the same as Eq. 2-9.

The use of Eq. 2-24 can be extended to bar systems consisting of several members. Since internal strain energy is a positive *scalar* quantity, the energies for the several members can be simply added arithmetically. This total strain energy $U$ can then be equated to the external work $W_e$ caused by one force for finding the deflection in the direction of that force. To illustrate, for the bracket shown in Fig. 2-24 for Example 2-4,

$$U = \frac{1}{2} \frac{P_{AB}^2 L_{AB}}{A_{AB}E} + \frac{1}{2} \frac{P_{BC}^2 L_{BC}}{A_{BC}E} = \frac{1}{2} P\Delta$$

where the subscripts refer to members. A solution of this equation gives deflection $\Delta$ of force $P$.

This method is extended in Chapter 12 to solution of problems with any number of applied forces for finding the deflections at any point in any direction.

## \*\*[23]2-13. Dynamic and Impact Loads

A freely falling weight, or a moving body, that strikes a structure delivers what is called a *dynamic* or *impact* load or force. Problems involving such forces may be analyzed rather simply on the basis of the following idealizing assumptions:

1. Materials behave elastically, and no dissipation of energy takes place at the point of impact or at the supports owing to local inelastic deformation of materials.
2. The inertia of a system resisting an impact may be neglected.
3. The deflection of a system is directly proportional to the magnitude of the applied force whether a force is dynamically or statically applied.

Then, using the principle of conservation of energy, it may be further assumed that at the *instant* a moving body is stopped, its kinetic energy is completely transformed into the internal strain energy of the resisting system. At this instant, the maximum deflection of a resisting system occurs and vibrations begin. However, since only maximum stresses and deflections are of primary interest, this subject will not be pursued.

As an example of a dynamic force applied to an elastic system, consider a falling weight striking a spring. This situation is illustrated in Fig. 2-39(a), where a weight $W$ falls from a height $h$ above the free length of a spring. *This system represents a very general case, since conceptually,*

---

[23] This is an optional section.

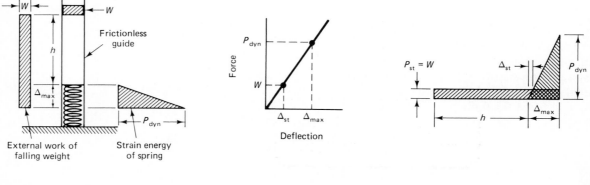

(a)                                            (b)                                            (c)

**Fig. 2-39** Behavior of an elastic system under an impact force.

*every elastic system may be treated as an equivalent spring.* Using the spring constant $k$, the static deflection $\Delta_{st}$ of the spring due to the weight $W$ is $\Delta_{st} = W/k$. Similarly, the maximum dynamic deflection $\Delta_{max} = P_{dyn}/k$, where $P_{dyn}$ is the maximum dynamic force experienced by the spring. Therefore, the dynamic force in terms of the weight $W$ and the deflections of the spring is

$$P_{dyn} = \frac{\Delta_{max}}{\Delta_{st}} W \qquad (2\text{-}25)$$

This relationship is shown in Fig. 2-39(b).

At the instant the spring deflects its maximum amount, all energy of the falling weight is transformed into the strain energy of the spring. Therefore, an equation representing the equality of external work to internal strain energy may be written as

$$W(h + \Delta_{max}) = \frac{1}{2} P_{dyn} \Delta_{max} \qquad (2\text{-}26)$$

A graphical interpretation of this equation is shown in Fig. 2-39(c). Note that a factor of one-half appears in front of the strain-energy expression, since the spring takes on the load *gradually*. Then, from Eq. 2-25,

$$W(h + \Delta_{max}) = \frac{1}{2} \frac{(\Delta_{max})^2}{\Delta_{st}} W$$

or

$$(\Delta_{max})^2 - 2\Delta_{st}\Delta_{max} - 2h\Delta_{st} = 0$$

hence

$$\Delta_{max} = \Delta_{st} + \sqrt{(\Delta_{st})^2 + 2h\Delta_{st}}$$

or

$$\Delta_{max} = \Delta_{st}\left(1 + \sqrt{1 + \frac{2h}{\Delta_{st}}}\right) \qquad (2\text{-}27)$$

and again using Eq. 2-25,

$$\boxed{P_{dyn} = W\left(1 + \sqrt{1 + \frac{2h}{\Delta_{st}}}\right)} \qquad (2\text{-}28)$$

Equation 2-27 gives the maximum deflection occurring in a spring struck by a weight $W$ falling from a height $h$, and Eq. 2-28 gives the maximum force experienced by the spring for the same condition. To apply these equations, the static deflection $\Delta_{st}$ caused by the gradually applied known weight $W$ is computed by the formulas derived earlier.

After the effective dynamic force $P_{dyn}$ is found, it may be used in computations as a static force. The magnification effect of a static force when dynamically applied is termed the *impact factor* and is given by the expression in parentheses appearing in Eqs. 2-27 and 2-28. The impact factor is surprisingly large in most cases. For example, if a force is applied to an elastic system *suddenly,* i.e., $h = 0$, it is equivalent to *twice* the same force *gradually* applied. If $h$ is large compared to $\Delta_{st}$, the impact factor is approximately equal to $\sqrt{2h/\Delta_{st}}$.

Similar equations may be derived for the case where a weight $W$ is moving horizontally with a velocity $v$ and is suddenly stopped by an elastic body. For this purpose, it is necessary to replace the external work done by the falling weight in the preceding derivation by the kinetic energy of a moving body, *using a consistent system of units.* Therefore, since the kinetic energy of a moving body is $Wv^2/2g$, where $g$ is the acceleration of gravity, it can be shown that

$$P_{dyn} = W \sqrt{\frac{v^2}{g\Delta_{st}}} \quad \text{and} \quad \Delta_{max} = \Delta_{st} \sqrt{\frac{v^2}{g\Delta_{st}}} \quad (2\text{-}29)$$

where $\Delta_{st}$ is the static deflection caused by $W$ acting in the horizontal direction. In Eq. 2-29, $W$ is in U.S. customary units.

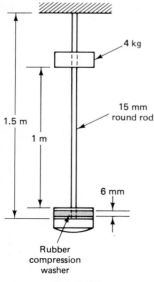

**Fig. 2-40**

### EXAMPLE 2-11

Determine the maximum stress in the steel rod shown in Fig. 2-40 caused by a mass of 4 kg falling freely through a distance of 1 m. Consider two cases: one as shown in the figure, and another when the rubber washer is removed. For the steel rod, assume $E = 200$ GPa, and for the washer, take $k = 4.5$ N/mm.

### Solution

The 4-kg mass applies a static force $P = ma = 4 \times 9.81 = 39.2$ N. The rod area $A = \pi \times 15^2/4 = 177$ mm$^2$. Note that the rod length is 1500 mm.
*Solution for rod with washer:*

$$\Delta_{st} = \frac{PL}{AE} + \frac{P}{k} = \frac{39.2 \times 1500}{177 \times 200 \times 10^3} + \frac{39.2}{4.5}$$
$$= 1.66 \times 10^{-3} + 8.71 = 8.71 \text{ mm}$$

$$\sigma_{max-dyn} = \frac{P_{dyn}}{A} = \frac{39.2}{177}\left(1 + \sqrt{1 + \frac{2 \times 1}{8.71 \times 10^{-3}}}\right) = 3.58 \text{ MPa}$$

*Solution for rod without washer:*

$$\sigma_{max-dyn} = \frac{39.2}{177} \left( 1 + \sqrt{1 + \frac{2 \times 1}{1.66 \times 10^{-6}}} \right) = 243 \text{ MPa}$$

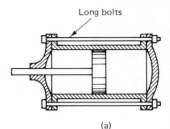

(a)

The large difference in the stresses for the two solutions suggests the need for flexible systems for resisting dynamic loads. A further study of this problem, and taking into account the results obtained in Example 2-9, leads to the conclusion that for obtaining the smallest dynamic stresses for the same system, one should:

1. select a material with a small elastic modulus;
2. make the total volume of the member large;
3. stress the material uniformly, and avoid stress concentrations.

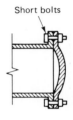

(b)

Several cases can be cited as illustrations of practical situations where these principles are used. Wood is used in railroad ties since its $E$ is low, and the cost per unit volume of the material is small. In pneumatic cylinders and jackhammers, Fig. 2-41, very long bolts are used to attach the ends to the tube. Long bolts provide a large volume of material, which, in operation, is uniformly stressed in tension. In the early stages of the development of this equipment, short bolts were used, and frequent failures occurred.

**Fig. 2-41** (a) Good design and (b) bad design of a pneumatic cylinder.

## Part B    STATICALLY INDETERMINATE SYSTEMS

## 2-14. General Considerations

As pointed out in Section 1-9, for some structural systems, the equations for static equilibrium are insuffficient for determining reactions. In such cases, some of the reactions are superfluous or redundant for maintaining equilibrium. In some other situations, redundancy may also result if some of the internal forces cannot be determined using the equations of statics alone. Both cases of such statical indeterminacy can arise in axially loaded systems. Two simple idealized examples are shown in Fig. 2-42. For the system shown in Fig. 2-42(a), reactions $R_1$ and $R_2$ cannot be determined using equations of statics alone. However, for the system shown in Fig. 2-42(b), whereas the reaction can be readily found, the distribution of forces between the two springs requires additional consideration. In both instances, the deformation characteristics of the system components must be considered.

There are various procedures for resolving structural indeterminacy in order to reduce a problem to statical determinacy such that the internal

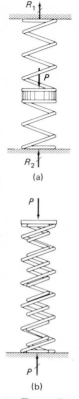

**Fig. 2-42** Examples of (a)
external statical
indeterminancy, and (b)
internal statical
indeterminancy.

forces can readily be found. Common to all of these procedures, the same three basic concepts encountered before are applied, and must be satisfied:

1. *Equilibrium conditions* for the system must be assured both in the local and global sense.
2. *Geometric compatibility* among the deformed parts of a body and at the boundaries must be satisfied.
3. *Constitutive relations* (stress-strain relations) for the materials of the system must be complied with.

Two general methods for solving simpler problems will be presented. The approach in one of these methods consists of first removing and then restoring a redundant reaction such that the compatibility condition at the boundaries is satisfied. This is the *force method* of analysis, since solution is obtained directly for the unknown reaction forces. Alternatively, the compatibility of displacements of adjoining members and at the boundaries is maintained throughout the loading process, and solution for displacements are obtained from equilibrium equations. This is the *displacement method* of analysis.

It is important to reiterate that in any one of these methods, the fundamental problem consists of fulfilling the three basic requirements: *equilibrium, compatibility,* and *conformity with constitutive relations.* The sequence in which they are applied is immaterial.

## 2-15. Force Method of Analysis

As an example of the force method of analysis, consider the linearly elastic axially loaded bar system shown in Fig. 2-43. The initially undeformed bars are shown in Fig. 2-43(a) with zig-zag lines as a reminder that they can be treated as springs. On applying force $P$ at $B$, reactions $R_1$ and $R_2$ develop at the ends and the system deforms, as shown in Fig. 2-43(b). Since only one nontrivial equation of statics is available for determining the two reactions, this system is statically indeterminate to the first degree. Here the upward direction of the applied force $P$, as well as that assumed for $R_1$ and $R_2$, coincides with the positive direction of the $x$ axis. For this reason, these quantities will be treated as positive. With this sign convention, if an applied force acts downward, it would be taken as negative. A calculated reaction with a negative sign signifies that it acts in the opposite direction from the assumed. Adherence to this sign convention is desirable, although in axially loaded bar problems, it is not essential since the directions of deflections and reactions can be usually seen by inspection. However, for computer solutions, as well as for the more complex problems discussed in Chapter 13, a strict adherence to a selected sign convention becomes necessary.

In applying the force method to axially loaded bars, one of the reactions

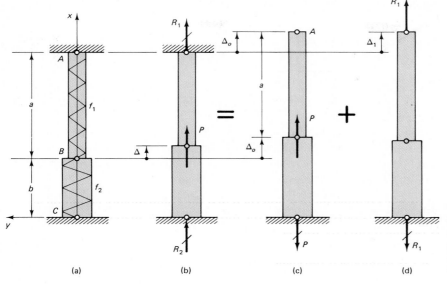

(a)          (b)          (c)          (d)

**Fig. 2-43** Force (flexibility) method of elastic analysis for a statically indeterminate axially loaded bar. Deformations are greatly exaggerated.

is temporarily removed, making the system statically determinate. Here an arbitrary choice is made to remove the upper reaction $R_1$, permitting the system of two bar segments to deform, as shown in Fig. 2-43(c). Such a simplified structural system is referred to as the *primary system*, since, from the point of view of statics, it can, by itself, carry the applied load. (However, from the point of view of strength, the redundant reaction may be necessary and, thus, in the actual field situation, cannot be removed.) Note that only the bottom bar segment is stressed here. Therefore, the same axial deformation $\Delta_0$, occurs at $A$, at the top of bar, as at point $B$. Then if the flexibility of the lower elastic bar is $f_2$, the deflection

$$\Delta_0 = f_2 P \qquad (2\text{-}30)$$

This result, shown in Fig. 2-43(c), violates the geometric boundary condition at $A$. In order to comply, the deflection $\Delta_1$ caused by $R_1$ acting on the *unloaded* bar $ABC$ is found next; see Fig. 2-43(d). This deflection is caused by the stretching of both bars. Therefore, if the flexibilities of these bars are $f_1$ and $f_2$, Fig. 2-43(a), the deflection

$$\Delta_1 = (f_1 + f_2)R_1 \qquad (2\text{-}31)$$

The compatibility of deformations at $A$ is then achieved by requiring that

$$\boxed{\Delta_0 + \Delta_1 = 0} \qquad (2\text{-}32)$$

P. 78

$$f = \frac{1}{K} = \frac{\Delta}{P} = \frac{L_i}{A_i E_i}$$

DEFLECTION
RESULTING FROM
THE APPLICATION
OF A UNIT FORCE

By substituting Eqs. 2-30 and 2-31 into Eq. 2-32 and solving for $R_1$, one has

$$R_1 = -\frac{f_2}{f_1 + f_2} P \tag{2-33}$$

The negative sign of the result indicates that $R_1$ acts in the opposite direction from the assumed. As to be expected, according to Eq. 2-31, this also holds true for $\Delta_1$.

The complete solution of this statically indeterminate problem is the algebraic sum of the solutions shown in Figs. 2-43(c) and (d). After the reactions become known, the previously discussed procedures for determining the internal forces and deflections apply.

Inasmuch as member flexibilities are particularly useful in formulating solutions by the force method, this approach is also known as the *flexibility method* of analysis.

The algebraic sum of the two solutions, as before, is an application of the *principle of superposition,* and will be frequently encountered in this text. This principle is based upon the premise that the resultant stress or strain in a system due to several forces is the algebraic sum of their effects when separately applied. *This assumption is true only if each effect is linearly related to the force causing it*. It is only approximately true when the deflections or deformations due to one force cause an abnormal change in the effect of another force. Fortunately, the magnitudes of deflections are relatively small in most engineering structures. In that regard, it is important to note that the deformation shown in Figs. 2-43(b) to (d) are *greatly exaggerated*. Moreover, since the deformations are very small, the *undeformed,* i.e., the initial, *bar lengths are used in calculating throughout*.

An illustration of force-deformation relationships for linear and nonlinear systems is shown in Fig. 2-44. For the linear systems considered, here

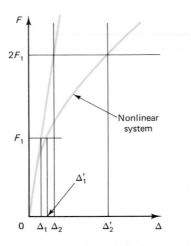

**Fig. 2-44** Comparison of force-displacement relationships between linear and nonlinear systems.

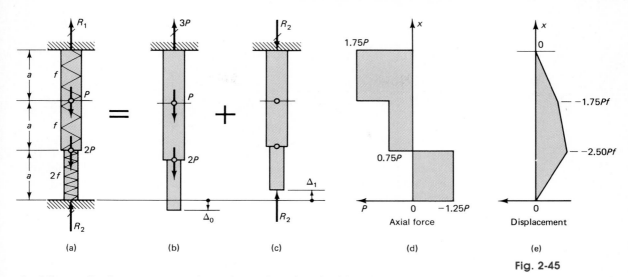

**Fig. 2-45**

doubling a displacement, say from $\Delta_1$ to $\Delta_2$, also doubles the load. This is not so for a nonlinear system. Therefore, for linear systems experiencing small deformations, the sequence or number of loads is immaterial.

The procedure just described is very general for linear systems and any number of axial loads, bar cross sections, different material properties, as well as thermal effects on the length of a bar system can be included in the analysis. However, the force method is not favored in practice because a systemic selection of the redundants for *large* problems is difficult.

Three examples follow illustrating applications of the force method to axially loaded elastic bar systems.

### EXAMPLE 2-12

An elastic bar at both ends is loaded as shown in Fig. 2-45. The known flexibility coefficients $f$ and $2f$ for each of the three bar segments are shown in the figure. Determine the reactions and plot the axial force and the axial displacement diagrams for the bar.

### Solution

Remove the lower support to obtain the free-body diagram shown in Fig. 2-45(b) and calculate $\Delta_0$. Since the applied forces act downward, because of the sign convention adopted in Fig. 2-43(b), they carry negative signs. The deflection caused by $R_2$ on an unloaded system is calculated next. Then, on solving Eq. 2-32, the reaction $R_2$ is determined. The remainder of the solution follows the same procedure as that described in Example 2-2.

$$\Delta_0 = \sum_i f_i P_i = -2fP - f(2P + P) = -5fP$$

and $$\Delta_1 = (2f + f + f)R_2 = 4fR_2$$

Since $$\Delta_1 + \Delta_2 = 0, \quad R_2 = 1.25P$$

Note that the applied forces are supported by a compressive reaction at the bottom and a tensile reaction at the top. In problems where the bar lengths and the cross-sectional areas, together with the elastic moduli $E$ for the materials, are given, the flexibilities are determined using Eq. 2-14.

The axial force diagram is plotted in Fig. 2-45(d). The compressive force in the bottom third of the bar causes a *downward* deflection of $1.25P \times 2f = 2.5Pf$. The tensile forces stretch the remainder of the bar $0.75Pf + 1.75Pf$ such that displacement at the top is zero. In this manner, the kinematic boundary conditions are satisfied at both ends of the bar.

## EXAMPLE 2-13

An elastic bar is held at both ends, as shown in Fig. 2-46. If the bar temperature increases by $\delta T$, what axial force develops in the bar? $AE$ for the bar is constant.

### Solution

First, the upper support is removed and $\Delta_0$ is determined using Eq. 2-18. The raising of the temperature causes no axial force in the bar. Thus, by using Eq. 2-13, $\Delta_1$ is calculated. By applying Eq. 2-32, the axial force in the bar, $R_1$, caused by the rise in temperature is found.

$$\Delta_0 = \alpha(\delta T)L$$

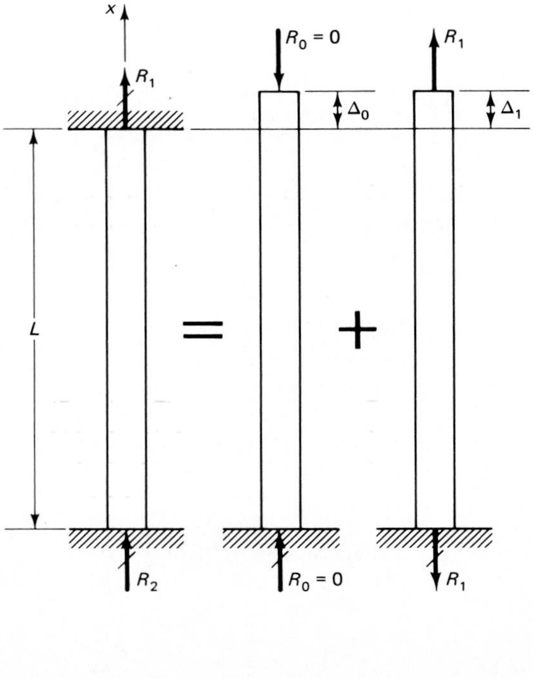

Fig. 2-46

and
$$\Delta_1 = R_1 f = \frac{R_1 L}{AE}$$

Since $\qquad \Delta_0 + \Delta_1 = 0, \qquad R_1 = -\alpha(\delta T)AE$

---

## EXAMPLE 2-14

For the planar system of the three elastic bars shown in Fig. 2-47(a), determine the forces in the bars caused by applied force $P$. The cross-sectional area $A$ of each bar is the same, and their elastic modulus is $E$.

### Solution

A free-body diagram of the assumed primary system with the support from the middle bar removed by cutting it at point $B$ is shown in Fig. 2-47(b). Then, by using statics, the forces in the bars are determined, and the deflection of point $D$ is calculated using the procedure illustrated in Example 2-4. Since bar $BD$ carries no force, deflection $\Delta_0$ at point $B$ is the same as it is at point $D$. Recognizing symmetry,

$$F_{10} = 0 \qquad \text{and} \qquad 2F_{20} \cos \alpha = P$$

Therefore,
$$F_{20} = \frac{P}{2 \cos \alpha}$$

Since $\qquad L_{AD} \cos \alpha = L, \qquad L_{AD} = L/\cos \alpha$

**Fig. 2-47**

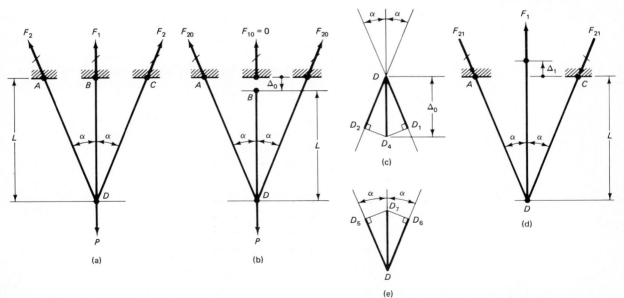

Hence, per Eq. 2-9, the stretch of bar $AD$ in the primary system is

$$(\Delta_{AD})_0 = \frac{PL}{2AE \cos^2 \alpha}$$

However, since $\Delta_0$ equals $DD_4$ in Fig. 2-47(c),

$$\Delta_0 \cos \alpha = (\Delta_{AD})_0 \quad \text{and} \quad \Delta_0 = - \frac{PL}{2AE \cos^3 \alpha}$$

where the negative sign signifies that the deflection is downward.

The same kind of relationship applies to the upward deflection of point $D$ caused by the force $F_1$; see Figs. 2-47(d) and (e). However, the deflection of point $B$ is increased by the stretch of the bar $BD$. The latter quantity is calculated using Eq. 2-9 again. On this basis,

$$\Delta_1 = \frac{F_1 L}{AE} + \frac{F_1 L}{2AE \cos^3 \alpha}$$

By applying Eq. 2-32, i.e., $\Delta_0 + \Delta_1 = 0$, and noting from statics that $F_1 + 2F_2 \cos \alpha = P$, on simplification,

$$F_1 = \frac{P}{2 \cos^3 \alpha + 1} \quad \text{and} \quad F_2 = \frac{P}{2 \cos^3 \alpha + 1} \cos^2 \alpha \quad (2\text{-}34)$$

## 2-16. Introduction to the Displacement Method

Another well-organized procedure for analyzing statically indeterminate problems is based on determining the displacements at selected points and providing information for finding the reactions and internal forces. As an example of this *displacement method* of analysis, consider the elastic axially loaded bar system shown in Fig. 2-48. The stiffnesses, $k_i = A_i E_i / L_i$, Eq. 2-12, for the bar segments are indicated in the figure as $k_1$ and $k_2$. An applied force $P$ at point $B$ causes reactions $R_1$ and $R_2$. These forces and the displacement $\Delta$ at $B$ are considered positive when they act in the positive direction of the $x$ axis. This problem is statically indeterminate to the first degree.

The main objective in this method of analysis is to determine the displacement $\Delta$, the principal parameter of the problem. In this example, there is only one such quantity and therefore the problem is said to have *one degree of kinematic indeterminacy,* or *one degree of freedom.* This is the only class of problems that is discussed in this section. More complex cases with several axial loads and changes in the cross sections of the bars, giving rise to several degrees of freedom, are considered in the next section.

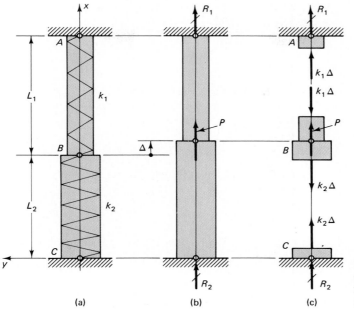

(a)　　　　　(b)　　　　　(c)

**Fig. 2-48** Displacement (stiffness) method of analysis for a statically indeterminate axially loaded bar.

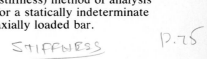

In this illustrative problem, it can be seen that the displacement $\Delta$ at $B$ causes compression in the upper bar $AB$ and tension in the lower bar $BC$. Therefore, if $k_1$ and $k_2$ are the respective stiffnesses for the bars, the respective internal forces are $k_1 \Delta$ and $k_2 \Delta$. These *internal* forces and reactions are shown on isolated free-bodies at points $A$, $B$, and $C$ in Fig. 2-48(c). These points are referred to as the node points. The sense of the internal forces is known since the upper bar is in compression and the lower one is in tension. By writing an equilibrium equation for the free body at node $B$, one has

$$-k_1 \Delta - k_2 \Delta + P = 0 \qquad (2\text{-}35)$$

and

$$\Delta = \frac{P}{k_1 + k_2} \qquad (2\text{-}36)$$

The equilibrium equations for the free-bodies at $i$ nodes $A$ and $C$ are

$$R_1 = -k_1 \Delta \quad \text{and} \quad R_2 = -k_2 \Delta \qquad (2\text{-}37)$$

Hence, with the aid of Eq. 2-36,

$$R_1 = -\frac{k_1}{k_1 + k_2} P \quad \text{and} \quad R_2 = -\frac{k_2}{k_1 + k_2} P \qquad (2\text{-}38)$$

The negative signs in Eq. 2-38 indicate that the reactions act in the opposite direction from the assumed.

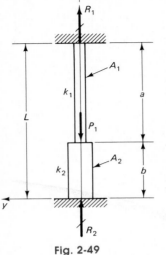

**Fig. 2-49**

Since in this solution bar stiffnesses are employed, this procedure is often called the *stiffness method*.

## EXAMPLE 2-15

An elastic stepped bar is loaded as shown in Fig. 2-49. Using the displacement method find the reactions. The bar segment stiffnesses $k_1$ and $k_2$, as well as their areas $A_1$ and $A_2$, and $E$ are given.

### Solution

According to Eq. 2-12 the stiffnesses $k$'s for the upper and lower bar segments, respectively, are

$$k_1 = A_1 E/a \qquad \text{and} \qquad k_2 = A_2 E/b$$

Therefore, per Eq. 2-36, the deflection $\Delta$ at $B$ due to *downward* force $P_1$ is

$$\Delta = -\frac{P_1}{k_1 + k_2} = -\frac{P_1}{A_1 E/a + A_2 E/b}$$

According to Eqs. 2-37, $R_1 = -k_1 \Delta$ and $R_2 = -k_2 \Delta$. By substituting the previous expressions for $\Delta$, $k_1$ and $k_2$, one obtains

$$R_1 = \frac{P_1}{1 + aA_2/bA_1} \qquad \text{and} \qquad R_2 = \frac{P_1}{1 + bA_1/aA_2} \qquad (2\text{-}39)$$

## **★★[24]2-17. Displacement Method with Several Degrees of Freedom**

In this section the displacement method is extended for axially loaded bars to include several degrees of freedom (d.o.f.). This method is the most widely used approach for solving both linear and nonlinear problems. However, the discussion will be limited to linearly elastic problems. As already noted in the previous section, solution of nonlinear problems using this method is beyond the scope of this text.

The displacement method is perfectly general and can be used for the analysis of statically determinate as well as indeterminate problems. With this in mind, consider a bar system consisting of three segments of variable stiffness defined by their respective spring constants $k_i$'s, as shown in Fig. 2-50(a). Each one of these segments terminates at a node point, some of which are common to the two adjoining bar segments. Each node, marked in the figure from 1 to 4, is permitted to displace vertically in

[24] This section is more advanced and can be omitted.

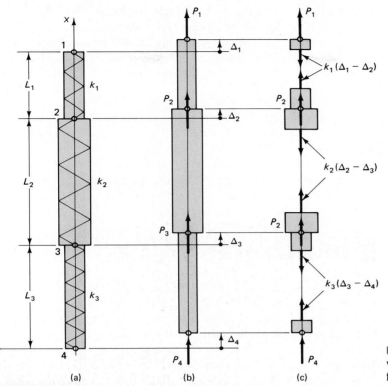

Fig. 2-50 Axially loaded bar with four degrees of freedom.

either direction. Therefore, this bar system has four degrees of freedom, i.e., one d.o.f. per node.

An application of forces at the nodes causes the bar system to displace in a geometrically compatible manner, as shown in Fig. 2-50(b). Here both the applied forces and the node displacements are shown with the positive sense coinciding with the positive direction of the x axis. Possible displacements at the nodes give rise to several special cases. With no deflection at the ends, one has a statically indeterminate problem. If, however, only one node point is held and forces or displacements are applied at the other nodes, the problem is statically determinate. However, if a displacement is specified at a node, it is not possible to also specify an applied force and vice versa.

With imposition of the applied forces and/or displacements, internal forces develop in the bar system. The magnitude and sense of these forces can be arrived at in the following manner. With the adopted sign convention, the bar segment extension[25] between the $i$th and the $(i + 1)$th nodes is $\Delta_i - \Delta_{i+1}$. By multiplying this stretch by the spring constant for the bar segment, the internal tensile force $(\Delta_i - \Delta_{i+1})k_i$ is determined.

[25] This can be clarified by noting the effect on a bar segment of node displacements taken one at a time.

Free-body diagrams for isolated nodes showing these internal as well as applied node forces are shown in Fig. 2-50(c).

The problem is resolved by writing equilibrium equation $\sum F_x = 0$ for each node. Thus, beginning with node 1, the following set of equations is obtained:

$$
\begin{aligned}
P_1 - k_1(\Delta_1 - \Delta_2) &&&= 0 \\
P_2 + k_1(\Delta_1 - \Delta_2) - k_2(\Delta_2 - \Delta_3) &&&= 0 \\
P_3 &&+ k_2(\Delta_2 - \Delta_3) - k_3(\Delta_3 - \Delta_4) &= 0 \\
P_4 &&&+ k_3(\Delta_3 - \Delta_4) = 0
\end{aligned}
\tag{2-40}
$$

It is to customary recast these equations into the following form

$$
\begin{aligned}
k_1 \Delta_1 \quad - k_1 \Delta_2 &&&= P_1 \\
- k_1 \Delta_1 + (k_1 + k_2) \Delta_2 \quad - k_2 \Delta_3 &&&= P_2 \\
- k_2 \Delta_2 \quad + (k_2 + k_3) \Delta_3 - k_3 \Delta_4 &= P_3 \\
- k_3 \Delta_3 \quad + k_3 \Delta_4 &= P_4
\end{aligned}
\tag{2-41}
$$

In most problems, the applied forces $P_i$'s are known, and the remaining $P_i$'s occurring at nodes of zero displacement are reactions. However, these equations can be applied to a broader range of problems by specifying displacements instead of applied forces. In such cases, at least one node must have a known (often zero) displacement where a reaction would develop. As noted earlier, at any one node, one can specify either an applied force or a displacement, but not both. These equations are solved simultaneously for the unknown quantities.

In typical applications of the displacement method, either the deflections $\Delta_i$'s or reactions $P_i$'s are the unknowns, and for clarity, it is customary to recast Eq. 2-41 in the following matrix form:

$$
\begin{bmatrix}
k_1 & -k_1 & 0 & 0 \\
-k_1 & k_1 + k_2 & -k_2 & 0 \\
0 & -k_2 & k_2 + k_3 & -k_3 \\
0 & 0 & -k_3 & k_3
\end{bmatrix}
\begin{Bmatrix}
\Delta_1 \\ \Delta_2 \\ \Delta_3 \\ \Delta_4
\end{Bmatrix}
=
\begin{Bmatrix}
P_1 \\ P_2 \\ P_3 \\ P_4
\end{Bmatrix}
\tag{2-42}
$$

This equation shows how the system *symmetric stiffness matrix* is built up from the member stiffnesses. The pattern of this matrix repeats for any number of node points. This formulation more clearly than the earlier case of single d.o.f. system shows why this approach is often referred to as the *stiffness method*. Excellent computer programs are available for solving these equations simultaneously.[26]

[26] E. L. Wilson, CAL-86, *Computer Assisted Learning of Structural Analysis and the CAL/SAP Development System*, Report No. UCB/SESM-86/05, Department of Civil Engineering, University of California, Berkeley, California, August 1986.

The displacement method is very extensively used in practice in the analysis of large complex problems with the aid of computers. Two simple examples follow.

## EXAMPLE 2-16

For the elastic weightless bar held at both ends, as shown in Fig. 2-51, determine the node displacements and the reactions using the displacement method. The cross section of the bar is constant throughout.

### Solution

Here only $\Delta_2$ and $\Delta_3$ have to be found as $\Delta_1 = \Delta_4 = 0$. Therefore, the system has two degrees of kinematic freedom. The stiffness coefficient $k$ is the same for each segment of the bar. Applying Eqs. 2-41 and setting $\Delta_1 = \Delta_4 = 0$, one obtains

$$
\begin{aligned}
-k\,\Delta_2 &&&= R_1 \\
2k\,\Delta_2 &- k\,\Delta_3 &&= -P \\
-k\,\Delta_2 &+ 2k\,\Delta_3 &&= -P \\
&- k\,\Delta_3 &&= R_2
\end{aligned}
$$

By solving the second and third equations simultaneously, $\Delta_2 = \Delta_3 = -P/k$, then from the first and the last equations, $R_1 = R_2 = P$. This result, which could be anticipated, means that, in effect, the upper load is hung from the top and the bottom one is supported at the base. The middle segment of the bar does not distort and deflects as a rigid body through a distance of $\Delta_2 = \Delta_3$.

In this problem, the force method would be simpler to apply than the displacement method since there is only one degree of static indeterminacy.

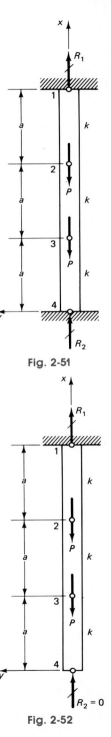

**Fig. 2-51**

## EXAMPLE 2-17

(a) Consider the same loaded bar as in Example 2-16 supported only at the top and free at the bottom; see Fig. 2-52. Determine the node displacements and the reaction. For this case, $R_2 = 0$. (b) Rework part (a) if the free end is displaced $3P/k$ upwards.

### Solution

(a) Here $\Delta_1 = 0$, and three nodal displacements, $\Delta_2$, $\Delta_3$, and $\Delta_4$, must be determined. Therefore, this statically determinate problem has three degrees of freedom. Applying Eqs. 2-41, one has

$$
\begin{aligned}
-k\,\Delta_2 &&&= R_1 \\
2k\,\Delta_2 &- k\,\Delta_3 &&= -P \\
-k\,\Delta_2 &+ 2k\,\Delta_3 &- k\,\Delta_4 &= -P \\
&- k\,\Delta_3 &+ k\,\Delta_4 &= 0
\end{aligned}
$$

By solving the last three equations simultaneously, $\Delta_2 = 2P/k$, $\Delta_3 = \Delta_4 =$

**Fig. 2-52**

$-3P/k$, and then from the first equation, $R_1 = 2P$. These results can be easily checked by the procedures discussed in Part A of this chapter.

(b) In this case, a force $R_2$ of unknown magnitude must be applied at the free end to cause the specified displacement $\Delta_4 = 3P/k$. As before, $\Delta_1 = 0$. Therefore, whereas the first three equations established for part (a) apply, the fourth equation must be revised to read

$$-k\,\Delta_3 + k\,\Delta_4 = R_2$$

After substituting the given value for $\Delta_4$ and solving the four applicable equations simultaneously, $\Delta_2 = 0$, $\Delta_3 = P/k$, and $R_1 = 0$.

### 2-18. Introduction to Statically Indeterminate Nonlinear Problems

The procedures discussed in the preceding three sections are very effective for solution of *linearly elastic* statically indeterminate axially loaded bar problems. By limiting the problems to one degree of kinematic indeterminacy, the procedure can be extended to include cases of inelastic material behavior. In this approach, the stepped bar in Fig. 2-53 or the symmetric bars in Fig. 2-47 can be analyzed regardless of the mechanical properties in each part of a two-part system. On the other hand, the bar in Fig. 2-45(a), having two degrees of kinematic indeterminacy and three distinctly differently stressed segments, is not susceptible to this kind of analysis.

In this extended approach, the forces remain the unknowns and are related at the juncture of the two systems by a compatibility condition. In such problems, a global *equilibrium* equation can always be written for a system. For example, for the bar in Fig. 2-53, such an equation is

$$R_1 + R_2 + P = 0 \tag{2-43}$$

Then, to assure *compatibility* at the juncture of the two bar segments, the deflections at $B$ are determined using two different paths. Therefore, since ends $A$ and $C$ are held, the deflection of bar $AB$ at $B$ is $\Delta_{AB}$ and that for bar $BC$ is $\Delta_{BC}$; and it follows that

$$\Delta_{AB} = \Delta_{BC} \tag{2-44}$$

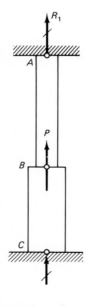

**Fig. 2-53** A bar of nonlinear material.

In calculating these quantities, it is usually convenient to visualize the bars cut and separated at $B$, and to determine $\Delta$'s for each part of the system maintained in equilibrium by the forces at the cut.

Any appropriate constitutive laws, including thermal effects and movement of supports, can be included in formulating the last equation. If the bar behavior is *linearly elastic*, with the aid of Eq. 2-9, the specialized equation becomes

$$\frac{R_1 L_1}{A_1 E_1} = \frac{R_2 L_2}{A_2 E_2} \tag{2-45}$$

Since no restrictions are placed on the constitutive relations for calculating deflections in Eq. 2-44, numerous nonlinear problems are tractable. Problems with internal statical indeterminacy can be solved in a similar manner. It must be emphasized, however, that, except for continuous members of linearly elastic material, *superposition cannot be used with the described procedure*. Several examples using the just-described procedure, as well as some other variations, follow.

## EXAMPLE 2-18

A stepped bar is held at both ends at immovable supports; see Fig. 2-54(a). The upper part of the bar has a cross-sectional area $A_1$; the area of the lower part is $A_2$. (a) If the material of the bar is elastic with an elastic modulus $E$, what are the reactions $R_1$ and $R_2$ caused by the application of an axial force $P_1$ at the point of discontinuity of the section? Use Eqs. 2-43 and 2-45. (b) If $A_1 = 600$ mm$^2$, $A_2 = 1200$ mm$^2$, $a = 750$ mm, $b = 500$ mm, and the material is linearly elastic–perfectly plastic, as shown in Fig. 2-54(d), determine the displacement $\Delta_1$ of the step as a function of the applied force $P_1$. Let $E = 200$ GPa. (c) Assuming that at the instant of impending yield in the whole bar, the applied force $P_1$ is removed, determine the residual force in the bar and the residual deflection at the bar step. (d) Using a stress-strain diagram for the material, show the strain history for each of the two bar parts during application and removal of force $P_1$.

## Solution

(a) In this approach, it is convenient to visualize the bar to be divided in two, as shown in Figs. 2-54(b) and (c). The upper part is subjected throughout its length to a tensile force $R_1$ and elongates an amount $\Delta_1$. The lower part contracts an amount $\Delta_2$ under the action of a compressive force $R_2$. These deflections must be equal. Therefore, using Eqs. 2-43 and 2-44 or its equivalent, Eq. 2-45, one has the following:

*From statics:*

$$R_1 + R_2 = P_1$$

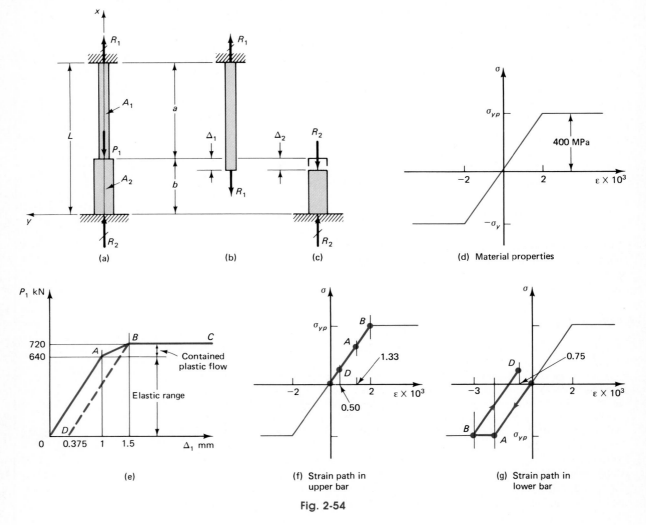

(a)  (b)  (c)  (d)  Material properties

(e)

(f)  Strain path in upper bar

(g)  Strain path in lower bar

**Fig. 2-54**

*From compatibility:*

$$\Delta_1 = \Delta_2 \quad \text{or} \quad \frac{R_1 a}{A_1 E} = \frac{R_2 b}{A_2 E}$$

By solving these two equations simultaneously,

$$R_1 = \frac{P_1}{1 + aA_2/bA_1} \quad \text{and} \quad R_2 = \frac{P_1}{1 + bA_1/aA_2} \qquad (2\text{-}39)$$

yielding the same result as found in Example 2-15.

(b) By direct substitution of data into Eqs. 2-39,

$$R_1 = \frac{P_1}{1 + 750 \times 1200/(500 \times 600)} = \frac{P_1}{4} \quad \text{and} \quad R_2 = \frac{3P_1}{4}$$

Hence, the normal stresses are

$$\sigma_1 = R_1/A_1 = P/2400 \quad \text{and} \quad \sigma_2 = R_2/A_2 = -P/1600$$

As $|\sigma_2| > \sigma_1$, the load at impending yield is found by setting $\sigma_2 = -400$ MPa. At this load, the lower part of the bar just reaches yield, and the strain attains the magnitude of $\varepsilon_{yp} = \sigma_{yp}/E = 2 \times 10^{-3}$. Therefore, from the previous relationship between $\sigma_2$ and $P$,

$$P_{yp} = 1600 \, \sigma_{yp} = 640 \times 10^3 \text{ N} = 640 \text{ kN}$$

and $\qquad \Delta_2 = \Delta_1 = \varepsilon_{yp}b = 2 \times 10^{-3} \times 500 = 1 \text{ mm}$

These quantities locate point $A$ in Fig. 2-54(e).

On increasing $P_1$ above 640 kN, the lower part of the bar continues to yield, carrying a compressive force $R_2 = \sigma_{yp}A_2 = 480$ kN. At the point of impending yield for the whole bar, the upper part just reaches yield. This occurs when $R_1 = \sigma_{yp}A_1 = 240$ kN and the strain in the upper part just reaches $\varepsilon_{yp} = \sigma_{yp}/E$. Therefore,

$$P_1 = R_1 + R_2 = 720 \text{ kN}$$

and $\qquad \Delta_1 = \varepsilon_{yp}a = 2 \times 10^{-3} \times 750 = 1.5 \text{ mm}$

These quantities locate point $B$ in Fig. 2-54(e). Beyond this point, the plastic flow is uncontained and $P_1 = 720$ kN is the *ultimate* or *limit* load of the rod.

Note the simplicity of calculating the limit load, which, however, provides no information on the deflection characteristics of the system. In general, plastic limit analysis is simpler than elastic analysis, which in turn is simpler than tracing the elastic-plastic load-deflection relationship.

**(c) According to the solution in part (b), when the applied force $P_1$ just reaches 720 kN and deflects 1.5 mm, point $B$ in Fig. 2-54(e), the whole bar becomes plastic. At this instant, $R_1 = 240$ kN and $R_2 = 480$ kN. On removing this force, the bar rebounds elastically (see Section 2-6). In the elastic equations, such a force must be treated with an opposite sign from that of the initially applied force. Therefore, per the solution found for part (b) based on Eqs. 2-39, the upper and lower reactions caused by the removal of the force $P_1$ are, respectively, $-P_1/4$ and $-3P_1/4$.

The residual force $R_{res}$ in the bar is equal to the initial force in either one of the bar parts less the reduction in these forces caused by the removal of the applied force. Hence, for the upper part of the bar,

$$R_{res} = R_1 - P_1/4 = 240 - 720/4 = 60 \text{ kN}$$

Likewise, for the lower part of the bar,

$$R_{res} = R_2 - 3P_1/4 = 480 - 3 \times 720/4 = 60 \text{ kN}$$

Both results are the same, as they should be, as no applied force remains at the bar discontinuity.

The residual deflection at the bar discontinuity can be determined using either part of the bar. For example, since the upper part loses $P_1/4 = 180$ kN of the tensile force, based on Eq. 2-9, it contracts $aP_1/(4A_1E) = 1.125$ mm. Hence, the residual deflection is $1.5 - 1.125 = 0.375$ mm, as shown in Fig. 2-49(e). The elastic rebound shown in this figure by the dashed line $BD$ is parallel to $OA$.

*(d) The strain histories for the two parts of the bar are given in Figs. 2-54(f) and (g). As shown in part (b), the lower segment begins to yield first. At that instant, $\Delta_1 = 1$ mm and the strain in the lower bar is $\Delta_1/b = 2 \times 10^{-3}$, whereas in the upper bar it is $\Delta_1/a = 1.33 \times 10^{-3}$. These results are identified by points $A$ in the figures. The instant when the upper bar begins to yield occurs at $\Delta_1 = 1.5$ mm. Therefore, the strains in both parts of the bar have increased by a factor of 1.5 and are so shown in the figures by their respective points $B$. No increase in the stress can occur in the lower bar during this time, as it is in a state of pure plastic deformation. When the applied load is completely removed, the residual deflection $\Delta_1 = 0.375$ mm. Hence, the corresponding residual strains $\Delta_1/a$ and $\Delta_1/b$ are, respectively, $0.50 \times 10^{-3}$ and $0.75 \times 10^{-3}$ m/m. The corresponding points are identified by points $D$ in Figs. 2-54(f) and (g).

## EXAMPLE 2-19

A 30-in long aluminum rod is enclosed within a steel-alloy tube; see Figs. 2-55(a) and (b). The two materials are bonded together. If the stress-strain diagrams for the two materials can be idealized as shown, respectively, in Fig. 2-55(d), what end deflection will occur for $P_1 = 80$ kips and for $P_2 = 125$ kips? The cross-sectional areas of steel $A_s$ and of aluminum $A_a$ are the same and equal to 0.5 in$^2$.

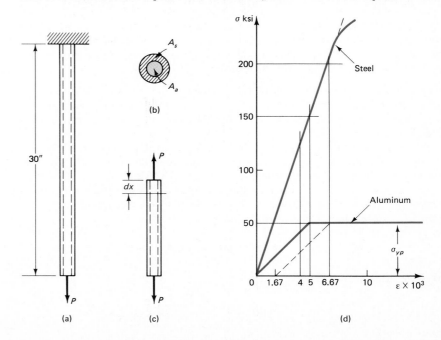

Fig. 2-55        (a)        (c)        (d)

## Solution

This problem is internally statically indeterminate since the manner in which the resistance to the force $P$ is distributed between the two materials is unknown. However, the total axial force at an arbitrary section can easily be determined; see Fig. 2-55(c). For an internal statically indeterminate problem, the requirements of equilibrium remain valid, but an additional condition is necessary to solve the problem. This auxiliary condition comes from the requirements of compatibility of deformations. However, since the requirements of statics involve forces and deformations involve displacements, a connecting condition based on the property of materials must be added.

Let subscripts a and s on $P$, $\varepsilon$, and $\sigma$ identify these quantities as being for aluminum and steel, respectively. Then, noting that the applied force is supported by a force developed in steel and aluminum and that at every section, the displacement or the strain of the two materials is the same, and tentatively assuming elastic response of both materials, one has the following:

*From equilibrium:*

$$P_a + P_s = P_1 \text{ or } P_2$$

*From compatibility:*

$$\Delta_a = \Delta_s \quad \text{or} \quad \varepsilon_a = \varepsilon_s$$

*From material properties:*

$$\varepsilon_a = \sigma_a/E_a \quad \text{and} \quad \varepsilon_s = \sigma_s/E_s$$

By noting that $\sigma_a = P_a/A_a$ and $\sigma_s = P_s/A_s$, one can solve the three equations. From the diagram the elastic moduli are $E_s = 30 \times 10^6$ psi and $E_a = 10 \times 10^6$ psi. Thus,

$$\varepsilon_a = \varepsilon_s = \frac{\sigma_a}{E_a} = \frac{\sigma_s}{E_s} = \frac{P_a}{A_a E_a} = \frac{P_s}{A_s E_s}$$

Hence, $P_s = [A_s E_s/(A_a E_a)]P_a = 3P_a$, and $P_a + 3P_a = P_1 = 80$ k; therefore, $P_a = 20$ k, and $P_s = 60$ k.

By applying Eq. 2-9 to either material, the tip deflection for 80 kips will be

$$\Delta = \frac{P_s L}{A_s E_s} = \frac{P_a L}{A_a E_a} = \frac{20 \times 10^3 \times 30}{0.5 \times 10 \times 10^6} = 0.120 \text{ in}$$

This corresponds to a strain of $0.120/30 = 4 \times 10^{-3}$ in/in. In this range, both materials respond elastically, which satisfies the material-property assumption made at the beginning of this solution. In fact, as may be seen from Fig. 2-55(d), since for the linearly elastic response, the strain can reach $5 \times 10^{-3}$ in/in for both materials, by direct proportion, the applied force $P$ can be as large as 100 kips.

At $P = 100$ kips, the stress in aluminum reaches 50 ksi. According to the idealized stress-strain diagram, no higher stress can be resisted by this material, although the strains may continue to increase. Therefore, beyond $P = 100$ kips, the aluminum rod can be counted upon to resist only $P_a = A_a\sigma_{yp} = 0.5 \times 50 = 25$ kips. The remainder of the applied load must be carried by the steel tube. Therefore for $P_2 = 125$ kips, 100 kips must be carried by the steel tube. Hence, $\sigma_s = 100/0.5 = 200$ ksi. At this stress level, $\varepsilon_s = 200/(30 \times 10^3) = 6.67 \times 10^{-3}$ in/in. Therefore, the tip deflection

$$\Delta = \varepsilon_s L = 6.67 \times 10^{-3} \times 30 = 0.200 \text{ in}$$

Note that it is not possible to determine $\Delta$ from the strain in aluminum, since no unique strain corresponds to the stress of 50 ksi, which is all that the aluminum rod can carry. However, in this case, the elastic steel tube contains the plastic flow. Thus, the strains in both materials are the same, i.e., $\varepsilon_s = \varepsilon_a = 6.67 \times 10^{-3}$ in/in; see Fig. 2-55(d).

If the applied force $P_2 = 125$ kips were removed, both materials in the rod would rebound elastically. Thus, if one imagines the bond between the two materials broken, the steel tube would return to its initial shape. But a permanent set (stretch) of $(6.67 - 5) \times 10^{-3} = 1.67 \times 10^{-3}$ in/in would occur in the aluminum rod. This incompatibility of strain cannot develop if the two materials are bonded together. Instead, residual stresses develop, which maintain the same axial deformations in both materials. In this case, the aluminum rod remains slightly compressed and the steel tube is slightly stretched. The procedure for the solution of this kind of problem is illustrated in the next example.

### EXAMPLE 2-20

A steel rod with a cross-sectional area of 2 in² and a length of 15.0025 in is loosely inserted into a copper tube, as shown in Fig. 2-56. The copper tube has a cross-sectional area of 3 in² and is 15.0000 in long. If an axial force $P = 25$ kips is applied through a rigid cap, what stresses will develop in the two materials? Assume that the elastic moduli of steel and copper are $E_s = 30 \times 10^6$ psi and $E_{cu} = 17 \times 10^6$ psi, respectively.

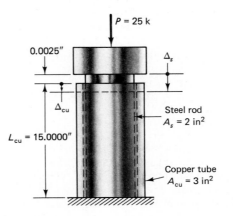

Fig. 2-56

## Solution

If the applied force $P$ is sufficiently large to close the small gap, a force $P_s$ will be developed in the steel rod and a force $P_{cu}$ in the copper tube. Moreover, upon loading, the steel rod will compress axially $\Delta_s$, which is as much as the axial deformation $\Delta_{cu}$ of the copper tube plus the initial gap. Hence,

*From statics:*

$$P_s + P_{cu} = 25{,}000 \text{ lb}$$

*From compatibility:*

$$\Delta_s = \Delta_{cu} + 0.0025$$

By applying Eq. 2-9, $\Delta = PL/AE$, substituting, and simplifying,

$$\frac{P_s L_s}{A_s E_s} = \frac{P_{cu} L_{cu}}{A_{cu} E_{cu}} + 0.0025$$

$$\frac{15.0025}{2 \times 30 \times 10^6} P_s - \frac{15}{3 \times 17 \times 10^6} P_{cu} = 0.0025$$

$$P_s - 1.176 P_{cu} = 10{,}000 \text{ lb}$$

Solving the two equations simultaneously,

$$P_{cu} = 6900 \text{ lb} \qquad \text{and} \qquad P_s = 18{,}100 \text{ lb}$$

and dividing these forces by the respective cross-sectional areas gives

$$\sigma_{cu} = 6900/3 = 2300 \text{ psi} \qquad \text{and} \qquad \sigma_s = 18{,}100/2 = 9050 \text{ psi}$$

If either of these stresses were above the proportional limit of its material or if the applied force were too small to close the gap, the above solution would not be valid. Also note that since the deformations considered are small, it is sufficiently accurate to use $L_s = L_{cu}$.

## Alternative Solution

The force $F$ necessary to close the gap may be found first, using Eq. 2-9. In developing this force, the rod acts as a "spring" and resists a part of the applied force. The remaining force $P'$ causes equal deflections $\Delta'_s$ and $\Delta'_{cu}$ in the two materials.

$$F = \frac{\Delta A_s E_s}{L_s} = \frac{0.0025 \times 2 \times 30 \times 10^6}{15.0025} = 10{,}000 \text{ lb} = 10 \text{ kips}$$

$$P' = P - F = 25 - 10 = 15 \text{ kips}$$

Then if $P'_s$ is the force resisted by the steel rod, in addition to the force $F$, and $P'_{cu}$ is the force carried by the copper tube,

*From statics:*

$$P'_s + P'_{cu} = P' = 15$$

*From compatibility:*

$$\Delta'_s = \Delta'_{cu} \quad \text{or} \quad \frac{P'_s L_s}{A_s E_s} = \frac{P'_{cu} L_{cu}}{A_{cu} E_{cu}}$$

$$\frac{15.0025}{2 \times 30 \times 10^6} P'_s = \frac{15}{3 \times 17 \times 10^6} P'_{cu} \qquad P'_{cu} = \frac{17}{20} P'_s$$

By solving the two appropriate equations simultaneously, it is found that $P'_{cu} = 6.9$ kips and $P'_s = 8.1$ kips, or $P_s = P'_s + F = 18.1$ kips.

If $(\sigma_{yp})_s = 40$ ksi and $(\sigma_{yp})_{cu} = 10$ ksi, the limit load for this assembly can be determined as follows:

$$P_{ult} = (\sigma_{yp})_s A_s + (\sigma_{yp})_{cu} A_{cu} = 110 \text{ kips}$$

At the ultimate load, both materials yield, therefore, the small discrepancy in the initial lengths of the parts is of no consequence.

## EXAMPLE 2-21

A copper tube 12-in long and having a cross-sectional area of 3 in² is placed between two very rigid caps made of Invar[27]; see Fig. 2-57(a). Four ¾-in steel bolts are symmetrically arranged parallel to the axis of the tube and are lightly

[27] Invar is a steel alloy which at ordinary temperatures has an $\alpha \approx 0$ and for this reason is used in the best grades of surveyor's tapes and watch springs.

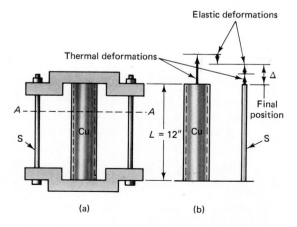

Fig. 2-57          (a)          (b)

tightened. Find the stress in the tube if the temperature of the assembly is raised from 60°F to 160°F. Let $E_{cu} = 17 \times 10^6$ psi, $E_s = 30 \times 10^6$ psi, $\alpha_{cu} = 9.1 \times 10^{-6}$ per °F, and $\alpha_s = 6.5 \times 10^{-6}$ per °F.

## Solution

If the copper tube and the steel bolts were free to expand, the axial thermal elongations shown in Fig. 2-57(b) would take place. However, since the axial deformation of the tube must be the same as that of the bolts, the copper tube will be pushed back and the bolts will be pulled out so that the net deformations will be the same. Moreover, as can be established by considering a free body of the assembly above some arbitrary section such as $A-A$ in Fig. 2-57(a), the compressive force $P_{cu}$ in the copper tube and the tensile force $P_s$ in the steel bolts are equal. Hence,

*From statics:*

$$P_{cu} = P_s = P$$

*From compatibility:*

$$\Delta_{cu} = \Delta_s = \Delta$$

This kinematic relation, on the basis of Fig. 2-57(b) with the aid of Eqs. 2-18 and 2-9, becomes

$$\alpha_{cu}(\delta T)L_{cu} - \frac{P_{cu}L_{cu}}{A_{cu}E_{cu}} = \alpha_s(\delta T)L_s + \frac{P_s L_s}{A_s E_s}$$

or, since $L_{cu} = L_s$, $\delta T = 100°$ and 0.442 in$^2$ is the cross section of one bolt,

$$9.1 \times 10^{-6} \times 100 - \frac{P_{cu}}{3 \times 17 \times 10^6}$$

$$= 6.5 \times 10^{-6} \times 100 + \frac{P_s}{4 \times 0.442 \times 30 \times 10^6}$$

By solving the two equations simultaneously, $P = 6750$ lb. Therefore, the stress in the copper tube is $\sigma_{cu} = 6750/3 = 2250$ psi.

The kinematic expression just used may also be set up on the basis of the following statement: the differential expansion of the two materials due to the change in temperature is accommodated by or is equal to the elastic deformations that take place in the two materials.

## EXAMPLE 2-22

A steel bolt having a cross-sectional area $A_1 = 1$ in$^2$ is used to grip two steel washers of total thickness, $L$, each having the cross-sectional area $A_2 = 9$ in$^2$; see Fig. 2-58(a). If the bolt in this assembly is tightened initially so that its stress

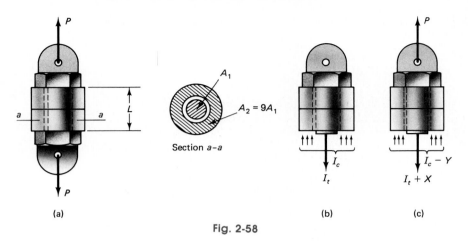

Section a–a

(a)

(b)        (c)

**Fig. 2-58**

is 20 ksi, what will be the final stress in this bolt after a force $P = 15$ kips is applied to the assembly?

## Solution

A free-body corresponding to the initial conditions of the assembly is in Fig. 2-58(b), where $I_t$ is the initial tensile force in the bolt, and $I_c$ is the initial compressive force in the washers. From statics, $I_t = I_c$. A free-body of the assembly after the force $P$ is applied is shown in Fig. 2-58(c), where $X$ designates the increase in the tensile force in the bolt, and $Y$ is the decrease in the compressive force on the washers due to $P$. As a result of these forces, $X$ and $Y$, if the adjacent parts remain in contact, the bolt elongates the same amount as the washers expand elastically. Hence, the final conditions are as follows:

*From statics:*

$$P + (I_c - Y) = (I_t + X)$$

or since $I_c = I_t$,

$$X + Y = P$$

*From compatibility:*

$$\Delta_{\text{bolt}} = \Delta_{\text{washers}}$$

By applying Eq. 2-9,

$$\frac{XL}{A_1E} = \frac{YL}{A_2E} \quad \text{i.e.,} \quad Y = \frac{A_2}{A_1}X$$

By solving the two equations simultaneously,

$$X = \frac{P}{1 + A_2/A_1} = \frac{P}{1 + 9} = 0.1P = 1500 \text{ lb}$$

Therefore, the increase of the stress in the bolt is $X/A_1 = 1500$ psi, and the stress in the bolt after the application of the force $P$ becomes 21,500 psi. This remarkable result indicates that most of the applied force is carried by decreasing the initial compressive force on the assembled washers since $Y = 0.9P$.

The solution is not valid if one of the materials ceases to behave elastically or if the applied force is such that the initial precompression of the assembled parts is destroyed.

Situations approximating the above idealized problem are found in many practical applications. A hot rivet used in the assembly of plates, upon cooling, develops within it enormous tensile stresses. Thoroughly tightened bolts, as in a head of an automobile engine or in a flange of a pressure vessel, have high initial tensile stresses; so do the steel tendons in a prestressed concrete beam. It is crucially important that on applying the working loads, only a small increase occurs in the initial tensile stresses.

## EXAMPLE 2-23

Extend the solution of Example 2-14 for the frame shown in Fig. 2-59(a) into the plastic range of material behavior and plot a force-displacement diagram. The cross-sectional area $A$ of each bar is the same. Assume ideal elastic-plastic behavior with the material yielding at $\sigma_{yp}$.

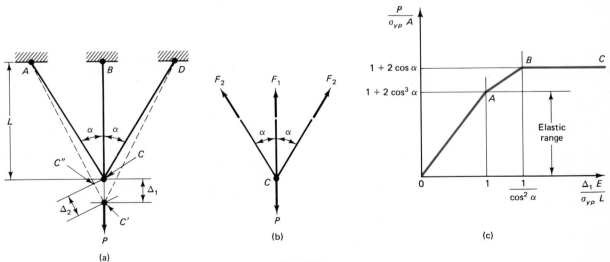

Fig. 2-59

### Solution

The *equilibrium equation* for forces at joint $C$, Fig. 2-59(c), recognizing symmetry, is

$$F_1 + 2F_2 \cos \alpha = P$$

The *compatibility equation* at joint $C$, Fig. 2-59(a), relating the elongations in bars $AC'$ and $DC'$ with that of bar $BC'$ is

$$\Delta_2 = \Delta_1 \cos \alpha$$

In both of these equations, it is assumed that the deformations are small. However, these equations hold true whether the bar material behaves elastically or plastically.

By noting that the inclined bars are $L/(\cos \alpha)$ long, using Eq. 2-9 and the established compatibility equation,

$$\frac{F_2[L/\cos \alpha]}{AE} = \frac{F_1 L}{AE} \cos \alpha \quad \text{or} \quad F_2 = F_1 \cos^2 \alpha$$

By substituting the last expression into the equilibrium equation at joint $C$, and simplifying leads to the same results as found in Example 2-14:

$$F_1 = \frac{P}{1 + 2 \cos^3 \alpha} \quad \text{and} \quad F_2 = \frac{P}{1 + 2 \cos^3 \alpha} \cos^2 \alpha \quad (2\text{-}34)$$

It is seen from this solution that the maximum force occurs in the vertical bar. At the impending yield $F_1 = \sigma_{yp}A$, and, per Eq. 2.9, $\Delta_1 = \sigma_{yp}L/E$. By substituting $F_1 = \sigma_{yp}A$ into the left side of Eq. 2-34, the force $P = \sigma_{yp}A(1 + 2 \cos^3 \alpha)$ at the limit of elastic behavior is obtained. This value of $P$ occurring at $\Delta_1 = \sigma_{yp}L/E$ is identified by point $A$ in Fig. 2-59(c).

By increasing force $P$ above the first yield in the vertical bar, force $F_1 = \sigma_{yp}A$ remains constant, and the equation of statics at joint $C$ is sufficient for determining force $F_2$ until the stress in the inclined bars reaches $\sigma_{yp}$. This occurs when $F_2 = \sigma_{yp}A$. At the impending yield in the inclined bars, and the vertical bar already yielding, the joint $C$ equilibrium equation gives $P = \sigma_{yp}A(1 + 2 \cos \alpha)$. This condition corresponds to the *plastic limit load* for the system. Note that the procedure of finding this load is rather simple, as the system is statically determinate when the limit load is reached. In Chapter 13, such a limit load is associated with the concept of the *collapse mechanism*.

At the impending yield in the inclined bars, per Eq. 2-9, $\Delta_2 = (\sigma_{yp}/E)[L/\cos \alpha]$ and $\Delta_1 = \Delta_2/\cos \alpha = \sigma_{yp}L/(E \cos^2 \alpha)$. This value of $\Delta_1$ locates the abscissa for point $B$ in Fig. 2-59(c). Beyond this point, all bars continue to yield without bound based on ideal plasticity.

## **28**2-19. Alternative Differential Equation Approach for Deflections

In Section 2-7, the axial deflection $u$ of a bar was in essence determined by solving a first-order differential $\varepsilon_x = du/dx$, Eq. 2-6. It is instructive to reformulate this problem as a second-order equation. Such an equation for linearly elastic materials follows from two observations. First, since, in general, $du/dx = \varepsilon = \sigma/E = P/AE$, one has

$$P = AE \frac{du}{dx} \qquad (2\text{-}46)$$

The second relation is based on the equilibrium requirements for an infinitesimal element of an axially loaded bar. For this purpose, consider a typical element such as that in Fig. 2-60, where all forces are shown with a positive sense according to the previously adopted sign convention. Since $\sum F_x = 0$ or $dP + p_x \, dx = 0$, and

$$\frac{dP}{dx} = -p_x \qquad \left[\frac{\text{lb}}{\text{in}}\right] \qquad (2\text{-}47)$$

This equation states that the rate of change with $x$ of the internal axial force $P$ is equal to the negative of the applied force $p_x$. On this basis, assuming $AE$ constant,

$$\frac{d}{dx}\left(\frac{du}{dx}\right) = \frac{1}{AE}\frac{dP}{dx} \qquad \text{or} \qquad AE \frac{d^2u}{dx^2} = -p_x \qquad (2\text{-}48)$$

[28] This section is optional; can be studied after Section 2-7 before the accompanying examples.

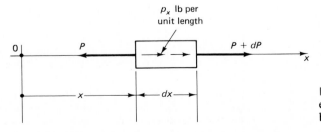

Fig. 2-60 Infinitesimal element of an axially loaded bar.

It is important to note that the three basic concepts of engineering mechanics of solids are included in deriving this governing differential equation. The requirements of *statics* are satisfied by making use of Eq. 2-47, and that of *kinematics* through the use of Eq. 2-6. The *constitutive relation* is defined by Eq. 2-3. A solution of Eq. 2-48 *subject to the prescribed boundary conditions constitutes a solution of any given axially loaded elastic bar problem.* Equation 2-48 is equally applicable to statically determinate *and* statically indeterminate problems. However for ease of solution $p_x/AE$ should be a continuous function. When the function is discontinuous, several alternatives are possible. One of them consists of obtaining solutions for each segment of a bar and enforcing continuity conditions at the junctures.[29] This is related to the statically determinate procedure discussed in Section 2-7, and to the statically indeterminate procedure considered in Sections 2-16 and 2-17. For concentrated forces, singularity functions, discussed in Section 5-16, can be used to advantage. However direct use of Eq. 2-48 for bars where several axial loads are applied and/or cross sections change becomes cumbersome. Therefore the procedures discussed before, including the scheme for dividing problems into statically determinate and indeterminate ones, are more useful in practical applications.

The example that follows illustrates the procedure when $p_x$ is a continuous function.

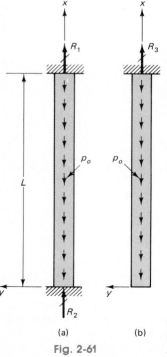

(a)          (b)

Fig. 2-61

### EXAMPLE 2-24

(a) Consider a bar of uniform cross section held between two rigid supports spun in a centrifuge such that an approximately uniformly distributed axial force $p_o$ N/m develops in the bar, as shown in Fig. 2-61(a). Determine the reactions at the ends. (b) If the same bar is supported only at one end, Fig. 2-61(b), what will the displacements $u(x)$ be along the bar?

### Solution

(a) Using Eq. 2-48, and noting Eq. 2-46, on integrating twice:

$$AE \frac{d^2u}{dx^2} = -(-p_o) = p_o$$

$$AE \frac{du}{dx} = p_o x + C_1 = P$$

$$AEu = \frac{p_o x^2}{2} + C_1 x + C_2$$

[29] This requires the displacements of the abutting bar segments at a discontinuity to be equal, and that the axial forces acting on an isolated infinitesimal element at the discontinuity be in equilibrium. (See, for example, the element at $B$ in Fig. 2-48(c) where at a discontinuity the force $P$ may also be zero.)

The constants of integration $C_1$ and $C_2$ can be found by noting that the deflection $u$ is zero at both ends, i.e., $u(0) = 0$ and $u(L) = 0$. Hence, from the last equation,

$$AEu(0) = 0 \quad \text{and} \quad C_2 = 0$$
$$AEu(L) = p_o L^2/2 + C_1 L = 0 \quad \text{and} \quad C_1 = -p_o L/2$$

Since $u'(x) = du/dx$, from Eq. 2-46,

$$R_2 = P(0) = AE\, u'(0) = -p_o L/2$$

The negative sign shows that this force is generated by compressive stresses. Similarly,

$$R_1 = P(L) = AE\, u'(L) = p_o L/2 = p_o L/2$$

These results indicate that the applied forces are shared equally by the two supports.

(b) The general solution for the problem found in (a) remains applicable. However, different constants of integration must be determined from the two boundary conditions. These are $P(0) = 0$ and $u(L) = 0$; hence, $AE\, u'(0) = 0$ and $C_1 = 0$. Similarly,

$$AEu(L) = p_o L^2/2 + C_2 = 0 \quad \text{and} \quad C_2 = -p_o L^2/2$$

Therefore,

$$AEu = -\frac{p_o}{2}(L^2 - x^2)$$

As is to be expected,

$$R_3 = AE\, u'(L) = p_o L$$

---

# Problems

## Section 2-4

✓ **2-1.** A standard steel specimen of $\frac{1}{2}$ in diameter is elongated 0.0087 in in an 8-in gage length when it was subjected to a tensile force of 6250 lb. If the specimen was known to be in the elastic range, what is the elastic modulus of the steel?

✓ **2-2.** The axial strain for an aluminum rod due to an applied force is $10^{-3}$ m/m. If the rod is 400 mm long and 12 mm in diameter, what axial stress is caused by the applied force? Assume elastic behavior and let $E = 75$ GPa.

## Section 2-7

✓ **2-3.** A steel rod 10 m long used in a control mechanism must transmit a tensile force of 5 kN without stretching more than 3 mm, nor exceeding an allowable stress of 150 MPa. (a) What is the diameter of the rod? Give the answer to the nearest millimeter. $E = 210$ GPa. Does strength or stiffness of the rod control the design? (b) Find the spring constant for the rod.

**2-4.** Revise the data in Example 2-2 to read as follows: $P_1 = 10$ kips, $P_3 = 100$ kips, and $P_4 = 30$ kips, and the bar segments $AB$, $BC$, and $CD$ are, respectively, 4-, 2-, and 3-ft long. Then find (a) the force $P_2$ necessary for equilibrium and (b) the total elongation of rod $AD$. The cross-sectional area of the rod from $A$ to $B$ is 1 in², from $B$ to $C$ is 4 in², and from $C$ to $D$ is 2 in². Let $E = 30 \times 10^3$ ksi. (c) Plot the axial displacement diagram along the bar.

**2-5.** Find the axial spring constant for the bar of variable cross section in Example 2-2.

**2-6.** Assume that segments $L_1$, $L_2$, and $L_3$ of the circular member of variable cross section in Problem 1-13 are, respectively, 600, 500, and 400 mm long. Plot the axial force, the axial strain, and the axial displacement diagrams along the bar length $E = 200$ GPa.

**2-7.** Find the axial spring constant for the bar in Problem 2-6.

**2-8.** A solid bar 50 mm in diameter and 2000 mm long consists of a steel and an aluminum part fastened together as shown in the figure. When axial force $P$ is applied to the system, a strain gage attached to the aluminum indicates an axial strain of 873 μm/m. (a) Determine the magnitude of applied force $P$. (b) If the system behaves elastically, find the total elongation of the bar. Let $E_{St} = 210$ GPa, and $E_{Al} = 70$ GPa.

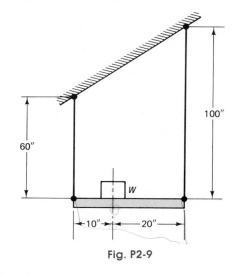

**Fig. P2-9**

**2-10.** In a California oil field, a very long steel drill pipe got stuck in hard clay (see figure). It was necessary to determine at what depth this occurred. The engineer on the job ordered the pipe subjected to a large upward tensile force. As a result of this operation, the pipe came up elastically 2 ft. At the same time, the pipe elongated 0.0014 in in an 8-in gage length. Approximately where was the pipe stuck? Assume that the cross-sectional area of the pipe was constant and that the media surrounding the pipe hindered the elastic deformation of the pipe very little in a static test.

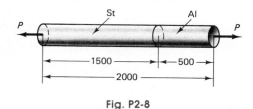

**Fig. P2-8**

**2-9.** Two wires are connected to a rigid bar, as shown in the figure. The wire on the left is of steel, having $A = 0.10$ in² and $E = 30 \times 10^6$ psi. The aluminum-alloy wire on the right has $A = 0.20$ in² and $E = 10 \times 10^6$ psi. (a) If a weight $W = 2000$ lb is applied as shown, how much will it deflect due to the stretch in the wires? (b) Where should the weight be located such that the bar would remain horizontal?

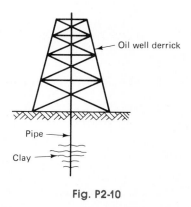

**Fig. P2-10**

**2-11.** A wall bracket is constructed as shown in the figure. All joints may be considered pin-connected. Steel rod $AB$ has a cross-sectional area of 5 mm². Member $BC$ is a rigid beam. If a 1000-mm diameter fric-

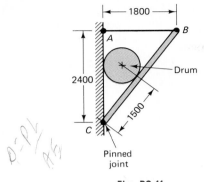

**Fig. P2-11**

tionless drum weighing 500 kg is placed in the position shown, what will be the elongation of rod $AB$? Let $E = 200$ GN/m$^2$.

**2-12.** Determine the shortening of steel tubular spreader bar $AB$ due to application of tensile forces at $C$ and $D$. The cross-sectional area of the tube is 100 mm$^2$. Let $E = 200$ GPa.

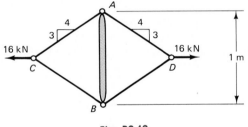

**Fig. P2-12**

***2-13.** Determine the elongation in rod $AB$ in Problem 1-43 if it is made of 0.125-in-diameter aluminum-alloy wire. Let $E = 10 \times 10^3$ ksi.

***2-14.** Determine the elongation in the 20-mm-diameter high-strength steel rod $CD$ for the frame in Problem 1-44. Let $E = 200$ GPa.

**2-15.** A rigid machine part $AD$ is suspended by double hangers $AE$ and $BF$, as shown in the figure. The hangers are made of cold-worked Monel Alloy (Ni-Cu) whose elastic modulus $E = 180$ GPa. This material yields at approximately 600 MPa. The cross-sectional area is 50 mm$^2$ for hanger pair $AE$ and 100 mm$^2$ for hanger pair $BF$. Determine the deflection that would occur at $D$ by applying a downward force of 10 kN at $C$. Check hanger stresses to assure that an elastic solution is applicable. Sketch deflected member $AD$, greatly exaggerating the vertical displacements (since

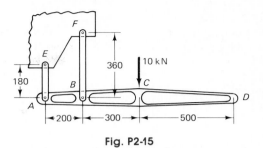

**Fig. P2-15**

vertical displacements are small, the horizontal displacements are negligibly small).

**2-16.** A planar mechanical system consists of two rigid bars, $BD$ and $EG$, and three rods, $AB$, $CF$, and $EH$, as shown in the figure. On application of force $P$ at $G$ the stress in all rods is 15 ksi. Each rod is 20 in. long. (a) Determine the vertical deflection of points $B$, $D$, $E$, and $G$ caused by the application of force $P = 300$ lb. (Since vertical displacements are small, the horizontal displacements are negligibly small.) (b) Show the deflected shape for the system, greatly exaggerating the vertical displacements. Let $E = 30 \times 10^3$ ksi.

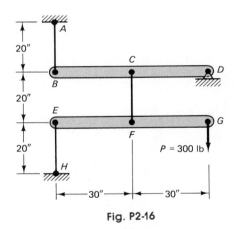

**Fig. P2-16**

**2-17.** If in Example 2-3, the rod is a 1 in$^2$ aluminum bar weighing 1.17 lb/ft, what should its length be for the free end to elongate 0.250 in under its own weight? $E = 10 \times 10^6$ psi.

**2-18.** What will be the deflection of the free end of the rod in Example 2-3 if, instead of Hooke's law, the stress-strain relationship is $\sigma = E\varepsilon^n$, where $n$ is a number dependent on the properties of the material?

**2-19.** A rod of two different cross-sectional areas is made of soft copper and is subjected to a tensile load

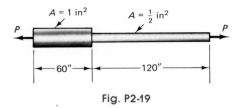

Fig. P2-19

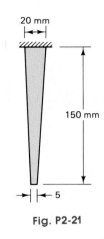

Fig. P2-21

as shown in the figure. (a) Determine the elongation of the rod caused by the application of force $P = 5$ kips. Assume that the axial stress-strain relationship is

$$\varepsilon = \sigma/16{,}000 + (\sigma/165)^3$$

where $\sigma$ is in ksi. (b) Find the residual bar elongation upon removal of force $P$. Assume that during unloading, copper behaves as a linearly elastic material with an $E$ equal to the tangent to the virgin $\sigma$-$\varepsilon$ curve at the origin.

$\checkmark$ **2-20.** A two-bar system has the configuration shown in the figure. The cross-sectional area for bar $AB$ is 0.200 in$^2$ and for bar $BC$ is 0.150 in$^2$. If the stress-strain diagram for the rods is bilinear as shown, how much would each wire elongate due to the application of vertical force $P = 4$ kips?

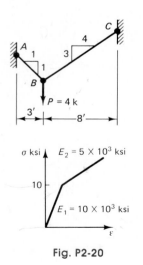

Fig. P2-20

**2-21.** The small tapered symmetric piece shown in the figure is cut from a 4-mm-thick plate. Determine the increase in length of this piece caused by its own

weight when hung from the top. The mass per unit volume for this material is $\gamma$ and the elastic modulus is $E$.

**2-22.** Two bars are to be cut from a 1-in-thick metal plate so that both bars have a constant thickness of 1 in. Bar $A$ is to have a constant width of 2 in throughout its entire length. Bar $B$ is to be 3 in wide at the top and 1 in wide at the bottom. Each bar is to be subjected to the same load $P$. Determine the ratio $L_A/L_B$ so that both bars will stretch the same amount. Neglect the weight of the bar.

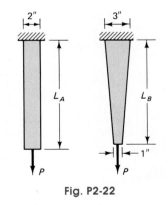

Fig. P2-22

**2-23.** The dimensions of a frustum of a right circular cone supported at the large end on a rigid base are shown in the figure. Determine the deflection of the top due to the weight of the body. The unit weight of material is $\gamma$; the elastic modulus is $E$. *Hint*: Consider the origin of the coordinate axes at the vertex of the extended cone.

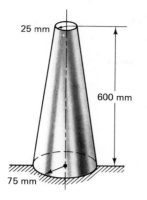

Fig. P2-23

**\*2-24.** Find the total elongation $\Delta$ of a slender elastic bar of constant cross-sectional area $A$, such as shown in the figure, if it is rotated in a horizontal plane with an angular velocity of $\omega$ radians per second. The unit weight of the material is $\gamma$. Neglect the small amount of extra material by the pin. *Hint*: First find the stress at a section a distance $r$ from the pin by integrating the effect of the inertial forces between $r$ and $L$. See Example 1-6.

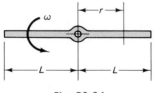

Fig. P2-24

**\*2-25.** An elastic rod having a cross-sectional area $A$ is bonded to the surrounding material, which has a thickness $a$, as shown in the figure. Determine the

change in the length of the rod due to the application of force $P$. Assume that the support provided for the rod by the surrounding material varies linearly as shown. Express the answer in terms of $P$, $A$, $a$, and $E$, where $E$ is the elastic modulus of the rod.

**2-26.** For the same frame as in Example 2-4, Fig. 2-24, find the horizontal and vertical deflections at point $B$ caused by applying a horizontal force of 3 kips at $B$. Assume linearly elastic behavior of the material.

**2-27.** Determine horizontal and vertical elastic displacements of load point $B$ for the two-bar system having the dimensions shown in the figure. Assume that for each bar, $AE = 10^4$ kips.

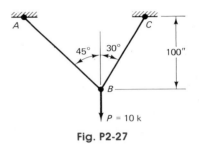

Fig. P2-27

**2-28.** For the data given in Problem 2-20, assuming linearly elastic behavior, find the horizontal and vertical displacements of load point $B$. Let $E = 10 \times 10^3$ ksi.

**2-29.** A jib crane has the dimensions shown in the figure. Rod $AB$ has a cross-sectional area of 300 mm$^2$ and tube $BC$, 320 mm$^2$. (a) Find the vertical stiffness of the crane at point $B$. (b) Determine the vertical deflection caused by the application of force $P = 16$ kN. Let $E = 200$ GPa.

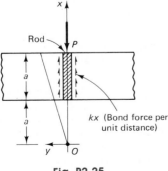

Fig. P2-25

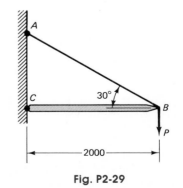

Fig. P2-29

## Section 2-8

**2-30.** A steel bar 2 in wide and 0.5 in thick is 25 in long, as shown in the figure. On application of force $P$, the bar width becomes narrower by $0.5 \times 10^{-3}$ in. Estimate the magnitude of applied force $P$ and the axial elongation of the bar. Assume elastic behavior and take $E = 30 \times 10^3$ ksi and $\nu = 0.25$.

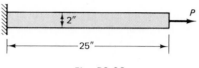

Fig. P2-30

**2-31.** A 10-mm-thick low-alloy-steel plate 150 mm wide and 2000 mm long is subjected to a set of uniformly distributed frictional forces along its two edges, as shown in the figure. If the total decrease in the transverse 150-mm dimension at section $a$–$a$ due to the applied forces is $15 \times 10^{-3}$ mm, what is the total elongation of the bar in the longitudinal direction. Let $E = 200$ GPa and $\nu = 0.25$. Assume that the steel behaves as a linearly elastic material.

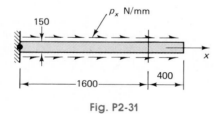

Fig. P2-31

## Section 2-9

**2-32.** A rigid bar rests on aluminum-alloy and steel uprights, as shown in the figure. (a) Determine the inclination of the horizontal bar after a raise in temperature of 100 °C. Assume the coefficients of thermal expansion for aluminum alloy and steel to be, respectively, $23.2 \times 10^{-6}$/°C and $11.7 \times 10^{-6}$/°C. To a greatly exaggerated scale, sketch the position of the bar after the raise in temperature. (b) What stresses would develop in the upright members if their tops were prevented from expanding? Let the elastic moduli for aluminum alloy and steel be, respectively, 75 GPa and 200 GPa. Compare the obtained stresses with those given in Table 1 of the Appendix. *Hint*: The ten-

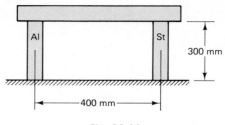

Fig. P2-32

dency for thermal expansion is counteracted by elastic contraction.

**2-33.** For the data given in Problem 2-20, find the vertical and horizontal displacement of point $B$ caused by a rise in temperature of 100 °F in the rod. Assume elastic behavior and use $\alpha$ and $E$ given in Table 1 of the Appendix for 6061-T6 aluminum alloy.

**2-34.** For the data given in Problem 2-29, find the vertical and horizontal displacements of point $B$ caused by a rise in temperature of 80 °C only in the rod. Let $\alpha = 11.7 \times 10^{-6}$/°C.

## Section 2-10

**2-35.** A 6 by 75 mm plate 600 mm long has a circular hole of 25 mm diameter located in its center. Find the axial tensile force that can be applied to this plate in the longitudinal direction without exceeding an allowable stress of 220 MPa.

**2-36.** Determine the extent by which a machined flat tensile bar used in a mechanical application is weakened by having an enlarged section, as shown in the figure. Since the bar is to be loaded cyclically, consider stress concentrations.

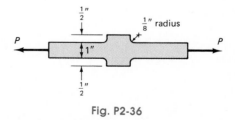

Fig. P2-36

**2-37.** A machine part 10 mm thick, having the dimensions shown in the figure, is to be subjected to cyclic loading. If the maximum stress is limited to 60 MPa, determine allowable force $P$. Approximate the

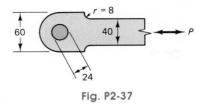

Fig. P2-37

stress concentration factors from Fig. 2-32. Where might a potential fracture occur?

**2-38.** A machine part of constant thickness for transmitting cyclical axial loading should have the dimensions shown in the figure. (a) Select the thickness needed in the member for transmitting an axial force of 12 kN in order to limit the maximum stress to 80 MPa. Approximate the stress concentration factors from Fig. 2-32. (b) Where might a potential fracture occur?

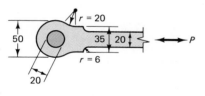

Fig. P2-38

**2-39.** A long slot is cut out from a 1 by 6 in steel bar 10 ft long, as shown in the figure. (a) Find the maximum stress if axial force $P = 50$ kips is applied to the bar. Assume that the upper curve in Fig. 2-32 is applicable. (b) For the same case, determine the total elongation of the rod. Neglect local effects of stress concentrations and assume that the reduced cross-sectional area extends for 24 in. (c) Estimate the elongation of the same rod if $P = 160$ kips. Assume that steel yields 0.020 in per inch at a stress of 40 ksi. (d) On removal of the load in part (c), what is the residual deflection? Let $E = 30 \times 10^6$ psi.

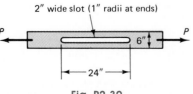

Fig. P2-39

## Sections 2-11 and 2-12

**2-40.** Verify the vertical deflection of point $B$ caused by applied force $P = 3$ kips in Example 2-4 using Eq. 2-24.

**2-41.** By applying Eq. 2-24, find the deflection of point $G$ in Problem 2-16.

**2-42.** Find the vertical deflection of point $B$ caused by the applied load in Problem 2-27 using Eq. 2-24.

**2-43.** Find the vertical deflection of point $B$ caused by the applied force $P$ in Problem 2-29 using Eq. 2-24.

**2-44.** A mechanical system consisting of a steel spreader bar $AB$ and four high-strength steel rods, $AC$, $CB$, $AD$, and $DB$, is subjected to forces at $C$ and $D$, as shown in the figure. Determine the increase in distance $CD$ that would occur on applying the two 8-kN forces. Both bars $AC$ and $CB$ have a cross-sectional area of 20 mm², and both bars $AD$ and $DB$, 40 mm². The cross-sectional area of the spreader bar is 100 mm². Let $E = 200$ GPa.

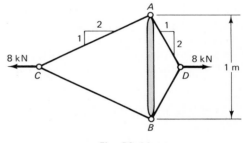

Fig. P2-44

## Section 2-13

**2-45.** Compare the dynamic stresses in the three steel bars of different diameters shown in the figure in their

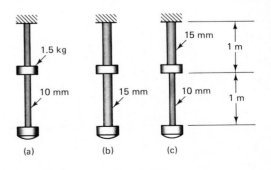

(a)          (b)          (c)

response to 1.5-kg masses falling freely through a distance of 1 m. Let $E = 200$ GPa. Assume no energy is dissipated through plastic deformation of the impact surfaces, nor at points of high local stresses occurring at supports.

**2-46.** Determine the stiffness required in the spring, for the system shown in the figure, for stopping a mass of 1 kg moving at a velocity of 3 m/sec such that, during impact, the spring deflection would not exceed 20 mm. Neglect frictional effects.

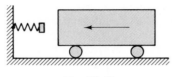

**Fig. P2-46**

## Section 2-15

**2-47.** An elastic bar of variable cross section, held at both ends, is loaded as shown in the figure. The flexibilities of the bar segments are $f/2$, $f$, and $f$. Determine the reactions, and plot the axial-force and axial-displacement diagrams.

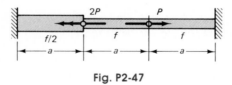

**Fig. P2-47**

**2-48.** Consider the same elastic bar of variable cross-sectional area shown in the two alternative figures. Determine deflection $\Delta_{ab}$ at $a$ caused by the application of a unit force at $b$, and show that it is equal to

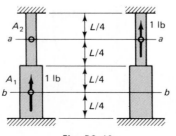

**Fig. P2-48**

$\Delta_{ba}$, the deflection at $b$ due to the application of a unit force at $a$. Let $A_1 = 2A_2$. (In Section 13-4, it is shown that this relationship is true in general for elastic systems. It is widely used in analysis. This conclusion can be reached by inspection for statically determinate bars.)

**2-49.** Consider the bar given in Example 2-2 and assume that ends $A$ and $D$ are held and that $P_2 = 390$ kN and $P_3 = 200$ kN act in the directions shown. (a) Determine the reactions. (b) Plot the axial force, axial strain, and axial displacement diagrams.

**2-50.** If in Problem 2-49, in addition to the applied forces, there is a drop in temperature of 100 °F, what reactions would develop at the supports? Let $\alpha = 13 \times 10^{-6}$/°F.

✓ **2-51.** For the 2-in² constant cross-sectional elastic bar shown in the figure, (a) determine the reactions, and (b) plot the axial-force, axial-strain, and axial-displacement diagrams. Let $E = 10 \times 10^3$ ksi.

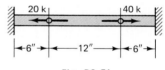

**Fig. P2-51**

✓ **2-52.** If a load of 1 kip is applied to a rigid bar suspended by three wires as shown in the figure, what force will be resisted by each wire? The outside wires are aluminum ($E = 10^7$ psi). The inside wire is steel ($E = 30 \times 10^6$ psi). Initially, there is no slack in the wire.

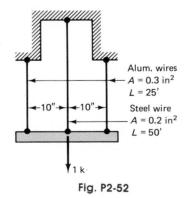

**Fig. P2-52**

**2-53.** Three identical equally spaced steel wires attached to a rigid bar support a mass $M$ developing a downward force of 5 kN, as shown in the figure. Initially, this force is equally distributed among the three wires. The stresses in the wires are well within the linearly elastic range of material behavior. (a) Determine the forces in the wires caused by a temperature drop of 50 °C in the right wire. Properties of the wires: $A = 10$ mm$^2$, $L = 2000$ mm, $E = 200 \times 10^3$ N/mm$^2$, $\alpha = 12.5 \times 10^{-6}$/°C. (b) At what change in temperature in the middle wire would it become slack?

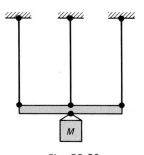

Fig. P2-53

**2-54.** Initially, on applying a 3-kN force to a rigid bar hung by three parallel wires (see the figure), all three wires become taut. What additional forces would develop in the wires if the left wire slips out 3 mm from its support. Each of the steel wires is 2000 mm long, has a cross-sectional area of 10 mm$^2$, and an elastic modulus of 200 GPa.

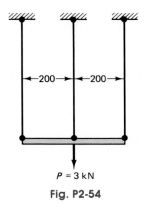

Fig. P2-54

**2-55.** Rework Example 2-14 by changing the bar inclination angles $\alpha$ to 30° and taking the cross-sectional area of bar *BD* as 2*A*. The cross sections of bars *AD* and *DC* remain equal to *A*.

## Section 2-16

**2-56.** An elastic bar held at both ends is loaded by an axial force $P$, as shown in the figure. Cross section $A$ of the bar is constant. (a) Determine the reactions and interpret the results in relation to the position of the applied force. (b) Plot the axial displacement diagram assuming that $E$ is known.

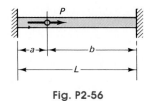

Fig. P2-56

**2-57.** For symmetrically arranged springs in parallel, the combined spring constant $k = \sum_n k_i$; see figure (a). Justify that for the springs in series, as in figure (b), the system spring constant $k$ follows from $1/k = \sum_n 1/k_i$, or, alternatively, $f = \sum_n f_i$, where $f$ is system flexibility, and $f_i$ the flexibility of an $i$th spring.

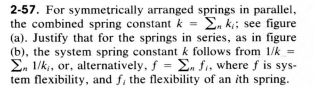

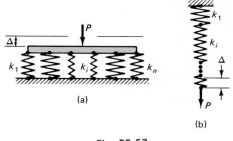

Fig. P2-57

**2-58.** A symmetrical arrangement of springs is attached to a rigid bar and carries an applied force $P$, as shown in the figure. (a) Find the reactions. *Hint:* Use

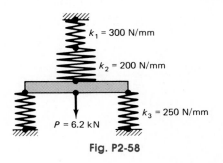

Fig. P2-58

the relationships given in Problem 2-57. (b) How is the total deflection distributed between the upper two springs?

**2-59.** Rework Problem 2-52 using the displacement method.

**2-60.** An elastic bar of variable cross section and held at both ends is axially loaded, as shown in the figure. The cross-sectional area of the small part is $A$ and of the larger, $2A$. (a) Using the displacement method, find the reactions. (b) Plot a qualitative axial-displacement diagram. *Hint:* Use the relationship given in Problem 2-57 for determining the combined stiffness of the bar segments to the left of $P$.

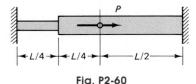

**Fig. P2-60**

**\*2-61.** A bar of constant thickness and held at both ends has the geometry shown in the figure. Determine the reactions caused by the axially applied force $P$. *Hint:* First find the stiffness for the tapered part of the bar.

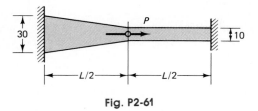

**Fig. P2-61**

**2-62.** A rigid bar is hinged at end $A$ and, in addition, is supported on three identical springs, each having stiffness $k$. (a) What is the degree of statical indeterminacy of this system? (b) How many degrees of free-

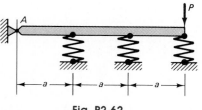

**Fig. P2-62**

dom are there? (c) Find the forces acting on the springs.

**2-63.** A rigid bar is supported by a pin at $A$ and two linearly elastic wires at $B$ and $C$, as shown in the figure. The area of the wire at $B$ is 60 mm$^2$ and for the one at $C$ is 120 mm$^2$. Determine the reactions at $A$, $B$, and $C$ caused by applied force $P = 6$ kN.

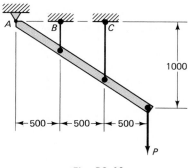

**Fig. P2-63**

**2-64.** Five steel rods, each having a cross-sectional area of 500 mm$^2$, are assembled in a symmetrical manner, as shown in the figure. Assume that the steel behaves as a linearly elastic material with $E = 200$ GPa. Determine the deflection of joint $A$ due to downward force $P = 2$ MN. Assume that, initially, the rods are taut.

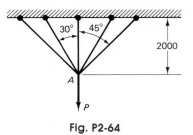

**Fig. P2-64**

## Section 2-17

**2-65.** An elastic bar of variable cross section and held at both ends is axially loaded at several points, as shown in the figure. The cross section for the larger area is $2A$, and for the smaller, $A$. (a) Compare the degrees of kinematic and static indeterminacies. (b) Determine the reactions if $P_1 = 3P$, $P_2 = 2P$, and $P_3 = P$. (c) Plot axial-force diagram.

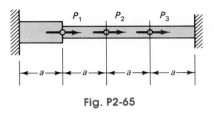

Fig. P2-65

**2-66.** Rework Problem 2-65 after removing force $P_1$. *Hint:* The degree of kinematic indeterminacy can be reduced by using a relationship given in Problem 2-57.

## Section 2-18

**2-67.** A material possesses a nonlinear stress-strain relationship given as $\sigma = K\varepsilon^n$, where $K$ and $n$ are material constants. If a rod made of this material and of constant area $A$ is initially fixed at both ends and is then loaded as shown in the figure, how much of applied force $P$ is carried by the left support?

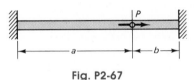

Fig. P2-67

**2-68.** A rod is fixed at $A$ and loaded with an axial force $P$, as shown in the figure. The material is elastic-perfectly plastic, with $E = 200$ GPa and a yield stress of 200 MPa. Prior to loading, a gap of 2 mm exists between the end of the rod and fixed support $C$. (a) Plot the load-displacement diagram for the load point assuming $P$ increases from zero to its ultimate value for the rod. The cross section from $A$ to $B$ is 200 mm² and that from $B$ to $C$ is 100 mm². (b) What will be the residual displacement of point $B$ upon release of the applied force?

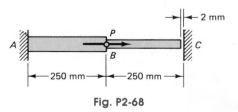

Fig. P2-68

**2-69.** The cross section of a short reinforced concrete column is as shown in the figure. Four 1-in round bars

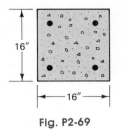

Fig. P2-69

serve as reinforcement. (a) Determine the instantaneous elastic strength of the column based on allowable stresses. (b) Estimate the ultimate (plastic) strength of the column. Assume that both materials are elastic-perfectly plastic. For steel, let $\sigma_{allow} = 24$ ksi, $\sigma_{yp} = 60$ ksi and $E = 30 \times 10^6$ psi, and for concrete, $\sigma_{allow} = 2000$ psi, $\sigma_{yp} = 3600$ psi and $E = 2 \times 10^6$ psi. (It has been shown experimentally that when steel yields, the concrete "yield" strength is approximately $0.85\sigma_{ult}$, where $\sigma_{ult}$ is the ultimate compressive strength of an unreinforced cylindrical specimen of the same material, age, and curing conditions. In order to achieve ductile behavior of columns, the use of lateral ties or spiral reinforcement is essential.)

**2-70.** A rigid platform rests on two aluminum bars ($E = 10^7$ psi) each 10.000 in long. A third bar made of steel ($E = 30 \times 10^6$ psi) and standing in the middle is 9.995 in long. (a) What will be the stress in the steel bar if a force $P$ of 100 kips is applied on the platform? (b) How much do the aluminum bars shorten? (c) What will be the ultimate (plastic) strength for the system if $(\sigma_{yp})_{Al} = 40$ ksi and $(\sigma_{yp})_{St} = 60$ ksi?

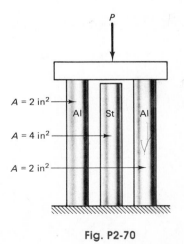

Fig. P2-70

**2-71.** A force $P = 1$ kN is applied to a rigid bar suspended by three wires, as shown in the figure. All wires are of equal size and the same material. For each wire, $A = 80$ mm$^2$, $E = 200$ GPa, and $L = 4$ m. If, initially, there were no slack in the wires, how will the applied load distribute between the wires?

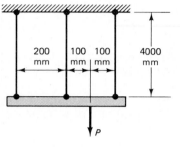

**Fig. P2-71**

✓ **2-72.** An aluminum rod 7 in long, having two different cross-sectional areas, is inserted into a steel link, as shown in the figure. If at 60 °F no axial force exists in the aluminum rod, what will be the magnitude of this force when the temperature rises to 160 °F? $E_{Al} = 10^7$ psi and $\alpha_{Al} = 12.0 \times 10^{-6}/°$F; $E_{St} = 30 \times 10^6$ psi and $\alpha_{St} = 6.5 \times 10^{-6}/°$F.

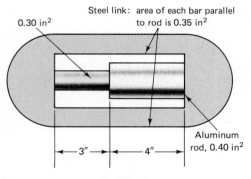

**Fig. P2-72**

**2-73.** An aluminum tube is axially compressed between the two heavy nuts of a steel bolt, as shown in the figure. If it is known that the axial stress in the sleeve at 80 °C is 20 MPa, at what temperature does this prestress become zero? For the aluminum tube: $A = 1000$ mm$^2$, $E = 70 \times 10^3$ MPa, and $\alpha = 23.2 \times 10^{-6}$ per °C. For the steel bolt: $A = 500$ mm$^2$, $E = 200 \times 10^3$ MPa, and $\alpha = 11.7 \times 10^{-6}$ per °C.

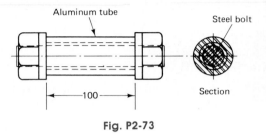

**Fig. P2-73**

**2-74.** Rework Example 2-23 after assuming that the elastic modulus $E_1$ for the middle bar is three times smaller than that for the outside bars, i.e., $E_2 = E_3 = 3E_1$.

**2-75.** Plot the load-deflection diagram for joint $A$ in Problem 2-64 caused by the applied force $P$, assuming that the yield stress for the material $\sigma_{yp} = 250$ MPa.

## Section 2-19

**\*2-76.** Rework Problem 2-25 using Eq. 2-48.

**2-77.** Rework Problem 2-24 using Eq. 2-48.

**\*2-78.** Rework Problem 2-56 using Eq. 2-48 and continuity conditions or singularity functions.

## chapter 3

# Generalized Hooke's Law, Pressure Vessels, and Thick-Walled Cylinders

## 3-1. Introduction

In addition to the normal strain discussed in the previous chapter in connection with axially loaded bars, in general, a body may also be subjected to shear strains. For the purposes of deformation analysis, such strains must be related to the applied shear stress. This topic is discussed in Part A of this chapter. In Part B, general mathematical definitions for normal and shear strains are given. Then, by employing the method of superposition, the generalized Hooke's law is synthesized, relating stresses and strains for a three-dimensional state of stress. Next, in Part C, thin-walled pressure vessels and shells of revolution are considered. The generalized Hooke's law is employed for the deformation analysis of these important elements of construction. In the concluding part, Part D, a solution for thick-walled cylinders is developed. This illustrates a solution of a typical boundary-value problem in the mathematical theory of elasticity, and, at the same time, provides bounds on the applicability of the equations established for thin-walled pressure vessels using engineering solid mechanics.

## Part A    CONSTITUTIVE RELATIONSHIPS FOR SHEAR

## 3-2. Stress-Strain Relationships for Shear

In addition to the normal strains related to the axial strains in bars discussed in Chapter 2, a body may be subjected to shear stresses that cause shear deformations. An example of such deformations is shown in

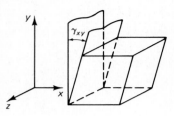

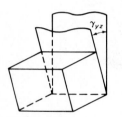

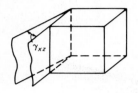

**Fig. 3-1** Possible shear deformations of an element.

Fig. 3-1. The change in the initial right angle between any two imaginary planes in a body defines *shear strain* γ (gamma). For infinitesimal elements, these small angles are measured in *radians*. The γ subscripts shown in Fig. 3-1 associate a particular shear strain with a pair of coordinate axes. Transformation of shear strain to any other mutually perpendicular set of planes will be discussed in Section 8-13.

For the purposes of deformation analysis, it is essential to establish a relationship between shear strain and shear stress based on experiments. As will become apparent in the next chapter, such experiments are most conveniently performed on thin-walled circular tubes in torsion. The elements of such tubes are essentially in a state of pure shear stress. An illustration of the conditions prevailing in a tube wall are shown in Fig. 3-2. The corresponding shear strains can be determined from the appropriate geometric measurements.

Note that per Section 1-4, the shear stresses on mutually perpendicular planes are equal; see Fig. 3-2(a). Moreover, since in this discussion, the stresses and strains are limited to a planar case, the subscripts for both can be omitted; see Fig. 3-2(b). By using experiments with thin-walled tubes, the generated shear stress-strain diagrams, except for their scale, greatly resemble those usually found for tension specimens (See Figs. 2-5, 2-6, and 2-13).

Two τ–γ diagrams are shown in Fig. 3-3. In the idealized diagram of elastic-perfectly plastic behavior, Fig. 3-3(b), $\tau_{yp}$ and $\gamma_{yp}$ designate, respectively, the shear yield stress and the shear yield strain.

In numerous technical problems, the shear stresses do not exceed the yield strength of the material. For most materials in this range of stress, just as for axially loaded bars, a *linear* relationship between pure shear stress and the angle γ it causes can be postulated. Therefore, mathematically, extension of Hooke's law for shear stress and strain reads

$$\tau = G\gamma \tag{3-1}$$

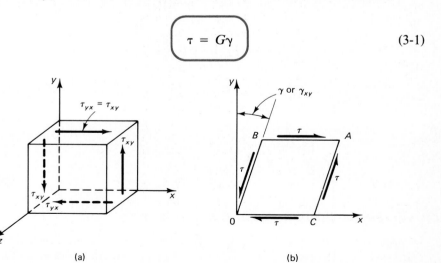

**Fig. 3-2** Element in pure shear.

(a)                     (b)

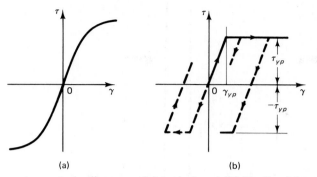

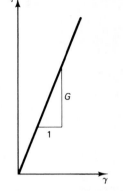

**Fig. 3-3** Shear stress-strain diagrams; (a) typical and (b) idealized for a ductile material.

**Fig. 3-4** Linear or Hookean relation between pure shear stress and strain.

where $G$ is a constant of proportionality called the *shear modulus of elasticity,* or the *modulus of rigidity.* Like $E$, $G$ is a constant for a given material. For emphasis, the relationship given by Eq. 3-1 is shown in Fig. 3-4.

## EXAMPLE 3-1

One of the shear mountings for a small piece of vibrating mechanical equipment has the dimensions shown in Fig. 3-5. The 8-mm thick pad of Grade 50 rubber[1] has $G = 0.64$ N/mm². Determine the shear spring constant $k_s$ for this mounting. Neglect the stiffness of the outer metal plates to which the rubber is bonded.

Solution

Here $\gamma \approx \dfrac{\Delta}{t}$; hence from Eq. 3-1, $\tau = G\gamma = \dfrac{G\,\Delta}{t}$

Further, $$F = \tau ab = \frac{G\,\Delta\,ab}{t}$$

Therefore, $$k_s = \frac{F}{\Delta} = \frac{Gab}{t} = \frac{0.64 \times 20 \times 40}{8} = 64 \text{ N/mm}$$

This solution neglects small local effects at the ends since no shear stresses act at the two boundaries.

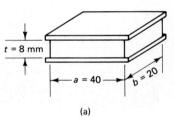

(a)

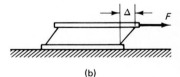

(b)

**Fig. 3-5**

## 3-3. Elastic Strain Energy for Shear Stresses

An expression for the elastic strain energy for an infinitesimal element in pure shear may be established in a manner analogous to that for one in uniaxial stress. Thus, consider an element in a state of shear, as shown

[1] P. B. Lindley, *Engineering Design with Natural Rubber* (Hertford, England: Malaysian Rubber Producers' Research Association, 1978).

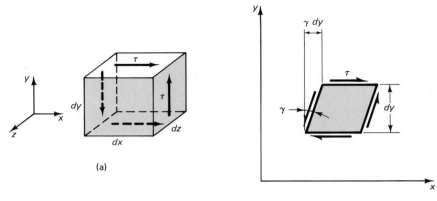

**Fig. 3-6** An element for deriving strain energy due to pure shear stresses.

in Fig. 3-6(a). The deformed shape of this element is shown in Fig. 3-6(b), where it is assumed that the bottom plane of the element is fixed in position.[2] As this element is deformed, the force on the top plane reaches a final value of $\tau\,dx\,dz$. The total displacement of this force for a small deformation of the element is $\gamma\,dy$; see Fig. 3-6(b). Therefore, since the external work done on the element is equal to the internal recoverable elastic strain energy,

$$dU_{\text{shear}} = \underbrace{\frac{1}{2}\,\tau\,dx\,dz}_{\text{average force}} \times \underbrace{\gamma\,dy}_{\text{distance}} = \frac{1}{2}\,\tau\gamma\,dx\,dy\,dz = \frac{1}{2}\,\tau\gamma\,dV \qquad (3\text{-}2)$$

where $dV$ is the volume of the infinitesimal element.

By recasting Eq. 3-2, the strain-energy density for shear becomes

$$(U_o)_{\text{shear}} = \left(\frac{dU}{dV}\right)_{\text{shear}} = \frac{\tau\gamma}{2} \qquad (3\text{-}3)$$

By using Hooke's law for shear stresses, $\tau = G\gamma$, Eq. 3-3 may be recast as

$$(U_o)_{\text{shear}} = \left(\frac{dU}{dV}\right)_{\text{shear}} = \frac{\tau^2}{2G} \qquad (3\text{-}4)$$

or

$$U_{\text{shear}} = \int_{\text{vol}} \frac{\tau^2}{2G}\,dV \qquad (3\text{-}5)$$

Note the similarity of Eqs. 3-2–3-5 to Eqs. 2-20–2-23 for elements in a state of uniaxial stress.

Applications of these equations are given in Chapters 4, 10, and 12.

---

[2] This assumption does not make the expression less general.

# Part B  GENERALIZED CONCEPTS OF STRAIN AND HOOKE'S LAW

## **\*\*³3-4. Mathematical Definition of Strain**

Since strains generally vary from point to point, the definitions of strain must relate to an infinitesimal element. With this in mind, consider an extensional strain taking place in one direction, as shown in Fig. 3-7(a). Some points like $A$ and $B$ move to $A'$ and $B'$, respectively. During straining, point $A$ experiences a displacement $u$. The displacement of point $B$ is $u + \Delta u$, since in addition to the rigid-body displacement $u$, common to the whole element $\Delta x$, a stretch $\Delta u$ takes place within the element. On this basis, the definition of the extensional or normal strain is[4]

$$\varepsilon = \lim_{\Delta x \to 0} \frac{\Delta u}{\Delta x} = \frac{du}{dx} \tag{3-6}$$

[3] This and the next section can be omitted without loss of continuity in the text.

[4] A more fundamental definition of extensional strain, more amenable to the more general concepts of *stretching* or *extending,* can be expressed, using Fig. 3-7(c), as

$$\varepsilon_x = \lim_{\Delta x \to 0} \frac{D'C' - DC}{DC} \tag{3-6a}$$

where the vectorial displacements of points $C$ and $D$ are $\mathbf{u}_C = CC'$ and $\mathbf{u}_D = DD'$. For the small deformations considered here, Eq. 3-6a reduces to Eq. 3-6. Also see Sections 8-11 and 12.

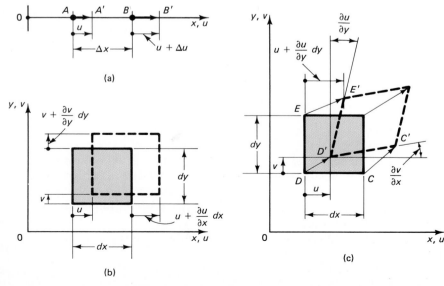

**Fig. 3-7** One and two-dimensional strained elements in initial and final positions.

If a body is strained in orthogonal directions, as shown for a two-dimensional case in Fig. 3-7(b), subscripts must be attached to ε to differentiate between the directions of the strains. For the same reason, it is also necessary to change the ordinary derivatives to partial ones. Therefore, if at a point of a body, $u$, $v$, and $w$ are the three displacement components occurring, respectively, in the $x$, $y$, and $z$ directions of the coordinate axes, the basic definitions of *normal strain* become

$$\varepsilon_x = \frac{\partial u}{\partial x} \qquad \varepsilon_y = \frac{\partial v}{\partial y} \qquad \varepsilon_z = \frac{\partial w}{\partial z} \qquad\qquad (3\text{-}7)$$

Note that double subscripts, analogously to those of stress, can be used for these strains. Thus,

$$\varepsilon_x \equiv \varepsilon_{xx} \qquad \varepsilon_y \equiv \varepsilon_{yy} \qquad \varepsilon_z \equiv \varepsilon_{zz} \qquad\qquad (3\text{-}8)$$

where one of the subscripts designates the direction of the line element, and the other, the direction of the displacement. Positive signs apply to elongations.

In addition to normal strains, an element can also experience a shear strain as shown for example in the $x$-$y$ plane in Fig. 3-7(c). This inclines the sides of the deformed element in relation to the $x$ and the $y$ axes. Since $v$ is the displacement in the $y$ direction, as one moves in the $x$ direction, $\partial v / \partial x$ is the slope of the initially horizontal side of the infinitesimal element. Similarly, the vertical side tilts through an angle $\partial u / \partial y$. On this basis, the initially right angle $CDE$ is reduced by the amount $\partial v / \partial x + \partial u / \partial y$. Therefore, for small angle changes, the definition of the *shear strain* associated with the $xy$ coordinates is

$$\gamma_{xy} = \gamma_{yx} = \frac{\partial v}{\partial x} + \frac{\partial u}{\partial y} \qquad\qquad (3\text{-}9)$$

To arrive at this expression, it is assumed that tangents of small angles are equal to the angles themselves in radian measure. Positive sign for the shear strain applies when the element is deformed, as shown in Fig. 3-7(c). (This deformation corresponds to the positive directions of the shear stresses; see Fig. 1-4.)

The definitions for the shear strains for the $xz$ and $yz$ planes are similar to Eq. 3-9:

$$\gamma_{xz} = \gamma_{zx} = \frac{\partial w}{\partial x} + \frac{\partial u}{\partial z} \qquad \gamma_{yz} = \gamma_{zy} = \frac{\partial w}{\partial y} + \frac{\partial v}{\partial z} \qquad (3\text{-}10)$$

In Eqs. 3-9 and 3-10, the subscripts on $\gamma$ can be permuted. This is permissible since no meaningful distinction can be made between the two sequences of each alternative subscript.

In examining Eqs. 3-7, 3-9, and 3-10, note that these six strain-displacement equations depend only on three displacement components $u$, $v$, and $w$. Therefore, these equations cannot be independent. Three independent equations can be developed showing the interrelationships among $\varepsilon_{xx}$, $\varepsilon_{yy}$, $\varepsilon_{zz}$, $\gamma_{xy}$, $\gamma_{yz}$, and $\gamma_{zx}$. The number of such equations reduces to one for a two-dimensional case. The derivation and the application of these equations, known as the *equations of compatibility,* are given in texts on the theory of elasticity.

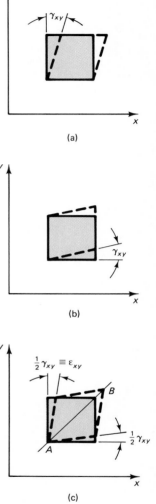

## **\*\*3-5. Strain Tensor**

The normal and the shear strains defined in the preceding section together express the strain tensor, which is highly analogous to the stress tensor already discussed. It is necessary, however, to modify the relations for the shear strains in order to have a tensor, an entity which must obey certain laws of transformation.[5] Thus, the physically attractive definition of the shear strain as the change in angle $\gamma$ is not acceptable when the shear strain is a component of a tensor. This heuristically may be attributed to the following. In Fig. 3-8(a), positive $\gamma_{xy}$ is measured from the vertical direction. The same positive $\gamma_{xy}$ is measured from the horizontal direction in Fig. 3-8(b). In Fig. 3-8(c), the same amount of shear deformation is shown to consist of two $\gamma_{xy}/2$'s. The deformed elements in Figs. 3-8(a) and (b) can be obtained by rotating the element in Fig. 3-8(c) as a rigid body through an angle of $\gamma_{xy}/2$. The scheme shown in Fig. 3-8(c) is the correct one for defining the shear-strain component as an element of a tensor. Since in this definition, the element is not rotated as a rigid body, the strain is said to be *pure* or *irrotational.* Following this approach, one redefines the shear strains as

Fig. 3-8 Shear deformations.

$$\varepsilon_{xy} = \varepsilon_{yx} = \frac{\gamma_{xy}}{2} = \frac{\gamma_{yx}}{2}$$

$$\varepsilon_{yz} = \varepsilon_{zy} = \frac{\gamma_{yz}}{2} = \frac{\gamma_{zy}}{2} \qquad (3\text{-}11)$$

$$\varepsilon_{zx} = \varepsilon_{xz} = \frac{\gamma_{zx}}{2} = \frac{\gamma_{xz}}{2}$$

[5] Rigorous discussion of this question is beyond the scope of this text. A better appreciation of it will develop, however, after the study of Chapter 8, where strain transformation for a two-dimensional case is considered.

From these equations, the strain tensor in matrix representation can be assembled as follows:

$$
\begin{pmatrix}
\varepsilon_x & \dfrac{\gamma_{xy}}{2} & \dfrac{\gamma_{xz}}{2} \\[2mm]
\dfrac{\gamma_{yx}}{2} & \varepsilon_y & \dfrac{\gamma_{yz}}{2} \\[2mm]
\dfrac{\gamma_{zx}}{2} & \dfrac{\gamma_{zy}}{2} & \varepsilon_z
\end{pmatrix}
\equiv
\begin{pmatrix}
\varepsilon_{xx} & \varepsilon_{xy} & \varepsilon_{xz} \\
\varepsilon_{yx} & \varepsilon_{yy} & \varepsilon_{yz} \\
\varepsilon_{zx} & \varepsilon_{zy} & \varepsilon_{zz}
\end{pmatrix}
\tag{3-12}
$$

The strain tensor is symmetric. Mathematically, the notation employed in the last expression is particularly attractive and has wide acceptance in continuum mechanics (elasticity, plasticity, rheology, etc.). Just as for the stress tensor, using indicial notation, one can write $\varepsilon_{ij}$ for the strain tensor.

Analogously to the stress tensor, the strain tensor can be diagonalized, having only $\varepsilon_1$, $\varepsilon_2$, and $\varepsilon_3$ as the surviving components. For a two-dimensional problem, $\varepsilon_3 = 0$; and one has the case of *plane strain*. The tensor for this situation is

$$
\begin{pmatrix}
\varepsilon_{xx} & \varepsilon_{xy} & 0 \\
\varepsilon_{yx} & \varepsilon_{yy} & 0 \\
0 & 0 & 0
\end{pmatrix}
\quad \text{or} \quad
\begin{pmatrix}
\varepsilon_1 & 0 & 0 \\
0 & \varepsilon_2 & 0 \\
0 & 0 & 0
\end{pmatrix}
\quad \text{or} \quad
\begin{pmatrix}
\varepsilon_1 & 0 \\
0 & \varepsilon_2
\end{pmatrix}
\tag{3-13}
$$

The transformation of strain suggested by Eq. 3-13 will be considered in Chapter 8.

The similarities and differences between plane strain and plane stress, defined in Section 1-4, will be discussed in the next section after the introduction of the generalized Hooke's law.

The reader should note that in discussing the concept of strain, the mechanical properties of the material were not involved. The equations are applicable whatever the mechanical behavior of the material. However, only small strains are defined by the presented equations. Also note that strains give only the relative displacement of points; rigid-body displacements do not affect the strains.

## 3-6. Generalized Hooke's Law for Isotropic Materials

In this article, six basic relationships between a general state of stress and strain are synthesized using the principle of superposition from the previously established simpler stress-strain equations. This set of equations is referred to as the *generalized Hooke's law*. These equations are applicable only to homogeneous *isotropic materials,* i.e., materials having the same properties in all directions. Hooke's law becomes more complex for anisotropic materials. For example, wood has decidedly different properties in the longitudinal, radial, and transverse directions, i.e., in the three orthogonal directions. Such materials, referred to as *orthotropic,* have nine independent material constants, whereas, as it will be shown in the next section, isotropic materials have only two. For fully *aniso-tropic* crystalline materials the number of independent material constants can be as large as 21.[6] In this book consideration is basically limited to isotropic materials, although by properly selecting the directions of axes, the developed procedures can be applied to orthotropic problems. Notable examples of these are wood and man-made materials, such as corrugated sheets or filament-reinforced plastics.

According to the basic concept of Hooke's law, a linear relationship exists between the applied stress and the resulting strain, such as shown in Fig. 3-9. During this process, a lateral contraction or expansion of a body takes place, depending on whether a body is being stretched or compressed. The extent of the lateral deformation is analytically formulated using Poisson's ratio (see Section 2-8). Qualitative illustrations of deformations caused by stresses applied along the coordinate axes are shown in Fig. 3-10.

Consider first that the element shown in Fig. 3-10(a) is subjected only to a tensile stress $\sigma_x$, as shown in Fig. 3-10(b). For this case, from $\sigma = E\varepsilon$, Eq. 2-3, one has $\varepsilon_x' = \sigma_x/E$, where $\varepsilon_x'$ is the strain in the $x$ direction. The corresponding lateral strains $\varepsilon_y'$ and $\varepsilon_z'$ along the $y$ and $z$ axes, re-

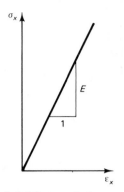

**Fig. 3-9** Linear relation between uniaxial stress and extensional strain.

[6] A. P. Boresi and O. M. Sidebottom, Advanced Mechanics of Materials, 4th ed., (New York: Wiley, 1985). I. S. Sokolnikoff, *Mathematical Theory of Elasticity* (New York: McGraw-Hill, 1956). L. E. Malvern, *Introduction to the Mechanics of a Continuous Medium,* (Englewood Cliffs, NJ: Prentice-Hall, 1969).

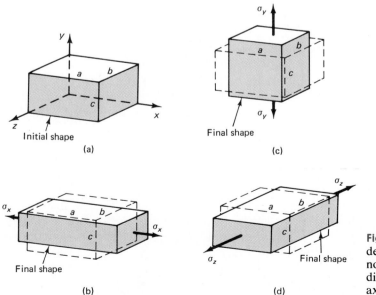
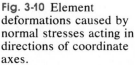

**Fig. 3-10** Element deformations caused by normal stresses acting in directions of coordinate axes.

spectively, follow, using Poisson's ratio, Eq. 2-16, and are $\varepsilon_y' = \varepsilon_x' = -\nu\sigma_x/E$. Similar expressions for strains $\varepsilon_x''$, $\varepsilon_y''$, and $\varepsilon_z''$ apply when the element is stressed, as shown in Fig. 3-10(c), and again for strains $\varepsilon_x'''$, $\varepsilon_y'''$, and $\varepsilon_z'''$, when stressed, as shown in Fig. 3-10(d). By *superposing* these strains, complete expressions for normal strains $\varepsilon_x$, $\varepsilon_y$, and $\varepsilon_z$ are obtained.

Since shear strains for the Cartesian axes can be treated as illustrated in Fig. 3-1, for the general problem only, the introduction of the appropriate subscripts into Eq. 3-1 are needed.

Based on the above, six equations for the generalized Hooke's law for *isotropic linearly elastic materials* for use with Cartesian coordinates can be written as

$$
\begin{aligned}
\varepsilon_x &= \frac{\sigma_x}{E} - \nu\frac{\sigma_y}{E} - \nu\frac{\sigma_z}{E} \\
\varepsilon_y &= -\nu\frac{\sigma_x}{E} + \frac{\sigma_y}{E} - \nu\frac{\sigma_z}{E} \\
\varepsilon_z &= -\nu\frac{\sigma_x}{E} - \nu\frac{\sigma_y}{E} + \frac{\sigma_z}{E} \\
\gamma_{xy} &= \frac{\tau_{xy}}{G} \\
\gamma_{yz} &= \frac{\tau_{yz}}{G} \\
\gamma_{zx} &= \frac{\tau_{zx}}{G}
\end{aligned}
\tag{3-14}
$$

These six equations have an inverse, i.e., they can be solved simultaneously to express stresses in terms of strains. This is left for the reader as an exercise.

If normal stresses are compressive, the signs of the corresponding terms change in the previous equations for the normal strains. The positive sense of the shear strains corresponding to the positive direction of the shear stresses (Fig. 1-3) is shown in Figs. 3-1 and 3-2. In the next section, it will be shown that in Eq. 3-14, the three elastic constants, $E$, $v$, and $G$ are not independent of each other, and that for *isotropic* materials, there are only two constants.

If a body experiences a change in temperature, the three normal strain equations should be modified by adding to each the expression given by Eq. 2-17. No changes in shear strains due to a change in temperature take place in isotropic materials since such materials have the same properties in all directions.

It should be clearly understood that Eq. 3-14 gives *strains*, i.e., *deformations per unit length*. If the strain is constant along the length of a member, in order to determine the deformation of such a member, the strain must be multiplied by the member's length. For example, the normal deformation $\Delta_x$ in the $x$ direction is given as

$$\Delta_x = \varepsilon_x L_x \qquad (3\text{-}15)$$

where $L_x$ is the member's length in the $x$ direction. Similar relations apply for $\Delta_y$ and $\Delta_z$. An integration process is used when strains vary along the length.

From the generalized Hooke's law equations, some useful comments can be made to clarify the distinction between *plane stress* and *plane strain* problems. An examination of Eq. 1-3 for the plane stress problem shows that $\sigma_x$ and $\sigma_y$ may exist. If either one or both of these stresses are present, according to the third Eq. 3-14, a normal strain $\varepsilon_z$ will develop. Conversely, in the plane strain problem, defined by Eq. 3-13, the normal strain $\varepsilon_z$ must be zero. Therefore, in this case, if either $\sigma_x$ and/or $\sigma_y$ are present, it can be concluded from the third Eq. 3-14 that $\sigma_z$ should not be zero. The similarity and the difference between the two kinds of problems can be further clarified from the table, where the stresses and strains are shown in matrix form.

| Plane Stress | Plane Strain |
|---|---|
| $\begin{pmatrix} \sigma_x & \tau & 0 \\ \tau & \sigma_y & 0 \\ 0 & 0 & 0 \end{pmatrix}$ | $\begin{pmatrix} \varepsilon_x & \gamma/2 & 0 \\ \gamma/2 & \varepsilon_y & 0 \\ 0 & 0 & 0 \end{pmatrix}$ |
| $\begin{pmatrix} \varepsilon_x & \gamma/2 & 0 \\ \gamma/2 & \varepsilon_y & 0 \\ 0 & 0 & \varepsilon_z \end{pmatrix}$ | $\begin{pmatrix} \sigma_x & \tau & 0 \\ \tau & \sigma_y & 0 \\ 0 & 0 & \sigma_z \end{pmatrix}$ |

## EXAMPLE 3-2

A 50 mm cube of steel is subjected to a uniform pressure of 200 MPa acting on all faces. Determine the change in dimension between two parallel faces of the cube. Let $E = 200$ GPa and $\nu = 0.25$.

## Solution

Using the first expression in Eq. 3-14 and Eq. 3-15, and noting that pressure is a compressive stress,

$$\varepsilon_x = \frac{(-200)}{200 \times 10^3} - \left(\frac{1}{4}\right)\frac{(-200)}{200 \times 10^3} - \left(\frac{1}{4}\right)\frac{(-200)}{200 \times 10^3}$$
$$= -5 \times 10^{-4} \text{ mm/mm}$$
$$\Delta_x = \varepsilon_x L_x = -5 \times 10^{-4} \times 50 = -0.025 \text{ mm (contraction)}$$

In this case $\Delta_x = \Delta_y = \Delta_z$.

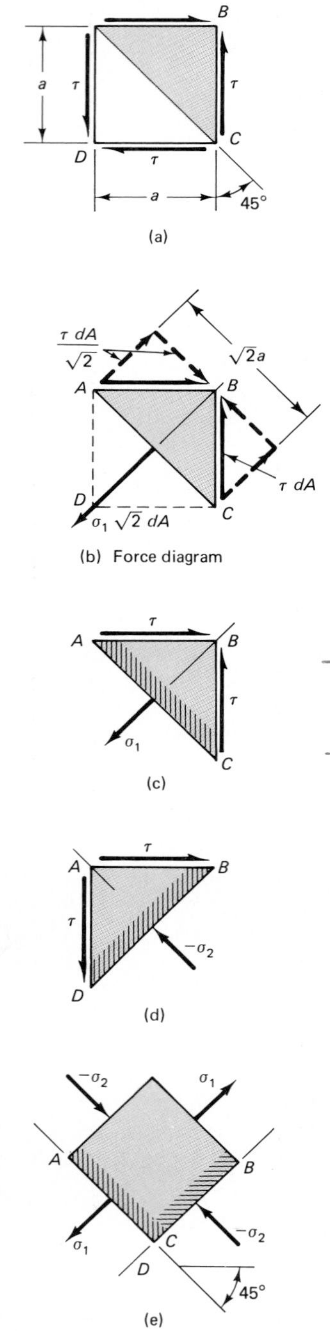

(a)

(b) Force diagram

(c)

(d)

(e)

Fig. 3-11 Transformation of pure shear stress into equivalent normal stresses.

## 3-7. *E, G*, and $\nu$ Relationships

In order to demonstrate the relationship among $E$, $G$, and $\nu$, first it must be shown that a state of pure shear, such as shown in Fig. 3-11(a), can be *transformed* into an *equivalent system of normal stresses*. This can be shown in the following manner.

Bisect square element *ABCD* by diagonal *AC* and isolate a triangular element, as shown in Fig. 3-11(b). If this element is *dz* thick, then each area associated with sides *AB* or *BC* is *dA*, and that associated with the diagonal *AC* is $\sqrt{2}\ dA$. Since the shear stress acting on the areas *dA* is $\tau$, the **forces** acting on these areas are $\tau\ dA$. The components of these forces acting *toward* diagonal *BD* are in equilibrium. On the other hand, the *components* parallel to diagonal *BD* develop a resultant $\tau\sqrt{2}\ dA$ acting normal to *AC*. This **force** is equilibrated by the *normal* stresses $\sigma_1$ acting on area $\sqrt{2}\ dA$ associated with diagonal *AC*. This gives rise to a **force** $\sigma_1\sqrt{2}\ dA$ shown in the figure. Since the shear stress resultant and this force must be equal, it follows that $\sigma_1 = \tau$. These *stresses* are shown in Eq. 3-11(c) and **cannot** be treated as forces.

By isolating an element with a side *BD,* as shown in Eq. 3-11(d), and proceeding in the same manner as before, a conclusion is reached that $\sigma_2 = -\tau$. The results of the two analyses are displayed in Fig. 3-11(e). This representation of stress is completely *equivalent* to that shown in Fig. 3-11(a). Therefore, *a pure shear stress at a point can be alternatively represented by the normal stresses at 45° with the directions of the shear stresses,* as shown in Fig. 3-11(e), and numerically,

$$\sigma_1 = -\sigma_2 = \tau \tag{3-16}$$

This important stress transformation enables one to proceed in establishing the relationship among $E$, $G$ and $\nu$. For this purpose, consider the deformed element shown in Fig. 3-12, and determine the strain in diagonal *DB* on two different bases. In one approach, determine strain from shear stresses; in the other, from the equivalent normal stresses.

Considering only infinitesimal deformations, and letting $\sin \gamma \approx \tan \gamma \approx \gamma$ and $\cos \gamma \approx 1$, it follows that displacement *BB'* due to shear is $a\gamma$. The projection of this displacement onto diagonal *DB'*, which, to the order of the approximation adopted, is equal to the stretch of *DB*, is $a\gamma/\sqrt{2}$. Therefore, since the length of *DB* is $\sqrt{2}a$, its normal strain $\varepsilon_{45°}$ is $\gamma/2$. Hence, recalling that $\tau = G\gamma$, Eq. 3-1, one has

$$\varepsilon_{45°} = \frac{\tau}{2G} \tag{3-17}$$

However, the shear stresses causing the deformation shown in Fig. 3-12 are equivalent to the normal stresses represented in Fig. 3-11(e). Therefore, if the x axis is directed along diagonal *DB*, the first Eq. 3-14 can be applied by taking $\sigma_x = \sigma_1$, $\sigma_y = -\sigma_2$, and $\sigma_z = 0$. In this manner an alternative expression for the normal strain in diagonal *DB* is found.

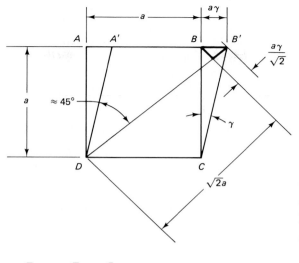

Fig. 3-12 Kinematics of element deformation for establishing a relationship between shear and extensional strains.

$$\varepsilon_{45°} = \frac{\sigma_1}{E} - v\frac{\sigma_2}{E} = \frac{\tau}{E}(1 + v) \qquad (3\text{-}18)$$

Equating the two alternative relations for the strain along the shear diagonal and simplifying,

$$G = \frac{E}{2(1 + v)} \qquad (3\text{-}19)$$

This is the basic relation between $E$, $G$, and $v$; it shows that these quantities are not independent of one another. If any two of these are determined experimentally, the third can be computed. Note that the shear modulus $G$ is always less than the elastic modulus $E$, since the Poisson ratio $v$ is a positive quantity. For most materials, $v$ is in the neighborhood of $\frac{1}{4}$.

## **7 3-8. Dilatation and Bulk Modulus

By extending some of the established concepts, one can derive an equation for volumetric changes in elastic materials subjected to stress. In the process of doing this, two new terms are introduced and defined.

The sides $dx$, $dy$, and $dz$ of an infinitesimal element after straining become $(1 + \varepsilon_x)\,dx$, $(1 + \varepsilon_y)\,dy$, and $(1 + \varepsilon_z)\,dz$, respectively. After subtracting the initial volume from the volume of the strained element, the change in volume is determined. This is

[7] Study of this section is optional.

$$(1 + \varepsilon_x) \, dx \, (1 + \varepsilon_y) \, dy \, (1 + \varepsilon_z) \, dz - dx \, dy \, dz$$
$$\approx (\varepsilon_x + \varepsilon_y + \varepsilon_z) \, dx \, dy \, dz$$

where the products of strain $\varepsilon_x \varepsilon_y + \varepsilon_y \varepsilon_z + \varepsilon_z \varepsilon_x + \varepsilon_x \varepsilon_y \varepsilon_z$, being small, are neglected. Therefore, in the infinitesimal (small) strain theory, $e$, the change in volume per unit volume, often referred to as *dilatation*, is defined as

$$e = \varepsilon_x + \varepsilon_y + \varepsilon_z \tag{3-20}$$

The shear strains cause no change in volume.

Based on the generalized Hooke's law, the dilatation can be found in terms of stresses and material constants. For this purpose, the first three Eqs. 3-14 must be added together. This yields

$$e = \varepsilon_x + \varepsilon_y + \varepsilon_z = \frac{1 - 2v}{E} (\sigma_x + \sigma_y + \sigma_z) \tag{3-21}$$

which means that dilatation is proportional to the algebraic sum of all normal stresses.

If an elastic body is subjected to hydrostatic pressure of uniform intensity $p$, so that $\sigma_x = \sigma_y = \sigma_z = -p$, then from Eq. 3-21,

$$e = -\frac{3(1 - 2v)}{E} p \quad \text{or} \quad \frac{-p}{e} = k = \frac{E}{3(1 - 2v)} \tag{3-22}$$

The quantity $k$ represents the ratio of the hydrostatic compressive stress to the decrease in volume and is called the *modulus of compression,* or *bulk modulus.*

## Part C    THIN-WALLED PRESSURE VESSELS

### 3-9. Cylindrical and Spherical Pressure Vessels

In this section, attention is directed toward two types of thin-walled pressure vessels: cylindrical and spherical. Both of these types of vessels are very widely used in industry; hence, this topic is of great practical importance. In analyzing such vessels for elastic deformations, an application of the generalized Hooke's law is required.

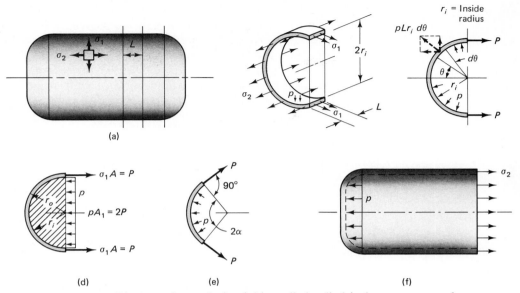

**Fig. 3-13** Diagrams for analysis of thin-walled cylindrical pressure vessels.

The walls of an ideal thin-walled pressure vessel act as a membrane, i.e., no bending of the walls takes place. A sphere is an ideal shape for a closed pressure vessel if the contents are of negligible weight; a cylindrical vessel is also good with the exception of the junctures with the ends, a matter to be commented on in more detail in the next section.

The analysis of pressure vessels will begin by considering a cylindrical pressure vessel such as a boiler, as shown in Fig. 3-13(a). A segment is isolated from this vessel by passing two planes perpendicular to the axis of the cylinder and one additional longitudinal plane *through* the same axis, shown in Fig. 3-13(b). The conditions of symmetry exclude the presence of any shear stresses in the planes of the sections, as shear stresses would cause an incompatible distortion of the tube. Therefore, the stresses that can exist on the sections of the cylinder can only be the normal stresses, $\sigma_1$ and $\sigma_2$, shown in Fig. 3-13(b). These stresses, multiplied by the respective areas on which they act, maintain the element of the cylinder in equilibrium against the internal pressure.

Let the internal pressure in excess of the external pressure be $p$ psi or Pa (gage pressure), and let the internal radius of the cylinder be $r_i$. Then the *force* on an infinitesimal area $Lr_i \, d\theta$ (where $d\theta$ is an infinitesimal angle) of the cylinder caused by the internal pressure acting *normal* thereto is $pLr_i \, d\theta$; see Fig. 3-13(c). The component of this force acting in the horizontal direction is $(pLr_i \, d\theta) \cos \theta$; hence, the total resisting force of $2P$ acting on the cylindrical segment is

$$2P = 2 \int_0^{\pi/2} pLr_i \cos d\theta = 2pr_iL \qquad (3\text{-}23)$$

Again from symmetry, half of this total force is resisted at the top cut through the cylinder and the other half is resisted at the bottom. The normal stresses $\sigma_2$ acting in a direction parallel to the axis of the cylinder do not enter into the above integration.

Instead of obtaining the force $2P$ caused by the internal pressure by integration, as above, a simpler equivalent procedure is available. From an alternate point of view, the two forces $P$ resist the force developed by the internal pressure $p$, which acts perpendicular to the *projected area* $A_1$ of the cylindrical segment onto the diametral plane; see Fig. 3-13(d). This area in Fig. 3-13(b) is $2r_iL$; hence, $2P = A_1p = 2r_iLp$. This force is resisted by the forces developed in the material in the longitudinal cuts, and since the outside radius of the cylinder is $r_o$, the area of *both* longitudinal cuts is $2A = 2L(r_o - r_i)$. Moreover, if the *average* normal stress acting on the longitudinal cut is $\sigma_1$, the force resisted by the walls of the cylinder is $2L(r_o - r_i)\sigma_1$. Equating the two forces, $2r_iLp = 2L(r_o - r_i)\sigma_1$.

Since $r_o - r_i$ is equal to $t$, the thickness of the cylinder wall, the last expression simplifies to

$$\sigma_1 = \frac{pr_i}{t} \tag{3-24}$$

The normal stress given by Eq. 3-24 is often referred to as the *circumferential* or the *hoop stress*. Equation 3-24 is valid only for thin-walled cylinders, as it gives the *average* stress in the hoop. However, as is shown in Example 3-6, the wall thickness can reach one-tenth of the internal radius and the error in applying Eq. 3-24 will still be small. Since Eq. 3-24 is used primarily for *thin-walled* vessels, where $r_i \approx r_o$, the subscript for the radius is usually omitted.

Equation 3-24 can also be derived by passing two longitudinal sections, as shown in Fig. 3-13(e). Because of the assumed membrane action, the forces $P$ in the hoop must be considered acting tangentially to the cylinder. The horizontal components of the forces $P$ maintain the horizontal component of the internal pressure in a state of static equilibrium.

The other normal stress $\sigma_2$ acting in a cylindrical pressure vessel acts *longitudinally,* Fig. 3-13(b), and it is determined by solving a simple axial-force problem. By passing a section through the vessel perpendicular to its axis, a free-body as shown in Fig. 3-13(f) is obtained. The force developed by the internal pressure is $p\pi r_i^2$, and the force developed by the longitudinal stress $\sigma_2$ in the walls is $\sigma_2(\pi r_o^2 - \pi r_i^2)$. Equating these two forces and solving for $\sigma_2$,

$$p\pi r_i^2 = \sigma_2(\pi r_o^2 - \pi r_i^2)$$

$$\sigma_2 = \frac{pr_i^2}{r_o^2 - r_i^2} = \frac{pr_i^2}{(r_o + r_i)(r_o - r_i)}$$

However, $r_o - r_i = t$, the thickness of the cylindrical wall, and since this development is restricted to *thin-walled* vessels, $r_o \approx r_i \approx r$; hence,

$$\sigma_2 = \frac{pr}{2t} \qquad (3\text{-}25)$$

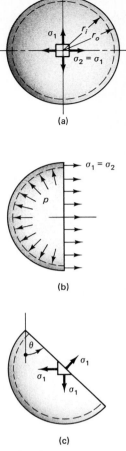

(a)

Note that for *thin-walled cylindrical* pressure vessels, $\sigma_2 \approx \sigma_1/2$.

An analogous method of analysis can be used to derive an expression for *thin-walled* spherical pressure vessels. By passing a section through the center of the sphere of Fig. 3-14(a), a hemisphere shown in Fig. 3-14(b) is isolated. By using the same notation as before, an equation identical to Eq. 3-25 can be derived. However, for a sphere, *any section that passes through the center of the sphere yields the same result* whatever the inclination of the element's side; see Fig. 3-14(c). Hence, the maximum membrane stresses for thin-walled *spherical pressure vessels* are

(b)

$$\sigma_1 = \sigma_2 = \frac{pr}{2t} \qquad (3\text{-}26)$$

(c)

Infinitesimal elements for the vessels analyzed showing the normal stresses $\sigma_1$ and $\sigma_2$ viewed from the outside are indicated in Figs. 3-13(a), 3-14(a), and 3-14(c). According to Eq. 1-10, the maximum shear stresses associated with these normal stresses are half as large. The planes on which these shear stresses act may be identified on elements viewed toward a section through the wall of a vessel. Such a section is shown in Fig. 3-15. The stress $\sigma_2$ acts perpendicularly to the plane of the figure.

**Fig. 3-14** Thin-walled spherical pressure vessel.

## EXAMPLE 3-3

Consider a closed cylindrical steel pressure vessel, as shown in Fig. 3-13(a). The radius of the cylinder is 1000 mm and its wall thickness is 10 mm. (a) Determine the hoop and the longitudinal stresses in the cylindrical wall caused by an internal pressure of 0.80 MPa. (b) Calculate the change in diameter of the cylinder caused by pressurization. Let $E = 200$ GPa, and $\nu = 0.25$. Assume that $r_i \approx r_o \approx r$.

### Solution

The stresses follow by direct application of Eqs. 3-24 and 3-25:

$$\sigma_1 = \frac{pr}{t} = \frac{0.8 \times 1}{10 \times 10^{-3}} = 80 \text{ MPa}$$

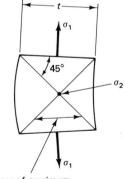

Planes of maximum shear stresses

**Fig. 3-15** In yielded steel pressure vessels shear slip planes at 45° can be observed on etched specimens.

**155**

and
$$\sigma_2 = \frac{pr}{2t} = \frac{0.8 \times 1}{2 \times 10 \times 10^{-3}} = 40 \text{ MPa}$$

The stress perpendicular to the cylinder wall, $\sigma_3 = p = 0.80$ MPa, on the inside decreases to zero on the outside. Being small, it can be neglected. Hence, on setting $\sigma_x = \sigma_1$, $\sigma_y = \sigma_2$, and $\sigma_z = 0$ in the first expression in Eq. 3-14, one obtains the hoop strain $\varepsilon_1$:

$$\varepsilon_1 = \frac{\sigma_1}{E} - v\frac{\sigma_2}{E} = \frac{80}{200 \times 10^3} - \frac{40}{4 \times 200 \times 10^3} = 0.35 \times 10^{-3} \text{ mm/mm}$$

On pressurizing the cylinder, the radius $r$ increases by an amount $\Delta$. For this condition, the hoop strain $\varepsilon_1$ can be found by calculating the difference in the strained and the unstrained hoop circumferences and dividing this quantity by the initial hoop length. Therefore,

$$\varepsilon_1 = \frac{2\pi(r + \Delta) - 2\pi r}{2\pi r} = \frac{\Delta}{r} \tag{3-27}$$

By recasting this expression and substituting the numerical value for $\varepsilon_1$ found earlier,

$$\Delta = \varepsilon_1 r = 0.35 \times 10^{-3} \times 10^3 = 0.35 \text{ mm}$$

---

## EXAMPLE 3-4

Consider a steel spherical pressure vessel of radius 1000 mm having a wall thickness of 10 mm. (a) Determine the maximum membrane stresses caused by an internal pressure of 0.80 MPa. (b) Calculate the change in diameter in the sphere caused by pressurization. Let $E = 200$ GPa, and $v = 0.25$. Assume that $r_i \approx r_o \approx r$.

## Solution

The maximum membrane normal stresses follow directly from Eq. 3-26.

$$\sigma_1 = \sigma_2 = \frac{pr}{2t} = \frac{0.80 \times 1}{2 \times 10 \times 10^{-3}} = 40 \text{ MPa}$$

The same procedure as in the previous example can be used for finding the expansion of the sphere due to pressurization. Hence, if $\Delta$ is the increase in the radius $r$ due to this cause, $\Delta = \varepsilon_1 r$, where $\varepsilon_1$ is the membrane strain on the great circle. However, from the first expression in Eq. 3-14, one has

$$\varepsilon_1 = \frac{\sigma_1}{E} - v\frac{\sigma_2}{E} = \frac{40}{200 \times 10^3} - \frac{40}{4 \times 200 \times 10^3} = 0.15 \times 10^{-3} \text{ mm/mm}$$

Hence, $\Delta = \varepsilon_1 r = 0.15 \times 10^{-3} \times 10^3 = 0.15$ mm

## EXAMPLE 3-5

For an industrial laboratory a pilot unit is to employ a pressure vessel of the dimensions shown in Fig. 3-16. The vessel will operate at an internal pressure of 0.7 MPa. If for this unit 20 bolts are to be used on a 650 mm bolt circle diameter, what is the required bolt diameter at the root of the threads? Set the allowable stress in tension for the bolts at 125 MPa; however, assume that at the root of the bolt threads the stress concentration factor is 2.

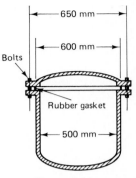

**Figure 3-16**

## Solution

The vertical force $F$ acting on the cover is caused by the internal pressure $p$ of 0.7 MPa acting on the horizontal projected area within the self-sealing rubber gasket, i.e.,

$$F = 0.7 \times 10^6 \times \pi(600/2)^2 = 198 \times 10^9 \text{ N}$$

Assuming that this force is equally distributed among the 20 bolts, the force $P$ per bolt is $198 \times 10^9/20 = 9.90 \times 10^9$ N. Using the given stress-concentration factor $K = 2$ and applying Eq. 2-19, the required bolt area $A$ at the root of the threads,

$$A = K \frac{P}{\sigma_{allow}} = \frac{2 \times 9.90 \times 10^9}{125 \times 10^6} = 158 \text{ mm}^2$$

Hence the required bolt diameter $d$ at the root of the threads d $= 2\sqrt{A/\pi} = 14.2$ mm. Note from Example 2-22 that initial tightening of the bolts results in a relatively small increase in total bolt stress when the vessel is pressurized.

## *[8]3.10. Remarks on Thin-Walled Pressure Vessels

It is instructive to note that for comparable size and wall thickness, the maximum normal stress in a spherical pressure vessel is only about one-half as large as that in a cylindrical one. The reason for this can be clarified by making reference to Figs. 3-17 and 3-18. In a cylindrical pressure vessel, the longitudinal stresses, $\sigma_2$, parallel to the vessel's axis, do not contribute to maintaining the equilibrium of the internal pressure $p$ acting on the curved surface; whereas in a spherical vessel, a system of equal stresses resists the applied internal pressure. These stresses, given by Eqs. 3-24–3-26, are treated as biaxial, although the internal pressure $p$ acting on the wall causes local compressive stresses on the inside equal

[8] Study of this section is optional.

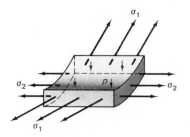

Fig. 3-17 An element of a thin-walled cylindrical pressure vessel.

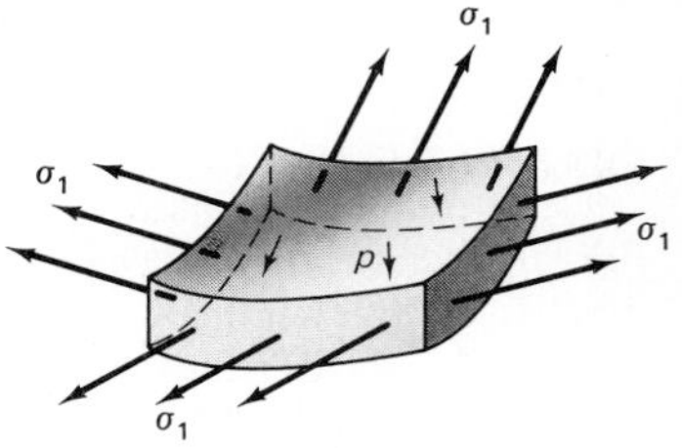

Fig. 3-18 An element of a thin-walled spherical pressure vessel.

to this pressure. As already pointed out in Example 3-3, such stresses are small in comparison with the membrane stresses $\sigma_1$ and $\sigma_2$, and are generally ignored for thin-walled pressure vessels. A more complete discussion of this problem is given in Section 3-13 and Example 3-6. A much more important problem arises at geometrical changes in the shape of a vessel. These can cause a disturbance in the membrane action. An illustration of this condition is given in Fig. 3-19 using the numerical results found in Examples 3-3 and 3-4.

If a cylindrical pressure vessel has hemispherical ends, as shown in Fig. 3-19(a), and if initially the cylinder and the heads were independent of each other, under pressurization they would tend to expand, as shown by the dashed lines. In general, the cylinder and the ends would expand by different amounts and would tend to create a discontinuity in the wall, as shown at $A$. However, physical continuity of the wall must be maintained by local bending and shear stresses in the neighborhood of the juncture, as shown in Fig. 3-19(b). If instead of relatively flexible hemi-

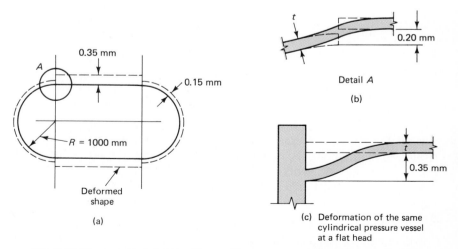

Fig. 3-19 Exaggerated deformations of pressure vessels at discontinuities.

spherical ends, thick end plates are used, the local bending and shear stresses increase considerably; see Fig. 3-19(c). For this reason, the ends (heads) of pressure vessels must be very carefully designed.[9] Flat ends are very undesirable.

A majority of pressure vessels are manufactured from curved sheets that are joined together by means of welding. Examples of welds used in pressure vessels are shown in Fig. 3-20, with preference given to the different types of butt joints. Some additional comments on welded joints may be found in Section 1-14.

In conclusion, it must be emphasized that the formulas derived for thin-walled pressure vessels in the preceding section should be used only for cases of *internal pressure*. If a vessel is to be designed for external pressure, as in the case of a vacuum tank or a submarine, *instability* (buckling) of the walls may occur, and stress calculations based on the previous formulas can be meaningless.

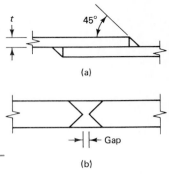

Fig. 3-20 Examples of welds used in pressure vessels. (a) Double-fillet lap joint, and (b) double-welded butt joint with V-grooves.

## Part D    THICK-WALLED CYLINDERS

## **[10]3-11. Introduction

Analysis of thick-walled cylinders under internal and external pressure is discussed in this part. This problem is related to the thin-walled cylindrical pressure vessel problem treated earlier. In order to solve the posed problem, a characteristic method of the *mathematical theory of elasticity* is employed. This consists of assuring equilibrium for each infinitesimal element, and, through the use of geometric relations, allowing only their compatible (possible) deformations. The equilibrium conditions are related to those of deformation using the generalized Hooke's law. Then the governing differential equation established on the preceding bases is solved *subject to the prescribed boundary conditions*. This approach differs from that used in engineering mechanics of solids, where the internal statical indeterminancy is resolved by means of a plausible kinematic assumption in each particular case. Occasionally, in engineering mechanics of solids, it becomes necessary to draw upon the solutions obtained using the methods of the mathematical theory of elasticity. This, for example, was already resorted to in treating stress concentrations at discontinuities in axially loaded bars. Solutions of two- and three-dimensional problems using the finite-element approach, philosophically, are in

---

[9] The ASME Unfired Pressure Vessel Code gives practical information on the design of ends; the necessary theory is beyond the scope of this text. In spite of this limitation, the elementary formulas for thin-walled cylinders developed here are suitable in the majority of cases.

[10] The remainder of this chapter can be omitted in a first course.

many respects similar to the methods of the mathematical theory of elasticity. In both cases, one seeks solutions to *boundary-value problems*.

Mathematically, the problem of thick-walled cylinders is rather simple, yet it clearly displays the characteristic method used in elasticity. Here the solution is carried further by including inelastic behavior of thick-walled cylinders. Both the elastic-plastic and the plastic states are examined.

The solution of the problem of thick-walled cylinders under internal pressure provides bounds on the applicability of the equations developed earlier for thin-walled cylinders. This solution is also useful for the design of extrusion molds and other mechanical equipment.

## **\*\*3-12.** Solution of the General Problem

Consider a long cylinder with axially restrained ends whose cross section has the dimensions shown in Fig. 3-21(a).[11] The inside radius of this cylinder is $r_i$; the outside radius is $r_o$. Let the internal pressure in the cylinder be $p_i$ and the outside, or external, pressure be $p_o$. Stresses in the wall of the cylinder caused by these pressures are sought.

This problem can be conveniently solved by using cylindrical coordinates. Since the cylinder is long, every ring of unit thickness measured perpendicular to the plane of the paper is stressed alike. A typical infinitesimal element of unit thickness is defined by two radii, $r$ and $r + dr$, and an angle $d\phi$, as shown in Fig. 3-21(b).

If the normal *radial* stress acting on the infinitesimal element at a distance $r$ from the center of the cylinder is $\sigma_r$, this variable stress at a distance $r + dr$ will be $\sigma_r + (d\sigma_r/dr)\, dr$. Both normal *tangential* stresses acting on the other two faces of the element are $\sigma_r$. These stresses, analogous to the hoop stresses in a thin cylinder, are equal. Moreover, since from the condition of symmetry, every element at the same radial distance from the center must be stressed alike, *no shear stresses act on the ele-*

[11] This problem was originally solved by Lamé, a French engineer, in 1833 and is sometimes referred to as the Lamé problem.

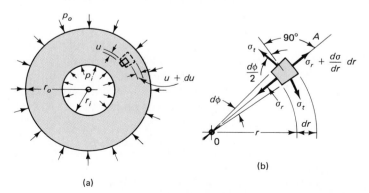

**Fig. 3-21** Thick-walled cylinder.

(a)

(b)

*ment shown.* Further, the axial stresses $\sigma_x$ on the two faces of the element are equal and opposite normal to the plane of the paper.

The nature of the stresses acting on an infinitesimal element having been formulated, a characteristic elasticity solution proceeds along the following pattern of reasoning.

## Static Equilibrium

The element chosen must be in static equilibrium. To express this mathematically requires the evaluation of *forces* acting on the element. These forces are obtained by multiplying stresses by their respective areas. The area on which $\sigma_r$ acts is $1 \times r\, d\phi$; that on which $\sigma_r + d\sigma_r$ acts is $1 \times (r + dr)\, d\phi$; and each area on which $\sigma_t$ acts is $1 \times dr$. The weight of the element itself is neglected. Since the angle included between the sides of the element is $d\phi$, both tangential stresses are inclined $\frac{1}{2}\, d\phi$ to the line perpendicular to $OA$. Then, summing the forces along a radial line, $\sum F_r = 0$,

$$\sigma_r r\, d\phi + 2\sigma_t\, dr\left(\frac{d\phi}{2}\right) - \left(\sigma_r + \frac{d\sigma_r}{dr}\, dr\right)(r + dr)\, d\phi = 0$$

Simplifying, and neglecting the infinitesimals of higher order,

$$\sigma_t - \sigma_r - r\frac{d\sigma_r}{dr} = 0 \qquad \text{or} \qquad \frac{d\sigma_r}{dr} + \frac{\sigma_r - \sigma_t}{r} = 0 \qquad (3\text{-}28)$$

This one equation has two unknown stresses, $\sigma_t$ and $\sigma_r$. Intermediate steps are required to express this equation in terms of one unknown so that it can be solved. This is done by introducing the geometry of deformations and properties of materials into the problem.

## Geometric Compatibility

The deformation of an element is described by its strains in the radial and tangential directions. If $u$ represents the *radial displacement* or *movement* of a cylindrical surface of radius $r$, Fig. 3-21(a), $u + (du/dr)\, dr$ is the radial displacement or movement of the adjacent surface of radius $r + dr$. Hence, the strain $\varepsilon_r$ of an element in the radial direction is

$$\varepsilon_r = \frac{\left(u + \dfrac{du}{dr}\, dr\right) - u}{dr} = \frac{du}{dr} \qquad (3\text{-}29)$$

The strain $\varepsilon_t$ in the tangential direction follows by subtracting from the length of the circumference of the deformed cylindrical surface of radius

$r + u$ the circumference of the unstrained cylinder of radius $r$ and dividing the difference by the latter length. Hence,

$$\varepsilon_t = \frac{2\pi(r + u) - 2\pi r}{2\pi r} = \frac{u}{r} \tag{3-30}$$

Note that Eqs. 3-29 and 3-30 give strains expressed in terms of *one* unknown variable $u$.

### Properties of Material

The generalized Hooke's law relating strains to stresses is given by Eq. 3-14, and can be restated here in the form[12]

$$\varepsilon_r = \frac{1}{E}(\sigma_r - v\sigma_t - v\sigma_x) \tag{3-31}$$

$$\varepsilon_t = \frac{1}{E}(-v\sigma_r + \sigma_t - v\sigma_x) \tag{3-32}$$

$$\varepsilon_x = \frac{1}{E}(-v\sigma_r - v\sigma_t + \sigma_x) \tag{3-33}$$

However, in the case of the thick-walled cylinder with axially restrained deformation, the problem is one of *plane strain,* i.e., $\varepsilon_x = 0$. The last equation then leads to a relation for the axial stress as

$$\sigma_x = v(\sigma_r + \sigma_t) \tag{3-34}$$

Introducing this result into Eqs. 3-31 and 3-32 and solving them simultaneously gives expressions for stresses $\sigma_r$ and $\sigma_t$ in terms of strains:

$$\sigma_r = \frac{E}{(1 + v)(1 - 2v)}[(1 - v)\varepsilon_r + v\varepsilon_t] \tag{3-35}$$

$$\sigma_t = \frac{E}{(1 + v)(1 - 2v)}[v\varepsilon_r + (1 - v)\varepsilon_t] \tag{3-36}$$

These equations bring the plane strain condition into the problem for elastic material.

### Formation of the Differential Equation

Now the equilibrium equation, Eq. 3-28, can be expressed in terms of one variable $u$. Thus, one eliminates the strains $\varepsilon_r$ and $\varepsilon_t$ from Eqs. 3-35

---

[12] Since an infinitesimal cylindrical element includes an *infinitesimal* angle between two of its sides, it can be treated as if it were an element in a Cartesian coordinate system.

and 3-36 by expressing them in terms of the displacement $u$, as given by Eqs. 3-29 and 3-30; then the radial and tangential stresses are

$$\sigma_r = \frac{E}{(1 + v)(1 - 2v)} \left[ (1 - v) \frac{du}{dr} + v \frac{u}{r} \right]$$

(3-37)

and

$$\sigma_t = \frac{E}{(1 - v)(1 - 2v)} \left[ v \frac{du}{dr} + (1 - v) \frac{u}{r} \right]$$

and, by substituting these values into Eq. 3-28 and simplifying, the desired governing differential equation is obtained:

$$\frac{d^2u}{dr^2} + \frac{1}{r} \frac{du}{dr} - \frac{u}{r^2} = 0$$

(3-38)

### Solution of the Differential Equation

As can be verified by substitution, the general solution of Eq. 3-38, which gives the radial displacement $u$ of any point on the cylinder, is

$$u = A_1 r + A_2/r$$

(3-39)

where the constants $A_1$ and $A_2$ must be determined from the conditions at the *boundaries* of the body.

Unfortunately, for the determination of the constants $A_1$ and $A_2$, the displacement $u$ is not known at either the inner or the outer boundary of the cylinder's wall. However, the known pressures are equal to the radial stresses acting on the elements at the respective radii. Hence,

$$\sigma_r(r_i) = -p_i \quad \text{and} \quad \sigma_r(r_o) = -p_o$$

(3-40)

where the minus signs are used to indicate compressive stresses. Moreover, since $u$ as given by Eq. 3-39 and $du/dr = A_1 - A_2/r^2$ can be substituted into the expression for $\sigma_r$ given by Eq. 3-37, the boundary conditions given by Eqs. 3-37 become

$$\sigma_r(r_i) = -p_i = \frac{E}{(1 + v)(1 - 2v)} \left[ A_1 - (1 - 2v) \frac{A_2}{r_i^2} \right]$$

(3-41)

$$\sigma_r(r_o) = -p_o = \frac{E}{(1 + v)(1 - 2v)} \left[ A_1 - (1 - 2v) \frac{A_2}{r_o^2} \right]$$

Solving these equations simultaneously for $A_1$ and $A_2$ yields

$$
\begin{aligned}
A_1 &= \frac{(1 + v)(1 - 2v)}{E} \frac{p_i r_i^2 - p_o r_o^2}{r_o^2 - r_i^2} \\
A_2 &= \frac{1 + v}{E} \frac{(p_i - p_o) r_i^2 r_o^2}{r_o^2 - r_i^2}
\end{aligned}
\tag{3-42}
$$

*These constants, when used in Eq. 3-39, permit the determination of the radial displacement of any point on the elastic cylinder subjected to the specified pressures.* Thus, displacements of the inner and outer boundaries of the cylinder can be computed.

If Eq. 3-39 and its derivative, together with the constants given by Eqs. 3-42, are substituted into Eqs. 3-37, and the results are simplified, general equations for the radial and tangential stresses at any point of an elastic cylinder are obtained. These are

$$
\sigma_r = C_1 - \frac{C_2}{r^2} \quad \text{and} \quad \sigma_t = C_1 + \frac{C_2}{r^2} \tag{3-43}
$$

where $\quad C_1 = \dfrac{p_i r_i^2 - p_o r_o^2}{r_o^2 - r_i^2} \quad$ and $\quad C_2 = \dfrac{(p_i - p_o) r_i^2 r_o^2}{r_o^2 - r_i^2}$

Note that $\sigma_r + \sigma_t$ is constant over the whole cross-sectional area of the cylinder. This means that the axial stress $\sigma_x$ as given by Eq. 3-34 is also constant over the entire cross-sectional area of the thick-walled cylinder.

### Remarks on the Thin-Disc Problem

The stress-strain relations used for a thick-walled cylinder corresponded to a *plane strain* condition. If, on the other hand, an annular thin disc were to be considered, the *plane stress* condition (i.e., $\sigma_x = 0$ and $\varepsilon_x = -v(\sigma_x + \sigma_y)/E$) governs. (See the discussion at the end of Section 3-6.) For this case, the stress-strain Eqs. 3-31 and 3-32 reduce to

$$
\varepsilon_r = \frac{1}{E}(\sigma_r - v\sigma_t) \quad \text{and} \quad \varepsilon_t = \frac{1}{E}(-v\sigma_r + \sigma_t) \tag{3-44}
$$

and by solving these equations simultaneously,

$$
\sigma_r = \frac{E}{1 - v^2}(\varepsilon_r + v\varepsilon_t) \quad \text{and} \quad \sigma_t = \frac{E}{1 - v^2}(\varepsilon_t + v\varepsilon_r) \tag{3-45}
$$

It is these stress-strain relations that must be used in the solution process. However, the resulting differential equation remains the same as Eq. 3-38, and the radial and tangential stresses are also identical to those in the thick-walled cylinder and are given by Eq. 3-43. The only difference is that a different constant $A_1$ must be used in Eq. 3-39 for determining the radial displacement $u$. The constant $A_2$ remains the same as in Eq. 3-42, whereas $A_1$ becomes

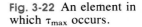

$$A_1 = \frac{1 - \nu}{E} \frac{p_i r_i^2 - p_o r_o^2}{r_o^2 - r_i^2} \qquad (3\text{-}46)$$

**Fig. 3-22** An element in which $\tau_{max}$ occurs.

## **3-13. Special Cases

*Internal pressure only*, i.e., $p_i \neq 0$ and $p_o = 0$, Fig. 3-22. For this case, Eqs. 3-43 simplify to

$$\sigma_r = \frac{p_i r_i^2}{r_o^2 - r_i^2} \left( 1 - \frac{r_o^2}{r^2} \right)$$
$$\sigma_t = \frac{p_i r_i^2}{r_o^2 - r_i^2} \left( 1 + \frac{r_o^2}{r^2} \right) \qquad (3\text{-}47)$$

Since $r_o^2/r^2 \geq 1$, $\sigma_r$ is always a compressive stress and is maximum at $r = r_i$. Similarly, $\sigma_t$ is always a tensile stress, and its maximum also occurs at $r = r_i$.

For brittle materials, the second Eq. 3-47 generally governs the design. However, for ductile materials, such as mild steel, it is more appropriate to adopt the criterion for the initiation of yielding due to shear rather than the material's capacity for resisting normal stress. This issue does not arise for thin-walled cylinders. In such problems, the maximum radial stress, equal to $p_i$, is negligible in comparison with $\sigma_1$. Therefore, according to Eq. 1-10, the relationship between the maximum normal and shear stresses is simple and direct, being $\tau_{max} = \sigma_1/2$, and either the normal or shear yield can be used as a criterion. However, for thick-walled cylinders, the radial stress $\sigma_r$ may be of the same order of magnitude as $\sigma_t$. For such a case, the maximum shear stress must be found by superposing the effects from both of the large normal stresses[13] in the manner shown in Fig. 3-23. Both of these stresses reach their maximum values at the inner surface of the cylinder. The maximum shear stress found in this manner should be compared with the maximum shear stress that a material can attain. Such a value can be taken as $\sigma_{yp}/2$, where $\sigma_{yp}$ is the normal yield stress in uniaxial tension. On this basis,

$$\tau_{max} = \frac{(\sigma_t)_{max} - (\sigma_r)_{max}}{2} = \frac{p_i r_o^2}{r_o^2 - r_i^2} = \frac{\sigma_{yp}}{2} \qquad (3\text{-}48)$$

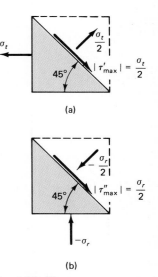

**Fig. 3-23** Stress transformations for obtaining maximum shear stresses.

[13] The axial stress $\sigma_z$ given by Eq. 3-34 does not enter the problem, since for $0 < \nu \leq 0.5$, this stress has an intermediate value between $\sigma_t$ and $\sigma_r$.

and
$$p_i = p_{yp} = \frac{\sigma_{yp}(r_o^2 - r_i^2)}{2r_o^2} \qquad (3\text{-}49)$$

*External pressure only*, i.e., $p_i = 0$ and $p_o \neq 0$. For this case, Eqs. 3-43 simplify to

$$\sigma_r = -\frac{p_o r_o^2}{r_o^2 - r_i^2}\left(1 - \frac{r_i^2}{r^2}\right)$$
$$\sigma_t = -\frac{p_o r_o^2}{r_o^2 - r_i^2}\left(1 + \frac{r_i^2}{r^2}\right) \qquad (3\text{-}50)$$

Since $r_i^2/r^2 \leq 1$, both stresses are always compressive. The maximum compressive stress is $\sigma_t$ and occurs at $r = r_i$.

Equations 3-50 must not be used for very thin-walled cylinders. Buckling of the walls may occur and strength formulas give misleading results.

## EXAMPLE 3-6

Make a comparison of the tangential stress distribution caused by the internal pressure $p_i$ as given by the Lamé formula in Section 3-12 with the distribution given by the approximate formula for thin-walled cylinders of Section 3-9 if (*a*) $r_o = 1.1r_i$, and if (*b*) $r_o = 4r_i$; see Fig. 3-24.

Solution

(*a*) Using Eq. 3-47b for $\sigma_t$,

$$(\sigma_t)_{r=r_i} = (\sigma_t)_{max} = \frac{p_i r_i^2}{(1.1r_i)^2 - r_i^2}\left[1 + \frac{(1.1r_i)^2}{r_i^2}\right] = 10.5p_i$$
$$(\sigma_t)_{r=r_o} = (\sigma_t)_{min} = \frac{p_i r_i^2}{(1.1r_i)^2 - r_i^2}\left[1 + \left(\frac{1.1r_i}{1.1r_i}\right)^2\right] = 9.5p_i$$

while, since the wall thickness $t = 0.1r_i$, the *average* hoop stress given by Eq. 3-24 is

$$(\sigma_t)_{avg} = \frac{p_i r_i}{t} = \frac{p_i r_i}{0.1r_i} = 10p_i$$

These results are shown in Fig. 3-24(a). Note particularly that in using Eq. 3-24, no appreciable error is involved.

(*b*) By using Eq. 3-47b for $\sigma_t$, the tangential stresses are obtained as before. These are

$$(\sigma_t)_{r=r_i} = (\sigma_t)_{max} = \frac{p_i r_i^2}{(4r_i)^2 - r_i^2}\left[1 + \frac{(4r_i)^2}{r_i^2}\right] = \frac{17}{15}p_i$$
$$(\sigma_t)_{r=r_o} = (\sigma_t)_{min} = \frac{p_i r_i^2}{(4r_i)^2 - r_i^2}\left[1 + \left(\frac{4r_i}{4r_i}\right)^2\right] = \frac{2}{15}p_i$$

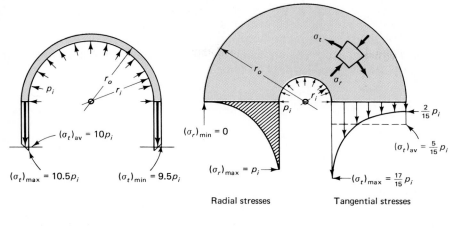

(a)  $r_o = 1.1r_i$ or $t = 0.1r_i$        (b)  $r_o = 4r_i$ or $t_i = 3r_i$        Fig. 3-24

The tangential stress is plotted in Fig. 3-28(b). A striking variation of the tangential stress can be observed from this figure. The average tangential stress given by Eq. 3-24, using $t = 3r_i$, is

$$(\sigma_t)_{av} = \frac{p_i r_i}{t} = \frac{5}{15} p_i = \frac{1}{3} p_i$$

The stress is nowhere near the true maximum stress.

The radial stresses were also computed by using Eq. 3-47a for $\sigma_r$, and the results are shown by the shaded area in Fig. 3-24(b).

It is interesting to note that no matter how thick a cylinder is made to resist internal pressure, the maximum tangential stress[14] will not be smaller than $p_i$. In practice, this necessitates special techniques to reduce the maximum stress. For example, in gun manufacture, instead of using a single cylinder, another cylinder is shrunk onto the smaller one, which sets up initial *compressive stresses* in the inner cylinder and tensile stresses in the outer one. In operation, the compressive stress in the inner cylinder is released first, and only then does this cylinder begin to act in tension. A greater range of operating pressures is obtained thereby.

## **\*\*3-14.** Behavior of Ideally Plastic Thick-Walled Cylinders

The case of a thick-walled cylinder under internal pressure alone was considered in the previous section, and Eq. 3-49 was derived for the onset of yield at the inner surface of the cylinder due to the maximum shear. Upon subsequent increase in the internal pressure, the yielding progresses toward the outer surface, and an elastic-plastic state prevails in the cyl-

[14] See Problem 3-21.

inder with a limiting radius $c$ beyond which the cross section remains elastic. As the pressure increases, the radius $c$ also increases until, eventually, the entire cross section becomes fully plastic at the ultimate load.

In the following discussion, as before, the maximum shear criterion for ideally plastic material will be assumed as

$$\tau_{max} = \frac{\sigma_t - \sigma_r}{2} = \frac{\sigma_{yp}}{2} \tag{3-51}$$

As noted earlier, this implies that $\sigma_x$ has an intermediate value between $\sigma_t$ and $\sigma_r$. A reexamination of Eqs. 3-34 and 3-47 shows this to be true in the elastic range, provided that $0 < v < 0.5$, but in the plastic range, this applies only if the ratio of outer to inner radius, $r_o/r_i$, is less than a certain value.[15] For $v = 0.3$, this ratio can be established to be 5.75; hence, the solutions to be obtained in this section will be valid only as long as $r_o < 5.75r_i$ (with $v = 0.3$). The task of finding the stress distribution is more complicated when this condition is not satisfied and is beyond the scope of this book.

### Plastic Behavior of Thick-Walled Cylinders

The equations of static equilibrium are applicable, regardless of whether the elastic or plastic state is considered. Hence, Eq. 3-28 is applicable, but must be supplemented by a yield condition.

*Static equilibrium,* Eq. 3-28:

$$\frac{d\sigma_r}{dr} + \frac{\sigma_r - \sigma_t}{r} = 0$$

*Yield condition,* Eq. 3-51:

$$\frac{\sigma_t - \sigma_r}{2} = \frac{\sigma_{yp}}{2}$$

By combining these two equations, the basic differential equation becomes

$$\frac{d\sigma_r}{dr} - \frac{\sigma_{yp}}{r} = 0 \quad \text{or} \quad d\sigma_r = \frac{\sigma_{yp}}{r} dr \tag{3-52}$$

The solution of this can be written as

[15] See W. T. Koiter, "On Partially Plastic Thick-Walled Tubes," *Biezeno Anniversary Volume on Applied Mechanics* (Haarlem, Holland: H. Stam, 1953), 233–251.

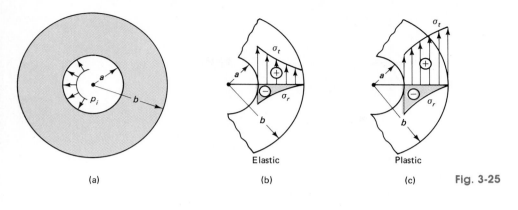

Elastic

Plastic

(a)  (b)  (c)  Fig. 3-25

$$\sigma_r = \sigma_{yp} \ln r + C \qquad (3\text{-}53)$$

For a cylinder with inner radius $a$ and outer radius $b$, the boundary condition (zero external pressure) can be expressed as

$$\sigma_r(b) = 0 = \sigma_{yp} \ln b + C \qquad (3\text{-}54)$$

Hence, the integration constant $C$ is given as

$$C = -\sigma_{yp} \ln b$$

The radial and tangential stresses are then obtained, using Eqs. 3-53 and 3-51, respectively. Thus,

$$\sigma_r = \sigma_{yp}(\ln r - \ln b) = \sigma_{yp} \ln r/b \qquad (3\text{-}55)$$
$$\sigma_t = \sigma_{yp} + \sigma_r = \sigma_{yp}(1 + \ln r/b) \qquad (3\text{-}56)$$

The stress distributions given by Eqs. 3-55 and 3-56 are shown in Fig. 3-25(c), whereas Fig. 3-25(b) shows the elastic stress distributions. Since the fully plastic state represents the ultimate collapse of the thick-walled cylinder, the ultimate internal pressure, using Eq. 3-55, is given as

$$p_{ult} = \sigma_r(a) = \sigma_{yp} \ln a/b \qquad (3\text{-}57)$$

### Elastic-Plastic Behavior of Thick-Walled Cylinders

For any value of $p_i$ that is intermediate to the yield and ultimate values given by Eq. 3-49 and Eq. 3-57, respectively, i.e., $p_{yp} < p_i < p_{ult}$, the cross section of the cylinder between the inner radius $a$ and an intermediate radius $c$ is fully plastic, whereas that between $c$ and the outer radius $b$ is in the elastic domain, Fig. 3-26. At the elastic-plastic interface, the yield condition is just satisfied, and the corresponding radial stress $X$ can be computed using Eq. 3-49 with $r_i = c$ and $r_o = b$; hence,

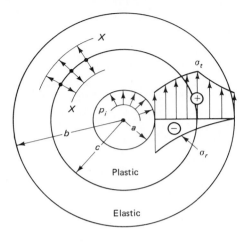

Fig. 3-26

$$X = \frac{\sigma_{yp}}{2} \frac{b^2 - c^2}{b^2} \tag{3-58}$$

This stress becomes the boundary condition to be used in conjunction with Eq. 3-53 for a fully plastic segment with inner radius $a$ and outer radius $c$. Hence,

$$\sigma_r(c) = -X = -\frac{\sigma_{yp}}{2} \frac{b^2 - c^2}{b^2} = \sigma_{yp} \ln c + C \tag{3-59}$$

and
$$C = -\frac{\sigma_{yp}}{2} \frac{b^2 - c^2}{b^2} - \sigma_{yp} \ln c \tag{3-60}$$

By substituting this value of $C$ into Eq. 3-55, the radial stress in the plastic region is obtained as

$$\sigma_r = \sigma_{yp} \ln \frac{r}{c} - \frac{\sigma_{yp}}{2} \frac{b^2 - c^2}{b^2} \tag{3-61}$$

and by using Eq. 3-51, the tangential stress in the plastic zone becomes

$$\sigma_t = \sigma_{yp} + \sigma_r = \sigma_{yp} \left( 1 + \ln \frac{r}{c} \right) - \frac{\sigma_{yp}}{2} \frac{b^2 - c^2}{b^2} \tag{3-62}$$

The internal pressure $p_i$ at which the plastic zone extends from $a$ to $c$ can be obtained, using Eq. 3-61, simply as $p_i = \sigma_r(a)$. Equations 3-47, with $r_i = c$ and $r_o = b$, provide the necessary relations for calculating the stress distributions in the elastic zone.

# Problems

## Section 3-2

**3-1.** Redesign the shear mounting in Example 3-1 to retain the same shear spring constant $k_s$, but changing its dimensions to a square pad with 10-mm-thick rubber.

## Section 3-6

**3-2.** Consider a 4-in square steel bar subjected to transverse biaxial tensile stresses of 20 ksi in the $x$ direction and 10 ksi in the $y$ direction. (a) Assuming the bar to be in a state of plane stress, determine the strain in the $z$ direction and the elongations of the plate in the $x$ and $y$ directions. (b) Assuming the bar to be in a state of plane strain, determine the stress in the $z$ direction and the elongations of the bar in the $x$ and $y$ directions. Let $E = 30 \times 10^3$ ksi and $\nu = 0.25$.

**3-3.** A piece of 50 by 250 by 10 mm steel plate is subjected to uniformly distributed stresses along its edges (see the figure). (a) If $P_x = 100$ kN and $P_y = 200$ kN, what change in thickness occurs due to the application of these forces? (b) To cause the same change in thickness as in part (a) by $P_x$ alone, what must be its magnitude? Let $E = 200$ GPa and $\nu = 0.25$.

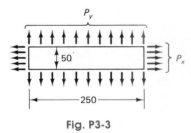

**Fig. P3-3**

**3-4.** A rectangular steel block, such as shown in Fig. 3-10(a), has the following dimensions: $a = 50$ mm, $b = 75$ mm, and $c = 100$ mm. The faces of this block are subjected to uniformly distributed forces of 180 kN (tension) in the $x$ direction, 200 kN (tension) in the $y$ direction, and 240 kN (compression) in the $z$ direction. Determine the magnitude of a single system of forces acting only in the $y$ direction that would cause the same deformation in the $y$ direction as the initial forces. Let $\nu = 0.25$.

## Sections 3-7 and 3-8

**3-5.** Using the values for $E$ and $G$ given in Table 1A of the Appendix, calculate Poisson's ratios for 2024-T4 aluminum alloy and steel.

**3-6.** Using Table 1A in the Appendix, calculate the bulk moduli for 6061-T6 alluminum alloy and steel in U.S. customary units.

## Section 3-9

**3-7.** A stainless-steel cylindrical shell has a 36-in inside diameter and is 0.5 in. thick. If the tensile strength of the material is 80 ksi and the factor of safety is 5, what is the allowable working pressure? Assume that appropriate hemispherical ends are provided. Also estimate the bursting pressure.

**3-8.** A "penstock," i.e., a pipe for conveying water to a hydroelectric turbine, operates at a head of 90 m. If the diameter of the penstock is 0.75 m and the allowable stress 50 MPa, what wall thickness is required?

**3-9.** A tank of butt-welded construction for the storage of gasoline is to be 40 ft in diameter and 16 ft high. (a) Select the plate thickness for the bottom row of plates. Allow 20 ksi for steel in tension and assume the efficiency of welds at 80%. Add approximately $\frac{1}{8}$ in to the computed wall thickness to compensate for corrosion. Neglect local stresses at the juncture of the vertical walls with the bottom. (Specific gravity of the gasoline to be stored is 0.721.) (b) Assuming that the bottom of the tank does not restrain the displacement of the tank walls, what increase in diameter would occur at the bottom? $E = 29 \times 10^3$ ksi and $\nu = 0.25$.

**3-10.** A cylindrical vessel is used for storing ammonia ($NH_3$) at the maximum temperature of 50 °C. The vapor pressure of $NH_3$ at 50 °C is 20 atm. The thickness of the vessel material is limited to 20 mm with a tensile strength of 400 MPa. (a) If the factor of safety is 5, assuming that all welds will be inspected with X-rays, what can be the maximum diameter of the vessel? (b) For the selected wall thickness, calculate the change in diameter that would occur with ammonia at 50 °C.

**3-11.** An air chamber for a pump, the sectional side view of which is shown in mm on the figure, consists of two pieces. Compute the number of 19-mm bolts (*net* area 195 mm$^2$) required to attach the chamber to the cylinder at plane $A–A$. The allowable tensile stress

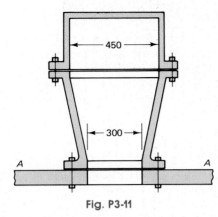

Fig. P3-11

in the bolts is 40 MPa, and the water and air pressure is 1.5 MPa.

**3-12.** A water tank made of wood staves is 5 m in diameter and 4 m high. Specify the spacing of 30 by 6 mm steel hoops if the allowable tensile stress for steel is set at 90 MPa. Use uniform hoop spacing within each meter of the tank's height.

**3-13.** A cylindrical pressure vessel of 120 in *outside* diameter, used for processing rubber, is 36 ft long. If the cylindrical portion of the vessel is made from 1-in thick steel plate and the vessel operates at 120-psi internal pressure, determine the total elongation of the circumference and the increase in the diameter's dimension caused by the operating pressure. $E = 29 \times 10^6$ psi and $\nu = 0.25$.

**3-14.** A thin ring is heated in oil 150 °C above room temperature. In this condition, the ring just slips on a solid cylinder, as shown in the figure. Assuming the

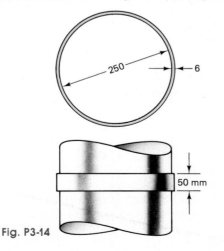

Fig. P3-14

cylinder to be completely rigid, (a) determine the hoop stress that develops in the ring upon cooling, and (b) determine what bearing develops between the ring and the cylinder. Let $\alpha = 2 \times 10^{-5}/°C$ and $E = 7 \times 10^7$ kN/m².

**3-15.** An aluminum alloy wire is stretched taut across the diameter of a cylindrical pressure vessel, as shown in the figure. For the wire: $A = 0.060$ mm², $E = 70 \times 10^3$ MPa, and $\alpha_{Al} = 23.4 \times 10^{-6}/°C$. The diameter of the steel pressure vessel is 2000 mm and the wall thickness is 10 mm. (In calculations, do not differentiate between the inside and mean diameters of the cylinder.) For steel, let $E = 200 \times 10^3$ MPa, $\alpha_{St} = 11.7 \times 10^{-6}/°C$, and Poisson's ratio $\nu = 0.30$. If this vessel is pressurized to 1 MPa and, at the same time, the temperature drops 50 °C, what stress would develop in the wire? Assume that the temperature of the wire as well as that of the cylinder simultaneously becomes lower and that the deformation of the cylinder caused by the pull of the wire can be neglected.

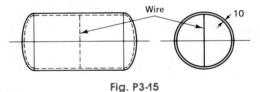

Fig. P3-15

**3-16.** A cylindrical pressure vessel shown in the figure is made by shrinking a brass tube over a mild steel tube. Both cylinders have a wall thickness of $\frac{1}{4}$ in. The nominal diameter of the vessel is 30 in and is to be used in all calculations involving the diameter. When the brass cylinder is heated 100 °F above room temperature, it exactly fits over the steel cylinder, which is at room temperature. What is the stress in the brass cylinder when the composite vessels cool to room temperature? For brass: $E_{Br} = 16 \times 10^6$ psi and $\alpha_{Br} = 10.7 \times 10^{-6}/°F$. For steel: $E_{St} = 30 \times 10^6$ psi and $\alpha_{St} = 6.7 \times 10^{-6}/°F$.

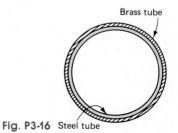

Fig. P3-16  Steel tube

**\*3-17.** An aluminum-alloy tube is shrunk onto a steel tube to form the pressure vessel illustrated in the figure. The wall thickness of each tube is 4 mm. The average diameter of the assembly to be used in calculations is 400 mm. If the composite tube is pressurized at 2 MPa, what additional hoop stress develops in the aluminum tube? Assume that the ends of the tube can freely expand, preventing the development of longitudinal stresses, i.e., $\sigma_x = 0$. Let $E_{Al} = 70 \times 10^3$ MPa and $E_{St} = 200 \times 10^3$ MPa. *Hint*: The interface pressure, say $\bar{p}$, between the two materials acts to cause hoop tension in the outer tube and hoop compression in the inner tube.

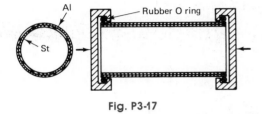

**Fig. P3-17**

**\*3-18.** Exceptionally light-weight pressure vessels have been developed by employing glass filaments for resisting the tensile forces and using epoxy resin as a binder. A diagram of a filament-wound cylinder is shown in the figure. If the winding is needed to resist only hoop stresses, the helix angle $\alpha = 90°$. If, however, the cylinder is closed, both hoop and longitudinal forces develop, and the required helix angle of the filaments $\alpha \approx 55°$ ($\tan^2 \alpha = 2$). Verify this result. (*Hint*: Isolate an element of unit width and a developed length of $\tan \alpha$ as in the figure. For such an element, the same number of filaments is cut by each section. Therefore, if $F$ is a force in a filament and $n$ is the number of filaments at a section, $P_y = Fn \sin \alpha$. Force $P_x$ can be found similarly. An equation based on the known ratio between the longitudinal and the hoop stress leads to the required result.)

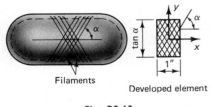

Filaments

Developed element

**Fig. P3-18**

## Sections 3-12 and 3-13

**3-19.** Verify the solution of Eq. 3-38.

**3-20.** Show that the ratio of the maximum tangential stress to the average tangential stress for a thick-walled cylinder subjected only to internal pressure is $(1 + \beta^2)/(1 + \beta)$, where $\beta = r_o/r_i$.

**3-21.** Show that no matter how large the outside diameter of a cylinder, subjected only to internal pressure, is made, the maximum tangential stress is not less than $p_i$. (*Hint*: Let $r_o \to \infty$.)

**3-22.** An alloy-steel cylinder has a 6-in ID (inside diameter) and a 18-in OD. If it is subjected to an internal pressure of $p_i = 24,000$ psi ($p_o = 0$), (a) determine the radial and tangential stress distributions and show the results on a plot. (b) Determine the maximum (principal) shear stress.·(c) Determine the change in external and internal diameters. $E = 30 \times 10^6$ psi and $\nu = 0.3$.

**3-23.** An alloy-steel cylinder has a 0.15-m ID and a 0.45 m OD. If it is subjected to an internal pressure of $p_i = 160$ MPa ($p_o = 0$), (a) determine the radial and tangential stress distributions and show the results on a plot. (b) Determine the maximum (principal) shear stress. (c) Determine the changes in external and internal diameters. $E = 200 \times 10^3$ MPa and $\nu = 0.3$.

**3-24.** Rework Problem 3-23 with $p_i = 0$ and $p_o = 80$ MPa.

**3-25.** Rework Problem 3-23 with $p_i = 160$ MPa and $p_o = 80$ MPa.

**3-26.** Isolate one-half of the cylinder of Problem 3-25 by passing a plane through the axis of the cylinder. Then, by integrating the tangential stresses over the respective areas, show that the isolated free body is in equilibrium.

**3-27.** Design a thick-walled cylinder of a 4-in internal diameter for an internal pressure of 8000 psi such as to provide: (a) a factor of safety of 2 against any yielding in the cylinder, and (b) a factor of safety of 3 against ultimate collapse. The yield stress of steel in tension is 36 ksi.

**3-28.** A 16-in OD steel cylinder with approximately a 10-in bore (ID) is shrunk onto another steel cylinder of 10-in OD with a 6-in ID. Initially, the internal diameter of the outer cylinder was 0.01 in smaller than the external diameter of the inner cylinder. The assembly was accomplished by heating the larger cylinder in oil. For both cylinders, $E = 30 \times 10^6$ psi and $\nu = 0.3$. (a) Determine the pressure at the boundaries between the two cylinders. (*Hint*: The elastic increase in the diameter of the outer cylinder with the elastic decrease in the diameter of the inner cyclinder accommodates the initial interference between the two cylinders.) (b) Determine the tangential and radial stresses caused by the pressure found in part (a). Show the results on a plot. (c) Determine the internal pressure to which the composite cylinder may be subjected without exceeding a tangential stress of 20,000 psi in the inner cylinder. (*Hint*: After assembly, the cylinders act as one unit. The initial compressive stress in the inner cylinder is released first.) (d) Superpose the tangential stresses found in part (b) with the tangential stresses resulting from the internal pressure found in part (c). Show the results on a plot.

**3-29.** Set up the differential equation for a thin disk rotating with an angular velocity of $\omega$ rad/s. The unit weight of the material is $\gamma$. *Hint*: Consider an element as in Fig. 3-21(b) and add an inertia term.

## Section 3-14

**3-30.** For a thick-walled cylinder of inner radius $a$ and outer radius $b = 2a$, (a) calculate the internal pressure at which the elastic-plastic boundary is at $r = 1.5\,a$, (b) determine the radial and tangential stress distributions due to the internal pressure found in part (a) and show them on a plot, and (c) calculate the ultimate collapse load. Assume the material to be elastic-perfectly plastic, with a yield stress of 250 MPa.

# chapter 4

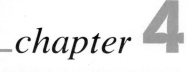

# Torsion

## 4-1. Introduction

Detailed methods of analysis for determining stresses and deformations in axially loaded bars were presented in the first two chapters. Analogous relations for members subjected to torques about their longitudinal axes are developed in this chapter. The constitutive relations for shear discussed in the preceding chapter will be employed for this purpose. *The investigations are confined to the effect of a single type of action, i.e., of a torque causing a twist or torsion in a member.* Members subjected simultaneously to torque and bending, frequently occurring in practice, are treated in Chapter 9.

By far, the major part of this chapter is devoted to the consideration of members having circular cross sections, either solid or tubular. Solution of such elastic and inelastic problems can be obtained using the procedures of engineering mechanics of solids. For the solution of torsion problems having noncircular cross sections, methods of the mathematical theory of elasticity (or finite elements) must be employed. This topic is briefly discussed in order to make the reader aware of the differences in such solutions from that for circular members. Further, to lend emphasis to the difference in the solutions discussed, this chapter is subdivided into four distinct parts. It should be noted, however, that in practice, members for transmitting torque, such as shafts for motors, torque tubes for power equipment, etc., are predominantly circular or tubular in cross section. Therefore, numerous applications fall within the scope of the formulas derived in this chapter.

## 4-2. Application of the Method of Sections

In engineering solid mechanics, in analyzing members for torque, regardless of the type of cross section, the basic method of sections (Section 1-2) is employed. For the torsion problems discussed here, there is *only*

**175**

*one* relevant equation of statics. Thus, if the $x$ axis is directed along a member, such an equation is $\sum M_x = 0$. Therefore, for statically determinate systems, there can only be one reactive torque. After determining this torque, an analysis begins by separating a member of a *section perpendicular to the axis of a member*. Then either side of a member can be isolated and the *internal* torque found. This internal torque must *balance* the externally applied torques, i.e., *the external and the internal torques are equal*, but have opposite sense. In statically determinate problems, the formal calculation of a reaction may be bypassed by isolating a bar segment with the unsupported end. Nevertheless, an equilibrium of the whole system must always be assured. In statically indeterminate problems, the reactions must always be found before one can calculate the internal torques. Some guidance on calculating reactions in statically indeterminate problems is provided in Section 4-9 of this chapter.

For simplicity, the members treated in this chapter will be assumed "weightless" or supported at frequent enough intervals to make the effect of bending negligible. Axial forces that may also act simultaneously on the bars are excluded for the present.

### EXAMPLE 4-1

Find the internal torque at section $K$–$K$ for the shaft shown in Fig. 4-1(a) and acted upon by the three torques indicated.

### Solution

The 30 N·m torque at $C$ is balanced by the two torques of 20 and 10 N·m at $A$ and $B$, respectively. Therefore, the body as a whole is in equilibrium. Next, by passing a section $K$–$K$ perpendicular to the axis of the rod *anywhere* between $A$ and $B$, a free body of a part of the shaft, shown in Fig. 4-1(b), is obtained. Whereupon, from $\sum M_x = 0$, or

$$\text{externally applied torque} = \text{internal torque}$$

the conclusion is reached that the internal or resisting torque developed in the shaft between $A$ and $B$ is 20 N·m. Similar considerations lead to the conclusion that the internal torque resisted by the shaft between $B$ and $C$ is 30 N·m.

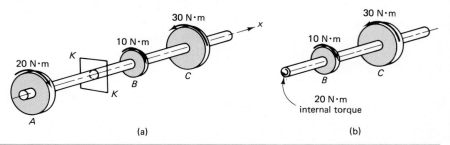

Fig. 4-1　　　　　　　　　　　　　　　　(a)　　　　　　　　　　　　　　　　(b)

It may be seen intuitively that for a member of constant cross section, the maximum internal torque causes the maximum stress and imposes the most severe condition on the material. Hence, in investigating a torsion member, several sections may have to be examined to determine the largest internal torque. A section where the largest internal torque is developed is the *critical section.* In Example 4-1, the critical section is anywhere between points $B$ and $C$. If the torsion member varies in size, it is more difficult to decide where the material is critically stressed. Several sections may have to be investigated and *stresses computed* to determine the critical section. These situations are analogous to the case of an axially loaded rod, and means must be developed to determine stresses as a function of the internal torque and the size of the member. In the next several sections, the necessary formulas are derived.

Instead of curved arrows as in Fig. 4-1, double-headed vectors following the right-hand screw rule sign convention will also be used in this text; see Fig. 4-2.

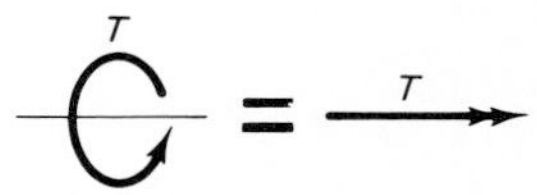

**Fig. 4-2** Alternative representations of torque.

# Part A TORSION OF CIRCULAR ELASTIC BARS

## 4-3. Basic Assumptions for Circular Members

To establish a relation between the internal torque and the stresses it sets up in members with *circular solid and tubular cross sections*, it is necessary to make two assumptions, the validity of which will be justified later. These, in addition to the homogeneity of the material, are as follows:

1. A plane section of material perpendicular to the axis of a circular member remains *plane* after the torques are applied, i.e., no *warpage* or distortion of parallel planes normal to the axis of a member takes place.[1]

2. In a circular member subjected to torque, *shear strains* $\gamma$ *vary linearly from the central axis* reaching $\gamma_{max}$ at the periphery. This assumption is illustrated in Fig. 4-3 and means that an imaginary plane such as $DO_1O_3C$ moves to $D'O_1O_3C$ when the torque is applied. Alternatively, if an imaginary radius $O_3C$ is considered fixed in direction, similar radii initially at $O_2B$ and $O_1D$ rotate to the respective new positions $O_2B'$ and $O_1D'$. These radii *remain* straight.

---

[1] Actually, it is also implied that parallel planes perpendicular to the axis *remain a constant* distance apart. This is not true if deformations are large. However, since the usual deformations are very small, stresses not considered here are negligible. For details, see S. Timoshenko, *Strength of Materials*, 3rd. ed., Part II, *Advanced Theory and Problems* (New York: Van Nostrand, 1956), Chapter VI.

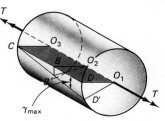

**Fig. 4-3** Variation of strain in circular member subjected to torque.

It must be emphasized that these assumptions *hold only for circular solid and tubular members*. For this class of members, these assumptions work so well that they *apply beyond the limit of the elastic behavior of a material*. These assumptions will be used again in Section 4-13, where stress distribution beyond the proportional limit is discussed.

3.  If attention is confined to the linearly *elastic* material, Hooke's law applies, and, it follows that shear stress is proportional to shear strain. For this case complete agreement between experimentally determined and computed quantities is found with the derived stress and deformation formulas based on these assumptions. Moreover, their validity can be rigorously demonstrated by the methods of the mathematical theory of elasticity.

## 4-4. The Torsion Formula

In the *elastic* case, on the basis of the previous assumptions, since stress is proportional to strain, and the latter varies linearly from the center, *stresses vary linearly from the central axis of a circular member.* The stresses induced by the assumed distortions are *shear* stresses and lie in the plane parallel to the section taken normal to the axis of a rod. The variation of the shear stress follows directly from the shear-strain assumption and the use of Hooke's law for shear, Eq. 3-1. This is illustrated in Fig. 4-4. Unlike the case of an axially loaded rod, this stress is *not* of uniform intensity. The maximum shear stress occurs at points most remote from the center $O$ and is designated $\tau_{max}$. These points, such as points $C$ and $D$ in Figs. 4-3 and 4-4, lie at the periphery of a section at a distance $c$ from the center. For linear shear stress variation, at *any* arbitrary point at a distance $\rho$ from $O$, the shear stress is $(\rho/c)\tau_{max}$.

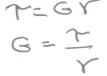

The resisting torque can be expressed in terms of stress once the stress distribution at a section is established. For equilibrium this internal resisting torque must equal the externally applied torque T. Hence,

$$\int_A \underbrace{\underbrace{\frac{\rho}{c}\tau_{max}}_{\text{stress}} \; \underbrace{dA}_{\text{area}}}_{\underbrace{\text{force}}} \quad \underbrace{\rho}_{\text{arm}} = T$$
$$\underbrace{\phantom{xxxxxxxxxxxxxxxxxxxxxx}}_{\text{torque}}$$

where the integral sums up all torques developed on the cut by the infinitesimal forces acting at a distance $\rho$ from a member's axis, $O$ in Fig. 4-4, over the whole area $A$ of the cross section, and where $T$ is the resisting torque.

At any given section, $\tau_{max}$ and $c$ are constant; hence, the previous relation can be written as

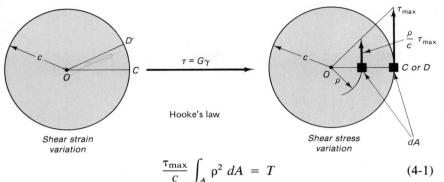

**Fig. 4-4** Shear strain assumption leading to elastic shear stress distribution in a circular member.

$$\frac{\tau_{max}}{c} \int_A \rho^2 \, dA = T \qquad (4\text{-}1)$$

However, $\int_A \rho^2 \, dA$, *the polar moment of inertia* of a cross-sectional area, is also a constant for a particular cross-sectional area. It will be designated by $J$ in this text. For a circular section, $dA = 2\pi\rho \, d\rho$, where $2\pi\rho$ is the circumference of an annulus[2] with a radius $\rho$ of width $d\rho$. Hence,

$$J = \int_A \rho^2 \, dA = \int_0^c 2\pi\rho^3 \, d\rho = 2\pi \left. \frac{\rho^4}{4} \right|_0^c = \frac{\pi c^4}{2} = \frac{\pi d^4}{32} \qquad (4\text{-}2)$$

i.e.,

$$J = \frac{\pi c^4}{2} = \frac{\pi d^4}{32} \qquad (4\text{-}2)$$

where $d$ is the diameter of a solid circular shaft. If $c$ or $d$ is measured in millimeters, $J$ has the units of $mm^4$; if in inches, the units become $in^4$.

By using the symbol $J$ for the polar moment of inertia of a circular area, Eq. 4-1 may be written more compactly as

$$\tau_{max} = \frac{Tc}{J} \qquad (4\text{-}3)$$

This equation is the well-known *torsion formula*[3] for circular shafts that expresses the maximum shear stress in terms of the resisting torque and the dimensions of a member. In applying this formula, the internal torque $T$ can be expressed[4] in newton-meters, N·m, or inch-pounds, $c$ in meters

[2] An annulus is an area contained between two concentric circles.
[3] It was developed by Coulomb, a French engineer, in about 1775 in connection with his work on electric instruments. His name has been immortalized by its use for a practical unit of quantity in electricity.
[4] 1 N·m is equal to 1 joule (J). However, in this text, the symbol $J$ is used only for the polar moment of inertia of a section.

or inches, and $J$ in $m^4$ or $in^4$. Such usage makes the units of the torsional shear stress

$$\frac{[N \cdot m][m]}{[m^4]} = \left[\frac{N}{m^2}\right]$$

or *pascals* (Pa) in SI units, or

$$\frac{[in\text{-}lb][in]}{[in^4]} = \left[\frac{lb}{in^2}\right]$$

or *psi* in the U.S. customary units.

A more general relation than Eq. 4-3 for a shear stress, $\tau$, at *any* point a distance $\rho$ from the center of a section is

$$\tau = \frac{\rho}{c}\tau_{max} = \frac{T\rho}{J} \tag{4-4}$$

Equations 4-3 and 4-4 *are applicable* with equal rigor *to circular tubes*, since the same assumptions as used in the previous derivation apply. It is necessary, however, to modify $J$. For a tube, as may be seen from Fig. 4-5, the limits of integration for Eq. 4-2 extend from $b$ to $c$. Hence, for a *circular tube*,

$$J = \int_A \rho^2 \, dA = \int_b^c 2\pi\rho^3 \, d\rho = \frac{\pi c^4}{2} - \frac{\pi b^4}{2} \tag{4-5}$$

or stated otherwise: $J$ for a circular tube equals $+J$ for a solid shaft using the outer diameter and $-J$ for a solid shaft using the inner diameter.

For very *thin* tubes, if $b$ is nearly equal to $c$, and $c - b = t$, the thickness of the tube, $J$ reduces to a simple approximate expression:

$$J \approx 2\pi R_{av}^3 t \tag{4-6}$$

where $R_{av} = (b + c)/2$, which is sufficiently accurate in some applications.

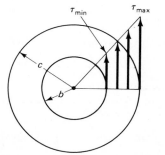

**Fig. 4-5** Variation of stress in an elastic circular tube.

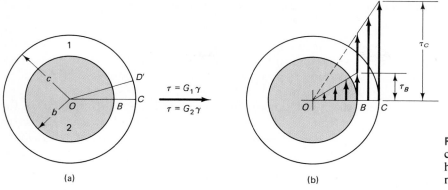

$$\tau = G_1 \gamma$$
$$\tau = G_2 \gamma$$

**Fig. 4-6** Elastic behavior of a circular member in torsion having an inner core of soft material.

If a circular bar is made from two different materials bonded together, as shown in Fig. 4-6(a), the same *strain* assumption applies as for a solid member. For such a case, through Hooke's law, the shear-*stress* distribution becomes as in Fig. 4-6(b). If the shear modulus for the outer stiffer tube is $G_1$ and that of the inner softer core is $G_2$, the ratio of the respective shear stresses on a ring of radius $OB$ is $G_1/G_2$.

### Procedure Summary

For the torsion problem of circular shafts the *three basic concepts* of engineering mechanics of solids as used above may be summarized in the following manner:

1. *Equilibrium conditions* are used for determining the internal resisting torques at a section.
2. *Geometry of deformation* (kinematics) is postulated such that shear strain varies linearly from the axis of a shaft.
3. *Material properties* (constitutive relations) are used to relate shear strains to shear stresses and permit calculation of shear stresses at a section.

Only a linear elastic case using Hooke's law is considered in the preceding discussion. This is extended to non-linear material behavior in Section 4-13.

These basic concepts are used for determining both stresses and angles-of-twist of circular shafts. However, similar to the case for axially loaded bars, large *local* stresses arise at points of application of concentrated torques or changes in cross section. According to *Saint-Venant's principle* the stresses and strains are accurately described by the developed theory only beyond a distance about equal to the diameter of a shaft from these locations. Typically local stresses are determined by using stress concentration factors.

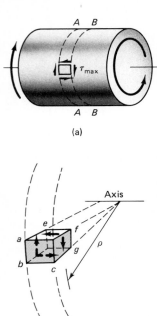

(a)

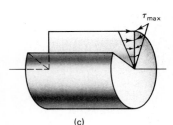

(b)

(c)

**Fig. 4-7** Existence of shear stresses on mutually perpendicular planes in a circular shaft subjected to torque.

## 4-5. Remarks on the Torsion Formula

So far the shear stresses as given by Eqs. 4-3 and 4-4 have been thought of as acting only in the plane of a cut perpendicular to the axis of the shaft. There indeed they are acting to form a couple resisting the externally applied torques. However, to understand the problem further, an infinitesimal cylindrical element,[5] shown in Fig. 4-7(b), is isolated.

The shear stresses acting in the planes perpendicular to the axis of the rod are known from Eq. 4-4. *Their directions coincide with the direction of the internal torque.* (This should be clearly visualized by the reader.) On adjoining parallel planes of a disc-like element, these stresses act in opposite directions. However, these shear stresses acting in the plane of the cuts taken normal to the axis of a rod *cannot exist alone*, as was shown in Section 1-4. Numerically, equal shear stresses must act on the axial planes (such as the planes *aef* and *bcg* in Fig. 4-7(b)) to fulfill the requirements of static equilibrium for an element.[6]

Shear stresses acting in the axial planes follow the same variation in intensity as do the shear stresses in the planes perpendicular to the axis of the rod. This variation of shear stresses on the mutually perpendicular planes is shown in Fig. 4-7(c), where a portion of the shaft has been removed for the purposes of illustration.

According to Section 3-7, such *shear* stresses can be *transformed* into an *equivalent* system of *normal* stresses acting at angles of 45° with the shear stresses (see Fig. 3-11). Numerically, these stresses are related to each other in the following manner: $\tau = \sigma_1 = -\sigma_2$. Therefore, if the shear strength of a material is less than its strength in tension, a shear failure takes place on a plane perpendicular to the axis of a bar; see Fig. 4-8. This kind of failure occurs gradually and exhibits *ductile* behavior.

[5] Two planes perpendicular to the axis of the rod, two planes through the axis, and two surfaces at different radii are used to isolate this element. Properties of such an element are expressible mathematically in cylindrical coordinates.

[6] Note that maximum shear stresses, as shown diagrammatically in Fig. 4-7(a), actually act on planes perpendicular to the axis of the rod and on planes passing through the axis of the rod. The representation shown is purely schematic. The free *surface* of a shaft is *free* of all stresses.

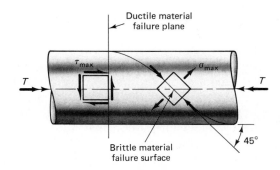

**Fig. 4-8** Potential torsional failure surfaces in ductile and brittle materials.

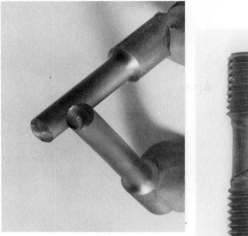

Fig. 4-9 Fractured torsion specimen of A322 steel.

Fig. 4-10 Fractured cast iron specimen in torsion. The photograph on the right shows the specimen more widely separated. (Threaded ends are normally not used for such specimens).

Alternatively, if the converse is true, i.e., $\sigma_1 < \tau$, a brittle fracture is caused by the tensile stresses along a helix forming an angle of 45° with the bar axis[7]; see Fig. 4-8. A photograph of a ductile fracture of a steel specimen is shown in Fig. 4-9, and that of a brittle fracture for cast iron in Fig. 4-10. Another examples of a brittle fracture for sandstone is shown in Fig. 4-11.

The stress transformation brought into the previous discussion, since it does not depend on material properties, is also applicable to anisotropic materials. For example, wood exhibits drastically different properties of strength in different directions. The shearing strength of wood on planes parallel to the grain is much less than on planes perpendicular to the grain. Hence, although equal intensities of shear stress exist on mutually perpendicular planes, wooden shafts of inadequate size fail longitudinally along axial planes. Such shafts are occasionally used in the process industries.

## EXAMPLE 4-2

Find the maximum torsional shear stress in shaft $AC$ shown in Fig. 4-1(a). Assume the shaft from $A$ to $C$ is 10 mm in diameter.

### Solution

From Example 4-1, the maximum internal torque resisted by this shaft is known to be 30 N·m. Hence, $T = 30$ N·m, and $c = d/2 = 5$ mm. From Eq. 4-2,

$$J = \frac{\pi d^4}{32} = \frac{\pi \times 10^4}{32} = 982 \text{ mm}^4$$

[7] Ordinary chalk behaves similarly. This may be demonstrated in the classroom by twisting a piece of chalk to failure.

Fig. 4-11 Part of fractured sandstone core specimen in torsion. (Experiment by D. Pirtz).

and from Eq. 4-3,

$$\tau_{max} = \frac{Tc}{J} = \frac{30 \times 10^3 \times 5}{982} = 153 \text{ MPa}$$

This maximum shear stress at 5 mm from the axis of the rod acts in the plane of a cut perpendicular to the axis of the rod *and* along the longitudinal planes passing through the axis of the rod (Fig. 4-7(c)). Just as for a Cartesian element, the shear stresses on mutually perpendicular planes for a cylindrical element are equal. It is instructive to note that the results of this solution can be represented in matrix form by two elements in a stress tensor as

$$\begin{pmatrix} 0 & \tau_{max} & 0 \\ \tau_{max} & 0 & 0 \\ 0 & 0 & 0 \end{pmatrix} = \begin{pmatrix} 0 & 153 & 0 \\ 153 & 0 & 0 \\ 0 & 0 & 0 \end{pmatrix} \text{ MPa} \qquad (4\text{-}7)$$

This is to be contrasted with the fully populated stress tensor given by Eq. 1-1b.

### EXAMPLE 4-3

Consider a long tube of 20 mm outside diameter, $d_o$, and of 16 mm inside diameter, $d_i$, twisted about its longitudinal axis with a torque $T$ of 40 N·m. Determine the shear stresses at the outside and the inside of the tube; see Fig. 4-12.

Solution

From Eq. 4-5,

$$J = \frac{\pi(c^4 - b^4)}{2} = \frac{\pi(d_o^4 - d_i^4)}{32} = \frac{\pi(20^4 - 16^4)}{32} = 9270 \text{ mm}^4$$

and from Eq. 4-3,

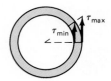

$$\tau_{max} = \frac{Tc}{J} = \frac{40 \times 10^3 \times 10}{9270} = 43.1 \text{ MPa}$$

Similarly from Eq. 4-4,

$$\tau_{min} = \frac{T\rho}{J} = \frac{40 \times 10^3 \times 8}{9270} = 34.5 \text{ MPa}$$

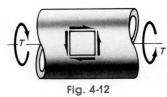

**Fig. 4-12**

In a thin-walled tube, all of the material works at approximately the same stress level. Therefore, thin-walled tubes are more efficient in transmitting torque than solid shafts. Such tubes are also useful for creating an essentially uniform "field" of pure shear stress needed for establishing $\tau$–$\gamma$ relationships (Section 3-2). To avoid local buckling, however, the wall thickness cannot be excessively thin.

## 4-6. Design of Circular Members in Torsion

In designing members for strength, allowable shear stresses must be selected. These depend on the information available from experiments and on the intended application. Accurate information on the capacity of materials to resist shear stresses comes from tests on thin-walled tubes. Solid shafting is employed in routine tests. Moreover, as torsion members are so often used in power equipment, many fatigue experiments are done. Typically, the shear strength of ductile materials is only about half as large as their tensile strength. The ASME (American Society of Mechanical Engineers) code of recommended practice for transmission shafting gives an allowable value in shear stress of 8000 psi for unspecified steel and 0.3 of yield, or 0.18 of ultimate, shear strength, whichever is smaller.[8] In practical designs, suddenly applied and shock loads warrant special considerations. (See Section 4-11.)

After the torque to be transmitted by a shaft is determined and the maximum allowable shear stress is selected, according to Eq. 4-3, the proportions of a member are given as

$$\frac{J}{c} = \frac{T}{\tau_{max}} \qquad (4\text{-}8)$$

where $J/c$ is the *parameter* on which the elastic strength of a shaft depends. For an axially loaded rod, such a parameter is the cross-sectional area of a member. For a *solid shaft*, $J/c = \pi c^3/2$, where $c$ is the outside radius. By using this expression and Eq. 4-8, the required radius of a shaft can be determined. Any number of *tubular* shafts can be chosen to satisfy Eq. 4-8 by varying the ratio of the outer radius to the inner radius, $c/b$, to provide the required value of $J/c$.

The reader should carefully note that large local stresses generally develop at changes in cross sections and at splines and keyways, where the torque is actually transmitted. These questions, of critical importance in the design of rotating shafts, are briefly discussed in the next section.

Members subjected to torque are very widely used as rotating shafts for transmitting power. For future reference, a formula is derived for the conversion of horsepower, the conventional unit used in the industry, into torque acting through the shaft. By definition, 1 hp does the work of 745.7 N·m/s. One N·m/s is conveniently referred to as a watt (W) in the SI units. Thus, 1 hp can be converted into 745.7 W. Likewise, it will be recalled from dynamics that power is equal to torque multiplied by the angle,

---

[8] Recommendations for other materials may be found in machine design books. For example, see J. E. Shigley, *Mechanical Engineering Design*, 3rd ed. (New York: McGraw-Hill, 1977) or R. C. Juvinal, *Stress, Strain, and Strength* (New York: McGraw-Hill, 1967).

measured in radians, through which the shaft rotates per unit of time. For a shaft rotating with a frequency of $f$ Hz,[9] the angle is $2\pi f$ rad/s. Hence, if a shaft were transmitting a constant torque $T$ measured in N·m, it would do $2\pi f T$ N·m of work per second. Equating this to the horsepower supplied

$$hp \times 745.7 = 2\pi f T \ [\text{N·m/s}]$$

or

$$T = \frac{119 \times hp}{f} \ [\text{N·m}] \tag{4-9}$$

or

$$T = \frac{159 \times kW}{f} \ [\text{N·m}] \tag{4-10}$$

where $f$ is the frequency in hertz of the shaft transmitting the horsepower, hp, or kilowatts, kW. These equations convert the applied power into applied torque.

In the U.S. customary system of units, 1 hp does work of 550 ft-lb per second, or $550 \times 12 \times 60$ in-lb per minute. If the shaft rotates at $N$ rpm (revolutions per minute), an equation similar to those above can be obtained:

$$T = \frac{63{,}000 \times hp}{N} \ [\text{in-lb}] \tag{4-11}$$

### EXAMPLE 4-4

Select a solid shaft for a 10-hp motor operating at 30 Hz. The maximum shear stress is limited to 55 MPa.

Solution

From Eq. 4-9,

$$T = \frac{119 \times hp}{f} = \frac{119 \times 10}{30} = 39.7 \ \text{N·m}$$

and from Eq. 4-8,

$$\frac{J}{c} = \frac{T}{\tau_{max}} = \frac{39.7 \times 10^3}{55} = 722 \ \text{mm}^3$$

$$\frac{J}{c} = \frac{\pi c^3}{2} \quad \text{or} \quad c^3 = \frac{2}{\pi}\frac{J}{c} = \frac{2 \times 722}{\pi} = 460 \ \text{mm}^3$$

---

[9] 1 hertz (Hz) = 1 cycle per second (cps).

Hence, $c = 7.72$ mm or $d = 2c = 15.4$ mm.
For practical purposes, a 16-mm shaft would probably be selected.

---

## EXAMPLE 4-5

Select solid shafts to transmit 200 hp each without exceeding a shear stress of 10,000 psi. One of these shafts operates at 20 rpm and the other at 20,000 rpm.

### Solution

Subscript 1 applies to the low-speed shaft and 2 to the high-speed shaft. From Eq. 4-11,

$$T_1 = \frac{\text{hp} \times 63,000}{N_1} = \frac{200 \times 63,000}{20} = 630,000 \text{ in-lb}$$

Similarly, $\qquad\qquad T_2 = 630$ in-lb

From Eq. 4-8,

$$\frac{J_1}{c} = \frac{T_1}{\tau_{max}} = \frac{630,000}{10,000} = 63 \text{ in}^3$$

$$\frac{J_1}{c} = \frac{\pi d_1^3}{16} \quad \text{or} \quad d_1^3 = \frac{16}{\pi}(63) = 321 \text{ in}^3$$

Hence, $\qquad\qquad d_1 = 6.85$ in $\qquad$ and $\qquad d_2 = 0.685$ in

This example illustrates the reason for the modern tendency to use high-speed machines in mechanical equipment. The difference in size of the two shafts is striking. Further savings in the weight of the material can be effected by using hollow tubes.

---

## 4-7. Stress Concentrations

Equations 4-3, 4-4, and 4-8 apply only to solid and tubular circular shafts while the material behaves elastically. Moreover, the cross-sectional areas along the shaft should remain reasonably constant. If a *gradual* variation in the diameter takes place, the previous equations give satisfactory solutions. On the other hand, for stepped shafts where the diameters of the adjoining portions change abruptly, large perturbations of shear stresses take place. High *local* shear stresses occur at points away from the center of the shaft. Methods of determining these local concentrations of stress are beyond the scope of this text. However, by forming a ratio of the true maximum shear stress to the maximum stress given by Eq. 4-3, a torsional stress-concentration factor can be obtained. An anal-

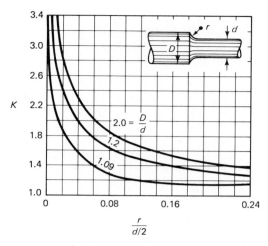

**Fig. 4-13** Torsional stress-concentration factors in circular shafts of two diameters.

ogous method was used for obtaining the stress-concentration factors in axially loaded members (Section 2-10). These factors depend only on the geometry of a member. Stress-concentration factors for various proportions of stepped round shafts are shown in Fig. 4-13.[10]

To obtain the actual stress at a geometrical discontinuity of a stepped shaft, a curve for a particular $D/d$ is selected in Fig. 4-13. Then, corresponding to the given $r/(d/2)$ ratio, the stress-concentration factor $K$ is read from the curve. Lastly, from the definition of $K$, the actual maximum shear stress is obtained from the modified Eq. 4-3

$$\tau_{\max} = K\frac{Tc}{J} \qquad (4\text{-}12)$$

where the shear stress $Tc/J$ is determined for the smaller shaft.

A study of stress-concentration factors shown in Fig. 4-13 emphasizes the need for a generous fillet radius $r$ at all sections where a transition in the shaft diameter is made.

Considerable stress increases also occur in shafts at oil holes and at keyways for attaching pulleys and gears to the shaft. A shaft prepared for a key, Fig. 4-14, is no longer a circular member. However, according to the procedures suggested by the ASME, in ordinary design, computations for shafts with keyways may be made using Eq. 4-3 or 4-8 , but the allowable shear stress must be *reduced* by 25%. This presumably compensates for the stress concentration and reduction in cross-sectional area.

Because of some inelastic or nonlinear response in real materials, for

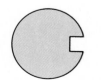

**Fig. 4-14** Circular shaft with a keyway.

[10] This figure is adapted from a paper by L. S. Jacobsen, "Torsional-Stress Concentrations in Shafts of Circular and Variable Diameter," *Trans. ASME* 47 (1925): 632.

reasons analogous to those pointed out in Section 2-10, the theoretical stress concentrations based on the behavior of linearly elastic material tend to be somewhat high.

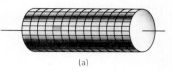

(a)

## 4-8. Angle-of-Twist of Circular Members

In this section, attention will be directed to a method for determining the angle-of-twist for solid and tubular circular *elastic shafts* subjected to torsional loading. The interest in this problem is at least threefold. First, it is important to predict the twist of a shaft per se since at times it is not sufficient to design it only to be strong enough; it also must not deform excessively. Then, magnitudes of angular rotations of shafts are needed in the torsional vibration analysis of machinery. Finally, the angular twist of members is needed in dealing with statically indeterminate torsional problems.

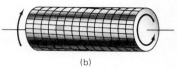

(b)

Fig. 4-15 Circular shaft (a) before and (b) after torque is applied.

According to assumption 1 stated in Section 4-3, planes perpendicular to the axis of a circular rod do not warp. The elements of a shaft undergo deformation of the type shown in Fig. 4-15(b). The shaded element is shown in its undistorted form in Fig. 4-15(a). From such a shaft, a typical element of length $dx$ is shown isolated in Fig. 4-16 similar to Fig. 4-3.

In the element shown, a line on its surface such as $CD$ is initially parallel to the axis of the shaft. After the torque is applied, it assumes a new position $CD'$. At the same time, by virtue of assumption 2, Section 4-3, radius $OD$ remains straight and rotates through a small angle $d\phi$ to a new position $OD'$.

Denoting the small angle $DCD'$ by $\gamma_{max}$, from geometry, one has two alternative expressions for the arc $DD'$:

$$\text{arc } DD' = \gamma_{max}\, dx \quad \text{ or } \quad \text{arc } DD' = d\phi\, c$$

where both angles are small and are measured in radians. Hence,

$$\gamma_{max}\, dx = d\phi\, c \tag{4-13}$$

$\gamma_{max}$ applies only in the zone of an infinitesimal "tube" of constant maximum shear stress $\tau_{max}$. Limiting attention to linearly elastic response makes Hooke's law applicable. Therefore, according to Eq. 3-1, the angle $\gamma_{max}$ is proportional to $\tau_{max}$, i.e., $\gamma_{max} = \tau_{max}/G$. Moreover, by Eq. 4-3, $\tau_{max} = Tc/J$. Hence, $\gamma_{max} = Tc/(JG)$.[11] By substituting the latter expression into Eq. 4-13 and simplifying, the governing differential equation for the angle-of-twist is obtained.

*(handwritten note: Y VARIES LINEARLY W/ DISTANCE FROM AXIS.)*

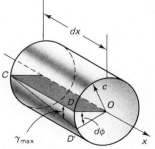

Fig. 4-16 Deformation of a circular bar element due to torque.

[11] The foregoing argument can be carried out in terms of any $\gamma$, which progressively becomes smaller as the axis of the rod is approached. The only difference in derivation consists in taking an arc corresponding to $BD$ an arbitrary distance $\rho$ from the center of the shaft and using $T\rho/J$ instead of $Tc/J$ for $\tau$.

$$\frac{d\phi}{dx} = \frac{T}{JG} \quad \text{or} \quad d\phi = \frac{T\,dx}{JG} \tag{4-14}$$

This gives the relative angle-of-twist of two adjoining sections an infinitesimal distance $dx$ apart. To find the total angle-of-twist $\phi$ between any two sections $A$ and $B$ on a shaft a finite distance apart, the rotations of all elements must be summed. Hence, a general expression for the angle-of-twist between any two sections of a shaft of a linearly elastic material is

$$\phi = \phi_B - \phi_A = \int_A^B d\phi = \int_A^B \frac{T_x\,dx}{J_xG} \tag{4-15}$$

where $\phi_B$ and $\phi_A$ are, respectively, the global shaft rotations at ends $B$ and $A$. The rotation at $A$ may not necessarily be zero. In this equation, the internal torque $T_x$ and the polar moment of inertia $J_x$ may vary along the length of a shaft. The direction of the angle of twist $\phi$ coincides with the direction of the applied torque $T$.

Equation 4-15 is valid for both solid and hollow circular shafts, which follows from the assumptions used in the derivation. The angle $\phi$ is measured in radians. Note the great similarity of this relation to Eq. 2-7 for the deformation of axially loaded rods. The following three examples illustrate applications of these concepts.

### EXAMPLE 4-6

Find the relative rotation of section $B$–$B$ with respect to section $A$–$A$ of the solid elastic shaft shown in Fig. 4-17 when a constant torque $T$ is being transmitted through it. The polar moment of inertia of the cross-sectional area $J$ is constant.

### Solution

In this case, $T_x = T$ and $J_x = J$; hence, from Eq. 4-15,

$$\phi = \int_A^B \frac{T_x\,dx}{J_xG} = \int_0^L \frac{T\,dx}{JG} = \frac{T}{JG}\int_0^L dx = \frac{TL}{JG}$$

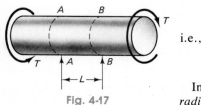

i.e.,

$$\phi = \frac{TL}{JG} \tag{4-16}$$

Fig. 4-17

In applying Eq. 4-16, note particularly that the angle $\phi$ must be expressed in *radians*. Also observe the great similarity of this relation to Eq. 2-9, $\Delta = PL/AE$, for axially loaded bars. Here $\phi \Leftrightarrow \Delta$, $T \Leftrightarrow P$, $J \Leftrightarrow A$, and $G \Leftrightarrow E$. Analogous to Eq. 2-9, Eq. 4-16 can be recast to express the *torsional spring constant*, or *torsional stiffness*, $k_t$ as

Fig. 4-18 Schematic representation of a torsion spring.

$$k_t = \frac{T}{\phi} = \frac{JG}{L} \quad \left[\frac{\text{in-lb}}{\text{rad}}\right] \quad \text{or} \quad \left[\frac{\text{N·m}}{\text{rad}}\right] \qquad (4\text{-}17)$$

This constant represents the torque required to cause a rotation of 1 radian, i.e., $\phi = 1$. It depends only on the material properties and size of the member. As for axially loaded bars, one can visualize torsion members as springs; see Fig. 4-18.

The reciprocal of $k_t$ defines the *torsional flexibility* $f_t$. Hence, for a circular solid or hollow shaft,

$$f_t = \frac{1}{k_t} = \frac{L}{JG} \quad \left[\frac{\text{rad}}{\text{in-lb}}\right] \quad \text{or} \quad \left[\frac{\text{rad}}{\text{N·m}}\right] \qquad (4\text{-}18)$$

This constant defines the rotation resulting from application of a unit torque, i.e., $T = 1$. On multiplying by the torque $T$, one obtains Eq. 4-16.

If in the analysis, a shaft must be subdivided into a number of regions, appropriate identifying subscripts should be attached to the definitions given by Eqs. 4-17 and 4-18. For example, for the $i$th segment of a bar, one can write $(k_t)_i = J_i G_i / L_i$ and $(f_t)_i = L_i / J_i G_i$.

The previous equations are widely used in mechanical vibration analyses of transmission shafts, including crank shafts.[12] These equations are also useful for solving statically indeterminate problems, considered in the next section. These equations are required in the design of members for torsional stiffness when it is essential to limit the amount of twist. For such applications, note that $J$, rather than the $J/c$ used in strength calculations, is the governing parameter. In axially loaded bar problems, the cross-sectional area $A$ serves both purposes.

Lastly, it should be noted that since in a torsion test, $\phi$, $T$, $L$, and $J$ can be measured or calculated from the dimensions of a specimen, the shear modulus of elasticity for a specimen can be determined from Eq. 4-16 since $G = TL/J\phi$.

---

## EXAMPLE 4-7

Consider the stepped shaft shown in Fig. 4-19(a) rigidly attached to a wall at $E$, and determine the angle-of-twist of the end $A$ when the two torques at $B$ and at $D$ are applied. Assume the shear modulus $G$ to be 80 GPa, a typical value for steels.

[12] According to S. P. Timoshenko, *Vibration Problems in Engineering*, 2d ed. (New York: Van Nostrand, 1937), in 1902, H. Frahm, a German engineer, was the first to recognize and study this important problem.

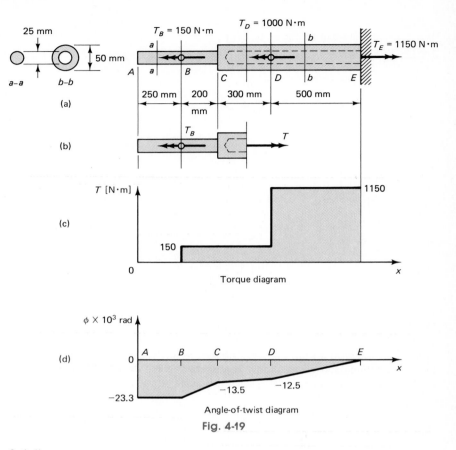

(a)

(b)

(c)

Torque diagram

(d)

Angle-of-twist diagram

**Fig. 4-19**

## Solution

Except for the difference in parameters, the solution of this problem is very similar
to that of Example 2-2 for an axially loaded bar. First, the torque at $E$ is determined
to assure equilibrium. Then internal torques at arbitrary sections, isolating the
*left* segment of a shaft, such as shown in Fig. 4-19(b), are examined. If the direction
of the torque vector $T$ coincides with that of the positive $x$ axis, it is taken as
positive, or vice versa. This leads to the conclusion that between $A$ and $B$ there
is no torque, whereas between $B$ and $D$ the torque is $+150$ Nm. The torque
between $D$ and $E$ is $+1150$ Nm. The torque diagram is drawn in Fig. 4-19(c). The
internal torques, identified by subscripts for the various shaft segments, are:

$$T_{AB} = 0, \ T_{BD} = T_{BC} = T_{CD} = 150 \text{ N·m, and } \ T_{DE} = 1150 \text{ N·m}$$

The polar moments of inertia for the two kinds of cross sections occurring in
this problem are found using Eqs. 4-2 and 4-5 giving

$$J_{AB} = J_{BC} = \frac{\pi d^4}{32} = \frac{\pi \times 25^4}{32} = 38.3 \times 10^3 \text{ mm}^4$$

$$J_{CD} = J_{DE} = \frac{\pi}{32}(d_o^4 - d_i^4) = \frac{\pi}{32}(50^4 - 25^4) = 575 \times 10^3 \text{ mm}^4$$

To find the angle-of-twist of the end $A$, Eq. 4-15 is applied for each segment and the results summed. The limits of integration for the segments occur at points where the values of $T$ or $J$ change abruptly.

$$\phi = \int_A^E \frac{T_x\,dx}{J_x G} = \int_A^B \frac{T_{AB}\,dx}{J_{AB}G} + \int_B^C \frac{T_{BC}\,dx}{J_{BC}G} + \int_C^D \frac{T_{CD}\,dx}{J_{CD}G} + \int_D^E \frac{T_{DE}\,dx}{J_{DE}G}$$

In the last group of integrals, $T$'s and $J$'s are constant between the limits considered, so each integral reverts to a known solution, Eq. 4-16. Hence,

$$\phi = \sum_i \frac{T_i L_i}{J_i G_i} = \frac{T_{AB}L_{AB}}{J_{AB}G} + \frac{T_{BC}L_{BC}}{J_{BC}G} + \frac{T_{CD}L_{CD}}{J_{CD}G} + \frac{T_{DE}L_{DE}}{J_{DE}G}$$

$$= 0 + \frac{150 \times 10^3 \times 200}{38.3 \times 10^3 \times 80 \times 10^3} + \frac{150 \times 10^3 \times 300}{575 + 10^3 \times 80 \times 10^3}$$

$$+ \frac{1150 \times 10^3 \times 500}{575 \times 10^3 \times 80 \times 10^3}$$

$$= 0 + 9.8 \times 10^{-3} + 1.0 \times 10^{-3} + 12.5 \times 10^{-3} = 23.3 \times 10^{-3} \text{ rad}$$

As can be noted from the above, the angles-of-twist for the four shaft segments starting from the left end are: 0 rad, $9.8 \times 10^{-3}$ rad, $1.0 \times 10^{-3}$ rad, and $12.5 \times 10^{-3}$ rad. Summing these quantities beginning from $A$, in order to obtain the function for the angle-of-twist along the shaft, gives the broken line from $A$ to $E$, shown in Fig. 4-19(d). Since no shaft twist can occur at the built-in end, this function must be zero at $E$, as required by the boundary condition. Therefore, according to the adopted sign convention, the angle-of-twist at $A$ is $-23.3 \times 10^{-3}$ rad occurring in the direction of applied torques.

No doubt local disturbances in stresses and strains occur at the applied concentrated torques and the change in the shaft size, as well as at the built-in end. However, these are local effects having limited influence on the overall behavior of the shaft.

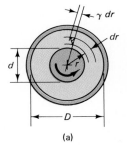

(a)

## **EXAMPLE 4-8

Determine the torsional stiffness $k_t$ for the rubber bushing shown in Fig. 4-20. Assume that the rubber is bonded to the steel shaft and the outer steel tube, which is attached to a machine housing. The shear modulus for the rubber is $G$. Neglect deformations in the metal parts of the assembly.

### Solution

Due to the axial symmetry of the problem, on every imaginary cylindrical surface of rubber of radius $r$, the applied torque $T$ is resisted by constant shear stresses $\tau$. The area of the imaginary surface is $2\pi r L$. On this basis, the equilibrium equa-

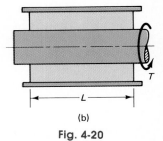

(b)

**Fig. 4-20**

tion for the applied torque $T$ and the resisting torque developed by the shear stresses $\tau$ acting at a radius $r$ is

$$T = (2\pi rL)\tau r \qquad [\text{area} \times \text{stress} \times \text{arm}]$$

From this relation, $\tau = T/2\pi r^2 L$. Hence, by using Hooke's law given by Eq. 3-1, the shear strain $\gamma$ can be determined for an infinitesimal tube of radius $r$ and thickness $dr$, Fig. 4-20(a), from the following relation:

$$\gamma = \frac{\tau}{G} = \frac{T}{2\pi LGr^2}$$

This shear strain in an infinitesimal tube permits the shaft to rotate through an infinitesimal angle $d\phi$. Since in the limit $r + dr$ is equal to $r$, the magnitude of this angle is

$$d\phi = \frac{\gamma \, dr}{r}$$

The total rotation $\phi$ of the shaft is an integral, over the rubber bushings, of these infinitesimal rotations, i.e.,

$$\phi = \int d\phi = \frac{T}{2\pi LG} \int_{d/2}^{D/2} \frac{dr}{r^3} = \frac{T}{\pi LG}\left(\frac{1}{d^2} - \frac{1}{D^2}\right)$$

From which

$$k_t = \frac{T}{\phi} = \frac{\pi LG}{1/d^2 - 1/D^2} \qquad (4\text{-}19)$$

## *4-9. Statically Indeterminate Problems

The analysis of statically indeterminate members subjected to twist parallels the procedures discussed earlier in Part B of Chapter 2 in connection with axially loaded bars. In considering *linearly elastic* problems with *one* degree of *external* indeterminacy, i.e., cases where there are two reactions, the *force* (flexibility) method is particularly advantageous. Such problems are reduced to statical determinacy by removing one of the redundant reactions and calculating the rotation $\phi_0$ at the released support. The required boundary conditions are then restored by twisting the member at the released end through an angle $\phi_1$ such that

$$\phi_0 + \phi_1 = 0 \qquad (4\text{-}20)$$

Such problems remain simple to analyze regardless of the number and kinds of applied torques or variations in the shaft size or material.

Torsion problems also occur with *internal* statical indeterminacy in composite shafts built up from two or more tubes or materials, such as shown in Fig. 4-6. In such cases, the angle-of-twist $\phi$ is the same for each constituent part of the member. Therefore, the *displacement* (stiffness) method is particularly simple to apply to linearly elastic problems. In such problems, the torque $T_i$ for each $i$th part of the shafts is $T_i = (k_t)_i\phi$, Eqs. 4-16 and 4-17. The total applied torque $T$ is then the sum of its parts, i.e.,

$$T = \sum_i (k_t)_i \phi \qquad (4\text{-}21)$$

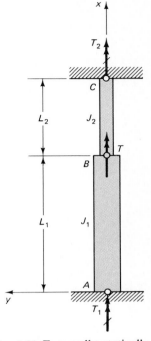

For complex *externally* statically indeterminate elastic problems with several kinematic degrees of freedom, the general displacement method similar to that given in Section 2-18 can be used. Here, however, the discussion is limited to the case of *one d.o.f.* Such cases can be analyzed using the procedure described in Section 2-16. Applying this approach to the shaft in Fig. 4-21, one can write the following two basic equations:

*For global equilibrium:*

$$T_1 + T_2 + T = 0 \qquad (4\text{-}22)$$

**Fig. 4-21** Externally statically indeterminate bar in torsion.

*For geometric compatibility:*

$$\phi_{AB} = \phi_{BC} \qquad (4\text{-}23)$$

where $\phi_{AB}$ and $\phi_{BC}$ are, respectively, the twists at $B$ of the bar segments $AB$ and $BC$, assuming that ends $A$ and $C$ are fixed.

According to Eq. 4-16, for linearly *elastic* behavior, Eq. 4-23 becomes

$$\frac{T_1 L_1}{J_1 G_1} = \frac{T_2 L_2}{J_2 G_2} \qquad (4\text{-}24)$$

where the shear moduli are given as $G_1$ and $G_2$ to provide for the possibility of different materials in the two parts of the shaft.

Solutions for one d.o.f. statically indeterminate *inelastic* problems closely follow the procedure given in Example 2-18 for axially loaded bars.

The previous procedures can be applied to the analysis of statically indeterminate bars having cross sections other than circular, such as discussed in Sections 4-14 and 4-16.

An example of an application of the force method for a statically indeterminate elastic problem follows.

**Torsion**

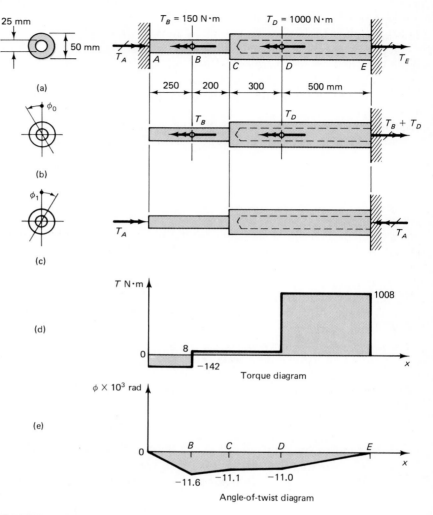

Fig. 4-22

## EXAMPLE 4-9

Assume that the stepped shaft of Example 4-7, while loaded in the same manner, is now built-in at both ends, as shown in Fig. 4-22. Determine the end reactions and plot the torque diagram for the shaft. Apply the force method.

### Solution

There are two unknown reactions, $T_A$ and $T_E$. One of them can be considered as redundant, and, arbitrarily, reaction $T_A$ is removed. This results in the free-body diagram shown in Fig. 4-22(b). The solution to Example 4-7 gives the end rotation $\phi_0 = 23.3 \times 10^{-3}$ rad.

From Example 4-7, $J_{AC} = 38.3 \times 10^3$ mm$^2$ and $J_{CE} = 575 \times 10^3$ mm$^2$. By applying $T_A$ to the *unloaded* bar, as shown in Fig. 4-22(c), end rotation $\phi_1$ at end $A$ is found using Eq. 4-16.

$$\phi_1 = \sum_i \frac{T_i L_i}{J_i G_i} = T_A \times 10^3 \left( \frac{450}{38.3 \times 10^3 \times 80 \times 10^3} + \frac{800}{575 \times 10^3 \times 80 \times 10^3} \right)$$

$$= (147 \times 10^{-6} + 17 \times 10^{-6}) T_A = 164 \times 10^{-6} T_A \text{ rad}$$

where $T_A$ has the units of N·m.

Using Eq. 4-20 and defining rotation in the direction of $T_A$ as positive, one has

$$-23.3 \times 10^{-3} + 164 \times 10^{-6} T_A = 0$$

Hence, $T_A = 142$ N·m and $T_B = 1150 - 142 = 1008$ N·m

The torque diagram for the shaft is shown in Fig. 4-22(d). As in Fig. 4-19(c) of Example 4-7, if the direction of the internal torque vector $T$ on the *left* part of an isolated shaft segment coincides with that of the positive $x$ axis, it is taken as positive. Note that most of the applied torque is resisted at the end $E$. Since the shaft from $A$ to $C$ is more flexible than from $C$ to $E$, only a small torque develops at $A$.

Calculating the angles-of-twist for the four segments of the shafts, as in Example 4-7, the angle-of-twist diagram along the shaft, Fig. 4-22(e), can be obtained. (Verification of this diagram is left as an exercise for the reader.) The angle-of-twist at $A$ and $E$ *must be zero from the prescribed boundary conditions*. As to be expected, the shaft twists in the direction of the applied torques.

Whereas this problem is indeterminate only to the first degree, it has three kinematic degrees of freedom. Two of these are associated with the applied torques and one with the change in the shaft size. Therefore, an application of the displacement method would be more cumbersome, requiring three simultaneous equations.

## **13**4-10. Alternative Differential Equation Approach for Torsion Problems

For constant $JG$, Eq. 4-14 can be recast into a second-order differential equation. Preliminary to this step, consider an element, shown in Fig. 4-23, subjected to the end torques $T$ and $T + dT$ and to an applied distributed torque $t_x$ having the units of in-lb/in or N·m/m. By using the right-hand

[13] This and the next two sections can be omitted.

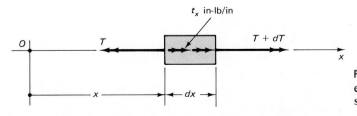

**Fig. 4-23** Infinitesimal element of a circular bar subjected to torque.

screw rule for the torques, all these quantities are shown in the figure as having a positive sense. For equilibrium of this infinitesimal element,

$$t_x \, dx + dT = 0 \qquad \text{or} \qquad \frac{dT}{dx} = -t_x \tag{4-25}$$

On differentiating Eq. 4-14 with respect to $x$,

$$JG \frac{d^2\phi}{dx^2} = \frac{dT}{dx} = -t_x \tag{4-26}$$

The constants appearing in the solution of this differential equation are determined from the boundary conditions at the ends of a shaft, and either the rotation $\phi$ or the torque $T$ must be specified. The rotation boundary conditions for $\phi$ should be evident from the problem, whereas those for the torque $T$ follow from Eq. 4-14 since $T = JG \, d\phi/dx$.

Equation 4-26 can be used for solution of statically determinate and indeterminate problems. By making use of singularity functions, discussed in Section 5-16, this equation can be employed for problems with concentrated moments.

The following example illustrates the application of Eq. 4-26 when the applied torque is a continuous function.

### **EXAMPLE 4-10

Consider an elastic circular bar having a constant $JG$ subjected to a uniformly varying torque $t_x$, as shown in Fig. 4-24. Determine the rotation of the bar along its length and the reactions at ends $A$ and $B$ for two cases: (*a*) Assume that end $A$ is free and that end $B$ is built-in, (*b*) assume that both ends of the bar are fixed.

### Solution

(*a*) By integrating Eq. 4-26 twice and determining the constants of integration $C_1$ and $C_2$ from the boundary conditions, the required solution is determined.

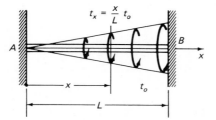

**Fig. 4-24**

$$JG \frac{d^2\phi}{dx^2} = -t_x = -\frac{x}{L}t_o$$

$$JG \frac{d\phi}{dx} = T = -\frac{t_o x^2}{2L} + C_1$$

$$T_A = T(0) = 0 \qquad \text{hence, } C_1 = 0$$

$$T_B = T(L) = -\frac{t_o L}{2}$$

$$JG\phi = -\frac{t_o x^3}{6L} + C_2$$

$$\phi_B = \phi(L) = 0 \qquad \text{hence, } C_2 = \frac{t_o L^2}{6}$$

$$JG\phi = \frac{t_o L^2}{6} - \frac{t_o x^3}{6L}$$

The negative sign for $T_B$ means that the torque vector acts in the direction opposite to that of the positive $x$ axis.

(b) Except for the change in the boundary conditions, the solution procedure is the same as in part (a).

$$JG \frac{d^2\phi}{dx^2} = -t_x = -\frac{x}{L}t_o$$

$$JG \frac{d\phi}{dx} = T = -\frac{t_o x^2}{2L} + C_1$$

$$JG\phi = -\frac{t_o x^3}{6L} + C_1 x + C_2$$

$$\phi_A = \phi(0) = 0 \qquad \text{hence, } C_2 = 0$$

$$\phi_B = \phi(L) = 0 \qquad \text{hence, } C_1 = \frac{t_o L}{6}$$

$$JG\phi = \frac{t_o L x}{6} - \frac{t_o x^3}{6L}$$

$$T_A = T(0) = \frac{t_o L}{6}$$

$$T_B = T(L) = -\frac{t_o L}{2} + \frac{t_o L}{6} = -\frac{t_o L}{3}$$

## **4-11. Energy and Impact Loads

The concepts of elastic strain energy and impact loads discussed in Sections 2-12 and 2-13 for axially loaded members, as well as those in Section 3-3 for pure shear, transfer directly to the torsion problem. For example, the deflection of a member can be determined by equating the internal

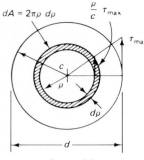

$dA = 2\pi\rho\, d\rho$    $\dfrac{\rho}{c}\tau_{\text{max}}$

$\tau_{\text{max}}$

**Fig. 4-25**

shear strain energy $U_{\text{sh}}$ for a member to the external work $W_e$ due to the applied force, Eq. 2-24. This concept can be applied to static problems (Example 2-10), as well as for elementary solutions of dynamic problems.

### EXAMPLE 4-11

(*a*) Find the energy absorbed by an elastic circular shaft subjected to a constant torque in terms of maximum shear stress and the volume of material; see Fig. 4-25. (*b*) Find the rotation of the end of an elastic circular shaft with respect to the built-in end when a torque $T$ is applied at the free end.

### Solution

(*a*) The shear stress in an elastic circular shaft subjected to a torque varies linearly from the longitudinal axis. Hence, the shear stress acting on an element at a distance $\rho$ from the center of the cross section is $\tau_{\text{max}}\rho/c$. Then, using Eq. 3-5 and integrating over the volume $V$ of the rod $L$ inches long, one obtains

$$
\begin{aligned}
U_{\text{sh}} &= \int_V \frac{\tau^2}{2G}\, dV = \int_V \frac{\tau_{\text{max}}^2 \rho^2}{2Gc^2}\, 2\pi\rho\, d\rho\, L \\
&= \frac{\tau_{\text{max}}^2}{2G}\frac{2\pi L}{c^2}\int_0^c \rho^3\, d\rho = \frac{\tau_{\text{max}}^2}{2G}\frac{2\pi L}{c^2}\frac{c^4}{4} \\
&= \frac{\tau_{\text{max}}^2}{2G}\left(\frac{1}{2}\,\text{vol}\right)
\end{aligned}
$$

If there were uniform shear stress throughout the member, a more efficient arrangement for absorbing energy would be obtained. Rubber bushings (Example 4-8) with their small $G$ values provide an excellent device for absorbing shock torques from a shaft.

(*b*) If torque $T$ is gradually applied to the shaft, the external work $W_e = \frac{1}{2}T\phi$, where $\phi$ is the angular rotation of the free end in radians. The expression for the internal train energy $U_{\text{sh}}$, which was found in part (*a*), may be written in a more convenient form by noting that $\tau_{\text{max}} = Tc/J$, the volume of the rod $\pi c^2 L$, and $J = \pi c^4/2$. Thus,

$$
U_{\text{sh}} = \frac{\tau_{\text{max}}^2}{2G}\left(\frac{1}{2}\,\text{vol}\right) = \frac{T^2 c^2}{2J^2 G}\frac{1}{2}\pi c^2 L = \frac{T^2 L}{2JG}
$$

Then, from $W_e = U_{\text{sh}}$

$$
\frac{T\phi}{2} = \frac{T^2 L}{2JG} \quad \text{and} \quad \phi = \frac{TL}{JG}
$$

which is the same as Eq. 4-16.

## **4-12. Shaft Couplings

Frequently, situations arise where the available lengths of shafting are not long enough. Likewise, for maintenance or assembly reasons, it is often desirable to make up a long shaft from several pieces. To join the pieces of a shaft together, the so-called flanged shaft couplings of the type shown in Fig. 4-26 are used. When bolted together, such couplings are termed *rigid*, to differentiate them from another type called *flexible* that provides for misalignment of adjoining shafts. The latter type is almost universally used to join the shaft of a motor to the driven equipment. Here only rigid-type couplings are considered. The reader is referred to machine-design texts and manufacturer's catalogues for the other type.

For rigid couplings, it is customary to assume that shear strains in the bolts vary directly (linearly) as their distance from the axis of the shaft. Friction between the flanges is neglected. Therefore, analogous to the torsion problem of circular shafts, if the bolts are of the same material, elastic shear *stresses* in the bolts also vary linearly as their respective distances from the center of a coupling. The shear stress in any one bolt is assumed to be *uniform* and is governed by the distance from its center to the center of the coupling. Then, if the shear stress in a bolt is multiplied by its cross-sectional area, the force in a bolt is found. On this basis, for example, for bolts of *equal size* in two "bolt circles," the forces on the bolts located by the respective radii $a$ and $b$ are as shown in Fig. 4-26(c). The moment of the forces developed by the bolts around the axis of a shaft gives the torque capacity of a coupling.

The previous reasoning is the same as that used in deriving the torsion formula for circular shafts, except that, instead of a continuous cross section, a discrete number of points is considered. This analysis is crude, since stress concentrations are undoubtedly present at the points of contact of the bolts with the flanges of a coupling.

The outlined method of analysis is valid only for the case of a coupling in which the bolts act primarily in shear. However, in some couplings, the bolts are tightened so much that the coupling acts in a different fashion. The initial tension in the bolts is great enough to cause the entire coupling to act in friction. Under these circumstances, the suggested analysis is

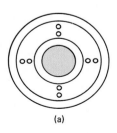
(a)

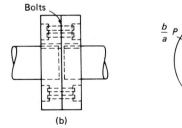
(b)

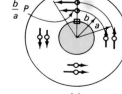
(c)

Fig. 4-26 Flanged shaft coupling.

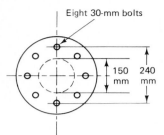

Eight 30-mm bolts

150 mm   240 mm

Fig. 4-27

not valid, or is valid only as a measure of the ultimate strength of the coupling should the stresses in the bolts be reduced. However, if high tensile strength bolts are used, there is little danger of this happening, and the strength of the coupling may be greater than it would be if the bolts had to act in shear.[14]

**EXAMPLE 4-12**

Estimate the torque-carrying capacity of a steel coupling forged integrally with the shaft, shown in Fig. 4-27, as controlled by an allowable shear stress of 40 MPa in the eight bolts. The bolt circle is diameter 240 mm.

Solution

Area of one bolt:

$$A = (1/4)\pi(30)^2 = 706 \text{ mm}^2$$

Allowable force for one bolt:

$$P_{\text{allow}} = A\tau_{\text{allow}} = 706 \times 40 = 28.2 \times 10^3 \text{ N}$$

Since eight bolts are available at a distance of 120 mm from the central axis,

$$T_{\text{allow}} = 28.2 \times 10^3 \times 120 \times 8 = 27.1 \times 10^6 \text{ N·mm} = 27.1 \times 10^3 \text{ N·m}$$

## Part B    TORSION OF INELASTIC CIRCULAR BARS

### 4-13. Shear Stresses and Deformations in Circular Shafts in the Inelastic Range

The torsion formula for circular sections previously derived is based on Hooke's law. Therefore, it applies only up to the point where the proportional limit of a material in shear is reached in the outer annulus of a shaft. Now the solution will be extended to include inelastic behavior of a material. As before, the equilibrium requirements at a section must be met. The deformation assumption of linear strain variation from the axis remains applicable. Only the difference in material properties affects the solution.

[14] See "Symposium on High-Strength Bolts," Part I, by L. T. Wyly, and Part II by E. J. Ruble, *Proc. AISC* (1950). Also see Section 1-13.

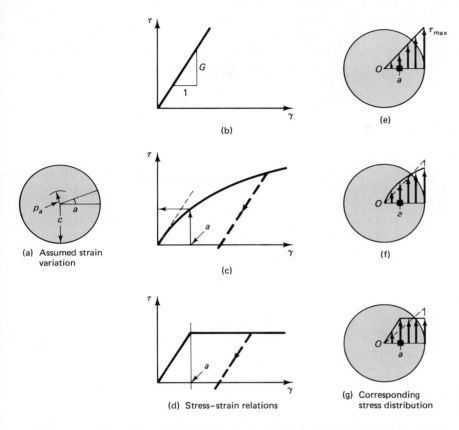

(b)

(e)

(a) Assumed strain variation

(c)

(f)

(d) Stress–strain relations

(g) Corresponding stress distribution

Fig. 4-28 Stresses in circular members due to torque.

A section through a shaft is shown in Fig. 4-28(a). The linear strain variation is shown schematically in the same figure. Some possible mechanical properties of materials in shear, obtained, for example, in experiments with thin tubes in torsion, are as shown in Figs. 4-28(b), (c), and (d). The corresponding shear-stress distribution is shown to the right in each case. The stresses are determined from the strain. For example, if the shear strain is $a$ at an interior annulus, Fig. 4-28(a), the corresponding stress is found from the stress-strain diagram. This procedure is appliable to solid shafts as well as to integral shafts made of concentric tubes of different materials, provided the corresponding stress-strain diagrams are used. The derivation for a linearly elastic material is simply a special case of this approach.

After the stress distribution is known, torque $T$ carried by these stresses is found as before, i.e.,

$$T = \int_A (\tau \, dA)\rho \tag{4-27}$$

This integral must be evaluated over the cross-sectional area of the shaft.

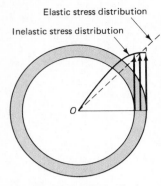

Elastic stress distribution

Inelastic stress distribution

**Fig. 4-29** For thin-walled tubes the difference between elastic and inelastic stresses is small.

Although the shear-stress distribution after the elastic limit is exceeded is nonlinear and the elastic torsion formula, Eq. 4-3, does not apply, it is sometimes used to calculate a fictitious stress for the ultimate torque. The computed stress is called the *modulus of rupture*; see the largest ordinates of the dashed lines on Figs. 4-28(f) and (g). It serves as a rough index of the ultimate strength of a material in torsion. For a thin-walled tube, the stress distribution is very nearly the same regardless of the mechanical properties of the material; see Fig. 4-29. For this reason, experiments with thin-walled tubes are widely used in establishing the shear stress-strain $\tau$-$\gamma$ diagrams.

If a shaft is strained into the inelastic range and the applied torque is then removed, every "imaginary" annulus rebounds elastically. Because of the differences in the strain paths, which cause permanent set in the material, residual stresses develop. This process will be illustrated in one of the examples that follow.

For determining the rate of twist of a circular shaft or tube, Eq. 4-13 can be used in the following form:

$$\frac{d\phi}{dx} = \frac{\gamma_{max}}{c} = \frac{\gamma_a}{\rho_a} \tag{4-28}$$

Here either the maximum shear strain at $c$ or the strain at $\rho_a$ determined from the stress-strain diagram must be used.

### EXAMPLE 4-13

A solid steel shaft of 24-mm diameter is so severely twisted that only an 8-mm diameter elastic core remains on the inside, Fig. 4-30(a). If the material properties can be idealized, as shown in Fig. 4-30(b), what residual stresses and residual rotation will remain upon release of the applied torque? Let $G = 80$ GPa.

### Solution

To begin, the magnitude of the initially applied torque and the corresponding angle of twist must be determined. The stress distribution corresponding to the given condition is shown in Fig. 4-30(c). The stresses vary linearly from 0 to 160 MPa when $0 \le \rho \le 4$ mm; the stress is a constant 160 MPa for $\rho > 4$ mm. Equation 4-27 can be used to determine the applied torque $T$. The release of torque $T$ causes elastic stresses, and Eq. 4-3 applies; see Fig. 4-30(d). The difference between the two stress distributions, corresponding to no external torque, gives the residual stresses.

$$T = \int_A \tau\rho \, dA = \int_0^c 2\pi\tau\rho^2 \, d\rho = \int_0^4 \left(\frac{\rho}{4} 160\right) 2\pi\rho^2 \, d\rho$$
$$+ \int_4^{12} (160) 2\pi\rho^2 \, d\rho$$
$$= (16 + 558) \times 10^3 \text{ N·mm} = 574 \times 10^3 \text{ N·mm}$$

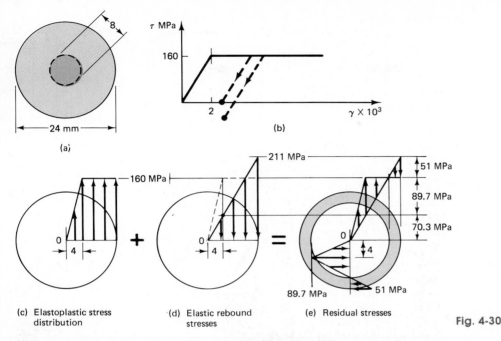

(a)

(b)

(c) Elastoplastic stress distribution

(d) Elastic rebound stresses

(e) Residual stresses

Fig. 4-30

Note the small contribution to the total of the first integral.

$$\tau_{max} = \frac{Tc}{J} = \frac{574 \times 10^3 \times 12}{(\pi/32) \times 24^4} = 211 \text{ MPa}$$

At $\rho = 12$ mm, $\tau_{residual} = 211 - 160 = 51$ MPa.

Two alternative residual stress diagrams are shown in Fig. 4-30(e). For clarity, the initial results are replotted from the vertical line. In the entire shaded portion of the diagram, the residual torque is clockwise; an exactly equal residual torque acts in the opposite direction in the inner portion of the shaft.

The initial rotation is best determined by calculating the twist of the elastic core. At $\rho = 4$ mm, $\gamma = 2 \times 10^{-3}$. The elastic rebound of the shaft is given by Eq. 4-16. The difference between the inelastic and the elastic twists gives the residual rotation per unit length of shaft. If the initial torque is reapplied in the same direction, the shaft responds elastically.

*Inelastic:*

$$\frac{d\phi}{dx} = \frac{\gamma_a}{\rho_a} = \frac{2 \times 10^{-3}}{4 \times 10^{-3}} = 0.50 \text{ rad/m}$$

*Elastic:*

$$\frac{d\phi}{dx} = \frac{T}{JG} = \frac{574 \times 10^3 \times 10^3}{(\pi/32) \times 24^4 \times 80 \times 10^3} = 0.22 \text{ rad/m}$$

*Residual:*

$$\frac{d\phi}{dx} = 0.50 - 0.22 = 0.28 \text{ rad/m}$$

## EXAMPLE 4-14

Determine the ultimate torque carried by a solid circular shaft of mild steel when shear stresses above the proportional limit are reached essentially everywhere. For mild steel, the shear stress-strain diagram can be idealized to that shown in Fig. 4-31(a). The shear yield-point stress, $\tau_{yp}$, is to be taken as being the same as the proportional limit in shear, $\tau_{pl}$.

## Solution

If a very large torque is imposed on a member, large strains take place everywhere except near the center. Corresponding to the large strains for the idealized material considered, the yield-point shear stress will be reached everywhere except near the center. However, the resistance to the applied torque offered by the material located near the center of the shaft is negligible as the corresponding $\rho$'s are small, Fig. 4-31(b). (See the contribution to torque $T$ by the elastic action in Example 4-13.) Hence, it can be assumed with a sufficient degree of accuracy that a constant shear stress $\tau_{yp}$ is acting everywhere on the section considered. The torque corresponding to this condition may be considered the *ultimate limit* torque. (Figure 4-31(c) gives a firmer basis for this statement.) Thus,

$$T_{\text{ult}} = \int_A (\tau_{yp} dA)\rho = \int_0^c 2\pi\rho^2 \tau_{yp}\, d\rho = \frac{2\pi c^3}{3}\tau_{yp}$$
$$= \frac{4}{3}\frac{\tau_{yp}}{c}\frac{\pi c^4}{2} = \frac{4}{3}\frac{\tau_{yp}J}{c} \qquad (4\text{-}29)$$

Since the maximum elastic torque capacity of a solid shaft is $T_{yp} = \tau_{yp}J/c$, Eq. 4-3, and $T_{\text{ult}}$ is $\frac{4}{3}$ times this value, the remaining torque capacity after yield is $\frac{1}{3}$ of that at yield. A plot of torque $T$ vs. $\theta$, the angle of twist per unit distance, as full plasticity develops is shown in Figure 4-31(c). Point $A$ corresponds to the

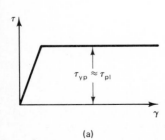

(a)

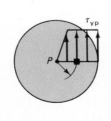

(b)

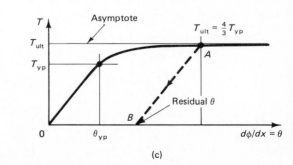

(c)

**Fig. 4-31**

results found in the preceding example, line $AB$ is the elastic rebound, and point $B$ is the residual $\theta$ for the same problem.

It should be noted that in machine members, because of the fatigue properties of materials, the ultimate static capacity of the shafts as evaluated here is often of minor importance.

## Part C    TORSION OF SOLID NONCIRCULAR MEMBERS

### *4-14. Solid Bars of Any Cross Section

The analytical treatment of solid noncircular members in torsion is beyond the scope of this book. Mathematically, the problem is complex.[15] The first two assumptions stated in Section 4-3 do not apply for noncircular members. Sections perpendicular to the axis of a member warp when a torque is applied. The nature of the distortions that take place in a rectangular section can be surmised from Fig. 4-32.[16] For a rectangular member, the corner elements do not distort at all. Therefore shear stresses at the corners are zero; they are maximum at the midpoints of the long sides. Figure 4-33 shows the shear-stress distribution along three radial lines emanating from the center. Note particularly the difference in this stress distribution compared with that of a circular section. For the latter, the stress is a maximum at the most remote point, but for the former, the stress is zero at the most remote point. This situation can be clarified by

[15] This problem remained unsolved until the famous French elastician B. de Saint Venant developed a solution for such problems in 1853. The general torsion problem is sometimes referred to as the St. Venant problem.

[16] An experiment with a rubber eraser on which a rectangular grating is ruled demonstrates this type of distortion.

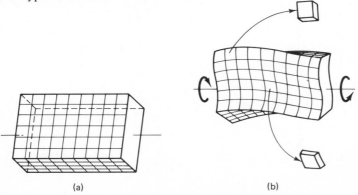

(a)                    (b)

**Fig. 4-32** Rectangular bar (a) before and (b) after a torque is applied.

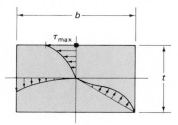

**Fig. 4-33** Shear stress distribution in a rectangular shaft subjected to a torque.

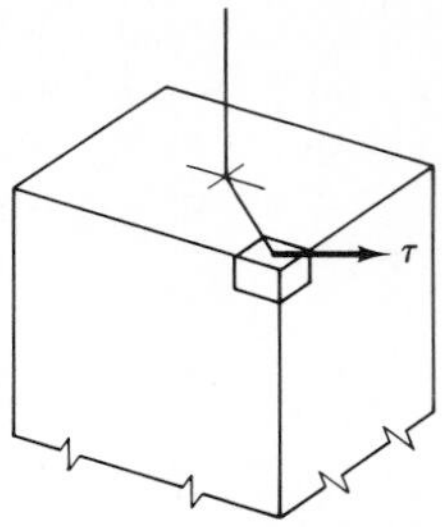

**Fig. 4-34** The shear stress shown cannot exist.

considering a corner element, as shown in Fig. 4-34. If a shear stress $\tau$ existed at the corner, it could be resolved into two components parallel to the edges of the bar. However, as shears always occur in pairs acting on mutually perpendicular planes, these components would have to be met by shears lying in the planes of the outside surfaces. The latter situation is impossible as outside surfaces are free of all stresses. Hence, $\tau$ must be zero. Similar considerations can be applied to other points on the boundary. All shear stresses in the plane of a cut near the boundaries act parallel to them.

Analytical solutions for torsion of rectangular, elastic members have been obtained.[17] The methods used are beyond the scope of this book. The final results of such analysis, however, are of interest. For the maximum shear stress (see Fig. 4-33) and the angle-of-twist, these results can be put into the following form:

$$\tau_{max} = \frac{T}{\alpha b t^2} \quad \text{and} \quad \phi = \frac{TL}{\beta b t^3 G} \tag{4-30}$$

where $T$ as before is the applied torque, $b$ is the length of the long side, and $t$ is the thickness or width of the short side of a rectangular section. The values of parameters $\alpha$ and $\beta$ depend upon the ratio $b/t$. A few of these values are recorded in the following table. For thin sections, where $b$ is much greater than $t$, the values of $\alpha$ and $\beta$ approach $\frac{1}{3}$.

**Table of Coefficients for Rectangular Bars[17]**

| $b/t$ | 1.00 | 1.50 | 2.00 | 3.00 | 6.00 | 10.0 | $\infty$ |
|---|---|---|---|---|---|---|---|
| $\alpha$ | 0.208 | 0.231 | 0.246 | 0.267 | 0.299 | 0.312 | 0.333 |
| $\beta$ | 0.141 | 0.196 | 0.229 | 0.263 | 0.299 | 0.312 | 0.333 |

[17] S. Timoshenko and J. N. Goodier, *Theory of Elasticity*, 3rd ed. (New York: McGraw-Hill, 1970), 312. The table is adapted from this source.

It is useful to recast the second Eq. 4-30 to express the torsional stiffness $k_t$ for a rectangular section, giving

$$k_t = \frac{T}{\phi} = \beta b t^3 \frac{G}{L} \qquad (4\text{-}31)$$

Formulas such as these are available for many other types of cross-sectional areas in more advanced books.[18]

For cases that cannot be conveniently solved mathematically, a remarkable method has been devised.[19] It happens that the solution of the partial differential equation that must be solved in the elastic torsion problem is mathematically identical to that for a thin membrane, such as a soap film, lightly stretched over a hole. This hole must be geometrically similar to the cross section of the shaft being studied. Light air pressure must be kept on one side of the membrane. Then the following can be shown to be true:

1. The shear stress at any point is proportional to the slope of the stretched membrane at the same point, Fig. 4-35(a).

2. The direction of a particular shear stress at a point is at right angles to the slope of the membrane at the same point, Fig. 4-35(a).

3. Twice the volume enclosed by the membrane is proportional to the torque carried by the section.

[18] R. J. Roark and W. C. Young, *Formulas for Stress and Strain*, 5th ed. (New York: McGraw-Hill, 1975). Finite-element analyses for solid bars of arbitrary cross section are also available. See, for example, L. R. Herrmann, "Elastic Torsional Analysis of Irregular Shapes," *J. Eng. Mech. Div., ASCE* (December 1965).

[19] This analogy was introduced by the German engineering scientist L. Prandtl in 1903.

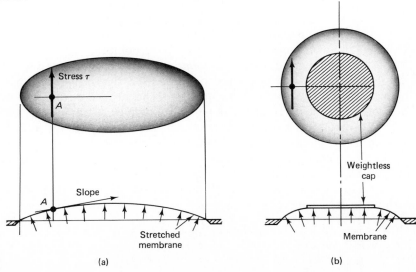

Fig. 4-35 Membrane analogy: (a) simply connected region, and (b) multiply connected (tubular) region.

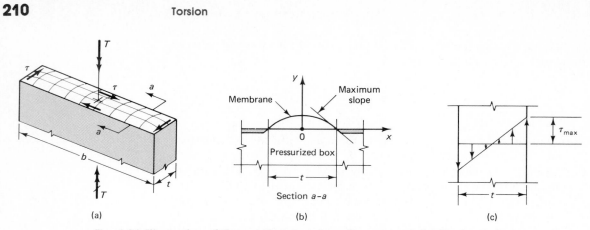

**Fig. 4-36** Illustration of the membrane analogy for a rectangular bar in torsion.

The foregoing analogy is called the *membrane analogy*. In addition to its value in experimental applications, it is a very useful mental aid for visualizing stresses and torque capacities of members. For example, consider a narrow rectangular bar subjected to torque $T$, as shown in Fig. 4-36. A stretched membrane for this member is shown in Fig. 4-36(a). If such a membrane is lightly stretched by internal pressure, a section through the membrane is a parabola, Fig. 4-36(b). For this surface, the maximum slope, hence maximum shear stress, occurs along the edges, Fig. 4-36(c). *No shear stress develops along a line bisecting the bar thickness t.* The maximum shear stresses along the short sides are small. The volume enclosed by the membrane is directly proportional to the torque the member can carry at a given maximum stress. For this reason, the sections shown in Fig. 4-37 can carry approximately the same torque at the same maximum shear stress (same maximum slope of the membrane) since the volume enclosed by the membranes would be approximately the same in all cases. (For all these shapes, $b = L$ and the $t$'s are equal.) However, use of a little imagination will convince the reader that the contour lines of a soap film will "pile up" at points $a$ of re-entrant corners. Hence, high local stresses will occur at those points.

Another analogy, the *sand-heap analogy*, has been developed for plastic torsion.[20] Dry sand is poured onto a raised flat surface having the shape of the cross section of the member. The surface of the sand heap so formed assumes a constant slope. For example, a cone is formed on a circular disc, or a pyramid on a square base. The constant maximum slope of the sand corresponds to the limiting surface of the membrane in the previous analogy. The volume of the sand heap, hence its weight, is proportional to the fully plastic torque carried by a section. The other items in connection with the sand surface have the same interpretation as those in the membrane analogy.

[20] A. Nadai, *Theory of Flow and Fracture of Solids*, Vol. 1, 2nd ed. (New York: McGraw-Hill, 1950).

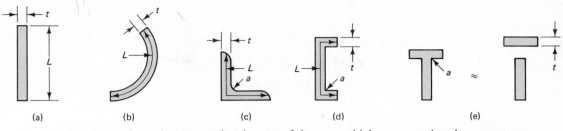

**Fig. 4-37** Members of equal cross-sectional areas of the same thickness carrying the same torque.

Statically indeterminate bars having any cross section are susceptible to the analysis procedures discussed in Section 4-9.

### EXAMPLE 4-15

By using the membrane analogy, determine an approximate value for the torsion constant $J_{equiv}$ for a W12 × 65 steel beam; see Fig. 4-38. Compare the calculated value with the 2.18 in$^4$ given in the *AISC Manual of Steel Construction.*

### Solution

By comparing the equations given for $\phi$ for a circular section, Eq. 4-16, with that for a rectangular bar, Eq. 4-30, it can be concluded that $J_{equiv} = \beta bt^3$. Further, a W12 × 65 section can be approximated, as implied in Fig. 4-37(e), by three separate narrow bars: two flanges and a web. Since $b/t$ for the flanges is 12/0.605 = 19.8 and that for the web is 10.91/0.390 = 28.0, from the table for both cases, $\beta \approx \frac{1}{3}$. Hence,

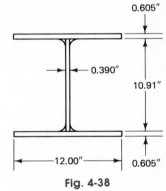

$$J_{equiv} = \tfrac{1}{3}(2 \times 12 \times 0.605^3 + 10.91 \times 0.390^3) = 1.99 \text{ in}^4$$

The value given in the *AISC Manual* is larger (2.18 in$^4$). The discrepancy can be attributed to neglecting the fillets at the four inside corners.

This problem can be solved from a different point of view using Eq. 4-21. The numerical work is identical.

**Fig. 4-38**

---

## **\*\*[21]4.15. Warpage of Thin-Walled Open Sections**

The solution of the general elastic torsion problem discussed in the preceding section is associated with the name of Saint-Venant. Solutions based on this rigorous approach (which includes membrane analogy,) for thin-walled open sections[22] may result in significant inaccuracies in some engineering applications. As pointed out in connection with the twist of

---

[21] This section presents only a qualitative discussion of this important topic.

[22] In mathematics, the boundaries of such sections are referred to as simply connected, i.e., such sections are neither tubular nor hollow.

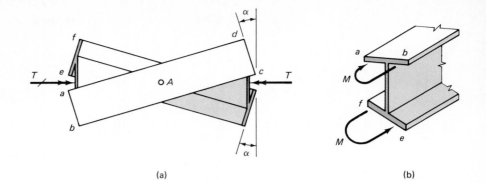

**Fig. 4-39** Cross-sectional warpage due to applied torque.

(a)

(b)

a narrow rectangular bar, Fig. 4-36, no shear stresses develop along a line bisecting thickness $t$. This means that *no in-plane deformation can take place along the entire width and length of the bar's middle surface.* The same holds true for middle surfaces of curved bars, as well as for an assembly of bars. In this sense, an $I$ section, shown in Fig. 4-39, consists of three flat bars, and, during twisting, the *three* middle surfaces of these bars *do not develop in-plane deformations.*

By virtue of symmetry, this $I$ section twists around its centroidal axis, which in this case is also the center of twist. During twisting, as the beam flanges displace laterally, the *undeformed* middle surface abcd rotates about point $A$, Fig. 4-39(a). Similar behavior is exhibited by the middle surface of the other flange. In this manner, plane sections of an $I$ beam warp, i.e., cease to be plane, during twisting. By contrast, for circular members, the sections perpendicular to the axis remain plane during twisting (See Section 4-3, assumption 1). Although warpage of the cross section does take place for other *thick* sections, including rectangular bars, this effect is negligible. On the other hand, for *thin-walled* torsion members, commonly employed in aircraft, automobiles, ships, bridges, etc., the cross-sectional warpage, or its restraint, may have an important effect[23] on member strength, and, particularly on its stiffness.

Warpage of cross sections in torsion is restrained in many engineering applications. For example, by welding an end of a steel $I$ beam to a rigid support, the attached cross section cannot warp. To maintain required compatibility of deformations, in-plane flange moments $M$, shown in Fig. 4-39(b),[24] must develop. Such an enforced restraint effectively stiffens a beam and reduces its twist. This effect is local in character and, at some distance from the support, becomes unimportant. Nevertheless, for short beams, cutouts, etc., the warpage-restraint effect is dominant. This important topic is beyond the scope of this text.[25]

[23] V. Z. Vlasov in a series of 1940 papers made basic contributions to this subject. See his book, *Thin-walled Elastic Beams*, 2nd ed. (Washington, DC: Israel Translations, Office of Technical Services, 1961).

[24] Shears that occur in the flanges and efficiently carry part of the applied torque are not shown in the diagram.

[25] For details, see, for example, J. T. Oden and E. A. Ripperger, *Mechanics of Elastic Structures*, 2nd ed. (New York: McGraw-Hill, 1981).

# Part D TORSION OF THIN-WALLED TUBULAR MEMBERS

## *4-16. Thin-Walled Hollow Members

Unlike solid noncircular members, *thin-walled* tubes of any shape can be rather simply analyzed for the magnitude of the shear stresses and the angle-of-twist caused by a torque applied to the tube. Thus, consider a tube of an arbitrary shape with varying wall thickness, such as shown in Fig. 4-40(a), subjected to torque $T$. Isolate an element from this tube, as shown enlarged in Fig. 4-40(b). This element must be in equilibrium under the action of forces $F_1$, $F_2$, $F_3$, and $F_4$. These forces are equal to the shear stresses acting on the cut planes multiplied by the respective areas.

From $\sum F_x = 0$, $F_1 = F_3$, but $F_1 = \tau_2 t_2 \, dx$, and $F_3 = \tau_1 t_1 \, dx$, where $\tau_2$ and $\tau_1$ are shear stresses acting on the respective areas $t_2 \, dx$ and $t_1 . dx$. Hence, $\tau_2 t_2 \, dx = \tau_1 t_1 \, dx$, or $\tau_1 t_1 = \tau_2 t_2$. However, since the longitudinal sections were taken an arbitrary distance apart, it follows from the previous relations that the product of the shear stress and the wall thickness is the same, i.e., constant, on any such planes. This constant will be denoted by $q$, which is measured in the units of force per unit distance along the perimeter. Therefore, its units are either N/m or lb/in.

In Section 1-4, Eq. 1-2, it was established that shear stresses on mutually perpendicular planes are equal at a corner of an element. Hence, at a corner such as $A$ in Fig. 4-40(b), $\tau_2 = \tau_3$; similarly, $\tau_1 = \tau_4$. Therefore, $\tau_4 t_1 = \tau_3 t_2$, or, in general, $q$ is constant in the plane of a section perpendicular to the axis of a member. On this basis, an analogy can be formulated. The inner and outer boundaries of the wall can be thought of as being the boundaries of a channel. Then one can imagine a constant quantity of water steadily circulating in this channel. In this arrangement, the

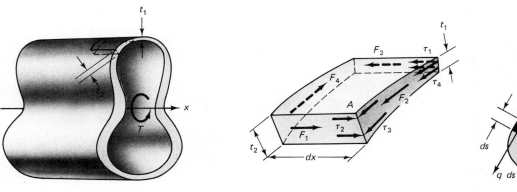

| (a) | (b) | (c) |

**Fig. 4-40** Thin-wall tubular member of variable thickness.

**214**    Torsion

quantity of water flowing through a plane across the channel is constant. Because of this analogy, the quantity $q$ has been termed the *shear flow*.

Next consider the cross section of the tube as shown in Fig. 4-40(c). The force per unit distance of the perimeter of this tube, by virtue of the previous argument, is constant and is the shear flow $q$. This shear flow multiplied by the length $ds$ of the perimeter gives a force $q\,ds$ per differential length. The product of this infinitesimal force $q\,ds$ and $r$ around some convenient point such as $O$, Fig. 4-40(c), gives the contribution of an element to the resistance of applied torque $T$. Adding or integrating this,

$$T = \oint rq\,ds \tag{4-32}$$

where the integration process is carried around the tube along the center line of the perimeter. Since for a tube, $q$ is a constant, this equation may be written as

$$T = q \oint r\,ds \tag{4-33}$$

Instead of carrying out the actual integration, a simple interpretation of the integral is available. It can be seen from Fig. 4-40(c) that $r\,ds$ is twice the value of the shaded area of an infinitesimal triangle of altitude $r$ and base $ds$. Hence, the complete integral is twice the whole area bounded by the center line of the perimeter of the tube. Defining this area by a special symbol $\circledA$, one obtains

$$T = 2\circledA q \quad \text{or} \quad q = \frac{T}{2\circledA} \tag{4-34}$$

This equation[26] applies only to *thin-walled* tubes. The area $\circledA$ is approximately an average of the two areas enclosed by the inside and the outside surfaces of a tube, or, as noted, it is an area enclosed by the center line of the wall's contour. Equation 4-34 is not applicable at all if the tube is slit, when Eqs. 4-30 should be used.

Since for any tube, the shear flow $q$ given by Eq. 4-34 is constant, from the definition of shear flow, the shear stress at any point of a tube where the wall thickness is $t$ is

$$\tau = \frac{q}{t} \tag{4-35}$$

[26] Equation 4-34 is sometimes called Bredt's formula in honor of the German engineer who developed it.

In the elastic range, Eqs. 4-34 and 4-35 are applicable to any shape of tube. For inelastic behavior, Eq. 4-35 applies only if thickness $t$ is constant. The analysis of tubes of more than one cell is beyond the scope of this book.[27]

For linearly *elastic* materials, the angle of twist for a hollow tube can be found by applying the principle of conservation of energy, Eq. 2-24. In this derivation, it is convenient to introduce the angle-of-twist per unit length of the tube defined as $\theta = d\phi/dx$. The elastic shear strain energy for the tube should also be per unit length of the tube. Hence, Eq. 3-5 for the elastic strain energy here reduces to $U_{sh} = \int_{vol} (\tau^2/2G)\, dV$, where $dV = 1 \times t\, ds$. By substituting Eq. 4-35 and then Eq. 4-34 into this relation and simplifying,

$$\overline{U}_{sh} = \oint \frac{T^2}{8\,\textcircled{A}^2 Gt}\, ds = \frac{T^2}{8\,\textcircled{A}^2 G} \oint \frac{ds}{t} \qquad (4\text{-}36)$$

where, in the last expression, the constants are taken outside the integral.

Equating this relation to the external work per unit length of member expressed as $\overline{W}_e = T\theta/2$, the governing differential equation becomes:

$$\theta = \frac{d\phi}{dx} = \frac{T}{4\,\textcircled{A}^2 G} \oint \frac{ds}{t} \qquad (4\text{-}37)$$

Here again it is useful to recast Eq. 4-37 to express the torsional stiffness $k_t$ for a thin-walled hollow tube. Since for a prismatic tube subjected to a constant torque, $\phi = \theta L$,

$$k_t = \frac{T}{\phi} = \frac{4\,\textcircled{A}^2}{\oint ds/t} \frac{G}{L} \qquad (4\text{-}38)$$

The cross-sectional warpage discussed in Section 4-15 is not very important for tubular members. Analysis of statically indeterminate tubular members follows the procedures discussed earlier.

## EXAMPLE 4-16

Rework Example 4-3 using Eqs. 4-34 and 4-35. The tube has outside and inside radii of 10 and 8 mm, respectively, and the applied torque is 40 N·m.

Solution

The mean radius of the tube is 9 mm and the wall thickness is 2 mm. Hence,

[27] J. T. Oden, and E. A. Ripperger, *Mechanics of Elastic Structures*, 2nd ed. (New York: McGraw-Hill, 1981).

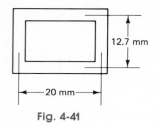

12.7 mm

20 mm

**Fig. 4-41**

$$\tau = \frac{q}{t} = \frac{T}{2\widehat{(A)}t} = \frac{40 \times 10^3}{2\pi \times 9^2 \times 2} = 39.3 \text{ MPa}$$

Note that by using Eqs. 4-34 and 4-35, only one shear stress is obtained and that it is just about the average of the two stresses computed in Example 4-3. The thinner the walls, the more accurate the answer, or vice versa.

It is interesting to note that a rectangular tube, shown in Fig. 4-41, with a wall thickness of 2 mm, for the same torque will have nearly the same shear stress as that of the circular tube. This is so because its enclosed area is about the same as the $\widehat{(A)}$ of the circular tube. However, some local stress concentrations will be present at the inside (reentrant) corners of a square tube.

### EXAMPLE 4-17

An aluminum extrusion has the cross section shown in Fig. 4-42. If torque $T = 300$ N·m is applied, (a) determine the maximum shear stresses that would develop in the three different parts of the member, and (b) find the torsional stiffness of the member. Neglect stress concentrations.

### Solution

The cross section consists essentially of three parts: a circular knob $\textcircled{1}$, a rectangular bar $\textcircled{2}$, and a rectangular hollow box with variable wall-thickness $\textcircled{3}$. During application of torque $T$, each one of these elements rotates through the same angle $\phi$, and therefore each element resists a torque $(k_t)_i\phi$. Hence, according to Eq. 4-21, the total torque resisted by the member is the sum of these quantities for the three parts. The expressions for $(k_t)_i$'s for the parts are given, respectively, by Eqs. 4-17, 4-31, and 4-38. These constants are

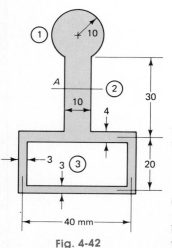

**Fig. 4-42**

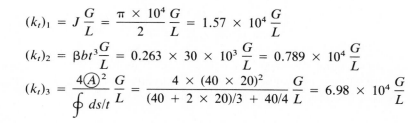

$$(k_t)_1 = J\frac{G}{L} = \frac{\pi \times 10^4}{2}\frac{G}{L} = 1.57 \times 10^4 \frac{G}{L}$$

$$(k_t)_2 = \beta bt^3\frac{G}{L} = 0.263 \times 30 \times 10^3 \frac{G}{L} = 0.789 \times 10^4 \frac{G}{L}$$

$$(k_t)_3 = \frac{4\widehat{(A)}^2}{\oint ds/t}\frac{G}{L} = \frac{4 \times (40 \times 20)^2}{(40 + 2 \times 20)/3 + 40/4}\frac{G}{L} = 6.98 \times 10^4 \frac{G}{L}$$

where all numerical values are in mm. In evaluating the integral in the last equation, it is assumed that the 4 mm thickness of the box extends for 40 mm.

By adding the stiffnesses for the parts, the member torsional stiffness $\sum (k_t)_i = 9.34 \times 10^4 G/L$.

The applied torque is distributed among the three parts in a ratio of $(k_t)_i/\sum (k_t)_i$. On this basis, the torques are $300 \times (1.57 \times 10^4 G/L)/(9.34 \times 10^4 G/L) = 50.4$ N·m for the knob, 25.3 N·m for the bar, and 224 N·m for the box. The maximum stresses in each of the parts are determined using, respectively, Eqs. 4-3, 4-30, and 4-34.

$$\tau_{1\text{-max}} = \frac{Tc}{J} = \frac{50.4 \times 10^3 \times 10}{\pi \times 10^4/2} = 32.1 \quad \text{MPa}$$

$$\tau_{2\text{-max}} = \frac{T}{\alpha b t^2} = \frac{25.3 \times 10^3}{0.267 \times 30 \times 10^2} = 31.6 \quad \text{MPa}$$

$$\tau_{3\text{-max}} = \frac{T}{2(A)t} = \frac{224 \times 10^3}{2 \times 40 \times 20 \times 3} = 46.7 \quad \text{MPa}$$

Stress $\tau_{1\text{-max}}$ occurs along the perimeter of the knob, $\tau_{2\text{-max}}$ at the midheight of the bar, and $\tau_{3\text{-max}}$ in the 3-mm walls of the tube. Due to the approximations made, these stresses cannot be considered precise. In mechanical applications, stress concentrations may be particularly important. Membrane analogy can be used to great advantage to determine the location of stress concentrations. Generous fillets at reentrant corners can be a remedy.

Member torsional stiffness found in this manner, such as needed for vibration analysis and for the solution of statically indeterminate elastic problems, would be sufficiently accurate since local effects such as stress concentrations play a minor role.

## Problems

### Sections 4-4 and 4-5

**4-1.** The solid cylindrical shaft of variable size, as shown in mm on the figure, is acted upon by the torques indicated. What is the maximum torsional stress in the shaft, and between what two pulleys does it occur?

**4-2.** A 6-in diameter core of 3 in radius is bored out from a 9-in diameter solid circular shaft. What percentage of the torsional strength is lost by this operation?

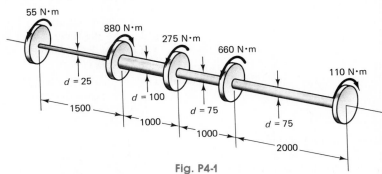

**Fig. P4-1**

### Section 4-6

**4-3.** A solid circular shaft of 2-in diameter is to be replaced by a hollow circular tube. If the outside diameter of the tube is limited to 3 in, what must be the thickness of the tube for the same linearly elastic material working at the same maximum stress? Determine the ratio of weights for the two shafts.

**4-4.** A 120-mm-diameter solid-steel shaft transmits 400 kW at 2 Hz. (a) Determine the maximum shear stress. (b) What would be the required shaft diameter to operate at 4 Hz at the same maximum stress?

**4-5.** A motor, through a set of gears, drives a line shaft, as shown in the figure, at 630 rpm. Thirty hp are delivered to a machine on the right; 90 hp on the left.

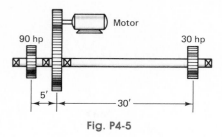

**Fig. P4-5**

Select a solid round shaft of the same size throughout. The allowable shear stress is 5750 psi.

**4-6.** (a) Design a hollow steel shaft to transmit 300 hp at 75 rpm without exceeding a shear stress of 6000 psi. Use 1.2:1 as the ratio of the outside diameter to the inside diameter. (b) What solid shaft could be used instead?

**4-7.** A 100-hp motor is driving a line shaft through gear $A$ at 26.3 rpm. Bevel gears at $B$ and $C$ drive rubber-cement mixers. If the power requirement of the mixer driven by gear $B$ is 25 hp and that of $C$ is 75 hp, what are the required shaft diameters? The allowable shear stress in the shaft is 6000 psi. A sufficient number of bearings is provided to avoid bending.

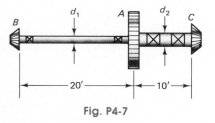

**Fig. P4-7**

## Section 4-7

**4-8.** A solid circular shaft of 150 mm diameter is machined down to a diameter of 75 mm along a part of the shaft. If, at the transition point of the two diameters, the fillet radius is 12 mm, what maximum shear stress is developed when a torque of 2700 N·m is applied to the shaft? What will the maximum shear stress be if the fillet radius is reduced to 3 mm?

**4-9.** Find the required fillet radius for the juncture of a 6-in diameter shaft with a 4-in diameter segment if the shaft transmits 110 hp at 100 rpm and the maximum shear stress is limited to 8000 psi.

## Section 4-8

**4-10.** What must be the length of a 5-mm diameter aluminum wire so that it could be twisted through one complete revolution without exceeding a shear stress of 42 MPa? $G = 27$ GPa.

**4-11.** The solid 50-mm-diameter steel line shaft shown in the figure is driven by a 30-hp motor at 3 Hz. (a) Find the maximum torsional stresses in sections $AB$, $BC$, $CD$, and $DE$ of the shaft. (b) Determine the total angle of twist between $A$ and $E$. Let $G = 84$ GPa.

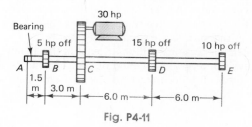

**Fig. P4-11**

**4-12.** A hollow steel rod 6 in long is used as a torsional spring. The ratio of inside to outside diameters is $\frac{1}{2}$. The required stiffness for this spring is $\frac{1}{12}$ of a degree per 1 in-lb of torque. (a) Determine the outside diameter of this rod. $G = 12 \times 10^6$ psi. (b) What is the torsional spring constant for this rod?

**4-13.** A solid aluminum-alloy shaft 50 mm in diameter and 1000 mm long is to be replaced by a tubular steel shaft of the same outer diameter such that the new shaft would neither exceed twice the maximum shear stress nor the angle of twist of the aluminum shaft. (a) What should be the inner radius of the tubular steel shaft? Let $G_{Al} = 28$ GPa and $G_{St} = 84$ GPa. (b) Which of the two criteria governs?

**4-14.** Two gears are attached to two 50-mm-diameter steel shafts, as shown in the figure. The gear at $B$ has

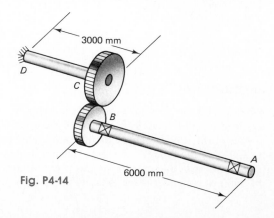

**Fig. P4-14**

a 200-mm pitch diameter; the gear at $C$, a 400-mm pitch diameter. Through what angle will end $A$ turn if at $A$ a torque of 560 N·m is applied and end $D$ of the second shaft is prevented from rotating? $G = 84$ GPa.

**4-15.** A circular steel shaft of the dimensions shown in the figure is subjected to three torques: $T_1 = 28$ k-in, $T_2 = -8$ k-in, and $T_3 = 10$ k-in. (a) What is the angle of twist of the right end due to the applied torques. (b) Plot the angle-of-twist diagram along the shaft. Let $G = 12 \times 10^6$ psi.

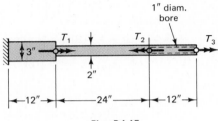

**Fig. P4-15**

**4-16.** A dynamometer is employed to calibrate the required power input to operate an exhaust fan at 20 Hz. The dynamometer consists of a 12-mm-diameter solid shaft and two disks attached to the shaft 300 mm apart, as shown in the figure. One disk is fastened through a tube at the input end; the other is near the output end. The relative displacement of these two disks as viewed in stroboscopic light was found to be 6° 0′. Compute the power input in hp required to operate the fan at the given speed. Let $G = 84$ GPa.

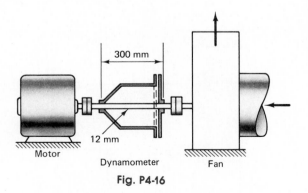

**Fig. P4-16**

**4-17.** A solid tapered steel shaft is rigidly fastened to a fixed support at one end and is subjected to a torque $T$ at the other end (see the figure). Find the angular

**Fig. P4-17**

rotation of the free end if $d_1 = 6$ in, $d_2 = 2$ in, $L = 20$ in, and $T = 27,000$ in-lb. Assume that the usual assumptions of strain in prismatic circular shafts subjected to torque apply, and let $G = 12 \times 10^6$ psi. (b) Determine the torsional flexibility of the shaft.

**4-18.** A thin-walled elastic frustum of a cone has the dimensions shown in the figure. (a) Determine the torsional stiffness of this member, i.e., the magnitude of torque per unit angle of twist. The shearing modulus for the material is $G$. (b) What is the torsional flexibility of this member?

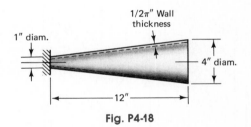

**Fig. P4-18**

**4-19.** The loading on a control torque tube for an aileron of an airplane may be idealized by a uniformly varying torque $t_x = kx$ in-lb/in, where $k$ is a constant (see the figure). Determine the angle of twist of the free end. Assume $JG$ to be constant.

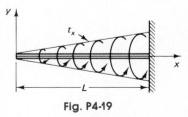

**Fig. P4-19**

***4-20.** A torque applied to a circular shaft is idealized as uniformly varying from the built-in end, see the fig-

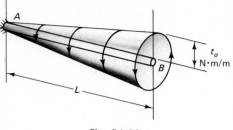

**Fig. P4-20**

ure. Determine the angle of twist of the right end. The torsional rigidity $JG$ of the shaft is constant.

**\*4-21.** A 2000-mm long circular shaft attached at one-end and free at the other is subjected to a linearly varying distributed torque along its length, as shown in the figure. The torsional rigidity $JG$ of the shaft is constant. Determine the angle of twist at the free end caused by the applied torque.

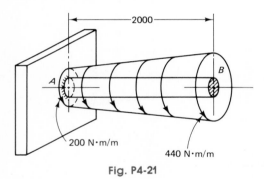

**Fig. P4-21**

## Section 4-9

**4-22.** An aluminum-alloy tube is shrunk onto a steel rod, forming a shaft that acts as a unit. This shaft is 40 in long and has the cross section shown in the figure. Assume elastic behavior and let $E_{St} = 3E_{Al} = 30 \times 10^3$ ksi. (a) What stresses would be caused by applying a torque $T = 200$ k-in? Show the shear stress distri-

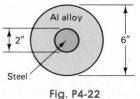

**Fig. P4-22**

bution on a graph. (b) Determine the torsional stiffness and flexibility of the shaft.

**4-23.** A tube of 50-mm outside diameter and 2-mm thickness is attached at the ends by means of rigid flanges to a solid shaft of 25-mm diameter, as shown in the figure. If both the tube and the shaft are made of the same linearly elastic material, what part of the applied torque $T$ is carried by the tube?

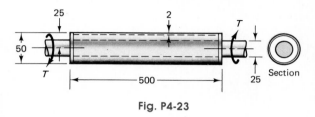

**Fig. P4-23**

**4-24.** Assume that in Problem 4-23, prior to welding the rigid end plates, the shaft is subjected to a torque of 200 N·m and maintained in this condition during the welding process. What residual torque will remain in the shaft upon release of the applied torque?

**4-25.** Using the displacement method, determine the reactions for the shaft shown in Fig. 4-21 for the following data: $T = 40$ k-in, $L_1 = 15$ in, $L_2 = 10$ in, $J_1 = 2\pi$ in$^4$, $J_2 = \pi/2$ in$^4$, and $G_1 = G_2 = G = 12 \times 10^3$ ksi. Also plot the angle-of-twist diagram for the shaft along its length.

**4-26.** Consider the same elastic stepped circular shaft shown in the two alternative figures. Using the force method, determine the angle of twist $\phi_{ab}$ at $a$ caused

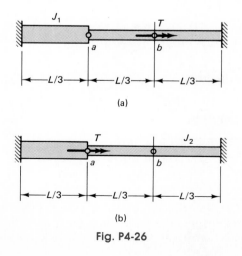

(a)

(b)

**Fig. P4-26**

by the application of a unit torque at *b*, and show that it is equal to $\phi_{ba}$, the angle-of-twist at *b* due to the application of a unit torque at *a*. Let $J_1 = 3J_2$. (See Prob. 2-48.)

**4-27.** (a) Using the force method, determine the reactions for the circular stepped shaft shown in the figure. The applied torques are $T_1 = 600$ lb-in, $T_2 = 500$ lb-in, and $T_3 = 200$ lb-in. The shaft diameters are $d_1 = 2.83$ in and $d_2 = 2.38$ in. (b) Plot the angle-of-twist diagram for the shaft along its length. Let $E = 10 \times 10^3$ ksi.

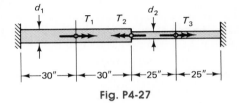

**Fig. P4-27**

**4-28.** An elastic circular shaft attached at both ends is subjected to a uniformly distributed torque $t_o$ per unit length along one-half of its length, as shown in the figure. (a) Using the force method, find the reactions. (b) Determine the angle of maximum twist and plot the angle-of-twist diagram along the shaft length. The torsional rigidity $JG$ of the shaft is constant.

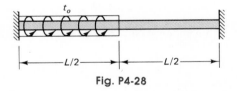

**Fig. P4-28**

**4-29.** Assume that the shaft in Problem 4-20 is attached at both ends. (a) Using the force method, determine the reactions. (b) Find the angle of maximum twist and plot the angle-of-twist diagram along the shaft length.

## Section 4-10

**4-30.** Rework Problem 4-20 using Eq. 4-26.

**4-31.** Rework Problem 4-21 using Eq. 4-26.

**\*4-32.** Using Eq. 4-26 and continuity conditions (see Section 2-19) or singularity functions, determine the reactions at the built-in ends caused by the application

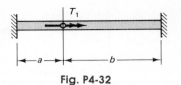

**Fig. P4-32**

of torque $T_1$; see the figure. Plot the torque $T(x)$ and the angle-of-twist $\varphi(x)$ diagrams.

**\*4-33.** Using Eq. 4-26 and continuity conditions (see Section 2-19) or singularity functions, determine the reactions caused by a uniformly distributed torque $t_o$ along one-half of the shaft length, as shown in the figure for Problem 4-28. Sketch the angle-of-twist diagram along the shaft length.

## Section 4-11

**4-34.** A circular stepped shaft has the dimensions shown in the figure. (a) Using an energy method, determine the angle of twist at the loaded end. $G$ is given. (b) Check the result using Eq. 4-16.

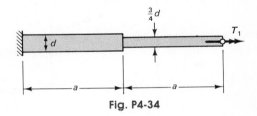

**Fig. P4-34**

## Section 4-12

**4-35.** A coupling is made with eight $\frac{3}{4}$-in-diameter high-strength bolts located on a 10-in-diameter bolt circle. (a) Calculate the torque that can be transmitted by this coupling if the allowable shear stress in the bolts is 10,500 psi. (b) Find the hp that can be transmitted when the shaft and couplings are rotating at 250 rpm.

**4-36.** A flange coupling has six bolts having a cross-sectional area of 0.2 in² each in a 8-in-diameter bolt circle, and six bolts having a cross sectional area of 0.5 in² each in a 5-in-diameter bolt circle. If the allowable shear stress in the bolt is 16 ksi, what is the torque capacity of this coupling?

## Section 4-13

**4-37.** A specimen of an SAE 1060 steel bar of 20-mm diameter and 450-mm length failed at a torque of 900

N·m. What is the modulus of rupture of this steel in torsion?

**4-38.** A solid steel shaft of 20-mm diameter and 1000 mm long is twisted such that a 16-mm-diameter core remains elastic; see the figure. (a) Determine the torque applied to cause the yield state. (b) Find the residual stress distribution that would occur on removing the torque. Draw the residual-stress pattern with the critical values. Assume the idealized mechanical properties for the material given in Fig. 4-30(b) of Example 4-13.

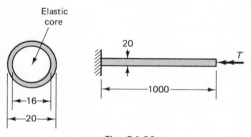

**Fig. P4-38**

**4-39.** If the shaft in Problem 4-38 is twisted at the free end through an angle $\phi = 0.25$ rad and then released, what will be the residual angle $\phi$? Also find the residual shear stresses. Draw the residual-stress pattern with the critical values.

**4-40.** A thin tube of nickel-alloy steel is shrunk onto a solid circular rod of mild streel. The cross-sectional dimensions of the composite shaft are shown in mm on the figure. Determine the torque developed by this shaft if the maximum shear stress measured on the surface is 480 MPa. For either steel, $G = 120$ GPa. However, the mild steel yields in shear at 120 MPa, whereas the alloy steel remains essentially linearly

elastic into the 600-MPa range. Idealized $\tau$-$\gamma$ diagrams for the two materials are illustrated in the figure.

**4-41.** If in Problem 4-40 the applied torque is released, (a) what will be the residual stress pattern? Draw the results with the critical values. (b) Determine the residual angle of twist per unit length of shaft.

## Section 4-14

**4-42.** Compare the maximum shear stress and angle of twist for members of equal length and cross-sectional areas for a square section, a rectangular section, and a circular section. All members are subjected to the same torque. The circular section is 100 mm in diameter and the rectangular section is 25 mm wide.

**4-43.** Compare the torsional strength and stiffness of thin-walled tubes of circular cross section of linearly elastic material with and without a longitudinal slot (see the figure).

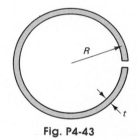

**Fig. P4-43**

**4-44.** An agitator shaft acting as a torsional member is made by welding four rectangular bars to a circular pipe, as shown in the figure. The pipe is of 4 in outside diameter and is $\frac{1}{2}$ in thick; each of the rectangular bars is $\frac{3}{4}$ by 2 in. If the maximum elastic shear stress, neglecting the stress concentrations, is limited to 8 ksi, what torque $T$ can be applied to this member?

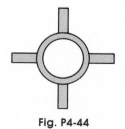

**Fig. P4-44**

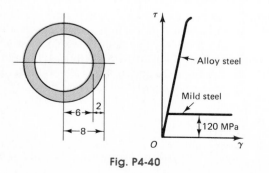

**Fig. P4-40**

**4-45.** A torsion member has the cross section shown in the figure. Estimate the torsion constant $J_{equiv}$.

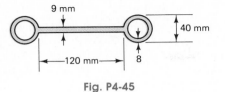

Fig. P4-45

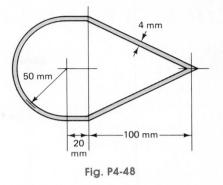

Fig. P4-48

**4-46.** Using the sand-heap analogy, determine the ultimate torsional moment of resistance for a rectangular section of $a$ by $2a$. (*Hint:* First, using the analogy, verify Eq. 4-29 for a solid circular shaft, where the height of the heap is $c\tau_{yp}$. Twice the volume included by the heap yields the required results.)

## Section 4-16

**4-47.** For a member having the cross section shown in the figure, find the maximum shear stresses and angles of twist per unit length due to an applied torque of 1000 in-lb. Neglect stress concentrations. Comment on the advantage gained by the increase in the wall thickness over part of the cross section.

MPa. Neglect the effect of stress concentrations. Is there any advantage to thicken the inclined plates? Use centerline dimensions.

**4-49.** A shaft having the cross section shown in the figure is subjected to a torque $T = 150$ N·m. (a) Estimate the percentage of torque carried by each of the two cross-sectional components, and calculate the maximum shear stresses in each part, neglecting stress concentrations. (b) Find the angle-of-twist per unit length caused by the applied torque. Let $G = 25 \times 10^3$ GPa.

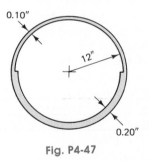

Fig. P4-47

**4-48.** A thin-walled cross section in the form of a simplified airfoil is shown in the figure. Determine the torque it would carry at a maximum shear stress of 20

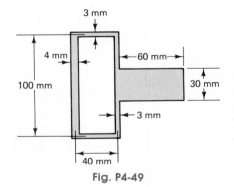

Fig. P4-49

# Axial Force, Shear, and Bending Moment

### 5-1. Introduction

The effect of axial forces and torsion on straight members was treated in the preceding chapters. There are other types of forces to which members may be subjected. In many instances in structural and machine design, members must resist forces applied laterally or transversely to their axes. Such members are called *beams*. The main members supporting floors of buildings are beams, just as an axle of a car is a beam. Many shafts of machinery act simultaneously as torsion members and as beams. With modern materials, the beam is a dominant member of construction. The determination of the system of internal forces necessary for equilibrium of any beam segment will be the main objective of this chapter.

For the axially or torsionally loaded members previously considered, only one internal force was required at an arbitrary section to satisfy the conditions of equilibrium. However, even for a beam with all forces in the same plane, i.e., a *planar* beam problem, a system of *three* internal force components can develop at a section. These are the axial force, the shear, and the bending moment. Determining these quantities is the focus of this chapter.

The chapter is divided into three parts. In Part A, methods for calculating reactions are reviewed; in Parts B and C, two different procedures for calculating the internal shear and bending moment and their graphic representations along a beam are discussed. At the end of Part C, an optional topic on singularity functions for solving such problems is introduced.

Attention will be largely confined to consideration of single beams, which, for convenience, will be shown in the horizontal position. Some discussion of related problems of planar frames resisting axial forces, shears, and bending moments is also given. Only statically determinate systems will be fully analyzed for these quantities. Special procedures to be developed in subsequent chapters are required for determining reactions in statically indeterminate problems for complete solutions. Extension to members in three-dimensional systems, where there are *six* possible internal force components, will be introduced in later chapters as needed, and will rely on the reader's knowledge of statics. In such problems at a section of a member there can be: an axial force, two shear components, two bending moment components, and a torque.

# Part A    CALCULATION OF REACTIONS

## *5-2. Diagrammatic Conventions for Supports[1]

In studying planar structures it is essential to adopt diagrammatic conventions for their supports and loadings inasmuch as several kinds of supports and a great variety of loads are possible. An *adherence* to such conventions avoids much confusion and minimizes the chances of making mistakes. These conventions form the pictorial language of engineers.

Three types of supports are recognized for planar structures. These are identified by the kind of resistance they offer to the forces. One type of support is physically realized by a *roller* or a *link*. It is capable of resisting a force in only *one specific line of action*. The link shown in Fig. 5-1(a) can resist a force only in the direction of line *AB*. The roller in Fig. 5-1(b) can resist only a vertical force, whereas the rollers in Fig. 5-1(c) can resist only a force that acts perpendicular to the plane *CD*. This type of support will be usually represented in this text by rollers as shown in Figs. 5-1(b) and (c), and it will be understood that *a roller support is capable of resisting a force in either direction*[2] along the line of action of the reaction. To avoid this ambiguity, a schematic link will be occasionally employed to indicate that the reactive force may act in either direction (see Fig. 5-4). A reaction of this type corresponds to a single unknown when equations of statics are applied. For inclined reactions, the *ratio* between the two components is fixed (see Example 1-3).

Another type of support that may be used is a *pin*. In construction, such a support is realized by using a detail shown in Fig. 5-2(a). In this text, such supports will be represented diagrammatically, as shown in

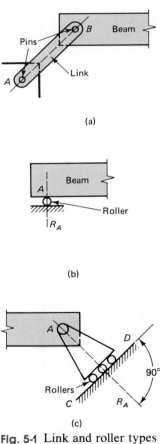

(a)

(b)

(c)

**Fig. 5-1** Link and roller types of supports. (The only possible lines of action of the reactions are shown by the dashed lines.)

---

[1] This and the next three sections are an informal review of statics.
[2] This imples that in the actual design, a link must be provided if the reaction acts away from the beam; in other words, the beam is not allowed to lift off from the support at *A* in Fig. 5-1(b). In this figure, it may be helpful to show the roller on top of the beam in the case of a downward reaction in order to make it clear that the beam is constrained against moving vertically at the support. This practice usually will be followed in the text.

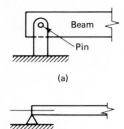

**Fig. 5-2** Pinned support: (a) actual, and (b) diagrammatic.

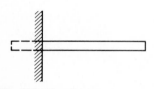

**Fig. 5-3** Fixed support.

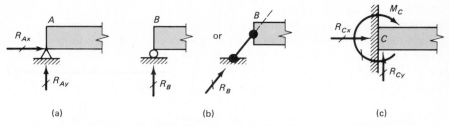

**Fig. 5-4** Three basic types of idealized supports for planar structural systems. Simple supports: (a) a pinned support resists two force components, and (b) a roller or a link resists only one directed force. Fixed support: (c) a fixed support resists two force components and a moment.

Fig. 5-2(b). A pinned support is capable of resisting a force acting in *any* direction of the plane. Hence, in general, the reaction at such a support may have two components, one in the horizontal and one in the vertical direction. Unlike the ratio applying to the roller or link support, that between the reaction components for the pinned support is *not fixed*. To determine these two components, two equations of statics must be used.

The third type of support is able to resist a force in any direction *and is also capable of resisting a moment or a couple*. Physically, such a support is obtained by building a beam into a wall, casting it into concrete, or welding the end of a member to the main structure. A system of *three* forces can exist at such a support, two components of force and a moment. Such a support is called a *fixed support*, i.e., the built-in end is fixed or prevented from rotating. The standard convention for indicating it is shown in Fig. 5-3.

To differentiate fixed supports from the roller and pin supports, which are not capable of resisting moment, the latter two are termed *simple supports*. Figure 5-4 summarizes the foregoing distinctions between the three types of supports and the kind of resistance offered by each type. In practice, engineers usually assume the supports to be of one of the three types by "judgment," although in actual construction, supports for beams do not always clearly fall into these classifications.

## *5-3. Diagrammatic Conventions for Loading

Structural members are called upon to support a variety of loads. For example, frequently a force is applied to a beam through a post, a hanger, or a bolted detail, as shown in Fig. 5-5(a). Such arrangements apply the force over a very limited portion of the beam and are idealized for the purposes of beam analysis as *concentrated* forces. These are shown diagrammatically in Fig. 5-5(b). On the other hand, in many instances, the forces are applied over a considerable portion of the beam. In a warehouse, for example, goods may be piled up along the length of a beam. Such *distributed* loads are defined by their load intensity at any point in force per unit length.

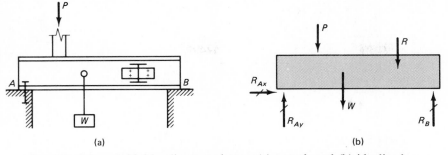

**Fig. 5-5** Concentrated loading on a beam, (a) actual, and (b) idealized.

Many types of distributed loads occur. Among these, two kinds are particularly important: the *uniformly distributed* loads and the *uniformly varying* loads. The first could easily be an idealization of the warehouse load just mentioned, where the same kind of goods are piled up to the same height along the beam. Likewise the beam itself, if of constant cross-sectional area, is an excellent illustration of the same kind of loading. A realistic situation and a diagrammatic idealization are shown in Fig. 5-6. This load is usually expressed as force per unit length of the beam, unless specifically noted otherwise. In SI units, it may be given as newtons per meter (N/m); in the U.S. customary units, as pounds per inch (lb/in), as pounds per foot (lb/ft), or as kilopounds per foot (k/ft).

Uniformly varying loads act on the vertical and inclined walls of a vessel containing liquid. This is illustrated in Fig. 5-7, where it is assumed that the vertical beam is *one* meter *wide* and $\gamma$ (N/m³) is the unit weight of the liquid. For this type of loading, it should be carefully noted that the maximum intensity of the load of $q_o$ N/m is applicable only to an *infinitesimal length* of the beam. It is twice as large as the average intensity of pressure. Hence, the total force exerted by such a loading on a beam is $(q_o h/2)$ N, and its resultant acts at a distance $h/3$ above the vessel's bottom. Horizontal bottoms of vessels containing liquid are loaded uniformly. Various aerodynamic loadings are of distributed type.

Finally, it is conceivable to load a beam with a concentrated moment applied to the beam essentially at a point. One of the possible arrangements for applying a concentrated moment is shown in Fig. 5-8(a), and

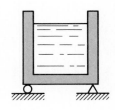

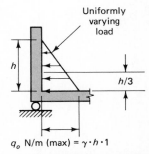

$$q_o \text{ N/m (max)} = \gamma \cdot h \cdot 1$$

**Fig. 5-7** Hydrostatic loading on a vertical wall.

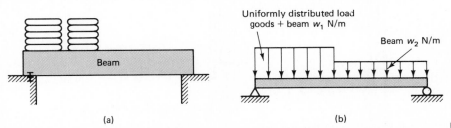

**Fig. 5-6** Distributed loading on a beam, (a) actual, and (b) idealized.

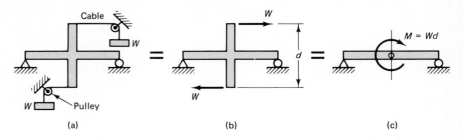

**Fig. 5-8** A method for applying a concentrated moment to a beam.

(a)                    (b)                    (c)

its diagrammatic representation to be used in this text is shown in Fig. 5-8(c).

A less artificial example of the application of a concentrated moment to a member, frequently occurring in the design of machine and structural elements, is illustrated in Fig. 5-9. In order to maintain the applied force $P$ in equilibrium at joint $C$, a shear $P$ and a moment $Pd$ must be developed at the support, Fig. 5-9(c). These forces apply a concentrated moment and an axial force, as shown in Fig. 5-9(b).

The necessity for a complete understanding of the foregoing symbolic representation for supports and forces cannot be overemphasized. Note particularly the kind of resistance offered by the different types of supports and the manner of representation of the forces at such supports. These notations will be used to construct free-body diagrams for beams.

## *5-4. Classification of Beams

Beams are classified into several groups, depending primarily on the kind of supports used. Thus, if the supports are at the ends and are either pins or rollers, the beams are *simply supported*, or *simple* beams, Figs. 5-10(a) and (b). The beam becomes a *fixed* beam, or *fixed-ended* beam, Fig. 5-

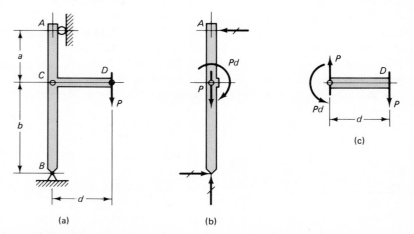

(a)                    (b)

**Fig. 5-9** Loaded horizontal member applies an axial force and a concentrated moment to the vertical member.

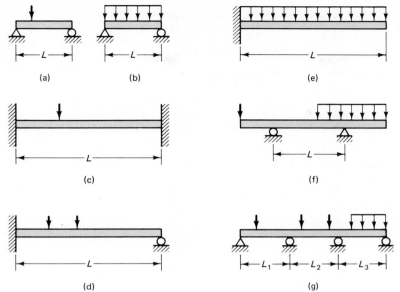

Fig. 5-10 Types of beams.

10(c), if the ends have fixed supports. Likewise, following the same scheme of nomenclature, the beam shown in Fig. 5-10(d) is a beam fixed at one end and simply supported at the other. Such beams are also called *restrained* beams, as one end is "restrained" from rotation. A beam fixed at one end and completely free at the other has a special name, a *cantilever* beam, Fig. 5-10(e).

If the beam projects beyond a support, the beam is said to have an *overhang*. Thus, the beam shown in Fig. 5-10(f) is an overhanging beam. If intermediate supports are provided for a physically continuous member acting as a beam, Fig. 5-10(g), the beam is termed a *continuous* beam.

For all beams, the distance $L$ between supports is called a *span*. In a continuous beam, there are several spans that may be of varying lengths.

In addition to classifying beams on the basis of supports, descriptive phrases pertaining to the loading are often used. Thus, the beam shown in Fig. 5-10(a) is a simple beam with a concentrated load, whereas the one in Fig. 5-10(b) is a simple beam with a uniformly distributed load. Other types of beams are similarly described.

For most of the work in engineering solid mechanics, it is also meaningful to further classify beams into statically determinate and statically indeterminate beams. If for a planar beam or a frame, the number of unknown reaction components, including a bending moment, does not exceed three, such a structural system is externally statically determinate. These unknowns can always be found from the equations of static equilibrium. The next section will briefly review the methods of statics for computing reactions for statically determinate beams. A procedure for determining reactions in indeterminate beams is given in Chapter 10.

## *5-5. Calculation of Beam Reactions

All subsequent work with beams in this chapter will begin with determination of the reactions. When all of the forces are applied in one plane, three equations of static equilibrium are available for the analysis. These are $\sum F_x = 0$, $\sum F_y = 0$, and $\sum M_z = 0$, and have already been discussed in Chapter 1. For straight beams in the horizontal position, the $x$ axis will be taken in a horizontal direction, the $y$ axis in the upward vertical direction, and the $z$ axis normal to the plane of the paper. The application of these equations to several beam problems is illustrated in the following examples and is intended to serve as a review of this important procedure. The deformation of beams, being small, is neglected when the equations of statics are applied. For stable beams, the small amount of deformation that does take place changes the points of application of the forces imperceptibly.

### EXAMPLE 5-1

Find the reactions at the supports for a simple beam loaded as shown in Fig. 5-11(a). Neglect the weight of the beam.

### Solution

The loading of the beam is already given in diagrammatic form. The nature of the supports is examined next, and the unknown components of these reactions are clearly indicated on the diagram. The beam, with the unknown reaction components and all the applied forces, is redrawn in Fig. 5-11(b) to emphasize this important step in constructing a free-body diagram. In order to differentiate among the applied forces and reactions, following the suggestion made in Section 1-5, slashes are drawn across the reaction force vectors.

At $A$, *two* unknown reaction components may exist, since the end is pinned. The reaction at $B$ can act only in a vertical direction since the end is on a roller. The points of application of all forces are carefully noted. After a free-body diagram of the beam is made, the equations of statics are applied to obtain the solution.

$$\sum F_x = 0 \qquad\qquad\qquad\qquad\qquad\qquad\qquad R_{Ax} = 0$$
$$\sum M_A = 0 \circlearrowleft + \quad 200 + 100 \times 0.2 + 160 \quad \times 0.3 - R_B \times 0.4 = 0$$
$$R_B = + 670 \text{ N} \uparrow$$
$$\sum M_B = 0 \circlearrowleft + \quad R_{Ay} \times 0.4 + 200 - 100 \quad \times 0.2 - 160 \times 0.1 = 0$$
$$R_{Ay} = -410 \text{ N} \downarrow$$

*Check:* $\quad \sum F_y = 0 \uparrow + \qquad\qquad\qquad -410 - 100 - 160 + 670 = 0$

Note that $\sum F_x = 0$ uses one of the three independent equations of statics; thus, only two additional reaction components can be determined from statics.

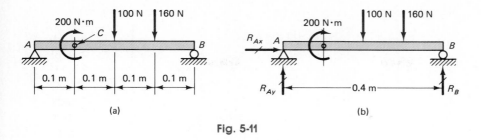

Fig. 5-11

If more unknown reaction components or moments exist at the support, the problem becomes statically indeterminate.

Note that the concentrated moment applied at $C$ enters only into the expressions for the summation of moments. The positive sign of $R_B$ indicates that its direction has been correctly assumed in Fig. 5-11(b). The opposite is the case of $R_{Ay}$, and the vertical reaction at $A$ acts downward. A check on the arithmetical work is available if the calculations are made as shown.

## Alternative Solution

In computing reactions, some engineers prefer to make calculations in the manner indicated in Fig. 5-12. Fundamentally, this involves the use of the same principles. Only the details are different. The reactions for every force are determined one at a time. The total reaction is obtained by summing these reactions. This procedure permits a running check of the computations as they are performed. For every force, the sum of its reactions is equal to the force itself. For example, for the 160-N force, it is easy to see that the upward forces of 40 N and 120 N total 160 N. On the other hand, the concentrated moment at $C$, being a couple, is resisted by a couple. It causes an *upward* force of 500 N at the right reaction and a *downward* force of 500 N at the left reaction.

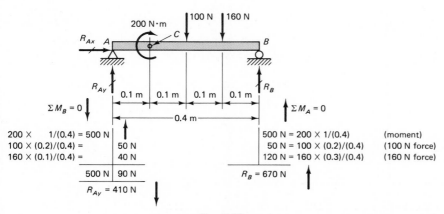

Fig. 5-12

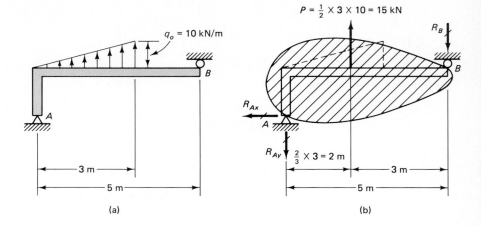

**Fig. 5-13**

(a)                                          (b)

### EXAMPLE 5-2

Find the reactions for the partially loaded beam with a uniformly varying load shown in Fig. 5-13(a). Neglect the weight of the beam.

### Solution

An examination of the supporting conditions indicates that there are three unknown reaction components; hence, the beam is statically determinate. These and the applied load are shown in Fig. 5-13(b). Note particularly that the configuration of the member is not important for computing the reactions. A crudely shaped outline, bearing no resemblance to the actual beam, is indicated to emphasize this point. However, this new body is supported at points $A$ and $B$ in the same manner as the original beam.

For calculating the reactions, the distributed load is replaced by an equivalent concentrated force $P$. It acts through the centroid of the distributed forces. These pertinent quantities are marked on the working sketch, Fig. 5-13(b). After a free-body diagram is prepared, the solution follows by applying the equations of static equilibrium.

$$\sum F_x = 0 \qquad\qquad\qquad\qquad\qquad\qquad\qquad R_{Ax} = 0$$

$$\sum M_A = 0\circlearrowleft + \qquad + 15 \times 2 - R_B \times 5 = 0 \qquad R_B = 6 \text{ kN} \downarrow$$

$$\sum M_B = 0\circlearrowright + \qquad - R_{Ay} \times 5 + 15 \times 3 = 0 \qquad R_{Ay} = 9 \text{ kN} \downarrow$$

*Check:* $\sum F_y = 0 \uparrow +$ $\qquad\qquad\qquad\qquad\qquad\qquad -9 + 15 - 6 = 0$

### EXAMPLE 5-3

Determine the reactions at $A$ and $B$ for the beam shown in Fig. 5-14(a) due to the applied force.

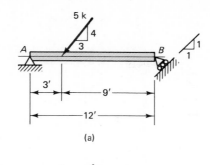

(a)

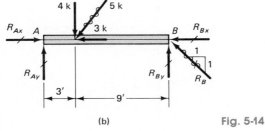

(b)

Fig. 5-14

## Solution

A free-body diagram is shown in Fig. 5-14(b). At $A$, there are two unknown reaction components, $R_{Ax}$ and $R_{Ay}$. At $B$, the reaction $R_B$ acts normal to the supporting plane and constitutes a single unknown. It is expedient to replace this force by the two components $R_{By}$ and $R_{Bx}$, which in this particular problem are numerically equal. Similarly, it is best to replace the inclined force with the two components shown. These steps reduce the problem to one where all forces are either horizontal or vertical. This is of great convenience in applying the equations of static equilibrium.

$$\sum M_A = 0 \circlearrowright + \qquad 4 \times 3 - R_{By} \times 12 = 0 \qquad R_{By} = 1\,\text{k} \uparrow \; = |\,R_{Bx}\,|$$
$$\sum M_B = 0 \circlearrowright + \qquad R_{Ay} \times 12 - 4 \times 9 = 0 \qquad R_{Ay} = 3\,\text{k} \uparrow$$
$$\sum F_x = 0 \rightarrow + \qquad R_{Ax} - 3 - 1 \qquad = 0 \qquad R_{Ax} = 4\,\text{k} \rightarrow$$
$$R_A \; = \sqrt{4^2 + 3^2} = 5\,\text{k}$$
$$R_B \; = \sqrt{1^2 + 1^2} = \sqrt{2}\,\text{k}$$

*Check:* $\sum F_y = 0 \uparrow +$ $\qquad\qquad\qquad\qquad\qquad + 3 - 4 + 1 = 0$

Occasionally, *hinges* or *pinned joints* are introduced into beams and frames. A hinge is capable of transmitting only horizontal and vertical forces. *No moment can be transmitted at a hinged joint.* Therefore, the point where a hinge occurs is a particularly convenient location for "separation" of the structure into parts for purposes of computing the reactions. This process is illustrated in Fig. 5-15. Each part of the beam so separated is treated independently. Each hinge provides an extra axis around which moments may be taken to determine reactions. The

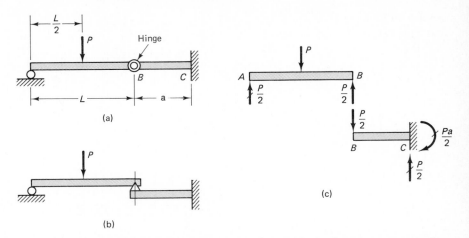

**Fig. 5-15** Structures "separated" at hinges to determine the reactions by statics.

introduction of a hinge or hinges into a continuous beam in many cases makes the system statically determinate. The introduction of a hinge into a determinate beam results in a beam that is not stable. Note that the reaction at the hinge for one beam acts in an *opposite direction* on the other beam.

## Part B     DIRECT APPROACH FOR AXIAL FORCE, SHEAR AND BENDING MOMENT

### 5-6. Application of the Method of Sections

The main objective of this chapter is to establish means for determining the forces that exist at a section of a beam or a frame. To obtain these forces, the method of sections, the basic approach of solid mechanics, will be applied. This procedure is referred to here as a direct approach.

The analysis of any beam or frame for determining the *internal forces* begins with the preparation of a free-body diagram showing both the applied and the reactive forces. The reactions can always be computed using the equations of equilibrium provided the system is statically determinate. If the system is statically indeterminate, the reactions are appropriately labeled and shown on the free-body. In this manner, for either case, the complete force system is identified. In the subsequent steps of analysis, *no distinction has to be made between the applied and reactive forces.* The method of sections can then be applied at any section of a structure by employing the previously used concept that <u>if a *whole body* is in equilibrium, *any part* of it is likewise in equilibrium.</u>

To be specific, consider a beam, such as shown in Fig. 5-16(a), with certain concentrated and distributed forces acting on it. The reactions are also presumed to be known, since they may be computed as in the examples considered earlier in Section 5-5. The externally applied forces and the reactions at the support keep the *whole body* in equilibrium. Now consider an imaginary cut *X–X normal* to the axis of the beam, which separates the beam into two segments, as shown in Figs. 5-16(b) and (c). Note particularly that the imaginary section goes through the distributed load and separates it too. Each of these beam segments is a free-body that must be in equilibrium. These conditions of equilibrium require the existence of a system of internal forces at the cut section of the beam. In general, at a section of such a member, a vertical force, a horizontal force, and a moment are necessary to maintain the isolated part in equilibrium. These quantities take on a special significance in beams and therefore will be discussed separately.

## 5-7. Axial Force in Beams

A horizontal force such as $P$, shown in Fig. 5-16(b) or (c), may be necessary at a section of a beam to satisfy the conditions of equilibrium. The magnitude and sense of this force follows from a particular solution of the equation $\sum F_x = 0$. If the horizontal force $P$ acts toward the section, it is called a *thrust*; if away, it is called *axial tension*. In referring to either of these forces, the term *axial force* is used. The effect of an axial force on a section of a member has already been discussed in Chapters 1 and 2. It was shown that it is imperative to apply this force through the *centroid*

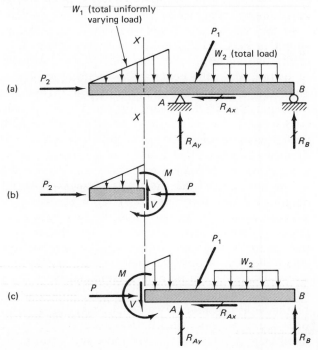

**Fig. 5-16** An application of the method of sections to a statically determinate beam.

of the cross-sectional area of a member to avoid bending. Similarly, here *the line of action of the axial force will always be directed through the centroid of the beam's cross-sectional area.*

Any section along a beam may be examined for the magnitude of the axial force in the previous manner. The tensile force at a section is customarily taken positive. The axial force (thrust) at section $X$–$X$ in Figs. 5-16(b) and (c) is equal to the horizontal force $P_2$.

## 5-8. Shear in Beams

In general, to maintain a segment of a beam, such as that shown in Fig. 5-16(b), in equilibrium, there must be an internal vertical force $V$ at the cut to satisfy the equation $\sum F_y = 0$. This internal force $V$, acting *at right angles* to the axis of the beam, is called the *shear*, or *shear force. The shear is numerically equal to the algebraic sum of all the vertical components of the external forces acting on the isolated segment*, but it is opposite in direction. Given the qualitative data shown in Fig. 5-16(b), $V$ is opposite in direction to the downward load to the left of the section. This shear may also be computed by considering the right-hand segment shown in Fig. 5-16(c). It is then equal numerically and is opposite in direction to the sum of all the vertical forces, including the vertical reaction components, to the right of the section. Whether the right-hand segment or the left is used to determine the shear at a section is immaterial—arithmetical simplicity governs. Shears at *any other section* may be computed similarly.

At this time, a significant observation must be made. The *same* shear shown in Figs. 5-16(b) and (c) at the section $X$–$X$ is opposite in direction in the two diagrams. For that *part* of the downward load $W_1$ to the left of section $X$–$X$, the beam at the section provides an upward support to maintain vertical forces in equilibrium. Conversely, the loaded portion of the beam exerts a downward force *on* the beam, as shown in Fig. 5-16(c). At a section, "two directions" of shear must be differentiated, depending upon *which segment* of the beam is considered. This follows from the familiar action-reaction concept of statics and has occurred earlier in the case of an axially loaded rod, and again in the torsion problem.

The direction of the shear at section $X$–$X$ would be reversed in *both* diagrams if the distributed load $W_1$ were acting upward. Frequently, a similar reversal in the direction of shear takes place at one section or another along a beam. Therefore, the adoption of a sign convention is necessary to differentiate between the two possible directions of shear. The definition of positive shear is illustrated in Fig. 5-17. A *downward* internal force $V$ acting at a section on an isolated *left segment* of the beam, as in Fig. 5-17(a), or an *upward* force $V$ acting at the same section on the *right segment* of the beam, as in Fig. 5-17(b), corresponds to positive shear. Positive shears are shown in Fig. 5-17(c) for an element isolated from a beam by two sections, and again in Fig. 5-17(d). The shear

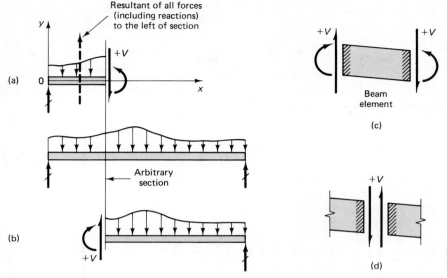

Resultant of all forces (including reactions) to the left of section

(a)

Arbitrary section

(b)

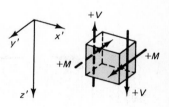

Beam element

(c)

(d)

Fig. 5-17 Definition of positive shear.

at section $X$–$X$ of Fig. 5-16(a) is a negative shear. Note that in addition to specifying the direction of a shear $V$, it is essential to associate it with a particular side of a section, Fig. 5-17(c). This is also true with stresses. (See discussion in Sections 1-3 and 1-4.)

The selected sign convention for shear in this book is the one generally used. Historically, it appears to be based on directing the coordinate axes as shown in Fig. 5-18(a). A few books[3] reverse the direction of positive shear to be consistent with the direction of axes in Fig. 5-18(b).

## 5-9. Bending Moment in Beams

The internal shear and axial forces at a section of a beam satisfy only two equations of equilibrium: $\sum F_x = 0$ and $\sum F_y = 0$. The remaining condition of static equilibrium for a planar problem is $\sum M_z = 0$. This, in general, can be satisfied only by developing a couple or an *internal resisting moment* within the cross-sectional area of the cut to counteract the moment caused by the external forces. The internal resisting moment must act in a direction opposite to the external moment to satisfy the governing equation $\sum M_z = 0$. It follows from the same equation that *the magnitude of the internal resisting moment equals the external moment*. These moments tend to bend a beam in the plane of the loads and are usually referred as *bending moments*.

To determine an internal bending moment maintaining a beam segment in equilibrium, either the left- or the right-hand part of a beam free-body

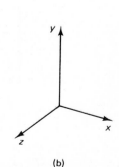

(a)

(b)

Fig. 5-18 Positive sense of shear and bending moment defined in (a) is used in this text with coordinates shown in (b).

[3] S. H. Crandall, N. C. Dahl, and T. J. Lardner, *An Introduction to the Mechanics of Solids*, 2nd ed. (New York: McGraw-Hill, 1978). J. L. Merriam, *Statics*, 2nd ed. (New York: Wiley, 1971). E. P. Popov, *Introduction to Mechanics of Solids*, (Englewood Cliffs, NJ: Prentice-Hall, 1968).

can be used, as shown in Figs. 5-16(b) and (c). The magnitude of the bending moment is found by the summation of the moments caused by all forces multiplied by their respective arms. The internal forces $V$ and $P$, as well as the applied couples, must be included in the sum. In order to exclude the moments caused by $V$ and $P$, it is advantageous to *select the point of intersection of these two internal forces as the point around which the moments are summed*. This point lies on the *centroidal axis* of the beam cross section. In Figs. 5-16(b) and (c), the internal bending moment may be physically interpreted as a pull on the top fibers of the beam and a push on the lower ones.

If the load $W_1$ in Fig. 5-16(a) were acting in the opposite direction, the resisting moments in Figs. 5-16(b) *and* (c) would reverse. This and similar situations require the adoption of a sign convention for the bending moments. This convention is associated with a definite physical action of the beam. For example, in Figs. 5-16(b) and (c), the internal moments shown cause tension in the upper part of the beam and compression the lower. This tends to increase the length of the top surface of the beam and to contract the lower surface. A continuous occurrence of such moments along the beam makes the beam deform convex upwards, i.e., "shed water." Such bending moments are assigned a *negative sign*. Conversely, a positive moment is defined as one that produces compression in the top part and tension in the lower part of a beam's cross section. Under such circumstances, the beam assumes a shape that "retains water." For example, a simple beam supporting a group of downward forces deflects down as shown in *exaggerated* form in Fig. 5-19(a), a fact suggested by physical intuition. Definitions for positive and negative bending moments are shown in Figs. 5-19(b) and (c). Note that, as for shears $V$, in addition to the sense of $M$, it is also *essential to associate the moment for a particular side of a section*.

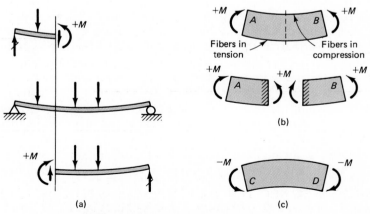

Fig. 5-19 Definition of bending moment signs.

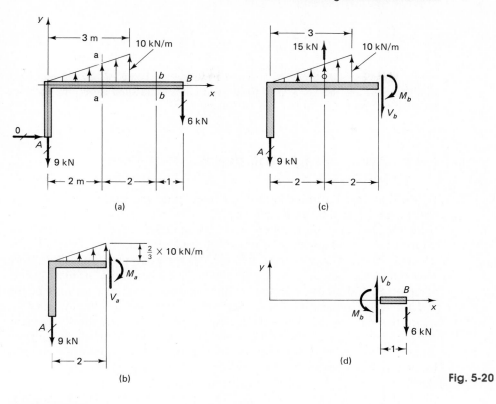

**Fig. 5-20**

## EXAMPLE 5-4

Consider earlier Example 5-2 and determine the internal system of forces at sections $a$–$a$ and $b$–$b$; see Fig. 5-20(a).

### Solution

A free-body for the member, including reactions, is shown in Fig. 5-20(a). A free-body to the left of section $a$–$a$ in Fig. 5-20(b) shows the maximum ordinate for the isolated part of the applied load. Using this information,

$$V_a = -9 + \frac{1}{2} \times 2 \times \frac{2}{3} \times 10 = -2.33 \text{ kN}$$

and

$$M_a = -9 \times 2 + \frac{1}{2} \times 2 \times \frac{2}{3} \times 10 \times \frac{1}{3} \times 2 = -13.6 \text{ kN·m}$$

These forces are shown with correct sense in the figure.

A free-body to the left of section $b$–$b$ is shown in Fig. 5-20(c), and to the right, in Fig. 5-20(d). It is evident that the second free-body is simpler for calculations, giving directly

$$V_b = +6 \text{ kN}$$

and
$$M_b = -6 \times 1 = -6 \text{ kN·m}$$

The same procedure can be used for frames consisting of several members rigidly joined together as well as for curved bars. In all such cases, the sections must be perpendicular to the axis of a member.

### 5-10. Axial-Force, Shear, and Bending-Moment Diagrams

By the methods discussed before, the magnitude and sense of axial forces, shears, and bending moments may be obtained at many sections of a beam. Moreover, with the sign conventions adopted for these quantities, a plot of their values may be made on *separate* diagrams. On such diagrams, ordinates may be laid off equal to the computed quantities from a base line representing the length of a beam. When these ordinate points are plotted and interconnected by lines, graphical representations of the functions are obtained. These diagrams, corresponding to the kind of quantities they depict, are called, respectively, *the axial-force diagram, the shear diagram, or the bending-moment diagram*. With the aid of such diagrams, the magnitudes and locations of the various quantities become immediately apparent. It is convenient to make these plots directly below the free-body diagram of the beam, using the same horizontal scale for the length of the beam. Draftsmanlike precision in making such diagrams is usually unnecessary, although the significant ordinates are generally marked with their numerical value.

The axial-force diagrams are not as commonly used as the shear and the bending-moment diagrams. This is so because the majority of beams investigated in practice are loaded by forces that act perpendicular to the axis of the beam. For such loadings of a beam, there are no axial forces at any section.

Shear and moment diagrams are exceedingly important. From them, a designer sees at a glance the kind of performance that is desired from a beam at every section. The procedure of sectioning a beam or a frame and finding the system of forces at the section is the most fundamental approach. It will be used in the following illustrative examples. In some of these examples, algebraic expressions for these functions along a beam will be given.

A systematic method for rapidly constructing shear and moment diagrams will be discussed in the next part of this chapter.

### EXAMPLE 5-5

Construct axial-force, shear, and bending-moment diagrams for the beam shown in Fig. 5-21(a) due to the inclined force $P = 5$ k.

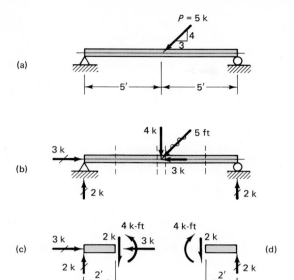

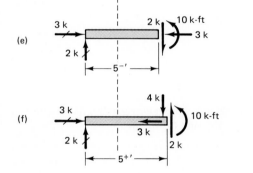

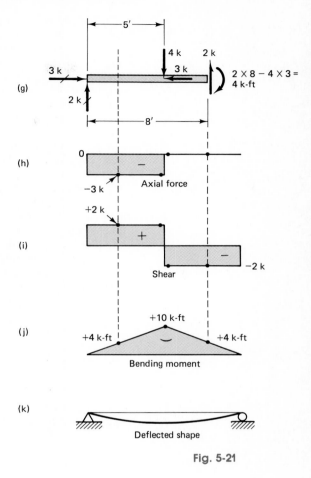

Fig. 5-21

## Solution

A free-body diagram of the beam is shown in Fig. 5-21(b). Reactions follow from inspection after the applied force is resolved into the two components. Then several sections through the beam are investigated, as shown in Figs. 5-21(c)–(g). In every case, the same question is posed: *What are the necessary internal forces to keep the segment of the beam in equilibrium?* The corresponding quantities are recorded on the respective free-body diagrams of the beam segment. The ordinates for these quantities are indicated by heavy dots in Figs. 5-21(h)–(j), with due attention paid to their signs.

Note that the free bodies shown in Figs. 5-21(d) and (g) are alternates, as they furnish the same information, and normally both would not be made. Note that a section *just to the left* of the applied force has one sign of shear, Fig. 5-21(e), whereas *just to the right*, Fig. 5-21(f), it has another. This indicates the importance of determining shears on either side of a concentrated force. For the condition shown, the beam *does not resist* a shear that is equal to the whole force. The bending moment in both cases is the same.

In this particular case, after a few individual points have been established on the three diagrams in Figs. 5-21(h)–(j), the behavior of the respective quantities across the whole length of the beam may be reasoned out. Thus, although the segment of the beam shown in Fig. 5-21(c) is 2 ft long, it may vary in length anywhere from zero to *just to the left* of the applied force, and *no change in the shear and the axial force occurs*. Hence, the ordinates in Figs. 5-21(h) and (i) *remain* constant for this segment of the beam. On the other hand, the bending moment depends directly on the distance from the support; hence, it varies linearly, as shown in Fig. 5-21(j). Similar reasoning applies to the segment shown in Fig. 5-21(d), enabling one to complete the three diagrams on the right-hand side. The use of the free-body of Fig. 5-21(g) for completing the diagram to the right of center yields the same result.

The sign of a bending moment, per Figs. 5-19(b) and (c), defines the sense in which a beam bends. Since, in this problem, throughout the beam length, the moments are positive, the beam curves to "retain water." In order to emphasize this physical behavior some analysts find it advantageous to draw a short curved line directly on the moment diagram, as shown in Fig. 5-21(j), to indicate the manner in which a beam or a beam segment curves.

Sometimes, in addition to or instead of the shear or moment diagrams, analytical expressions for these functions are necessary. For the origin of $x$ at the left end of the beam, the following relations apply:

$$V = +2 \text{ k} \qquad \text{for } 0 < x < 5$$
$$V = -2 \text{ k} \qquad \text{for } 5 < x < 10$$
$$M = +2x \text{ k-ft} \qquad \text{for } 0 \leq x \leq 5$$
$$M = +2x - 4(x - 5) = +20 - 2x \text{ k-ft} \qquad \text{for } 5 \leq x \leq 10$$

These expressions can be easily established by mentally replacing the distances of 2 ft and 8 ft, respectively, in Figs. 5-21(c) and (g) by an $x$.

## EXAMPLE 5-6

Determine axial-force, shear, and bending-moment diagrams for the cantilever loaded with an inclined force at the end; see Fig. 5-22(a).

### Solution

First, the inclined force is replaced by the two components shown in Fig. 5-22(b) and the reactions are determined. The *three* unknowns at the support follow from the familiar equations of statics. This completes the free-body diagram shown in Fig. 5-22(b). *Completeness in indicating all of these forces is of the utmost importance.*

A segment of the beam is shown in Fig. 5-22(c); from this segment, it may be seen that the axial force and the shear force remain the same regardless of the distance $x$. On the other hand, the bending moment is a variable quantity. A summation of moments around $C$ gives $PL - Px$ acting in the direction shown. This represents a *negative* moment. The moment at the support is likewise a *negative* bending moment as it tends to pull on the *upper* fibers of the beam. The three diagrams are plotted in Figs. 5-22(d)–(f).

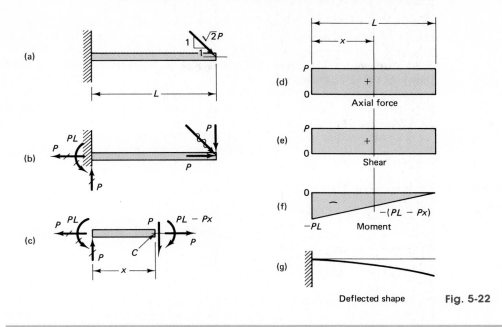

Fig. 5-22

---

## EXAMPLE 5-7

Construct shear and bending-moment diagrams for the beam loaded with the forces shown in Fig. 5-23(a).

### Solution

An arbitrary section at a distance $x$ from the left support isolates the beam segment shown in Fig. 5-23(b). This section is applicable for any value of $x$ just to the left

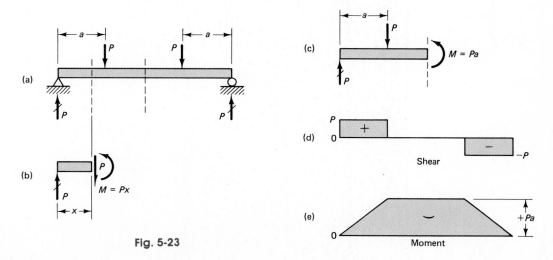

Fig. 5-23

of the applied force $P$. The shear, regardless of the distance from the support, remains constant and is $+P$. The bending moment varies linearly from the support, reaching a maximum of $+Pa$.

An arbitrary section applicable anywhere *between* the two applied forces is shown in Fig. 5-23(c). No shear force is necessary to maintain equilibrium of a segment in this part of the beam. Only a constant bending moment of $+Pa$ must be resisted by the beam in this zone. Such a state of bending or flexure is called *pure* bending.

Shear and bending-moment diagrams for this loading condition are shown in Figs. 5-23(d) and (e). No axial-force diagram is necessary, as there is no axial force at any section of the beam.

## EXAMPLE 5-8

Plot shear and a bending-moment diagrams for a simple beam with a uniformly distributed load; see Fig. 5-24.

### Solution

The best way of solving this problem is to write algebraic expressions for the quantities sought. For this purpose, an arbitrary section taken at a distance $x$ from the left support is used to isolate the segment shown in Fig. 5-24(b). Since the applied load is continuously distributed along the beam, this section is typical and applies to *any section* along the length of the beam.

The shear $V$ is equal to the left upward reaction *less* the load to the left of the section. The internal bending moment $M$ resists the moment caused by the reaction on the left *less* the moment caused by the forces to the left of the same section. The summation of moments is performed around an axis *at the section*.

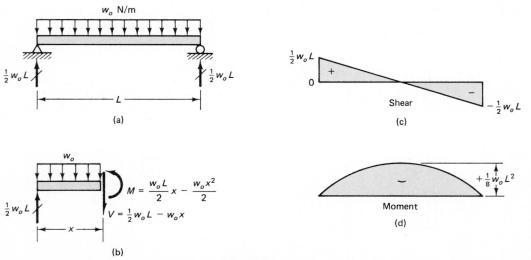

Fig. 5-24

Although it is customary to isolate the left-hand segment, similar expressions may be obtained by considering the right-hand segment of the beam, with due attention paid to sign conventions. The plot of the $V$ and $M$ functions is shown in Figs. 5-24(c) and (d).

## EXAMPLE 5-9

For the beam in Example 5-4, shown in Fig. 5-25(a), express the shear $V$ and the bending moment $M$ as a function of $x$ along the horizontal member.

### Solution

Unlike the preceding example, in this case, a load discontinuity occurs at $x = 3$ m. Therefore, the solution is determined in two parts for each of which the functions $V$ and $M$ are continuous. A free-body diagram for the beam segment under the load is shown in Fig. 5-25(b), and for the remainder, in Fig. 5-25(c). The required expressions for $0 < x < 3$ are

$$V(x) = -9 + \frac{1}{2}x\left(\frac{x}{3} \times 10\right) = -9 + \frac{5}{3}x^2 \text{ kN}$$

$$M(x) = -9x + \frac{1}{2}x\left(\frac{x}{3} \times 10\right)\left(\frac{x}{3}\right) = -9x + \frac{5}{9}x^3 \text{ kN·m}$$

For $3 < x < 5$,

$$V(x) = -9 + 15 = +6 \text{ kN}$$
$$M(x) = -9x + 15(x - 2) = 6x - 30 \text{ kN·m}$$

To obtain the last expression, it would have been a little simpler to use a free-body diagram similar to Fig. 5-20(d).

This problem can also be solved using the singularity functions discussed in Section 5-16.

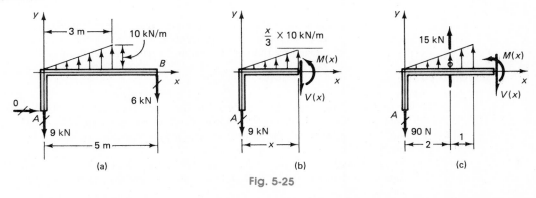

**Fig. 5-25**

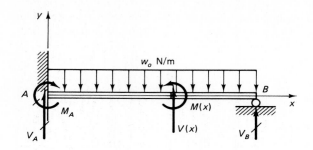

Fig. 5-26

## EXAMPLE 5-10

Write analytic expressions for $V$ and $M$ for the beam shown in Fig. 5-26.

### Solution

Unlike the preceding cases, this is a statically indeterminate problem to the first degree having one redundant reaction. There is no horizontal reaction at $A$. Except for carefully identifying the unknown reactions as $V_A$, $V_B$, and $M_A$, the procedure is the same as before, although numerical results cannot be obtained until the reactions are determined. On this basis, at a distance $x$ away from the origin,

$$V(x) = V_A - w_o x$$

and

$$\begin{aligned} M(x) &= M_A + V_A x - (w_o x)x/2 \\ &= M_A + V_A x - w_o x^2/2 \end{aligned}$$

Sometimes, it will be necessary to use such expressions in the process of solving for unknown reactions in Chapters 10 and 12.

## EXAMPLE 5-11

Consider a structural system of three interconnected straight bars, as shown in Fig. 5-27(a). At arbitrary sections, determine the internal forces $P$, $V$, and $M$ in the members caused by the application of a vertical force $P_1$ at $D$.

### Solution

The frame is conveniently analyzed by isolating the three straight members, as shown in Fig. 5-27(b). For each case, a different coordinate system is indicated, and sections through the members are shown at arbitrary distances from the origin.

The solution begins by calculating the reaction at $A$, which is then shown on beam segment $AB$. At an arbitrary section through this beam, the internal forces are seen to be

$$P(x_1) = +P_1, \quad V(x_1) = 0 \quad \text{and} \quad M(x_1) = +2P_1 a$$

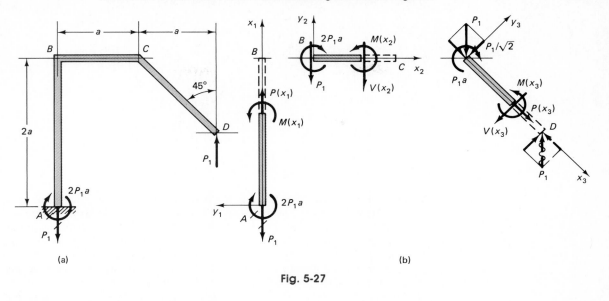

(a)                                    (b)

Fig. 5-27

These forces are constant throughout the length of the vertical bar and become the reactions at $B$ for the beam segment $BC$. It is important to note that the axial force in member $AB$ acts as shear in $BC$. After the reactions at $B$ for $BC$ are known, the usual procedure gives the following internal forces:

$$P(x_2) = 0, \; V(x_2) = -P_1 \quad \text{and} \quad M(x_2) = +2P_1a - P_1x_2$$

For member $CD$, except for the need for resolving the force $P_1$ at $C$, the procedure for determining the internal forces is the same as before, giving

$$P(x_3) = -P_1/\sqrt{2}, V(x_3) = -P_1/\sqrt{2} \quad \text{and} \quad M(x_3) = +P_1a - P_1x_3/\sqrt{2}$$

By substituting $x_3 = \sqrt{2}a$ into the last expression, it can be verified that the bending moment at $D$ is zero, as it should be.

Shear and bending-moment diagrams for this structural system can be plotted directly on the outline of the frame.

## EXAMPLE 5-12

Consider a curved beam whose centroidal axis is bent into a semicircle of 0.2 m radius, as shown in Fig. 5-28(a). If this member is being pulled by the 1000-N forces shown, find the axial force, the shear, and the bending moment at section $A-A$, $\alpha = 45°$. The centroidal axis and the applied forces all lie in the same plane.

## Solution

There is no essential difference in the method of attack in this problem compared with that in a straight-beam problem. The body as a whole is examined for con-

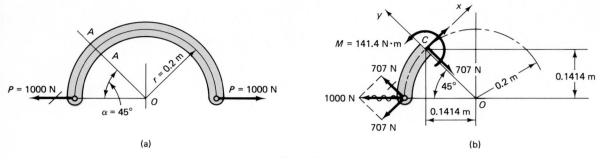

(a)                                                    (b)

Fig. 5-28

ditions of equilibrium. From the conditions of the problem here, such is already the case. Next, a segment of the beam is isolated; see Fig. 5-28(b). *Section A–A is taken perpendicular to the axis of the beam.* Before determining the quantities wanted at the cut, the applied force $P$ is resolved into components parallel and perpendicular to the cut. These directions are taken respectively as the $y$ and $x$ axes. This resolution replaces $P$ by the components shown in Fig. 5-28(b). From $\sum F_x = 0$, the axial force at the cut is $+707$ N. From $\sum F_y = 0$, the shear is 707 N in the direction shown. The bending moment at the cut can be determined in several different ways. For example, if $\sum M_o = 0$ is used, note that the lines of action of the applied force $P$ and the shear at the section pass through $O$. Therefore, only the axial force at the centroid of the cut times the radius has to be considered, and the *resisting* bending moment is $707(0.2) = 141.4$ N·m, acting in the direction shown. An alternative solution may be obtained by applying $\sum M_C = 0$. At $C$, a point lying on the centroid, the axial force and the shear intersect. The bending moment is then the product of the applied force $P$ and the 0.1414-m arm. In both of these methods of determining bending moment, use of the components of the force $P$ is avoided as this is more involved arithmetically.

It is suggested that the reader complete this problem in terms of a general angle $\alpha$. Several interesting observations may be made from such a general solution. The moments at the ends will vanish for $\alpha = 0°$ and $\alpha = 180°$. For $\alpha = 90°$, the shear vanishes and the axial force becomes equal to the applied force $P$. Likewise, the maximum bending moment is associated with $\alpha = 90°$.

## Part C    SHEAR AND BENDING MOMENTS BY INTEGRATION

### 5-11. Differential Equations of Equilibrium for a Beam Element

Instead of the direct approach of cutting a beam and determining shear and moment at a section by statics, an efficient alternative procedure can be used. For this purpose, certain fundamental differential relations must be derived. These can be used for the construction of shear and moment diagrams as well as for the calculation of reactions.

Consider a beam element $\Delta x$ long, isolated by two adjoining sections taken perpendicular to its axis, Fig. 5-29(b). Such an element is shown as a free-body in Fig. 5-29(c). All the forces shown acting on this element have positive sense. The positive sense of the distributed external force $q$ is taken to coincide with the direction of the positive $y$ axis. As the shear and the moment may each change from one section to the next, note that on the right side of the element, these quantities are, respectively, designated $V + \Delta V$ and $M + \Delta M$.

From the condition for equilibrium of vertical forces, one obtains[4]

$$\Sigma F_y = 0 \uparrow + \qquad V + q\,\Delta x - (V + \Delta V) = 0$$

or

$$\frac{\Delta V}{\Delta x} = q \qquad\qquad (5\text{-}1)$$

For equilibrium, the summation of moments around $A$ also must be zero. So, upon noting that from point $A$ the arm of the distributed force is $\Delta x/2$, one has

[4] No variation of $q(x)$ within $\Delta x$ need be considered, since, in the limit as $\Delta x \to 0$, the change in $q$ becomes negligibly small. This simplification is not an approximation.

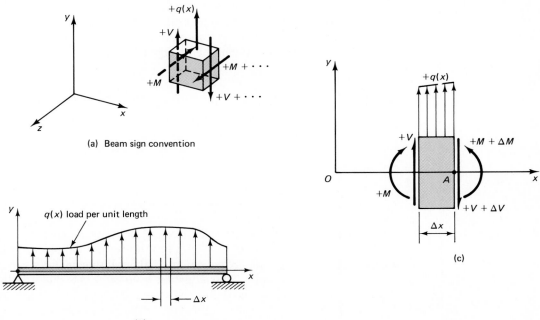

(a) Beam sign convention

(b)

(c)

Fig. 5-29 Beam and beam elements between adjoining sections.

$$\sum M_A = 0 \circlearrowleft + \qquad (M + \Delta M) - V\Delta x - M - (q\,\Delta x)(\Delta x/2) = 0$$

or

$$\frac{\Delta M}{\Delta x} = V + \frac{q\,\Delta x}{2} \tag{5-2}$$

Equations 5-1 and 5-2 in the limit as $\Delta x \to 0$ yield the following two basic differential equations:

$$\frac{dV}{dx} = q \tag{5-3}$$

and

$$\frac{dM}{dx} = V \tag{5-4}$$

By substituting Eq. 5-4 into Eq. 5-3, another useful relation is obtained:

$$\frac{d}{dx}\left(\frac{dM}{dx}\right) = \frac{d^2M}{dx^2} = q \tag{5-5}$$

This differential equation can be used for determining reactions of statically determinate beams from the boundary conditions, whereas Eqs. 5-3 and 5-4 are very convenient for construction of shear and moment diagrams. These applications will be discussed next.

## 5-12. Shear Diagrams by Integration of the Load

By transposing and integrating Eq. 5-3 gives the shear $V$:

$$V = \int_0^x q\,dx + C_1 \tag{5-6}$$

By assigning definite limits to this integral, it is seen that the shear at a section is simply an integral (i.e., a sum) of the vertical forces along the beam from the left end of the beam *to the section in question* plus a constant of integration $C_1$. This constant is equal to the shear on the left-hand end. Between any two definite sections of a beam, the shear changes by the amount of the vertical force included *between* these sections. If no force occurs between any two sections, no change in shear takes place. If a concentrated force comes into the summation, a discontinuity, or a

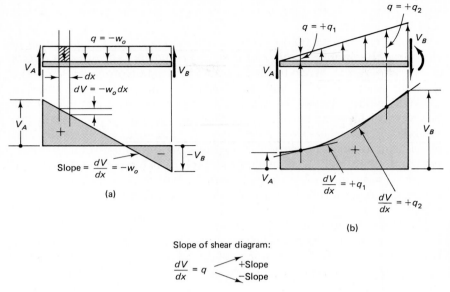

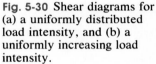

Slope of shear diagram:

$$\frac{dV}{dx} = q \qquad \begin{array}{l} +\text{Slope} \\ -\text{Slope} \end{array}$$

**Fig. 5-30** Shear diagrams for (a) a uniformly distributed load intensity, and (b) a uniformly increasing load intensity.

"jump," in the value of the shear occurs. The continuous summation process remains valid nevertheless, since a concentrated force may be thought of as being a distributed force extending for an infinitesimal distance along the beam.

On the basis of the above reasoning, a shear diagram can be established by the summation process. For this purpose, *the reactions must always be determined first*. Then the vertical components of forces *and reactions* are successively summed *from the left end* of the beam to preserve the mathematical sign convention for shear adopted in Fig. 5-17. The shear at a section is simply equal to the sum of *all* vertical forces to the left of the section.

When the shear diagram is constructed from the load diagram by the summation process, two important observations can be made regarding its shape. First, the sense of the applied load determines the sign of the slope of the shear diagram. If the applied load acts upward, the slope of the shear diagram is positive, and vice versa. Second, this slope is equal to the corresponding applied load intensity. For example, consider a segment of a beam with a uniformly distributed downward load $w_o$ and known shears at both ends, as shown in Fig. 5-30(a). Since here the applied load intensity $w_o$ is *negative* and *uniformly distributed*, i.e., $q = -w_o = $ constant, the slope of the shear diagram exhibits the same characteristics. Alternatively, the linearly varying load intensity acting upward on a beam segment with known shears at the ends, shown in Fig. 5-30(b), gives rise to a differently shaped shear diagram. Near the left end of this segment, the locally applied *upward* load $q_1$ is *smaller* than the corresponding one $q_2$ near the right end. Therefore, the *positive* slope of the shear diagram on the left is *smaller* than it is on the right, and the shear diagram is concave upward.

Do not fail to note that *a mere systematic consecutive summation of the vertical components of the forces is all that is necessary to obtain the shear diagram.* When the consecutive summation process is used, the diagram must end up with the previously calculated shear (reaction) at the right end of a beam. No shear acts through the beam just beyond the last vertical force or reaction. *The fact that the diagram closes in this manner offers an important check on the arithmetical calculations.* This check should never be ignored. It permits one to obtain solutions independently with almost complete assurance of being correct. The semi-graphical procedure of integration outlined before is very convenient in practical problems. It is the basis for sketching qualitative shear diagrams rapidly.

From the physical point of view, the shear sign convention is not completely consistent. Whenever beams are analyzed, a shear diagram drawn from one side of the beam is opposite in sign to a diagram constructed by looking at the same beam from the other side. The reader should verify this statement on some simple cases, such as a cantilever with a concentrated force at the end and a simply supported beam with a concentrated force in the middle. For design purposes, the sign of the shear is usually unimportant.

### 5-13. Moment Diagrams by Integration of the Shear

Transposing and integrating Eq. 5-4 gives the bending moment

$$M = \int_0^x V \, dx + C_2 \tag{5-7}$$

where $C_2$ is a constant of integration corresponding to boundary conditions at $x = 0$. This equation is analogous to Eq. 5-6 developed for the construction of shear diagrams. The meaning of the term $V \, dx$ is shown graphically by the hatched areas of the shear diagrams in Fig. 5-31. The summation of these areas between definite sections through a beam corresponds to an evaluation of the definite integral. If the ends of a beam are on rollers, pin-ended, or free, the starting and the terminal moments are zero. If the end is built-in (fixed against rotation), in statically determinate beams, the end moment is known from the reaction calculations. If the fixed end of a beam is on the left, this moment with the proper[5] sign is the *initial constant of integration* $C_2$.

[5] Bending moments carry signs according to the convention adopted in Fig. 5-19. Moments that cause *compression* in the top fibers of the beam are positive.

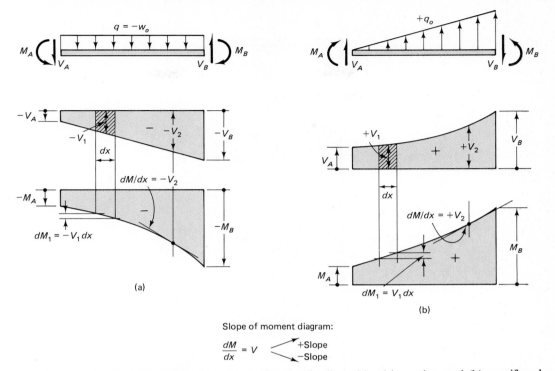

**Fig. 5-31** Shear and moment diagrams for (a) a uniformly distributed load intensity, and (b) a uniformly increasing load intensity.

By proceeding *continuously along the beam from the left-hand end* and summing up the areas of the shear diagram with due regard to their sign, the moment diagram is obtained. This process of obtaining the moment diagram from the shear diagram by summation is exactly the same as that employed earlier to go from loading to shear diagrams. *The change in moment in a given segment of a beam is equal to the area of the corresponding shear diagram.* Qualitatively, the shape of a moment diagram can be easily established from the slopes at some selected points along the beam. These slopes have the same sign and magnitude as the corresponding shears on the shear diagram, since according to Eq. 5-4, $dM/dx = V$. Alternatively, the change of moment $dM = V\,dx$ can be studied along the beam. Examples are shown in Fig. 5-31. According to these principles, variable shears cause nonlinear variation of the moment. A constant shear produces a uniform change in the bending moment, resulting in a straight line in the moment diagram. If no shear occurs along a certain portion of a beam, *no change in moment* takes place.

Since $dM/dx = V$, according to the fundamental theorem of calculus, the *maximum or minimum moment occurs where the shear is zero.*

In a bending-moment diagram obtained by summation, *at the right-hand end* of the beam, an invaluable check on the work is available again. *The*

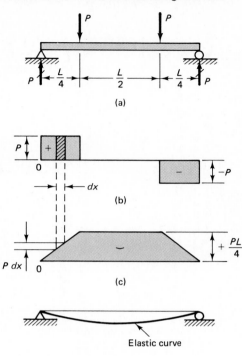

**Fig. 5-32**

(d)

*terminal conditions for the moment must be satisfied.* If the end is free or pinned, the computed sum must equal zero. If the end is built-in, the end moment computed by summation equals the one calculated initially for the reaction. These are the boundary conditions and must always be satisfied.

## EXAMPLE 5-13

Construct shear and moment diagrams for the symmetrically loaded beam shown in Fig. 5-32(a) by the integration process.

### Solution

The reactions are each equal to $P$. To obtain the shear diagram, Fig. 5-32(b), the summation of forces is started from the left end. The left reaction acts *up*, so an ordinate on the shear diagram at this force equal to $P$ is plotted *up*. Since there are no other forces until the quarter point, *no change in the magnitude of the shear ordinate is made until that point.* Then a downward force $P$ brings the ordinate back to the base line, and this zero ordinate remains until the next downward force $P$ is reached where the shear changes to $-P$. At the right end, the upward reaction closes the diagram and provides a check on the work. This shear diagram is *antisymmetrical.*

The moment diagram, Fig. 5-32(c), is obtained by summing up the area of the shear diagram. As the beam is simply supported, the moment at the left end is zero. The sum of the positive portion of the shear diagram *increases at a constant rate* along the beam until the quarter point, where the moment reaches a magnitude of $+PL/4$. This moment remains constant in the middle half of the beam. *No change in the moment can be made in this zone* as there is no corresponding shear area.

Beyond the second force, the moment decreases by $-P\,dx$ in *every dx*. Hence, the moment diagram in this zone has a constant, negative slope. Since the positive and the negative areas of the shear diagram are equal, at the right end, the moment is zero. This is as it should be, since the right end is on a roller. Thus, a check on the work is obtained. This moment diagram is *symmetrical*.

## EXAMPLE 5-14

Consider a simple beam with a uniformly increasing load intensity from an end, as shown in Fig. 5-33(a). The total applied load is $W$. (*a*) Construct shear and moment diagrams with the aid of the integration process. (*b*) Derive expressions for $V$ and $M$ using Eq. 5-5.

$$q = \frac{d^2 M}{dx^2}$$

Solution

(*a*) Since the total load $W = kL^2/2$, $k = 2W/L^2$. For the given load distribution, the downward reactions are $W/3$ and $2W/3$, as shown in Fig. 5-33(a). Therefore, the shear diagram given in Fig. 5-33(b) begins and ends as shown. Since the rate of applied load is smaller on the left end than on the right, the shear diagram is concave upward. The point of zero shear occurs where the reaction on the left is balanced by the applied load, i.e.,

$$\frac{W}{3} = \frac{1}{2}x_1\frac{2W}{L^2}x_1 \qquad \text{hence, } x_1 = \frac{L}{\sqrt{3}}$$

$$W = (kL)\frac{L}{2}$$

$$k = \frac{2W}{L^2}$$

At $x_1$, the bending moment is maximum; therefore,

$$M_{\max} = M\left(\frac{L}{\sqrt{3}}\right) = -\frac{W}{3}\frac{L}{\sqrt{3}} + \frac{1}{2}\frac{L}{\sqrt{3}}\frac{2W}{L^2}\frac{L}{\sqrt{3}}\left(\frac{1}{3}\frac{L}{\sqrt{3}}\right) = -\frac{2WL}{9\sqrt{3}}$$

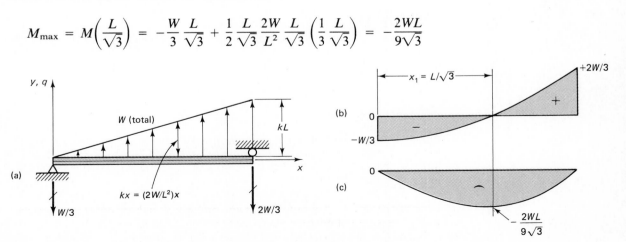

**Fig. 5-33**

By following the rules given in Fig. 5-31, the moment diagram has the shape shown in Fig. 5-33(c).

Although the shear and bending moment diagrams could be sketched qualitatively, it was necessary to supplement the results analytically for determining the critical values.

(*b*) Applying Eq. 5-5 and integrating it twice, one has

$$\frac{d^2M}{dx^2} = q = +kx = +\frac{2W}{L^2}x$$

$$\frac{dM}{dx} = \frac{kx^2}{2} + C_1 \quad \text{and} \quad M = \frac{kx^3}{6} + C_1x + C_2$$

However, the boundary conditions require that the moments at $x = 0$ and $x = L$ be zero, i.e., $M(0) = 0$ and $M(L) = 0$. Therefore, since

$$M(0) = 0 \qquad C_2 = 0$$

and, similarly, since $M(L) = 0$,

$$\frac{kL^3}{6} + C_1L = 0 \quad \text{or} \quad C_1 = -\frac{kL^2}{6}$$

With these constants,

$$V = \frac{dM}{dx} = \frac{kx^2}{2} - \frac{kL^2}{6} = \frac{Wx^2}{L^2} - \frac{W}{3}$$

and

$$M = \frac{kx^2}{6} - \frac{kL^2x}{6} = \frac{Wx^3}{3L^2} - \frac{Wx}{3}$$

These results agree with those found earlier.

The attractive features of the boundary-value approach used in this example for solving differential equations can be extended to situations of discontinuous loads using the singularity functions discussed in Section 5-16.

## EXAMPLE 5-15

Construct shear and bending-moment diagrams for loaded beam shown in Fig. 5-34(a) with the aid of the integration process.

### Solution

Reactions must be calculated first, and, before proceeding further, the inclined force is resolved into its horizontal and vertical components. The horizontal reaction at $A$ is 30 kips and acts to the right. From $\sum M_A = 0$, the vertical reaction at $B$ is found to be 37.5 kips (check this). Similarly, the reaction at $A$ is 27.5 kips.

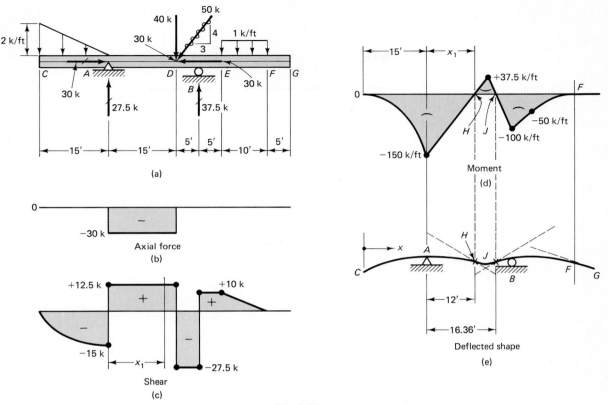

Fig. 5-34

The sum of the vertical reaction components is 65 kips and equals the sum of the vertical forces.

The diagram for the axial force is shown in Fig. 5-34(b). This compressive force only acts in the segment $AD$ of the beam.

With reactions known, the summation of forces is begun from the left end of the beam to obtain the shear diagram, Fig. 5-34(c). At first, the downward distributed load accumulates at a rapid rate. Then, as the load intensity decreases, for an equal increment of distance along the beam, a smaller change in shear occurs. Hence, the shear diagram in the zone $CA$ is a curved line, which is concave up. This is in accord with Eq. 5-3, illustrated in Fig. 5-30. Since $dV/dx = q = -w_0$, the negative slope of this shear diagram is large on the left, and gradually decreases to zero at $A$. The total downward force from $C$ to $A$ is 15 kips, and this is the negative ordinate of the shear diagram, *just to the left of the support* $A$. At $A$, the *upward* reaction of 27.5 kips moves the ordinate of the shear diagram to $+12.5$ kips. This value of the shear applies to a section through the beam *just to the right* of the support $A$. The abrupt *change* in the shear at $A$ is equal to the reaction, but this total does not represent the shear through the beam.

No forces are applied to the beam between $A$ and $D$; hence, there is no change in the value of the shear. At $D$, the 40-kip downward component of the concentrated force drops the value of the shear to $-27.5$ kips. Similarly, the value of the shear is raised to $+10$ kips at $B$. Since between $E$ and $F$, the uniformly distributed load acts downward, according to Eq. 5-3, and shown in Fig. 5-30, a decrease in shear takes place at a constant rate of 1 kip/foot. Thus, at $F$, the shear is zero, which serves as the final check.

To construct the moment diagram shown in Fig. 5-34(d) by the summation method, areas of the shear diagram in Fig. 5-34(c) must be continuously summed from the left end. In the segment $CA$, at first, less area is contributed to the sum in a distance $dx$ than a little farther along, so a line that is concave down appears in the moment diagram. This is in accord with Eq. 5-4, $dM/dx = V$, illustrated in Fig. 5-31. Here $V$, defining the slope of the moment diagram is negative, and progressively becomes larger to the right. The moment at $A$ is equal to the area of the shear diagram in the segment $CA$. This area is enclosed by a curved line, and it may be determined by integration,[6] since the shear along this segment may be expressed analytically. This procedure often is cumbersome, and instead, the bending moment at $A$ may be obtained from the fundamental definition of a moment at a section. By passing a section through $A$ and isolating the segment $CA$, the moment at $A$ is found. The other areas of the shear diagram in this example are easily determined. Due attention must be paid to the signs of these areas. It is convenient to arrange the work in tabular form. At the right end of the beam, the customary check is obtained.

$$
\begin{array}{lll}
M_A & & \\
& -\tfrac{1}{2}(15)2(10) = -150.0 \text{ k-ft} & \text{(moment around } A\text{)} \\
& +12.5(15) = +187.5 & \text{(shear area } A \text{ to } D\text{)} \\
M_D & \overline{+\phantom{0}37.5 \text{ k-ft}} & \\
& -27.5(5) = -137.5 & \text{(shear area } D \text{ to } B\text{)} \\
M_B & \overline{-100.0 \text{ k-ft}} & \\
& +10(5) = +\phantom{0}50.0 & \text{(shear area } B \text{ to } E\text{)} \\
M_E & \overline{-\phantom{0}50.0 \text{ k-ft}} & \\
& +\tfrac{1}{2}(10)10 = +\phantom{0}50.0 & \text{(shear area } E \text{ to } F\text{)} \\
M_F & \overline{\phantom{00}0.0 \text{ k-ft}} & (\textit{check})
\end{array}
$$

## 5-14. Effect of Concentrated Moment on Moment Diagrams

In the derivation for moment diagrams by summation of shear-diagram areas, no *external concentrated moment* acting on the infinitesimal element was included, yet such a moment may actually be applied. Hence, the summation process derived applies only up to the point of application of an external moment. *At a section just beyond an externally applied moment, a different bending moment is required to maintain the segment of a beam in equilibrium.* For example, in Fig. 5-35 an external clockwise moment $M_A$ is acting on the element of the beam at $A$. Then, if the internal clockwise moment on the left is $M_O$, for equilibrium of the element, the

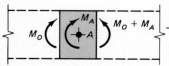

**Fig. 5-35** An external concentrated moment acting on an element of a beam.

[6] In this case, the shear curve is a second-degree parabola whose vertex is on a vertical line through $A$. For areas enclosed by various curves, see Table 2 of the Appendix.

resisting counterclockwise moment on the right must be $M_O + M_A$. At the point of the externally applied moment, a discontinuity, or a "jump," equal to the concentrated moment appears in the moment diagram. Hence, in applying the summation process, due regard must be given the concentrated moments as their effect is not apparent in the shear diagram. The conventional summation process may be applied up to the point of application of a concentrated moment. At this point, a vertical "jump" equal to the external moment must be made in the diagram. The direction of this vertical "jump" in the diagram depends upon the sense of the concentrated moment and is best determined with the aid of a sketch analogous to Fig. 5-35. After the discontinuity in the moment diagram is passed, the summation process of the shear-diagram areas may be continued over the remainder of the beam.

### EXAMPLE 5-16

Construct the bending-moment diagram for the horizontal beam loaded as shown in Fig. 5-36(a).

### Solution

By taking moments about either end of the beam, the vertical reactions are found to be $P/6$. At $A$, the reaction acts down; at $C$, it acts up. From $\sum F_x = 0$, it is known that at $A$, a horizontal reaction equal to $P$ acts to the left. The shear diagram is drawn next; see Fig. 5-36(b). It has a constant negative ordinate for the *whole* length of the beam. After this, by using the summation process, the moment diagram shown in Fig. 5-36(c) is constructed. The moment at the left end of the beam is zero, since the support is pinned. The total change in moment from $A$ to $B$ is given by the area of the shear diagram between these sections and equals $-2Pa/3$. The moment diagram in zone $AB$ has a constant negative slope. For further analysis, an element is isolated from the beam, as shown in Fig. 5-36(d). The moment on the left-hand side of this element is *known to be* $-2Pa/3$, and the concentrated moment caused by the applied force $P$ about the neutral axis

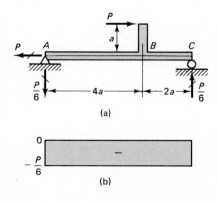

(a)

(b)

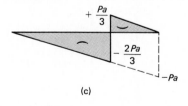

(c)

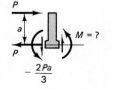

(d)

**Fig. 5-36**

of the beam is $Pa$; hence, for equilibrium, the moment on the right side of the element must be $+Pa/3$. At $B$, an upward "jump" of $+Pa$ is made in the moment diagram, and just to the right of $B$, the ordinate is $+Pa/3$. Beyond point $B$, the summation of the shear diagram area is continued. The area between $B$ and $C$ is equal to $-Pa/3$. This value *closes* the moment diagram at the right end of the beam, and thus the boundary conditions are satisfied. Note that the lines in the moment diagram that are inclined downward to the right are parallel. This follows because the shear everywhere along the beam is negative and constant.

### EXAMPLE 5-17

Construct shear and moment diagrams for the member shown in Fig. 5-37(a). All dimensions are shown in mm. Neglect the weight of the beam.

### Solution

In this case, unlike all cases considered so far, definite dimensions are assigned for the *depth* of the beam. The beam, for simplicity, is assumed to be rectangular in its cross-sectional area; consequently, the *centroidal axis* lies 80 mm below the top of the beam. Note carefully that this beam is not supported at the centroidal axis.

A free-body diagram of the beam with the applied force resolved into components is shown in Fig. 5-37(b). Reactions are computed in the usual manner. Moreover, since the shear diagram is concerned only with the vertical forces, it is easily constructed and is shown in Fig. 5-37(c).

In constructing the moment diagram shown in Fig. 5-37(d), particular care must be exercised. As was emphasized earlier, the bending moments may always be determined by considering a segment of a beam, and they are most conveniently computed by taking moments of external forces *around a point on the centroidal axis of the beam*. Thus, by passing a section just to the right of $A$ and considering the left-hand segment, it can be seen that a positive moment of 48 N·m is resisted

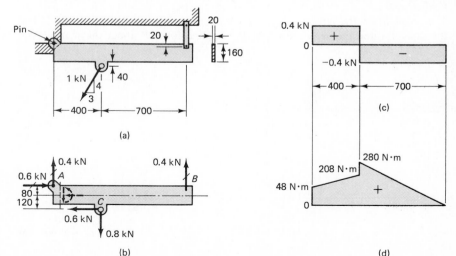

Fig. 5-37                    (a)                    (b)                    (c)                    (d)

by the beam at this end. Hence, the plot of the moment diagram must *start* with an ordinate of $+48$ N·m. The other point on the beam where a concentrated moment occurs is $C$. Here the horizontal component of the applied force induces a clockwise moment of $0.6 \times 120 = 72$ N·m around the neutral axis. Just to the right of $C$, this moment must be resisted by an additional positive moment. This causes a discontinuity in the moment diagram. The summation process of the shear-diagram areas applies for the segments of the beam where no external moments are applied. The necessary calculations are carried out in tabular form.

$$
\begin{array}{llll}
M_A & +0.6 \times 80 & = & +\ 48\ \text{N·m} \\
 & +0.4 \times 400 & = & +160 & \text{(shear area } A \text{ to } C) \\
\text{Moment just to left of } C & & = & \overline{+208\ \text{N·m}} \\
 & +0.6 \times 120 & = & +\ 72 & \text{(external moment at } C) \\
\text{Moment just to right of } C & & = & \overline{+280\ \text{kN·m}} \\
 & -0.4 \times 700 & = & -280 & \text{(shear area } C \text{ to } B) \\
M_B & & = & \overline{\quad 0\quad} & (check)
\end{array}
$$

Note that in solving this problem, the forces were considered *wherever they actually act on the beam*. The investigation for shear and moments at a section of a beam determines what the beam is actually experiencing. At times, this differs from the procedure of determining reactions, where the actual framing or configuration of a member is not important.

Again, it must be emphasized that if a moment or a shear is needed at a *particular* section through any member, *the basic method of sections may always be used*. For inclined members, the shear acts *normal to the axis of the beam*.

## 5-15. Moment Diagram and the Elastic Curve

As defined in Section 5-9, a positive moment causes a beam to deform concave upwards or to "retain water," and vice versa. Hence, the shape of the deflected axis of a beam can be *definitely* established from the *sign of the moment diagram*. The trace of this axis of a loaded elastic beam in a deflected position is known as the *elastic curve*. It is customary to show the elastic curve on a sketch, where the actual small deflections tolerated in practice are greatly *exaggerated*. A sketch of the elastic curve clarifies the physical action of a beam. It also provides a useful basis for quantitative calculations of beam deflections to be discussed in Sections 10-13 and 10-14. Some of the preceding examples for which bending-moment diagrams were constructed will be used to illustrate the physical action of a beam.

An inspection of Fig. 5-32(c) shows that the bending moment throughout the length of the beam is *positive*. Accordingly, the elastic curve shown in Fig. 5-32(d) is *concave up at every point*. Correct representation of convexity or concavity of the elastic curve is important. In this case, the ends of the beam rest on supports.

In a more complex moment diagram, Fig. 5-34(d), zones of positive and negative moment occur. Corresponding to the zones of negative moment, a *definite* curvature of the elastic curve that is concave down takes place; see Fig. 5-34(e). On the other hand, for the zone *HJ*, where the positive moment occurs, the concavity of the elastic curve is upward. Where curves join, as at *H* and *J*, there are lines that are *tangent* to the two joining curves since the beam is physically *continuous*. Also note that the free end *FG* of the beam is tangent to the elastic curve at *F*. There is no curvature in *FG*, since the moment is zero in that segment of the beam.

If the suggestion made in Example 5-5, indicating the curvature of beam segments by means of short curved lines on the moment diagram is followed, as in Fig. 5-34(d), the elastic curve is simply an assembly of such curves drawn to a proper scale.

The point of transition on the elastic curve into reverse curvature is called the *point of inflection* or contraflexure. At this point, the moment changes its sign, and the beam is not called upon to resist any moment. This fact often makes these points a desirable place for a field connection of large members, and their location is calculated. A procedure for determining points of inflection will be illustrated in the next example.

### EXAMPLE 5-18

Find the location of the inflection points for the beam analyzed in Example 5-15; see Fig. 5-34(a).

Solution

By definition, an inflection point corresponds to a point on a beam where the bending moment is zero. Hence, an inflection point can be located by setting up an algebraic expression for the moment in a beam for the segment where such a point is anticipated, and solving this relation equated to zero. By measuring $x$ from end $C$ of the beam, Fig. 5-34(e), the bending moment for segment $AD$ of the beam is $M = -\frac{1}{2}(15)(2)(x - 5) + (27.5)(x - 15)$. By simplifying and setting this expression equal to zero, a solution for $x$ is obtained.

$$M = 12.5x - 337.5 = 0 \qquad x = 27 \text{ ft}$$

Therefore, the inflection point occurring in segment $AD$ of the beam is $27 - 15 = 12$ ft from support $A$.

Similarly, by writing an algebraic expression for the bending moment for segment $DB$ and setting it equal to zero, the location of inflection point $J$ is found.

$$M = -\tfrac{1}{2}(15)(2)(x - 5) + 27.5(x - 15) - 40(x - 30) = 0$$

where $x = 31.36$ ft; hence, the distance $AJ = 16.36$ ft.

Often a more convenient method for finding the inflection points consists of utilizing the known relations between the shear and moment diagrams. Thus, since the moment at $A$ is $-150$ kip-ft, the point of zero moment occurs when the positive portion of the shear-diagram area from $A$ to $H$ equals this moment, i.e., $-150 + 12.5x_1 = 0$. Hence, distance $AH = 150/12.5 = 12$ ft as before.

Similarly, by beginning with a known positive moment of $+37.5$ kip-ft at $D$, the second inflection point is known to occur when a portion of the negative shear-diagram area between $D$ and $J$ reduces this value to zero. Hence, distance $DJ = 37.5/27.5 = 1.36$ ft, or distance $AJ = 15 + 1.36 = 16.36$ ft, Fig. 5-34(e), as before.

Just as any infinitesimal beam element must be in equilibrium, so must also any corner element in a continuous frame with rigid joints. Therefore, the bending moments at a corner can act only either as shown in Fig. 5-38(a) or 5-38(b). The associated parts of elastic curves are shown in these figures.

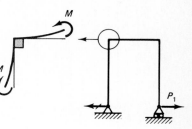

(a)

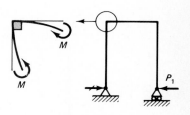

(b)

**Fig. 5-38** Elastic curves at corners of planar rigid frames.

## **\*\*<sup>7</sup>5-16. Singularity Functions**

As was pointed out earlier, analytical expressions for the shear $V(x)$ and the moment $M(x)$ of a given beam may be needed in an analysis. If the loading $q(x)$ is a continuous function between the supports, solution of the differential equation $d^2M/dx^2 = q(x)$ is a convenient approach for determining $V(x)$ and $M(x)$ (see Example 5-14). Here this will be extended to situations in which the loading function is discontinuous. For this purpose, the notation of operational calculus will be used. The functions $q(x)$ considered here are polynomials with integral powers of $x$. The treatment of other functions is beyond the scope of this text. For the functions considered, however, the method is perfectly general. Further applications of this approach will be given in Chapter 10 for calculating deflections of beams.

Consider a beam loaded as in Fig. 5-39. Since the applied loads are point (concentrated) forces, four distinct regions exist to which different bending moment expressions apply. These are

$$M = R_1 x \qquad\qquad \text{when } 0 \le x \le d$$
$$M = R_1 x - P_1(x - d) \qquad\qquad \text{when } d \le x < b$$
$$M = R_1 x - P_1(x - d) + M_b \qquad\qquad \text{when } b < x \le c$$
$$M = R_1 x - P_1(x - d) + M_b + P_2(x - c) \qquad \text{when } c \le x \le L$$

[7] This Section can be omitted.

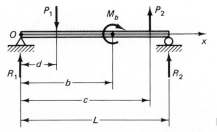

**Fig. 5-39** A loaded beam.

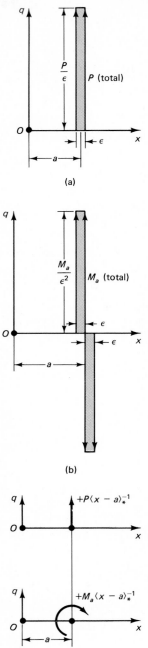

All four equations can be written as one, providing one defines the following symbolic function:

$$\langle x - a \rangle^n = \begin{cases} 0 & \text{for } 0 < x < a \\ (x - a)^n & \text{for } a < x < \infty \end{cases} \qquad (5\text{-}8)$$

where $n \geq 0$ $(n = 0, 1, 2, \ldots)$.

The expression enclosed by the pointed brackets does not exist until $x$ reaches $a$. For $x$ beyond $a$, the expression becomes an ordinary binomial. For $n = 0$ and for $x > a$, the function is unity. On this basis, the four separate functions for $M(x)$ given for the beam of Fig. 5-39 can be combined into one expression that is appliable across the whole span:[8]

$$M = R_1 \langle x - 0 \rangle^1 - P_1 \langle x - d \rangle^1 + M_b \langle x - b \rangle^0 + P_2 \langle x - c \rangle^1$$

Here the values of $a$ are 0, $d$, $b$, and $c$, respectively.

To work with this function further, it is convenient to introduce two additional symbolic functions. One is for the concentrated force, treating it as a degenerate case of a distributed load. The other is for the concentrated moment, treating it similarly. Rules for integrating all these functions must be also established. In this discussion, the heuristic (nonrigorous) approach will be followed.

A concentrated (point) force may be considered as an enormously strong distributed load acting over a small interval $\epsilon$, Fig. 5-40(a). By treating $\epsilon$ as a constant, the following is true

$$\lim_{\epsilon \to 0} \int_{a - \epsilon/2}^{a + \epsilon/2} \frac{P}{\epsilon} \, dx = P \qquad (5\text{-}9)$$

Here it can be noted that $P/\epsilon$ has the dimensions of force per unit distance such as lb/in, and corresponds to the distributed load $q(x)$ in the earlier treatment. Therefore, as $\langle x - a \rangle^1 \to 0$, by an analogy of $\langle x - a \rangle^1$ to $\epsilon$, for a concentrated force at $x = a$,

$$q = P \langle x - a \rangle_*^{-1} \qquad (5\text{-}10)$$

**Fig. 5-40** Concentrated force $P$ and moment $M_a$: (a) and (b) considered as distributed load, and (c) symbolic notation for $P$ and $M$ at $a$.

[8] This approach was first introduced by A. Clebsch in 1862. O. Heaviside in his *Electromagnetic Theory* initiated and greatly extended the methods of operational calculus. In 1919, W. H. Macaulay specifically suggested the use of special brackets for beam problems. The reader interested in further and/or more rigorous development of this topic should consult texts on mathematics treating Laplace transforms.

For $q$, this expression is dimensionally correct, although $\langle x - a \rangle_*^{-1}$ at $x = a$ becomes infinite and by definition is zero everywhere else. Thus, it is a *singular function*. In Eq. 5-10, the asterisk subscript of the bracket is a reminder that according to Eq. 5-9, the integral of this expression extending over the range $\epsilon$ remains bounded and upon integration, yields the point force itself. Therefore, a special symbolic rule of integration must be adopted:

$$\int_0^x P \langle x - a \rangle_*^{-1} \, dx = P \langle x - a \rangle^0 \qquad (5\text{-}11)$$

The coefficient $P$ in the previous functions is known as the *strength* of singularity. For $P$ equal to unity, the *unit point load function* $\langle x - a \rangle_*^{-1}$ is also called the *Dirac delta* or the *unit impulse function*.

By analogous reasoning, see Fig. 5-40(b), the loading function $q$ for concentrated moment at $x = a$ is

$$q = M_a \langle x - a \rangle_*^{-2} \qquad (5\text{-}12)$$

This function in being integrated twice defines two symbolic rules of integration. The second integral, except for the exchange of $P$ by $M$, has already been stated as Eq. 5-11.

$$\int_0^x M_a \langle x - a \rangle_*^{-2} \, dx = M_a \langle x - a \rangle_*^{-1} \qquad (5\text{-}13)$$

$$\int_0^x M_a \langle x - a \rangle_*^{-1} \, dx = M_a \langle x - a \rangle^0 \qquad (5\text{-}14)$$

In Eq. 5-12, the expression is correct dimensionally since $q$ has the units of lb/in. For $M_a$ equal to unity, one obtains the *unit point moment function*, $\langle x - a \rangle_*^{-2}$, which is also called the *doublet* or *dipole*. This function is also singular being infinite at $x = a$ and zero elsewhere. However, after integrating twice, a bounded result is obtained. Equations 5-10, 5-12, and 5-13 are symbolic in character. The relation of these equations to the given point loads is clearly evident from Eqs. 5-11, and 5-14.

The integral of binomial functions in pointed brackets for $n \geq 0$ is given by the following rule:

$$\int_0^x \langle x - a \rangle_n \, dx = \frac{\langle x - a \rangle^{n+1}}{n + 1} \qquad \text{for } n \geq 0 \qquad (5\text{-}15)$$

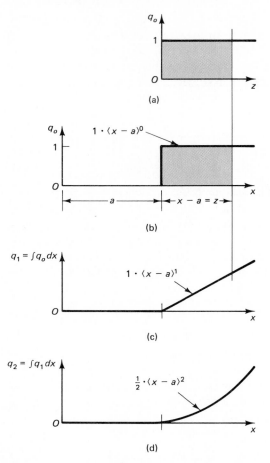

**Fig. 5-41** Typical integrations.

This integration process is shown in Fig. 5-41. If the distance $a$ is set equal to zero, one obtains conventional integrals.

### EXAMPLE 5-19

Using symbolic functional notation, determine $V(x)$ and $M(x)$ caused by the loading in Fig. 5-42(a).

### Solution

To solve this problem, Eq. 5-5 can be used. The applied load $q(x)$ acts downward and begins at $x = 0$. Therefore, a term $q = -w_o$ or $w_o\langle x - 0\rangle^0$, which means the same, must exist. This function, however, propagates across the whole span; see Fig. 5-42(b). To terminate the distributed load at $x = L/2$ as required in this problem, another function $+w_o\langle x - L/2\rangle^0$ must be added. The two expressions together represent correctly the applied load.

For this simply supported beam, the known boundary conditions are $M(0) = 0$ and $M(L) = 0$. These are used to determine the reactions:

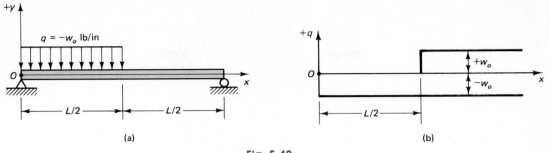

Fig. 5-42

$$\frac{d^2M}{dx^2} = q = -w_o\langle x - 0\rangle^0 + w_o\langle x - L/2\rangle^0$$

$$\frac{dM}{dx} = V = -w_o\langle x - 0\rangle^1 + w_o\langle x - L/2\rangle^1 + C_1$$

$$M(x) = -\tfrac{1}{2}\,w_o\langle x - 0\rangle^2 + \tfrac{1}{2}\,w_o\langle x - L/2\rangle^2 + C_1 x + C_2$$

$$M(0) = C_2 = 0$$

$$M(L) = -\tfrac{1}{2}\,w_o L^2 + \tfrac{1}{2}\,w_o(L/2)^2 + C_1 L = 0$$

Hence,
$$C_1 = +\tfrac{3}{8}\,p_o L$$

and
$$V(x) = -w_o\langle x - 0\rangle^1 + w_o\langle x - L/2\rangle^1 + \tfrac{3}{8}\,w_o L$$
$$M(x) = -\tfrac{1}{2}\,w_o\langle x - 0\rangle^2 + \tfrac{1}{2}\,w_o\langle x - L/2\rangle^2 + \tfrac{3}{8}\,w_o Lx$$

After the solution is obtained, these relations are more easily read by rewriting them in conventional form:

$$\left.\begin{array}{l} V = +\tfrac{3}{8}\,w_o L - w_o x \\ M = +\tfrac{3}{8}\,w_o Lx - \tfrac{1}{2}\,w_o x^2 \end{array}\right\} \quad \text{when } 0 < x \le L/2$$

$$\left.\begin{array}{l} V = +\tfrac{3}{8}\,w_o L - \tfrac{1}{2}\,w_o L = -\tfrac{1}{8}\,w_o L \\ M = +\tfrac{1}{8}\,w_o L^2 - \tfrac{1}{8}\,w_o Lx \end{array}\right\} \quad \text{when } L/2 \le x < L$$

The reactions can be checked by conventional statics. By setting $V = 0$, the location of maximum moment can be found. A plot of these functions is left for the reader to complete.

## EXAMPLE 5-20

Find $V(x)$ and $M(x)$ for a beam loaded as shown in Fig. 5-43. Use singularity functions and treat it as a boundary-value problem.

Solution

By making direct use of Eqs. 5-10 and 5-12, the function $q(x)$ can be written in symbolic form. From the conditions $M(0) = 0$ and $M(L) = 0$, with $L = 3a$, the constants of integration can be found:

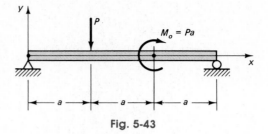

**Fig. 5-43**

$$d^2M/dx^2 = q = -P\langle x - a\rangle_*^{-1} + Pa\langle x - 2a\rangle_*^{-2}$$

$$dM/dx = V = -P\langle x - a\rangle^0 + Pa\langle x - 2a\rangle_*^{-1} + C_1$$

$$M = -P\langle x - a\rangle^1 + Pa\langle x - 2a\rangle^0 + C_1x + C_2$$

$$M(0) = C_2 = 0$$

and

$$M(3a) = -2Pa + Pa + 3C_1a = 0$$

Hence,

$$C_1 = +\tfrac{1}{3}P = \tfrac{1}{3}P\langle x - 0\rangle^0$$

and

$$V(x) = +\tfrac{1}{3}P\langle x - 0\rangle^0 - P\langle x - a\rangle^0 + Pa\langle x - 2a\rangle_*^{-1}$$

$$M(x) = +\tfrac{1}{3}P\langle x - 0\rangle^1 - P\langle x - a\rangle^1 + Pa\langle x - 2a\rangle^0$$

In the final expression for $V(x)$, the last term has no value if the expression is written in conventional form. Such terms are used only as tracers during the integration process.

It is suggested that the reader check the reactions by conventional statics, write out $V(x)$ and $M(x)$ for the three ranges of the beam within which these functions are continuous, and compare these with a plot of the shear and moment diagrams constructed by the summation procedure.

A suggestion of the manner of representing a uniformly varying load, Fig. 5-44(a), acting on a part of a beam is indicated in Fig. 5-44(b). Three separate functions are needed to define the given load completely.

In the previous discussion, it has been tacitly assumed that the reactions are at the ends of the beams. If such is not the case, the unknown constants $C_1$ and $C_2$ must be introduced into Eq. 5-5 as point loads, i.e., as

$$C_1\langle x - a\rangle_*^{-1} \qquad \text{and} \qquad C_2\langle x - b\rangle_*^{-1}$$

This is the condition shown in Fig. 5-44(c). No additional constants of integration are necessary in a solution obtained in this manner.

Singularity functions can be used to advantage in statically indeterminate problems for axially loaded bars, as well as for torsion members and beams. However, the solutions are *limited to prismatic members*. If the cross section varies along the length of a member, the procedure for using singularity functions becomes impractical.

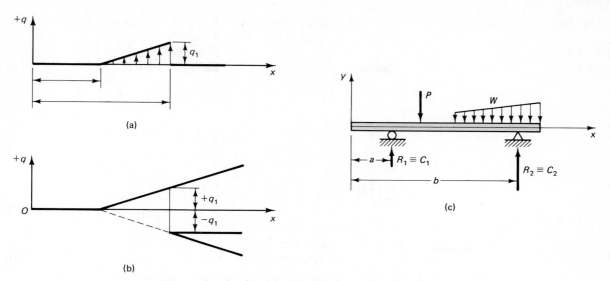

Fig. 5-44 Illustration for formulating singularity functions for reactions.

# Problems

**5-1 through 5-4.** Determine the reaction components caused by the applied loads for the planar framing shown in the figures. *Correctly drawn free-body dia-*

*grams are essential parts of solutions.* (*Hint for Prob. 5-1:* The effect on a structure of two cable forces acting over a frictionless pulley is the same as that of the same two forces applied at the center of the axle. Prove before using.)

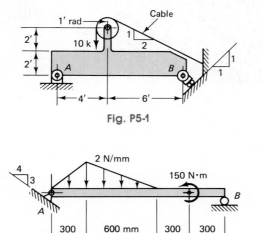

Fig. P5-1

Fig. P5-2

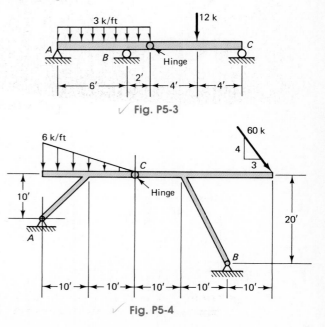

✓ Fig. P5-3

✓ Fig. P5-4

## Sections 5-6 through 5-9

**5-5 through 5-19.** For the planar structures shown in the figures, find the reactions and determine the axial forces *P*, the shears *V*, and the bending moments *M* caused by the applied loads at sections *a–a*, *b–b*, etc.

as specified. Magnitude and sense of calculated quantities should be shown on separate free-body diagrams. For simplicity, assume that members can be represented by lines. When sections such as *a–a* and *b–b* are shown close together, one section is just to the left of a given dimension and the other is just to the right.

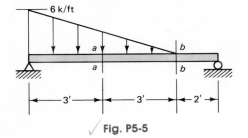

✓ **Fig. P5-5**

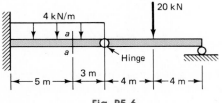

**Fig. P5-6**

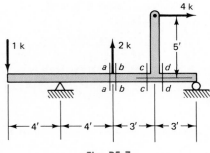

**Fig. P5-7**

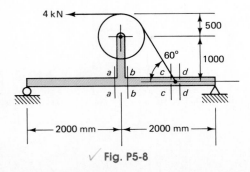

✓ **Fig. P5-8**

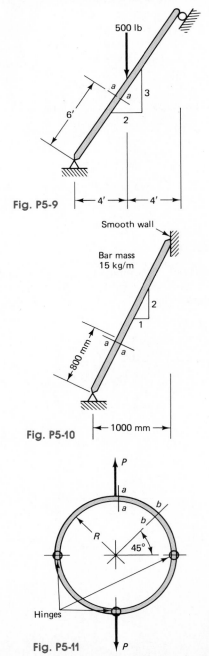

**Fig. P5-9**

**Fig. P5-10**

**Fig. P5-11**

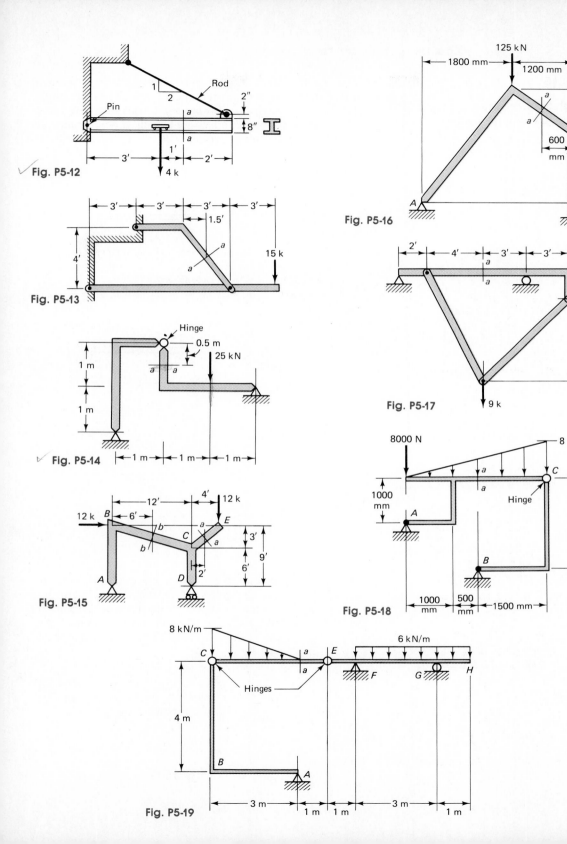

Fig. P5-12

Fig. P5-13

Fig. P5-14

Fig. P5-15

Fig. P5-16

Fig. P5-17

Fig. P5-18

Fig. P5-19

**271**

## Section 5-10

**5-20 through 5-24.** Plot shear and moment diagrams for the beams shown in the figures.

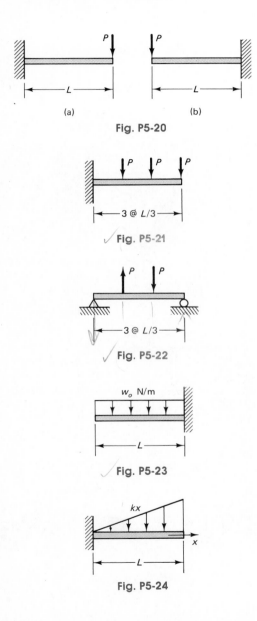

**(a)**     **(b)**

**Fig. P5-20**

**3 @ $L/3$**

**Fig. P5-21**

**3 @ $L/3$**

**Fig. P5-22**

$w_o$ N/m

**$L$**

**Fig. P5-23**

$kx$

$x$

**$L$**

**Fig. P5-24**

**5-25.** Plot shear and moment diagrams for the beam shown in Fig. 5-15.

**5-26 through 5-28.** For the beams loaded as shown in the figures, write explicit expressions for $M(x)$'s along the spans. Assume the origins of $x$ at $A$. Since the applied loads are discontinuous, different functions apply for regions $AC$ and $CB$.

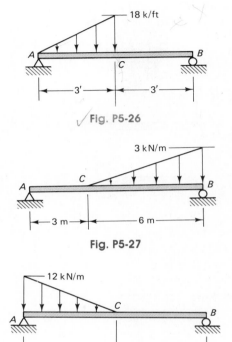

18 k/ft

**3'**     **3'**

**Fig. P5-26**

3 kN/m

**3 m**     **6 m**

**Fig. P5-27**

12 kN/m

**3 m**     **3 m**

**Fig. P5-28**

**5-29 through 5-31.** Write explicit expressions for $M(x)$ along the spans for the statically indeterminate beams loaded as shown in the figures. Assume the origins of $x$ at $A$. Consider the reactions on the left as unknowns. Take advantage of symmetry in Prob. 5-29.

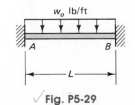

$w_o$ lb/ft

$A$     $B$

**$L$**

**Fig. P5-29**

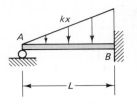

Fig. P5-30

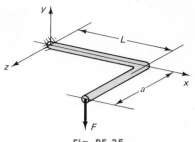

Fig. P5-35

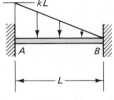

Fig. P5-31

**5-32.** Establish general expressions for the axial force $P(\theta)$, shear $V(\theta)$, and moment $M(\theta)$ for the curved bar in Example 5-12, Fig. 5-28. Angle $\theta$ is measured counterclockwise from the positive $x$ axis.

**5-33.** Establish general expressions for the axial force $P(\theta)$, shear $V(\theta)$, and moment $M(\theta)$ for the ring with three hinges of Prob. 5-11. Angle $\theta$ is measured counterclockwise from the positive $x$ axis.

**5-34.** A rectangular bar bent into a semicircle is built in at one end and is subjected to a radial pressure of $p$ lb per unit length (see figure). Write the general expressions for $P(\theta)$, $V(\theta)$, and $M(\theta)$, and plot the results on a polar diagram. Show positive directions assumed for $P$, $V$, and $M$ on a free-body diagram.

general expressions for $V$, $M$, and $T$ (torque) caused by the application of a force $F$ normal to the plane of the bent bar. Plot the results. (b) If in addition to the applied force $F$, the weight of the bar $w$ lb per unit length is also to be considered, what system of internal force components develops at the fixed end?

## Section 5-11

**5-36.** Using the differential equation, Eq. 5-5, determine $V(x)$ and $M(x)$ for the beam loaded as shown in Prob. 5-24. Verify the reactions using conventional statics. (*Hint:* The constants of integration can be found from the boundary conditions $V(L) = 0$ and $M(L) = 0$.)

**5-37 through 5-39.** Using Eq. 5-5 for the statically determinate beams shown in the figures, find $V(x)$ and $M(x)$ and the reactions at the supports. Plot the shear and moment diagrams. (*Hint:* The constants of integration are found from the boundary conditions for $V$ and $M$. This approach cannot be extended to statically indeterminate beams, which require the use of a higher-order differential equation, discussed in Chapter 10.)

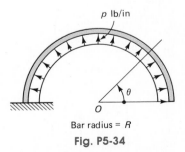

Bar radius = $R$

Fig. P5-34

**5-35.** A bar in the shape of a right angle, as shown in the figure, is fixed at one of its ends. (a) Write the

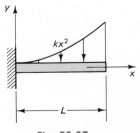

Fig. P5-37

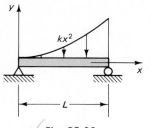

Fig. P5-38

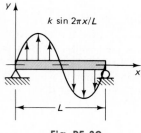

Fig. P5-39

## Sections 5-12 and 5-13

**Problems 5-20 through 5-31 can also be assigned for solution using the methods developed in these two sections.**

**5-40 through 5-66.** Plot shear and moment diagrams for the beams shown in the figures using the methods of Sections 5-12 and 5-13. It is also suggested to draw the deflected shapes of the beams using the criteria given in Fig. 5-19. (A more detailed discussion for drawing such shapes is given in Section 5-15.)

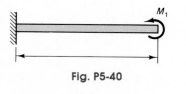

Fig. P5-40

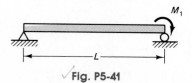

Fig. P5-41

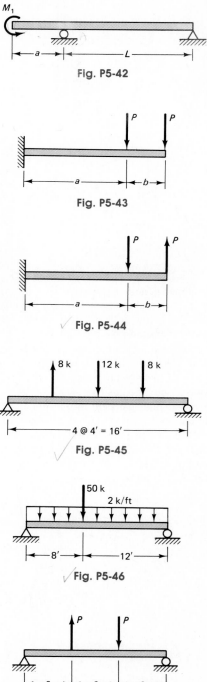

Fig. P5-42

Fig. P5-43

Fig. P5-44

Fig. P5-45

Fig. P5-46

Fig. P5-47

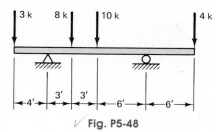

Fig. P5-48

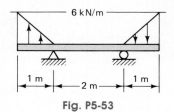

Fig. P5-53

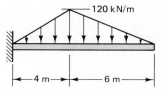

Fig. P5-49

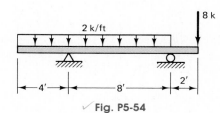

Fig. P5-54

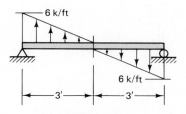

Fig. P5-50

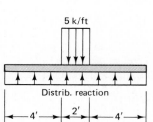

Fig. P5-55

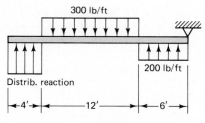

Fig. P5-51

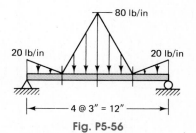

Fig. P5-56

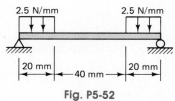

Fig. P5-52

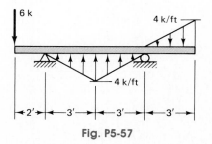

Fig. P5-57

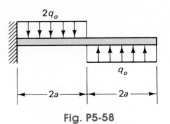

Fig. P5-58

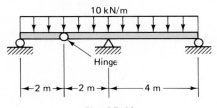

Fig. P5-59

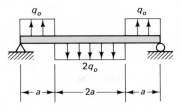

Fig. P5-60

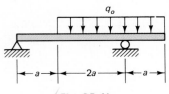

✓ Fig. P5-61

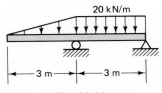

Fig. P5-62

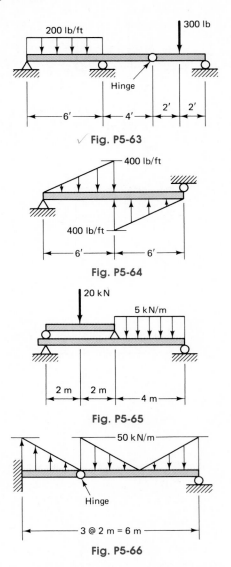

✓ Fig. P5-63

Fig. P5-64

Fig. P5-65

Fig. P5-66

**5-67.** A small narrow barge is loaded as shown in the figure. Plot shear and moment diagrams for the applied loading.

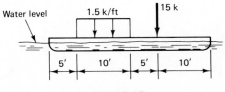

Fig. P5-67

**5-68.** The load distribution for a small single-engine airplane in flight may be idealized as shown in the figure. In this diagram, vector $A$ represents the weight of the engine; $B$, the uniformly distributed cabin weight; $C$, the weight of the aft fuselage; and $D$, the forces from the tail control surfaces. The upward forces $E$ are developed by the two longerons from the wings. Using this data, construct plausible, qualitative shear and moment diagrams for the fuselage.

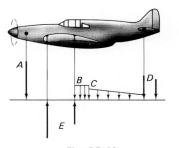

Fig. P5-68

**5-69.** The moment diagram for a beam supported at $A$ and $B$ is shown in the figure. How is the beam loaded?

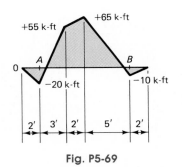

Fig. P5-69

**5-70.** The redundant moment over support $B$ for the beam shown in the figure can be shown to be $-400$ kN·m by the methods discussed in Chapter 10. Plot the shear and moment diagrams for this beam.

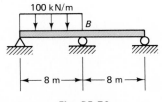

Fig. P5-70

**5-71 through 5-73.** For the structural systems shown in the figures, plot the axial force $P$, shear $V$, and moment $M$ diagrams. Note that the axial force and shear contribute to the equilibrium of forces at a joint in bent members (see Fig. 5-27).

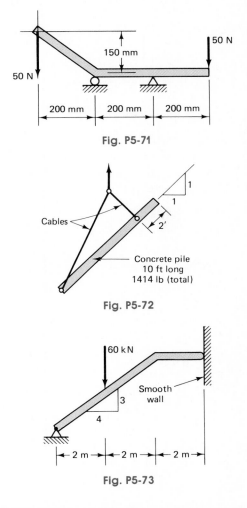

Fig. P5-71

Fig. P5-72

Fig. P5-73

**5-74.** For member $DF$ of the frame in Prob. 1-44, plot the axial force, shear, and moment diagrams caused by the applied force.

## Sections 5-14 and 5-15

**5-75 through 5-81.** For the structural systems shown in the figures, plot the axial force $P$, shear $V$, and moment $M$ diagrams due to the applied loads. These diagrams are to be confined only to the main horizontal members. Note that the beams in the last four problems have finite depth.

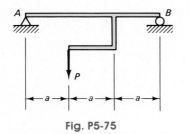

Fig. P5-75

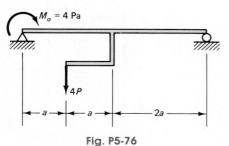

Fig. P5-76

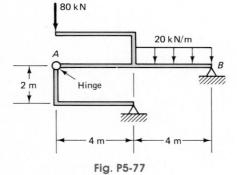

Fig. P5-77

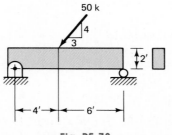

Fig. P5-78

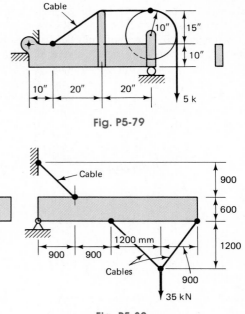

Fig. P5-79

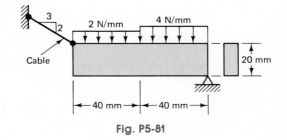

Fig. P5-80

Fig. P5-81

## Section 5-16

**5-82 through 5-87.** For the beams loaded as shown in the figures, using singularity functions and Eq. 5-5, (a) find $V(x)$ and $M(x)$. Check reactions by conventional statics. (b) Plot shear and moment diagrams.

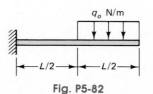

Fig. P5-82

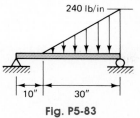

Fig. P5-83

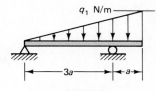

Fig. P5-85

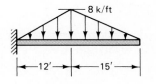

Fig. P5-84

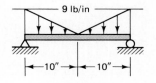

Fig. P5-86

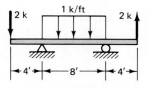

Fig. P5-87

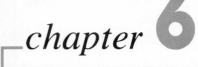

# chapter 6

# Pure Bending and Bending with Axial Forces

## 6-1. Introduction

In the previous chapter, it was shown that a system of internal forces consisting of an axial force, a shear force, and a bending moment may develop in planar frames and beams. The stresses caused by an axial force were already discussed in Chapter 1. The stresses due to bending are considered in this chapter. For this treatment, it is convenient to divide the chapter into two parts. In Part A, only members having *symmetric* cross sections and subjected to bending in the plane of symmetry are considered. Both elastic and inelastic stress distributions caused by bending are discussed. Stress distribution in curved bars is also included. In Part B, the problem is generalized to include *unsymmetric* bending of members with symmetric cross sections as well as bending of members of arbitrary cross section. Consideration is also given to problems where bending occurs in the presence of axial forces. For completeness, a discussion on area moments of inertia for arbitrary cross sections is included in Part C.

For simplicity, members will generally be shown as beams in a horizontal position. When a segment of a beam is in equilibrium under the action of bending moments alone, such a condition is referred to as *pure bending*, or *flexure*. A cantilever loaded with a concentrated moment at the end, or a segment of a beam between the concentrated forces, as shown in Fig. 5-23, are examples of pure bending. Studies in subsequent chapters will show that usually the *bending stresses* in slender beams are dominant. Therefore, the formulas derived in this chapter for pure bending are directly applicable in numerous design situations.

It is important to note that some beams by virtue of their slenderness or lack of lateral support may become unstable under an applied load and may buckle laterally and collapse. Such beams do not come within the scope of this chapter. A better appreciation of the instability phenomenon will result after the study of column buckling in Chapter 11.

## Part A  BENDING OF BEAMS WITH SYMMETRIC CROSS SECTIONS

### 6-2. The Basic Kinematic Assumption

In the simplified engineering theory of bending, to establish the relation among the applied bending moment, the cross-sectional properties of a member, and the internal stresses and deformations, the approach applied earlier in the torsion problem is again employed. This requires, first, that a plausible deformation assumption reduce the internally statically indeterminate problem to a determinate one; second, that the deformations causing strains be related to stresses through the appropriate stress-strain relations; and, finally, that the equilibrium requirements of external and internal forces be met. The key kinematic assumption for the deformation of a beam as used in the simplified theory is discussed in this section. A generalization of this assumption forms the basis for the theories of plates and shells.

For present purposes, consider a horizontal prismatic beam having a cross section with a vertical axis of symmetry; see Fig. 6-1(a). A horizontal line through the centroid of the cross section will be referred to as the axis of a beam. Next, consider a typical element of the beam between two planes perpendicular to the beam axis. In side view, such an element is identified in the figure as *abcd*. When such a beam is subjected to equal end moments $M_z$ acting around the $z$ axis, Fig. 6-1(b), this beam bends in the plane of symmetry, and the planes initially perpendicular to the beam axis slightly tilt. Nevertheless, the lines such as *ad* and *bc* becoming *a'd'* and *b'c'* remain straight.[1] This observation forms the basis for the fundamental hypothesis[2] of the flexure theory. It may be stated thus: *plane sections through a beam taken normal to its axis remain plane after the beam is subjected to bending*.

---

[1] This can be demonstrated by using a rubber model with a ruled grating drawn on it. Alternatively, thin vertical rods passing through the rubber block can be used. In the immediate vicinity of the applied moments, the deformation is more complex. However, in accord with the St. Venant's principle (Section 2-10), this is only a local phenomenon that rapidly dissipates.

[2] This hypothesis with an inaccuracy was first introduced by Jacob Bernoulli (1645–1705), a Swiss mathematician. At a later date a great Swiss mathematician, Leonard Euler (1707–1783), who largely worked in Russia and Germany, made important use of this concept. This assumption is often referred to as the Bernoulli-Euler hypothesis. In the correct final form, it dates back to the writings of the French engineering educator M. Navier (1785–1836).

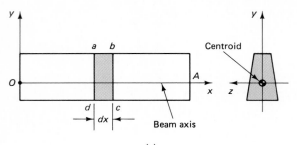
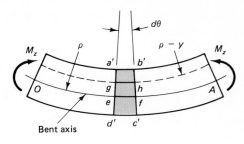

(a)                              (b)

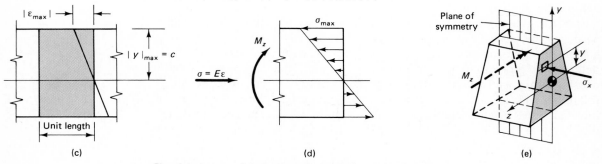

(c)                             (d)                            (e)

**Fig. 6-1** Assumed behavior of elastic beam in bending.

    As demonstrated in texts on the theory of elasticity, this assumption is completely true for elastic, rectangular members in pure bending. If shears also exist, a small error is introduced.[3] Practically, however, this assumption is generally applicable with a high degree of accuracy whether the material behaves elastically or plastically, providing the depth of the beam is small in relation to its span. In this chapter, the stress analysis of all beams is based on this assumption.

    In pure bending of a prismatic beam, the beam axis deforms into a part of a circle of radius $\rho$, (rho) as shown in Fig. 6-1(b). For an element defined by an infinitesimal angle $d\theta$, the fiber length $ef$ of the beam axis is given as $ds = \rho \, d\theta$. Hence,

$$\frac{d\theta}{ds} = \frac{1}{\rho} = \kappa \qquad (6\text{-}1)$$

where the reciprocal of $\rho$ defines the axis *curvature* $\kappa$ (kappa). In pure bending of prismatic beams, both $\rho$ and $\kappa$ are constant.

---

[3] See the discussion in Section 7-5.

The fiber length *gh* located on a radius $\rho - y$ can be found similarly. Therefore, the difference between fiber lengths *gh* and *ef* identified here as *dû* can be expressed as follows

$$d\hat{u} = (\rho - y)\, d\theta - \rho\, d\theta = -y\, d\theta \qquad (6\text{-}2)$$

By dividing by *ds* and using Eq. 6-1, the last term becomes $\kappa$. Moreover, since the deflection and rotations of the beam axis are very small, the cosines of the angles involved in making the projections of *dû* and *ds* onto the horizontal axis are very nearly unity. Therefore, in the development of the simplified beam theory, it is possible to replace *dû* by *du*, the axial fiber deformation, and *ds* by *dx*.[4] Hence, by dividing Eq. 6-2 by *ds* and approximating *dû/ds* by *du/dx*, which according to Eq. 2-6 is the normal strain $\varepsilon_x$, one has

$$\boxed{\varepsilon_x = -\kappa y} \quad = -\frac{1}{\rho} y \qquad (6\text{-}3)$$

This equation establishes the expression for the basic kinematic hypothesis for the flexure theory. However, although it is clear that the strain in a bent beam varies along the beam depth linearly with *y*, information is lacking for locating the origin of the *y* axis. With the aid of Hooke's law and an equation of equilibrium, this problem is resolved in the next section.

## 6-3. The Elastic Flexure Formula

By using Hooke's law, the expression for the normal strain given by Eq. 6-3 can be recast into a relation for the normal longitudinal stress $\sigma_x$:

$$\boxed{\sigma_x = E\,\varepsilon_x = -E\,\kappa\, y} \quad -\frac{E y}{\rho} \qquad (6\text{-}4)$$

In this equation, the variable *y* can assume both positive and negative values.

Two nontrivial equations of equilibrium are available to solve the beam flexure problem. One of these determines the origin for *y*; the second completes the solution for the flexure formula. Using the first one of these equations, requiring that in pure bending, the sum of all forces at a section in the *x* direction must vanish, one has

[4] A further discussion of the approximations involved may be found in Section 10-3.

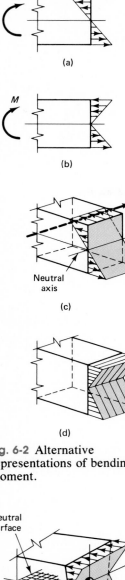

(a)

(b)

Neutral
axis

(c)

(d)

**Fig. 6-2** Alternative
representations of bending
moment.

Neutral
surface

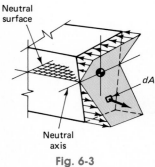

Neutral
axis

**Fig. 6-3**

$$\sum F_x = 0 \qquad \int_A \sigma_x \, dA = 0 \qquad (6\text{-}5)$$

where the subscript $A$ indicates that the summation of the infinitesimal forces must be carried out over the entire cross-sectional area $A$ of the beam. This equation with the aid of Eq. 6-4 can be rewritten as

$$\int_A -E \kappa y \, dA = -E \kappa \int_A y \, dA = 0 \qquad (6\text{-}6)$$

where the constants $E$ and $\kappa$ are taken outside the second integral. By definition, this integral $\int y \, dA = \bar{y} A$, where $\bar{y}$ is the distance from the origin to the centroid of an area $A$. Since here this integral equals zero and area $A$ is not zero, distance $\bar{y}$ must be set equal to zero. Therefore, the $z$ axis must pass through the *centroid* of a section. According to Eqs. 6-3 and 6-4, this means that along the $z$ axis so chosen, both the normal strain $\varepsilon_x$ and the normal stress $\sigma_x$ are zero. In bending theory, this axis is referred to as the *neutral axis* of a beam. The neutral axis for any *elastic* beam of homogeneous material can be easily determined by finding the centroid of a cross-sectional area.

Based on this result, linear variation in strain is schematically shown in Fig. 6-1(c). The corresponding *elastic* stress distribution in accordance with Eq. 6-4 is shown in Fig. 6-1(d). Both the absolute maximum strain $\varepsilon_x$ and the absolute maximum stress $\sigma_{\max}$ occur at the *largest* value of $y$.

Alternative representations of the elastic bending stress distribution in a beam are illustrated in Fig. 6-2. Note the need for awareness that the problem is *three-dimensional*, although for simplicity, two-dimensional representations are generally used. The locus of a neutral axis along a length of a beam defines the *neutral surface*, as noted in Fig. 6-3.

To complete the derivation of the elastic flexure formula, the second relevant equation of equilibrium must be brought in: the sum of the externally applied and the internal resisting moments must vanish, i.e., be in equilibrium. For the beam segment in Fig. 6-4(a), this yields

$$\sum M_O = 0 \circlearrowleft + \qquad M_z - \int_A \underbrace{E\kappa y}_{\text{stress}} \underbrace{dA}_{\text{area}} \qquad y = 0 \qquad (6\text{-}7)$$

$$\underbrace{\phantom{E\kappa y \quad dA}}_{\text{force}} \qquad \underbrace{\phantom{y}}_{\text{arm}}$$

A negative sign in front of the integral is necessary because the *compressive* stresses $\sigma_x$ develop a counterclockwise moment around the $z$ axis. The *tensile* stresses below the neutral axis, where $y$'s have a negative sign, contribute to this moment in the same manner. This sign also follows directly from Eq. 6-4. From a slightly different point of view, Eq. 6-7 states that the clockwise external moment $M_z$ is balanced by the counterclockwise moment developed by the internal stresses at a section. Recasting Eq. 6-7 into this form, and recognizing that $E$ and $\kappa$ are constants,

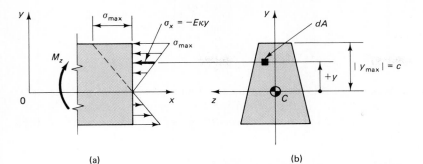
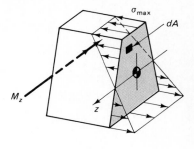

Fig. 6-4 Segment of a beam in pure flexure.

$$M_z = E\kappa \int_A y^2 \, dA \qquad (6\text{-}8)$$

In mechanics, the last integral, depending only on the geometrical properties of a cross-sectional area, is called the rectangular *moment of inertia* or *second moment of the area A* and will be designated in this text by $I$. It must be found with respect to the cross section's neutral (centroidal) axis. Since $I$ must always be determined with respect to a particular axis, it is often meaningful to identify it with a subscript corresponding to such an axis. For the case considered, this subscript is $z$, i.e.,

$$I_z = \int_A y^2 \, dA \qquad (6\text{-}9)$$

With this notation, Eq. 6-8 yields the following result:

$$\kappa = \frac{M_z}{E I_z} \qquad (6\text{-}10)$$

This is the basic relation giving the curvature of an elastic beam subjected to a specified moment.

By substituting Eq. 6-10 into Eq. 6-4, the elastic *flexure formula*[5] for beams is obtained:

[5] It took nearly two centuries to develop this seemingly simple expression. The first attempts to solve the flexure problem were made by Galileo in the seventeenth century. In the form in which it is used today, the problem was solved in the early part of the nineteenth century. Generally, Navier is credited for this accomplishment. However, some maintain that credit should go to Coulomb, who also derived the torsion formula.

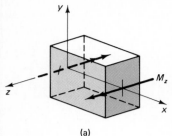

(a)

$$\sigma_x = -\frac{M_z}{I_z} y \qquad (6\text{-}11)$$

The derivation of this formula was carried out with the coordinate axes shown in Fig. 6-5(a). If the derivation for a member having a doubly symmetric cross section were done with the coordinates shown in Fig. 6-5(b), the expression for the longitudinal stress $\sigma_x$ would read

$$\sigma_x = +\frac{M_y}{I_y} z \qquad (6\text{-}12)$$

(b)

Fig. 6-5 Definitions of positive moments.

The sign reversal in relation to Eq. 6-11 is necessary because a positive $M_y$ causes tensile stresses for positive $z$'s.

Application of these equations to biaxial bending as well as an extension of the bending theory for beams with unsymmetric cross sections is considered in Sections 6-11 and 6-14. In this part of the chapter, attention is confined to beams having symmetric cross sections bent in the plane of symmetry. For such applications, it is customary to recast the flexure formula to give the *maximum* normal stress $\sigma_{\max}$ directly and to designate the value of $|y|_{\max}$ by $c$. It is also common practice to dispense with the sign as in Eq. 6-11 as well as with subscripts on $M$ and $I$. Since the normal stresses must develop a couple statically equivalent to the internal bending moment, their sense can be determined by inspection. On this basis, the flexure formula becomes

$$\sigma_{\max} = \frac{Mc}{I} \qquad (6\text{-}13)$$

In conformity with the above practice, in dealing with bending of *symmetric* beam sections, the simplified notation of leaving out $z$ subscripts in Eq. 6-11 on $M$ and $I$ will be employed often in this text.

The flexure formula and its variations discussed before are of unusually great importance in applications to structural and machine design. In applying these formulas, the internal bending moment can be expressed in newton-meters [N·m] or inch-pounds [in-lb], $c$ in meters [m] or inches [in], and $I$ in $m^4$ or $in^4$. The use of consistent units as indicated makes the units of $\sigma$: $[N\cdot m][m]/[m^4] = N/m^2 = Pa$, or $[in\text{-}lb][in]/[in^4] = [lb/in^2] = psi$, as to be expected.

It should be noted that $\sigma_x$ as given by Eqs. 6-11 or 6-12 is the only stress that results from pure bending of a beam. Therefore, in the matrix representation of the stress tensor, one has

$$\begin{pmatrix} \sigma_x & 0 & 0 \\ 0 & 0 & 0 \\ 0 & 0 & 0 \end{pmatrix}$$

As will be pointed out in Chapter 8, this stress may be transformed or resolved into stresses acting along different sets of coordinate axes.

In concluding this discussion, it is interesting to note that due to Poisson's ratio, the compressed zone of a beam expands laterally;[6] the tensile zone contracts. The strains in the $y$ and $z$ directions are $\varepsilon_y = \varepsilon_z = -\nu\varepsilon_x$, where $\varepsilon_x = \sigma_x/E$, and $\sigma_x$ is given by Eq. 6-11. This is in complete agreement with the rigorous solution. Poisson's effect, as may be shown by the methods of elasticity, deforms the neutral axis into a curve of large radius; and the neutral surface becomes curved in two opposite directions; see Fig. 6-6. In the previous treatment, the neutral surface was assumed to be curved in one direction only. These interesting details are not significant in most practical problems.

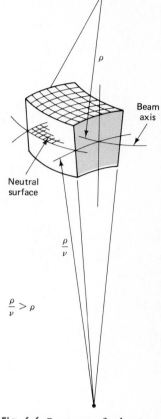

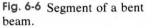

Fig. 6-6 Segment of a bent beam.

### Procedure Summary and Extensions

The same *three basic concepts* of engineering mechanics of solids that were used in developing the theories for axially loaded bars and circular shafts in torsion are used in the preceding derivation of flexure formulas. These may be summarized as follows:

1. *Equilibrium conditions* (statics) are used for determining the internal resisting bending moment at a section.
2. *Geometry of deformation* (kinematics) is used by assuming that plane sections through a beam remain plane after deformation. This leads to the conclusion that normal strains along a beam section vary linearly from the neutral axis.
3. *Properties of materials* (constitutive relations) in the form of Hooke's law are assumed to apply to the longitudinal normal strains. Poisson effect of transverse contraction and expansion is neglected.

In extending this approach to bending of beams of two and more materials (Section 6-8), as well as to inelastic bending of beams (Section 6-10), the first two of the enumerated concepts remain fully applicable. Only the third, dealing with the mechanical properties of materials must be modified. As an example of a change necessary for such cases consider the beam having the cross section shown in Fig. 6-7(a). This beam is made up of two materials, 1 and 2, bonded together at their interface. The elastic

---

[6] An experiment with an ordinary rubber eraser is recommended!

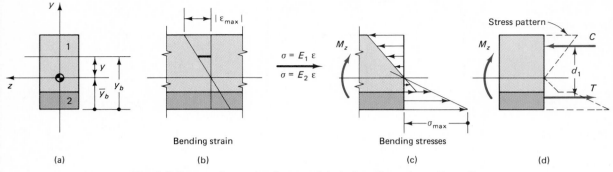

Bending strain

Bending stresses

(a)    (b)    (c)    (d)

**Fig. 6-7** Beam of two elastic materials in bending where $E_2 > E_1$.

moduli for the two materials are $E_1$ and $E_2$, where the subscripts identify the material. For the purposes of discussion assume that $E_2 > E_1$.

When such a composite beam is bent, as for a beam of one material, the strains vary linearly, as shown in Fig. 6-7(b). However, the longitudinal stresses depend on the elastic moduli and are as shown in Fig. 6-7(c). At the interface between the two materials, whereas the strain for both materials is the same, the stresses are different, and depend on the magnitudes of $E_1$ and $E_2$. The remaining issue in such problems consists of locating the neutral axis or surface. This can be easily done for beams having cross sections with symmetry around the vertical axes.

For beams of several different materials, the elastic moduli for each material must be identified. Let $E_i$ be such an elastic modulus for the $i$th material in a composite cross section. Then Eq. 6-4 can be generalized to read

$$\sigma_x = E_i \varepsilon_x = -E_i \kappa y \tag{6-14}$$

Where from Fig. 6-7(a), $y = y_b - \bar{y}_b$. In this relation $y_b$ is arbitrarily measured from the bottom of the section, and $\bar{y}_b$ locates the neutral axis as shown.

Since for pure bending the force $F_x$ at a section in the $x$ direction must vanish, following the same procedure as before, and substituting Eq. 6-14 into Eq. 6-5,

$$F_x = \int_A \sigma_x \, dA = -\kappa \int_A E_i y \, dA = 0 \tag{6-15}$$

The last expression differs from Eq. 6-6 only by not placing $E_i$ outside of the integral. By substituting $y = y_b - \bar{y}_b$ into Eq. 6-15, and recognizing that $\bar{y}_b$ is a constant,

$$-\kappa \int_A E_i y_b \, dA + \kappa \bar{y}_b \int_A E_i \, dA = 0$$

and
$$\bar{y}_b = \frac{\int_A E_i \, y_b \, dA}{\int_A E_i \, dA} \qquad\qquad (6\text{-}16)$$

where the integration must be carried out with appropriate $E_i$'s, for each material. This equation defines the *modulus-weighted centroid* and locates the neutral axis.

Essentially the same process is used for inelastic bending analysis of beams by changing the stress-strain relations. The first two of the enumerated basic concepts remain applicable.

The developed theory for elastic beams of *one* material is in complete agreement with the mathematically exact solution[7] based on the theory of elasticity for pure bending of an elastic rectangular bar. However, even for this limited case, the boundary conditions at the ends require the surface stresses $\sigma_x$ to be distributed over the ends as given by Eq. 6-11. For this case plane sections through a beam remain precisely plane after bending. However, in usual applications, per *Saint-Venant's principle*, it is generally assumed that the stresses, at a distance about equal to the depth of a member away from the applied moment, are essentially uniform and are given by Eq. 6-11. The local stresses at points of force application or change in cross section are calculated using stress concentration factors. In applications the theory discussed is routinely applied to any kind of cross section, whether a material is elastic or plastic.

In conclusion it should be noted that, in all cases in pure bending, the stresses acting on the area above the neutral axis develop a force of one sense, whereas those below the neutral axis develop a force acting in the opposite direction. An example is shown in Fig. 6-7(d) where the tension $T$ is equal to the compression $C$, and the $T - C$ couple is equal to the moment $M_z$. This method of reducing stresses to forces and a couple can be used to advantage in some problems.

## *[8]6-4. Computation of the Moment of Inertia

In applying the flexure formula, the rectangular moment of inertia $I$ of the cross-sectional area about the neutral axis must be determined. Its value is defined by the integral of $y^2 \, dA$ over the entire cross-sectional area of a member, and it must be emphasized that for the flexure formula, the moment of inertia must be computed around the neutral axis. This axis passes through the centroid of the cross-sectional area. It is shown

---

[7] S. Timoshenko, and J. N. Goodier, *Theory of Elasticity*, 3rd ed. (New York: McGraw-Hill, 1970), 284.

[8] This is a review section.

in Sections 6-15 and 6-16 that for symmetric cross sections, the neutral axis is perpendicular to the axis of symmetry. The moment of inertia around such an axis is either a maximum or a minimum, and for that reason, this axis is one of the *principal* axes for an area. The procedures for determining centroids and moments of inertia of areas are generally thoroughly discussed in texts on statics.[9] However, for completeness, they are reviewed in what follows.

The first step in evaluating $I$ for an area is to find its centroid. An integration of $y^2\, dA$ is then performed with respect to the horizontal axis passing through the area's centroid. In applications of the flexure formula, the actual integration over areas is necessary for only a few elementary shapes, such as rectangles, triangles, etc. Values of moments of inertia for some simple shapes may be found in texts on statics as well as in any standard civil or mechanical engineering handbook (also see Table 2 of the Appendix). Most cross-sectional areas used may be divided into a combination of these simple shapes. To find $I$ for an area composed of several simple shapes, the *parallel-axis theorem* (sometimes called the *transfer formula*) is necessary; its development follows.

Consider that the area $A$ shown in Fig. 6-8 is a *part* of a complex area of a cross section of a beam in flexure. The centroidal axis $z_c$ for *this* area is at a distance $d_z$ from the centroidal $z$ axis for the *whole* cross-sectional area. Then, by definition, the moment of inertia $I_{z_c}$ of the area $A$ around its $z_c$ axis is

$$I_{z_c} = \int_A y_c^2\, dA \qquad (6\text{-}17)$$

On the other hand, the moment of inertia $I_z$ of the *same* area $A$ around the $z$ axis is

$$I_z = \int_A (y_c + d_z)^2\, dA$$

By squaring the quantities in the parentheses and placing the constants outside the integrals,

$$I_z = \int_A y_c^2\, dA + 2d_z \int_A y_c\, dA + d_z^2 \int_A dA$$

Here the first integral according to Eq. 6-17 is equal to $I_{z_c}$, the second integral vanishes as $y_c$ passes through the centroid of $A$, and the last integral reduces to $A d_z^2$. Hence,

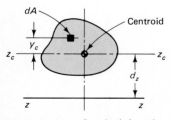

**Fig. 6-8** Area for deriving the parallel-axis theorem.

[9] For example, see J. L. Meriam and L. G. Kraige, *Engineering Mechanics*, Vol. 1, Statics, 2nd ed. (New York: Wiley, 1986).

$$I_z = I_{z_c} + Ad_z^2 \tag{6-18}$$

This is the *parallel-axis theorem*. It can be stated as follows: the moment of inertia of an area around any axis is equal to the moment of inertia of the same area around a parallel axis passing through the area's centroid, plus the product of the same area and the square of the distance between the two axes.

In calculations, Eq. 6-18 must be applied to *each* part into which a cross-sectional area has been subdivided and the results summed to obtain $I_z$ for the whole section, i.e.,

$$I_z \text{ (whole section)} = \sum(I_{z_c} + Ad_z^2) \tag{6-18a}$$

After this process is completed, the $z$ subscript may be dropped in treating bending of *symmetric* cross sections.

The following examples illustrate the method of computing $I$ directly by integration for two simple areas. Then an application of the parallel-axis theorem to a composite area is given. Values for $I$ for commercially fabricated steel beams, angles, and pipes are given in Tables 3 to 8 of the Appendix.

## EXAMPLE 6-1

Find the moment of inertia around the horizontal axis passing through the centroid for the rectangular area shown in Fig. 6-9.

## Solution

The centroid of this section lies at the intersection of the two axes of symmetry. Here it is convenient to take $dA$ as $b\,dy$. Hence,

$$I_z = I_o = \int_A y^2\,dA = \int_{-h/2}^{+h/2} y^2 b\,dy = b\left.\frac{y^3}{3}\right|_{-h/2}^{+h/2} = \frac{bh^3}{12}$$

Hence,

$$I_z = \frac{bh^3}{12} \quad \text{and} \quad I_y = \frac{b^3h}{12} \tag{6-19}$$

These expressions are used frequently, as rectangular beams are common.

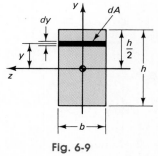

Fig. 6-9

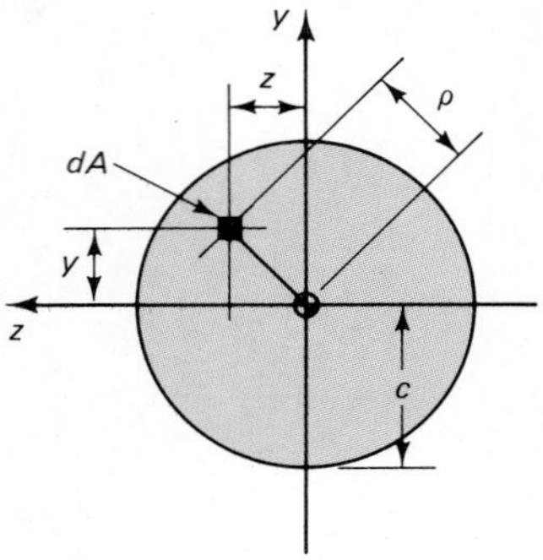

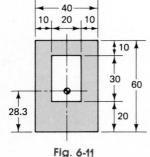

**Fig. 6-10**

## EXAMPLE 6-2

Find the moment of inertia about a diameter for a circular area of radius $c$; see Fig. 6-10.

### Solution

To find $I$ for a circle, first note that $\rho^2 = z^2 + y^2$, as may be seen from the figure. Then using the definition of $J$, noting the symmetry around both axes, and using Eq. 4-2,

$$J = \int_A \rho^2 \, dA = \int_A (y^2 + z^2) \, dA = \int_A y^2 \, dA + \int_A z^2 \, dA$$
$$= I_z + I_y = 2I_z$$

$$\boxed{I_z = I_y = \frac{J}{2} = \frac{\pi c^4}{4}} \qquad (6\text{-}20)$$

In mechanical applications, circular shafts often act as beams; hence, Eq. 6-20 will be found useful. For a tubular shaft, the moment of inertia of the hollow interior must be subtracted from the last expression.

## EXAMPLE 6-3

Determine the moment of inertia $I$ around the horizontal axis for the area shown in mm in Fig. 6-11 for use in the flexure formula.

### Solution

As the moment of inertia is for use in the flexure formula, it must be obtained around the axis through the centroid of the area. Hence, the centroid of the area must be found first. This is most easily done by treating the entire outer section and deducting the hollow interior from it. For convenience, the work is carried out in tabular form. Then the parallel-axis theorem is used to obtain $I$.

| Area | $A$ [mm$^2$] | $y$ [mm] (*from bottom*) | $Ay$ |
|---|---|---|---|
| Entire area | $40 \times 60 = 2400$ | 30 | 72 000 |
| Hollow interior | $-20 \times 30 = -600$ | 35 | $-21\,000$ |
| | $\sum A = 1800 \text{ mm}^2$ | | $\sum Ay = 51\,000 \text{ mm}^3$ |

$$\bar{y} = \frac{\sum Ay}{\sum A} = \frac{51\,000}{1800} = 28.3 \text{ mm from bottom}$$

**Fig. 6-11**

*For the entire area:*

$$I_{z_c} = \frac{bh^3}{12} = \frac{40 \times 60^3}{12} = 72 \times 10^4 \text{ mm}^4$$
$$Ad^2 = 2400(30 - 28.3)^2 = \underline{0.69 \times 10^4 \text{ mm}^4}$$
$$I_z = 72.69 \times 10^4 \text{ mm}^4$$

*For the hollow interior:*

$$I_{z_c} = \frac{bh^3}{12} = \frac{20 \times 30^3}{12} = 4.50 \times 10^4 \text{ mm}^4$$
$$Ad^2 = 600(35 - 28.3)^2 = \underline{2.69 \times 10^4 \text{ mm}^4}$$
$$I_z = 7.19 \times 10^4 \text{ mm}^4$$

*For the composite section:*

$$I_z = (72.69 - 7.19)10^4 = 65.50 \times 10^4 \text{ mm}^4$$

Note particularly that in applying the parallel-axis theorem, each element of the composite area contributes two terms to the total $I$. One term is the moment of inertia of an area around its own centroidal axis, the other term is due to the transfer of its axis to the centroid of the whole area. Methodical work is the prime requisite in solving such problems correctly.

## 6-5. Applications of the Flexure Formula

The largest stress at a section of a beam is given by Eq. 6-13, $\sigma_{max} = Mc/I$, and in most practical problems, it is this maximum stress that has to be determined. Therefore, it is desirable to make the process of determining $\sigma_{max}$ as simple as possible. This can be accomplished by noting that both $I$ and $c$ are constants for a given section of a beam. Hence, $I/c$ is a constant. Moreover, since this ratio is only a function of the cross-sectional dimensions of a beam, it can be uniquely determined for any cross-sectional area. This ratio is called the *elastic section modulus* of a section and will be designated by $S$. With this notation, Eq. 6-13 becomes

$$\boxed{\sigma_{max} = \frac{Mc}{I} = \frac{M}{I/c} = \frac{M}{S}} \qquad (6\text{-}21)$$

or stated otherwise

$$\text{maximum bending stress} = \frac{\text{bending moment}}{\text{elastic section modulus}}$$

If the moment of inertia $I$ is measured in $in^4$ (or $m^4$) and $c$ in in (or m), $S$ is measured in $in^3$ (or $m^3$). Likewise, if $M$ is measured in in-lb (or N·m), the units of stress, as before, become psi (or $N/m^2$). It bears repeating that the distance $c$ as used here is measured from the neutral axis to the most remote fiber of the beam. This makes $I/c = S$ a minimum, and consequently $M/S$ gives the maximum stress. The efficient sections for resisting elastic bending have as large an $S$ as possible for a given amount of material. This is accomplished by locating as much of the material as possible far from the neutral axis.

The use of the *elastic section modulus* in Eq. 6-21 corresponds somewhat to the use of the area term $A$ in Eq. 1-13 ($\sigma = P/A$). However, only the maximum flexural stress on a section is obtained from Eq. 6-21, whereas the stress computed from Eq. 1-13 holds true across the whole section of a member.

Equation 6-21 is widely used in practice because of its simplicity. To facilitate its use, section moduli for many manufactured cross sections are tabulated in handbooks. Values for a few steel sections are given in Tables 3 to 8 in the Appendix. Equation 6-21 is particularly convenient for the design of beams. Once the maximum bending moment for a beam is determined and an allowable stress is decided upon, Eq. 6-21 may be solved for the required section modulus. This information is sufficient to select a beam. However, a detailed consideration of beam design will be delayed until Chapter 9. This is necessary inasmuch as a shear force, which in turn causes stresses, usually also acts at a beam section. The interaction of the various kinds of stresses must be considered first to gain complete insight into the problem.

The following two examples illustrate calculations for bending stresses at specified sections, where, in addition to bending moments, shears are also required for equilibrium. As shown in the next chapter, the presence of small or moderate shears does not significantly affect the bending stresses in slender beams. Both moment and shear frequently occur at the same section simultaneously.

## EXAMPLE 6-4

A 300 by 400 mm wooden cantilever beam weighing 0.75 kN/m carries an upward concentrated force of 20 kN at the end, as shown in Fig. 6-12(a). Determine the maximum bending stresses at a section 2 m from the free end.

### Solution

A free-body diagram for a 2-m segment of the beam is shown in Fig. 6-12(c). To keep this segment in equilibrium requires a shear of $20 - (0.75 \times 2) = 18.5$ kN and a bending moment of $(20 \times 2) - (0.75 \times 2 \times 1) = 38.5$ kN·m at the section. Both of these quantities are shown with their proper sense in Fig. 6-12(c). The distance from the neutral axis to the extreme fibers $c = 200$ mm. This is applicable to both the tension and the compression fibers.

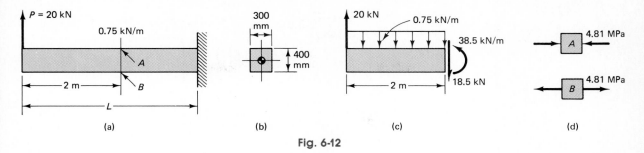

Fig. 6-12

From Eq. 6-19:

$$I_z = \frac{bh^3}{12} = \frac{300 \times 400^3}{12} = 16 \times 10^8 \text{ mm}$$

From Eq. 6-13:

$$\sigma_{max} = \frac{Mc}{I} = \frac{38.5 \times 10^6 \times 200}{16 \times 10^8} = \pm 4.81 \text{ MPa}$$

From the sense of the bending moment shown in Fig. 6-12(c), the top fibers of the beam are seen to be in compression and the bottom ones in tension. In the answer given, the positive sign applies to the tensile stress and the negative sign applies to the compressive stress. Both of these stresses decrease at a linear rate toward the neutral axis, where the bending stress is zero. The normal stresses acting on infinitesimal elements at $A$ and $B$ are shown in Fig. 6-12(d). It is important to learn to make such a representation of an element as it will be frequently used in Chapters 8 and 9.

## Alternative Solution

If only the maximum stress is desired, the equation involving the section modulus may be used. The section modulus for a rectangular section in algebraic form is

$$S = \frac{I}{c} = \frac{bh^3}{12} \frac{2}{h} = \frac{bh^2}{6} \tag{6-22}$$

In this problem, $S = 300 \times 400^2/6 = 8 \times 10^6 \text{ mm}^3$, and by Eq. 6-21,

$$\sigma_{max} = \frac{M}{S} = \frac{38.5 \times 10^6}{8 \times 10^6} = 4.81 \text{ MPa}$$

Both solutions lead to identical results.

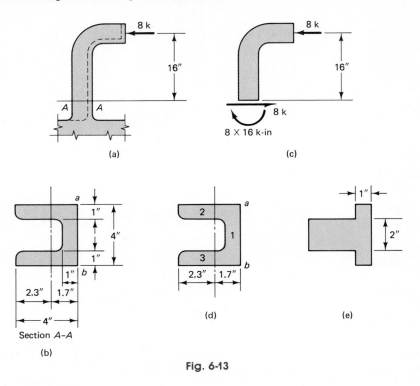

(a)

(c)

Section *A–A*

(b)

(d)

(e)

**Fig. 6-13**

## EXAMPLE 6-5

Find the maximum tensile and compressive stresses acting normal to section *A–A* of the machine bracket shown in Fig. 6-13(a) caused by the applied force of 8 kips.

### Solution

The shear and bending moment of proper magnitude and sense to maintain the segment of the member in equilibrium are shown in Fig. 6-13(c). Next the neutral axis of the beam must be located. This is done by locating the centroid of the area shown in Fig. 6-13(b); see also Fig. 6-13(d). Then the moment of inertia about the neutral axis is computed. In both these calculations, the legs of the cross section are assumed rectangular, neglecting fillets. Then, keeping in mind the sense of the resisting bending moment and applying Eq. 6-13, one obtains the desired values.

| Area Number | $A$ [in$^2$] | $y$ [in] (from $ab$) | $Ay$ |
|:---:|:---:|:---:|:---:|
| 1 | 4.0 | 0.5 | 2.0 |
| 2 | 3.0 | 2.5 | 7.5 |
| 3 | 3.0 | 2.5 | 7.5 |
| $\sum A = 10.0$ in$^2$ | | | $\sum Ay = 17.0$ in$^3$ |

$$\bar{y} = \frac{\sum Ay}{\sum A} = \frac{17.0}{10.0} = 1.70 \text{ in} \qquad \text{from line } ab$$

$$I = \sum (I_o + Ad^2) = \frac{4 \times 1^3}{12} + 4 \times 1.2^2$$

$$+ \frac{2 \times 1 \times 3^3}{12} + 2 \times 3 \times 0.8^2 = 14.43 \text{ in}^4$$

$$\sigma_{max} = \frac{Mc}{I} = \frac{8 \times 16 \times 2.3}{14.43} = 20.4 \text{ ksi} \qquad \text{(compression)}$$

$$\sigma_{max} = \frac{Mc}{I} = \frac{8 \times 16 \times 1.7}{14.43} = 15.1 \text{ ksi} \qquad \text{(tension)}$$

These stresses vary linearly toward the neutral axis and vanish there. The results obtained would be the same if the cross-sectional area of the bracket were made T-shaped, as shown in Fig. 6-13(e). The properties of this section about the significant axis are the same as those of the channel. Both these sections have an axis of symmetry.

---

The previous example shows that members resisting flexure may be proportioned so as to have a different maximum stress in tension than in compression. This is significant for materials having different strengths in tension and compression. For example, cast iron is strong in compression and weak in tension. Thus, the proportions of a cast-iron member may be so set as to have a low maximum tensile stress. The potential capacity of the material may thus be better utilized.

## *6-6. Stress Concentrations

The flexure theory developed in the preceding sections applies only to beams of constant cross section, i.e., *prismatic* beams. If the cross-sectional area of the beam varies gradually, no significant deviation from the stress pattern discussed earlier takes place. However, if notches, grooves, bolt holes, or an abrupt change in the cross-sectional area of the beam occur, high *local* stresses arise. This situation is analogous to the ones discussed earlier for axial and torsion members. Again, it is very difficult to obtain analytical expressions for the actual stress. In the past, most of the information regarding the actual stress distribution came from accurate photoelastic experiments. Numerical methods employing finite elements are now extensively used for the same purpose.

Fortunately, as in the other cases discussed, only the geometric proportions of the member affect the local stress pattern. Moreover, since interest generally is in the maximum stress, stress-concentration factors may be used to an advantage. The ratio $K$ of the actual maximum stress to the nominal maximum stress in the *minimum* section, as given by Eq.

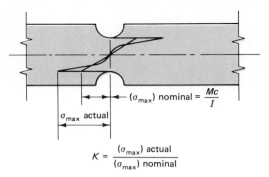

$(\sigma_{max})$ nominal $= \dfrac{Mc}{I}$

$\sigma_{max}$ actual

**Fig. 6-14** Meaning of stress-concentration factor in bending.

$$K = \frac{(\sigma_{max}) \text{ actual}}{(\sigma_{max}) \text{ nominal}}$$

6-13, is defined as the stress-concentration factor in bending. This concept is illustrated in Fig. 6-14. Hence, in general,

$$(\sigma_{max})_{actual} = K\frac{Mc}{I} \qquad (6\text{-}23)$$

In this equation $Mc/I$ is for the small width of a bar.

Figures 6-15 and 6-16 are plots of stress-concentration factors for two representative cases.[10] The factor $K$, depending on the proportions of the member, may be obtained from these diagrams. A study of these graphs indicates the desirability of generous fillets and the elimination of sharp notches to reduce local stress concentrations. These remedies are highly desirable in machine design. For ductile materials, where the applied forces are static, stress concentrations are less important.

Stress concentrations become particularly significant if the cross-sec-

[10] These figures are adapted from a paper by M. M. Frocht, "Factors of Stress Concentration Photoelastically Determined," *Trans. ASME* 57, (1935): A-67.

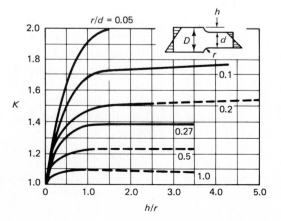

**Fig. 6-15** Stress-concentration factors in pure bending for flat bars with various fillets.

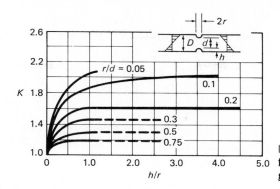

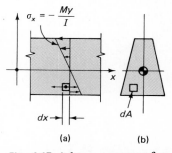

Fig. 6-16 Stess-concentration factors in bending for grooved flat bars.

tional area has reentrant angles. For example, high localized stresses may occur at the point where the flange[11] and the web of an I beam meet. To minimize these, commercially rolled or extruded shapes have a generous fillet at all such points.

In addition to stress concentrations caused by changes in the cross-sectional area of a beam, another effect is significant. Forces often are applied over a limited area of a beam. Moreover, the reactions act only locally on a beam at the points of support. In the previous treatment, all such forces were idealized as concentrated forces. In practice, the average bearing pressure between the member delivering such a force and the beam are computed at the point of contact of such forces with the beam. This bearing pressure, or stress, acts normal to the neutral surface of a beam and is at *right angles to the bending stresses discussed in this chapter*. A more detailed study of the effect of such forces shows that they cause a disturbance of all stresses on a local scale, and the bearing pressure as normally computed is a crude approximation. The stresses at right angles to the flexural stresses behave more nearly as shown in Fig. 2-30.

The reader must remember that the stress-concentration factors apply only while the material behaves elastically. Inelastic behavior of material tends to reduce these factors.

## *[12]6-7. Elastic Strain Energy in Pure Bending

In Section 2-11, the elastic strain energy for an infinitesimal element subjected to a normal stress was formulated. Using this as a basis, the elastic strain energy for beams in pure bending can be found. For this case, the normal stress varies linearly from the neutral axis, as shown in Fig. 6-17, and, according to Eq. 6-11, in simplified notation, this stress $\sigma = -My/I$. The volume of a typical infinitesimal beam element is $dx\,dA$, where $dx$ is its length, and $dA$ is its cross-sectional area. By substituting

---

[11] The *web* is a thin vertical part of a beam. Thin horizontal parts of a beam are called *flanges*.

[12] This section can be postponed until study of Chapter 12.

Fig. 6-17 A beam segment for deriving strain energy in bending.

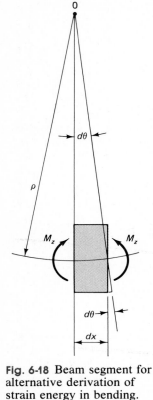

**Fig. 6-18** Beam segment for alternative derivation of strain energy in bending.

these expressions into Eq. 2-23 and integrating over the volume $V$ of the beam, the expression for the elastic strain energy $U$ in a beam in pure bending is obtained.

$$U = \int_V \frac{\sigma_x^2}{2E} dV = \int_V \frac{1}{2E} \left( -\frac{My}{I} \right)^2 dx\, dA$$

Rearranging terms and remembering that $M$ at a section of a beam is constant and that the order of performing the integration is arbitrarily,

$$U = \int_{\text{length}} \frac{M^2}{2EI^2} dx \int_{\text{area}} y^2\, dA = \int_0^L \frac{M^2\, dx}{2EI} \qquad (6\text{-}24)$$

where the last simplification is possible since, by definition, $I = \int y^2\, dA$. Equation 6-24 reduces the volume integral for the elastic energy of prismatic beams in pure flexure to a single integral taken over the length $L$ of a beam.

Alternatively, Eq. 6-24 can be derived from a different point of view, by considering an elementary segment of a beam $dx$ long, as is shown in Fig. 6-18. Before the application of bending moments $M$, the two planes perpendicular to the axis of the beam are parallel. After the application of the bending moments, extensions of the same two planes, which remain planes, intersect at $O$, and the angle included between these two planes is $d\theta$. Moreover, since the full value of the moment $M$ is attained *gradually*, the *average* moment acting through an angle $d\theta$ is $\frac{1}{2}M$. Hence, the external work $W_e$ done on a segment of a beam is $dW_e = \frac{1}{2}M\, d\theta$. Further, since for small deflections, $dx \approx \rho\, d\theta$, where $\rho$ is the radius of curvature of the elastic curve, per Eq. 6-10 $1/\rho = M/EI$. Hence, from the principle of conservation of energy, the internal strain energy of an element of a beam is

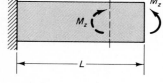

$$dU = dW_e = \frac{1}{2}M\, d\theta = \frac{1}{2}M \frac{dx}{\rho} = \frac{M^2\, dx}{2EI}$$

which has the same meaning as Eq. 6-24.

## EXAMPLE 6-7

Find the elastic strain energy stored in a rectangular cantilever beam due to a bending moment $M$ applied at the end; see Fig. 6-19.

Section

**Fig. 6-19**

## Solution

The bending moment at every section of this beam, as well as the flexural rigidity $EI$, is constant. By direct application of Eq. 6-24,

$$U = \int_0^L \frac{M^2\,dx}{2EI} = \frac{M^2}{2EI} \int_0^L dx = \frac{M^2 L}{2EI}$$

It is instructive to write this result in another form. Thus, since $\sigma_{\max} = Mc/I$, $M = \sigma_{\max}I/c = 2\sigma_{\max}I/h$, and $I = bh^3/12$,

$$U = \frac{(2\sigma_{\max}I/h)^2 L}{2EI} = \frac{\sigma_{\max}^2}{2E}\left(\frac{bh\,L}{3}\right) = \frac{\sigma_{\max}^2}{2E}\left(\frac{1}{3}\text{vol}\right)$$

For a given maximum stress, the volume of the material in this beam is only one-third as effective for absorbing energy as it would be in a uniformly stressed bar, where $U = (\sigma^2/2E)(\text{vol})$. This results from *variable* stresses in a beam. If the bending moment also varies along a prismatic beam, the volume of the material becomes even less effective.

---

# **\*\*[13]6-8. Beams of Composite Cross Section**

Important uses of beams made of different materials occur in practice. Wooden beams are sometimes reinforced by metal straps, plastics are reinforced with fibers, and reinforced concrete is concrete with steel reinforcing bars. The elastic bending theory discussed before can be readily extended to include such beams of composite cross section.

Consider an elastic beam of several materials bonded together with a vertical axis of symmetry as shown in Fig. 6-20(a). The elastic moduli $E_i$ for the different materials are given. As for a homogeneous material, the longitudinal extensional strains $\varepsilon_x$ are assumed to vary linearly as shown

[13] This is an optional section.

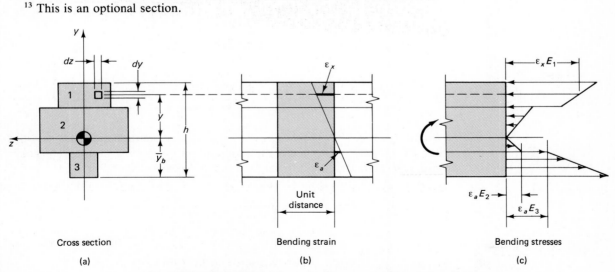

Cross section        Bending strain        Bending stresses

(a)             (b)             (c)

**Fig. 6-20** Elastic beam of composite cross section in bending.

$\sigma_x = E_i \, \varepsilon_x$

in Fig. 6-20(b). The neutral axis for this section, passing through the *modulus-weighted* centroid, is located by the distance $\bar{y}_b$ and can be calculated using Eq. 6-16. The stresses shown in Fig. 6-20(c) follow from Eq. 6-14. At the interfaces between two materials, depending on the relative values of their $E_i$'s, a sharp discontinuity in stress magnitudes arises.

Following the same procedure as in Eq. 6-7, the resisting bending moment

$$M_z = \kappa \int_A E_i y^2 \, dA = \kappa (EI)^* \tag{6-25}$$

where the curvature $\kappa$, being constant for the section, is taken outside the integral, and $(EI)^*$ defines symbolically the value of the integral in the middle expression. Hence

$$\kappa = \frac{M_z}{(EI)^*} \tag{6-26}$$

and by substituting this relation into Eqs. 6-3 and 6-14,

$$\varepsilon_x = -\frac{M_z}{(EI)^*}y \quad \text{and} \quad \sigma_x = -E_i\frac{M_z}{(EI)^*}y \tag{6-27}$$

where the last expression is an analogue to Eq. 6-11, and can be immediately specialized for a homogeneous beam.

In calculations of bending of composite cross sections, sometimes it is useful to introduce the concept of an *equivalent* or *transformed cross-sectional area* in *one* material. This requires arbitrary selection of a *reference* $E_i$, defined here as $E_{ref}$. Using this notation the integral in Eq. 6-15, for constant curvature $\kappa$, can be recast as follows:

$$\int_A E_i y \, dA = E_{ref} \int_a y\frac{E_i}{E_{ref}} \, dA = E_{ref} \int_A y(n_i \, dA) = 0 \tag{6-28}$$

$n = \dfrac{E_i}{E_{REF}}$

where $n_i \, dA = (E_i/E_{ref}) \, dA$. Therefore a beam of composite cross section can be considered to have the mechanical properties of the *reference material*, provided the differential areas $dA$ are multiplied by $n_i$, the ratio of $E_i$ to $E_{ref}$. After transforming a cross section in this manner, conventional elastic analysis is applicable. In transformed sections the stresses *vary linearly* from the neural axis in *all* materials. The actual stresses are obtained for the reference material, whereas the stresses in the other materials must be multiplied by $n_i$.

This procedure is illustrated on the two examples that follow.

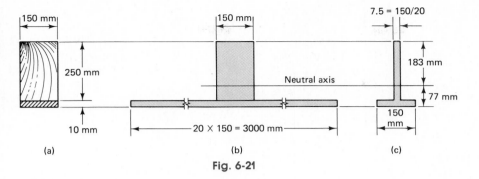

Fig. 6-21

## EXAMPLE 6-8

Consider a composite beam of the cross-sectional dimensions shown in Fig. 6-21(a). The upper 150 by 250 mm part is wood, $E_w = 10$ GPa; the lower 10 by 150 mm strap is steel, $E_s = 200$ GPa. If this beam is subjected to a bending moment of 30 kN·m around a horizontal axis, what are the maximum stresses in the steel and wood?

### Solution

Select $E_w$ as $E_{ref}$. Then $n_s = E_s/E_w = 20$. Hence the transformed cross section is as in Fig. 6-21(b) with the equivalent width of steel equal to $150 \times 20 = 3000$ mm. The centroid and moment of inertia around the centroidal axis for this transformed section are, respectively,

$$\bar{y} = \frac{150 \times 250 \times 125 + 10 \times 3000 \times 255}{150 \times 250 + 10 \times 3000} = 183 \text{ mm} \quad \text{(from the top)}$$

$$I_z = \frac{150 \times 250^3}{12} + 150 \times 250 \times 58^2 + \frac{3000 \times 10^3}{12} + 10 \times 3000 \times 72^2$$
$$= 478 \times 10^6 \text{ mm}^4$$

The maximum stress in the wood is

$$(\sigma_w)_{max} = \frac{Mc}{I} = \frac{0.03 \times 10^9 \times 183}{478 \times 10^6} = 11.5 \text{ MPa}$$

The maximum stress in the steel is

$$(\sigma_s)_{max} = n\sigma_w = 20 \times \frac{0.03 \times 10^9 \times 77}{478 \times 10^6} = 96.7 \text{ MPa}$$

### Alternative Solution

Select $E_s$ as $E_{ref}$. Then $n_w = E_w/E_s = 1/20$, and the transformed section is as in Fig. 6-21(c).

$$\bar{y} = \frac{7.5 \times 250 \times 135 + 150 \times 10 \times 5}{7.5 \times 250 + 150 \times 10}$$
$$= 77 \text{ mm} \quad \text{(from the bottom)}$$
$$I_z = \frac{7.5 \times 250^3}{12} + 7.5 \times 250 \times 58^2 + \frac{150 \times 10^3}{12}$$
$$+ 150 \times 10 \times 72^2 = 23.9 \times 10^6 \text{ mm}^4$$
$$(\sigma_s)_{\max} = \frac{0.03 \times 10^9 \times 77}{23.9 \times 10^6} = 96.7 \text{ MPa}$$
$$(\sigma_w)_{\max} = \frac{\sigma_s}{n} = \frac{1}{20} \times \frac{0.03 \times 10^9 \times 183}{23.9 \times 10^6} = 11.5 \text{ MPa}$$

Note that if the transformed section is an equivalent wooden section, the stresses in the actual wooden piece are obtained directly. Conversely, if the equivalent section is steel, stresses in steel are obtained directly. The stress in a material stiffer than the material of the transformed section is increased, since, to cause the same unit strain, a higher stress is required.

### EXAMPLE 6-9

Determine the maximum stress in the concrete and the steel for a reinforced-concrete beam with the section shown in Fig. 6-21(a) if it is subjected to a positive bending moment of 50,000 ft-lb. The reinforcement consists of two #9 steel bars. (These bars are $1\frac{1}{8}$ in in diameter and have a cross-sectional area of 1 in$^2$). Assume the ratio of $E$ for steel to that of concrete to be 15, i.e., $n = 15$.

### Solution

Plane sections are assumed to remain plane in an elastic reinforced-concrete beam. Strains vary linearly from the neutral axis, as shown in Fig. 6-22(b) by the line $ab$. A transformed section in terms of concrete is used to solve this problem. However, concrete is so weak in tension that there is no assurance that minute cracks will not occur in the tension zone of the beam. For this reason, no credit is given to concrete for resisting tension. On the basis of this assumption, concrete

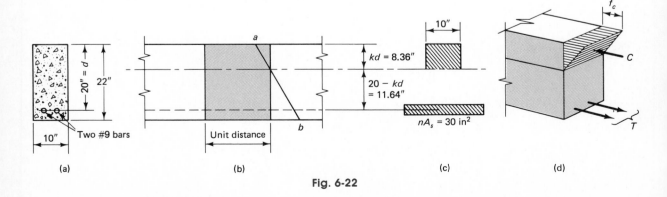

Fig. 6-22

in the tension zone of a beam only holds the reinforcing steel in place.[14] Hence, in this analysis, it virtually does not exist at all, and the transformed section assumes the form shown in Fig. 6-22(c). The cross section of concrete has the beam shape above the neutral axis; below it, no concrete is shown. Steel, of course, can resist tension, so it is shown as the transformed concrete area. For computation purposes, the steel is located by a single dimension from the neutral axis to its centroid. There is a negligible difference between this distance and the true distances to the various steel fibers.

So far, the idea of the neutral axis has been used, but its location is unknown. However, it is known that this axis coincides with the axis through the centroid of the transformed section. It is further known that the first (or statical) moment of the area on one side of a centroidal axis is equal to the first moment of the area on the other side. Thus, let $kd$ be the distance from the top of the beam to the centroidal axis, as shown in Fig. 6-22(c), where $k$ is an unknown ratio,[15] and $d$ is the distance from the top of the beam to the center of the steel. An algebraic restatement of the foregoing locates the neutral axis, about which $I$ is computed and stresses are determined as in the preceding example.

$$\underbrace{10(kd)}_{\substack{\text{concrete} \\ \text{area}}} \; \underbrace{(kd/2)}_{\text{arm}} = \underbrace{30}_{\substack{\text{transformed} \\ \text{steel area}}} \; \underbrace{(20 - kd)}_{\text{arm}}$$

$$5(kd)^2 = 600 - 30(kd)$$
$$(kd)^2 + 6(kd) - 120 = 0$$

Hence,

$$kd = 8.36 \text{ in} \quad \text{and} \quad 20 - kd = 11.64 \text{ in}$$

$$I = \frac{10(8.36)^3}{12} + 10(8.36)\left(\frac{8.36}{2}\right)^2 + 0 + 30(11.64)^2 = 6020 \text{ in}^4$$

$$(\sigma_c)_{max} = \frac{Mc}{I} = \frac{50{,}000 \times 12 \times 8.36}{6020} = 833 \text{ psi}$$

$$\sigma_s = n\frac{Mc}{I} = \frac{15 \times 50{,}000 \times 12 \times 11.64}{6020} = 17{,}400 \text{ psi}$$

## Alternate Solution

After $kd$ is determined, instead of computing $I$, a procedure evident from Fig. 6-22(d) may be used. The resultant force developed by the stresses acting in a "hydrostatic" manner on the compression side of the beam must be located $kd/3$ below the top of the beam. Moreover, if $b$ is the width of the beam, this resultant force $C = \frac{1}{2}(\sigma_c)_{max}b(kd)$ (average stress times area). The resultant tensile force $T$ acts at the center of the steel and is equal to $A_s\sigma_s$, where $A_s$ is the cross-sectional area of the steel. Then, if $jd$ is the distance between $T$ and $C$, and since $T = C$, the applied moment $M$ is resisted by a couple equal to $Tjd$ or $Cjd$.

---

[14] Actually, it is used to resist shear and provide fireproofing for the steel.
[15] This conforms with the usual notation used in books on reinforced concrete. In this text, $h$ is generally used to represent the height or depth of the beam.

$$jd = d - kd/3 = 20 - (8.36/3) = 17.21 \text{ in}$$

$$M = Cjd = \frac{1}{2}b(kd)(\sigma_c)_{\max}(jd)$$

$$(\sigma_c)_{\max} = \frac{2M}{b(kd)(jd)} = \frac{2 \times 50,000 \times 12}{10 \times 8.36 \times 17.21} = 833 \text{ psi}$$

$$M = Tjd = A_s\sigma_s jd$$

$$\sigma_s = \frac{M}{A_s(jd)} = \frac{50,000 \times 12}{2 \times 17.21} = 17,400 \text{ psi}$$

Both methods naturally give the same answer. The second method is more convenient in practical applications. Since steel and concrete have different allowable stresses, the beam is said to have balanced reinforcement when it is designed so that the respective stresses are at their allowable level simultaneously. Note that the beam shown would become virtually worthless if the bending moments were applied in the opposite direction.

## **[16]6-9. Curved Bars

The flexure theory for curved bars is developed in this section. Attention is confined to bars having an axis of symmetry of the cross section, with this axis lying in one plane along the length of the bar. Only the elastic case is treated,[17] with the usual proviso that the elastic modulus is the same in tension and compression.

Consider a curved member such as shown in Figs. 6-23(a) and (b). The outer fibers are at a distance of $r_o$ from the center of curvature $O$. The inner fibers are at a distance of $r_i$. The distance from $O$ to the centroidal axis is $\bar{r}$. The solution[18] of this problem is again based on the familiar assumption: Sections perpendicular to the axis of the beam remain plane after a bending moment $M$ is applied. This is diagrammatically represented by the line $ef$ in relation to an element of the beam $abcd$. The element is defined by the central angle $\phi$.

Although the basic deformation assumption is the same as for straight beams, and, from Hooke's law, the normal stress $\sigma = E\varepsilon$, a difficulty is encountered. The initial length of a beam fiber such as $gh$ depends upon the distance $r$ from the center of curvature. Thus, although the total deformation of beam fibers (described by the small angle $d\phi$) follows a linear law, strains do not. The elongation of a generic fiber $gh$ is $(R - r)\,d\phi$, where $R$ is the distance from $O$ to the neutral surface (not yet known),

[16] Study of this section is optional.
[17] For plastic analysis of curved bars, see, for example, H. D. Conway, "Elastic-Plastic Bending of Curved Bars of Constant and Variable Thickness," *J. Appl. Mech.* 27/4 (December 1960): 733–734.
[18] This approximate solution was developed by E. Winkler in 1858. The exact solution of the same problem by the methods of the mathematical theory of elasticity is due to M. Golovin, who solved it in 1881.

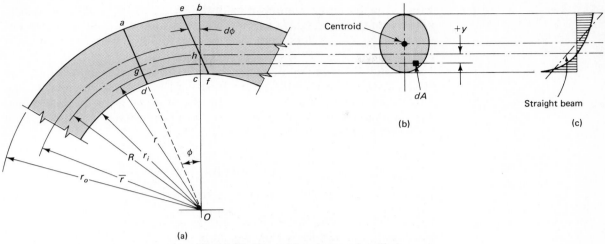

(a)

(b)

(c)

**Fig. 6-23** Curved bar in pure bending.

and its initial length is $r\phi$. The strain $\varepsilon$ of any arbitrary fiber is $(R - r)$ $(d\phi)/r\phi$, and the normal stress $\sigma$ on an element $dA$ of the cross-sectional area is

$$\sigma = E\varepsilon = E\frac{(R - r)\ d\phi}{r\phi} \qquad (6\text{-}29a)$$

For future use, note also that

$$\frac{\sigma r}{R - r} = \frac{E\ d\phi}{\phi} \qquad (6\text{-}29b)$$

Equation 6-29a gives the normal stress acting on an element of area of the cross section of a curved beam. The location of the neutral axis follows from the condition that the summation of the forces acting perpendicular to the section must be equal to zero, i.e.,

$$\sum F_n = 0 \qquad \int_A \sigma\ dA = \int_A \frac{E(R - r)\ d\phi}{r\phi}\ dA = 0$$

However, since $E$, $R$, $\phi$, and $d\phi$ are constant at any one section of a stressed bar, they may be taken outside the integral sign and a solution for $R$ obtained. Thus:

$$\frac{E\ d\phi}{\phi} \int_A \frac{R - r}{r}\ dA = \frac{E\ d\phi}{\phi} \left( R \int_A \frac{dA}{r} - \int_A dA \right) = 0$$

$$\boxed{R = \frac{A}{\displaystyle\int_A dA/r}} \qquad (6\text{-}30)$$

where $A$ is the cross-sectional area of the beam, and $R$ locates the neutral axis. Note that the neutral axis so found does not coincide with the centroidal axis. This differs from the situation found to be true for straight elastic beams.

Now that the location of the neutral axis is known, the equation for the stress distribution is obtained by equating the external moment to the internal resisting moment built up by the stresses given by Eq. 6-29a. The summation of moments is made around the $z$ axis, which is normal to the plane of the figure at $O$ in Fig. 6-23(a).

$$\sum M_z = 0 \qquad M = \int_A \sigma \, dA \underbrace{(R - r)}_{} = \int_A \frac{E(R - r)^2 \, d\phi}{r\phi} \, dA$$
$$\text{force} \qquad \text{arm}$$

Again, remembering that $E$, $R$, $\phi$, and $d\phi$ are constant at a section, using Eq. 6-29b, and performing the algebraic steps indicated, the following is obtained:

$$M = \frac{E \, d\phi}{\phi} \int_A \frac{(R - r)^2}{r} \, dA = \frac{\sigma r}{R - r} \int_A \frac{(R - r)^2}{r} \, dA$$
$$= \frac{\sigma r}{R - r} \int_A \frac{R^2 - Rr - Rr + r^2}{r} \, dA$$
$$= \frac{\sigma r}{R - r} \left( R^2 \int_A \frac{dA}{r} - R \int_A dA - R \int_A dA + \int_A r \, dA \right)$$

Here, since $R$ is a constant, the first two integrals vanish as may be seen from the expression in parentheses appearing just before Eq. 6-30. The third integral is $A$, and the last integral, by definition, is $\bar{r}A$ where $\bar{r}$ is the radius of the centroidal axis. Hence,

$$M = \frac{\sigma r}{R - r}(\bar{r}A - RA)$$

from where the normal stress acting on a curved beam at a distance $r$ from the center of curvature is

$$\boxed{\sigma = \frac{M(R - r)}{rA(\bar{r} - R)}} \tag{6-31}$$

If positive $y$ is measured toward the center of curvature from the neutral axis, and $\bar{r} - R = e$, Eq. 6-31 may be written in a form that more closely resembles the flexure formula for straight beams:

$$\sigma = \frac{My}{Ae(R - y)} \qquad (6\text{-}32)$$

These equations indicate that the stress distribution in a curved bar follows a hyperbolic pattern. A comparison of this result with the one that follows from the formula for straight bars is shown in Fig. 6-23(c). Note particularly that in the curved bar, the neutral axis is pulled toward the center of the curvature of the beam. This results from the higher stresses developed below the neutral axis. The theory developed applies, of course, only to elastic stress distribution and only to beams in pure bending. For a consideration of situations where an axial force is also present at a section, see Section 6-12.

## EXAMPLE 6-10

Compare stresses in a 50 by 50 mm rectangular bar subjected to end moments of 2083 N·m in three special cases: (a) straight beam, (b) beam curved to a radius of 250 mm along the centroidal axis, i.e., $\bar{r} = 250$ mm, Fig. 6-24(a), and (c) beam curved to $\bar{r} = 75$ mm.

## Solution

($a$) This follows directly by applying Eqs. 6-21 and 6-22.

$$S = bh^2/6 = 50 \times 50^2/6 = 20.83 \times 10^3 \text{ mm}^3$$

$$\sigma_{max} = \frac{M}{S} = \frac{2083 \times 10^3}{20.83 \times 10^3} = \pm 100 \text{ MPa}$$

This result is shown in Fig. 6-24(c). $\bar{r} = \infty$ since a straight bar has an infinite radius of curvature.

To solve parts ($b$) and ($c$) the neutral axis must be located first. This is found in general terms by integrating Eq. 6-30. For the rectangular section, the elementary area is taken as $b \, dr$, Fig. 6-24(b). The integration is carried out between the limits $r_i$ and $r_o$, the inner and outer radii, respectively.

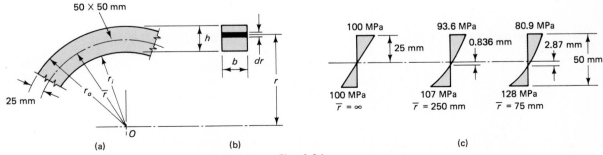

**Fig. 6-24**

$$R = \frac{A}{\displaystyle\int_A dA/r} = \frac{bh}{\displaystyle\int_{r_i}^{r_o} b\,dr/r} = \frac{h}{\displaystyle\int_{r_i}^{r_o} dr/r}$$

$$= \frac{h}{|\ln r|_{r_i}^{r_o}} = \frac{h}{\ln (r_o/r_i)} = \frac{h}{2.3026 \log (r_o/r_i)} \tag{6-33}$$

where $h$ is the depth of the section, ln is the natural logarithm, and log is a logarithm with a base of 10 (common logarithm).

(b) For this case, $h = 50$ mm, $\bar{r} = 250$ mm, $r_i = 225$ mm, and $r_o = 275$ mm. The solution is obtained by evaluating Eqs. 6-33 and 6-31. Subscript $i$ refers to the normal stress $\sigma$ of the inside fibers; $o$ of the outside fibers.

$$R = \frac{50}{\ln (275/225)} = 249.164 \text{ mm}$$

$$e = \bar{r} - R = 250 - 249.164 = 0.836 \text{ mm}$$

$$\sigma_i = \frac{M(R - r_i)}{r_i A(\bar{r} - R)} = \frac{2083 \times 10^3 \times (249.164 - 225)}{225 \times 50^2 \times 0.836}$$

$$= 107 \text{ MPa}$$

$$\sigma_o = \frac{M(R - r_o)}{r_o A(\bar{r} - R)} = \frac{2083 \times 10^3 \times (249.164 - 275)}{275 \times 50^2 \times 0.836}$$

$$= -93.6 \text{ MPa}$$

The negative sign of $\sigma_o$ indicates a compressive stress. These quantities and the corresponding stress distribution are shown in Fig. 6-24(c); $\bar{r} = 250$ mm.

(c) This case is computed in the same way. Here $h = 50$ mm, $\bar{r} = 75$ mm, $r_i = 50$ mm, and $r_o = 100$. Results of the computation as shown in Fig. 6-24(c).

$$R = \frac{50}{\ln (100/50)} = \frac{50}{\ln 2} = 72.13 \text{ mm}$$

$$e = \bar{r} - R = 75 - 73.13 = 2.87 \text{ mm}$$

$$\sigma_i = \frac{2083 \times 10^3 \times (72.13 - 50)}{50 \times 50^2 \times 2.87} = 128 \text{ MPa}$$

$$\sigma_o = \frac{2083 \times 10^3 \times (72.13 - 100)}{100 \times 50^2 \times 2.87} = -80.9 \text{ MPa}$$

Several important conclusions, generally true, may be reached from this example. First, *the usual flexure formula is reasonably good for beams of considerable curvature.* Only 7 percent error in the maximum stress occurs in part (b) for $\bar{r}/h = 5$, an error tolerable for most applications. For greater ratios of $\bar{r}/h$, this error diminishes. As the curvature of the beam increases, the stress on the concave side rapidly increases over the one given by the usual flexure formula. When $\bar{r}/h = 1.5$, a 28 percent error occurs. Second, the evaluation of the integral for $R$ over

Finding the location of the neutral axis such that $T = C$ may require a trial-and-error process, although direct procedures have been devised for some cross sections.[21] After the neutral axis is correctly located, the resisting bending moment $M_z$ at the same section is known to be $C(a + b)$ or $T(a + b)$, see Fig. 6-25(d). Alternatively, in the form of a general equation,

$$M_z = -\int \sigma y \, dA \qquad (6\text{-}37)$$

The problem is greatly simplified if the beam cross section is symmetric around the horizontal axis and material properties are the same in tension and compression. For these conditions it is known a priori that the neutral axis passes through the centroid of the section, and Eq. 6-37 can be directly applied. The behavior of such a beam in bending is shown qualitatively in Fig. 6-26. A sequence of progressively increasing strains associated with plane sections is shown in Fig. 6-26(b). These maximum strains define the maximum stresses in the outer fibers of the beam, Fig. 6-26(c), resulting in progressively increasing bending stresses.

As can be seen from Figs. 6-26(a) and (c), the maximum attainable stress is $\sigma_3$. The instantaneous stress distribution in the beam associated with $\sigma_3$, for this brittle material, is given by the curved line $AB$ in Fig. 6-26(c). However, in routine experiments the *nominal stress* in the extreme fibers is often computed by applying the elastic flexure formula, Eq. 6-13, using the experimentally determined ultimate bending moment. The stress so

[21] A. Nadai, *Theory of Flow and Fracture of Solids*, vol. I (New York: McGraw-Hill, 1950), 356.

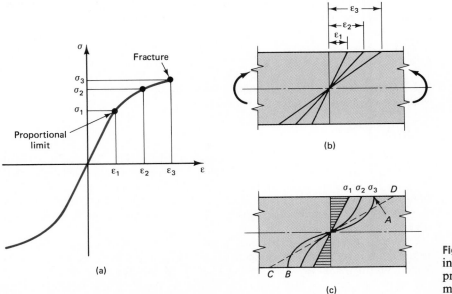

Fig. 6-26 Rectangular beam in bending exceeding the proportional limit of the material.

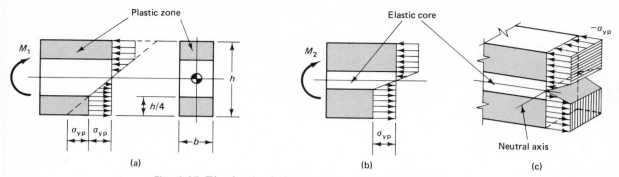

**Fig. 6-27** Elastic-plastic beam at large levels of straining.

found is called the *rupture modulus* of the material in bending. This stress is associated with the line $CD$ in Fig. 6-26(c) and is larger than the stress actually attained.

The elastic perfectly plastic idealization [Fig. 2-13(b)], for reasons of simplicity, is very frequently used for beams of ductile materials in determining their behavior in bending, and as an important example of inelastic bending, consider a rectangular beam of elastic-plastic material; see Fig. 6-27. In such an idealization of material behavior, a sharp separation of the member into distinct elastic and plastic zones is possible. For example, if the strain in the extreme fibers is double that at the beginning of yielding, only the middle half of the beam remains elastic; see Fig. 6-27(a). In this case, the outer quarters of the beam yield. The magnitude of moment $M_1$ corresponding to this condition can be readily computed (see Example 6-13). At higher strains, the elastic zone, or core, diminishes. Stress distribution corresponding to this situation is shown in Figs. 6-27(b) and (c).

## EXAMPLE 6-11

Determine the ultimate plastic capacity in flexure of a mild steel beam of rectangular cross section. Consider the material to be ideally elastic-plastic.

### Solution

The idealized stress-strain diagram is shown in Fig. 6-28(a). It is assumed that the material has the same properties in tension and compression. The strains that can take place in steel during yielding are much greater than the maximum elastic strain (15 to 20 times the latter quantity). Since unacceptably large deformations of a beam would occur at larger strains, the plastic moment may be taken as the ultimate moment.

The stress distribution shown in Fig. 6-28(b) applies after a large amount of deformation takes place. In computing the resisting moment, the stresses corresponding to triangular areas *abc* and *bde* may be neglected without unduly impairing the accuracy. They contribute little resistance to the applied bending moment because of their short moment arms. Hence, the idealization of the stress distribution to that shown in Fig. 6-28(c) is permissible and has a simple physical meaning. The whole upper half of the beam is subjected to a *uniform compressive* stress $-\sigma_{yp}$, whereas the lower half is all under a *uniform tension* $\sigma_{yp}$. That the beam is divided evenly into a tension and a compression zone follows from symmetry. Numerically,

$$C = T = \sigma_{yp}(bh/2) \qquad \text{i.e., stress} \times \text{area}$$

Each one of these forces acts at a distance $h/4$ from the neutral axis. Hence, the *plastic*, or ultimate resisting, moment of the beam is

$$M_p \equiv M_{ult} = C\left(\frac{h}{4} + \frac{h}{4}\right) = \sigma_{yp}\frac{bh^2}{4}$$

where $b$ is the breadth of the beam, and $h$ is its height.

The same solution may be obtained by directly applying Eqs. 6-36 and 6-37. Noting the sign of stresses, one can conclude that Eq. 6-36 is satisfied by taking the neutral axis through the middle of the beam. By taking $dA = b\,dy$ and noting the symmetry around the neutral axis, one changes Eq. 6-37 to

$$M_p \equiv M_{ult} = -2\int_0^{h/2} (-\sigma_{yp})yb\,dy = \sigma_{yp}bh^2/4 \qquad (6\text{-}38)$$

The resisting bending moment of a beam of rectangular section when the outer fibers just reach $\sigma_{yp}$, as given by the elastic flexure formula, is

$$M_{yp} = \sigma_{yp}I/c = \sigma_{yp}(bh^2/6) \qquad \text{therefore, } M_p/M_{yp} = 1.50$$

The ratio $M_p/M_{yp}$ depends only on the cross-sectional properties of a member and is called the *shape factor*. The shape factor just given for the rectangular beam shows that $M_{yp}$ may be exceeded by 50 percent before the ultimate plastic capacity of a rectangular beam is reached.

For static loads such as occur in buildings, ultimate capacities can be approximately determined using plastic moments. The procedures based on such concepts are referred to as *the plastic method of analysis* or *design*. For such work, *plastic section modulus Z* is defined as follows:

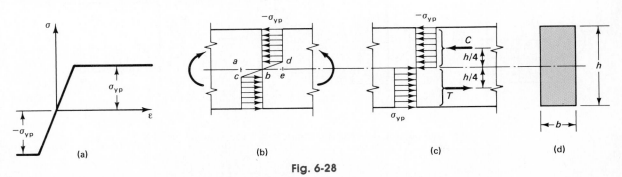

(a)    (b)    (c)    (d)

**Fig. 6-28**

$$M_p = \sigma_{yp}Z \qquad (6\text{-}39)$$

For the rectangular beam just analyzed, $Z = bh^2/4$.

The *Steel Construction Manual*[22] provides a table of plastic section moduli for many common steel shapes. An abridged list of these moduli for steel sections is given in Table 9 of the Appendix. For a given $M_p$ and $\sigma_{yp}$ the solution of Eq. 6-39 for $Z$ is very simple.

The method of limit or plastic analysis is unacceptable in machine design in situations where fatigue properties of the material are important.

### EXAMPLE 6-12

Find the residual stresses in a rectangular beam upon removal of the ultimate plastic bending moment.

### Solution

The stress distribution associated with an ultimate plastic moment is shown in Fig. 6-29(a). The magnitude of this moment has been determined in the preceding example and is $M_p = \sigma_{yp}bh^2/4$. Upon release of this plastic moment $M_p$, every fiber in the beam can rebound elastically. The material elastic range during the unloading is double that which could take place initially (see Fig. 2-13). Therefore, since $M_{yp} = \sigma_{yp}bh^2/6$ and the moment being released is $\sigma_{yp}(bh^2/4)$ or $1.5M_{yp}$, the maximum stress calculated on the basis of elastic action is $\frac{3}{2}\sigma_{yp}$, as shown in Fig. 6-29(b). Superimposing the initial stresses at $M_p$ with the elastic rebound stresses due to the release of $M_p$, one finds the residual stresses; see Fig. 6-29(c). Both tensile and compressive longitudinal residual stresses remain in the beam. The tensile zones are colored in the figure. If such a beam were machined by gradually reducing its depth, the release of the residual stresses would cause undesirable deformations of the bar.

[22] American Institute of Steel Construction, *AISC Steel Construction Manual*, 9th ed. (Chicago: AISC, 1989).

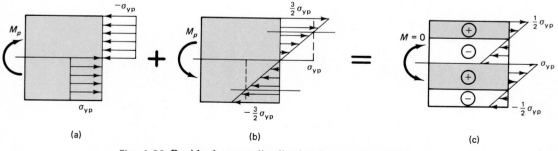

**Fig. 6-29** Residual stress distribution in a rectangular bar.

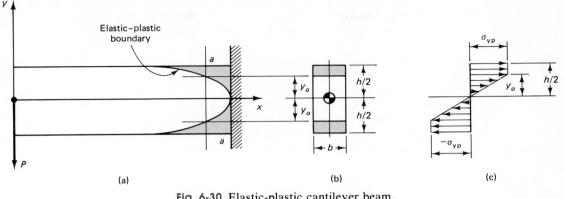

**Fig. 6-30** Elastic-plastic cantilever beam.

## EXAMPLE 6-13

Determine the moment resisting capacity of an elastic-plastic rectangular beam.

### Solution

To make the problem more definite, consider a cantilever loaded as shown in Fig. 6-30(a). If the beam is made of ideal elastic-plastic material and the applied force $P$ is large enough to cause yielding, plastic zones will be formed (shown shaded in the figure). At an arbitrary section $a$–$a$, the corresponding stress distribution will be as shown in Fig. 6-30(c). The elastic zone extends over the depth of $2y_o$. Noting that within the elastic zone the stresses vary linearly and that everywhere in the plastic zone the longitudinal stress is $\sigma_{yp}$, the resisting moment $M$ is

$$
M = -2 \int_0^{y_o} \left( -\frac{y}{y_o}\sigma_{yp} \right)(b \, dy)y - 2 \int_{y_o}^{h/2} (-\sigma_{yp})(b \, dy)y \tag{6-40}
$$

$$
= \sigma_{yp}\frac{bh^2}{4} - \sigma_{yp}\frac{by_o^2}{3} = M_p - \sigma_{yp}\frac{by_o^2}{3}
$$

where the last simplification is done in accordance with Eq. 6-38. In this general equation, if $y_o = 0$, the moment capacity becomes equal to the ultimate plastic moment. However, if $y_o = h/2$, the moment reverts to the limiting elastic case, where $M = \sigma_{yp}bh^2/6$. When the applied bending moment along the span is known, the elastic-plastic boundary can be determined by solving Eq. 6-40 for $y_o$. As long as an elastic zone or core remains, the plastic deformations cannot progress without a limit. This is a case of contained plastic flow.

After the applied force $P$ is released, along the length of the beam where plastic deformations occurred, residual stresses will remain. A typical residual stress distribution for this region is shown in Fig. 6-31. This is a more realistic stress distribution pattern than the one shown in Fig. 6-29(c), where the idealization of sharply dividing the tension and compression zones in the beam at the neutral axis in reality is impossible to attain. That pattern of stress distribution represents the limiting case of the stress distribution pattern given in Fig. 6-31. Many inelastic

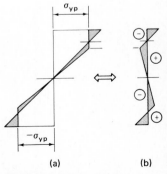

**Fig. 6-31** Residual stress distribution in the beam.

materials tend to have a stress-strain relationship such as shown in Fig. 6-26(a). The residual stress pattern for such materials would resemble the stress difference between curved line *AB* and straight line *CD* of Fig. 6-26(c).

### EXAMPLE 6-14

Determine the plastic moment strength for the reinforced concrete beam in Example 6-9. Assume that the steel reinforcement yields at 40,000 psi and that the ultimate strength of concrete $f'_c = 2500$ psi.

### Solution

When the reinforcing steel begins to yield, large deformations commence. This is taken to be the ultimate capacity of steel; hence, $T_{ult} = A_s\sigma_{yp}$.

At the ultimate or plastic moment, experimental evidence indicates that the compressive stresses in concrete can be approximated by the rectangular stress block shown in Fig. 6-32.[23] It is customary to assume the average stress in this compressive stress block to be $0.85f'_c$. On this basis, keeping in mind that $T_{ult} = C_{ult}$, one has

$$T_{ult} = \sigma_{yp}A_s = 40,000 \times 2 = 80,000 \text{ lb} = C_{ult}$$
$$k'd = \frac{C_{ult}}{0.85f'_c b} = \frac{80,000}{0.85 \times 2,500 \times 10} = 3.77 \text{ in}$$
$$M_{ult} = T_{ult}(d - k'd/2) = 80,000(20 - 3.77/2)/12 = 121,000 \text{ ft-lb}$$

[23] For further details, see P. M. Ferguson, J. E. Breen, and J. O. Jirsa, *Reinforced Concrete Fundamentals*, 5th ed. (New York: Wiley, 1988), or R. Park and T. Paulay, *Reinforced Concrete Structures* (New York: Wiley, 1975).

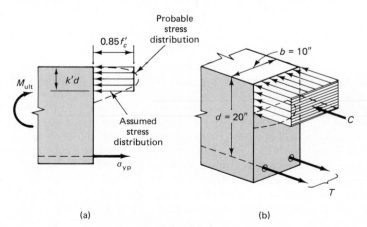

(a)                                        (b)

**Fig. 6-32**

## Part B  UNSYMMETRIC BENDING AND BENDING WITH AXIAL LOADS

### 6-11. Bending About Both Principal Axes[24]

As a simple example of skew or unsymmetrical pure bending, consider the rectangular beam shown in Fig. 6-33. The applied moments $M$ act in the plane *abcd*. By using the vector representation for **M** shown in Fig. 6-33(b), this vector forms an angle $\alpha$ with the $z$ axis and can be resolved into the two components, $M_y$ and $M_z$. Since the cross section of this beam has symmetry about both axes, the formulas derived in Section 6-3 are directly applicable. Because of symmetry, the product of inertia for this section is zero, and the orthogonal axes shown are the *principal* axes for the cross section. This also holds true for the centroidal axes of singly symmetric areas. (For details see Sections 6-15 and 6-16.)

By assuming *elastic* behavior of the material, a superposition of the stresses caused by $M_y$ and $M_z$ is the solution to the problem. Hence, using Eqs. 6-11 and 6-12,

$$\sigma_x = -\frac{M_z y}{I_z} + \frac{M_y z}{I_y} \tag{6-41}$$

where all terms have the previously defined meanings.

[24] Some readers may prefer to study Section 6-14 first, and then consider this section as a special case.

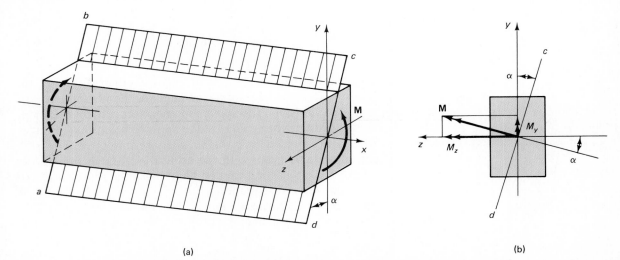

(a)                                                        (b)

**Fig. 6-33** Unsymmetrical bending of a beam with doubly symmetric cross section.

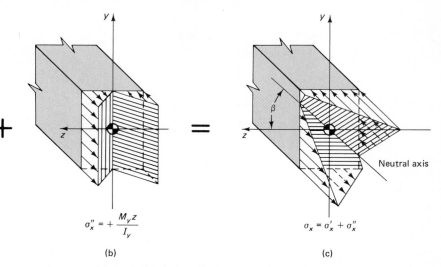

$$\sigma'_x = -\frac{M_z y}{I_z}$$

(a)

$$\sigma''_x = +\frac{M_y z}{I_y}$$

(b)

$$\sigma_x = \sigma'_x + \sigma''_x$$

(c)

**Fig. 6-34** Superposition of elastic bending stresses.

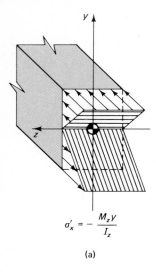

(a)

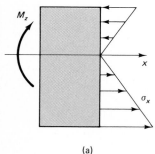

(b)

**Fig. 6-35** Pure bending around a principal axis.

A graphical illustration of superposition is given in Fig. 6-34. Note that a line of zero stress, i.e., a neutral axis, forms at an angle β with the z axis. Analytically, such an axis can be determined by setting the stress given by Eq. 6-41 to zero, i.e.,

$$-\frac{M_z y}{I_z} + \frac{M_y z}{I_y} = 0 \qquad \text{or} \qquad \tan \beta = \frac{y}{z} = \frac{M_y I_z}{M_z I_y} \qquad (6\text{-}42)$$

Since, in general, $M_y = M \sin \alpha$ and $M_z = M \cos \alpha$, this equation reduces to

$$\tan \beta = \frac{I_z}{I_y} \tan \alpha \qquad (6\text{-}43)$$

This equation shows that unless $I_z = I_y$, or α is either 0° or 90°, the angles α and β are not equal. Therefore, in general, the neutral axis and the normal to a plane in which the applied moment acts do not coincide. ✳

The results just given can be generalized to apply to beams having cross sections of any shape provided the *principal* axes are employed. To justify this statement, consider a beam with the arbitrary cross section shown in Fig. 6-35. Let such an elastic beam be bent about the *principal z* axis and assume that the stress distribution is given as $\sigma_x = -M_z y/I_z$, Eq. 6-11. If this stress distribution causes no bending moment $M_y$ around the y axis, this is the correct solution of the problem. Forming such an expression gives

$$M_y = \int_A -\frac{M_z}{I_z} yz \, dA = -\frac{M_z}{I} \int_A yz \, dA = 0 \qquad (6\text{-}44)$$

where the constants are placed in front of the second integral, which is equal to zero because by definition a product of inertia for a principal axis vanishes.

By virtue of the above, the restriction placed on the elastic flexure formula at the beginning of the chapter limiting it to applications for symmetric cross sections can be removed. However, in the application of Eq. 6-41, the _principal axes for a cross section must be used._ A procedure for bypassing this requirement is given in Section 6-14.

## EXAMPLE 6-15

The 100 by 150 mm wooden beam shown in Fig. 6-36(a) is used to support a uniformly distributed load of 4 kN (total) on a simple span of 3 m. The applied load acts in a plane making an angle of 30° with the vertical, as shown in Fig. 6-36(b) and again in Fig. 6-36(c). Calculate the maximum bending stress at midspan, and, for the same section, locate the neutral axis. Neglect the weight of the beam.

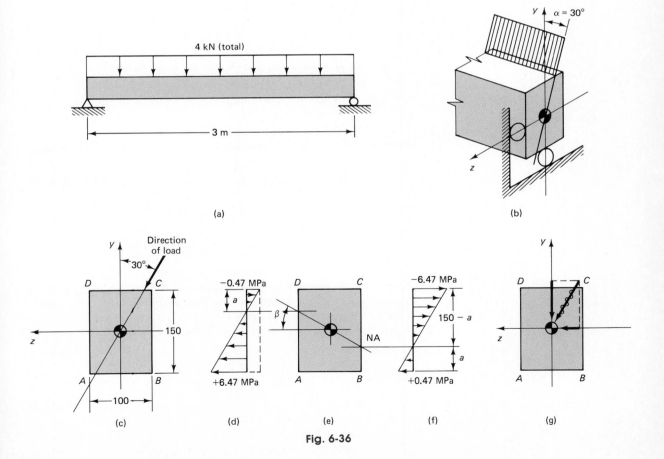

Fig. 6-36

Solution

The maximum bending *in the plane of the applied load* occurs at midspan, and according to Example 5-8, it is equal to $w_o L^2/8$ or $WL/8$, where $W$ is the total load on span $L$. Hence,

$$M = \frac{WL}{8} = \frac{4 \times 3}{8} = 1.5 \text{ kN·m}$$

Next, this moment is resolved into components acting around the respective axes, and $I_z$ and $I_y$ are calculated.

$$M_z = M \cos \alpha = 1.5 \times \sqrt{3}/2 = 1.3 \text{ kN·m}$$
$$M_y = M \sin \alpha = 1.5 \times 0.5 = 0.75 \text{ kN·m}$$
$$I_z = 100 \times 150^3/12 = 28.1 \times 10^6 \text{ mm}^4$$
$$I_y = 150 \times 100^3/12 = 12.5 \times 10^6 \text{ mm}^4$$

By considering the sense of the moment components, it can be concluded that the maximum tensile stress occurs at $A$. Similar reasoning applies when considering the other corner points. Alternatively, the values for the coordinate points can be substituted directly into Eq. 6-41. On either basis,

$$\sigma_A = -\frac{M_z(-c_1)}{I_z} + \frac{M_y c_2}{I_y} = \frac{1.3 \times 10^6 \times 75}{28.1 \times 10^6} + \frac{0.75 \times 10^6 \times 50}{12.5 \times 10^6}$$

$$= +3.47 + 3.00 = +6.47 \text{ MPa}$$
$$\sigma_B = +3.47 - 3.00 = +0.47 \text{ MPa}$$
$$\sigma_C = -3.47 - 3.00 = -6.47 \text{ MPa}$$
$$\sigma_D = -3.47 + 3.00 = -0.47 \text{ MPa}$$

Note that the stress magnitudes on diametrically opposite corners are numerically equal.

The neutral axis is located by the angle $\beta$, using Eq. 6-43:

$$\tan \beta = \frac{28.1 \times 10^6}{12.5 \times 10^6} \tan 30° = 1.30 \quad \text{or} \quad \beta = 52.4°$$

Alternatively, it can be found from the stress distribution, which varies linearly between any two points. For example, from similar triangles, $a/(150 - a) = 0.47/6.47$, giving $a = 10.2$ mm. This locates the neutral axis shown in Fig. 6-36(e) as it must pass through the section centroid. These results lead to the same $\beta$.

## *EXAMPLE 6-16

Determine the maximum tensile and compressive stresses caused by a bending moment of 10 kN·m acting around the horizontal axis for the angle shown in mm in Fig. 6-37.

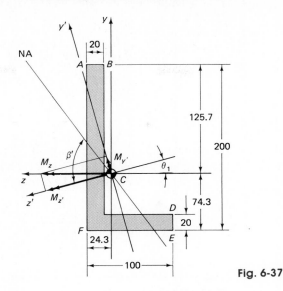

Fig. 6-37

## Solution

It is *incorrect* to solve this problem using the $y$ and $z$ coordinates shown with the flexure formulas developed so far in the text. The solution can be obtained, however, using the *principal axes* for the cross section. These are determined in Example 6-15, where it is found that the axes must be rotated counterclockwise through an angle $\theta_1 = 14.34°$ to locate such axes. For these principal axes, $I_{max} = I_{z'} = 23.95 \times 10^6$ mm$^4$ and $I_{min} = I_{y'} = 2.53 \times 10^6$ mm$^4$. For these axes,

$$M_{z'} = +M \cos \theta_1 = 10 \times 10^6 \cos 14.34° = 9.689 \times 10^6 \text{ N·mm}$$
$$M_{y'} = +M \sin \theta_1 = 10 \times 10^6 \sin 14.34° = 2.475 \times 10^6 \text{ N·mm}$$

The highest stressed points on the cross section lie at points farthest from the neutral axis. To locate this axis, the angle $\beta$ is given by Eq. 6-43. Hence, using the $y'$ and $z'$ coordinates,

$$\tan \beta' = \frac{I_{z'}}{I_{y'}} \tan \theta_1 = \frac{23.95 \times 10^6}{2.53 \times 10^6} \tan 14.34° = 2.42$$

and $\beta' = 67.5°$. Since this angle is measured from the $z'$ axis, it forms an angle of $67.5° - 14.3° = 53.2°$ with the $z$ axis. Note the large inclination of the neutral axis with respect to the $z$ axis, which is much larger than $\theta_1$.

Having established the neutral axis, by inspection of the sketch, it can be seen that the highest stressed point in compression is at $B$, whereas that in tension is at $F$. By locating these points in the $y'z'$ coordinate system of the *principal axes* and applying Eq. 6-41, the required stresses are found.

$$y'_B = z_B \sin \theta_1 + y_B \cos \theta_1 = +4.3 \sin \theta_1 + 125.7 \cos \theta_1 = 122.9 \text{ mm}$$
$$z'_B = z_B \cos \theta_1 - y_B \sin \theta_1 = +4.3 \cos \theta_1 - 125.7 \sin \theta_1 = -26.95 \text{ mm}$$

$$\sigma_B = -\frac{M_{z'}y_B'}{I_{z'}} + \frac{M_{y'}z_B'}{I_{y'}}$$

$$= -\frac{9.689 \times 10^6 \times 122.9}{23.95 \times 10^6} + \frac{2.475 \times 10^6 \times (-26.95)}{2.53 \times 10^6}$$

$$= -76.1 \text{ MPa}$$

Similarly,

$$y_F' = z_F \sin\theta_1 + y_F \cos\theta_1 = 24.3\sin\theta_1 - 74.3\cos\theta_1 = -65.97 \text{ mm}$$
$$z_F' = z_F \cos\theta_1 - y_F \sin\theta_1 = 24.3\cos\theta_1 + 74.3\sin\theta_1 = +41.93 \text{ mm}$$

$$\sigma_F = -\frac{M_{z'}y_F'}{I_{z'}} + \frac{M_{y'}z_F'}{I_{y'}}$$

$$= -\frac{9.689 \times 10^6 \times (-65.97)}{23.95 \times 10^6} + \frac{2.475 \times 10^6 \times 41.93}{2.53 \times 10^6}$$

$$= +67.7 \text{ MPa}$$

---

When unsymmetrical bending of a beam is caused by applied transverse forces, another procedure equivalent to that just given *is often more convenient*. The applied forces are first resolved into components that act parallel to the principal axes of the cross-sectional area. Then the bending moments caused by these components around the respective axes are computed for use in the flexure formula. In Example 6-15, such components of the applied load are shown in Fig. 6-36(g). To avoid *torsional stresses*, the applied transverse forces must act through the *shear center*, a concept discussed in the next chapter. For bilaterally symmetrical sections, e.g., a rectangle, a circle, an I beam, etc., *the shear center coincides with the geometric center (centroid) of the cross section*. For other cross sections, such as a channel, the shear center lies elsewhere, as at $S$ shown in Fig. 6-38, and it is at this point that the transverse force must be applied to prevent occurrence of torsional stresses. Single angles acting as beams must be treated similarly (see Fig. 7-24). For analysis of unsymmetrical bending, the applied forces must be resolved *at* the shear center parallel to the principal axes of the cross section.

## 6-12. Elastic Bending with Axial Loads

A solution for pure bending around both principal axes of a member can be extended to include the effect of axial loads by employing *superposition*. Such an approach is applicable only in the range of *elastic* behavior of members. Further, if an applied axial force causes compression, a member must be stocky, lest a buckling problem of the type considered in Chapter 11 arises. With these reservations, Eq. 6-41 can be generalized to read

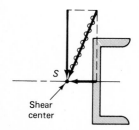

**Fig. 6-38** Lateral force through shear center $S$ causes no torsion.

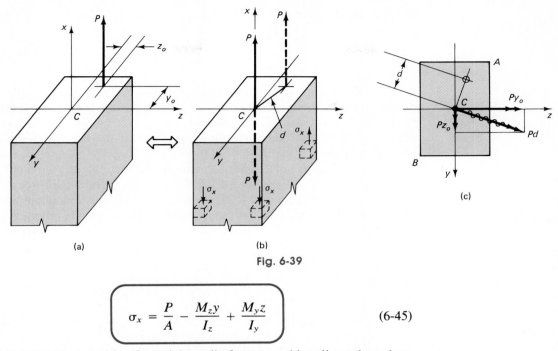

Fig. 6-39

$$\sigma_x = \frac{P}{A} - \frac{M_z y}{I_z} + \frac{M_y z}{I_y} \qquad (6\text{-}45)$$

where $P$ is taken positive for axial tensile forces, and bending takes place around the two *principal y* and $z$ axes.

For the particular case of an eccentrically applied axial force, consider the case shown in Fig. 6-39(a). By applying two equal but opposite forces $P$ at centroid $C$, as shown in Fig. 6-39(b), an *equivalent* problem is obtained. In this formulation, the applied axial force $P$ acting at $C$ gives rise to the term $P/A$ in Eq. 6-45; whereas a couple $Pd$ developed by the opposed forces $P$ a distance $d$ apart causes unsymmetrical bending. The moment $Pd$ applied by this couple can be resolved into two components along the principal axes, as shown in Fig. 6-39(c). These components are $M_y = Pz_o$ and $M_z = Py_o$. Since the sense of these moments coincides with the positive directions of the $y$ and $z$ axes, these moments in Eq. 6-45 are positive.

Provided the principal axes are used, Eq. 6-45 can be applied to members of any cross section. In some instances, however, it may be more advantageous to use an arbitrary set of orthogonal axes and to determine the bending stresses using Eq. 6-64 given in Section 6-14. To complete a solution, the normal stress caused by axial force must be superposed.

It is instructive to note that in calculus, the equation of a plane is given as

$$Ax + By + Cz + D = 0$$

where $A, B, C$, and $D$ are constants. By setting $A = 1$, $x = \sigma_x$, $B = M_z/I_z$, $C = -M_y/I_y$, and $D = -P/A$, it can be recognized that Eq. 6-45 defines a plane. Similarly, since $\varepsilon = \sigma/E$, Eq. 6-45 can be recast in terms of strain to read

$$\varepsilon_x = x = -(by + cz + d) \tag{6-47}$$

where $a = 1$, and $b$, $c$, and $d$ are constants. Since this equation also defines a plane, the basic strain assumption of the simplified engineering theory of flexure is verified. However, because of the presence of axial strain due to $P$, the "plane sections" not only rotate, but also translate an amount $P/AE$.

Based on the above discussion, it can be concluded that the longitudinal strain magnitudes in members subjected to bending and axial forces can be represented by distances from a reference plane to an inclined plane. The same is true for *elastic* stresses. These inclined planes intersect the reference plane in a line. This line of *zero stress* or *strain* is analogous to the neutral axis occurring in pure bending. Unlike the former case, however, when $P \neq 0$, this line does not pass through the centroid of a section. For large axial forces and small bending moments, the line of zero stress or strain may lie outside a cross section. The significance of this line is that the normal stresses or strains vary from it linearly.

It should be noted that in many instances, the bending moment in a member is caused by transverse forces rather than by an eccentrically applied axial force such as illustrated in Fig. 6-39. In such cases, Eq. 6-45 remains applicable.

Several illustrative examples follow, beginning with situations where bending takes place only around one of the principal axes.

### EXAMPLE 6-19

A 50 by 75 mm, 1.5 m long elastic bar of negligible weight is loaded as shown in mm in Fig. 6-40(a). Determine the maximum tensile and compressive stresses acting normal to the section through the beam.

### Solution

To emphasize the method of superposition, this problem is solved by dividing it into two parts. In Fig. 6-40(b), the bar is shown subjected only to the axial force, and in Fig. 6-40(c) the same bar is shown subjected only to the transverse force. For the axial force, the normal stress throughout the length of the bar is

$$\sigma = \frac{P}{A} = \frac{25 \times 10^3}{50 \times 75} = 6.67 \text{ MPa} \qquad \text{(tension)}$$

This result is indicated in Fig. 6-40(d). The normal stresses due to the transverse force depend on the magnitude of the bending moment, and the maximum bending moment occurs at the applied force. As the left reaction is 2.7 kN, $M_{\text{max}} = 2.7$

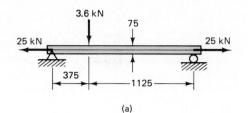

(a)

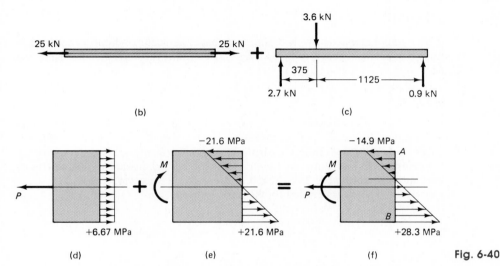

(b)　　　　　　　　　　　　　(c)

(d)　　　　　　　(e)　　　　　　　(f)　　　　**Fig. 6-40**

$\times\ 10^3 \times 375 = 1.013 \times 10^6$ N·mm. From the flexure formula, the maximum stresses at the extreme fibers caused by this moment are

$$\sigma = \frac{Mc}{I} = \frac{6M}{bh^2} = \frac{6 \times 1.013 \times 10^6}{50 \times 75^2} = \pm 21.6 \text{ MPa}$$

These stresses act normal to the section of the beam and decrease linearly toward the neutral axis as in Fig. 6-40(e). Then, to obtain the compound stress for any particular element, bending stresses must be added algebraically to the direct tensile stress. Thus, as may be seen from Fig. 6-40(f), at point $A$, the resultant normal stress is 14.9 MPa compression, and at $B$, it is 28.3 MPa tension. Side views of the stress vectors as commonly drawn are shown in the figure.

Although in this problem, the given axial force is larger than the transverse force, bending causes higher stresses. However, the reader is cautioned not to regard slender compression members in the same light.

Note that in the final result, the line of zero stress, which is located at the centroid of the section for flexure, moves upward. Also note that the local stresses, caused by the concentrated force, which act normal to the top surface of the beam, were not considered. Generally, these stresses are treated independently as local bearing stresses.

The stress distribution shown in Fig. 6-40(f) would change if instead of the axial tensile forces applied at the ends, compressive forces of the same magnitude were acting on the member. The maximum tensile stress would be reduced to 14.9 MPa from 28.3 MPa, which would be desirable in a beam made of a material weak in tension and carrying a transverse load. This idea is utilized in prestressed construction. Tendons made of high-strength steel rods or cable passing through a beam with anchorages at the ends are used to precompress concrete beams. Such artificially applied forces inhibit the development of tensile stresses. Prestressing also has been used in racing-car frames.

### **EXAMPLE 6-18

A 50 by 50 mm elastic bar bent into a U shape, as in Fig. 6-41(a), is acted upon by two opposing forces $P$ of 8.33 kN each. Determine the maximum normal stress occurring at section $A-B$.

### Solution

The section to be investigated is in the curved region of the bar, but this makes no essential difference in the procedure. First, a segment of the bar is taken as a free-body, as shown in Fig. 6-41(b). At section $A-B$, the axial force, applied at the centroid of the section, and the bending moment necessary to maintain equilibrium are determined. Then, each element of the force system is considered separately. The stress caused by the axial forces is

$$\sigma = \frac{P}{A} = \frac{8.33 \times 10^3}{50 \times 50} = 3.33 \text{ MPa} \qquad \text{(compression)}$$

and is shown in the first diagram of Fig. 6-41(c). The normal stresses caused by the bending moment may be obtained by using Eq. 6-31. However, for this bar, bent to a 75-mm radius, the solution is already known from Example 6-10. The stress distribution corresponding to this case is shown in the second diagram of Fig. 6-41(c). By superposing the results of these two solutions, the compound stress distribution is obtained. This is shown in the third diagram of Fig. 6-41(c). The maximum compressive stress occurs at $A$ and is 131 MPa. An isolated element for point $A$ is shown in Fig. 6-41(d). Shear stresses are absent at section $A-B$ as

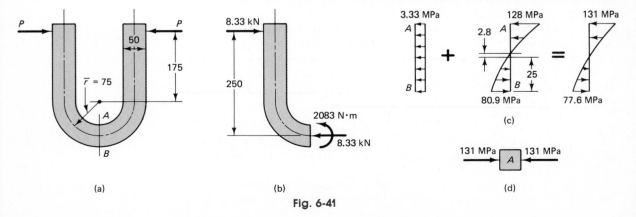

(a)                                    (b)                                    (c)

(d)

**Fig. 6-41**

no shear force is necessary to maintain equilibrium of the segment shown in Fig. 6-41(b). The relative insignificance of the stress caused by the axial force is striking.

Problems similar to the above commonly occur in machine design. Hooks, C clamps, frames of punch presses, etc. illustrate the variety of situations to which the foregoing methods of analysis must be applied.

### EXAMPLE 6-19

Consider a tapered block having a rectangular cross section at the base, as shown in Figs. 6-42(a) and (b). Determine the maximum eccentricity $e$ such that the stress at $B$ caused by the applied force $P$ is zero.

### Solution

In order to maintain applied force $P$ in equilibrium, there must be an axial compressive force $P$ and a moment $Pe$ at the base having the senses shown. The stress caused by the axial force is $\sigma = -P/A = -P/bh$, whereas the largest tensile stress caused by bending is $\sigma_{max} = Mc/I = M/S = 6Pe/bh^2$, where $bh^2/6$ is the elastic section modulus of the rectangular cross section. To satisfy the condition for having stress at $B$ equal to zero, it follows that

$$\sigma_B = -\frac{P}{bh} + \frac{6Pe}{bh^2} = 0 \quad \text{or} \quad e = \frac{h}{e}$$

which means that if force $P$ is applied at a distance of $h/6$ from the centroidal axis of the cross section, the stress at $B$ is just zero. Stress distributions across the base corresponding, respectively, to the axial force and bending moment are shown in Figs. 6-42(c) and (d), and their algebraic sum in Fig. 6-42(e).

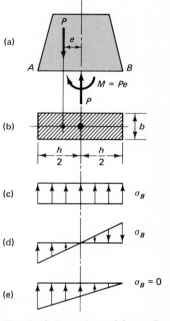

**Fig. 6-42** Location of force $P$ causing zero stress at B.

In the above problem, if force $P$ were applied closer to the centroid of the section, a smaller bending moment would be developed at section $A-B$, and there would be some compression stress at $B$. The same argument may be repeated for the force acting to the right of the centroidal axis. Hence, a practical rule, much used by the early designers of masonry structures, may be formulated thus: _if the resultant of all vertical forces acts within the middle third of the rectangular cross section, there is no tension in the material at that section_. It is understood that the resultant acts in a vertical plane containing one of the axes of symmetry of the rectangular cross-sectional area.

The foregoing discussion may be generalized in order to apply to any planar system of forces acting on a member. The resultant of these forces

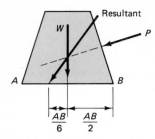

**Fig. 6-43** Resultant causing no tension at *B*.

may be made to intersect the plane of the cross section, as is shown in Fig. 6-43. At the point of intersection of this resultant with the section, it may be resolved into horizontal and vertical components. If the vertical component of the resultant fulfills the conditions of the former problem, no tension will be developed at point *B*, as the horizontal component causes only shear stresses. Hence, a more general "middle-third" rule may be stated thus: there will be no tension at a section of a member of a *rectangular* cross section if the resultant of the forces above this section *intersects* one of the axes of symmetry of the section within the middle third.

## EXAMPLE 6-20

Find the stress distribution at section *ABCD* for the block shown in mm in Fig. 6-44(a) if $P = 64$ kN. At the same section, locate the line of zero stress. Neglect the weight of the block.

### Solution

In this problem, it is somewhat simpler to recast Eq. 6-45 with the aid of Eq. 6-22, defining the elastic section modulus $S = I/c$ as $bh^2/6$. The normal stress at

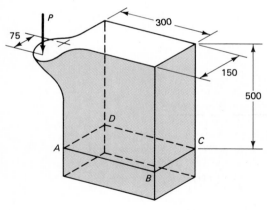

(a)

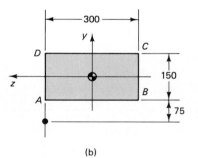

(b)

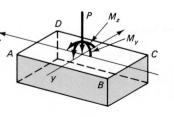

(c)

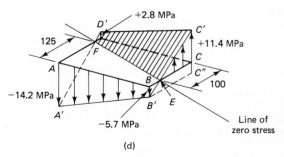

(d)

**Fig. 6-44**

any $I$th corner of the block can be found directly from such a transformed equation. This equation reads

$$\sigma_I = \frac{P}{A} - \frac{M_z}{S_z} + \frac{M_y}{S_y} \qquad (6\text{-}48)$$

where $S_z = bh^2/6$, and $S_y = hb^2/6$.

The forces acting on section $ABCD$, Fig. 6-44(c), are $P = -64 \times 10^3$ N, $M_y = -64 \times 10^3 \times 150 = -9.6 \times 10^6$ N·mm, and $M_z = -64 \times 10^3 \times (75 + 75) = -9.6 \times 10^6$ N·mm. The cross-sectional area has the following properties: $A = 150 \times 300 = 45 \times 10^3$ mm$^2$, $S_z = 300 \times 150^2/6 = 1.125 \times 10^6$ mm$^3$, and $S_y = 150 \times 300^2/6 = 2.25 \times 10^6$ mm$^3$.

The normal stresses at the corners are found using Eq. 6-48, assigning signs for the stresses caused by moments by inspection. For example, from Fig. 6-44(c), it can be seen that due to $M_y$, the stresses at corners $A$ and $D$ are compressive. Other cases are treated similarly. Using this approach,

$$\sigma_A = -\frac{64 \times 10^3}{45 \times 10^3} - \frac{9.6 \times 10^6}{1.125 \times 10^6} - \frac{9.6 \times 10^6}{2.25 \times 10^6}$$
$$= -1.42 - 8.53 - 4.27 = -14.2 \text{ MPa}$$
$$\sigma_B = -1.42 - 8.53 + 4.27 = -5.7 \text{ MPa}$$
$$\sigma_C = -1.42 + 8.53 + 4.27 = +11.4 \text{ MPa}$$
$$\sigma_D = -1.42 + 8.53 - 4.27 = +2.8 \text{ MPa}$$

These stresses are shown in Fig. 6-44(d). The ends of these four stress vectors at $A'$, $B'$, $C'$, and $D'$ lie in the plane $A'B'C'D'$. The vertical distance between planes $ABCD$ and $A'B'C'D'$ defines the total stress at any point on the cross section. The intersection of plane $A'B'C'D'$ with plane $ABCD$ locates the line of zero stress $FE$.

By drawing a line $B'C''$ parallel to $BC$, similar triangles $C'B'C''$ and $C'EC$ are obtained; thus, the distance $CE = [11.4/(11.4 + 5.7)]150 = 100$ mm. Similarly, distance $AF$ is found to be 125 mm. Points $E$ and $F$ locate the line of zero stress. If the weight of the block is neglected, the stress distribution on any other section parallel to $ABCD$ is the same.

## EXAMPLE 6-21

Find the zone over which the vertical downward force $P_o$ may be applied to the rectangular weightless block shown in Fig. 6-45(a) without causing any tensile stresses at the section $A$–$B$.

### Solution

The force $P = -P_o$ is placed at an arbitrary point in the first quadrant of the $yz$ coordinate system shown. Then the same reasoning used in the preceding example shows that with this position of the force, the greatest tendency for a tensile stress exists at $A$. With $P = -P_o$, $M_z = +P_o y$, and $M_y = -P_o z$, setting the stress at

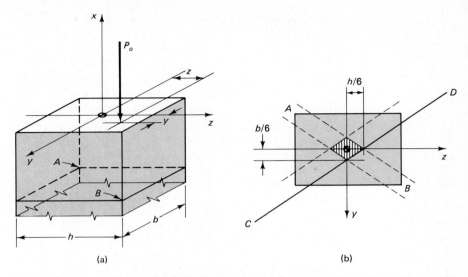

**Fig. 6-45**

(a)                                                      (b)

A equal to zero fulfills the limiting condition of the problem. Using Eq. 6-45 allows the stress at A to be expressed as

$$\sigma_A = 0 = \frac{-P_o}{A} - \frac{(P_o y)(-b/2)}{I_{zz}} + \frac{(-P_o z)(-h/2)}{I_{yy}}$$

or

$$-\frac{P_o}{A} + \frac{P_o y}{b^2 h/6} + \frac{P_o z}{b h^2/6} = 0$$

Simplifying,

$$\frac{z}{h/6} + \frac{y}{b/6} = 1$$

which is an equation of a straight line. It shows that when $z = 0$, $y = b/6$; and when $y = 0$, $z = h/6$. Hence, this line may be represented by line *CD* in Fig. 6-45(b). A vertical force may be applied to the block anywhere on this line and the stress at A will be zero. Similar lines may be established for the other three corners of the section; these are shown in Fig. 6-45(b). If force *P* is applied on any one of these lines or on any line parallel to such a line toward the centroid of the section, there will be no tensile stress at the corresponding corner. Hence, force *P* may be applied anywhere within the ruled area in Fig. 6-45(b) without causing tensile stress at any of the four corners or anywhere else. This zone of the cross-sectional area is called the *kern* of a section. By limiting the possible location of the force to the lines of symmetry of the rectangular cross section, the results found in this example verify the "middle-third" rule discussed in Example 6-19.

## EXAMPLE 6-22

Consider a "weightless" rigid block resting on a linearly elastic foundation not capable of transmitting any tensile stresses, as shown in Fig. 6-46(a). Determine the stress distribution in the foundation when applied force *P* is so placed that a part of the block lifts off.

## Solution

Assume that only a portion $AB$ of the foundation of length $x$ and width $b$ is effective in resisting applied force $P$. This corresponds to the colored area in Fig. 6-46(c). The stress along line $B–B$ is zero by definition. Hence, the following equation for the stress at $B$ may be written.

$$\sigma_B = -\frac{P}{xb} + P\left(\frac{x}{2} - k\right)\frac{6}{bx^2} = 0$$

where $x/2 - k$ is the eccentricity of the applied force with respect to the centroidal axis of the shaded contact area, and $bx^2/6$ is its section modulus. By solving for $x$, it is found that $x = 3k$ and the pressure distribution will be "triangular," as shown in Fig. 6-46(b) (why?). As $k$ decreases, the intensity of pressure on line $A–A$ increases; when $k$ is zero, the block becomes unstable.

Problems such as this arise, for example, in the design of foundations for chimneys, as no tensile stresses can develop at the contact surface of a concrete pad with soil. Similar problems arise in foundations for heavy machinery. Similar reasoning can be applied where a number of forces are acting on a member and the contact area is of any shape.

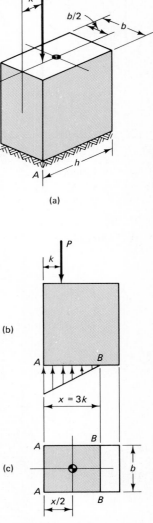

(a)

(b)

(c)

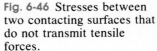

Fig. 6-46 Stresses between two contacting surfaces that do not transmit tensile forces.

## *6-13. Inelastic Bending with Axial Loads

In Section 6-10, it is pointed out that the basic kinematic assumption that plane sections through a beam taken normal to its axis remain plane after a beam is bent remains valid even if the material behaves inelastically. Similarly, plane sections perpendicular to a beam axis move along it parallel to themselves when an inelastic member is loaded axially. For small deformations, the *normal strains* corresponding to these actions can be *superposed*. As a result of such superposition, a plane defined by Eq. 6-47 can be formulated. Such general analysis of inelastic beams is rather cumbersome and is not considered in this text.[25] Here attention is confined to a planar case.

The superposition of strains for a planar member simultaneously subjected to an axial force $P$ and a bending moment $M$ is shown schematically in Fig. 6-47. For clarity, the strains are greatly exaggerated. Superposition of strains due to $P$ and $M$ moves a plane section axially and rotates it as shown. If axial force $P$ causes strain larger than any strain of opposite sign that is caused by $M$, the combined strains will not change their sign within a section.

By supplementing these basic kinematic assumptions with the stress-strain relations and conditions of equilibrium, one can solve either elastic or inelastic problems. It is important to note, however, that *superposition*

[25] M. S. Aghbabian and E. P. Popov, "Unsymmetrical Bending of Rectangular Beams Beyond the Elastic Limit," *Proceedings, First U.S. National Congress of Applied Mechanics* (Michigan: Edwards Bros., 1951), 579–584.

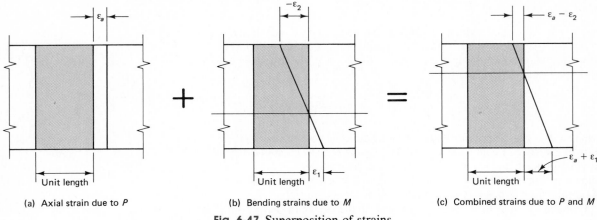

(a) Axial strain due to $P$      (b) Bending strains due to $M$      (c) Combined strains due to $P$ and $M$

**Fig. 6-47** Superposition of strains.

*of stresses is applicable only in elastic problems where deformations are small.*

An example follows illustrating an elastic as well as an inelastic solution for a member simultaneously subjected to bending and axial forces.

### EXAMPLE 6-23

Consider a rectangular elastic-plastic beam bent around the horizontal axis and simultaneously subjected to an axial tensile force. Determine the magnitudes of the axial forces and moments associated with the stress distributions shown in Figs. 6-48(a), (b), and (e).

### Solution

The stress distribution shown in Fig. 6-48(a) corresponds to the limiting elastic case, where the maximum stress is at the point of impending yielding. For this case, the stress-superposition approach can be used. Hence,

$$\sigma_{\max} = \sigma_{yp} = \frac{P_1}{A} + \frac{M_1 c}{I} \qquad (6\text{-}49)$$

Force $P$ at yield can be defined as $P_{yp} = A\sigma_{yp}$; from Eq. 6-21, the moment at yield is $M_{yp} = (I/c)\sigma_{yp}$. Dividing Eq. 6-49 by $\sigma_{yp}$ and substituting the relations for $P_{yp}$ and $M_{yp}$, after simplification,

$$\frac{P_1}{P_{yp}} + \frac{M_1}{M_{yp}} = 1 \qquad (6\text{-}50)$$

This establishes a relationship between $P_1$ and $M_1$ so that the maximum stress just equals $\sigma_{yp}$. A plot of this equation corresponding to the case of impending yield is represented by a straight line in Fig. 6-49. Plots of such relations are called *interaction curves* or *diagrams*.

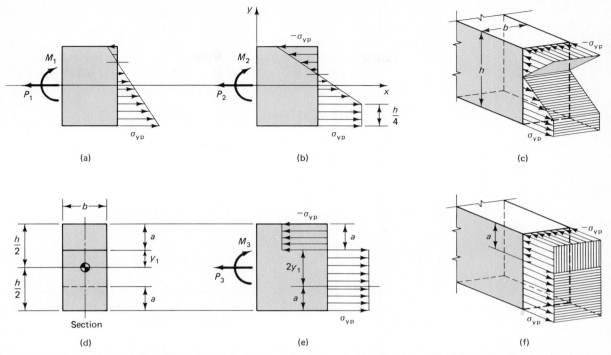

**Fig. 6-48** Combined axial and bending stresses: (a) elastic stress distribution, (b) and (c) elastic-plastic stress distribution, and (e) and (f) fully plastic stress distribution.

The stress distribution shown in Figs. 6-48(b) and (c) occurs after yielding has taken place in the lower quarter of the beam. With this stress distribution given, one can determine directly the magnitudes of $P$ and $M$ from the conditions of equilibrium. If on the other hand, $P$ and $M$ were given, since superposition does not apply, a cumbersome process would be necessary to determine the stress distribution.

For the stresses given in Figs. 6-48(b) and (c), one simply applies Eqs. 6-36 and 6-37 developed for inelastic bending of beams, except that in Eq. 6-36, the sum of the normal stresses must equal axial force $P$. Noting that in the elastic zone, the stress can be expressed algebraically as $\sigma = \sigma_{yp}/3 - \sigma_{yp}y/(3h/8)$ and that in the plastic zone $\sigma = \sigma_{yp}$, one has

$$P_2 = \int_A \sigma \, dA = \int_{-h/4}^{+h/2} \frac{\sigma_{yp}}{3}\left(1 - \frac{8y}{h}\right)b \, dy + \int_{-h/2}^{-h/4} \sigma_{yp}b \, dy = \sigma_{yp}\frac{bh}{4}$$

$$M_2 = -\int_A \sigma y \, dA = -\int_{-h/4}^{+h/2} \frac{\sigma_{yp}}{3}\left(1 - \frac{8y}{h}\right)yb \, dy - \int_{-h/2}^{-h/4} \sigma_{yp}yb \, dy$$

$$= \frac{3}{16}\sigma_{yp}bh^2$$

Note that the axial force just found exactly equals the force acting on the plastic area of the section. Moment $M_2$ is greater than $M_{yp} = \sigma_{yp}bh^2/6$ and less than $M_{ult} = M_p = \sigma_{yp}bh^2/4$; see Eq. 6-38.

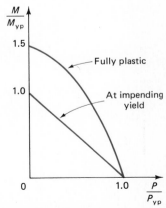

**Fig. 6-49** Interaction curves for $P$ and $M$ for a rectangular member.

The axial force and moment corresponding to the fully plastic case shown in Figs. 6-48(e) and (f) are simple to determine. As may be seen from Fig. 6-48(e), the axial force is developed by $\sigma_{yp}$ acting on the area $2y_1b$. Because of symmetry, these stresses make no contribution to the moment. Forces acting on the top and the bottom areas $ab = [(h/2) - y_1]b$, Fig. 6-48(d), form a couple with a moment arm of $h - a = h/2 + y_1$. Therefore,

$$P_3 = 2y_1 b\sigma_{yp} \qquad \text{or} \qquad y_1 = P_3/2b\sigma_{yp}$$

and

$$M_3 = ab\sigma_{yp}(h - a) = \sigma_{yp}b(h^2/4 - y_1^2) = M_p - \sigma_{yp}by_1^2$$

$$= \frac{3M_{yp}}{2} - \frac{P_3^2}{4b\sigma_{yp}}$$

Then dividing by $M_p = 3M_{yp}/2 = \sigma_{yp}bh^2/4$ and simplifying, one obtains

$$\frac{2M_3}{3M_{yp}} + \left(\frac{P_3}{P_{yp}}\right)^2 = 1 \tag{6-51}$$

This is a general equation for the interaction curve for $P$ and $M$ necessary to achieve the fully plastic condition in a rectangular member (see Fig. 6-49). Unlike the equation for the elastic case, the relation is nonlinear.

## **\*\*[26]6-14. Bending of Beams with Unsymmetric (Arbitrary) Cross Section**

A general equation for *pure* bending of *elastic* members of arbitrary cross section whose reference axes are not the principal axes can be formulated using the same approach as for the symmetrical cross sections considered earlier. Again, it is assumed that any plane section through a beam, taken normal to its axis, remains plane after the beam is subjected to bending. Then two basic requirements for equilibrium are enforced: (1) the total axial force on any cross section of a beam must be zero, and (2) the external bending moment at a section must be developed by the internal stresses acting on the cross section. Hooke's law is postulated for uniaxial normal strain.

In order to derive the required equation, consider a beam having an arbitrary cross section such as that shown in Fig. 6-50. The orientation of the $y$ and $z$ orthogonal axes is chosen arbitrarily. Let this beam be subjected to a pure bending moment **M** having the components $M_y$ and $M_z$, respectively, around the $y$ and $z$ axes; see Fig. 6-50(a).

According to the fundamental hypothesis, during bending, a *plane* section through a beam would rotate and intersect the $yz$ plane at an angle $\beta$ with the $z$ axis, as shown in the figure. A generic infinitesimal area $dA$

---

[26] This section is of an advanced character and can be omitted.

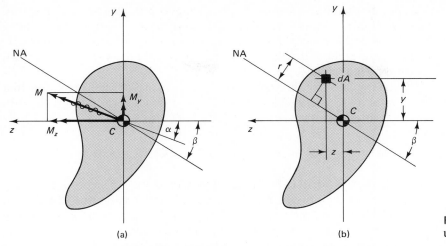

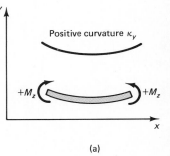

**Fig. 6-50** Bending of unsymmetric cross section.

in the positive quadrant of the $y$ and $z$ axes is located by the perpendicular distance $r$ from this line. Then analogous to Eq. 6-3, the longitudinal normal strain $\varepsilon_x$ is assumed to be

$$\varepsilon_x = -\kappa r \qquad (6\text{-}52)$$

where in the chosen coordinates,

$$r = y \cos \beta - z \sin \beta \qquad (6\text{-}53)$$

Then, by analogy to Eq. 6-4, the longitudinal *elastic* stress $\sigma_x$ acting on the cross section is

$$\sigma_x = E\varepsilon_x = -E\kappa r \qquad (6\text{-}54)$$

and using Eq. 6-53, this relation becomes

$$\sigma_x = -E\kappa y \cos \beta + E\kappa z \sin \beta \qquad (6\text{-}55)$$

where $\kappa \cos \beta$ is the projected curvature $\kappa_y$ in the $xy$ plane, as may be seen from the limiting case of setting $\beta$ equal to zero. Similarly, $\kappa \sin \beta$ is the projected curvature $\kappa_z$ in the $xz$ plane. By adopting this notation, Eq. 6-55 can be recast to read

$$\sigma_x = -E\kappa_y y + E\kappa_z z \qquad (6\text{-}56)$$

The difference in signs in the two expressions on the right side of the equations arises from the adopted sign convention and can be clarified by making reference to Fig. 6-51. Here it can be noted that a mathematically defined positive curvature, causing an increase in the slope of

**Fig. 6-51** Relationships between positive moments and curvatures in $xy$- and $xz$-planes.

a bent beam with an increase in the distance from the origin, gives rise to two different cases. In the $xy$ plane, positive curvature and positive bending moments have the same sense. The opposite is true in the $xz$ plane. Hence, the normal stresses $\sigma_x$ due to these two curvatures must be of opposite sign.

By having an analytic expression for $\sigma_x$, Eq. 6-56, the condition that the sum of all forces in the $x$ direction must equal zero, i.e., $\sum F_x = 0$, can be written as

$$\int \sigma_x \, dA = -E\kappa_y \int y \, dA + E\kappa_z \int z \, dA = 0 \qquad (6\text{-}57)$$

This equation is identically satisfied provided that the coordinate axes are taken with their origin at the *centroid* of the cross section. This result was anticipated and the arbitrary orthogonal axes in Fig. 6-50 are shown passing through the centroid $C$ of the cross section.

By imposing the conditions of moment equilibrium at a section, two moment component equations can be written requiring that the externally applied moment around either axis is balanced by the internal system of stresses. One of these equations pertains to the moments around the $z$ axis; the other, around the $y$ axis. Hence, as previously defined, if $M_z$ is the known applied moment component around the $z$ axis and $M_y$ is the known applied moment component around the $y$ axis, one has the following two equations:

$$M_z = \int -\sigma_x y \, dA = E\kappa_y \int y^2 \, dA - E\kappa_z \int yz \, dA \qquad (6\text{-}58)$$

and

$$M_y = \int +\sigma_x z \, dA = -E\kappa_y \int yz \, dA + E\kappa_z \int z^2 \, dA \qquad (6\text{-}59)$$

where the constants are taken outside the integrals in the expressions on the right. The meaning of these integrals is discussed in Section 6-15. According to Eq. 6-66, these integrals define the moments and product of inertia for a cross sectional area as $I_z$, $I_y$, and $I_{yz}$, permitting the recasting of the last two equations as

$$EI_z\kappa_y - EI_{yz}\kappa_z = M_z \qquad (6\text{-}60)$$

and

$$-EI_{yz}\kappa_y + EI_y\kappa_z = M_y \qquad (6\text{-}61)$$

Solving these two equations simultaneously gives

$$E\kappa_y = +\frac{M_zI_y + M_yI_{yz}}{I_yI_z - I_{yz}^2} \qquad (6\text{-}62)$$

and

$$E\kappa_z = -\frac{M_yI_z + M_zI_{yz}}{I_yI_z - I_{yz}^2} \qquad (6\text{-}63)$$

By substituting these constants in Eq. 6-56, the expression for the *elastic* bending stress $\sigma_x$ for *any* beam cross section with arbitrarily directed orthogonal coordinate axes is

$$\sigma_x = -\frac{M_z I_y + M_y I_{yz}}{I_y I_z - I_{yz}^2} y + \frac{M_y I_z + M_z I_{yz}}{I_y I_z - I_{yz}^2} z \qquad (6\text{-}64)$$

This is the *generalized flexure formula*. If the principal axes for a cross section are used, where $I_{yz}$ is zero, this equation simplifies to Eq. 6-41.

By setting Eq. 6-64 equal to zero, the angle $\beta$ for locating the neutral axis in the *arbitrary* coordinate system is obtained, giving

$$\tan \beta = \frac{y}{z} = \frac{M_y I_z + M_z I_{yz}}{M_z I_y + M_y I_{yz}} \qquad (6\text{-}65)$$

For the principal axes, this equation reverts to Eq. 6-43.


## EXAMPLE 6-24

Using the general equation for elastic bending stress, verify the stresses found at points $B$ and $F$ for the angle of Example 6-16 shown in mm in Fig. 6-52. Show that these stresses are, respectively, the minimum and the maximum. The applied moment $M_z = 10$ kN·m.

Solution

In Example 6-25, it is found that $I_z = 22.64 \times 10^6$ mm$^4$, $I_y = 3.84 \times 10^6$ mm$^4$, and $I_{yz} = 5.14 \times 10^6$ mm$^4$. Substituting these values and $M_z = +10$ kN·m into Eq. 6-64, and defining, respectively, the coordinates of points $B$ and $F$ as (125.7, 4.3) and (−74.3, 24.3), one has

$$\sigma_B = -\frac{10 \times 10^6 \times 3.84 \times 10^6}{3.84 \times 22.64 \times 10^{12} - 5.14^2 \times 10^{12}} \times 125.7$$
$$+ \frac{10 \times 10^6 \times 5.14 \times 10^6}{3.84 \times 22.64 \times 10^{12} - 5.14^2 \times 10^{12}} \times 4.3$$
$$= -0.6345 \times 125.7 + 0.8493 \times 4.3 = -76.1 \text{ MPa}$$

and

$$\sigma_F = -0.6345 \times (-74.3) + 0.8943 \times 24.3 = +67.8 \text{ MPa}$$

To show that these stresses are the minimum and the maximum, respectively, locate the neutral axis using Eq. 6-65, giving

$$\tan \beta = \frac{10 \times 10^6 \times 5.14 \times 10^6}{10 \times 10^6 \times 3.84 \times 10^6} = 1.34 \quad \text{or} \quad \beta = 53.3°$$

By sketching this line on the given cross section, it is evident by inspection

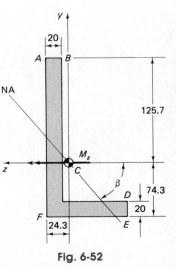

Fig. 6-52

that the farthest distances measured perpendicular to NA are associated with points $B$ and $F$. Therefore, the largest stresses occur at these points.

Some minor discrepancies among the results given in Example 6-16 and in this example are due to roundoff errors.

---

# Part C　AREA MOMENTS OF INERTIA

## *[27]6-15.　Area Moments and Products of Inertia

Moments of inertia, or second moments of area around the $z$ axis were already encountered in connection with symmetric cross sections. Here this concept is generalized for two orthogonal axes for any cross-sectional shape. With the $yz$ coordinates chosen as shown in Fig. 6-53, by definition, the moments and product of inertia of an area are given as

$$I_z = \int y^2 \, dA \qquad I_y = \int z^2 \, dA \qquad \text{and} \qquad I_{yz} = \int yz \, dA \qquad (6\text{-}66)$$

Note that these axes are chosen to pass through the centroid $C$ of the area. The use of such *centroidal* axes is essential in the solution of bending problems. It is also important to note that the product of inertia vanishes either for *doubly* or *singly symmetric* areas; see Fig. 6-54. This can be seen by referring to Fig. 6-54(b), where, due to symmetry, for each $y(+z) \, dA$, there is a $y(-z) \, dA$, and their sum vanishes.

In Section 6-4, it was shown that in calculating moments of inertia for symmetric cross sections having complex areas, it is advantageous to subdivide such areas into simple parts for which the moments of inertia are available in formulas. Then by applying the parallel-axis theorem to each part and adding, Eq. 6-18a, the moment of inertia for the whole section is obtained. By making reference to the general case shown in Fig. 6-55, it can be concluded that the previously developed formula, Eq. 6-18, for the transfer of a moment of inertia for an area from the $z_c$ to the $z$ axis remains applicable. Moreover, except for a change in notation, a similar formula applies for transferring a moment of inertia from the $y_c$ to the $y$ axis. Therefore, the following two formulas for the transfer of axes are available for the moments of inertia:

$$I_z = I_{z_c} + Ad_z^2 \qquad (6\text{-}18)$$

and

$$I_y = I_{y_c} + Ad_y^2 \qquad (6\text{-}67)$$

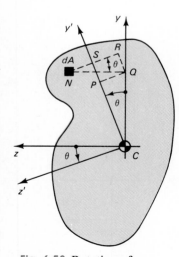

**Fig. 6-53** Rotation of coordinate axes.

[27] This is an optional part of the chapter.

where $I_{z_c}$ and $I_{y_c}$ are, respectively, moments of inertia around the $z_c$ and $y_c$ axes, $A$ is the area considered, and $d_z$ and $d_y$ are, respectively, the distances from $C$ to the axes $z$ and $y$.

By starting with the definition for the product of inertia, Eq. 6-66, and following the same procedure as before for $I_z$ and $I_y$, the transfer-of-axis formula for the product of inertia, after simplifications, becomes

$$I_{yz} = \int (y_c + d_z)(z_c + d_y)\, dA = I_{y_c z_c} + A d_y d_z \qquad (6\text{-}68)$$

where $I_{y_c z_c}$ is the product of inertia of the area $A$ around the centroidal $y_c$ and $z_c$ axes.

As noted earlier, the respective expressions given by Eqs. 6-18, 6-67, and 6-68 for all parts of a complex area should be summed to obtain $I_y$, $I_z$, and $I_{yz}$ for the whole cross section.

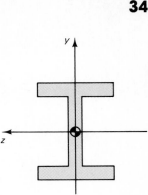

(a)

## *6-16. Principal Axes of Inertia

In the previous discussion, the $yz$ centroidal axes for an area of a general shape were chosen arbitrarily. Therefore, it is important to investigate how the moments and product of inertia change if these orthogonal axes are rotated. This is shown in Fig. 6-53, where the axes are rotated through an angle $\theta$, forming a new set of $y'z'$ coordinates. Generally, the moments and product of inertia corresponding to these axes are different from the values of $I_y$, $I_z$, and $I_{yz}$. In order to transform these quantities from one set of coordinates to another, one notes that

$$y' = CP + PS = y \cos \theta + z \sin \theta$$
$$z' = NR - RS = z \cos \theta - y \sin \theta$$

Then, based on the definitions for moments and product of inertia given in Eqs. 6-66,

$$
\begin{aligned}
I_{z'} &= \int (y')^2\, dA = \int (y \cos \theta + z \sin \theta)^2\, dA \\
&= \cos^2 \theta \int y^2\, dA + \sin^2 \theta \int z^2\, dA + 2 \sin \theta \cos \theta \int yz\, d\theta \\
&= I_z \cos^2 \theta + I_y \sin^2 \theta + 2 I_{yz} \sin \theta \cos \theta \\
&= I_z \frac{1 + \cos 2\theta}{2} + I_y \frac{1 - \cos 2\theta}{2} + I_{yz} \sin 2\theta
\end{aligned}
$$

Hence, on using trigonometric identities,

$$I_{z'} = \frac{I_z + I_y}{2} + \frac{I_z - I_y}{2} \cos 2\theta + I_{yz} \sin 2\theta \qquad (6\text{-}69)$$

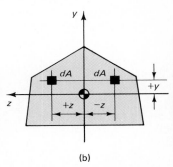

(b)

Fig. 6-54 (a) Doubly and (b) singly symmetric cross sections.

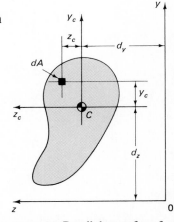

Fig. 6-55 Parallel transfer of axes.

Similarly,     $I_{y'} = \dfrac{I_z + I_y}{2} - \dfrac{I_z - I_y}{2} \cos 2\theta - I_{yz} \sin 2\theta$     (6-70)

and     $I_{y'z'} = -\dfrac{I_z - I_y}{2} \sin 2\theta + I_{yz} \cos 2\theta$     (6-71)

These equations relate the moments and the product of inertia of an area (second moments) in the new $y'z'$ coordinates to the initial ones in the $yz$ coordinates through the angle $\theta$. Note that $I_{y'} + I_{z'} = I_y + I_z$, i.e., the sum of the moments of inertia around two mutually perpendicular axes remains the same, i.e., *invariant*, regardless of the angle $\theta$. As noted earlier, the product of inertia $I_{yz}$ vanishes for doubly and singly symmetric sections.

A maximum or a minimum value of $I_{z'}$ or $I_{y'}$ can be found by differentiating either Eq. 6-69 or 6-70 with respect to $\theta$ and setting the derivative equal to zero, i.e.,

$$\frac{dI_{z'}}{d\theta} = -(I_z - I_y)\sin 2\theta + 2I_{yz}\cos 2\theta = 0$$

Hence,     $\tan 2\theta_1 = \dfrac{2I_{yz}}{I_z - I_y}$     (6-72)

This equation gives two roots within 360° that are 180° apart. Since this is for a double angle $2\theta_1$, the roots for $\theta_1$ are 90° apart. One of these roots locates an axis around which the moment of inertia is a maximum; the other locates the conjugate axis for the minimum moment of inertia. These two centroidal axes are known as the *principal axes of inertia*. As can be noted from Eq. 6-71, the same angles define the axes for which the product of inertia is zero. This means that *the product of inertia for the principal axes is zero*.

By defining sines and cosines in terms of the double angle roots of Eq. 6-72 (see Fig. 8-5), substituting these into Eq. 6-69, or Eq. 6-70, and simplifying, expressions for the *principal moments of inertia* are found:

$$I_{\substack{\max \\ \min}} = I_1 \text{ or } I_2 = \frac{I_z + I_y}{2} \pm \sqrt{\left(\frac{I_z - I_y}{2}\right)^2 + I_{yz}^2}$$     (6-73)

where, by definition, $I_1 = I_{\max}$, and $I_2 = I_{\min}$. The axes for which these maximum and minimum moments of inertia apply are defined by Eq. 6-72. By directly substituting one of the roots of this equation into Eq. 6-69, one can determine whether the selected root gives a maximum or a minimum value of the moment of inertia.

## EXAMPLE 6-15

For an angle having the cross section shown in mm in Fig. 6-56, find the principal axes and the principal moments of inertia.

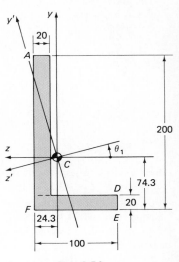

Fig. 6-56

### Solution

It can be verified by the procedure discussed earlier that the centroid of the area lies 74.3 mm from the bottom and 24.3 mm from the left side. The moments and product of inertia about the $y$ and $z$ axes can be calculated by dividing the angle into two rectangles and using the transfer-of-axes Eqs. 6-18, 6-67, and 6-68. Due to the symmetry of the two rectangles into which the angle is divided, there are no product of inertia terms for these parts around their own centroidal axes. For rectangles around their centroidal axes, $I = bh^3/12$, Eq. 6-19.

$$I_z = 20 \times 180^3/12 + 20 \times 180 \times (125.7 - 90)^2$$
$$+ 100 \times 20^3/12 + 100 \times 20 \times (-74.3 + 10)^2 = 22.64 \times 10^6 \, \text{mm}^4$$
$$I_y = 180 \times 20^3/12 + 180 \times 20 \times (24.3 - 10)^2$$
$$+ 20 \times 100^3/12 + 20 \times 100 \times (-50 + 24.3)^2 = 3.84 \times 10^6 \, \text{mm}^4$$
$$I_{yz} = 0 + 20 \times 180 \times (125.7 - 90)(24.3 - 10)$$
$$+ 0 + 100 \times 20(-74.3 + 10)(-50 + 24.3) = 5.14 \times 10^6 \, \text{mm}^4$$

By substituting these values into Eq. 6-73,

$$I_{max} = I_1 = 23.95 \times 10^6 \, \text{mm}^4 \text{ and } I_{min} = I_2 = 2.53 \times 10^6 \, \text{mm}^4$$

From Eq. 6-72,

$$\tan 2\theta_1 = \frac{2 \times 5.14 \times 10^6}{(22.64 - 3.84) \times 10^6} = 0.547 \qquad \text{hence, } \theta_1 = 14.34°$$

From inspection of Fig. 6-56, this angle is seen to define an axis for the maximum moment of inertia. A substitution of this value of $\theta_1$ into Eq. 6-69 can confirm this conclusion. In this case, $I_{max}$ is associated with the $z'$ axis at $\theta_1 = 14.34°$, i.e., $I_{max} = I_{z'}$; conversely, $I_{min} = I_{y'}$.

---

# Problems

## Sections 6-3 through 6-5

**6-1 through 6-4.** Determine bending moment capacities around the horizontal axes for the cross-sectional areas with the dimensions shown in the figures. The allowable elastic stress is either 165 MPa or 24 ksi. For properties of W steel shapes, channels, and angles, see Tables 4, 5, and 7, respectively, in the Appendix.

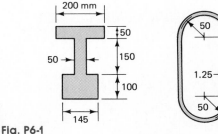

Fig. P6-1

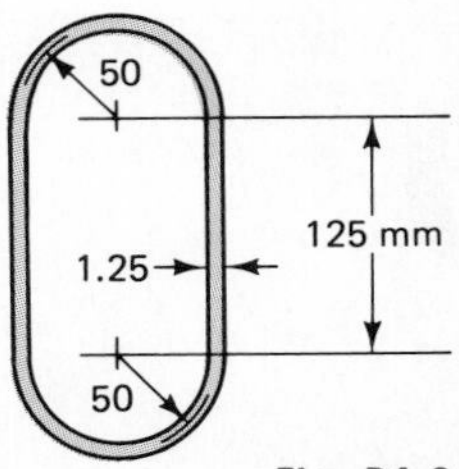

Fig. P6-2

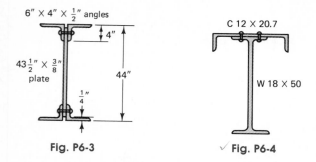

**Fig. P6-3**                    ✓ **Fig. P6-4**

**6-5.** Verify the section moduli given in the Appendix tables for S 12 × 40.8, W 10 × 112, and C 12 × 20.7.

**6-6.** If the applied moment is 40 k-ft, and the allowable elastic stress is 24 ksi, (a) what W section should be used for bending around the horizontal axis, and (b) around the vertical axis?

✓ **6-7.** A W 16 × 100 steel beam is supported at $A$ and $B$ as shown in the figure. What is the magnitude of the uniformly distributed load if a strain gage attached to the top of the upper flange measures 0.0002 in/in when the load is applied? $E = 29 \times 10^3$ ksi.

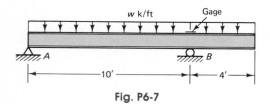

**Fig. P6-7**

**6-8.** A small steel T beam is used in an inverted position to span 400 mm. If, due to the application of the three forces shown in the figure, the longitudinal gage at $A$ registers a compressive strain of $50 \times 10^{-3}$, how large are the applied forces? $E = 200$ GPa.

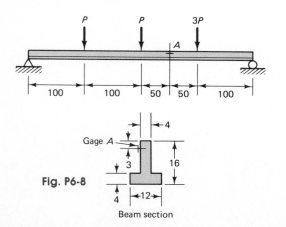

**Fig. P6-8**

**6-9 and 6-10.** Determine *elastic* positive bending-moment capacities around the horizontal axes for beams having the cross sections shown in the figures. The maximum elastic stress in tension for Prob. 6-9 is 10 ksi, and in compression, 15 ksi; the corresponding stresses for Prob. 6-10 are 100 MPa and 150 MPa.

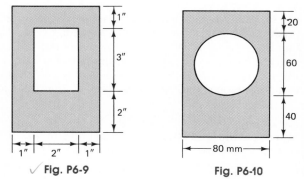

✓ **Fig. P6-9**                    **Fig. P6-10**

**6-11.** A beam having a solid rectangular cross section with the dimensions shown in the figure is subjected to a positive bending moment of 16 000 N·m acting around the horizontal axis. (a) Find the compressive force acting on the shaded area of the cross section developed by the bending stresses. (b) Find the tensile force acting on the cross-hatched area of the cross section.

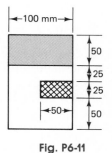

**Fig. P6-11**

**6-12.** Consider a linearly elastic beam subjected to a bending moment $M$ around its principal axis $z$ for which the moment of inertia of the cross-sectional area is $I$. Show that for such a beam, the normal force $F$ acting on any part of the cross-sectional area $A_1$ is

$$F = MQ/I$$

where

$$Q = \int_{A_1} y \, dA = \bar{y} A_1$$

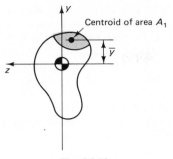

Fig. P6-12

and $\overline{y}$ is the distance from the neutral axis of the cross section to the centroid of the area $A_1$, as shown in the figure.

**6-13.** Determine the magnitude and position of the total tensile force $T$ acting on this section when a positive moment of 100 kN·m is applied. Since the magnitude of this tensile force $T$ equals the compressive force $C$ acting on the section, verify that the $T–C$ couple is equal to the applied moment.

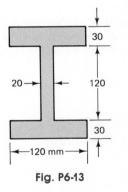

Fig. P6-13

**6-14.** Two 2 × 6 in full-sized wooden planks are glued together to form a T section, as shown in the figure.

Fig. P6-14

If a positive bending moment of 2270 ft-lb is applied to such a beam acting around a horizontal axis, (a) find the stresses at the extreme fibers, (b) calculate the total compressive force developed by the normal stresses above the neutral axis because of the bending of the beam, and (c) find the total force due to the tensile bending stresses at a section and compare it with the result found in (b).

**\*6-15.** By integration, determine the force developed by the bending stresses and its position acting on the shaded area of the cross section of the beam shown in the figure if the beam is subjected to a negative bending moment of 3500 N·m acting around the horizontal axis.

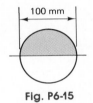

Fig. P6-15

**\*6-16.** A beam has the cross section of an isosceles triangle, as shown in the figure, and is subjected to a negative bending moment of 4000 N·m around the horizontal axis. (a) Show by integration that $I_o = bh^3/36$. (b) Determine the location and magnitude of the resultant tensile and compressive forces acting on a section if $b = h = 150$ mm.

Fig. P6-16

**6-17.** For a linearly elastic material, at the same maximum stress for a square member in the two different positions shown in the figure, determine the ratio of the bending moments. Bending takes place around the horizontal axis.

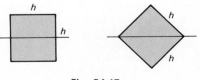

Fig. P6-17

**\*\*6-18.** Show that the elastic stress in a rectangular beam bent around its diagonal can be reduced by removing the small triangular areas, as shown in the figure. This is referred to as the Emerson paradox.[†] (*Hint:* Let the sides of the removed triangular areas be $ka$, where $k$ is a constant. In calculating $I$ for the section, treat it as consisting of two rectangles, the large one having sides $(1 - k)a$, and the small one having the width $ka\sqrt{2}$.)

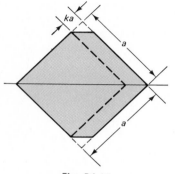

**Fig. P6-18**

**6-19.** A channel-shaped member, as shown in the figure, acts as a horizontal beam in a machine. When vertical forces are applied to this member the distance *AB* increases by 0.0010 in and the distance *CD* decreases by 0.0090 in. What is the sense of the applied moment, and what normal stresses occur in the extreme fibers? $E = 15 \times 10^6$ psi.

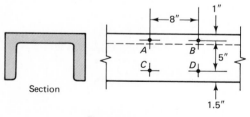

**Fig. P6-19**

[†] In 1864, in Saint-Venant's additions to Navier's book, he calls the removed fibers useless. However, he, as well as Emerson, recognized that the *elastic* failure of these fibers does not indicate that the truncated section possesses greater static strength than the complete section. However, in machine design, for members subjected to fatigue, the removal of sharp corners may be advantageous. See I. Todhunter and K. Pearson, *A History of the Theory of Elasticity and of the Strength of Materials* (New York: Dover, 1960), Vol. II, Part I, p. 109.

**6-20.** A solid steel beam having the cross-sectional dimensions partially shown in the figure was loaded in the laboratory in pure bending. Bending took place around a horizontal neutral axis. Strain measurements showed that the top fibers contracted 0.0003 m/m longitudinally; the bottom fibers elongated 0.0006 m/m longitudinally. Determine the total normal force that acted on the shaded area indicated in the figure at the time the strain measurements were made. $E = 200$ GPa. All dimensions are in mm.

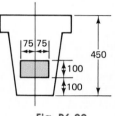

**Fig. P6-20**

**6-21.** As the screw of a large steel C clamp, such as shown in the figure, is tightened upon an object, the strain in the horizontal direction due to bending only is being measured by a strain gage at point *B*. If a strain of $900 \times 10^{-6}$ in/in is noted, what is the force on the screw corresponding to the value of the observed strain? $E = 30 \times 10^6$ psi.

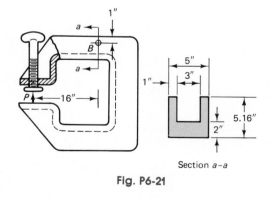

**Fig. P6-21**

**6-22.** A T beam shown in the figure is made of a material the behavior of which may be idealized as having a tensile proportional limit of 20 MPa and a compressive proportional limit of 40 MPa. With a factor of safety of $1\frac{1}{2}$ on the initiation of yielding, find the magnitude of the largest force $F$ that may be applied to

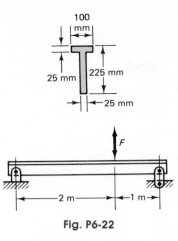

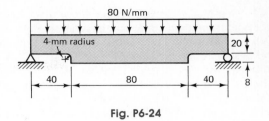

Fig. P6-24

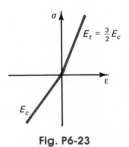

Fig. P6-22

ciently accurate. Do not consider the stress concentrations at the supports.

**6-25.** Considering the beam of a 160-mm span and the loading conditions given in the preceding problem, determine the distances from the supports such that the stresses at midspan and at the depth transition points are the same.

this beam in a downward direction as well as in an upward direction. Base answers only on the consideration of the maximum bending stresses caused by $F$.

***6-23.** A $150 \times 300$ mm rectangular section is subjected to a positive bending moment of 240 000 N·m around the "strong" axis. The material of the beam is nonisotropic and is such that the modulus of elasticity in tension is $1\frac{1}{2}$ times as great as in compression; see the figure. If the stresses do not exceed the proportional limit, find the maximum tensile and compressive stresses in the beam.

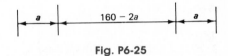

Fig. P6-25

## Section 6-7

**6-26.** Show that the elastic strain energy due to bending for a simple uniformly loaded beam of rectangular cross section is $(\sigma_{\max}^2/2E)(\frac{8}{45}AL)$, where $\sigma_{\max}$ is the maximum bending stress, $A$ is the cross-sectional area, and $L$ is the length of the beam.

**6-27.** Show that $U_{\text{bending}} = (\sigma_{\max}^2/2E)(\text{Vol}/9)$ for a cantilever of rectangular cross section supporting a concentrated load $P$ at the end.

Fig. P6-23

## Section 6-6

**6-24.** A small beam, shown in the figure, is to carry a cyclically applied load of 80 N/mm. The beam is 12-mm thick, and spans 160 mm. Determine the maximum stress at midspan and at depth transition points. Assume that the factors given in Fig. 6-15 are suffi-

## Section 6-8

**6-28.** A composite beam of two different materials has the cross section shown in Fig. 6-7(a). For the upper $50 \times 80$ mm bar, the elastic modulus $E_1 = 15$ GPa, and for the lower $50 \times 20$ mm bar, $E_2 = 40$ GPa. Find the maximum bending stresses in both materials caused by an applied positive moment of 12 kN·m acting around the $z$ axis. Do not use the method of transformed sections. (*Hint:* Use Eq. 6-16 to locate the neutral axis and the direct procedure shown in Figs. 6-7 and 6-20.)

**6-29.** Consider a composite beam whose cross section is made from three different materials bonded together, as shown in Fig. 6-20(a). Bar 1 is 40 × 20 mm and has an elastic modulus $E_1 = 15$ GPa; bar 2 is 60 × 40 mm with $E_2 = 10$ GPa; and bar 3 is 20 × 20 mm with $E_3 = 30$ GPa. Determine the maximum bending stresses in each of the three materials caused by an applied moment of 10 kN·m acting around the $z$ axis. Do not use the method of transformed sections; see the hint in the preceding problem.

**6-30 and 6-31.** Using transformed sections, determine the maximum bending stresses in each of the two materials for the composite beams shown in the figures when subjected to positive bending moments of 80 kN·m each. $E_{St} = 210$ GPa and $E_{Al} = 70$ GPa. (*Hint for Prob. 6-31:* For an ellipse with semiaxes $a$ and $b$, $I = \pi a b^3/4$ around the major centroidal axis.)

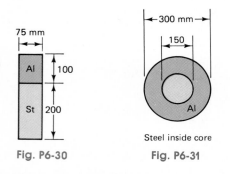

Fig. P6-30          Fig. P6-31

Steel inside core

**6-32 and 6-33.** Determine the allowable bending moment around horizontal neutral axes for the composite beams of wood and steel plates having the cross-sectional dimensions shown in the figures. Materials are fastened so that they act as a unit. $E_{St} = 30 \times 10^6$ psi and $E_w = 1.2 \times 10^6$ psi. The allowable bending stresses are $\sigma_{St} = 20$ ksi and $\sigma_w = 1.2$ ksi.

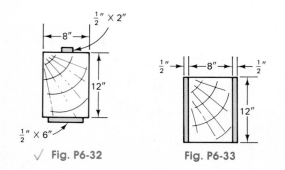

Fig. P6-32          Fig. P6-33

**6-34.** A 150-mm thick concrete slab is longitudinally reinforced with steel bars, as shown in the figure. Determine the allowable bending moment per 1-m width of this slab. Assume $n = 12$ and the allowable stresses for steel and concrete as 150 MN/m² and 8 MN/m², respectively.

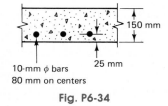

10-mm $\phi$ bars
80 mm on centers

Fig. P6-34

**6-35.** A beam has the cross section shown in the figure, and is subjected to a positive bending moment that causes a tensile stress in the steel of 20 ksi. If $n = 12$, what is the value of the bending moment?

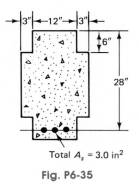

Total $A_s = 3.0$ in²

Fig. P6-35

## Section 6-9

**6-36.** Rework Example 6-10 by changing $h$ to 100 mm.

**6-37.** Derive Eq. 6-35.

**6-38.** What is the largest bending moment that may be applied to a curved bar, such as shown in Fig. 6-23(a), with $\bar{r} = 3$ in, if it has a circular cross-sectional area of 2-in diameter and the allowable stress is 12 ksi?

## Section 6-10

**6-39 through 6-43.** Find the ratios $M_{ult}/M_{yp}$ for beams having the cross sections shown in the figures. Bending occurs around the horizontal axes. Assume idealized elastic-plastic behavior as in Example 6-11.

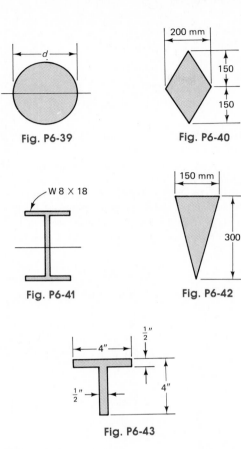

Fig. P6-39

Fig. P6-40

Fig. P6-41

Fig. P6-42

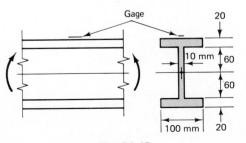

Fig. P6-43

**6-44.** Find the ultimate moment capacity for a beam having the cross section shown for Prob. 6-1. Assume that the material yields in tension and in compression at 200 MPa.

**6-45.** A steel I beam subjected to pure bending develops a longitudinal strain of $-1.6 \times 10^{-3}$ in the top flange in the location shown on the figure. (a) What

bending moment causes this strain? Assume ideal elastic-plastic material behavior with $E = 200$ GPa and $\sigma_{yp}$ = 240 MPa. (b) What residual strain would remain in the gage upon release of the applied load? (c) Draw the residual stress pattern.

**6-46.** An I beam is made up from three steel plates welded together as shown in the figure. The flanges are of stronger steel than the web. (a) What bending moment would the section develop when the largest stresses in the flanges just reach yield? The stress-strain properties of the two steels can be idealized as shown on the diagram. (b) Draw the residual stress pattern.

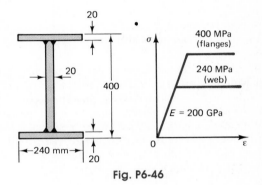

Fig. P6-46

**6-47.** A small sandwich beam spanning 400 mm is made up by bonding two aluminum alloy strips to an alloy steel bar, as shown in the figure. The idealized stress-strain diagrams are shown in the figure. What is the magnitude of the applied bending moment if it causes $-7.5 \times 10^{-3}$ longitudinal strain in the gage glued to the top of the aluminum alloy strip?

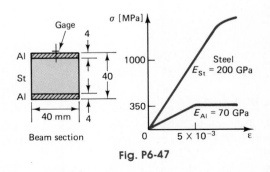

Fig. P6-47

**6-48.** On applying a bending moment around the horizontal axis to the T beam having the dimensions

Fig. P6-45

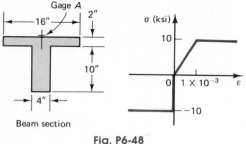

Gage A

16″    2″

10″

4″

Beam section

$\sigma$ (ksi)

10

0   $1 \times 10^{-3}$   $\varepsilon$

−10

**Fig. P6-48**

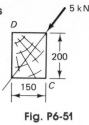

D   5 kN

200

150   C

**Fig. P6-51**

shown in the figure, the measured longitudinal strain at gage A is $-2 \times 10^{-3}$. Determine the magnitude of the applied bending moment if the stress-strain relation for the material can be idealized as shown on the diagram.

**6-49.** A $100 \times 180$ mm rectangular beam is of a material with the stress-strain characteristics shown in the figure. (a) Find the largest moment for which the entire cross section remains elastic. (b) Determine the ultimate moment capacity, and draw the resulting stress distribution. (c) What is the residual stress distribution after a release of the ultimate bending moment? (d) Show that the residual stresses are self-equilibrating.

shown in the figure. Determine the largest bending stresses and locate the neutral axis.

**6-52.** A 10-ft cantilever made up from the standard steel shape S $12 \times 50$ has its web in a vertical position, as shown in the figure. Determine the maximum bending stresses 2 ft from the support caused by the application of the variously inclined force $P$ acting through the centroid of the section at the free end. Let $\alpha$ be 0°, 1°, and 5°.

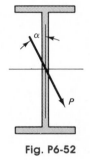

**Fig. P6-52**

**6-53.** A beam having the cross-sectional dimensions, in mm, shown in the figure is subjected to a bending moment of 500 N·m around its horizontal axis. Determine the maximum bending stresses.

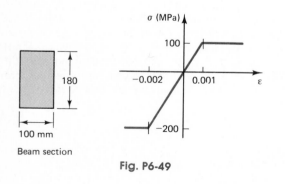

$\sigma$ (MPa)

100

−0.002   0.001   $\varepsilon$

−200

180

100 mm

Beam section

**Fig. P6-49**

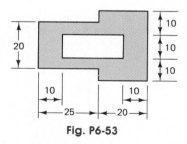

20

10   10   10

10   10

25   20

**Fig. P6-53**

## Section 6-11

**6-50.** Rework Example 6-15 by assuming that the span is 6000 mm, the beam is $150 \times 200$ mm, and $\alpha$ is 20°.

**6-51.** A $150 \times 200$ mm beam spanning 6000 mm is loaded in the middle of the span with an inclined force of 5 kN along the diagonal of the cross section, as

**6-54.** A biaxially symmetric cruciform aluminum extrusion has the cross-sectional dimensions, in mm, shown in the figure. It is used in a tilted position as a cantilever to carry an applied force $P = 100$ N at the end. (a) Determine the maximum flexural tensile stress

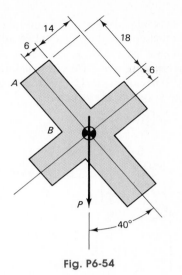

Fig. P6-54

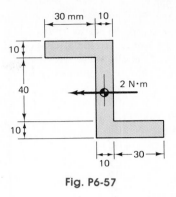

Fig. P6-57

200 mm from the loaded end of the cantilever. Assume linearly elastic behavior of the material. (b) Locate a point of zero stress on line $AB$.

**6-55.** Determine the bending stresses at the corners in the cantilever loaded, as shown in the figure, at a section 500 mm from the free end. Also locate the neutral axis.

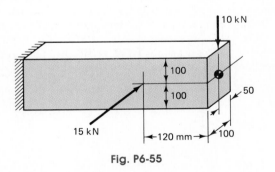

Fig. P6-55

**6-56.** Rework Example 6-16 by assuming that the angle is subjected to a 4-kN·m bending moment around the vertical axis.

**6-57.** Determine the maximum stresses in the Z section caused by a 2-N·m bending moment acting around the $z$ axis. As found in Prob. 6-83, the principal moments of inertia are $I_1 = I_{z'} = 753.9757 \times 10^3$ mm$^4$, $I_2 = I_{y'} = 96.0243 \times 10^3$ mm$^4$, and $\theta_1 = 32.8862°$. (*Hint:* Locating the neutral axis gives an indication as to where the largest stresses occur.)

## Section 6-12

**6-58.** AW 10 × 49 beam 8 ft long is subjected to a pull $P$ of 100 k, as shown in the figure. At the ends, where the pin connections are made, the beam is reinforced with doubler plates. Determine the maximum flange stress in the middle of the member caused by the applied forces $P$. Qualitatively, briefly discuss the load transfer at the ends. Most likely, where are the highest stressed regions in this member?

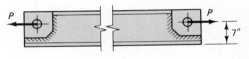

Fig. P6-58

**6-59.** For the machine link shown in the figure, determine the offset distance $e$ such that the tensile and compressive stresses in the T section are equal.

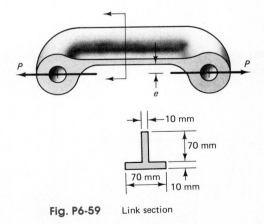

Fig. P6-59    Link section

**6-60.** A frame for a punch press has the proportions shown in the figure. What force $P$ can be applied to this frame controlled by the stresses in the sections such as $a$–$a$, if the allowable stresses are 4,000 psi in tension and 12,000 psi in compression?

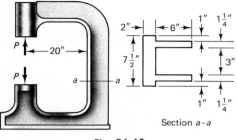

**Fig. P6-60**

**6-61.** A force of 169.8 k is applied to bar $BC$ at $C$, as shown in the figure. Find the maximum stress acting normal to section $a$–$a$. Member $BC$ is made from a piece of 6 by 6 in steel bar. Neglect the weight of the bar.

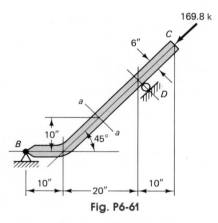

**Fig. P6-61**

**6-62.** Calculate the maximum compressive stress acting on section $a$–$a$ caused by the applied load for the

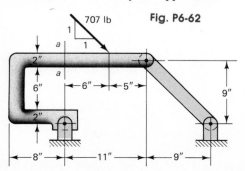

**Fig. P6-62**

structure shown in the figure. The cross section at section $a$–$a$ is that of a solid circular bar of 2-in diameter.

**6-63.** Compute the maximum compressive stress acting normal to section $a$–$a$ for the structure shown in the figure. Post $AB$ has a 12 by 12 in cross section. Neglect the weight of the structure.

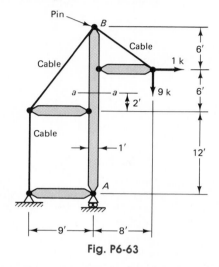

**Fig. P6-63**

**6-64.** In order to obtain the magnitude of an eccentric vertical force $F$ on a tee-shaped steel column, strain gages are attached at $A$ and $B$, as shown in the figure. Determine the force $F$ if the longitudinal strain at $A$ is $-100 \times 10^{-6}$ in/in and at $B$ is $-800 \times 10^{-6}$ in/in. $E = 30 \times 10^{6}$ psi and $G = 12 \times 10^{6}$ psi. The cross-sectional area of the column is 24 in$^2$.

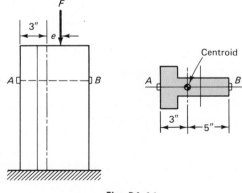

**Fig. P6-64**

**\*6-65.** A bar having a $100 \times 100$ mm cross section is subjected to a force $F$, as shown in the figure. The

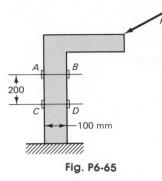

**Fig. P6-65**

longitudinal stresses on the extreme fibers at two sections 200 mm apart are determined experimentally to be $\sigma_A = 0$; $\sigma_B = -30$ MPa; $\sigma_C = -24$ MPa; and $\sigma_D = -6$ MPa. Determine the magnitude of the vertical and horizontal components of force $F$.

**6-66.** A rectangular vertical member fixed at the base is loaded as shown in the figure. Find the location for a gage on member face $AB$ such that no longitudinal strain would occur due to the application of force $P = 6$ kN. Does the answer depend on the magnitude of force $P$? Assume elastic behavior. All dimensions are given in mm.

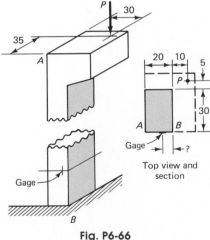

**Fig. P6-66**

**6-67.** An inclined tensile force $F$ is applied to an aluminum alloy bar such that its line of action goes through the centroid of the bar, as shown in mm in the figure. (The detail of the attachment is not shown.) What is the magnitude of force $F$ if it causes a longi-

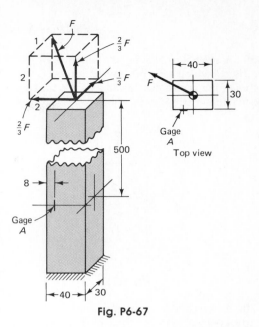

**Fig. P6-67**

tudinal strain of $+20 \times 10^{-6}$ in the gage at $A$? Assume that the bar behaves as a linearly elastic material and let $E = 70$ GPa.

**6-68.** A magnesium alloy bar is bonded to a steel bar of the same size forming a beam having the cross-sectional dimensions in mm shown in the figure. (a) If on application of an eccentric axial force $P$, the upper longitudinal gage measures a compressive strain of $2 \times 10^{-3}$, and the lower one, a tensile strain of $2 \times 10^{-3}$, what is the magnitude of applied force $P$? Assume elastic behavior of the materials with $E_{Mg} = 45$ GPa and $E_{St} = 200$ GPa. (b) Where would one have to apply axial force $P$ to cause no bending? (It is interesting to note that this locates the neutral axis for this beam.)

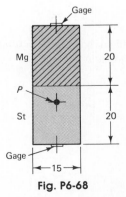

**Fig. P6-68**

**6-69.** A steel hook, having the proportions in the figure, is subjected to a downward force of 19 k. The radius of the centroidal curved axis is 6 in. Determine the maximum stress in this hook.

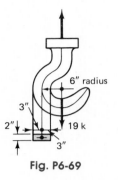

**Fig. P6-69**

**6-70.** A steel bar of 50-mm diameter is bent into a nearly complete circular ring of 300-mm outside diameter, as shown in the figure. (a) Calculate the maximum stress in this ring caused by applying two 10-kN forces at the open end. (b) Find the ratio of the maximum stress found in (a) to the largest compressive stress acting normal to the same section.

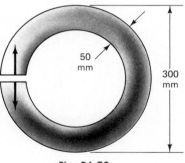

**Fig. P6-70**

**6-71.** A short block has cross-sectional dimensions in plan view as shown in the figure. Determine the range

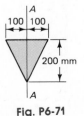

**Fig. P6-71**

along the line $A$–$A$ over which a downward vertical force could be applied to the top of the block without causing any tension at the base. Neglect the weight of the block.

**6-72.** The cross-sectional area in plan view of a short block is in the shape of an "arrow," as shown in the figure. Find the position of the vertical downward force on the line of symmetry of this section so that the stress at $A$ is just zero.

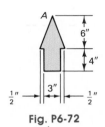

**Fig. P6-72**

**6-73.** Determine the kern for a member having a solid circular cross section.

**6-74.** For a small triangular dam of concrete weighing approximately 2550 kg/m³, as shown in the figure, find the approximate normal stress distribution at section $A$–$B$ using elementary methods for prismatic members when the water behind the dam is at the level indicated. For the purpose of calculation, consider one linear meter of the dam in the direction perpendicular to the plane of the paper as an isolated beam. The dimensions shown are in meters.

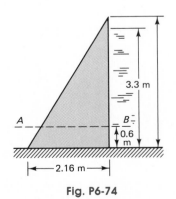

**Fig. P6-74**

**6-75.** What should the total height $h$ of the dam shown in the cross-sectional view be so that the foundation

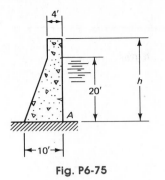

**Fig. P6-75**

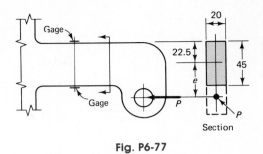

**Fig. P6-77**

What will be the reading of the gages when applied force $P$ is released?

pressure at $A$ is just zero? Assume that water weighs 62.5 lb/ft$^3$ and concrete 150 lb/ft$^3$.

## Section 6-13

**6-76.** A T beam of perfectly elastic-plastic material has the dimensions shown in the figure. (a) If the longitudinal strain at the bottom of the flange is $-\varepsilon_{yp}$ and is known to be zero at the juncture of the web and the flange, what axial force $P$ and bending moment $M$ act on the beam? (b) What would the strain reading be after the applied forces causing $P$ and $M$ in (a) are removed? Let $\sigma_{yp} = 200$ MPa.

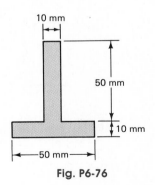

**Fig. P6-76**

**6-77.** A magnesium alloy casting has the dimensions given in the figure in mm. During application of force $P$, the upper gage recorded a tensile strain of $3 \times 10^{-3}$, and the lower one, a compressive strain of $6 \times 10^{-3}$. (a) Estimate the magnitude of applied force $P$ and its eccentricity $e$ assuming idealized behavior for the material. Let $\sigma_{yp} = 135$ MPa and $\varepsilon_{yp} = 3 \times 10^{-3}$. (b)

## Section 6-14

**6-78.** Rework Example 6-24 for an applied moment $M_y = 4$ kN·m.

**6-79.** Using the generalized flexure formula, find the largest stresses in a beam with a Z cross section, having the dimensions shown in the figure for Prob. 6-57, due to a pure bending moment $M_z$ of 2 N·m. Also locate the neutral axis. See answers to Prob. 6-83 for area moments of inertia for the cross section.

**6-80.** Rework the preceding problem for an applied moment $M_y = 6$ kN·m.

## Section 6-15

**6-81.** (a) Find the product of inertia for the triangular area shown in the figure with respect to the given axes. (b) For the same area, determine the product of inertia with respect to the vertical and horizontal axes through the centroid.

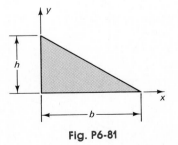

**Fig. P6-81**

**6-82.** (a) Find the principal axes and principal moments of inertia for the cross-sectional area of the angle shown in the figure. (b) The given dimensions of the

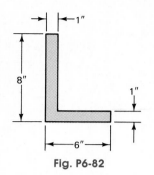

**Fig. P6-82**

condition, $I_{min} + I_{max} = I_{x'} + I_{y'} = I_x + I_y$, hence one can readily solve for $I_{max}$.)

**6-83.** For the Z cross section shown in the figure, first determine area moments of inertia $I_y$, $I_z$, and $I_{yz}$; then obtain the directions of the principal axes and principal moments of inertia.

cross section, except for small radii at the ends and a fillet, correspond to the cross-sectional dimensions of an $8 \times 6 \times 1$ in angle listed in Table 7 of the Appendix. Using the information given in that table, calculate the principal moments of inertia and compare with the results found in (a). (*Hint:* Note that per Section 11-6 and Example 11-2, $I_{min} = Ar^2_{min}$. The $r$ listed in Table 7 for the $z$ axis is $r_{min}$. Further, from the invariance

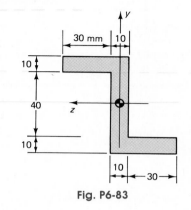

**Fig. P6-83**

# chapter 7

# Shear Stresses in Beams and Related Problems

## 7-1. Introduction

This chapter is divided into two parts. Part A, the major part, is devoted to a study of shear stresses in beams caused by transverse shear. The related problem of attaching separate longitudinal parts of a beam by means of bolts, gluing, or welding is also considered. In Part B, *superposition* of direct shear stresses of the type discussed in Part A with those caused by torque, as in springs, is treated. This problem is analogous to that encountered in the previous chapter in the study of beams simultaneously subjected to bending and axial forces.

The discussion in this chapter is largely limited to *elastic* analyses, the most widely used approach in the solution of the type of problems considered.

## Part A    SHEAR STRESSES IN BEAMS

## 7-2. Preliminary Remarks

In deriving the torsion and the flexure formulas, the same sequence of reasoning was employed. First, a strain distribution was assumed across the section; next, properties of the material were brought in to relate these strains to stresses; and, finally, the equations of equilibrium were used to establish the desired relations. However, the development of the expression linking the shear force and the cross-sectional area of a beam to the stress follows a different path. The previous procedure cannot be employed, as no simple assumption for the strain distribution due to the shear force can be made. Instead, an indirect approach is used. *The stress distribution caused by flexure, as determined in the preceding chapter, is assumed, which, together with the equilibrium requirements, resolves the problem of the shear stresses.*

First, it will be necessary to recall that the shear force is *inseparably* linked with a *change* in the bending moment at adjoining sections through a beam. Thus, if a shear and a bending moment are present at one section through a beam, a *different* bending moment will exist at an adjoining section, although the shear may remain constant. This will lead to the establishment of the shear stresses on the imaginary longitudinal planes through the members that are parallel to its axis. Then, since at a point, equal shear stresses exist on the mutually perpendicular planes, the shear stresses whose direction is coincident with the shear force at a section will be determined. Initially, only beams having symmetrical cross sections with applied forces acting in the plane of symmetry will be considered. The related problem of determining interconnection requirements for fastening together several longitudinal elements of built-up or composite beams will also be discussed.

In order to gain some insight into the problem, recall Eq. 5-4. Writing it in two alternative forms,

$$dM = V\,dx \quad \text{or} \quad \frac{dM}{dx} = V \qquad (7\text{-}1)$$

Equation 7-1 means that if shear $V$ is acting at a section, there will be a *change* in the bending moment $M$ on an adjoining section. The *difference* between the bending moments on the adjoining sections is equal to $V\,dx$. If no shear is acting, *no change in the bending moment occurs.* Alternatively, the rate of change in moment along a beam is equal to the shear. Therefore, although shear is treated in this chapter as an independent action on a beam, it is *inseparably* linked with the change in the bending moment along the beam's length.

As an example of the above, consider the shear and moment diagrams from Example 5-7, shown in Fig. 7-1. Here at any two sections such as $A$ and $B$ taken through the beam anywhere between applied forces $P$, the bending moment is the same. *No shear* acts at these sections. On the other hand, between any two sections such as $C$ and $D$ near the support, a change in the bending moment does take place. Shear forces act at these sections. These shears are shown acting on an element of the beam in Fig. 7-1(d). Note that in this zone of the beam, the *change* in the bending moment in a distance $dx$ is $P\,dx$ as shear $V$ is equal to $P$. In subsequent discussion, the possibility of equal, as well as of different, bending moments on two adjoining sections through a beam is of great importance.

Before a detailed analysis is given, a study of a sequence of photographs of a model (Fig. 7-2) may prove helpful. The model represents a segment

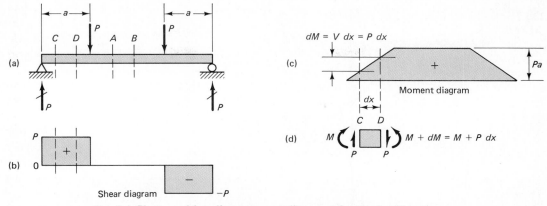

**Fig. 7-1** Shear and bending moment diagrams for the loading shown.

of an I beam. In Fig. 7-2(a), in addition to the beam itself, blocks simulating stress distribution caused by bending moments may be seen. The moment on the right is assumed to be larger than the one on the left. This system of forces is in equilibrium providing vertical shears $V$ (not seen in this view) also act on the beam segment. By separating the model along the neutral surface, one obtains two separate parts of the beam segment, as shown in Fig. 7-2(b). Again, either one of these parts alone must be in equilibrium.

If the upper and the lower segments of Fig. 7-2(b) are connected by a dowel or a bolt in an actual beam, the axial forces on either the upper or the lower part caused by the bending moment stresses must be maintained in equilibrium by a force in the dowel. The force that must be resisted can be evaluated by summing the forces in the axial direction caused by bending stresses. In performing such a calculation, either the upper or the lower part of the beam segment can be used. The horizontal force transmitted by the dowel is the force needed to balance the net force caused by the bending stresses acting on the two adjoining sections. Alternatively, by subtracting the same bending stress on both ends of the segment, the same results can be obtained. This is shown schematically in Fig. 7-2(c), where assuming a zero bending moment on the left, only the normal stresses due to the increment in moment within the segment need be shown acting on the right.

If, initially, the I beam considered is one piece requiring no bolts or dowels, an imaginary longitudinal plane can be used to separate the beam segment into two parts; see Fig. 7-2(d). As before, the net force that must be developed across the cut area to maintain equilibrium can be determined. Dividing this force by the area of the imaginary horizontal cut gives average shear stresses acting in this plane. In the analysis, it is again expedient to work with the change in bending moment rather than with the total moments on the end sections.

After the shear stresses on one of the planes are found (i.e., the hor-

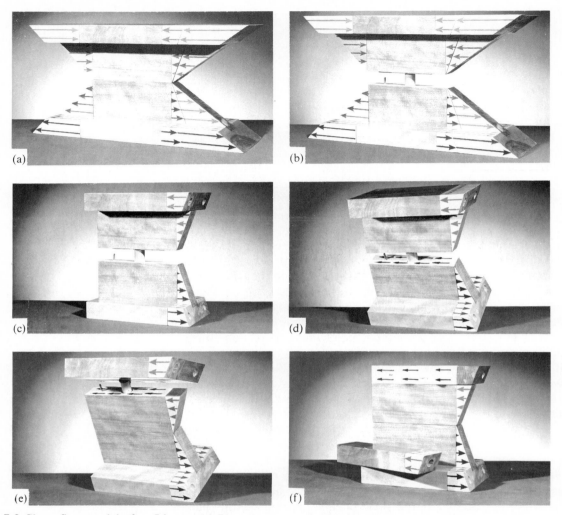

**Fig. 7-2** Shear flow model of an I beam. (a) Beam segment with bending stresses simulated by blocks. (b) Shear force transmitted through a dowel. (c) For determining the force on a dowel only a change in moment is needed. (d) The longitudinal shear force divided by the area of the imaginary cut yields shear stress. (e) Horizontal cut below the flange for determining the shear stress. (f) Vertical cut through the flange for determining the shear stress.

izontal one in Fig. 7-2(d)), shear stresses on mutually perpendicular planes of an infinitesimal element also become known since they must be numerically equal, Eq. 1-2. This approach establishes the shear stresses in the plane of the beam section taken normal to its axis.

The process discussed is quite general; two additional illustrations of separating the segment of the beam are in Figs. 7-2(e) and (f). In Fig. 7-2(e), the imaginary horizontal plane separates the beam just below the flange. Either the upper or the lower part of this beam can be used in calculating the shear stresses in the cut. The imaginary vertical plane cuts off a part of the flange in Fig. 7-2(f). This permits calculation of shear stresses lying in a vertical plane in the figure.

Before finally proceeding with the development of equations for determining the shear stresses in connecting bolts and in beams, an intuitive example is worthy of note. Consider a wooden plank placed on top of another, as shown in Fig. 7-3. If these planks act as a beam and are not interconnected, sliding at the surfaces of their contact will take place. The interconnection of these planks with nails or glue is necessary to make them act as an integral beam. In the next section, an equation will be derived for determining the required interconnection between the component parts of a beam to make them act as a unit. In the following section, this equation will be specialized to yield shear stresses in solid beams.

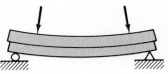

**Fig. 7-3** Sliding between planks not fastened together.

## 7-3. Shear Flow

Consider an elastic beam made from several continuous longitudinal planks whose cross section is shown in Fig. 7-4(a). For simplicity, the beam has a rectangular cross section, but such a limitation is not necessary. To make this beam act as an integral member, it is assumed that

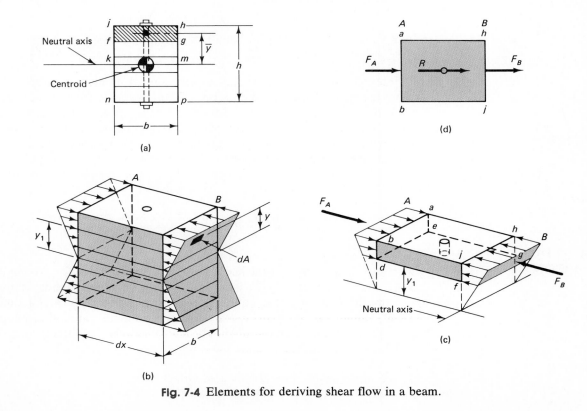

**Fig. 7-4** Elements for deriving shear flow in a beam.

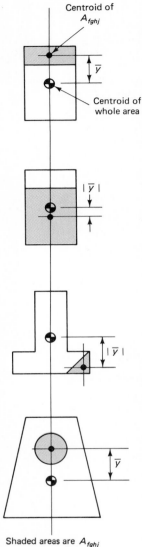

Centroid of
$A_{fghj}$

$\bar{y}$

Centroid of
whole area

$|\bar{y}|$

$|\bar{y}|$

$\bar{y}$

Shaded areas are $A_{fghj}$

**Fig. 7-5** Procedure for
determining $|Q|$.

the planks are fastened at intervals by vertical bolts. An element of this beam isolated by two parallel sections, both of which are perpendicular to the axis of the beam, is shown in Fig. 7-4(b).

If the element shown in Fig. 7-4(b) is subjected to a bending moment $+M_A$ at end $A$ and to $+M_B$ at end $B$, bending stresses that act normal to the sections are developed. These bending stresses vary linearly from their respective neutral axes, and at any point at a distance $y$ from the neutral axis are $-M_By/I$ on the $B$ end and $-M_Ay/I$ on the $A$ end.

From the beam element, Fig. 7-4(b), isolate the top plank, as shown in Fig. 7-4(c). The fibers of this plank nearest the neutral axis are located by the distance $y_1$. Then, since stress times area is equal to force, the forces acting perpendicular to ends $A$ and $B$ of this plank may be determined. At end $B$, the force acting on an infinitesimal area $dA$ at a distance $y$ from the neutral axis is $(-M_By/I)\,dA$. The total force acting on the area $fghj$, $A_{fghj}$, is the sum, or the integral, of these elementary forces over this area. Denoting the total force acting normal to the area $fghj$ by $F_B$ and remembering that, at section $B$, $M_B$ and $I$ are constants, one obtains the following relation:

$$F_B = \int_{\substack{\text{area}\\ fghj}} -\frac{M_By}{I}\,dA = -\frac{M_B}{I}\int_{\substack{\text{area}\\ fghj}} y\,dA = -\frac{M_BQ}{I} \quad (7\text{-}2)$$

where

$$Q = \int_{\substack{\text{area}\\ fghj}} y\,dA = A_{fghj}\bar{y} \quad (7\text{-}3)$$

The integral defining $Q$ is the first or the statical moment of area $fghj$ around the neutral axis. By definition, $\bar{y}$ is the distance from the neutral axis to the centroid of $A_{fghj}$.[1] Illustrations of the manner of determining $Q$ are in Fig. 7-5. Equation 7-2 provides a convenient means of calculating the longitudinal force acting normal to any selected part of the cross-sectional area.

Next consider end $A$ of the element in Fig. 7-4(c). One can then express the total force acting normal to the area $abde$ as

$$F_A = -\frac{M_A}{I}\int_{\substack{\text{area}\\ abde}} y\,dA = -\frac{M_AQ}{I} \quad (7\text{-}4)$$

where the meaning of $Q$ is the same as that in Eq. 7-2 since for prismatic beams, an area such as $fghj$ is equal to the area $abde$. Hence, if the moments at $A$ and $B$ were equal, it would follow that $F_A = F_B$, and the bolt shown in the figure would perform a nominal function of keeping the planks together and would not be needed to resist any known longitudinal forces.

[1] Area $fgpn$ and its $\bar{y}$ may also be used to find $|Q|$.

On the other hand, if $M_A$ is not equal to $M_B$, which is always the case when shears are present at the adjoining sections, $F_A$ is not equal to $F_B$. More push (or pull) develops on one end of a "plank" than on the other, as different normal stresses act on the section from the two sides. Thus, if $M_A \neq M_B$, equilibrium of the horizontal forces in Fig. 7-4(c) may be attained only by developing a horizontal resisting force $R$ in the bolt. If $M_B > M_A$, then $|F_B| > |F_A|$, and $|F_A| + R = |F_B|$, Fig. 7-4(d). The force $|F_B| - |F_A| = R$ tends to shear the bolt in the plane of the plank $edfg$.[2] If the shear force acting across the bolt at level $km$, Fig. 7-4(a), were to be investigated, the two upper planks should be considered as one unit.

If $M_A \neq M_B$ and the element of the beam is only $dx$ long, the bending moments on the adjoining sections change by an infinitesimal amount. Thus, if the bending moment at $A$ is $M_A$, the bending moment at $B$ is $M_B = M_A + dM$. Likewise, in the same distance $dx$, the longitudinal forces $F_A$ and $F_B$ change by an infinitesimal force $dF$, i.e., $|F_B| - |F_A| = dF$. By substituting these relations into the expression for $F_B$ and $F_A$ found above, with areas $fghj$ and $abde$ taken equal, one obtains an expression for the differential longitudinal push (or pull) $dF$:

$$dF = |F_B| - |F_A| = \left(\frac{M_A + dM}{I}\right)Q - \left(\frac{M_A}{I}\right)Q = \frac{dM}{I}Q$$

In the final expression for $dF$, the actual bending moments at the adjoining sections are eliminated. Only the difference in the bending moments $dM$ at the adjoining sections remains in the equation.

Instead of working with a force $dF$, which is developed in a distance $dx$, it is more significant to obtain a similar force per unit of beam length. This quantity is obtained by dividing $dF$ by $dx$. Physically, this quantity represents the difference between $F_B$ and $F_A$ for an element of the beam of unit length. The quantity $dF/dx$ will be designated by $q$ and will be referred to as the *shear flow*. Since force is measured in newtons or pounds, shear flow $q$ has units of newtons per meter or pounds per inch. Then, recalling that $dM/dx = V$, one obtains the following expression for the shear flow in beams:

$$q = \frac{dF}{dx} = \frac{dM}{dx}\frac{1}{I}\int_{\substack{\text{area} \\ fghj}} y\, dA = \frac{VA_{fghj}\bar{y}}{I} = \frac{VQ}{I} \qquad (7\text{-}5)$$

[2] The forces $(|F_B| - |F_A|)$ and $R$ are not collinear, but the element shown in Fig. 7-4(c) is in equilibrium. To avoid ambiguity, shear forces acting in the vertical cuts are omitted from the diagram.

In this equation, $I$ is the moment of interia of the *entire* cross-sectional area around the neutral axis, just as it does in the flexure formula from which it came. The total shear force at the section investigated is represented by $V$, and the integral of $y\, dA$ for determining $Q$ extends only over the cross-sectional area of the beam to one side of this area at which $q$ is investigated.

In retrospect, note carefully that Eq. 7-5 *was derived on the basis of the elastic flexure formula,* but no term for a bending moment appears in the final expressions. This resulted from the fact that only the change in the bending moments at the adjoining sections had to be considered, and the latter quantity is linked with shear $V$. Shear $V$ was substituted for $dM/dx$, and this masks the origin of the established relations. Equation 7-5 is very useful in determining the necessary interconnection between the elements making up a beam. This will be illustrated by examples.

### EXAMPLE 7-1

Two long wooden planks form a T section of a beam, as shown in mm in Fig. 7-6(a). If this beam transmits a constant vertical shear of 3000 N, find the necessary spacing of the nails between the two planks to make the beam act as a unit. Assume that the allowable shear force per nail is 700 N.

### Solution

In attacking such problems, the analyst must ask: What part of a beam has a tendency to slide longitudinally from the remainder? Here this occurs in the plane of contact of the two planks; Eq. 7-5 must be applied to determine the shear flow in this plane. To do this, the neutral axis of the whole section and its moment of inertia around the neutral axis must be found. Then as $V$ is known and $Q$ is defined as the statical moment of the area of the upper plank around the neutral axis, $q$ may be determined. The distance $y_c$ from the top to the neutral axis is

$$y_c = \frac{50 \times 200 \times 25 + 50 \times 200 \times 150}{50 \times 200 + 50 \times 200} = 87.5 \text{ mm}$$

$$I = \frac{200 \times 50^3}{12} + 50 \times 200 \times 62.5^2 + \frac{50 \times 200^3}{12} + 50 \times 200 \times 62.5^2$$

$$= 113.54 \times 10^6 \text{ mm}^4$$

$$Q = A_{fghj}\bar{y} = 50 \times 200 \times (87.5 - 25) = 625 \times 10^3 \text{ mm}^3$$

$$q = \frac{VQ}{I} = \frac{3000 \times 625 \times 10^3}{113.54 \times 10^6} = 16.5 \text{ N/mm}$$

Thus, a force of 16.5 N/mm must be transferred from one plank to the other along the length of the beam. However, from the data given, each nail is capable of resisting a force of 700 N; hence, one nail is adequate for transmitting shear along $700/16.5 = 42$ mm of the beam length. As shear remains constant at the consecutive sections of the beam, the nails should be spaced throughout at 42-mm intervals.

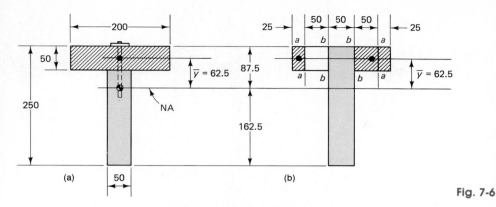

Fig. 7-6

## Solution for an Alternative Arrangement of Planks

If, instead of using the two planks as before, a beam of the same cross section were made from five pieces, Fig. 7-6(b), a different nailing schedule would be required.

To begin, the shear flow between one of the outer 25 by 50 mm planks and the remainder of the beam is found, and although the contact surface *a–a* is vertical, the procedure is the same. The push or pull on an element is built up in the same manner:

$$Q = A_{fghj}\bar{y} = 25 \times 50 \times 62.5 = 78.1 \times 10^3 \text{ mm}^3$$

$$q = \frac{VQ}{I} = \frac{3000 \times 78.1 \times 10^3}{113.5 \times 10^6} = 2.06 \text{ N/mm}$$

If the same nails as before are used to join the 25 by 50 mm piece to the 50 by 50 mm piece, they may be 700/2.06 = 340 mm apart. This nailing applies to both sections *a–a*.

To determine the shear flow between the 50 by 250 mm vertical piece and either one of the 50 by 50 mm pieces, the whole 75 by 50 mm area must be used to determine *Q*. It is the difference of pushes (or pulls) on this whole area that causes the unbalanced force that must be transferred at the surface *b–b*:

$$Q = A_{fghj}\bar{y} = 75 \times 50 \times 62.5 = 234 \times 10^3 \text{ mm}^3$$

$$q = \frac{VQ}{I} = \frac{3000 \times 234 \times 10^3}{113.4 \times 10^6} = 6.19 \text{ N/mm}$$

Nails should be spaced at 700/6.19 = 113 mm, intervals along the length of the beam in both sections *b–b*. These nails should be driven in first, then the 25 by 50 mm pieces put on.

## EXAMPLE 7-2

A simple beam on a 6-m span carries a load of 3 kN/m including its own weight. The beam's cross section is to be made from several wooden pieces, as is shown

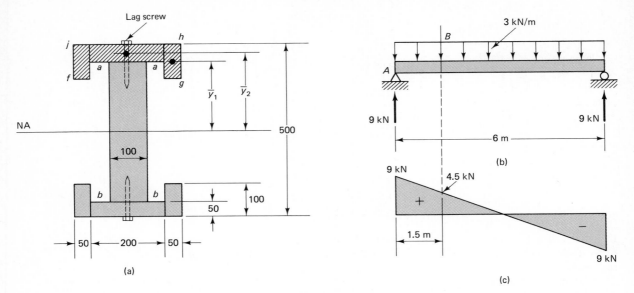

**Fig. 7-7**

in mm in Fig. 7-7(a). Specify the spacing of the 10-mm lag screws shown that is necessary to fasten this beam together. Assume that one 10-mm lag screw, as determined by laboratory tests, is good for 2 kN when transmitting lateral load parallel to the grain of the wood. For the *entire* section, $I$ is equal to $2.36 \times 10^9$ mm².

### Solution

To find the spacing of the lag screws, the shear flow at section $a–a$ must be determined. The loading on the given beam is shown in Fig. 7-7(b), and to show the variation of the shear along the beam, the shear diagram is constructed in Fig. 7-7(c). Next, to apply the shear flow formula, $\int_{\substack{\text{area} \\ fghj}} y \, dA = Q$ must be determined. This is done by considering the *hatched* area to one side of the cut $a–a$ in Fig. 7-7(a). The statical moment of this area is most conveniently computed by multiplying the areas of the *two* 50 by 100 mm pieces by the distances from their centroids to the neutral axis of the beam and adding to this product a similar quantity for the 50 by 200 mm piece. The largest shear flow occurs at the supports, as the largest vertical shears $V$ of 9 kN act there:

$$
\begin{aligned}
Q &= A_{fghj}\bar{y} = 2A_1\bar{y}_1 + A_2\bar{y}_2 \\
&= 2 \times 50 \times 100 \times 200 + 50 \times 200 \times 225 = 4.25 \times 10^6 \text{ mm}^3 \\
q &= \frac{VQ}{I} = \frac{9 \times 4.25 \times 10^9}{2.36 \times 10^9} = 16.2 \text{ N/mm}
\end{aligned}
$$

At the supports, the spacing of the lag screws must be $2 \times 10^3/16.2 = 123$ mm apart. This spacing of the lag screws applies only at a section where shear $V$ is equal to 9 kN. Similar calculations for a section where $V = 4.5$ kN gives $q = 8.1$ N/mm; and the spacing of the lag screws becomes $2 \times 10^3/8.1 = 246$ mm.

Thus, it is proper to specify the use of 10-mm lag screws on 120-mm centers for a distance of 1.5 m nearest both of the supports and 240-mm spacing of the same lag screws for the middle half of the beam. A greater refinement in making the transition from one spacing of fastenings to another may be desirable in some problems. The *same* spacing of lag screws should be used at section *b–b* as at section *a–a*.

In numerous practical applications, beams are made up by bolting or riveting longitudinal pieces, as shown in Fig. 7-8(a), or welding them, as shown in Fig. 7-8(b). Spacing of selected bolts or rivets, as well as sizing of welds, is determined using procedures analogous to those described before. The strength of individual bolts or rivets is discussed in Section 1-13 and that of welds is treated in Section 1-14. Note that the bolts may be staggered along the length of a beam, and that some may act in double shear. The welds may be either continuous or intermittent.

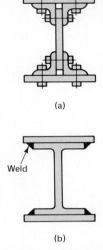

(a)

(b)

**Fig. 7-8** Composite beam sections: (a) plate girder, (b) I beam reinforced with plates.

## 7-4. The Shear-Stress Formula for Beams

The shear-stress formula for beams may be obtained by modifying the shear flow formula. Thus, analogous to the earlier procedure, an element of a beam may be isolated between two adjoining sections taken perpendicular to the axis of the beam. Then by passing *another imaginary longitudinal section* through this element parallel to the axis of the beam, a new element is obtained, which corresponds to the element of one "plank" used in the earlier derivations. A side view of such an element is shown in Fig. 7-9(a), where the imaginary longitudinal cut is made at a distance $y_1$ from the neutral axis. The cross-sectional area of the beam is shown in Fig. 7-9(c).

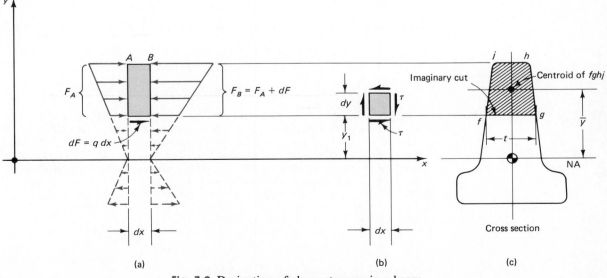

(a)        (b)        (c)

**Fig. 7-9** Derivation of shear stresses in a beam.

If shear forces exist at the sections through the beam, a different bending moment acts at section $A$ than at $B$. Hence, more push or pull is developed on one side of the *partial area fghj* than on the other, and, as before, this *longitudinal* force in a distance $dx$ is

$$dF = \frac{dM}{I} \int_{\substack{\text{area} \\ fghj}} y \, dA = \frac{dM}{I} A_{fghj}\bar{y} = \frac{dM}{I}Q$$

In a *solid* beam, the force resisting $dF$ may be developed only *in the plane* of the longitudinal cut taken parallel to the axis of the beam. Therefore, assuming that the shear stress $\tau$ is *uniformly distributed*[3] across the section of width $t$, the shear stress in the *longitudinal plane* may be obtained by dividing $dF$ by the area $t \, dx$. This yields the horizontal shear stress $\tau$. For an *infinitesimal* element, however, numerically equal shear stresses[4] act on the mutually perpendicular planes; see Fig. 7-9(b). Hence, the same relation gives *simultaneously* the longitudinal shear stress and the shear stress *in the plane of the vertical section at the longitudinal cut*.[5]

$$\tau = \frac{dF}{dx \, t} = \frac{dM}{dx} \frac{A_{fghj}\bar{y}}{It}$$

This equation may be simplified, since according to Eq. 7-1, $dM/dx = V$, and by Eq. 7-5, $q = VQ/I$. Hence,

$$\boxed{\tau = \frac{VA_{fghj}\bar{y}}{It} = \frac{VQ}{It} = \frac{q}{t}} \qquad (7\text{-}6)$$

Equation 7-6 is the important formula for the shear stresses in a beam.[6]

---

[3] This procedure is best suited to situations where the section sides are parallel and are away from significant changes in the shape of the cross section. For limitations see Section 7-6.

[4] Note that the sense of positive $\tau$ agrees with the positive sense for $V$ in beams adopted in Section 5-8.

[5] The presence of $\bar{y}$ in this relation may be explained differently. If the shear is present at a section through a beam, the moments at the adjoining sections are $M$ and $M + dM$. The magnitude of $M$ is irrelevant for determining the shear stresses. Hence, alternately, *no* moment need be considered at one section *if at the adjoining section, a bending moment dM is assumed to act*. Then on a partial area of the section, such as the shaded area in Fig. 7-9(c), this bending moment $dM$ will cause an *average normal stress* $(dM)\bar{y}/I$, as given by the flexure formula. In the latter relation, $\bar{y}$ locates the fiber that is at an *average* distance from the neutral axis in the *partial* area of a section. Multiplying $(dM)\bar{y}/I$ by the partial area of the section leads to the same expression for $dF$ as before.

[6] This formula was derived by D. I. Jouravsky in 1855. Its development was prompted by observing horizontal cracks in wood ties on several of the railroad bridges between Moscow and St. Petersburg.

It gives the shear stresses *at* the longitudinal cut. As before, $V$ is the *total* shear force at a section, and $I$ is the moment of inertia of the *whole* cross-sectional area about the neutral axis. Both $V$ and $I$ are constant at a section through a beam. Here $Q$ is the statical moment around the neutral axis of the *partial* area of the cross section to one side of the imaginary longitudinal cut, and $\bar{y}$ is the distance from the neutral axis of the beam to the centroid of the partial area $A_{fghj}$. Finally, $t$ is the width of the imaginary longitudinal cut, which is usually equal to the thickness or width of the member. The shear stress at different longitudinal cuts through the beam assumes different values as the values of $Q$ and $t$ for such sections differ.

Care must be exercised in making the longitudinal cuts preparatory for use in Eq. 7-6. The proper sectioning of some cross-sectional areas of beams is shown in Figs. 7-10(a), (b), (d), and (e). The use of inclined cutting planes should be avoided *unless* the section is made across a small thickness. When the axis of symmetry of the cross-sectional area of the beam is vertical and in the plane of the applied forces, the longitudinal cuts are usually made horizontally. In such cases, the solution of Eq. 7-6 gives simultaneous values of *horizontal and vertical* shear stresses, as such planes are mutually perpendicular, Eq. 1-2. The latter stresses act in the plane of the transverse section through the beam. Collectively, these shear stresses resist the shear force at the same section, thus satisfying the relation of statics, $\sum F_y = 0$. The validity of this statement for a special case will be proved in Example 7-3.

For thin members only, Eq. 7-6 may be used to determine the shear stresses with a cut such as $f$–$g$ of Fig. 7-10(b). These shear stresses act in a vertical plane and are directed perpendicularly to the plane of the paper. Matching shear stresses act horizontally; see Fig. 7-10(c). These shear stresses act in *entirely different directions* than those obtained by making horizontal cuts, such as $f$–$g$ in Figs. 7-10(a) and (d). As these shear stresses do not contribute directly to the resistance of vertical shear $V$, their significance will be discussed in Section 7-7.

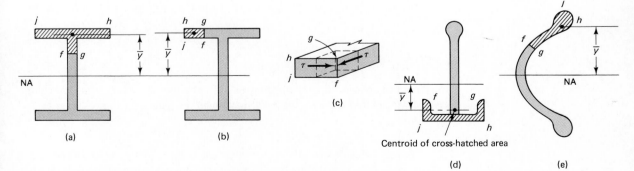

**Fig. 7-10** Sectioning for partial areas of cross sections for computing shear stresses.

### Procedure Summary

The same *three basic concepts* of engineering mechanics of solids as before are used in developing the formula for shear stresses in beams. However, their use is less direct.

1. *Equilibrium conditions* are used
   (a) for determining the shear at a section,
   (b) by using the relationship between the shear and the rate of change in bending moment along a span, and
   (c) by determining the force at a longitudinal section of a beam element for obtaining the average shear stress.

2. *Geometry of deformation*, as in pure bending, is assumed such that plane sections remain plane after deformation, leading to the conclusion that normal strains in a section vary linearly from the neutral axis. Since, due to shear, the cross sections do not remain plane, but warp, this assumption is less accurate than for pure bending. However, for small and moderate magnitudes of shear, and slender members, this assumption is satisfactory.

3. *Material properties* are considered to obey Hooke's law, although extension to other constitutive relations is possible for elementary solutions.

These conditions treat the problem as one-dimensional, and the assumed geometry of deformation is insensitive to the effects of concentrated forces and/or changes in the cross-sectional areas of beams. Therefore again reliance is largely placed on *Saint-Venant's principle*. In other words, only at distances beyond the member depth from such disturbances are the solutions accurate. Therefore solutions are best suited for slender members; see Section 7-5. Further, rigorous solutions show that for wide longitudinal sections, solutions are somewhat inaccurate due to complex warpage of their cross sections near the sides.

An application of Eq. 7-6 for determining shear stresses in a rectangular beam is given next. Based on the results obtained in this example, a general discussion follows of the effect of shear on warpage of initially plane sections in beams. Then two additional examples on the application of Eq. 7-6 are provided.

### EXAMPLE 7-3

Derive an expression for the shear-stress distribution in a beam of solid rectangular cross section transmitting a vertical shear $V$.

#### Solution

The cross-sectional area of the beam is shown in Fig. 7-11(a). A longitudinal cut through the beam at a distance $y_1$ from the neutral axis isolates the partial area $fghj$ of the cross section. Here $t = b$ and the infinitesimal area of the cross section

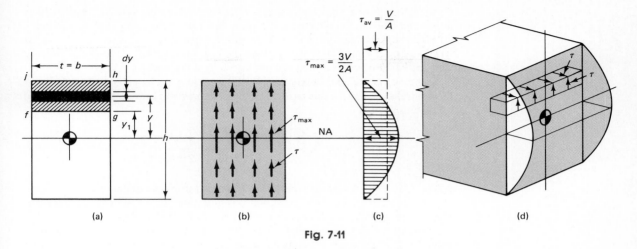

**Fig. 7-11**

may be conveniently expressed as $b\,dy$. By applying Eq. 7-6, the horizontal shear stress is found *at* level $y_1$ of the beam. At the same cut, numerically equal vertical shear stresses act *in the plane of the cross section*, Eq. 1-2.

$$\tau = \frac{VQ}{It} = \frac{V}{It} \int_{\substack{\text{area}\\ fghj}} y\,dA = \frac{V}{Ib} \int_{y_1}^{h/2} by\,dy$$

$$= \frac{V}{I} \left. \frac{y^2}{2} \right|_{y_1}^{h/2} = \frac{V}{2I} \left[ \left(\frac{h}{2}\right)^2 - y_1^2 \right] \qquad (7\text{-}7)$$

This equation shows that in a beam of rectangular cross section, both the horizontal and the vertical shear stresses vary parabolically. The maximum value of the shear stress is obtained when $y_1$ is equal to zero. *In the plane of the cross section*, Fig. 7-11(b), this is diagrammatically represented by $\tau_{\max}$ *at* the neutral axis of the beam. At increasing distances from the neutral axis, the shear stresses gradually diminish. At the upper and lower boundaries of the beam, the shear stresses cease to exist as $y_1 = \pm h/2$. These values of the shear stresses at the various levels of the beam may be represented by the parabola shown in Fig. 7-11(c). An isometric view of the beam with horizontal and vertical shear stresses is shown in Fig. 7-11(d).

To satisfy the condition of statics, $\sum F_y = 0$, at a section of the beam, the sum of all the vertical shear stresses $\tau$ times their respective areas $dA$ must be equal to the vertical shear $V$. That this is the case may be shown by integrating $\tau\,dA$ over the *whole* cross-sectional area $A$ of the beam, using the general expression for $\tau$ found before.

$$\int_A \tau\,dA = \frac{V}{2I} \int_{-h/2}^{+h/2} \left[ \left(\frac{h}{2}\right)^2 - y_1^2 \right] b\,dy_1 = \frac{Vb}{2I} \left[ \left(\frac{h}{2}\right)^2 y_1 - \left(\frac{y_1^3}{3}\right) \right]_{-h/2}^{+h/2}$$

$$= \frac{Vb}{2bh^3/12} \left[ \left(\frac{h}{2}\right)^2 h - \frac{2}{3}\left(\frac{h}{2}\right)^3 \right] = V$$

As the derivation of Eq. 7-6 was indirect, this proof showing that the shear stresses integrated over the section equal the vertical shear is reassuring. Moreover, since an agreement in signs is found, this result indicates that *the direction of the shear stresses at the section through a beam is the same as that of the shear force* V. This fact may be used to determine the sense of the shear stresses.

As noted before, the maximum shear stress in a rectangular beam occurs at the neutral axis, and for this case, the general expression for $\tau_{max}$ may be simplified by setting $y_1 = 0$.

$$\tau_{max} = \frac{Vh^2}{8I} = \frac{Vh^2}{8bh^3/12} = \frac{3}{2}\frac{V}{bh} = \frac{3}{2}\frac{V}{A} \tag{7-8a}$$

where $V$ is the total shear, and $A$ is the *entire* cross-sectional area. The same result may be obtained more directly if it is noted that to make $VQ/It$ a maximum, $Q$ must attain its largest value, as in this case $V$, $I$, and $t$ are constants. From the property of the statical moments of areas around a centroidal axis, the maximum value of $Q$ is obtained by considering one-half the cross-sectional area around the neutral axis of the beam. Hence, alternately,

$$\tau_{max} = \frac{VQ}{It} = \frac{V\left(\frac{bh}{2}\right)\left(\frac{h}{4}\right)}{\left(\frac{bh^3}{12}\right)b} = \frac{3}{2}\frac{V}{A} \tag{7-8b}$$

Since beams of rectangular cross-sectional area are used frequently in practice, Eq. 7-8b is very useful. It is widely used in the design of wooden beams since the shear strength of wood on planes parallel to the grain is small. Thus, although equal shear stresses exist on mutually perpendicular planes, wooden beams have a tendency to split longitudinally along the neutral axis. Note that the maximum shear stress is $1\frac{1}{2}$ times as great as the *average shear stress* $V/A$. Nevertheless, in the analysis of bolts and rivets, it is customary to determine their shear strengths by dividing the shear force $V$ by the cross-sectional area $A$ (see Section 1-8). Such practice is considered justified since the allowable and ultimate strengths are initially determined in this manner from tests. For beams, on the other hand, Eq. 7-6 is generally applied.

### **Alternative Solution

From the point of view of elasticity, internal stresses and strains in beams are statically indeterminate. However, in the engineering theory discussed here, the introduction of a kinematic hypothesis that plane sections remain plane after bending changes this situation. Here, in Eq. 6-11, it is asserted that in a beam, $\sigma_x = -My/I$. Therefore, one part of Eq. 1-5—that giving the differential equation of equilibrium for a two-dimensional problem with a body force $X = 0$—suffices to solve for the unknown shear stress. From the conditions of no shear stress at the top and the bottom boundaries, $\tau_{yx} = 0$ at $y = \pm h/2$, the constant of integration is found.

From Eq. 1-5:

$$\frac{\partial \sigma_x}{\partial x} + \frac{\partial \tau_{xy}}{\partial y} = 0$$

But $\qquad \sigma_x = -\dfrac{My}{I} \qquad$ hence,[7] $\qquad \dfrac{\partial \sigma_x}{\partial x} = -\dfrac{\partial M}{\partial x}\dfrac{y}{I} = \dfrac{Vy}{I}$

and Eq. 1-5 becomes $\qquad \dfrac{Vy}{I} + \dfrac{d\tau_{xy}}{dy} = 0$

Upon integrating, $\qquad \tau_{xy} = -\dfrac{Vy^2}{2I} + C_1$

Since $\qquad \tau_{xy}(\pm h/2) = 0 \qquad$ one has $\qquad C_1 + \dfrac{Vh^2}{8I}$

and $\qquad \tau_{xy} = \tau_{yx} = \dfrac{V}{2I}\left[\left(\dfrac{h}{2}\right)^2 - y^2\right]$

This agrees with the result found earlier, since here $y = y_1$.

---

## *7.5. Warpage of Plane Sections Due to Shear

A solution based on the mathematical theory of elasticity for a rectangular beam subjected simultaneously to bending *and* shear shows that plane sections perpendicular to the beam axis warp, i.e., they do not remain plane. This can also be concluded from Eq. 7-7 derived in the preceding example.

According to Hooke's law, shear strains must be associated with shear stresses. Therefore, the shear stresses given by Eq. 7-7 give rise to shear strains. According to this equation, the maximum shear stress, hence, maximum shear strain, occurs at $y = 0$; conversely, no shear strain takes place at $y = \pm h/2$. This behavior warps the initially plane sections through a beam, as shown qualitatively in Fig. 7-12, and contradicts the fundamental assumption of the simplified bending theory for pure flexure. However, based on rigorous analysis, warpage of the sections is known to be important only for very short members and is negligibly small for slender members. This can be substantiated by the two-dimensional finite-element studies for rectangular cantilevers shown in Figs. 7-13 and 7-14. In both

[7] In the elasticity sign convention used here, positive shear stress acts upward on the right face of an element as shown in Fig. 1-5. By analogy this requires that $\partial M/\partial x = -V$.

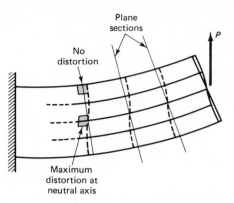

**Fig. 7-12** Shear distortions in a beam.

instances, the beams are fixed along lines *AB* at the nodal points indicated by dots in the figures. To avoid local disturbances of the type shown earlier in Fig. 2-31, in each case, applied forces *P* are distributed parabolically per Eq. 7-7 to the nodal points along lines *CD*.

The displacements of the nodal points of the elements for both beams shown in the figures are *greatly exaggerated*. For the numerical values used, they are increased by a factor of 3000 compared with the linear dimensions of the members. Considerable warpage of the initially plane sections can be clearly observed for the short cantilever in Fig. 7-13. By contrast, for the longer member in Fig. 7-14, the warpage of the sections is imperceptible. This study together with an examination of analytical results as well as experimental measurements on beams suggests that the assumption of "plane sections" is reasonable. It should also be noted that if shear force *V* along a beam is constant and the boundaries provide no restraint, the warping of all cross sections is the same. Therefore, the strain distribution caused by bending remains the same as in pure bending. Based on these considerations, a far-reaching conclusion can be made that the presence of shear at a section does not invalidate the expressions for bending stresses derived earlier.

It is cautioned, however, that local disturbances of stresses occur at the points of load applications and the use of the elementary elastic theory for short beams is questionable.

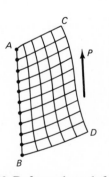

**Fig. 7-13** Deformed mesh for a short cantilever from a finite element solution.

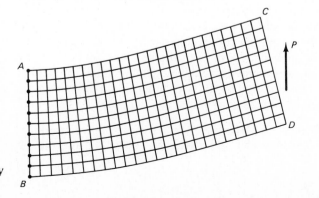

**Fig. 7-14** Finite element solution showing deformation of a moderately long cantilever.

## EXAMPLE 7-4

Using the simplified theory, determine the shear-stress distribution due to shear $V$ in the elastic-plastic zone of a rectangular beam.

### Solution

This situation occurs, for example, in a cantilever loaded as shown in Fig. 7-15(a). In the elastic-plastic zone, the external bending moment $M = -Px$, whereas, according to Eq. 6-40, the internal resisting moment $M = M_p - \sigma_{yp}by_o^2/3$. Upon noting that $y_o$ varies with $x$ and differentiating the above equations, one notes the following equality:

$$\frac{dM}{dx} = -P = -\frac{2by_o\sigma_{yp}}{3}\frac{dy_o}{dx}$$

This relation will be needed later. First, however, proceeding as in the elastic case, consider the equilibrium of a beam element, as shown in Fig. 7-15(b). Larger longitudinal forces act on the right side of this element than on the left. By separating the beam at the neutral axis and equating the force at the cut to the difference in the longitudinal force, one obtains

$$\tau_o\,dx\,b = \sigma_{yp}\,dy_o\,b/2$$

where $b$ is the width of the beam. After substituting $dy_o/dx$ from the relation found earlier and eliminating $b$, one finds the maximum horizontal shear stress $\tau_o$:

$$\tau_o = \frac{\sigma_{yp}}{2}\frac{dy_o}{dx} = \frac{3P}{4by_o} = \frac{3}{2}\frac{P}{A_o} \tag{7-9}$$

where $A_o$ is the cross-sectional area of the elastic part of the cross section. The shear-stress distribution for the elastic-plastic case is shown in Fig. 7-15(c). This can be contrasted with that for the elastic case, shown in Fig. 7-15(d). Since equal and opposite normal stresses occur in the plastic zones, no unbalance in longitudinal forces occurs and no shear stresses are developed.

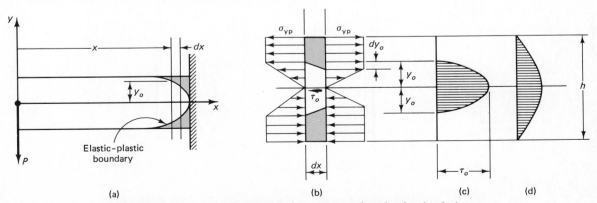

**Fig. 7-15** Shear stress distribution in a rectangular elastic-plastic beam.

This elementary solution has been refined by using a more carefully formulated criterion of yielding caused by the simultaneous action of normal and shear stresses.[8]

### EXAMPLE 7-5

An I beam is loaded as shown in Fig. 7-16(a). If it has the cross section shown in Fig. 7-16(c), determine the shear stresses at the levels indicated. Neglect the weight of the beam.

### Solution

From the free-body diagram of the beam segment in Fig. 7-16(b), it is seen that the vertical shear at all sections is 50 kips. Bending moments do not enter directly into the present problem. The shear flow at the various levels of the beam is computed in the following table using Eq. 7-5. Since $\tau = q/t$, Eq. 7-6, the shear stresses are obtained by dividing the shear flows by the respective widths of the beam.

$$I = 6 \times 12^3/12 - 5.5 \times 11^3/12 = 254 \text{ in}^4$$

[8] D. C. Drucker, "The Effect of Shear on the Plastic Bending of Beams," *J. Appl. Mech.* 23 (1956):509–514.

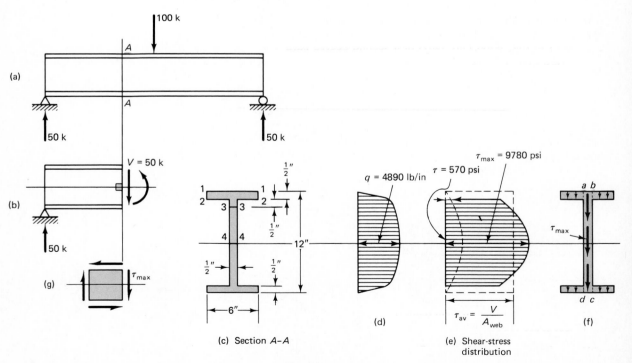

**Fig. 7-16**

For use in Eq. 7-5, the ratio $V/I = 50,000/254 = 197$ lb/in$^4$.

| Level | $A_{fghj}$[a] | $\bar{y}$[b] | $Q = A_{fghj}\bar{y}$ | $q = VQ/I$ | $t$ | $\tau$ (psi) |
|-------|------------|----------|------------------|----------|-----|-----------|
| 1–1 | 0 | 6 | 0 | 0 | 6.0 | 0 |
| 2–2 | $0.5 \times 6 = 3.00$ | 5.75 | 17.25 | 3400 | 6.0<br>0.5 | 570<br>6800 |
| 3–3 | $\begin{cases} 0.5 \times 6 = 3.00 \\ 0.5 \times 0.5 = 0.25 \end{cases}$ | 5.75<br>5.25 | $\left.\begin{matrix} 17.25 \\ 1.31 \end{matrix}\right\}18.56$ | 3650 | 0.5 | 7300 |
| 4–4 | $\begin{cases} 0.5 \times 6 = 3.00 \\ 0.5 \times 5.5 = 2.75 \end{cases}$ | 5.75<br>2.75 | $\left.\begin{matrix} 17.25 \\ 7.56 \end{matrix}\right\}24.81$ | 4890 | 0.5 | 9780 |

[a] $A_{fghj}$ is the partial area of the cross section above a given level in in$^2$.
[b] $\bar{y}$ is distance in mm from the neutral axis to the centroid of the partial area.

The positive signs of $\tau$ show that, for the section considered, the stresses act downward on the right face of the elements. This sense of the shear stresses coincides with the sense of shear force $V$. For this reason, a strict adherence to the sign convention is often unnecessary. It is always true that $\int_A \tau\, dA$ is equal to $V$ and has the same sense.

Note that at level 2–2, two widths are used to determine the shear stress—one just above the line 2–2, and one just below. A width of 6 in corresponds to the first case, and 0.5 in to the second. This transition point will be discussed in the next section. The results obtained, which by virtue of symmetry are also applicable to the lower half of the section, are plotted in Figs. 7-16(d) and (e). By a method similar to the one used in the preceding example, it may be shown that the curves in Fig. 7-16(e) are parts of a second-degree parabola.

The variation of the shear stress indicated by Fig. 7-16(e) may be interpreted as is shown in Fig. 7-16(f). The maximum shear stress occurs at the neutral axis, and the vertical shear stresses throughout the web of the beam are nearly of the same magnitude. The vertical shear stresses occurring in the flanges are very small. For this reason, the maximum shear stress in an I beam is often approximated by dividing the total shear $V$ by the cross-sectional area of the web with the web height assumed equal to the beam overall height, area *abcd* in Fig. 7-16(f). Hence,

$$(\tau_{max})_{approx} = \frac{V}{A_{web}} \tag{7-10}$$

In the example considered, this gives

$$(\tau_{max})_{approx} = \frac{50,000}{0.5 \times 12} = 8330 \text{ psi}$$

This stress differs by about 15 percent from the one found by the accurate formula. For most cross sections, a much closer approximation to the true maximum shear stress may be obtained by dividing the shear by the web area between the flanges only. For the above example, this procedure gives a stress of 9091 psi, which is an error of only about 8 percent. It should be clear from the above that division of $V$ by the whole cross-sectional area of the beam to obtain the shear stress is not permissible.

An element of the beam at the neutral axis is shown in Fig. 7-16(g). At levels 3–3 and 2–2, bending stresses, in addition to the shear stresses, act on the vertical faces of the elements. No shear stresses and only bending stresses act on the elements at level 1–1.

---

The sides of cross sections were assumed to be parallel in all the preceding examples. If they are not parallel, both $Q$ and $t$ vary with the section level, and the maximum shear stress may not occur at the neutral axis. However, using Eq. 7-6, the maximum *average* shear stress can always be found. For example it can be shown that for a symmetric triangular cross section such a maximum shear stress is midway between the apex and the base. For such cross sections, the stresses vary across a longitudinal section, and are particularly inaccurate near the sloping sides; see Fig. 7-18. Similar results may develop at longitudinal sections taken at an angle with the axes.

The same procedures as described before are used for determining longitudinal shear stresses in composite beams at bonded or glued surfaces.

### *7-6. Some Limitations of the Shear-Stress Formula

The shear-stress formula for beams is based on the flexure formula. Hence, all of the limitations imposed on the flexure formula apply. The material is assumed to be elastic with the same elastic modulus in tension as in compression. The theory developed applies only to straight beams. Moreover, there are additional limitations that are not present in the flexure formula. Some of these will be discussed now.

Consider a section through the I beam analyzed in Example 7-5. Some of the results of this analysis are reproduced in Fig. 7-17. The shear stresses computed earlier for level 1–1 apply to the infinitesimal element *a*. The vertical shear stress is zero for this element. Likewise, *no* shear stresses exist *on* the top plane of the beam. This is as it should be, since the top *surface* of the beam is a *free* surface. In mathematical phraseology, this means that the conditions at the boundary are satisfied. For beams of rectangular cross section, the situation at the boundaries is correct.

A different condition is found when the shear stresses determined for the I beam at levels 2–2 are scrutinized. The shear stresses were found to be 570 psi for the elements such as *b* or *c* shown in the figure. This requires matching horizontal shear stresses on the inner surfaces of the flanges. However, the latter surfaces *must be free* of the shear stresses as they are *free boundaries* of the beam. This leads to a contradiction that cannot be resolved by the methods of engineering mechanics of solids. The more advanced techniques of the mathematical theory of elas-

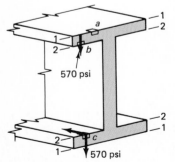

**Fig. 7-17** Boundary conditions are not satisfied at the levels 2-2.

570 psi

570 psi

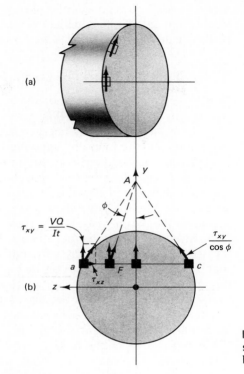

**Fig. 7-18** Modification of shear stresses to satisfy the boundary conditions.

ticity or three-dimensional finite-element analysis must be used to obtain an accurate solution.

Fortunately, the above defect of the shear-stress formula for beams is not serious. The vertical shear stresses in the flanges are small. The large shear stresses occur in the web and, for all practical purposes, are correctly given by Eq. 7-6. *No appreciable error is involved by using the relations derived in this chapter for thin-walled members*, and the majority of beams belong to this group. Moreover, as stated earlier, the solution for the shear stresses for a beam with a rectangular cross section is correct.

In mechanical applications, circular shafts frequently act as beams. Hence, beams having a solid circular cross section form an important class. These beams are not "thin-walled." An examination of the boundary conditions for circular members, Fig. 7-18(a), leads to the conclusion that when shear stresses are present, they must act parallel to the boundary. As no matching shear stress can exist *on* the free surface of a beam, no shear stress component can act normal to the boundary. However, according to Eq. 7-6, *vertical* shear stresses of *equal* intensity act at every level, such as $ac$ in Fig. 7-18(b). This is incompatible with the boundary conditions for elements $a$ and $c$, and the solution indicated by Eq. 7-6 is inconsistent.[9] Fortunately, the *maximum* shear stresses occurring at the neutral axis satisfy the boundary conditions and are within about 5 percent of their true value.[10]

## *7-7. Shear Stresses in Beam Flanges

In an I beam, the existence of shear stresses acting in a vertical longitudinal cut as $c–c$ in Fig. 7-19(a) was indicated in Fig. 7-2(f) and Section 7-4. These shear stresses act perpendicular to the plane of the paper. Their magnitude may be found by applying Eq. 7-6, and their sense follows by considering the bending moments at the adjoining sections through the beam. For example, if for the beam shown in Fig. 7-19(b), *positive* bending moments increase toward the reader, larger *normal* forces act on the *near* section. For the elements shown, $\tau t\, dx$ or $q\, dx$ must aid the smaller force acting on the partial area of the cross section. This fixes the sense of the shear stresses in the longitudinal cuts. However, numerically equal shear stresses act on the mutually perpendicular planes of an *infinitesimal* element, and the shear stresses on such planes either meet or part with their directional arrowheads at a corner. Hence, the sense of the shear stresses in the plane of the section becomes known also.

The magnitude of the shear stresses varies for the different vertical cuts. For example, if cut $c–c$ in Fig. 7-19(a) is at the edge of the beam, the *hatched* area of the beam's cross section is zero. However, if the thickness of the flange is constant, and cut $c–c$ is made progressively closer to the web, this area increases from zero at a linear rate. Moreover, as $\bar{y}$ remains constant for any such area, $Q$ also increases linearly from zero toward the web. Therefore, since $V$ and $I$ are constant at any section through the beam, shear flow $q_c = VQ/I$ follows the same variation. If the thickness of the flange remains the same, the shear stress $\tau_c = VQ/It$ varies similarly. The same variation of $q_c$ and $\tau_c$ applies on both sides of the axis of symmetry of the cross section. However, as may be seen from Fig. 7-19(b), these quantities in the plane of the cross section act in *opposite* directions on the two sides. The variation of these shear stresses or shear flows is represented in Fig. 7-19(c), where for simplicity, it is assumed that the web has zero thickness.

---

[9] The exact elastic solution of this problem is beyond the scope of this text. However, a better approximation of the true stresses may be obtained rather simply. First, an assumption is made that the shear stress as found by Eq. 7-6 gives a true *component* of the shear stress acting in the *vertical direction*. Then, since at every level, the shear stresses at the boundary must act tangent to the boundary, the lines of action of these shear stresses intersect at some point, as $A$ in Fig. 7-18(b). Thus, a second assumption is made that all shear stresses at a given level act in a direction toward a single point, as $A$ in Fig. 7-18(b). Therefore, the shear stress at any point such as $F$ becomes equal to $\tau_{xy}/\cos\phi$. The stress system found in the above manner is consistent.

[10] A. E. H. Love, *Mathematical Theory of Elasticity*, 4th ed. (New York: Dover, 1944), 348.

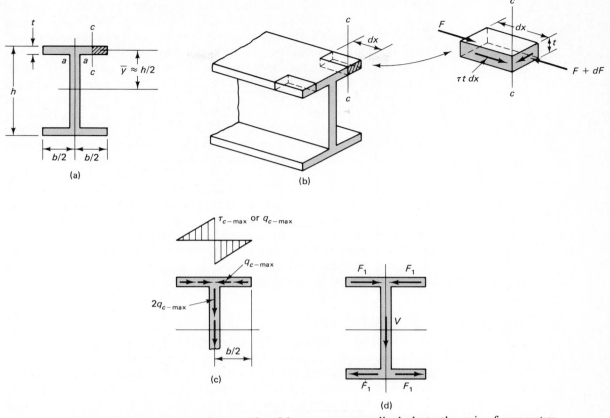

**Fig. 7-19** Shear forces in the flanges of an I beam act perpendicularly to the axis of symmetry.

In common with all stresses, the shear stresses shown in Fig. 7-19(c), when integrated over the area on which they act, are equivalent to a force. The magnitude of the horizontal force $F_1$ for *one-half* of the flange, Fig. 7-19(d), is equal to the *average* shear stress multiplied by *one-half of the whole area of the flange*, i.e.,

$$F_1 = \left(\frac{\tau_{c\text{-max}}}{2}\right)\left(\frac{bt}{2}\right) \quad \text{or} \quad F_1 = \left(\frac{q_{c\text{-max}}}{2}\right)\left(\frac{b}{2}\right) \quad (7\text{-}11)$$

If an I beam transmits a vertical shear, these horizontal forces act in the upper and lower flanges. However, because of the *symmetry* of the cross section, these equal forces occur in pairs and *oppose* each other, and cause no apparent external effect.

To determine the shear flow at the juncture of the flange and the web, cut $a$–$a$ in Fig. 7-19(a), the *whole* area of the flange times $\bar{y}$ must be used in computing the value of $Q$. However, since in finding $q_{c\text{-max}}$, *one-half* the flange area times the same $\bar{y}$ has already been used, the *sum* of the *two horizontal shear flows* coming in from opposite sides gives the *vertical* shear flow[11] at cut *a–a*. Hence, figuratively speaking, the horizontal shear flows "turn through 90° and merge to become the vertical shear flow." Thus, the shear flows at the various horizontal cuts through the web may be determined in the manner explained in the preceding sections. Moreover, as the resistance to the vertical shear *V* in thin-walled I beams is developed mainly in the web, it is so shown in Fig. 7-19(d). The sense of the shear stresses and shear flows *in the web* coincides with the direction of the shear *V*. Note that the vertical shear flow "splits" upon reaching the lower flange. This is represented in Fig. 7-19(d) by the two forces $F_1$ that are the result of the horizontal shear flows in the flanges.

The shear forces that act at a section of an I beam are shown in Fig. 7-19(d), and, for equilibrium, *the applied vertical forces must act through the centroid of the cross-sectional area* to be coincident with *V*. If the forces are so applied, *no torsion* of the member will occur. This is true for all sections having cross-sectional areas with an axis of symmetry. To avoid torsion of such members, the applied forces must act in the plane of symmetry of the cross section and the axis of the beam. A beam with an unsymmetrical section will be discussed next.

## 7-8. Shear Center

Consider a beam having the cross section of a channel; see Fig. 7-20(a). The walls of this channel are assumed to be sufficiently thin that the computations may be based on *center line* dimensions. Bending of this channel takes place around the horizontal axis and although this cross section does not have a vertical axis of symmetry, it will be *assumed* that the bending stresses are given by the usual flexure formula. Assuming further that this channel resists a vertical shear, the bending moments will vary from one section through the beam to another.

By taking an arbitrary vertical cut as *c–c* in Fig. 7-20(a), *q* and *τ* may

---

[11] The same statement *cannot* be made with regard to the shear stresses, as the thickness of the flange may differ from that of the web.

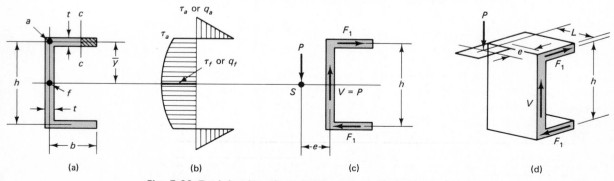

(a)          (b)          (c)          (d)

**Fig. 7-20** Deriving location of shear center for a channel.

be found in the usual manner. Along the horizontal legs of the channel, these quantities vary linearly from the free edge, just as they do for one side of the flange in an I beam. The variation of $q$ and $\tau$ is parabolic along the web. The variation of these quantities is shown in Fig. 7-20(b), where they are plotted along the center line of the channel's section.

The *average* shear stress $\tau_a/2$ multiplied by the areas of the flange gives a force $F_1 = (\tau_a/2)bt$, and the sum of the vertical shear stresses over the area of the web is the shear $V = \int_{-h/2}^{+h/2} \tau t \, dy$.[12] These shear forces acting in the plane of the cross section are shown in Fig. 7-20(c) and indicate that a force $V$ *and a couple $F_1h$* are developed at the section through the channel. Physically, there is a tendency for the channel to twist around some longitudinal axis. To prevent twisting and thus maintain the applicability of the initially assumed bending-stress distribution, the externally applied forces must be applied in such a manner as to *balance the internal couple $F_1h$.* For example, consider the segment of a cantilever beam of negligible weight, shown in Fig. 7-20(d), to which a vertical force $P$ is applied parallel to the web at a distance $e$ from the web's *center line.* To maintain this applied force in equilibrium, an *equal and opposite* shear force $V$ must be developed in the web. Likewise, to cause *no twisting of the channel*, couple *Pe* must *equal* couple $F_1h$. At the same section through the channel, bending moment $PL$ is resisted by the *usual* flexural stresses (these are not shown in the figure).

An expression for distance $e$, locating the plane in which force $P$ must be applied so as to cause *no twist* in the channel, may now be obtained. Thus, remembering that $F_1h = Pe$ and $P = V$,

$$e = \frac{F_1h}{P} = \frac{(1/2)\tau_a bth}{P} = \frac{bth}{2P}\frac{VQ}{It} = \frac{bth}{2P}\frac{Vbt(h/2)}{It} = \frac{b^2h^2t}{4I} \quad (7\text{-}12)$$

Note that distance $e$ is independent of the magnitude of applied force $P$, as well as of its location along the beam. Distance $e$ is a property of a section and is measured outward from the *center* of the web to the applied force.

A similar investigation may be made to locate the plane in which the horizontal forces must be applied so as to cause no twist in the channel. However, for the channel considered, by virtue of symmetry, it may be seen that this plane coincides with the neutral plane of the former case. The intersection of these two mutually perpendicular planes with the plane of the cross section locates a point that is called the *shear center*.[13] The shear center is designated by the letter $S$ in Fig. 7-20(c). The shear center for any cross section lies on a longitudinal line parallel to the axis of the

---

[12] When the thickness of a channel is variable, it is more convenient to find $F_1$ and $V$ by using the respective shear flows, i.e., $F_1 = (q_a/2)b$ and $V = \int_{-h/2}^{+h/2} q \, dy$. Since the flanges are thin, the vertical shear force carried by them is negligible.

[13] A. Eggenschwyler and R. Maillart of Switzerland clarified this concept only in 1921.

beam. _Any transverse force applied through the shear center causes no torsion of the beam._ A detailed investigation of this problem shows that when a member of any cross-sectional area _is_ twisted, the twist takes place around the shear center, which remains fixed. For this reason, the shear center is sometimes called the _center of twist._

For cross-sectional areas having one axis of symmetry, the shear center is always located on the axis of symmetry. For those that have two axes of symmetry, the shear center coincides with the centroid of the cross-sectional area. This is the case for the _I_ beam that was considered in the previous section.

The exact location of the shear center for unsymmetrical cross sections of thick materials is difficult to obtain and is known only in a few cases. If the material _is thin_, as has been assumed in the preceding discussion, relatively simple procedures may always be devised to locate the shear center of the cross section. The usual procedure consists of determining the shear forces, as $F_1$ and $V$ before, at a section, and then finding the location of the external force necessary to keep these forces in equilibrium.

### EXAMPLE 7-6

Find the approximate location of the shear center for a beam with the cross section of the channel shown in Fig. 7-21.

### Solution

Instead of using Eq. 7-12 directly, some further simplifications may be made. The moment of inertia of a thin-walled channel around its neutral axis may be found with sufficient accuracy by neglecting the moment of inertia of the flanges _around their own axes_ (only!). This expression for $I$ may then be substituted into Eq. 7-12 and, after simplifications, a formula for $e$ of channels is obtained.

$$I \approx I_{web} + (Ad^2)_{flanges} = th^3/12 + 2bt(h/2)^2 = th^3/12 + bth^2/2$$

$$e = \frac{b^2h^2t}{4I} = \frac{b^2h^2t}{4(bth^2/2 + th^3/12)} = \frac{b}{2 + h/3b} \tag{7-13}$$

Equation 7-13 shows that when the width of flanges $b$ is very large, $e$ approaches its maximum value of $b/2$. When $h$ is very large, $e$ approaches its minimum value of zero. Otherwise, $e$ assumes an intermediate value between these two limits. For the numerical data given in Fig. 7-21,

$$e = \frac{5}{2 + 10/(3 \times 5)} = 1.87 \text{ in}$$

Hence, the shear center $S$ is $1.87 - 0.05 = 1.82$ in from the outside vertical face of the channel.

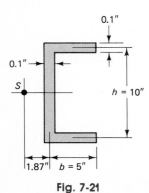

0.1″

0.1″

$S$

$h = 10″$

1.87″　$b = 5″$

**Fig. 7-21**

## EXAMPLE 7-7

Find the approximate location of the shear center for the cross section of the I beam shown in Fig. 7-22(a). Note that the flanges are unequal.

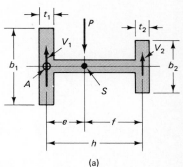

(a)

### Solution

This cross section has a horizontal axis of symmetry and the shear center is located on it; where it is located remains to be answered. Applied force $P$ causes significant bending and shear stresses *only in the flanges*, and the contribution of the web to the resistance of applied force $P$ is negligible.

Let the shear force resisted by the left flange of the beam be $V_1$, and by the right flange, $V_2$. For equilibrium, $V_1 + V_2 = P$. Likewise, to have no twist of the section, from $\sum M_A = 0$, $Pe = V_2h$ (or $Pf = V_1h$). Thus, only $V_2$ remains to be determined to solve the problem. This may be done by noting that the right flange is actually an ordinary rectangular beam. The shear stress (or shear flow) in such a beam is distributed parabolically, Fig. 7-22(b), and since the area of a parabola is two-thirds of the base times the maximum altitude, $V_2 = \frac{2}{3}b_2(q_2)_{\max}$. However, since the total shear $V = P$, by Eq. 7-5, $(q_2)_{\max} = VQ/I = PQ/I$, where $Q$ is the statical moment of the *upper half of the right-hand flange*, and $I$ is the moment of inertia of the *whole* section. Hence,

$q_{2-\max}$

Shear flow in right flange

(b)

**Fig. 7-22**

$$Pe = V_2h = \frac{2}{3}b_2(q_2)_{\max}h = \frac{\frac{2}{3}hb_2PQ}{I}$$

$$e = \frac{2hb_2}{3I}Q = \frac{2hb_2}{3I}\frac{b_2t_2}{2}\frac{b_2}{4} = \frac{h}{I}\frac{t_2b_2^3}{12} = \frac{hI_2}{I} \qquad (7\text{-}14)$$

where $I_2$ is the moment of inertia of the *right-hand flange* around the neutral axis. Similarly, it may be shown that $f = hI_1/I$, where $I_1$ applies to the *left flange*. If the web of the beam is thin, as originally assumed, $I \approx I_1 + I_2$, and $e + f = h$, as is to be expected.

---

A similar analysis leads to the conclusion that the shear center for a symmetrical angle is located at the intersection of the center lines of its legs, as shown in Figs. 7-23(a) and (b). This follows since the shear flow at every section, as c–c, is directed along the center line of a leg. These shear flows yield two identical forces, $F_1$, in the legs. The vertical components of these forces equal the vertical shear applied through $S$. An analogous situation is also found for any angle or T section, as shown in Figs. 7-24(a) and (b). The location of the shear center for various members is particularly important in aircraft applications.[14]

[14] For further details, see E. F. Bruhn, *Analysis and Design of Flight Vehicle Structures* (Cincinnati: Tri-State, 1965). See also P. Kuhn, *Stresses in Aircraft and Shell Structures* (New York: McGraw-Hill, 1956).

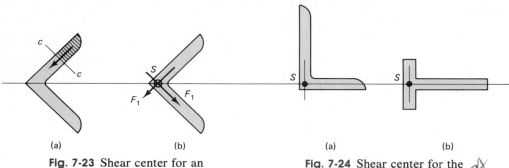

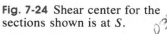

**Fig. 7-23** Shear center for an equal leg angle is at $S$.

**Fig. 7-24** Shear center for the sections shown is at $S$.

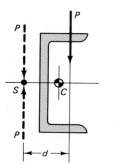

**Fig. 7-25** Torsion-bending of a channel.

As remarked earlier, in order to prevent torsion of a beam, the applied force must act through the shear center. When such a force forms an angle with the vertical, it is best to resolve it into components along the principal axes of the cross section, as shown in Fig. 6-38. If force $P$ is applied outside shear center $S$, as shown in Fig. 7-25, two equal but opposite forces $P$ can be introduced at $S$ without changing the problem. Then, in addition to the stresses caused by $P$ applied at $S$, the torsional stresses caused by the torque equal to $Pd$ must be considered, as described in Chapter 4.

It is to be noted that generally, in addition to the shear stresses discussed in this chapter, bending stresses usually also act on the elements considered. Transformation of this kind of state of stress is discussed in Chapter 8. In the remainder of this chapter, only superposition of the shear stresses is considered.

---

## Part B        SUPERPOSITION OF SHEAR STRESSES

### 7-9. Combined Direct and Torsional Shear Stresses

The analysis for combined direct and torsional shear stresses consists of two parts that are then superposed. In one of these parts, the direct shear stresses are determined using the procedures of Part A of this chapter; in the second, the shear stresses caused by torques susceptible to the methods of analysis treated in Chapter 4 are used.

The two analyses for combined shear stresses must be determined for the *same elementary area* regardless of cause. Multiplying these stresses by the respective area gives *forces*. Since these forces can be added vectorially, on reversing the process, i.e., on dividing the vector sum by the initial area, one obtains the combined shear stress. Such being the case, the shear stresses acting on the *same plane* of an infinitesimal element can be combined vectorially.[15] Generally, the maximum torsional shear stresses as well as the maximum direct shear stress for beams occur at the boundaries of cross sections and are collinear. Therefore, an algebraic sum of these stresses gives the combined shear stress at a point. However, on the interior of such members, a vectorial sum of the direct and torsional stresses is necessary.

In treating beam problems, as noted earlier, it must be recognized that in addition to the shear stresses discussed before, generally, normal stresses caused by bending also act on the elements considered. Procedures for combining such normal stresses with shear stresses are discussed in the next chapter.

## EXAMPLE 7-8

Find the maximum shear stress due to the applied forces in plane $A-B$ of the 10-mm diameter high-strength steel shaft shown in Fig. 7-26(a).

[15] The inverse problem of resolving a shear stress was considered in connection with Fig. 4-34.

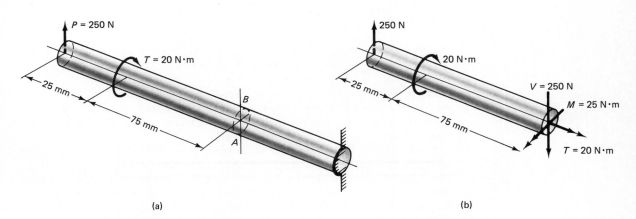

(a)    (b)

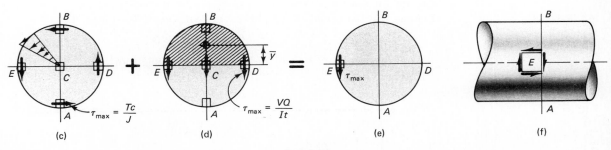

(c)    (d)    (e)    (f)

**Fig. 7-26**

### Solution

Since only the stresses due to the applied forces are required, the weight of the shaft need not be considered. The free-body of a segment of the shaft is shown in Fig. 7-26(b). The system of forces at the cut necessary to keep this segment in equilibrium consists of a torque, $T = 20$ N·m, a shear, $V = 250$ N, and a bending moment, $M = 25$ N·m.

Due to torque $T$, the shear stresses in cut $A–B$ vary linearly from the axis of the shaft and reach the maximum value given by Eq. 4-3, $\tau_{max} = Tc/J$. These maximum shear stresses, agreeing in sense with the *resisting* torque $T$, are shown at points $A$, $B$, $D$, and $E$ in Fig. 7-26(c).

The direct shear stresses caused by shear force $V$ may be obtained by using Eq. 7-6, $\tau = VQ/It$. For elements $A$ and $B$, Fig. 7-26(d), $Q = 0$; hence, $\tau = 0$. The shear stress reaches its maximum value at level $ED$. For this, $Q$ is equal to the cross-hatched area shown in Fig. 7-26(d) multiplied by the distance from its centroid to the neutral axis. The latter quantity is $\bar{y} \doteq 4c/3\pi$, where $c$ is the radius of the cross-sectional area; see Table 2 in the Appendix. Hence,

$$Q = \frac{\pi c^2}{2} \frac{4c}{3\pi} = \frac{2c^3}{3}$$

Moreover, since $t = 2c$, and $I = \pi c^4/4$, the maximum direct shear stress is

$$\tau_{max} = \frac{VQ}{It} = \frac{V}{2c} \frac{2c^3}{3} \frac{4}{\pi c^4} = \frac{4V}{3\pi c^2} = \frac{4V}{3A} \qquad (7\text{-}15)$$

where $A$ is the *entire* cross-sectional area of the rod. (A similar expression was derived in Example 7-3 for a beam of rectangular section.) In Fig. 7-26(d), this shear stress is shown acting down on the elementary areas at $E$, $C$, and $D$. This direction agrees with the direction of shear $V$.

To find the maximum combined shear stress in plane $A–B$, the stresses shown in Figs. 7-26(c) and (d) are superposed. Inspection shows that the maximum shear stress is at $E$, since in the two diagrams, the shear stresses at $E$ act in the same direction. There are no direct shear stresses at $A$ and $B$, while at $C$ there is no torsional shear stress. The two shear stresses act in *opposite* directions at $D$.

The combined shear stresses at the five points, $A$, $B$, $C$, $D$, and $E$, unlike most of the interior points, require no formal vectorial addition for determining their magnitudes. Since the torsional shear stresses at the interior points are smaller than those at the boundary, the maximum combined shear occurs at $E$.

$$J = \frac{\pi d^4}{32} = \frac{\pi \times 10^4}{32} = 982 \text{ mm}^4$$

$$I = \frac{J}{2} = 491 \text{ mm}^4$$

$$A = \frac{1}{4}\pi d^2 = 78.5 \text{ mm}^2$$

$$(\tau_{max})_{torsion} = \frac{Tc}{J} = \frac{20 \times 10^3 \times 5}{982} = 102 \text{ MPa}$$

$$(\tau_{max})_{direct} = \frac{VQ}{It} = \frac{4V}{3A} = \frac{4 \times 250}{3 \times 78.5} = 4 \text{ MPa}$$

$$\tau_E = 102 + 4 = 106 \text{ MPa}$$

A planar representation of the shear stress at $E$ with the matching stresses on the longitudinal planes is shown in Fig. 7-26(f). No normal stress acts on this element as it is located on the neutral axis.

## **\*\*[16]7-10. Stresses in Closely Coiled Helical Springs**

Helical springs, such as the one shown in Fig. 7-27(a), are often used as elements of machines. With certain limitations, these springs may be analyzed for stresses by a method similar to the one used in the preceding example. The discussion will be limited to springs manufactured from rods or wires of circular cross section.[17] Moreover, *any one coil of such a spring will be assumed to lie in a plane that is nearly perpendicular to the axis of the spring*. This requires that the adjoining coils be close together. With this limitation, a section taken perpendicular to the axis of the spring's rod becomes *nearly vertical*.[18] Hence, to maintain equilibrium of a segment of the spring, only a shear force $V = F$ and a torque $T = F\bar{r}$ are required at *any* section through the rod; see Fig. 7-27(b). Note that

---

[16] This section is on a specialized topic, and is optional.

[17] For an extensive discussion on springs, see A. M. Wahl, *Mechanical Springs* (Cleveland: Penton, 1944).

[18] This eliminates the necessity of considering an axial force and a bending moment at the section taken through the spring.

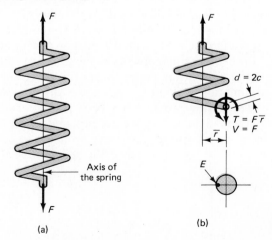

(a)

(b)

**Fig. 7-27** Closely coiled helical spring.

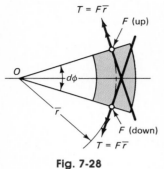

$T = F\bar{r}$

$F$ (up)

$O$

$d\phi$

$\bar{r}$

$F$ (down)

$T = F\bar{r}$

**Fig. 7-28**

$\bar{r}$ is the distance from the axis of the spring to the *centroid of the rod's cross-sectional area.*

Here it should be noted that in previous work, it has been reiterated that if a shear is present at a section, a change in the bending moment must take place along the member. Here a shear acts at every section of the rod, yet no bending moment nor a change in it appears to occur. This is so only because the rod is *curved.* Such an element of the rod viewed from the top is shown in Fig. 7-28. At both ends of the element, the torques are equal to $F\bar{r}$, and, using vectorial representation, act in the directions shown. The component of these vectors toward the axis of the spring $O$, resolved at the point of intersection of the vectors, $2F\bar{r}\,d\phi/2 = F\bar{r}\,d\phi$, opposes the couple developed by the vertical shears $V = F$, which are $\bar{r}\,d\phi$ apart.

The maximum shear stress at an arbitrary section through the rod could be obtained as in the preceding example, by superposing the torsional and the direct shearing stresses. This maximum shear stress occurs at the inside of the coil at point $E$, Fig. 7-27(b). However, in the analysis of springs, it has become *customary to assume that the shear stress caused by the direct shear force is uniformly distributed over the cross-sectional area of the rod.* Hence, the nominal direct shear stress for any point on the cross section is $\tau = F/A$. Superposition of this *nominal* direct and the torsional shear stress at $E$ gives the maximum combined shear stress. Thus, since $T = F\bar{r}$, $d = 2c$, and $J = \pi d^4/32$,

$$\tau_{\max} = \frac{F}{A} + \frac{Tc}{J} = \frac{Tc}{J}\left(\frac{FJ}{ATc} + 1\right) = \frac{16F\bar{r}}{\pi d^3}\left(\frac{d}{4\bar{r}} + 1\right) \qquad (7\text{-}16)$$

It is seen from this equation that as the diameter of the rod $d$ becomes small in relation to the coil radius $\bar{r}$, the effect of the direct shear stress also becomes small. On the other hand, if the reverse is true, the first term in the parentheses becomes important. However, in the latter case, the results indicated by Eq. 7-16 are considerably in error, and Eq. 7-16 should not be used, as it is based on the torsion formula for *straight rods.* As $d$ becomes numerically comparable to $\bar{r}$, the length of the inside fibers of the coil differs greatly from the length of the outside fibers, and the assumptions of strain used in the torsion formula are not applicable.

The spring problem has been solved exactly[19] by the methods of the mathematical theory of elasticity, and while these results are complicated, for any one spring, they may be made to depend on a single parameter $m = 2\bar{r}/d$, which is called the *spring index.* Thus, Eq. 7-16 may be rewritten as

$$\tau_{\max} = K\frac{16F\bar{r}}{\pi d^3} \qquad (7\text{-}17)$$

[19] O. Goehner, "Die Berechnung Zylindrischer Schraubenfedern," *Zeitschrift des Vereins deutscher Ingenieure* 76/1 (March 1932): 269.

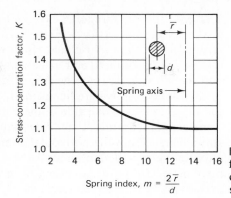

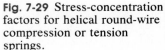

Fig. 7-29 Stress-concentration factors for helical round-wire compression or tension springs.

where $K$ may be interpreted as a stress-concentration factor for closely coiled helical springs made from circular rods. A plot of $K$ vs. the spring index is shown[20] in Fig. 7-29. For heavy springs, the spring index is small; hence, the stress-concentration factor $K$ becomes important. For all cases, factor $K$ accounts for the correct amount of direct shear stress. Very high stresses are commonly allowed in springs because high-strength materials are used in their fabrication. For good-quality spring steel, working shear stresses range anywhere from 200 to 700 MPa (30 to 100 ksi).

## **7-11. Deflection of Closely Coiled Helical Springs

For completeness, the deflection of closely coiled helical springs will be discussed in this section. Attention will be confined to closely coiled helical springs with a large spring index, i.e., the diameter of the wire will be assumed small in comparison with the radius of the coil. This permits the treatment of an element of a spring between two closely adjoining sections through the wire as a *straight circular bar in torsion*. The effect of direct shear on the deflection of the spring will be ignored. This is usually permissible as the latter effect is small.

Consider a helical spring such as shown in Fig. 7-30. A typical element $AB$ of this spring is subjected throughout its length to a torque $T = F\bar{r}$. This torque causes a relative rotation between the two adjoining planes, $A$ and $B$, and with sufficient accuracy, the amount of this rotation may be obtained by using Eq. 4-14, $d\phi = T\,dx/JG$, for straight circular bars. For this equation, the applied torque $T = F\bar{r}$, $dx$ is the length of the element, $G$ is the shear modulus of elasticity, and $J$ is the polar moment of inertia of the *wire's cross-sectional area*.

If the plane $A$ of the wire is imagined fixed, the rotation of the plane

[20] An analytical expression that gives the value of $K$ within 1 or 2 percent of the true value is frequently used. This expression in terms of spring index $m$ is $K_1 = (4m - 1)/(4m - 4) + 0.615/m$. It was derived by A. M. Wahl in the 1940s on the basis of some simplifying assumptions and is known as the *Wahl correction factor* for curvature in helical springs.

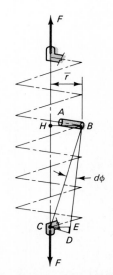

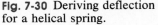

Fig. 7-30 Deriving deflection for a helical spring.

*B* is given by the foregoing expression. The contribution of this element to the movement of force *F* at *C* is equal to distance *BC* multiplied by angle $d\phi$, i.e., $CD = BC\ d\phi$. However, since element *AB* is small, distance *CD* is also small, and this distance may be considered perpendicular (although it is an arc) to line *BC*. Moreover, only the vertical component of this deflection is significant, as in a spring consisting of many coils, for any element on one side of the spring, there is a corresponding equivalent element on the other. The diametrically opposite elements of the spring balance out the horizontal component of the deflection and permit only the vertical deflection of force *F*. Therefore, by finding the *vertical* increment *ED* of the deflection of force *F* due to an element of spring *AB* and summing such increments for *all* elements of the spring, the deflection of the whole spring is obtained.

From similar triangles *CDE* and *CBH*,

$$\frac{ED}{CD} = \frac{HB}{BC} \quad \text{or} \quad ED = \frac{CD}{BC}\,HB$$

However, $CD = BC\ d\phi$, $HB = \bar{r}$, and *ED* may be denoted by $d\Delta$, as it represents an infinitesimal vertical deflection of the spring due to rotation of an element *AB*. Thus, $d\Delta = \bar{r}\ d\phi$ and

$$\Delta = \int d\Delta = \int \bar{r}\ d\phi = \int_0^L \bar{r}\,\frac{T\ dx}{JG} = \frac{TL\bar{r}}{JG}$$

However, $T = F\bar{r}$, and for a closely coiled spring, the *length L of the wire* may be taken with sufficient accuracy as $2\pi\bar{r}N$, where *N* is the number of *live* or active coils of the spring. Hence, the deflection $\Delta$ of the spring is

$$\Delta = \frac{2\pi F\bar{r}^3 N}{JG} \tag{7-18a}$$

or if the value of *J* for the wire is substituted,

$$\Delta = \frac{64F\bar{r}^3 N}{Gd^4} \tag{7-18b}$$

Equations 7-18a and 7-18b give the deflection of a closely coiled helical spring along its axis when such a spring is subjected to either a tensile or compressive force *F*. In these formulas, the effect of the direct shear stress on the deflection is neglected, i.e., they give only the effect of torsional deformations.

The behavior of a spring may be conveniently defined by its *spring constant k*. From Eq. 7-18b, the spring constant for a helical spring made from a wire with a circular cross section is

$$k = \frac{F}{\Delta} = \frac{Gd^4}{64\bar{r}^3N} \qquad \left[\frac{N}{m}\right] \qquad \text{or} \qquad \left[\frac{lb}{in}\right]$$

**Fig. 7-31**

### **EXAMPLE 7-9

Find the maximum stress in the 15-mm diameter steel rod shown in Fig. 7-31 caused by a 3-kg mass freely falling through 0.5 m. The steel helical spring of 35 mm outside diameter inserted into the system is made of 5-mm round wire and has 10 live coils. Let $E = 200$ GPa and $G = 80$ GPa.

### Solution

The static deflection of the 3-kg mass exerting a force of $3g = 29.4$ N on the spring is computed first. It consists of two parts: the deflection of the rod given by Eq. 2-9, and the deflection of the spring given by Eq. 7-18b. For use in Eq. 7-18b, $\bar{r} = 15$ mm. Then, from Eq. 2-28, the dynamic force acting on the spring and the rod is found. This force is used for finding the stress in the rod. Here the rod cross-sectional area $A = \pi \times 15^2/4 = 177$ mm$^2$.

$$\Delta_{st} = \Delta_{rod} + \Delta_{spr} = \frac{PL}{AE} + \frac{64F\bar{r}^3N}{Gd^4}$$

$$= \frac{29.4 \times 750}{177 \times 200 \times 10^3} + \frac{64 \times 29.4 \times 15^3 \times 10}{80 \times 10^3 \times 5^4} = 1.27 \text{ mm}$$

$$P_{dyn} = W\left(1 + \sqrt{1 + \frac{2h}{\Delta_{st}}}\right) = 29.4\left(1 + \sqrt{1 + \frac{2 \times 500}{1.27}}\right) = 855 \text{ N}$$

$$\sigma_{dyn} = \frac{P_{dyn}}{A} = \frac{855}{177} = 4.8 \text{ MPa}$$

For a free fall of the mass of 0.5 m without the spring, an elastic rod stress would be 210 MPa. For the system with the spring, most of the reduction in stress is due to $\Delta_{spr}$.

# Problems

## Section 7-3

✓ **7-1.** The cross section of a beam made up of a full-sized 6 × 6 in member reinforced with a 2 × 6 in plank is shown in the figure. What forces are exerted on 20d (20-penny) common nails, spaced 6 in apart and staggered, when force $P = 500$ lb is applied to the middle of the span? Calculate the shear flow two ways: using the cross section of the plank and then using the cross section of the larger member.

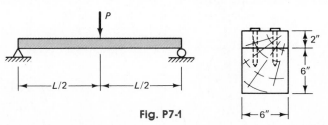

**Fig. P7-1**

**7-2.** The shear diagram for the box beam supporting a uniformly distributed load is conservatively approximated for design by the stepped diagram shown in the figure. If the beam is nailed together with 16d (16-penny) box nails from four full-sized pieces, as shown in the cross section, what nail spacing should be used along the span? Assume that each nail is good for 75 lb in shear.

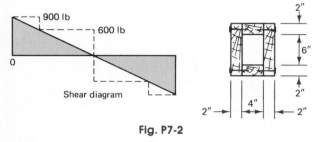

**Fig. P7-2**

**7-3.** A 10-in square box beam is to be made from four 2-in thick wood pieces. Two possible designs are considered, as shown in the figure. Moreover, the design shown in (a) can be turned 90° in the application. (a) Select the design requiring the minimum amount of nailing for transmitting shear. (b) If the shear to be transmitted by this member is 620 lb, what is the nail spacing for the best design? The nailing is to be done with 16d (16-penny) box nails that are good for 50 lb each in shear.

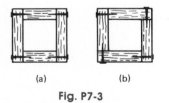

(a)  (b)

**Fig. P7-3**

**7-4.** A beam is loaded so that the moment diagram varies, as shown in the figure. For the cross section

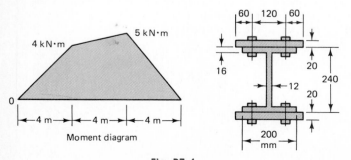

**Fig. P7-4**

shown, determine the bolt spacing for the critical region of the span. The bolts are arranged in pairs and the allowable shear force per high-strength bolt is 120 kN.

**7-5.** A wooden box beam, made up from 2-in thick boards, has the dimensions shown in the figure. If the beam transmits a vertical shear of 760 lb, what should be the longitudinal spacing of the nails (a) for connecting board $A$ with boards $B$ and $C$, and (b) for connecting board $D$ with boards $B$ and $C$?

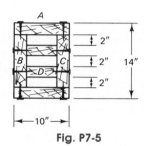

**Fig. P7-5**

**7-6.** Two W 8 × 67 beams are arranged as shown in the figure. Determine the bending and shear capacities of this member if the allowable bending stress is 24 ksi and the shear capacity of each bolt is 20 k. The bolts are arranged in pairs and are spaced 6 in on center.

**Fig. P7-6**

**7-7.** A plate girder is made up from two 14 × ½ in cover plates, four 6 × 4 × ½ in angles, and a 39½ × ⅜ in web plate, as shown in the figure. If at the section consid-

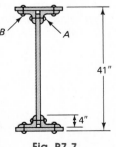

**Fig. P7-7**

ered, a total vertical shear of 150 k is transmitted, what must be the spacing of rivets *A* and *B*? For the girder around the neutral axis, *I* is 14,560 in⁴. Assume $\frac{3}{4}$-in rivets and note that one rivet is good for 6.63 k in single shear, 13.25 k in double shear, and 11.3 k in bearing on a $\frac{3}{8}$-in plate.

**7-8.** A simply supported beam has a cross section consisting of a C 12 × 20.7 and a W 18 × 50 fastened together by $\frac{3}{4}$-in-diameter bolts spaced longitudinally 6 in apart in each row, as shown in the figure. If this beam is loaded with a downward concentrated force of 112 k in the middle of the span, what is the shear stress in the bolts? Neglect the weight of the beam. The moment of inertia *I* of the whole member around the neutral axis is 1120 in⁴.

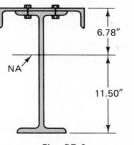

**Fig. P7-8**

**7-9.** Machined channel-like caps are attached to a plate to form the beam cross section shown in the figure. The interconnecting 10-mm-diameter rivets are placed longitudinally 80 mm on center. If the allowable shear stress for the rivets is 50 MPa, what is the allowable shear for this section?

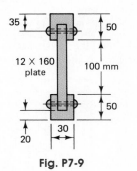

**Fig. P7-9**

**7-10.** A T-flange girder is used to support a 900-kN load in the middle of a 7-m simple span. The dimensions of the girder are given in the figure in a cross-

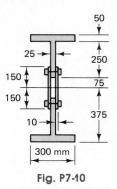

**Fig. P7-10**

sectional view. If the 22-mm-diameter rivets are spaced 125 mm apart longitudinally, what shear stress will be developed in the rivets by the applied loading? The moment of inertia of the girder around the neutral axis is approximately 4300 × 10⁶ mm⁴.

## Sections 7-4 and 7-5

**7-11.** Show that a formula, analogous to Eq. 7-8a, for beams having a solid circular cross section of area *A* is $\tau_{max} = \frac{4}{3}V/A$.

**7-12.** Show that a formula, analogous to Eq. 7-8a, for thin-walled circular tubes acting as beams having a net cross-sectional area *A* is $\tau_{max} = 2V/A$.

**7-13.** A T beam has the cross section shown in the figure. Calculate the shear stresses for the indicated six horizontal sections when the beam transmits a vertical shear of 240 kN. Plot the results as in Fig. 7-16(e).

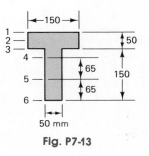

**Fig. P7-13**

**7-14.** A box beam has the cross section shown in the figure. Calculate shear stresses at several horizontal sections when the beam bending moment changes along the beam at the rate of 500 kN·m/m and plot the results as in Fig. 7-16(e).

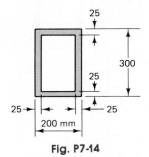

**Fig. P7-14**

**7-15.** A thin-walled extrusion has a cross section in the form of an isosceles triangle, as shown in the figure. Using Eq. 7-6, determine the shear stresses at the midheight and centroidal levels of the cross section corresponding to the vertical shear $V = 100$ kN. Calculate the approximate section properties for the member using the *centerline dimensions* shown on the detail. (*Hint:* For a thin inclined rectangular area, $I = bLh^2/12$, where $b$ is its width, $L$ its length, and $h$ its vertical height. Justify before using.)

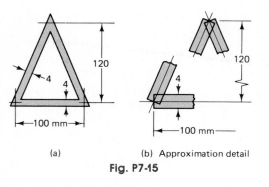

(a)                     (b) Approximation detail

**Fig. P7-15**

**7-16.** A beam has a cross-sectional area in the form of an isosceles triangle for which the base $b$ is equal to one-half its height $h$. (a) Using calculus and the conventional stress-analysis formula, determine the location of the maximum shear stress caused by a vertical shear $V$. Draw the manner in which the shear stress varies across the section. (b) If $b = 3$ in, $h =$

**Fig. P7-16**

6 in, and $\tau_{max}$ is limited to 100 psi, what is the maximum vertical shear $V$ that this section may carry?

**7-17.** A beam has a rhombic cross section, as shown in the figure. Assume that this beam transmits a vertical shear of 5000 N, and investigate the shear stresses at levels 50 mm apart, beginning with the apex. Report the results on a plot similar to the one shown in Fig. 7-16(e).

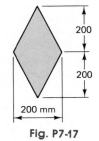

**Fig. P7-17**

**7-18.** A beam is loaded such that the moment diagram varies as shown in the figure. (a) Find the maximum longitudinal shear force acting on the $\frac{1}{2}$-in-diameter bolts spaced 12 in apart. (b) Find the shear stress in the glued joint.

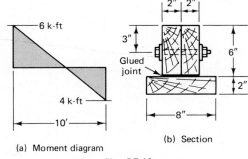

(a) Moment diagram                     (b) Section

**Fig. P7-18**

**7-19.** A beam has the cross-sectional dimensions shown in the figure. If the allowable stresses are 7 ksi

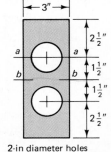

**Fig. P7-19**     2-in diameter holes

in tension, 30 ksi in compression, and 8 ksi in shear, what is the maximum allowable shear and the maximum allowable bending moment for this beam? Consider only the vertical loading of the beam and confine calculations for shear to sections $a$–$a$ and $b$–$b$.

**7-20.** A wooden I beam is made up with a narrow lower flange because of space limitations, as shown in the figure. The lower flange is fastened to the web with nails spaced longitudinally 6 in apart, and the vertical boards in the lower flange are glued in place. Determine the stress in the glued joints and the force carried by each nail in the nailed joint if the beam is subjected to a vertical shear of 400 lb. The moment of inertia for the whole section around the neutral axis is 2640 in⁴.

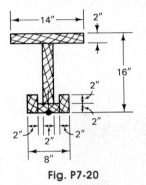

**Fig. P7-20**

**7-21.** A steel cantilever beam is fabricated from two structural tees welded together as shown in the figure. Determine the allowable force $P$ that the beam can carry if the allowable stress in bending is 150 MPa; in shear, 100 MPa; and along the weld, 2 MN/m. Neglect the weight of the beam.

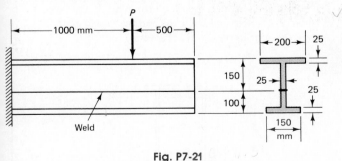

**Fig. P7-21**

**7-22.** A box beam is fabricated by nailing plywood sides to two longitudinal wooden pieces, as shown in the figure. If the allowable shear stress for plywood is 1.5 MPa and the allowable shear strength per nail is

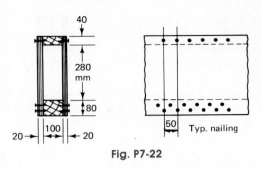

**Fig. P7-22**

500 N, determine the maximum allowable vertical shear for this member.

**7-23.** A wooden joist having the cross-sectional dimensions, in mm, shown in the figure is to be made from Douglas Fir lumber flanges and structural-grade plywood web. If the allowable shear stress on plywood is 2 MPa, what strength glue must be specified for the interfaces between the flanges and the web for a balanced design in shear?

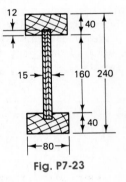

**Fig. P7-23**

**7-24.** A beam is made up of four 50 × 100 mm full-sized Douglas Fir pieces that are glued to a 25 × 450 mm Douglas Fir plywood web, as shown in the figure. Determine the maximum allowable shear and the maximum allowable bending moment that this section can

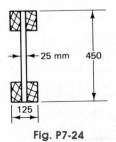

**Fig. P7-24**

carry if the allowable bending stress is 10 MPa; the allowable shear stress in plywood is 600 kN/m² and the allowable shearing stress in the glued joints is 300 kN/m². All dimensions in the figure are in mm.

**7-25.** Calculate the bending and the shear stresses due to the applied force $P$ acting on element $A$ for the cantilever shown in the figure, and show these stresses acting on an isolated sketch of the element.

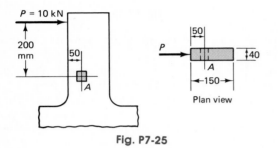

**Fig. P7-25**

**7-26.** A W 14 × 90 beam supports a uniformly distributed load of 4 k/ft, including its own weight, as shown in the figure. Determine the bending and the shear stresses acting on elements $A$ and $B$. Show the magnitude and sense of the computed quantities on infinitesimal elements.

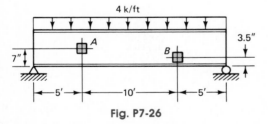

**Fig. P7-26**

**7-27.** Isolate the 50 × 150 × 200 mm shaded element from the rectangular beam having a 200 × 300 mm cross section and loaded as shown in the figure. On a free-body diagram, indicate the location, magnitude,

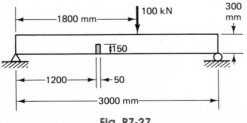

**Fig. P7-27**

and sense of all resultant forces due to the bending and shear stresses acting on this segment. Neglect the weight of the beam.

**\*7-28.** Two steel bars are bonded to an aluminum alloy core, making up a sandwich beam having the cross section shown in the figure. If this beam is loaded so that the bending moment changes at the rate of 5 kN·m/m, what maximum shear stress develops in the member? For steel, $E_{St}$ = 210 GPa; for aluminum alloy, $E_{Al}$ = 70 GPa. (*Hint:* The stress distribution pattern can be established, as shown in Fig. 6-7(c), and only the change in bending moment per unit distance along the beam need be considered, as shown in Fig. 7-2(d). Alternatively, a transformed section, discussed in Section 6-8, can be used.)

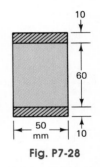

**Fig. P7-28**

**\*7-29.** The cross section of a beam of two different materials has the dimensions shown in the figure. The elastic modulus for the vertical web members is $E_w$ = 30 × 10³ ksi and that for the five pieces for the flange material is $E_f$ = 15 × 10³ ksi. If the vertical shear transmitted by this member is 4 k, (a) what is the maximum shear stress in the web? (b) What is the largest shear stress between the webs and the flange material? (See the hint in preceding problem.)

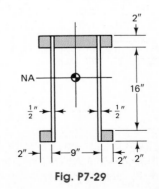

**Fig. P7-29**

**\*7-30.** An aluminum alloy extrusion for use as a beam has the cross-sectional dimensions in mm shown in the figure. Due to a vertical shear $V = 10$ kN, determine the shear stresses in the vertical walls of the member at the horizontal section passing through the section centroid.

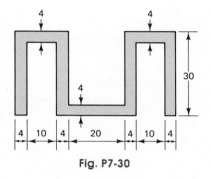

Fig. P7-30

**\*7-31.** The cross-sectional dimensions of a beam of a synthetic thermoplastic material are given in mm in the figure. The member material is 3 mm thick throughout. (a) Calculate the moment of inertia $I$ for the entire cross-sectional area around the horizontal centroidal axis. Use centerline dimensions as shown in Prob. 7-15. (b) Determine the magnitudes of the shear stresses at sections $a$–$a$, $b$–$b$, and $c$–$c$ due to the vertical shear $V = 10$ kN.

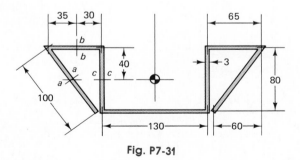

Fig. P7-31

**7-32.** A metal box beam has the cross section in mm shown in the figure. If $V/I$ is 0.006 N/mm$^4$, what shear stresses occur at sections $a$–$a$ and $b$–$b$? The centroid for the cross section is given. Use centerline dimensions for calculating area properties (see Prob. 7-15). (*Hint:* Take advantage of symmetry.)

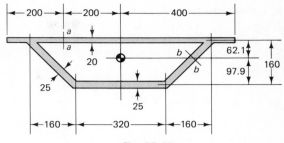

Fig. P7-32

**7-33.** A metal beam is made up of four $30 \times 24 \times 4$ mm angles attached with an adequate number of bolts to a $100 \times 4$ mm web plate, as shown in mm in the figure. Determine the shear stresses at section $a$–$a$. Let $V/I = 0.01$ N/mm$^4$.

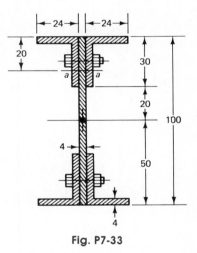

Fig. P7-33

**7-34.** A beam is fabricated by slotting 4-in standard steel pipes longitudinally and then securely welding them to a $23 \times \frac{3}{8}$ in web plate, as shown in the figure.

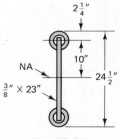

Fig. P7-34

$I$ of the composite section around the neutral axis is 1018 in⁴. If at a certain section, this beam transmits a vertical shear of 40 k, determine the shear stress in the pipe and in the web plate at a level 10 in above the neutral axis.

## Sections 7-7 and 7-8

**7-35.** A beam having a cross section with the dimensions in mm shown in the figure is in a region where there is a constant, positive vertical shear of 100 kN. (a) Calculate the shear flow $q$ acting at each of the four sections indicated in the figure. (b) Assuming a positive bending moment of 27 kN·m at one section and a larger moment at the adjoining section 10 mm away, draw isometric sketches of each segment of the beam isolated by the sections 10 mm apart and the four sections shown in the figure, and on the sketches indicate all forces acting on the segments. Neglect vertical shear stresses in the flanges.

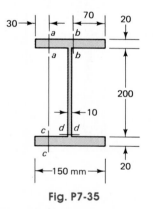

Fig. P7-35

**7-36.** A beam having the cross section with the dimensions shown in the figure transmits a vertical shear

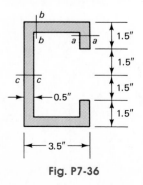

Fig. P7-36

$V = 7$ k applied through the shear center. (a) Determine the shear stresses at sections $a$–$a$, $b$–$b$, and $c$–$c$. $I$ around the neutral axis is 35.7 in⁴. The thickness of the material is $\frac{1}{2}$ in throughout. (b) Sketch the shear stress distribution along the centerline of the member.

**7-37 through 7-40.** Determine the location of the shear center for the beams having the cross-sectional dimensions shown in the figures. All members are to be considered thin-walled, and calculations should be based on the centerline dimensions.

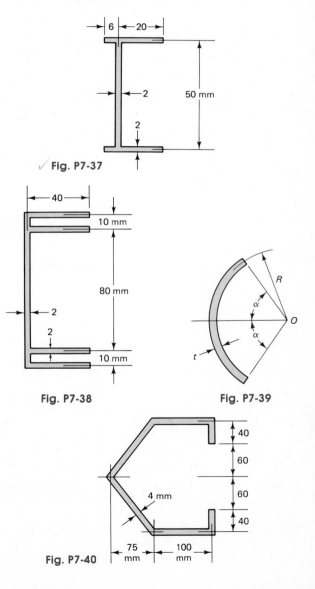

✓ Fig. P7-37

Fig. P7-38

Fig. P7-39

Fig. P7-40

**\*\*7-41.** Show that for the Z cross section shown in part (a) of the figure, the shear center lies on the vertical axis passing through the section's centroid. Demonstrating that the force resultant due to the internal shear flow is zero in each flange constitutes a proof. (*Hint:* Apply vertical shear force $V_y$ shown in the figure. Since the position of the shear center is independent of the magnitude of shear $V_y$, it can be chosen arbitrarily. Likewise, it is the rate of change in $M$, as in Fig. 7-2(d), rather than its magnitude that is of importance. Therefore, let the change in moment $\Delta M$ in a unit span distance be $V_y \times 1 = 10$ lb-in. Using this $\Delta M$, calculate the stresses along the centerline of the Z section using the generalized flexure formula, Eq. 6-64. Such calculations should verify the normal stress distribution shown in part (b) of the figure. By integrating these stresses along the section centerline, as in Fig. 7-20(b), the shear flow in the plane of the section is found, part (c) of the figure. The force resultants based on these shears vanish in both flanges.) Note that for the given section, $I_z = 2.133$ in$^4$, $I_y = 0.533$ in$^4$, and $I_{yz} = 0.800$ in$^4$.

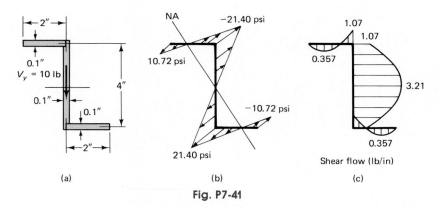

(a)    (b)    (c)

Shear flow (lb/in)

**Fig. P7-41**

**\*\*7-42.** Show that for the Z cross section in the preceding problem, the shear center lies on the horizontal line passing through the section's centroid. Demonstrating that an applied horizontal force is equally divided between the two flanges constitutes a solution of the problem. (*Hint:* Apply a horizontal force $V_z$, say equal to 10 lb, through the section's centroid, and as in the preceding problem, calculate the normal stresses and the shear stresses in the plane of the cross section. An auxiliary plot of the shear stresses along the section centerline is useful in the solution. Note that the resultant shear force in the web is zero.)

**7-43 and 7-44.** Determine the location of the shear center for the beams with idealized cross sections shown in the figures. Neglect the areas of the plates connecting the longitudinal stringers, each one of which has an effective area $A$ concentrated at a point for resisting longitudinal forces. There are two such areas in Fig. P7-43 and eight in Fig. P7-44. (This kind of idealization is often used in aircraft design.)

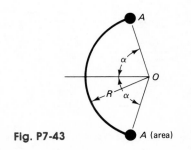

**Fig. P7-43**

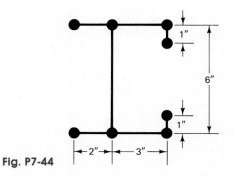

**Fig. P7-44**

## Section 7-9

**7-45 through 7-47.** Cantilevers of the kind shown in the figure for Prob. 7-45 are subjected to horizontal forces $P$, causing bending, direct shear, and torsion. Determine the stresses at the surfaces due to these actions at points $A$ and $B$, and show the results on isolated infinitesimal elements. Elements $A$ should be viewed from the top, and elements $B$ from the left. Use centerline dimensions for the box in calculating the torsional stresses in Prob. 7-45. The details for applying forces $P$ are not shown.

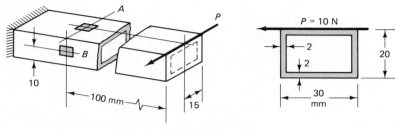

**Fig. P7-45**

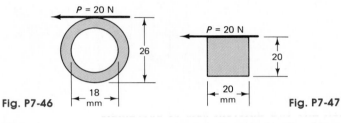

**Fig. P7-46**

**Fig. P7-47**

## Sections 7-10 and 7-11

**7-48.** A helical valve spring is made of $\frac{1}{4}$-in-diameter steel wire and has an outside diameter of 2 in. In operation, the compressive force applied to this spring varies from 20 lb minimum to 70 lb maximum. If there are eight active coils, what is the valve lift (or travel), and what is the maximum shear stress in the spring when in operation? $G = 11.6 \times 10^6$ psi.

**7-49.** If a helical tension spring consisting of 12 live coils of 6-mm steel wire and of 30 mm outside diameter is attached to the end of another helical tension spring of 18 live coils of 8-mm steel wire and of 40 mm outside diameter, what is the spring constant for this two-spring system? What is the largest force that may be applied to these springs without exceeding a shear stress of 480 MPa? $G = 82$ GN/m$^2$.

**7-50.** A heavy helical steel spring is made from a 1-in-diameter rod and has an outside diameter of 9 in. As originally manufactured, it has the pitch $p = 3\frac{1}{2}$ in; see the figure. If a force $P$, of such magnitude that the rod's $\frac{1}{8}$-in thick outer annulus becomes plastic, is applied to this spring, estimate the reduction of the pitch of the spring on removal of the load. Assume linearly elastic-plastic material with $\tau_{yp} = 50$ ksi, and $G = 12 \times 10^3$ ksi. Neglect the effects of stress concentration and of the direct shear on deflection. (*Hint:* See Examples 4-13 and 4-14.)

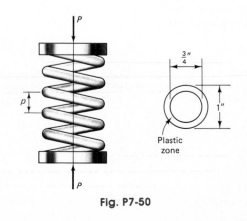

**Fig. P7-50**

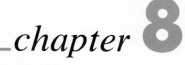

## Transformation of Stress and Strain; Yield and Fracture Criteria

### 8-1. Introduction

In the first two parts of this chapter, a formal treatment for changing the components of the state of stress or strain given in one set of coordinate axes to any other set of rotated axes is discussed. This transformation of stress or strain between any two different sets of coordinate axes is a mathematical process and does not necessarily require the use of formulas derived earlier. The connection between the established stress-analysis formulas and stress transformation is considered in the next chapter. Transformation of stress is discussed in Part A of this chapter; strain transformation in Part B. In both instances, *the discussion is largely confined to problems in two dimensions*. The possibility of transforming a given state of stress involving both normal and shear stresses to any other set of rotated coordinate axes permits an examination of the effect of such stresses on a material. In this manner, criteria for the onset of yield or the occurrence of fracture can be hypothesized. This important topic is treated in Part C.

## Part A    TRANSFORMATION OF STRESS

### 8-2. The Basic Problem

In several of the preceding chapters, stresses caused by separate actions causing either normal and/or shear stresses were considered. The superposition of normal stresses acting on the same element, when axial forces and bending occur simultaneously, was discussed in Chapter 6. Similarly,

**403**

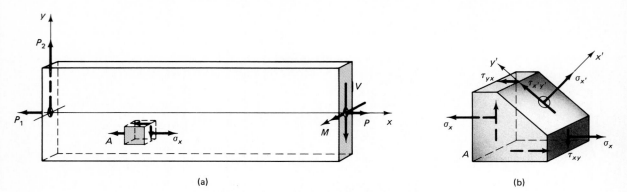

**Fig. 8-1** State of stress at a point on different planes.

the superposition of shear stresses caused by torque and direct shear was considered in Chapter 7. Moreover, in Chapter 3, Section 3-7, it was demonstrated that a state of pure shear can be transformed into an *equivalent* state of normal stresses (Fig. 3-11). Often in stress analysis, a more general problem arises, such as shown in Fig. 8-1(a). In the illustrated case, element $A$ is subjected to a normal stress $\sigma_x$ due to axial pull *and* bending, and *simultaneously* experiences a direct shear stress $\tau_{xy}$. The combined normal stress $\sigma_x$ follows by superposition. However, the combination of the normal stress $\sigma_x$ with the shear stress $\tau_{xy}$ requires special treatment. Essentially, this requires a consideration of stresses on an inclined plane, such as shown in Fig. 8-1(b). Since an inclined plane may be chosen arbitrarily, the state of stress at a point can be described in an *infinite number of ways,* which are all *equivalent.*

Stress has a magnitude and a sense, and is *also* associated with an area over which it acts. Such mathematical entities are *tensors* and are of a higher order than vectors.[1] However, the components of the stress on the *same area* are vectors. These stress components can be superposed by *vector addition.* As noted earlier, this was used in Chapter 6 and 7. Therefore the stresses can be referred to as *stress vectors* or *tractions,* provided they act on the same surface in or on a body. Only a change in the orientation of an area displays the non-vectorial character of the stress as a whole.

In the discussion that follows, direct use of stress vectors is avoided by multiplying stresses by their respective areas to obtain forces, which are vectors, and then adding them vectorially. On obtaining the force components on an inclined plane the process is reversed by dividing these force components by the inclined area to obtain the stresses on such planes.

This procedure will be first illustrated by a numerical example. Then the developed approach will be generalized to obtain algebraic relations for a stress transformation for finding stresses on any inclined plane from

---

[1] *Scalars* are tensors of rank zero, *vectors* are tensors of the first rank, and stresses and strains are second-rank tensors. See Section 1–4.

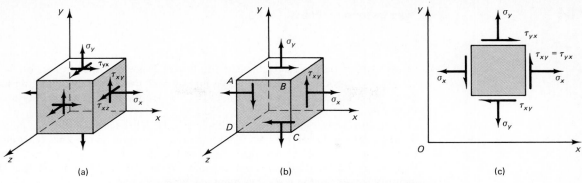

Fig. 8-2 Representations of stresses acting on an element.

a given state of stress. The methods used in these derivations do not involve properties of a material. Therefore, provided the initial stresses are given, the derived relations are applicable whether the material behaves elastically or plastically or even if it is not a solid. However, the planes on which the normal or the shear stresses reach their *maximum intensity* have a particularly significant effect on materials.

The general state of stress shown in Fig. 8-2(a) will not be considered in deriving the laws of stress transformation at a point. Instead, the two-dimensional stress problem indicated in Fig. 8-2(b) will be studied. In practical applications, this is a particularly important case as it is usually possible to select a critical element at an outer boundary of a member. The outside face of such an element, as *ABCD* in the figure, is generally free of significant surface stresses, whereas the stresses right at the surface acting parallel to it are usually the largest. Planar representation of the stresses, as shown in Fig. 8-2(c), will be used in most derivations and examples.

## EXAMPLE 8-1

Let the state of stress for an element of unit thickness be as shown in Fig. 8-3(a). An alternative representation of the state of stress at the same point may be given on an infinitesimal wedge with an angle of $\alpha = 22\frac{1}{2}°$, as in Fig. 8-3(b). Find the stresses that must act on plane *AB* of the wedge to keep the element in equilibrium.

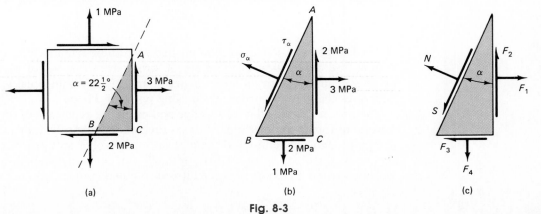

Fig. 8-3

**405**

### Solution

Wedge $ABC$ is part of the element in Fig. 8-3(a); therefore, the stresses on faces $AC$ and $BC$ are known. The unknown normal and shear stresses acting on face $AB$ are designated in the figure by $\sigma_\alpha$ and $\tau_\alpha$, respectively. Their sense is assumed arbitrarily.

To determine $\sigma_\alpha$ and $\tau_\alpha$, for convenience only, let the area of the face defined by line $AB$ be unity such as m². Then the area corresponding to line $AC$ is equal to $1 \times \cos \alpha = 0.924$ m² and that to $BC$ is equal to $1 \times \sin \alpha = 0.383$ m². (More rigorously, the area corresponding to line $AB$ should be taken as $dA$, but this quantity cancels out in the subsequent algebraic expressions.) Forces $F_1, F_2, F_3$, and $F_4$, Fig. 8-3(c), can be obtained by multiplying the stresses by their respective areas. The unknown equilibrant forces $N$ and $S$ act, respectively, normal and tangential to plane $AB$. Then applying the equations of static equilibrium to the forces acting on the wedge gives forces $N$ and $S$.

$$F_1 = 3 \times 0.924 = 2.78 \text{ MN} \qquad F_2 = 2 \times 0.924 = 1.85 \text{ MN}$$
$$F_3 = 2 \times 0.383 = 0.766 \text{ MN} \qquad F_4 = 1 \times 0.383 = 0.383 \text{ MN}$$

$$\sum F_N = 0 \qquad N = F_1 \cos \alpha - F_2 \sin \alpha - F_3 \cos \alpha + F_4 \sin \alpha$$
$$= 2.78(0.924) - 1.85(0.383) - 0.766(0.924)$$
$$+ 0.383(0.383)$$
$$= 1.29 \text{ MN}$$

$$\sum F_S = 0 \qquad S = F_1 \sin \alpha + F_2 \cos \alpha - F_3 \sin \alpha - F_4 \cos \alpha$$
$$= 2.78(0.383) + 1.85(0.924) - 0.766(0.383)$$
$$- 0.383(0.924)$$
$$= 2.12 \text{ MN}$$

Forces $N$ and $S$ act on the plane defined by $AB$, which was initially assumed to be 1 m². Their positive signs indicate that their assumed directions were chosen correctly. By dividing these forces by the area on which they act, the stresses acting on plane $AB$ are obtained. Thus, $\sigma_\alpha = 1.29$ MPa and $\tau_\alpha = 2.12$ MPa and act in the direction shown in Fig. 8-3(b).

---

The foregoing procedure accomplished something remarkable. It transformed the *description* of the state of stress from one set of planes to another. Either system of stresses pertaining to an infinitesimal element describes the state of stress at the same point of a body.

The procedure of isolating a wedge and using the equations of the equilibrium of forces to determine stresses on inclined planes is fundamental. Ordinary sign conventions of statics suffice to solve any problem. The reader is urged to return to this approach whenever questions arise regarding the more advanced procedures developed in the remainder of this chapter.

## 8.3. Transformation of Stresses in Two-Dimensional Problems

By following the same procedure as in the last example, equations for the normal and shear stresses acting on an inclined plane can be derived in algebraic form. Such expressions are called stress-transformation equations. These equations are based on the initially given stresses acting on an element of known orientation and the plane being investigated defined by a normal to it. The dependence of the stresses on the inclination of the plane is clearly apparent.

Algebraic equations are developed using an element of unit thickness in Fig. 8-4(a) in a state of two-dimensional stress initially referred to the $xy$ axes. If normal stresses $\sigma_x$ and $\sigma_y$ are tensile stresses, they are taken as positive, and are negative if compressive. *Positive shear stress is defined as acting upward in the positive direction of the y-axis on the right (positive) face DE of the element.* This sign convention for shear stresses was introduced in Chapter 1 (see Fig. 1-3) and is generally used in continuum mechanics (elasticity, plasticity, rheology). However it *differs* from the conventional *beam shear sign convention* used in Chapters 5 and 7. Here the stress transformation is sought from the $xy$ coordinate axes to the $x'y'$ axes. The angle $\theta$, which locates the $x'$ axis, is positive when measured from the $x$ axis toward the $y$ axis in a counterclockwise direction.

The *outward normal* to the section forms an angle $\theta$ with the $x$-axis. If an area of the wedge isolated by this section is $dA$, the areas associated with the faces $AC$ and $AB$ are $dA \cos\theta$ and $dA \sin\theta$, respectively. By multiplying the stresses by their respective areas, a diagram with the forced acting on the wedge is constructed, Fig. 8-4(c). Then, by applying the equations of static equilibrium to the forces acting on the wedge, stresses $\sigma_{x'}$, and $\tau_{x'y'}$ are obtained:

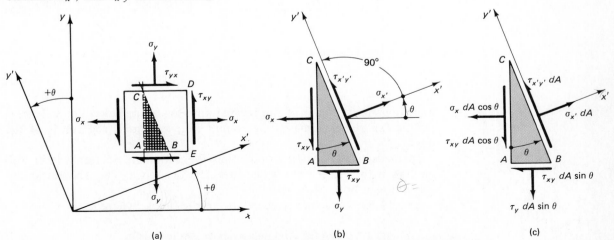

(a) (b) (c)

**Fig. 8-4** Derivation of stress transformation on an inclined plane.

$$\sum F_{x'} = 0 \qquad \sigma_{x'}\, dA = \sigma_x\, dA \cos\theta \cos\theta + \sigma_y\, dA \sin\theta \sin\theta$$
$$+ \tau_{xy}\, dA \cos\theta \sin\theta + \tau_{xy}\, dA \sin\theta \cos\theta$$
$$\sigma_{x'} = \sigma_x \cos^2\theta + \sigma_y \sin^2\theta + 2\tau_{xy} \sin\theta \cos\theta$$
$$= \sigma_x \frac{1 + \cos 2\theta}{2} + \sigma_y \frac{1 - \cos 2\theta}{2} + \tau_{xy} \sin 2\theta$$

$$\sigma_{x'} = \frac{\sigma_x + \sigma_y}{2} + \frac{\sigma_x - \sigma_y}{2} \cos 2\theta + \tau_{xy} \sin 2\theta \tag{8-1}$$

Similarly, from $\sum F_{y'} = 0$,

$$\tau_{x'y'} = -\frac{\sigma_x - \sigma_y}{2} \sin 2\theta + \tau_{xy} \cos 2\theta \tag{8-2}$$

Equations 8-1 and 8-2 are the general expressions for the normal and the shear stress, respectively, on any plane located by the angle $\theta$ and caused by a known system of stresses. These relations are the equations for transformation of stress from one set of coordinate axes to another. Note particularly that $\sigma_x$, $\sigma_y$, and $\tau_{xy}$ are initially known stresses.

Replacing $\theta$ in Eq. 8-1 by $\theta + 90°$ gives the normal stress in the direction of the $y'$ axis. This stress can be designated as $\sigma_{y'}$; see Fig. 1-3(b). Hence, on noting that $\cos(2\theta + 180°) = -\cos 2\theta$, and $\sin(2\theta + 180°) = -\sin 2\theta$, one has

$$\sigma_{y'} = \frac{\sigma_x + \sigma_y}{2} - \frac{\sigma_x - \sigma_y}{2} \cos 2\theta - \tau_{xy} \sin 2\theta \tag{8-3}$$

By adding Eqs. 8-1 and 8-3,

$$\sigma_{x'} + \sigma_{y'} = \sigma_x + \sigma_y \tag{8-4}$$

meaning that the sum of the normal stresses on any two mutually perpendicular planes remains the same,[2] i.e., *invariant*, regardless of the angle $\theta$.

Mathematically analogous equations were found in Section 6-16 in connection with area moments and products of inertia. The transformation equations in both cases are alike.

It is to be noted that in *plane strain* problems, where $\varepsilon_z = \gamma_{zx} = \gamma_{zy}$

---

[2] A similar relation for three-dimensional problems is $\sigma_{x'} + \sigma_{y'} + \sigma_{z'} = \sigma_x + \sigma_y + \sigma_z$.

= 0, a normal stress $\sigma_x$ can also develop. From Eq. 3-14c, this stress is given as

$$\sigma_z = \nu(\sigma_x + \sigma_y) \tag{8-5}$$

The forces resulting from this stress do not enter the relevant equilibrium equations used in deriving stress-transformation expressions. Moreover, by virtue of Eq. 8-4, the $\sigma_x + \sigma_y$ term in Eq. 8-5 remains constant regardless of $\theta$. Therefore, the derived equations for stress transformation are applicable for problems of *plane stress* as well as *plane strain*.

## 8-4. Principal Stresses in Two-Dimensional Problems

Interest often centers on the determination of the largest possible stress, as given by Eqs. 8-1 and 8-2, and the planes on which such stresses occur is found first. To find the plane for a maximum or a minimum normal stress, Eq. 8-1 is differential with respect to $\theta$ and the derivative set equal to zero, i.e.,

$$\frac{d\sigma_{x'}}{d\theta} = -\frac{\sigma_x - \sigma_y}{2} 2 \sin 2\theta + 2\tau_{xy} \cos 2\theta = 0$$

Hence,

$$\tan 2\theta_1 = \frac{\tau_{xy}}{(\sigma_x - \sigma_y)/2} \tag{8-6}$$

where the subscript of the angle $\theta$ is used to designate the angle that defines the plane of the maximum or minimum normal stress. Equation 8-6 has two roots, since the value of the tangent of an angle in the diametrically opposite quadrant is the same, as may be seen from Fig. 8-5. These roots are 180° apart, and, as Eq. 8-6 is for a double angle, the roots

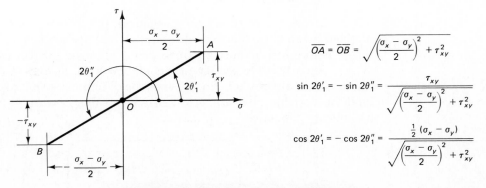

$$\overline{OA} = \overline{OB} = \sqrt{\left(\frac{\sigma_x - \sigma_y}{2}\right)^2 + \tau_{xy}^2}$$

$$\sin 2\theta_1' = -\sin 2\theta_1'' = \frac{\tau_{xy}}{\sqrt{\left(\frac{\sigma_x - \sigma_y}{2}\right)^2 + \tau_{xy}^2}}$$

$$\cos 2\theta_1' = -\cos 2\theta_1'' = \frac{\frac{1}{2}(\sigma_x - \sigma_y)}{\sqrt{\left(\frac{\sigma_x - \sigma_y}{2}\right)^2 + \tau_{xy}^2}}$$

**Fig. 8-5** Angle functions for principal stresses.

of $\theta_1$ are 90° apart. One of these roots locates a plane on which the maximum normal stress acts; the other locates the corresponding plane for the minimum normal stress. To distinguish between these two roots, a prime and double prime notation is used.

Before evaluating these stresses, carefully observe that if the location of planes on which no shear stresses act is wanted, Eq. 8-2 must be set equal to zero. This yields the same relation as that in Eq. 8-6. Therefore, an important conclusion is reached: on planes on which maximum or minimum normal stresses occur, there are no shear stresses. These planes are called the *principal planes* of stress, and the stresses acting on these planes—the maximum and minimum normal stresses—are called the *principal stresses*.

The magnitudes of the principal stresses can be obtained by substituting the values of the sine and cosine functions corresponding to the double angle given by Eq. 8-6 into Eq. 8-1. Then the results are simplified, and the expression for the maximum normal stress (denoted by $\sigma_1$) and the minimum normal stress (denoted by $\sigma_2$) becomes

$$(\sigma_{x'})_{\substack{\max \\ \min}} = \sigma_{1 \text{ or } 2} = \frac{\sigma_x + \sigma_y}{2} \pm \sqrt{\left(\frac{\sigma_x - \sigma_y}{2}\right)^2 + \tau_{xy}^2} \qquad (8\text{-}7)$$

where the positive sign in front of the square root must be used to obtain $\sigma_1$ and the negative sign to obtain $\sigma_2$. The planes on which these stresses act can be determined by using Eq. 8-6. A particular root of Eq. 8-6 substituted into Eq. 8-1 will check the result found from Eq. 8-4 and at the same time will locate the plane on which this principal stress acts.

## 8-5. Maximum Shear Stresses in Two-Dimensional Problems

If $\sigma_x$, $\sigma_y$, and $\tau_{xy}$ are known for an element, the shear stress on any plane defined by an angle $\theta$ is given by Eq. 8-2, and a study similar to the one made before for the normal stresses may be made for the shear stress. Thus, similarly, to locate the planes on which the maximum or the minimum shear stresses act, Eq. 8-2 must be differentiated with respect to $\theta$ and the derivative set equal to zero. When this is carried out and the results are simplified,

$$\tan 2\theta_2 = -\frac{(\sigma_x - \sigma_y)/2}{\tau_{xy}} \qquad (8\text{-}8)$$

where the subscript 2 is attached to $\theta$ to designate the plane on which the shear stress is a maximum or a minimum. Like Eq. 8-6, Eq. 8-8 has two roots, which again may be distinguished by attaching to $\theta_2$ a prime or a double prime notation. The two planes defined by this equation are mutually perpendicular. Moreover, the value of $\tan 2\theta_2$ given by Eq. 8-8 is a negative reciprocal of the value of $\tan 2\theta_1$ in Eq. 8-6. Hence, the roots for the double angles of Eq. 8-8 are 90° away from the corresponding roots of Eq. 8-6. This means that the angles that locate the planes of maximum or minimum shear stress form angles of 45° with the planes of the principal stresses. A substitution into Eq. 8-2 of the sine and cosine functions corresponding to the double angle given by Eq. 8-8 and determined in a manner analogous to that in Fig. 8-5 gives the maximum and the minimum values of the shear stresses. These, after simplifications, are

$$\tau_{\substack{\max \\ \min}} = \pm \sqrt{\left(\frac{\sigma_x - \sigma_y}{2}\right)^2 + \tau_{xy}^2} \qquad (8\text{-}9)$$

Thus, the maximum shear stress differs from the minimum shear stress only in sign. Moreover, since the two roots given by Eq. 8-8 locate planes 90° apart, this result also means that the numerical values of the shear stresses on the mutually perpendicular planes are the same. This concept was repeatedly used after being established in Section 1-4. In this derivation, the difference in sign of the two shear stresses arises from the convention for locating the planes on which these stresses act. From the physical point of view, these signs have no meaning, and for this reason, the largest shear stress regardless of sign will often be called the *maximum shear stress.*

The definite sense of the shear stress can always be determined by direct substitution of the particular root of $\theta_2$ into Eq. 8-2. A positive shear stress indicates that it acts in the direction assumed in Fig. 8-4(b), and vice versa. The determination of the maximum shear stress is of utmost importance for materials that are weak in shear strength. This will be discussed further in the Part C of this chapter.

Unlike the principal stresses for which no shear stresses occur on the principal planes, the maximum shear stresses act on planes that are usually not free of normal stresses. Substitution of $\theta_2$ from Eq. 8-8 into Eq. 8-1 shows that the normal stresses that act on the planes of the maximum shear stresses are

$$\sigma' = \frac{\sigma_x + \sigma_y}{2} \qquad (8\text{-}10)$$

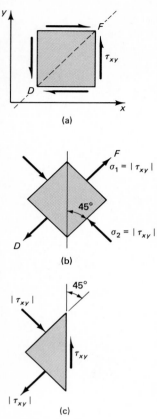

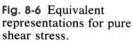

**Fig. 8-6** Equivalent representations for pure shear stress.

Therefore, a normal stress acts simultaneously with the maximum shear stress unless $\sigma_x + \sigma_y$ vanishes.

If $\sigma_x$ and $\sigma_y$ in Eq. 8-9 are the principal stress, $\tau_{xy}$ is zero and Eq. 8-9 simplifies to

$$\tau_{max} = \frac{\sigma_1 - \sigma_2}{2} \tag{8-11}$$

Here it is useful to recall a relationship between pure shear and the principal stresses discussed earlier in connection with Fig. 3-11. The results of this analysis are displayed in Fig. 8-6. Equation 8-7 clearly shows that in the absence of normal stresses, the principal stresses are numerically equal to the shear stress. The sense of the normal stresses follows from Eq. 8-6. The shear stresses act toward the diagonal $DF$ in the direction of the principal tensile stresses; see Fig. 8-6(a).

### EXAMPLE 8-2

For the state of stress in Example 8-1, reproduced in Fig. 8-7(a), (a) rework the previous problem for $\theta = -22\frac{1}{2}°$, using the general equations for the transformation of stress; (b) find the principal stresses and show their sense on a properly oriented element; and (c) find the maximum shear stresses with the associated normal stresses and show the results on a properly oriented element.

### Solution

(a) By directly applying Eqs. 8-1 and 8-2 for $\theta = -22\frac{1}{2}°$, with $\sigma_x = +3$ MPa, $\sigma_y = +1$ MPa, and $\tau_{xy} = +2$ MPa, one has

$$\begin{aligned}
\sigma_{x'} &= \frac{3+1}{2} + \frac{3-1}{2}\cos(-45°) + 2\sin(-45°) \\
&= 2 + 1 \times 0.707 - 2 \times 0.707 = +1.29 \text{ MPa} \\
\tau_{x'y'} &= -\frac{3-1}{2}\sin(-45°) + 2\cos(-45°) \\
&= +1 \times 0.707 + 2 \times 0.707 = +2.12 \text{ MPa}
\end{aligned}$$

The positive sign of $\sigma_{x'}$ indicates tension; whereas the positive sign of $\tau_{x'y'}$ indicates that the shear stress acts in the $+y'$ direction, as shown in Fig. 8-4(b). These results are shown in Fig. 8-7(b) as well as in Fig. 8-7(c).

(b) The principal stresses are obtained by means of Eq. 8-7. The planes on which the principal stresses act are found by using Eq. 8-6.

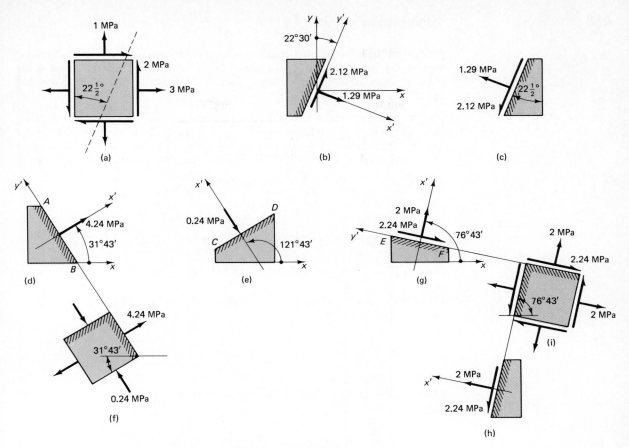

Fig. 8-7

$$\sigma_{1\text{ or }2} = \frac{3+1}{2} \pm \sqrt{\left(\frac{3-1}{2}\right)^2 + 2^2} = 2 \pm 2.24$$

$$\sigma_1 = +4.24 \text{ MPa (tension)} \qquad \sigma_2 = -0.24 \text{ MPa (compression)}$$

$$\tan 2\theta_1 = \frac{\tau_{xy}}{(\sigma_x - \sigma_y)/2} = \frac{2}{(3-1)/2} = 2$$

$$2\theta_1 = 63°26' \qquad \text{or} \qquad 63°26' + 180° = 243°26'$$

Hence, $\qquad \theta_1' = 31°43' \qquad$ and $\qquad \theta_1'' = 121°43'$

This locates the two principal planes, $AB$ and $CD$, Figs. 8-7(d) and (e), on which $\sigma_1$ and $\sigma_2$ act. On which one of these planes the principal stresses act is unknown. So, Eq. 8-1 is solved by using, for example, $\theta_1' = 31°43'$. The stress found by this calculation is the stress that acts on plane $AB$. Then, since $2\theta_1' = 63°26'$,

$$\sigma_{x'} = \frac{3+1}{2} + \frac{3-1}{2} \cos 63°26' + 2 \sin 63°26' = +4.24 \text{ MPa} = \sigma_1$$

This result, besides giving a check on the previous calculations, shows that the maximum principal stress acts on plane *AB*. The complete state of stress at the given point in terms of the principal stresses is shown in Fig. 8-7(f). Note that the results satisfy Eq. 8-4.

(*c*) The maximum shear stress is found by using Eq. 8-9. The planes on which these stresses act are defined by Eq. 8-8. The sense of the shear stresses is determined by substituting one of the roots of Eq. 8-8 into Eq. 8-2. Normal stresses associated with the maximum shear stress are determined by using Eq. 8-11.

$$\tau_{max} = \sqrt{[(3-1)/2]^2 + 2^2} = \sqrt{5} = 2.24 \text{ MPa}$$

$$\tan 2\theta_2 = -\frac{(3-1)/2}{2} = -0.500$$

$$2\theta_2 = 153°26' \quad \text{or} \quad 153°26' + 180° = 333°26'$$

Hence,     $\theta_2' = 76°43'$     and     $\theta_2'' = 166°43'$

These planes are shown in Figs. 8-7(g) and (h). Then, by using $2\theta_2' = 153°26'$ in Eq. 8-2,

$$\tau_{x'y'} = -\frac{3-1}{2}\sin 153°26' + 2\cos 153°26' = -2.24 \text{ MPa}$$

which means that the shear along plane *EF* has an opposite sense to that of the *y'* axis. From Eq. 8-7,

$$\sigma' = \frac{3+1}{2} = 2 \text{ MPa}$$

The complete results are shown in Fig. 8-7(i). Note again that Eq. 8-4 is satisfied.

The description of the state of stress can now be exhibited in three alternative forms: as the originally given data, and in terms of the stresses found in parts (*b*) and (*c*) of this problem. All these descriptions of the state of stress at the given point are equivalent. In matrix representation, this yields

$$\begin{pmatrix} 3 & 2 \\ 2 & 1 \end{pmatrix} \quad \text{or} \quad \begin{pmatrix} 4.24 & 0 \\ 0 & -0.24 \end{pmatrix} \quad \text{or} \quad \begin{pmatrix} 2 & -2.24 \\ -2.24 & 2 \end{pmatrix} \text{ MPa}$$

## 8.6. Mohr's Circle of Stress for Two-Dimensional Problems

In this section, the basic Eqs. 8-1 and 8-2 for the stress transformation at a point will be reexamined in order to interpret them graphically. In doing this, two objectives will be pursued. First, by graphically interpreting these equations, a greater insight into the general problem of stress transformation will be achieved. This is the main purpose of this section. Second, with the aid of graphical construction, a quicker solution of stress-transformation problems can often be obtained. This will be discussed in the following section.

A careful study of Eqs. 8-1 and 8-2 shows that they represent a circle written in parametric form. That they do represent a circle is made clearer by first rewriting them as

$$\sigma_{x'} - \frac{\sigma_x + \sigma_y}{2} = \frac{\sigma_x - \sigma_y}{2} \cos 2\theta + \tau_{xy} \sin 2\theta \qquad (8\text{-}12)$$

$$\tau_{x'y'} = -\frac{\sigma_x - \sigma_y}{2} \sin 2\theta + \tau_{xy} \cos 2\theta \qquad (8\text{-}13)$$

Then by squaring both these equations, adding, and simplifying,

$$\left(\sigma_{x'} - \frac{\sigma_x + \sigma_y}{2}\right)^2 + \tau_{x'y'}^2 = \left(\frac{\sigma_x - \sigma_y}{2}\right)^2 + \tau_{xy}^2 \qquad (8\text{-}14)$$

In a given problem, $\sigma_x$, $\sigma_y$, and $\tau_{xy}$ are the three known constants, and $\sigma_{x'}$ and $\tau_{x'y'}$ are the variables. Hence, Eq. 8-14 may be written in more compact form as

$$(\sigma_{x'} - a)^2 + \tau_{x'y'}^2 = b^2 \qquad (8\text{-}15)$$

where $a = (\sigma_x + \sigma_y)/2$, and $b^2 = [(\sigma_x - \sigma_y)/2]^2 + \tau_{xy}^2$ are constants.

This equation is the familiar expression of analytical geometry, $(x - a)^2 + y^2 = b^2$, for a circle of radius $b$ with its center at $(+a,0)$. Hence, if a circle satisfying this equation is plotted, the simultaneous values of a point $(x,y)$ on this circle correspond to $\sigma_{x'}$ and $\tau_{x'y'}$ for a particular orientation of an inclined plane. The ordinate of a point on the circle is the shear stress $\tau_{x'y'}$; the abscissa is the normal stress $\sigma_{x'}$. The circle so constructed is called a *circle of stress* or *Mohr's circle of stress*.[3]

By using the previous interpretation, a Mohr's circle for the stresses given in Fig. 8-8(a) is plotted in Fig. 8-8(c) with $\sigma$ and $\tau$ as the coordinate axes. The center $C$ is at $(a,0)$, and the circle radius $R = b$. Hence,

$$a = OC = \frac{\sigma_x + \sigma_y}{2} \qquad (8\text{-}16)$$

and

$$b = R = \sqrt{\left(\frac{\sigma_x - \sigma_y}{2}\right)^2 + \tau_{xy}^2} \qquad (8\text{-}17)$$

The coordinates for point $A$ on the circle correspond to the stresses in Fig. 8-8(a) on the *right face of the element*. For this face of the element,

---

[3] It is so named in honor of Otto Mohr of Germany, who in 1895 suggested its use in stress-analysis problems.

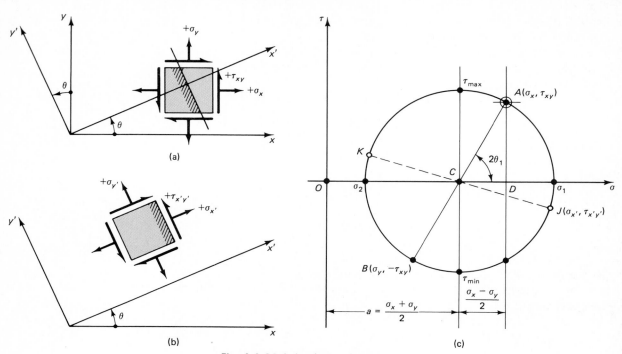

**Fig. 8-8** Mohr's circle of stress.

$\theta = 0°$, i.e., the $xy$ and the $x'y'$ axes coincide, and $\sigma_{x'} = \sigma_x$, and $\tau_{x'y'} = \tau_{xy}$. The positive directions for these stresses coincide with the positive directions of the axes. Since $AD/CD = \tau_{xy}/[(\sigma_x - \sigma_y)/2]$, according to Eq. 8-6, the angle $ACD$ is equal to $2\theta_1$. The coordinates for the conjugate point $B$ correspond to the stresses in Fig. 8-8(a) on the upper face of the element. This follows from Eqs. 8-1 and 8-2 with $\theta = 90°$ or alternatively for $\sigma_y'$ from Eq. 8-3 with $\theta = 0°$.

The same reasoning applies to any other orientation of an element, such as shown in Fig. 8-8(b). A pair of conjugate points $J$ and $K$ can always be found on the circle to give the corresponding stresses, Fig. 8-8(c). An infinity of possible states of stress dependent on the angle $\theta$ are defined by the stress circle. Therefore, the following important observations regarding the state of stress at a point can be made based on the Mohr's circle.

1. The largest possible normal stress is $\sigma_1$; the smallest is $\sigma_2$. No shear stresses exist together with either one of these principal stresses.

2. The largest shear stress $\tau_{max}$ is numerically equal to the radius of the circle, also to $(\sigma_1 - \sigma_2)/2$. A normal stress equal to $(\sigma_1 + \sigma_2)/2$ acts on each of the planes of maximum shear stress.

3. If $\sigma_1 = \sigma_2$, Mohr's circle degenerates into a point, and no shear stresses at all develop in the $xy$ plane.

4. If $\sigma_x + \sigma_y = 0$, the center of Mohr's circle coincides with the origin of the $\sigma\tau$ coordinates, and the state of pure shear exists.

5. The sum of the normal stresses on any two mutually perpendicular planes is invariant, i.e.,

$$\sigma_x + \sigma_y = \sigma_1 + \sigma_2 = \sigma_{x'} + \sigma_{y'} = \text{constant}$$

## *8-7. Construction of Mohr's Circles for Stress Transformation

The transformation of two-dimensional state of stress from one set of coordinates to another can always be made by direct application of statics as in Example 8-1, or, using the derived equations in Sections 8-3, 8-4, and 8-5. The latter equations can readily be programmed for a computer. However, the graphical display of stress transformations using a Mohr's circle offers a comprehensive view of a solution and is useful in some applications. Two alternative techniques for achieving such solutions are given in what follows. The physical planes on which the transformed stresses act are clearly displayed in the first method; in the second, the derivation for stress transformation is simpler, although determining the direction of the transformed stress is a little less convenient. The choice of method is a matter of preference.

### Method 1

The basic problem consists of constructing the circle of stress for given stresses $\sigma_x$, $\sigma_y$, and $\tau_{xy}$, such as shown in Fig. 8-9(a), and then determining the state of stress on an *arbitrary* plane $a$–$a$. A procedure for determining the stresses on any inclined plane requires justification on the basis of the equations derived in Section 8-3.

According to Eq. 8-16 the center $C$ of a Mohr's circle of stress is located on the $\sigma$ axis at a distance $(\sigma_x + \sigma_y)/2$ from the origin. Point $A$ on the circle has the coordinates $(\sigma_x, \tau_{xy})$ corresponding to the *stresses acting on the right-hand face of the element* in the positive direction of the coordinate axes, Fig. 8-9(a). Point $A$ will be referred to as the *origin of planes*. This information is sufficient to draw a circle of stress, Fig. 8-9(b).

The next step consists of drawing on the circle of stress a line through $A$ parallel to plane $a$–$a$ in the physical plane of Fig. 8-9(a). The intersection of this line at $J$ with the stress circle gives the stresses acting on plane $a$–$a$. This requires some justification. For this purpose, the indicated geometric construction must be reviewed in detail.

According to the previous derivation shown in Fig. 8-8(c), angle $ACF$ in Fig. 8-9(b) is equal to $2\theta_1$. Further, since line $CH$ is drawn perpendicular to line $AJ$, angle $ACJ$ is bisected, and $\alpha = 2\theta_1 - \theta$. Hence, angle $JCF$ is $\theta - \alpha = 2\theta - 2\theta_1$, and it remains to be shown that the coordinates of point $J$ define the stresses acting on inclined plane $a$–$a$. For this purpose,

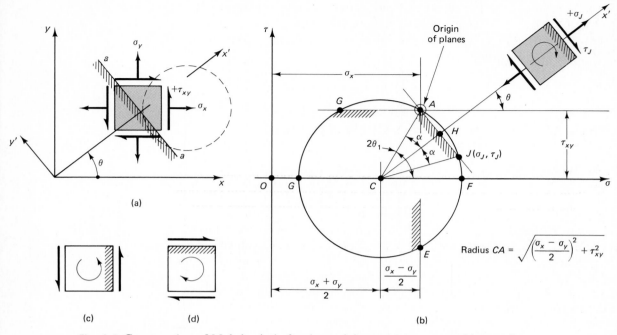

**Fig. 8-9** Construction of Mohr's circle for determining stresses on an arbitrary plane.

one notes from Fig. 8-9(b) that if $R$ is the radius of a circle, $R \cos 2\theta_1 = (\sigma_x - \sigma_y)/2$ and $R \sin 2\theta_1 = \tau_{xy}$. Then, forming expressions for the normal and shear stresses at $J$ based on the construction of the circle in Fig. 8-9(b) and making use of trigonometric identities for double angles, one has

$$
\begin{aligned}
\sigma_J &= \frac{\sigma_x + \sigma_y}{2} + R \cos (2\theta - 2\theta_1) \\
&= \frac{\sigma_x + \sigma_y}{2} + R (\cos 2\theta \cos 2\theta_1 + \sin 2\theta \sin 2\theta_1) \\
&= \frac{\sigma_x + \sigma_y}{2} + \frac{\sigma_x - \sigma_y}{2} \cos 2\theta + \tau_{xy} \sin 2\theta
\end{aligned} \tag{8-18}
$$

and

$$
\begin{aligned}
\tau_J &= R \sin(2\theta - 2\theta_1) = R \sin 2\theta \cos 2\theta_1 - R \cos 2\theta \sin 2\theta_1 \\
&= + \frac{\sigma_x - \sigma_y}{2} \sin 2\theta - \tau_{xy} \cos 2\theta
\end{aligned} \tag{8-19}
$$

Except for the sign of $\tau_J$, the last expressions are identical to Eqs. 8-1 and 8-2, and, therefore, define the stresses acting on the element shown in the upper right quadrant of Fig. 8-9(b). The hatched side of this element is parallel to line $AJ$ on the stress circle, which is parallel to line $a$–$a$ in

Fig. 8-9(a). However, since the sign of $\tau_J$ is opposite to that in the basic transformation, Eq. 8-2, a special rule for the direction of shear stress has to be introduced.

For this purpose, consider the initial data for the element shown in Fig. 8-9(a), where all stresses are shown with positive sense. By isolating the shear stresses acting on the vertical faces, Fig. 8-9(c), it can be seen that *these stresses alone* cause a *counterclockwise* couple. By considering lines emanating from the origin of planes $A$, for the first case, Fig. 8-9(c), the circle is intersected at $E$, whereas for the second case, Fig. 8-9(d), it is intersected at $G$. This can be generalized into a rule: if the point of intersection of a line emanating from the origin of planes $A$ intersects the circle *above* the $\sigma$ axis, the shear stresses on the opposite sides of an element cause a *clockwise* couple. Conversely, if the point of intersection lies *below* the $\sigma$ axis, the shear stresses on the opposite sides cause a *counterclockwise* couple. According to this rule, the shear stresses at $J$ in Fig. 8-9(b) act with a clockwise sense.

This general procedure is illustrated for two particularly important cases. For the data given in Fig. 8-10(a), the principal stresses are found in Fig. 8-10(b), and the maximum shear stresses are found in Fig. 8-10(c).

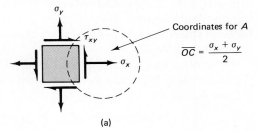

(a)

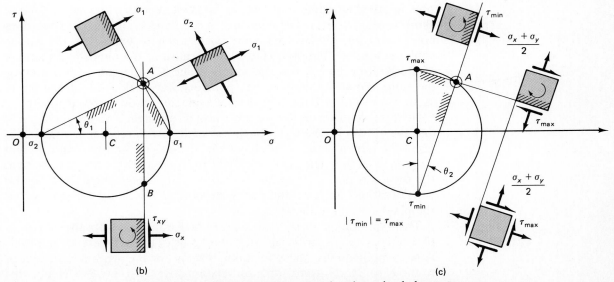

(b)                                                      (c)

**Fig. 8-10** Determining principal normal and maximal shear stresses.

For the first case, it is known that the extreme values on the abscissa, $\sigma_1$ and $\sigma_2$, give the principal stresses. Connecting these points with the origin of planes $A$ locates the planes on which these stresses act. Angle $\theta_1$ can be determined by trigonometry. Either one of the two solutions is sufficient to obtain the complete solution shown on the element on the right.

The magnitudes of the maximum absolute shear stresses are known to be given by the radius of the Mohr's circle. As shown in Fig. 8-10(c), these stresses are located above and below $C$. Connecting these points with the origin of planes $A$ determines the planes on which these stresses act. The corresponding elements are shown in the upper two diagrams of the elements, where the associated mean normal stresses are also indicated. Either one of these solutions with the aid of equilibrium concepts is sufficient for the complete solution shown on the bottom element in the figure.

### *Method 2*

The state of stress in the $xy$ coordinate system is shown in Fig. 8-11(a). The origin for these coordinates is arbitrarily chosen at the center of the infinitesimal element. The objective is to transform the given stresses to those in the rotated set of $x'y'$ axes as shown in Figs. 8-11(a) and (b) by using Mohr's circle.

As before, the center $C$ of the Mohr's circle is located at $(\sigma_x + \sigma_y)/2$. Again the right hand face of the element defines $\sigma_x$ and $\tau_{xy}$ used to locate a point on the circle. However,

if $\tau_{xy} > 0$, it is plotted *downwards* at $\sigma_x$, and

if $\tau_{xy} < 0$, it is plotted *upwards* at $\sigma_x$.

This in effect amounts to directing the positive $\tau$ axis downward, and is shown in Fig. 8-11(c). The coordinates of $\sigma_x$ and $\tau_{xy}$ locate the governing point $A_c$ on the circle. This point corresponds to point $A$ in the earlier construction; see Fig. 8-8. However, because of the opposite directions of the positive $\tau$ axes, whereas points $A$ and $A_c$ are related, they are not the same. Point $B_c$, conjugate to point $A_c$, can be located on the circle as shown in Fig. 8-11(c). The double angle $2\theta$, follows from geometry.

Next the diameter $A_c B_c$ is rotated through an angle $2\theta$ in the *same sense* that the $x'$ axis is rotated through the angle $\theta$ with respect to the $x$ axis. Then the new point $A_c'$ determines the stresses $\sigma_{x'}$ and $\tau_{x'y'}$ on the right hand face of the element in Fig. 8-11(b). Note that for the case shown, the shear stress $\tau_{x'y'}$ is negative, since at $\sigma_{x'}$ it is above the $\sigma$ axis. Similar considerations apply to the conjugate point $B_c'$ defining the stresses on the plane normal to the $y'$ axis.

The expressions for $\sigma_{x'}$ and $\tau_{x'y'}$ can be formulated from the construction of the Mohr's circle shown in Fig. 8-11(c) using Eqs. 8-16 and 8-17. After simplfications, these relations, except for the sign of $\tau_{x'y'}$, reduce to the basic stress transformation relations, Eqs. 8-1 and 8-2. Hence this

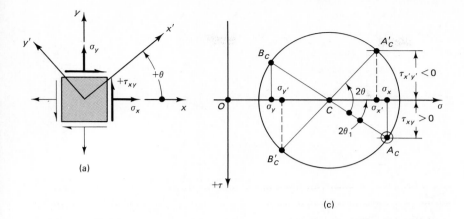

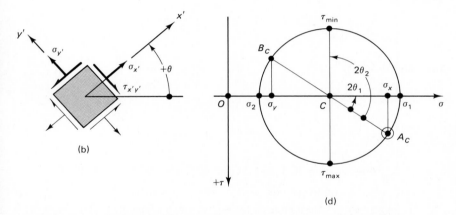

Fig. 8-11 Alternative construction of Mohr's circle of stress. Stresses on arbitrary and principal normal and shear planes are shown in (c) and (d), respectively.

construction of the Mohr's circle is justified. For proof modify Eqs. 8-18 and 8-19.

Procedure for determining the principal normal stress is shown in Fig. 8-11(d). After drawing a Mohr's circle the principal stresses $\sigma_1$ and $\sigma_2$ are known. The required rotation $\theta_1$ of the axes in the direction of these stresses is obtained by calculating the double angle $2\theta_1$ from the diagram. Similarly, the principal shear stresses are given by the coordinates of the points on a circle at their extreme values on the $\tau$ axis. The required rotation $\theta_2$ of these axes is obtained by calculating the double angle $2\theta_2$ from the diagram.

Method 1 is used in the two examples that follow.

## EXAMPLE 8-3

Given the state of stress shown in Fig. 8-12(a), transform it (a) into the principal stresses, and (b) into the maximum shear stresses and the associated normal stresses. Show the results for both cases on properly oriented elements. Use Method 1.

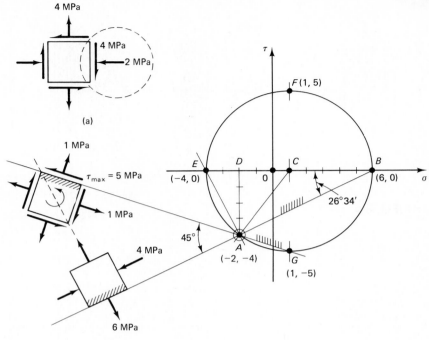

**Fig. 8-12**                    (b)

### Solution

To construct Mohr's circle of stress, the following quantities are required.

1.  Center of circle on the $\sigma$ axis: $(-2 + 4)/2 = +1$ MPa.
2.  Origin of planes $A$ from data on the right face of element: $(-2, -4)$ MPa.
3.  Radius of circle: $CA = \sqrt{CD^2 + DA^2} = 5$ MPa.

After drawing the circle, one obtains $\sigma_1 = +6$ MPa, $\sigma_2 = -4$ MPa, and $\tau_{max} = 5$ MPa.

Line $AB$ on the stress circle locates the principal plane for $\sigma_1 = 6$ MPa. The angle $\theta_1$ is 26°34′, since $\tan \theta_1 = AD/DB = 4/8 = 0.5$. The other principal stress, $\sigma_2 = -4$ MPa, acts at right angle to the above plane. These results are shown on a properly oriented element.

Line $AG$ on the circle at 45° with the principal planes determines the planes for maximum shear, $\tau_{max} = 5$ MPa, and the associated mean normal stress $\sigma' = 1$ MPa. The latter stress corresponds to $\sigma$ at the circle center. Complete results are shown on a properly oriented element.

It is worthy to note that the directions of the principal stresses can be anticipated and can be used in calculations as a check. A suitable inspection procedure is shown in Fig. 8-13. To begin with, it is known that tensile stresses of equal magnitude to the shear stress develop along a diagonal as shown in Fig. 8-6. Therefore, the maximum tensile stress $\sigma_1$, which is the result of all stresses, must act as shown in Fig. 8-13(b). Situations with compressive stresses can be treated similarly, Fig. 8-13(d).

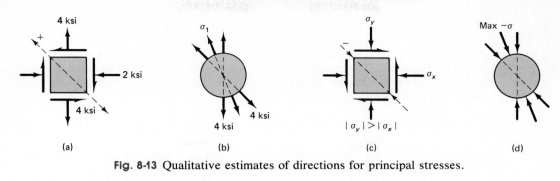

**Fig. 8-13** Qualitative estimates of directions for principal stresses.

## EXAMPLE 8-4

Using Mohr's circle, transform the stresses shown in Fig. 8-14(a) into stresses acting on the plane at an angle of $22\frac{1}{2}°$ with the vertical axis. Use Method 1.

### Solution

For this case, the center of the circle is at $(1 + 3)/2 = +2$ MPa on the $\sigma$ axis. The origin of planes $A$ is at $(3,3)$, and the radius $R = \sqrt{1^2 + 3^2} = 3.16$ MPa. By using these data, a stress circle is plotted in Fig. 8-14(b) on which an inclined line at 22.5° locating point $J$ is drawn.

Angle $\beta$ is 71.57°, since $\tan \beta = AD/CD = 3$. A normal to $AJ$ forms an angle of 22.5° with the $\sigma$ axis. Therefore, $\alpha = 71.57° - 22.5° = 49.07°$, and angle $FCJ$ is $\alpha - 22.5° = 26.57°$. This locates $J$ on the circle. Hence, $\sigma_J = 2 + R \cos(-26.57°) = 2 + 3.16(0.894) = 4.83$ MPa, and $\tau_J = R \sin(-26.57°) = 3.16 (-0.447) = -1.41$ MPa.

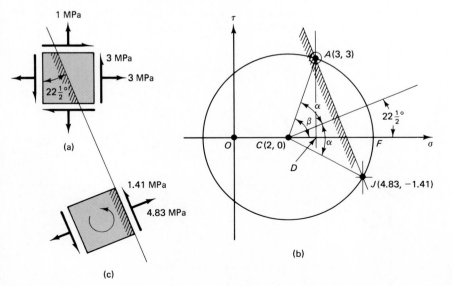

(a)

(b)

(c)

**Fig. 8-14**

These results are shown on a properly oriented element in Fig. 8-14(c). Since $\tau_J$ is negative, the shear stresses are shown acting counterclockwise.

Again it should be remarked that the equations for stress transformation are identical in form to the equations for determining the principal axes and moments of inertia of areas (Section 6-12). Therefore, Mohr's circle can be constructed for finding these equations.[4]

---

## **\*\*[5]8-8. Principal Stresses for a General State of Stress**

Traditionally, in an introductory text on solid mechanics, attention is largely confined to stresses in two dimensions. Since, however, the physical elements studied are always three-dimensional, for completeness, it is desirable to consider the consequences of three-dimensionality on stress transformations. The concepts developed in this section have an impact on the discussion that follows in this chapter, as well as on some issues considered in the next chapter.

Consider a general state of stress and define an infinitesimal tetrahedron[6] as shown in Fig. 8-15(a). Instead of considering an inclined plane in the $xy$ coordinate system, as before for a wedge, the unknown stresses are sought on an arbitrary oblique plane $ABC$ in the three dimensional $xyz$ coordinate system. A set of known stresses on the other three faces of the mutually perpendicular planes of the tetrahedron is given. These stresses are the same as shown earlier in Fig. 1-3(a).

A unit normal **n** to the oblique plane defines its orientation. This unit

[4] See J. L. Meriam, and L. G. Kraige, *Statics,* 2nd ed. (New York: Wiley, 1986).

[5] This section is more advanced and can be omitted.

[6] A tetrahedron was first introduced in the study of stress transformations by the great French mathematician A. L. Cauchy in the 1820s.

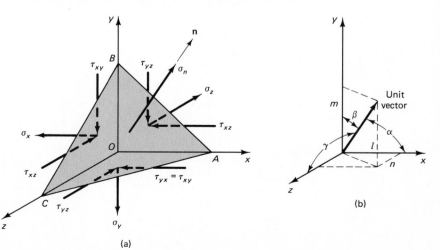

**Fig. 8-15** Tetrahedron for deriving a principal stress on an oblique plane.

vector is identified by its direction cosines $l$, $m$, and $n$, where $\cos \alpha = l$, $\cos \beta = m$, and $\cos \gamma = n$. The meaning of these quantities is illustrated in Fig. 8-14(b). From the same figure, it can be noted that since $l^2 + n^2 = d^2$ and $d^2 + m^2 = 1$,

$$l^2 + m^2 + n^2 = 1 \qquad (8\text{-}20)$$

Further, if the infinitesimal area $ABC$ is defined as $dA_{ABC} \equiv dA$, then the three areas of the tetrahedron along the coordinate axes, identified by their subscripts, are $dA_{BOC} = dA\, l$, $dA_{AOC} = dA\, m$, and $dA_{AOB} = dA\, n$.[7]

Force equilibrium equations for the tetrahedron can now be written by multiplying the stresses given in Fig. 8-15(a) by the respective areas established. For simplicity, it will be assumed that only a *normal stress* $\sigma_n$, i.e., a *principal stress,* is acting on face $ABC$. The components of the corresponding normal force $(\sigma_n\, dA)$ are obtained by resolving it along the coordinate axes using the direction cosines, Fig. 8-15(b). On this basis,

$$\sum F_x = 0 \qquad (\sigma_n\, dA)l - \sigma_x\, dA\, l - \tau_{xy}\, dA\, m - \tau_{xz}\, dA\, n = 0$$
$$\sum F_y = 0 \qquad (\sigma_n\, dA)m - \sigma_y\, dA\, m - \tau_{yz}\, dA\, n - \tau_{xy}\, dA\, l = 0 \qquad (8\text{-}21)$$
$$\sum F_z = 0 \qquad (\sigma_n\, dA)n - \sigma_z\, dA\, n - \tau_{xz}\, dA\, l - \tau_{yz}\, dA\, m = 0$$

Simplifying, changing signs, and regrouping terms,

$$(\sigma_x - \sigma_n)l + \tau_{xy}m + \tau_{xz}n = 0$$
$$\tau_{xy}l + (\sigma_y - \sigma_n)m + \tau_{yz}n = 0 \qquad (8\text{-}22)$$
$$\tau_{xz}l + \tau_{yz}m + (\sigma_z - \sigma_n)n = 0$$

By virtue of Eq. 8-20, all three direction cosines cannot be zero. However, the system of linear homogeneous equations has a nontrivial solution if and only if the determinant of the coefficients of $l$, $m$, and $n$ vanishes. Hence,

$$\begin{vmatrix} \sigma_x - \sigma_n & \tau_{xy} & \tau_{xz} \\ \tau_{xy} & \sigma_y - \sigma_n & \tau_{yz} \\ \tau_{xz} & \tau_{yz} & \sigma_z - \sigma_n \end{vmatrix} = 0 \qquad (8\text{-}23)$$

[7] These areas are the projection of $dA$ on the respective coordinate planes. To clarify, consider the two-dimensional wedge shown in Fig. 8-A and compare the volumes using two different paths. Let the area associated with side $AC$ be $A_{AC}$, and the corrresponding wedge height be $AB$. Then the wedge volume is $A_{AC}AB/2$. On the other hand, if the area for the side $CB$ is $A_{CB}$, and the wedge height is $AB \cos \theta = ABl$, the volume is $A_{CB}ABl/2$. By equating the volumes and simplifying, $A_{AC} = A_{CB}l$. By carrying out this procedure in three dimensions, the relations given above can be justified.

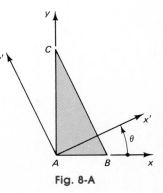

Fig. 8-A

Expansion of this determinant gives

$$\sigma_n^3 - I_\sigma \sigma_n^2 + II_\sigma \sigma_n - III_\sigma = 0 \qquad (8\text{-}24)$$

$$I_\sigma = \sigma_x + \sigma_y + \sigma_z \qquad (8\text{-}25)$$

where $\quad II_\sigma = (\sigma_x \sigma_y + \sigma_y \sigma_z + \sigma_z \sigma_x) - (\tau_{xy}^2 + \tau_{yz}^2 + \tau_{zx}^2) \qquad (8\text{-}26)$

$$III_\sigma = \sigma_x \sigma_y \sigma_z + 2\tau_{xy}\tau_{yz}\tau_{xz} - (\sigma_x \tau_{yz}^2 + \sigma_y \tau_{xz}^2 + \sigma_z \tau_{xy}^2) \qquad (8\text{-}27)$$

Here if the initial coordinate system is changed, thereby changing the three mutually perpendicular planes of the tetrahedron, the $\sigma_n$ on the inclined plane must remain the same. Therefore the constants $I_\sigma$, $II_\sigma$, and $III_\sigma$ in Eq. 8-24 must also remain the same, and hence they are *invariant*. Moreover, since the matrix of Eq. 8-23 is symmetric, and all of its elements are real, according to the Descartes rule of signs, in general, Eq. 8-24 has *three real root*.[8] These roots are the *eigenvalues* of the determinental Eq. 8-23 and are the *principal stresses* of the problem.

The three roots giving the principal stresses can be found from Eq. 8-24 using synthetic division or Newton's method of tangents.[9] It is customary to order the principal stress such that $\sigma_1 > \sigma_2 > \sigma_3$. Any one of these roots can be substituted into any two of Eqs. 8-22, and together with Eq. 8-20 form a set of three simultaneous equations. A solution of these equations gives the direction cosines for the selected principal stress. The three principal directions for the principal stresses are *orthogonal*. The planes normal to the principal directions are the *principal planes of stress*. If two or three of the principal stresses are equal, there are an infinite number of principal directions. This is further commented upon in the next section.

## 8-9. Mohr's Circle for a General State of Stress

In the preceding section, it was shown that for a general state of stress, there are three orthogonal principal stresses, $\sigma_1$, $\sigma_2$, and $\sigma_3$, provided $\sigma_1 \neq \sigma_2 \neq \sigma_3$. These stresses act along the principal axes. Plane stress problems fall within the scope of the general theory when one of the principal stresses is zero. So do the plane strain problems when one of the principal stresses is given by Eq. 8-5. However, degenerate cases arise requiring special treatment.

If only two of the principal stresses are equal, the remaining principal stress has a unique direction. Any other two orthogonal directions of an

[8] I. S. Sokolnikoff, *Mathematical Theory of Elasticity,* 2nd ed. (New York: McGraw-Hill, 1956), 47.

[9] A. Ralston, *A First Course in Numerical Analysis* (New York: McGraw-Hill, 1965). M. G. Salvadori, and M. L. Baron, *Numerical Method in Engineering* (Englewood Cliffs, NJ: Prentice-Hall, 1952).

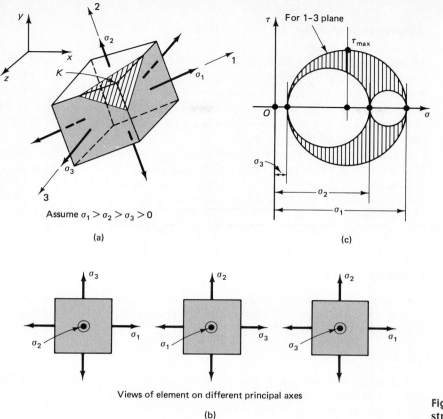

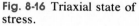

Assume $\sigma_1 > \sigma_2 > \sigma_3 > 0$

(a)

Views of element on different principal axes

(b)

**Fig. 8-16** Triaxial state of stress.

orthogonal triad are the principal directions. This case may be referred to as a *cylindrical* or *axisymmetric* state of stress.[10] If all three principal stresses are equal, the state of stress is said to be *spherical*. Any triad of orthogonal axes for this case gives the principal axes.

For the general case, consider the illustration given in Fig. 8-16, where the ordered principal stresses are $\sigma_1 > \sigma_2 > \sigma_3$. Suppose further that, after an appropriate stress transformation, principal axes 1, 2, and 3 and the corresponding principal stresses are oriented as shown on an element in Fig. 8-16(a). By viewing this element along the three principal axes, 3 two-dimensional diagrams, shown in Fig. 8-16(b), are obtained. For each of these diagrams, one can draw a Mohr's circle of stress. This is shown in Fig. 8-16(c) with a cluster of three circles. As far as the stress magnitudes are concerned, the outer circle is the most important one.

Although by definition (Section 8-4), the principal stresses are the maximum and the minimum ones, it is of interest as to where on a plot, such as Fig. 8-16(c), the stresses on all arbitrary oblique planes lie. Such a plane is designated by $K$ in Fig. 8-16(a). The results of this study, considering stresses in three dimensions, show[11] that the coordinate points

[10] O. Hoffman and G. Sachs, *Introduction to the Theory of Plasticity for Engineers* (New York: McGraw-Hill, 1953).

[11] *Ibid.*, 13.

**427**

for all possible planes lie either on one of the three circles or in an area between them, shown hatched in Fig. 8-16(c). A series of circles is defined within this area having their centers on the $\sigma$ axis by holding any one of the direction cosines constant. Therefore, it is convenient to refer to the three circles drawn as the *principal stress circles*. The largest of these is the *major* principal stress circle.

Inasmuch as all stresses in their various transformations may play a role in causing either yield or breakdown of a material, it is often instructive to plot all three principal circles of stress, as shown in Fig. 8-16(c). Two examples of this kind follow. In making such plots, the degenerate cases, when two or all principal stresses are equal, must be kept in mind. For such cases, a Mohr's circle becomes a point.

### EXAMPLE 8-5

For the data of Example 8-3, repeated in Fig. 8-17(a), construct three Mohr's principal circles of stress by viewing an element from three principal directions. Assume that this is a plane stress problem.

#### Solution

The principal stresses for this problem in two dimensions have already been determined in Example 8-3. The results are repeated in Fig. 8-17(b). Since this is a plane stress problem, the stress in the direction normal to the paper is zero. The complete state of stress showing all principal stresses is in Fig. 8-17(c). The 3 two-dimensional diagrams of the element viewed from different directions are in Figs. 8-17(d)–(f). The cluster of three Mohr's principal circles is shown in Fig. 8-17(g).

If the given stresses were for a *plane strain* problem, the middle principal stress, instead of being zero, per Eq. 8-5, would be $\sigma_2 = \nu(6 - 4) = +2\nu$, where $\nu$ is Poisson's ratio.

### EXAMPLE 8-6

For the plane stress shown in Fig. 8-18(a), draw the three Mohr's principal circle diagrams and determine the state of stress for maximum shear.

#### Solution

Two of the principal stresses are given; the third is zero, as this is a plane stress problem. The three principal stress circles are shown in Fig. 8-18(b). The maximum shear stress occurs in the planes shown in Fig. 8-18(c). This stress is associated with point $D$ on the major principal circle, and in physical orientation is given in Fig. 8-18(d).

This type of problem occurs in pressure-vessel analyses, where it is important to recognize that large shear stresses may arise.

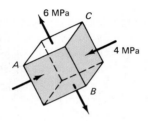

4 MPa
4 MPa
2 MPa

(a)

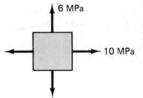

6 MPa
4 MPa
26°34′

(b)

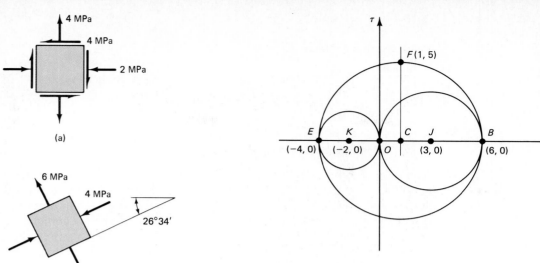

F(1, 5)

E
(−4, 0)

K
(−2, 0)

O

C

J
(3, 0)

B
(6, 0)

$\tau$

$\sigma$

(g)

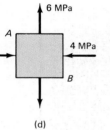

6 MPa
C
4 MPa
A
B

(c)

6 MPa
A
4 MPa
B

(d)

6 MPa
C
B
4 MPa

(e)

C
4 MPa
A
6 MPa

(f)

**Fig. 8-17**

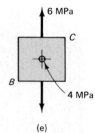

6 MPa
10 MPa

(a)

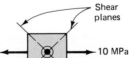

Shear
planes
10 MPa

(c)

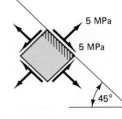

5 MPa
5 MPa
45°

(d)

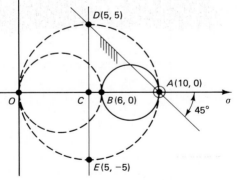

$\tau$

D(5, 5)

O

C

B(6, 0)

A(10, 0)

45°

$\sigma$

E(5, −5)

(b)

**Fig. 8-18**

**429**

# Part B TRANSFORMATION OF STRAIN

## 8-10. Strains in Two Dimensions

In the following four sections, study is directed toward strain transformation in two dimensions. This includes consideration of plane stress and plain strain problems. It will be shown that the transformation of normal and shear strains from one set of rotated axes to another is completely analogous to the transformation of normal and shear stresses presented earlier. Therefore, after establishing the strain transformation equations, it is possible to introduce Mohr's circle of strain. A procedure for reducing surface strain measurements made by means of strain gages into principal stresses completes this part of the chapter.

In studying the strains at a point, only the relative displacement of the adjoining points is of importance. Translation and rotation of an element as a whole are of no consequence since these displacements are *rigid-body displacements*. For example, if the extensional strain of a diagonal $ds$ of the original element in Fig. 8-19(a) is being studied, the element in its deformed condition can be brought back for comparison purposes, as shown in Fig. 8-19(c). It is immaterial whether the horizontal (dashed) or the vertical (dotted) sides of the deformed and the undeformed elements are matched to determine $d\Delta$. For the small strains and rotations considered, the relevant quantity, elongation $d\Delta$ in the direction of the diagonal, is essentially the same regardless of the method of comparison employed. In treating strains in this manner, only kinematic questions have relevance. The mechanical properties of material do not enter the problem.

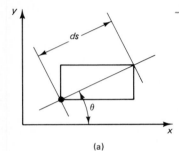

(a)

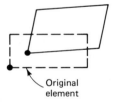

Original
element

(b)

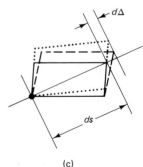

(c)

**Fig. 8-19** Strains are determined from relative deformations.

## 8-11. Transformation of Strain in Two Dimensions

In establishing the equations for the transformation of strain, strict adherence to a sign convention is necessary. The sign convention used here is related to the one chosen for the stresses in Section 8-3. The normal strains $\varepsilon_x$ and $\varepsilon_y$ corresponding to elongations in the $x$ and $y$ directions, respectively, are taken positive. The shear strain is considered positive if the 90° angle between the $x$- and the $y$-axes becomes smaller. For convenience in deriving the strain transformation equations, the element distorted by positive shear strain will be taken as that shown in Fig. 8-20(a).

Next, suppose that the strains $\varepsilon_x$, $\varepsilon_y$, and $\gamma_{xy}$ associated with the $xy$ axes are known and extensional strain along some new $x'$ axis is required. The new $x'y'$ system of axes is related to the $xy$ axes as in Fig. 8-20(b). In these new coordinates, a length $OA$, which is $dx'$ long, may be thought of as being a diagonal of a rectangular differential element $dx$ by $dy$ in the initial coordinates.

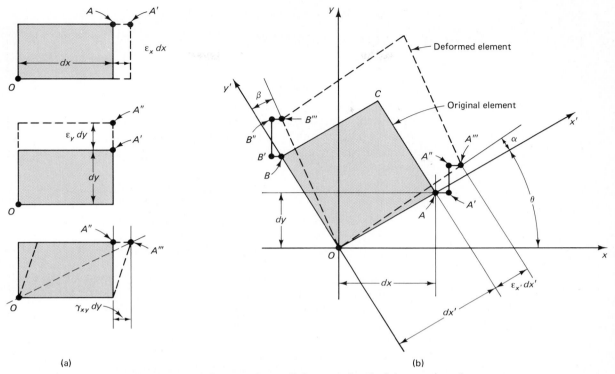

**Fig. 8-20** Exaggerated deformations of elements for deriving strains along new axes.

By considering point $O$ fixed, one can compute the displacements of point $A$ caused by the imposed strains on a different basis in the two coordinate systems. The displacement in the $x$ direction is $AA' = \varepsilon_x\, dx$; in the $y$ direction, $A'A'' = \varepsilon_y\, dy$. For the shear strain, assuming it causes the horizontal displacement shown in Fig. 8-20(a), $A''A''' = \gamma_{xy}\, dy$. The order in which these displacements occur is arbitrary. In Fig. 8-20(b), displacement $AA'$ is shown first, then $A'A''$, and finally $A''A'''$. By projecting these displacements onto the $x'$ axis, one finds the displacement of point $A$ along the $x'$ axis. Then, recognizing that by definition, $\varepsilon_{x'}\, dx'$ in the $x'y'$ coordinate system is also the elongation of $OA$, one has the following equality:

$$\varepsilon_{x'}\, dx' = AA' \cos\theta + A'A'' \sin\theta + A''A''' \cos\theta$$

By substituting the appropriate expressions for the displacements and dividing through by $dx'$, one has

$$\varepsilon_{x'} = \varepsilon_x \frac{dx}{dx'} \cos\theta + \varepsilon_y \frac{dy}{dx'} \sin\theta + \gamma_{xy} \frac{dy}{dx'} \cos\theta$$

Since, however, $dx/dx' = \cos\theta$ and $dy/dx' = \sin\theta$,

$$\boxed{\varepsilon_{x'} = \varepsilon_x \cos^2\theta + \varepsilon_y \sin^2\theta + \gamma_{xy} \sin\theta \cos\theta} \tag{8-28}$$

Equation 8-28 is the basic expression for normal strain transformation in a plane in an arbitrary direction defined by the $x'$ axis.[12] In order to apply this equation, $\varepsilon_x$, $\varepsilon_y$, and $\gamma_{xy}$ must be known. By use of trigonometric identities already encountered in deriving Eq. 8-1, the last equation can be rewritten also as

$$\boxed{\varepsilon_{x'} = \frac{\varepsilon_x + \varepsilon_y}{2} + \frac{\varepsilon_x - \varepsilon_y}{2}\cos 2\theta + \frac{\gamma_{xy}}{2}\sin 2\theta} \tag{8-29}$$

To complete the study of strain transformation at a point, shear-strain transformation must also be established. For this purpose, consider an element $OACB$ with sides $OA$ and $OB$ directed along the $x'$ and the $y'$ axes, as shown in Fig. 8-20(b). By definition, the shear strain for this element is the change in angle $AOB$. From the figure, the change of this angle is $\alpha + \beta$.

For small deformations, the small angle $\alpha$ can be determined by projecting the displacements $AA'$, $A'A''$, and $A''A'''$ onto a normal to $OA$ and dividing this quantity by $dx'$. In applying this approach, the tangent of the angle is assumed equal to the angle itself. This is acceptable as the strains are small. Thus,

$$\begin{aligned}
\alpha \approx \tan\alpha &= \frac{-AA'\sin\theta + A'A''\cos\theta - A''A'''\sin\theta}{dx'} \\
&= -\varepsilon_x \frac{dx}{dx'}\sin\theta + \varepsilon_y \frac{dy}{dx'}\cos\theta - \gamma_{xy}\frac{dy}{dx'}\sin\theta \\
&= -(\varepsilon_x - \varepsilon_y)\sin\theta\cos\theta - \gamma_{xy}\sin^2\theta
\end{aligned}$$

---

[12] Using direction cosines $l$, $m$, and $n$ (see Section 8-8), Eq. 8-28 can be rewritten as

$$\varepsilon_{x'} = \varepsilon_x l^2 + \varepsilon_y m^2 + \gamma_{xy} lm \tag{8-28a}$$

As is shown in books on the theory of elasticity or continuum mechanics, this normal strain transformation in three dimensions becomes

$$\varepsilon_{x'} = \varepsilon_x l^2 + \varepsilon_y m^2 + \varepsilon_z n^2 + \gamma_{xy} lm + \gamma_{yz} mn + \gamma_{zx} ln$$

Therefore, Eq. 8-28 can be applied *only* for strain transformation in two dimensions.

By analogous reasoning,

$$\beta \approx -(\varepsilon_x - \varepsilon_y) \sin \theta \cos \theta + \gamma_{xy} \cos^2 \theta$$

Therefore, since the shear strain $\gamma_{x'y'}$ of an angle included between the $x'y'$ axes is $\beta + \alpha$, one has

$$\gamma_{x'y'} = -2(\varepsilon_x - \varepsilon_y) \sin \theta \cos \theta + \gamma_{xy}(\cos^2 \theta - \sin^2 \theta)$$

or

$$\gamma_{x'y'} = -(\varepsilon_x - \varepsilon_y) \sin 2\theta + \gamma_{xy} \cos 2\theta \qquad (8\text{-}30)$$

This is the second fundamental expression for the transformation of strain. Note that when $\theta = 0°$, the shear strain associated with the $xy$ axes is recovered.

The basic Eqs. 8-29 and 8-30 for strain transformation in a plane are analogous to Eqs. 8-1 and 8-2 for stress transformation in two dimensions. Fundamentally, this is because both stresses and strains are second-rank tensors and mathematically obey the same laws of transformation. This similarity will be emphasized in discussing Mohr's circle of strain.

## **8-12. Alternative Derivation for Strain Transformation in Two-Dimensions

An approach more suitable for deriving strain transformation in three dimensions is presented in this section.

Consider element $AB$ initially $ds$ long, as shown in Fig. 8-21. After straining, this element displaces to the position $A'B'$ and becomes $ds^*$ long. The initial length is $ds^2 = dx^2 + dy^2$ and the strained length of the element is $(ds^*)^2 = (dx^*)^2 + (dy^*)^2$, with $dx^* = dx + du$ and $dy^* = dy + dv$.

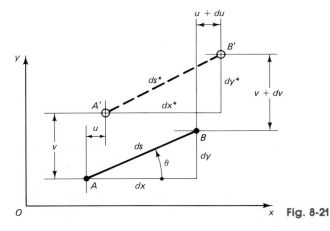

**Fig. 8-21**

The infinitesimal increments of strain, $du$ and $dv$ for the absolute displacements $u$ and $v$; can be found formally by applying the chain rule of differentiation to obtain total differentials, i.e.,

$$du = \frac{\partial u}{\partial x}\,dx + \frac{\partial u}{\partial y}\,dy \qquad \text{and} \qquad dv = \frac{\partial v}{\partial x}\,dx + \frac{\partial v}{\partial y}\,dy$$

By using these relations, the strain is most conveniently defined as the difference between $(ds^*)^2$ and $ds^2$. This difference is zero for unstrained bodies.

For the small-deformation theory, in the expression for $(ds^*)^2$, the squares of small quantities can be neglected in comparison with the quantities themselves. Thus, after some algebraic manipulations and simplifications,

$$(ds^*)^2 = \left(1 + 2\frac{\partial u}{\partial x}\right)dx^2 + 2\frac{\partial u}{\partial y}\,dx\,dy + \left(1 + 2\frac{\partial v}{\partial y}\right)dy^2 + 2\frac{\partial v}{\partial x}\,dx\,dy$$

Hence,

$$(ds^*)^2 - ds^2 = 2\frac{\partial u}{\partial x}\,dx^2 + 2\frac{\partial v}{\partial y}\,dy^2 + 2\left(\frac{\partial u}{\partial y} + \frac{\partial v}{\partial x}\right)dx\,dy$$

and, by recalling Eqs. 3-7 and 3-9, which define strains as derivatives of displacements, one has

$$(ds^*)^2 - ds^2 = 2\varepsilon_x\,dx^2 + 2\varepsilon_y\,dy^2 + 2\gamma_{xy}\,dy\,dx \qquad (8\text{-}31)$$

For small deformations, to a high degree of accuracy, one has

$$(ds^*)^2 - ds^2 = (ds^* + ds)\left(\frac{ds^* - ds}{ds}\right)ds \approx 2\varepsilon_\theta\,ds^2 \qquad (8\text{-}32)$$

where the normal strain $\varepsilon_\theta = (ds^* - ds)/ds$ from the classical definition of small strain, and, since $ds^*$ differs very little from $ds$, $ds^* + ds \approx 2\,ds$.

By equating Eqs. 8-31 and 8-32, dividing through by $ds^2$, and recognizing that $\cos\theta = dx/ds$ and $\sin\theta = dy/ds$, one obtains

$$\varepsilon_\theta = \varepsilon_x\cos^2\theta + \varepsilon_y\sin^2\theta + \gamma_{xy}\sin\theta\cos\theta \qquad (8\text{-}33)$$

This equation for normal strain is essentially identical to Eq. 8-28.

By taking two initially mutually perpendicular sides of an element and then forming the scalar product for the same two sides in the deformed state, Eq. 8-30 for the shear strain can be reproduced.

Extension of this procedure to a three dimensional case is direct, and is left for the reader to complete.

## *8-13. Mohr's Circle for Two-Dimensional Strain

The two basic equations for the transformations of strains in two dimensions derived in the preceding section mathematically resemble the equations for the transformation of stresses derived in Section 8-3. To achieve greater similarity between the appearances of the new equations and those of the earlier ones, Eq. 8-30 after division throughout by 2 is rewritten as Eq. 8-34.

$$\varepsilon_{x'} = \frac{\varepsilon_x + \varepsilon_y}{2} + \frac{\varepsilon_x - \varepsilon_y}{2} \cos 2\theta + \frac{\gamma_{xy}}{2} \sin 2\theta \qquad (8\text{-}29)$$

$$\frac{\gamma_{x'y'}}{2} = -\frac{\varepsilon_x - \varepsilon_y}{2} \sin 2\theta + \frac{\gamma_{xy}}{2} \cos 2\theta \qquad (8\text{-}34)$$

Since these strain-transformation equations with the shear strains divided by 2 are mathematically identical to the stress transformation Eqs. 8-1 and 8-2, Mohr's circle of strain can be constructed. In this construction, every point on the circle gives two values: one for the normal strain, the other for the shear strain *divided by 2*. (For further reasons, see Section 3-5.) Strains corresponding to elongation are positive; for contraction, they are negative. For positive shear strains the angle between the $x$- and the $y$-axes becomes smaller; see Fig. 8-20(a). In plotting the circle, the positive axes are taken *in accordance with the sign convention for Method 1 for Mohr's circle of stress,* upward and to the right. The vertical axis is measured in terms of $\gamma/2$.

As an illustration of Mohr's circle of strain, consider that $\varepsilon_x$, $\varepsilon_y$, and $+\gamma_{xy}$ are given. Then on the $\varepsilon - \gamma/2$ axes in Fig. 8-22 the center of the

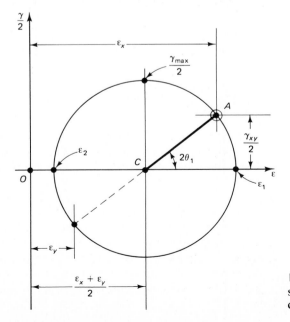

Fig. 8-22 Mohr's circle of strain using sign convention of Method 1, Section 8-7.

circle $C$ is at $[(\varepsilon_x + \varepsilon_y)/2,0]$ and, from the given data, the origin of planes $A$ on the circle is at $(\varepsilon_x, \gamma_{xy}/2)$. An examination of this circle leads to conclusions analogous to those reached before for the circle of stress.

1. The maximum normal strain is $\varepsilon_1$; the minimum is $\varepsilon_2$. These are the principal strains, and no shear strains are associated with them. The directions of the normal strains coincide with the directions of the principal stresses. As can be deduced from the circle, the analytical expression for the principal strains is

$$(\varepsilon_{x'})_{\substack{\max \\ \min}} = \varepsilon_{1\text{ or }2} = \frac{\varepsilon_x + \varepsilon_y}{2} \pm \sqrt{\left(\frac{\varepsilon_x - \varepsilon_y}{2}\right)^2 + \left(\frac{\gamma_{xy}}{2}\right)^2} \qquad (8\text{-}35)$$

where the positive sign in front of the square root is to be used for $\varepsilon_1$, the maximum principal strain in the algebraic sense. The negative sign is to be used for $\varepsilon_2$, the minimum principal strain. The planes on which the principal strains act can be defined analytically from Eq. 8-34 by setting it equal to zero. Thus,

$$\tan 2\theta_1 = \frac{\gamma_{xy}}{\varepsilon_x - \varepsilon_y} \qquad (8\text{-}36)$$

Since this equation has two roots, it is completely analogous to Eq. 8-6 and can be treated in the same manner.

2. The largest shear strain $\gamma_{\max}$ is equal to *two times* the radius of the circle. Normal strains of $(\varepsilon_1 + \varepsilon_2)/2$ in two mutually perpendicular directions are associated with the maximum shear strain.

3. The sum of normal strains in any two mutually perpendicular directions is *invariant*, i.e., $\varepsilon_1 + \varepsilon_2 = \varepsilon_x + \varepsilon_y = $ constant. Other properties of strains at a point can be established by studying the circle further.

Mathematically, in every respect, strain transformation is identical to stress transformation. Therefore, in a general three-dimensional strain problem, there are three principal directions in which principal normal strains develop. For *plane strain,* when $\varepsilon_z = \gamma_{zx} = \gamma_{zy} = 0$, besides the two principal strains $\varepsilon_1$ and $\varepsilon_2$, another principal strain $\varepsilon_3 = \varepsilon_z = 0$. By identifying the latter principal strain by a point on the $\varepsilon - \gamma/2$ plot, it is possible to draw a cluster of three principal strain circles just as before for the stress circles (Figs. 8-17 and 8-18). This procedure is illustrated in the next example. Mohr's strain circles degenerate to a point when two or three principal strains are equal.

For determining strain in the $z$ direction for *plane stress,* one must first

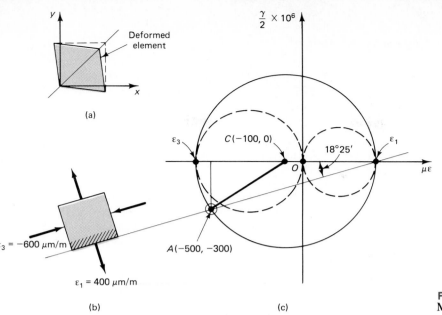

(a)

(b)

(c)

Fig. 8-23 (Sign convention of Method 1, Section 8-7.)

form an inverse of the first three of Eqs. 3-14, i.e., to solve them simultaneously to express stresses in terms of strain. For the stress in the $z$ direction, this gives

$$\sigma_z = \frac{E}{(1 + \nu)(1 - 2\nu)} [(1 - \nu)\varepsilon_z + \nu(\varepsilon_x + \varepsilon_y)] \qquad (8\text{-}37)$$

Then, since for plane stress $\sigma_z = 0$,

$$\varepsilon_z = -\frac{\nu}{1 - \nu}(\varepsilon_x + \varepsilon_y) \qquad (8\text{-}38)$$

Since $(\varepsilon_x + \varepsilon_y)$ is invariant, $\varepsilon_z$ remains constant for any planar coordinate transformation. Hence, at a point, either the Mohr's circle of strain, or its fundamental equivalent of algebraic transformations for the two-dimensional problem, is applicable.

## EXAMPLE 8-7

It is observed that an element of a body in a state of plane strain contracts 500 μm/m along the $x$ axis, elongates 300 μm/m in the $y$ direction, and distorts through an angle[13] of 600 μrad, as shown in Fig. 8-23(a). Using Mohr's circle, determine

---

[13] This measurement may be made by scribing a small square on a body, straining the body, and then measuring the change in angle that takes place. Photographic enlargements of grids, or photogrammetric procedures, have been used for this purpose.

**437**

the in-plane principal strains for the given data and show the directions in which they occur. On the same diagram, draw the remaining two principal strain circles.

### Solution

The given data are $\varepsilon_x = -500$ $\mu$m/m, $\varepsilon_y = +300$ $\mu$m/m, and $\gamma_{xy} = -600$ $\mu$m/m. Hence, on the $\varepsilon - \gamma/2$ system of axes, the center $C$ is at $(\varepsilon_x + \varepsilon_y)/2 = -100$ $\mu$m/m from $O$, Fig. 8-23(b). The origin of planes $A$ is at $(-500, -300)$. The circle radius $AC$ is 500 $\mu$m/m. Hence, $\varepsilon_1 = +400$ $\mu$m/m acts in the direction perpendicular to line $A\varepsilon_1$ and $\varepsilon_3 = -600$ $\mu$m/m acts in the direction perpendicular to the line $A\varepsilon_3$ (not shown). From geometry, $\theta = \tan^{-1} 300/900 = 18°25'$.

Since this is a *plane strain* problem, another principal strain, $\varepsilon_2 = 0$, is at the origin $O$ of the coordinate axes. Therefore, the two small dashed-line strain circles are shown on the figure to complete the problem.

---

## *8-14.  Strain Rosettes

Measurements of normal strain are particularly simple to make, and highly reliable techniques have been developed for this purpose. In such work, these strains are measured along several closely clustered gage lines, diagrammatically indicated in Fig. 8-24(a) by lines $a$–$a$, $b$–$b$, and $c$–$c$. These gage lines may be located on the member investigated with reference to some coordinate axes (such as $x$ and $y$) by the respective angles $\theta_1$, $\theta_2$, and $\theta_3$. By comparing the initial distance between any two corresponding gage points with the distance in the stressed member, the elongation in the gage length is obtained. Dividing the elongation by the gage length gives the strain in the $\theta_1$ direction, which will be designated $\varepsilon_{\theta_1}$. By performing the same operation with the other gage lines, $\varepsilon_{\theta_2}$ and $\varepsilon_{\theta_3}$ are obtained. If the distances between the gage points are small, measurements approximating the strains at a point are obtained.

Arrangements of gage lines at a point in a cluster, as shown in Fig. 8-24, are called *strain rosettes*. If three strain measurements are taken at

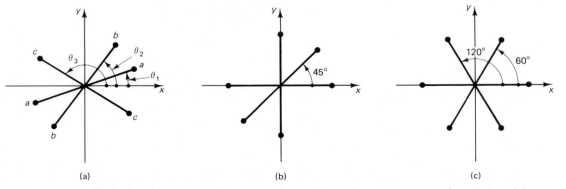

**Fig. 8-24**  (a) General strain rosette; (b) rectangular or 45° strain rosette; (c) equiangular or delta rosette.

a rosette, the information is sufficient to determine the complete state of plane strain at a point.

As already noted in Section 2-2, a particularly versatile and accurate method for measuring strain employs electric strain gages. These gages, made either of fine wire or foil glued to a member, are very sensitive for measuring the change in electrical resistance due to deformation in a member. An appropriate calibration[14] relates gage resistance to strain. Several types of rosettes are in general use. These usually consist of three single-element gages grouped together, as shown in Fig. 8-25. Metal-foil rosettes of this type are available in a wide range of sizes, with active gage lengths varying from 0.8 to 12 mm.

If angles $\theta_1$, $\theta_2$, and $\theta_3$, together with the corresponding strains $\varepsilon_{\theta_1}$, $\varepsilon_{\theta_2}$, and $\varepsilon_{\theta_3}$, are known from measurements, three simultaneous equations patterned after Eq. 8-28 can be written. In writing these equations, it is convenient to employ the following notation: $\varepsilon_{x'} \equiv \varepsilon_{\theta_1}$, $\varepsilon_{x''} \equiv \varepsilon_{\theta_2}$, and $\varepsilon_{x'''} \equiv \varepsilon_{\theta_3}$.

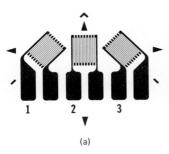

(a)

(b)

**Fig. 8-25** Three-element metal-foil electrical-resistance strain gages (Courtesy of Micro-Measurements Division, Measurements Group, Inc., Raleigh, North Carolina, U.S.A.).

$$\varepsilon_{\theta_1} = \varepsilon_x \cos^2 \theta_1 + \varepsilon_y \sin^2 \theta_1 + \gamma_{xy} \sin \theta_1 \cos \theta_1$$
$$\varepsilon_{\theta_2} = \varepsilon_x \cos^2 \theta_2 + \varepsilon_y \sin^2 \theta_2 + \gamma_{xy} \sin \theta_2 \cos \theta_2 \qquad (8\text{-}39)$$
$$\varepsilon_{\theta_3} = \varepsilon_x \cos^2 \theta_3 + \varepsilon_y \sin^2 \theta_3 + \gamma_{xy} \sin \theta_3 \cos \theta_3$$

This set of equations can be solved for $\varepsilon_x$, $\varepsilon_y$, and $\gamma_{xy}$, and the problem reverts to the cases already considered.

To minimize computational work, the gages in a rosette are usually arranged in an orderly manner. For example, in Fig. 8-24(b), $\theta_1 = 0°$, $\theta_2 = 45°$, and $\theta_3 = 90°$. This arrangement of gage lines is known as the *rectangular* or the 45° *strain rosette*. By direct substitution into Eq. 8-39, it is found that for this rosette

$$\varepsilon_x = \varepsilon_{0°} \qquad \varepsilon_y = \varepsilon_{90°} \qquad 2\varepsilon_{45°} = \varepsilon_x + \varepsilon_y + \gamma_{xy}$$

or

$$\boxed{\gamma_{xy} = 2\varepsilon_{45°} - (\varepsilon_{0°} + \varepsilon_{90°})} \qquad (8\text{-}40)$$

Thus, $\varepsilon_x$, $\varepsilon_y$, and $\gamma_{xy}$ become known. Variations of this arrangement are shown in Fig. 8-25.

Another arrangement of gage lines is shown in Fig. 8-24(c). This is known as the *equiangular*, or the *delta*, or the 60° *rosette*. Again, by substituting into Eq. 8-39 and simplifying,

$$\boxed{\varepsilon_x = \varepsilon_{0°}} \qquad \boxed{\varepsilon_y = (2\varepsilon_{60°} + 2\varepsilon_{120°} - \varepsilon_{0°})/3} \qquad (8\text{-}41a)$$

[14] See Society for Experimental Mechanics (SEM), A. S. Kobayashi (ed.), *Handbook on Experimental Mechanics* (Englewood Cliffs, NJ: 1987).

and
$$\gamma_{xy} = 2(\varepsilon_{60°} - \varepsilon_{120°})/\sqrt{3} \qquad (8\text{-}41b)$$

Other types of rosettes are occasionally used in experiments. The data from all rosettes can be analyzed by applying Eq. 8-39, solving for $\varepsilon_x$, $\varepsilon_y$, and $\gamma_{xy}$, and then either applying the strain-transformation equations or constructing Mohr's circle for finding the principal strains.

Sometimes rosettes with more than three lines are used. An additional gage line measurement provides a check on the experimental work. For these rosettes, the invariance of the strains in the mutually perpendicular directions can be used to check the data.

The application of the experimental rosette technique in complicated problems of stress analysis is almost indispensable.

In most problems where strain rosettes are used, it is necessary to determine the principal stresses at the point of strain measurement. In this problem, the surface where the strains are measured is generally free of significant normal surface stresses, i.e. $\sigma_z = 0$. Therefore, this is a *plane stress* problem. Hence, the relevant Eqs. 3-14 written in terms of principal stresses, $\sigma_1$ and $\sigma_2$, become

$$\varepsilon_1 = \frac{\sigma_1}{E} - v\frac{\sigma_2}{E} \quad \text{and} \quad \varepsilon_2 = \frac{\sigma_2}{E} - v\frac{\sigma_1}{E} \qquad (8\text{-}42)$$

Solving these equations simultaneously for the principal stresses, one obtains the required relations:

$$\sigma_1 = \frac{E}{1 - v^2}(\varepsilon_1 + v\varepsilon_2) \qquad \sigma_2 = \frac{E}{1 - v^2}(\varepsilon_2 + v\varepsilon_2) \qquad (8\text{-}43)$$

The elastic constants $E$ and $v$ must be determined from some appropriate experiments. With the aid of such experimental work, very complicated problems can be solved successfully.

### EXAMPLE 8-8

At a certain point on a steel machine part, measurements with an electric rectangular rosette indicate that $\varepsilon_{0°} = -500$ μm/m, $\varepsilon_{45°} = +200$ μm/m, and $\varepsilon_{90°} = +300$ μm/m. Assuming that $E = 200$ GPa and $v = 0.3$, find the principal stresses at the point investigated.

Solution

From Eq. 8-40,

$$\gamma_{xy} = 2\varepsilon_{45°} - (\varepsilon_{0°} + \varepsilon_{90°}) = 2 \times 200 - (-500 + 300) = 600 \text{ μm/m}$$

The principal strains for this data were determined in Example 8-7, and are $\varepsilon_1 = 400\ \mu m/m$ and $\varepsilon_2 = -600\ \mu m/m$. Hence, using Eqs. 8-43,

$$\sigma_1 = \frac{200 \times 10^3}{1 - 0.3^2}\,[400 + 0.3 \times (-600)] \times 10^{-6} = +48.4\ \text{MPa}$$

$$\sigma_2 = \frac{200 \times 10^3}{1 - 0.3^2}\,(-600 + 0.3 \times 400) \times 10^{-6} = 105\ \text{MPa}$$

Tensile stress $\sigma_1$ acts in the direction of $\varepsilon_1$; see Fig. 8-23. The compressive stress $\sigma_2$ acts in the direction of $\varepsilon_2$.

# Part C    YIELD AND FRACTURE CRITERIA

## 8-15. Introductory Remarks

From the preceding study of the text, it should be apparent that in numerous technical problems, the state of stress and strain at critical points may be very complex. Idealized mathematical procedures for determining those states, as well as their transformations to different coordinates, are available. However the precise response of real materials to such stresses and strains defies accurate formulations. A number of questions remain unsettled and are part of an active area of materials research. As yet no comprehensive theory can provide accurate predictions of material behavior under the multitude of static, dynamic, impact, and cyclic loading, as well as temperature effects. Only the classical idealizations of yield and fracture criteria for materials are discussed here. Of necessity, they are used in the majority of structural and machine design. These strength theories are structured to apply to particular classes of materials. The two most widely accepted criteria for the onset of inelastic behavior (yield) for ductile materials under combined stresses are discussed first. This is followed by presentation of a fracture criterion for brittle materials. It must be emphasized that, in classifying materials in this manner, strictly speaking, one refers to the brittle or ductile state of the material, as this characteristic is greatly affected by temperature as well as by the state of stress itself. For example, some low-carbon steels below their transition temperatures of about 10°C (+50°F) become brittle, losing their excellent ductile properties, and behave like different materials; see Fig. 8-26. More complete discussion of such issues are beyond the scope of this text.

Most of the information on yielding and fracture of materials under the action of biaxial stresses comes from experiments on thin-walled cylinders. A typical arrangement for such an experiment is shown in Fig. 8-27. The ends of a thin-walled cylinder of the material being investigated

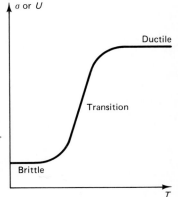

**Fig. 8-26** Typical transition curve for stress or energy to fracture vs. temperature for carbon steel.

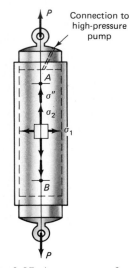

P

Connection to
high-pressure
pump

A

$\sigma''$

$\sigma_2$

$\sigma_1$

B

P

**Fig. 8-27** Arrangement for controlled ratios of principal stresses.

are closed by substantial caps. This forms the hollow interior of a cylindrical pressure vessel. By pressurizing the available space until the yielding or bursting occurs, the elements of the wall are subjected to biaxial stresses of a constant ratio $\sigma_1/\sigma_2 = 2$. By applying an additional tensile force $P$ to the caps, the $\sigma_2$ stress is increased to any predetermined amount $\sigma_2 + \sigma''$. By applying a compressive force, the $\sigma_2$ stress can be minimized or eliminated. Actual compressive stress in the longitudinal direction is undesirable, as the tube may buckle. By maintaining a fixed ratio between the principal stresses until the failure point is reached, the desired data on a material are obtained. Analogous experiments with tubes simultaneously subjected to torque, axial force, and pressure are also used. An interpretation of these data, together with all other related experimental evidence, including the simple tension tests, permits a formulation of theories of failure for various materials subjected to combined stresses.

## 8-16. Maximum Shear-Stress Theory

The maximum shear-stress theory,[15] or simply the maximum shear theory, results from the observation that in a *ductile* material slip occurs during yielding along critically oriented planes. This suggests that the maximum shear stress plays the key role, and it is assumed that yielding of the material depends only on the maximum shear stress that is attained within an element. Therefore, whenever a certain critical value $\tau_{cr}$ is reached, yielding in an element commences.[16] For a given material, this value usually is set equal to the shear stress at yield in simple tension or compression. Hence, according to Eq. 8-9, if $\sigma_x = \pm\sigma_1 \neq 0$, and $\sigma_y = \tau_{xy} = 0$,

$$\tau_{max} \equiv \tau_{cr} = \left| \pm \frac{\sigma_1}{2} \right| = \frac{\sigma_{yp}}{2} \qquad (8\text{-}44)$$

which means that if $\sigma_{yp}$ is the yield-point stress found, for example, in a simple tension test, the corresponding maximum shear stress is half as large. This conclusion also follows easily from Mohr's circle of stress.

In applying this criterion to a biaxial plane stress problem, two different cases arise. In one case, the signs of the principal stresses $\sigma_1$ and $\sigma_2$ are the same. Taking them, for example, to be tensile, Fig. 8-28(a), and setting $\sigma_3 = 0$, the resulting Mohr's principal stress circles are as shown in Fig.

[15] This theory appears to have been originally proposed by C. A. Coulomb in 1773. In 1868, H. Tresca presented the results of his work on the flow of metals under great pressures to the French Academy. Now this theory often bears his name.

[16] In single crystals, slip occurs along preferential planes and in preferential directions. In studies of this phenomenon, the effective component of the shear stress causing slip must be carefully determined. Here it is assumed that because of the random orientation of numerous crystals, the material has isotropic properties, and so by determining $\tau_{max}$, one finds the critical shear stress.

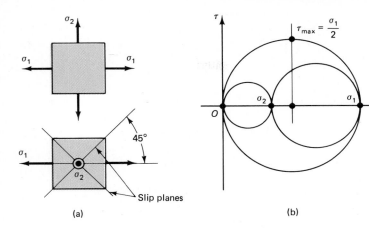

(a)

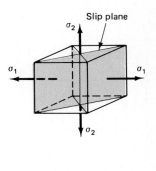

(b)

(c)

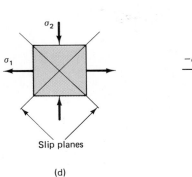

(d)

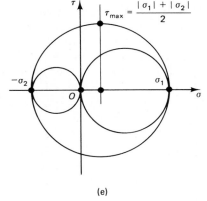

(e)

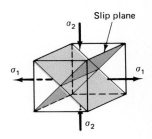

(f)

**Fig. 8-28** Planes of $\tau_{max}$ for biaxial stress.

8-28(b). Here the maximum shear stress is of the same magnitude as would occur in a simple uniaxial stress, Figs. 8-28(a) and (c). Therefore, if $|\sigma_1| > |\sigma_2|$, according to Eq. 8-44, $|\sigma_1|$ must not exceed $\sigma_{yp}$. Similarly, if $|\sigma_2| > |\sigma_1|$, $|\sigma_2|$ must not be greater than $\sigma_{yp}$. Therefore, the criteria corresponding to this case are

$$|\sigma_1| \le \sigma_{yp} \quad \text{and} \quad |\sigma_2| \le \sigma_{yp} \qquad (8\text{-}45)$$

The second case is considered in Figs. 8-28(d)–(f), where the signs of $\sigma_1$ and $\sigma_2$ are opposite, and $\sigma_3 = 0$. The largest Mohr's circle passes through $\sigma_1$ and $\sigma_2$, and the maximum shear stress $\tau_{max} = (|\sigma_1| + |\sigma_2|)/2$. The alternative possible slip planes are identified in Fig. 8-28(d) and (f). This maximum shear stress cannot exceed the shear yield criterion in simple tension, i.e., $\tau_{max} \le \sigma_{yp}/2$. Hence,

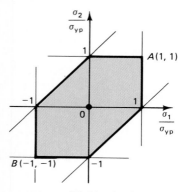

**Fig. 8-29** Yield criterion based on maximum shear stress.

$$\left| \pm \frac{\sigma_1 - \sigma_2}{2} \right| \leq \frac{\sigma_{yp}}{2} \tag{8-46}$$

or, for impending yield,

$$\frac{\sigma_1}{\sigma_{yp}} - \frac{\sigma_2}{\sigma_{yp}} = \pm 1 \tag{8-46a}$$

A plot of this equation gives the two sloping lines shown in Fig. 8-29. Dividing Eqs. 8-45 by $\sigma_{yp}$ puts them into the same form as Eq. 8-46a. These modified equations, $\sigma_1/\sigma_{yp} = \pm 1$, and $\sigma_2/\sigma_{yp} = \pm 1$, plot, respectively, in Fig. 8-29 as two vertical and two horizontal lines. Then, by treating $\sigma_1/\sigma_{yp}$ and $\sigma_2/\sigma_{yp}$ as coordinates of a point in this principal stress space, some important conclusions can be reached.

If a point defined by $\sigma_1/\sigma_{yp}$ and $\sigma_2/\sigma_{yp}$ falls on the hexagon shown in Fig. 8-29, a material begins and continues to yield. No such stress points can lie outside the hexagon because one of the three yield criteria equations given before for perfectly plastic material would be violated. The stress points falling within the hexagon indicate that a material behaves elastically.

Note that, according to the maximum shear theory, if hydrostatic tensile or compressive stresses are added, i.e., stresses such that $\sigma_1' = \sigma_2' = \sigma_3'$, no change in the material response is predicted. Adding these stresses merely shifts the Mohr's circles of stress along the $\sigma$ axis and $\tau_{max}$ remains the same. Also note that since the maximum shear stresses are defined on planes irrespective of material directional properties, it is implicit that the material is *isotropic*.

The derived yield criterion for perfectly plastic material is often referred to as the *Tresca yield condition* and is one of the widely used laws of plasticity.

## *8-17. Maximum Distortion-Energy Theory

Another widely accepted criterion of yielding for *ductile* isotropic materials is based on energy concepts.[17] In this approach, the total elastic energy is divided into two parts: one associated with the volumetric changes of the material, and the other causing shear distortions. By equating the shear distortion energy at yield point in simple tension to that under combined stress, the yield criterion for combined stress is established.

---

[17] The first attempt to use the total energy as the criterion of yielding was made by E. Beltrami of Italy in 1885. In its present form, the theory was proposed by M. T. Huber of Poland in 1904 and was further developed and explained by R. von Mises (1913) and H. Hencky (1925), both of Germany and the United States.

In order to derive the expression giving the yield condition for combined stress, the procedure of resolving the general state of stress must be employed. This is based on the concept of superposition. For example, it is possible to consider the stress tensor of the three principal stresses—$\sigma_1$, $\sigma_2$, and $\sigma_3$—to consist of two additive component tensors. The elements of one component tensor are defined as the mean "hydrostatic" stress:

$$\bar{\sigma} = \frac{\sigma_1 + \sigma_2 + \sigma_3}{3} \tag{8-47}$$

The elements of the other tensor are $(\sigma_1 - \bar{\sigma})$, $(\sigma_2 - \bar{\sigma})$, and $(\sigma_3 - \bar{\sigma})$. Writing this in matrix representation, one has

$$\begin{pmatrix} \sigma_1 & 0 & 0 \\ 0 & \sigma_2 & 0 \\ 0 & 0 & \sigma_3 \end{pmatrix} = \begin{pmatrix} \bar{\sigma} & 0 & 0 \\ 0 & \bar{\sigma} & 0 \\ 0 & 0 & \bar{\sigma} \end{pmatrix} + \begin{pmatrix} \sigma_1 - \bar{\sigma} & 0 & 0 \\ 0 & \sigma_2 - \bar{\sigma} & 0 \\ 0 & 0 & \sigma_3 - \bar{\sigma} \end{pmatrix} \tag{8-48}$$

This resolution of the general state of stress is shown schematically in Fig. 8-30. The special case of resolving the uniaxial state of the stress in the figure has been carried a step further. The sum of the stresses in Figs. 8-30(f) nd (g) corresponds to the last tensor of Eq. 8-48.

For the three-dimensional state of stress, the Mohr's circle for the first

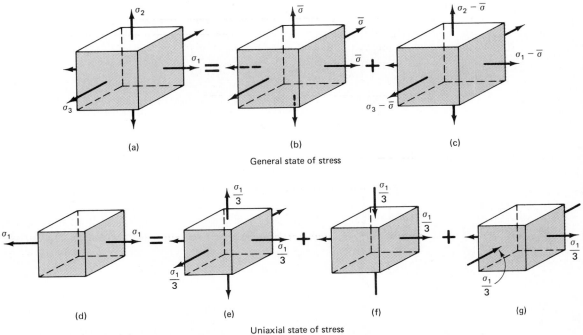

General state of stress

Uniaxial state of stress

**Fig. 8-30** Resolution of principal stresses into spherical (dilatational) and deviatoric (distortional) stresses.

tensor component of Eq. 8-48 degenerates into a point located at $\bar{\sigma}$ on the $\sigma$ axis. Therefore, the stresses associated with this tensor are the same in every possible direction. For this reason, this tensor is called the *spherical stress tensor*. Alternatively, from Eq. 3-21, which states that dilatation of an elastic body is proportional to $\bar{\sigma}$, this tensor is also called the *dilatational stress tensor*.

The last tensor of Eq. 8-48 is called the *deviatoric or distortional stress tensor*. A good reason for the choice of these terms may be seen from Figs. 8-30(f) and (g). The state of stress consisting of tension and compression on the mutually perpendicular planes is equivalent to pure shear stress. The latter system of stresses is known to cause no volumetric changes in isotropic materials, but instead, distorts or deviates the element from its initial cubic shape.

Having established the basis for resolving or decomposing the state of stress into dilatational and distortional components, one may find the strain energy due to distortion. For this purpose, first the strain energy per unit volume, i.e., strain density, for a three-dimensional state of stress must be found. Since this quantity does not depend on the choice of coordinate axes, it is convenient to express it in terms of principal stresses and strains. Thus, generalizing Eq. 2-21 for three-dimensions using superposition, one has

$$U_O = U_{\text{total}} = \frac{1}{2}\sigma_1\varepsilon_1 + \frac{1}{2}\sigma_2\varepsilon_2 + \frac{1}{2}\sigma_3\varepsilon_3 \qquad (8\text{-}49)$$

where, by substituting for strains, Eqs. 3-14, expressed in terms of principal stresses, after simplifications,

$$U_{\text{total}} = \frac{1}{2E}(\sigma_1^2 + \sigma_2^2 + \sigma_3^2) - \frac{v}{E}(\sigma_1\sigma_2 + \sigma_2\sigma_3 + \sigma_3\sigma_1) \quad (8\text{-}50)$$

The strain energy per unit volume due to the dilatational stresses can be determined from this equation by first setting $\sigma_1 = \sigma_2 = \sigma_3 = p$, and then replacing $p$ by $\bar{\sigma} = (\sigma_1 + \sigma_2 + \sigma_3)/3$. Thus,

$$U_{\text{dilatation}} = \frac{3(1 - 2v)}{2E}p^2 = \frac{1 - 2v}{6E}(\sigma_1 + \sigma_2 + \sigma_3)^2 \qquad (8\text{-}51)$$

By subtracting Eq. 8-51 from Eq. 8-50, simplifying, and noting from Eq. 3-19 that $G = E/2(1 + v)$, one finds the distortion strain energy for combined stress:

$$U_{\text{distortion}} = \frac{1}{12G}[(\sigma_1 - \sigma_2)^2 + (\sigma_2 - \sigma_3)^2 + (\sigma_3 - \sigma_1)^2] \quad (8\text{-}52)$$

According to the basic assumption of the distortion-energy theory, the expression of Eq. 8-52 must be equated to the maximum elastic distortion energy in simple tension. The latter condition occurs when one of the principal stresses reaches the yield point, $\sigma_{yp}$, of the material. The distortion strain energy for this is $2\sigma_{yp}^2/12G$. Equating this to Eq. 8-52, after minor simplifications, one obtains the basic law for yielding of an ideally plastic material:

$$(\sigma_1 - \sigma_2)^2 + (\sigma_2 - \sigma_3)^2 + (\sigma_3 - \sigma_1)^2 = 2\sigma_{yp}^2 \qquad (8\text{-}53)$$

For plane stress, $\sigma_3 = 0$, and Eq. 8-53 in dimensionless form becomes

$$\left(\frac{\sigma_1}{\sigma_{yp}}\right)^2 - \left(\frac{\sigma_1}{\sigma_{yp}}\frac{\sigma_2}{\sigma_{yp}}\right) + \left(\frac{\sigma_2}{\sigma_{yp}}\right)^2 = 1 \qquad (8\text{-}54)$$

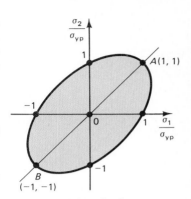

**Fig. 8-31** Yield criterion based on maximum distortion energy.

This is an equation of an ellipse, a plot of which is shown in Fig. 8-31. Any stress falling within the ellipse indicates that the material behaves elastically. Points on the ellipse indicate that the material is yielding. This is the same interpretation as that given earlier for Fig. 8-29. On unloading, the material behaves elastically.

This theory does not predict changes in the material response when hydrostatic tensile or compressive stresses are added. Since only differences of the stresses are involved in Eq. 8-53, adding a constant stress to each does not alter the yield condition. For this reason in the three-dimensional stress space, the yield surface becomes a cylinder with an axis having all three direction cosines equal to $1/\sqrt{3}$. Such a cylinder is shown in Fig. 8-32. The ellipse in Fig. 8-31 is simply the intersection of this cylinder with the $\sigma_1$–$\sigma_2$ plane. It can be shown also that the yield surface for the maximum shear stress criterion is a hexagon that fits into the tube, Fig. 8-31.

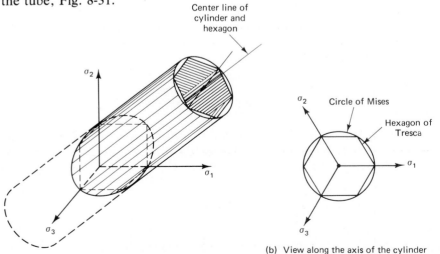

(a)

(b) View along the axis of the cylinder

**Fig. 8-32** Yield surfaces for triaxial state of stress.

The fundamental relation given by Eq. 8-53 may also be derived by formulating the second invariant, Eq. 8-26, of the deviatoric stresses given by the last matrix in Eq. 8-48. Such an approach is generally favored in the mathematical theory of plasticity. The derivation given before gives greater emphasis to physical behavior. As can be noted from the structure of Eq. 8-53 and the accompanying Figs. 8-31 and 8-32, it is a continuous function, making it attractive in analytical and numerical applications. This widely used constitutive equation for perfectly plastic material is often referred to as the *Huber–Hencky–Mises* or simply the *von Mises yield condition*.[18]

Both the maximum shear stress and the distortion energy yield conditions have been used in the study of viscoelastic phenomena under combined stress. Extension of these ideas to strain hardening materials is also possible. Such topics, however, are beyond the scope of this text.

### 8-18. Comparison of Maximum-Shear and Distortion-Energy Theories for Plane Stress

Plane stress problems occur especially frequently in practice and are largely emphasized in this text. Therefore, it is useful to make a comparison between the two most widely used yield criteria for ductile materials for this case. The maximum shear-stress criterion directs its attention to the maximum shear stress in an element. The distortion-energy criterion does this in a more comprehensive manner by considering in three dimensions the energy caused by shear deformations. Since shear stresses are the main parameters in both approaches, the difference between the two is not large. A comparison between them for plane stress is shown in Fig. 8-33. Here the Tresca hexagon for the maximum shear-stress theory and the von Mises ellipse for the maximum distortion-energy theory have the meanings already described. Either one of the lines gives a criterion for yield for a perfectly plastic material. Yield of a material is said to begin whenever either uniaxial or biaxial stresses reach the bounding lines. If a stress point for the principal stresses $\sigma_1$ and $\sigma_2$ falls within these curves, a material behaves elastically. Since no strain hardening behavior (see Fig. 2-13) is included in these mathematical models, no stress points can lie outside the curves, as yielding continued at the stress level given by the curves. More advanced theories are not considered in this text.[19]

It can be seen from Fig. 8-33 that the discrepancy between the two theories is not very large, the maximum shear-stress theory being in general more conservative. As to be expected, the uniaxial stresses given by

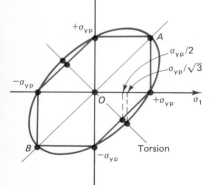

**Fig. 8-33** Comparison of Tresca and von Mises yield criteria.

[18] In the past, this condition has been also referred to as the *octahedral shearing stress theory*. See A. Nadai, *Theory of Flow and Fracture of Solids* (New York: McGraw-Hill, 1950), 104, or A. P. Boresi and O. M. Sidebottom, *Advanced Mechanics of Materials,* 4th ed. (New York: Wiley, 1985), 18.

[19] K. Washizu, *Variational Methods in Elasticity and Plasticity,* 2nd ed. (New York: Pergamon, 1975). L. E. Malvern, *Introduction to the Mechanics of a Continuous Medium* (New York: Prentice-Hall, 1969).

both are equal to those corresponding to simple tension or compression. It is assumed that these basic stresses are of *equal* magnitude. The yield criteria in the second and fourth quadrant indicate smaller strengths at yield than that for uniaxial stresses. The largest discrepancy occurs when two of the principal stresses are equal but of opposite sign. This condition develops, for example, in torsion of thin-walled tubes. According to the maximum shear-stress theory, when $\pm\sigma_1 = \mp\sigma_2$, these stresses at yield can reach only $\sigma_{yp}/2$. The maximum distortion-energy theory limits this stress to $\sigma_{yp}/\sqrt{3} = 0.577\sigma_{yp}$. Points corresponding to these stresses are identified in Fig. 8-33. These values of yield in shear stress are frequently used in design applications.

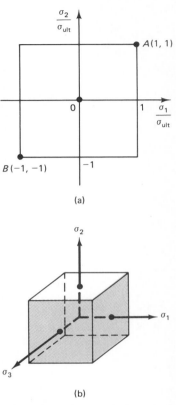

(a)

## 8-19. Maximum Normal Stress Theory

The maximum normal stress theory or simply the maximum stress theory[20] asserts that failure or fracture of a material occurs when the maximum normal stress at a point reaches a critical value regardless of the other stresses. Only the largest principal stress must be determined to apply this criterion. The critical value of stress $\sigma_{ult}$ is usually determined in a tensile experiment, where the failure of a specimen is defined to be either excessively large elongation or fracture. Usually, the latter is implied.

Experimental evidence indicates that this theory applies well to *brittle* materials in all ranges of stresses, providing a tensile principal stress exists. Failure is characterized by the separation, or the cleavage, fracture. This mechanism of failure differs drastically from the ductile fracture, which is accompanied by large deformations due to slip along the planes of maximum shear stress.

The maximum stress theory can be interpreted on graphs as the other theories. This is done in Fig. 8-34. Failure occurs if points fall on the surface. Unlike the previous theories, this stress criterion gives a bounded surface of the stress space.

(b)

**Fig. 8-34** Fracture envelope based on maximum stress criterion.

## 8-20. Comparison of Yield and Fracture Criteria

Comparison of some classical experimental results with the yield and fracture criteria presented before is shown in Fig. 8-35.[21] Note the particularly good agreement between the maximum distortion-energy theory and experimental results for ductile materials. However, the maximum

[20] This theory is generally credited to W. J. M. Rankine, an eminent British educator (1820–1872). An analogous theory based on the maximum strain, rather than stress, being the basic criterion of failure was proposed by the great French elastician, B. de Saint-Venant (1797–1886). Experimental evidence does not corroborate the latter approach.

[21] The experimental points shown on this figure are based on classical experiments by several investigators. The figure is adapted from a compilation made by G. Murphy, *Advanced Mechanics of Materials* (New York: McGraw-Hill, 1964), 83.

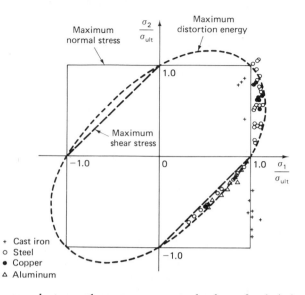

**Fig. 8-35** Comparison of yield and fracture criteria with test data.

normal stress theory appears to be best for brittle materials and can be unsafe for ductile materials.

All the theories for uniaxial stress agree since the simple tension test is the standard of comparison. Therefore, if one of the principal stresses at a point is large in comparison with the other, all theories give practically the same results. The discrepancy between the criteria is greatest in the second and fourth quadrants, when both principal stresses are numerically equal.

In the development of the theories discussed before, it has been assumed that the properties of material in tension and compression are alike—the plots shown in several of the preceding figures have two axes of symmetry. On the other hand, it is known that some materials such as rocks, cast iron, concrete, and soils have drastically different properties depending on the sense of the applied stress. This is the greatest flaw in applying the classical idealizations to materials having large differences in their mechanical behavior in tension and compression. An early attempt to adopt the maximum shear theory to achieve better agreement with experiments was made by Duguet in 1885.[22] The improved model recognizes the higher strengths of brittle materials in biaxial compression than in tension. Therefore, the region in biaxial tension in the principal stress space is made smaller than it is for biaxial compression; see Fig. 8-36. In the second and fourth quadrant, a linear change between the two of the above regions is assumed. A. A. Griffith,[23] in a sense, refined the

---

[22] A. Nadai, *Theory of Flow and Fracture of Solids* (New York: McGraw-Hill, 1950).

[23] A. A. Griffith, "The Phenomena of Rupture and Flow of Solids," *Philosophical Transactions of the Royal Society of London,* Series A, 1920, Vol. 221, 163–198.

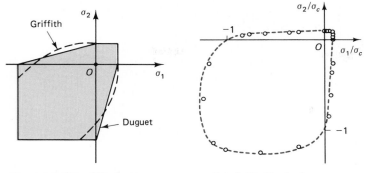

Fig. 8-36 Plausible fracture criteria for brittle materials.

Fig. 8-37 Dashed curve shows analytical fit for three different strength concretes.

explanation for the previous observations by introducing the idea of sur-face energy at microscopic cracks and showing the greater seriousness of tensile stresses compared with compressive ones with respect to fail-ure. According to this theory, an existing crack will rapidly propagate if the available elastic strain energy release rate is greater than the increase in the surface energy of the crack. The original Griffith concept has been considerably expanded by G. R. Irwin.[24] Careful recent experimental re-search on concrete specimens of different strengths strongly corroborates this approach, Fig. 8-37.[25] This work now has been extended to include strain hardening effects, and has been implemented for use with a com-puter.[26]

Another important attempt for rationalizing fracture of materials having different properties in tension and compression is due to Mohr.[27] In this approach, several different experiments must be conducted on the same material. For example, if the results of experiments in tension, compres-sion, and shear are available, the results can be represented on the same plot using their respective largest principal stress circles, as shown in Fig. 8-38(a). The points of contact of the *envelopes* with the stress circles define the state of stress at a fracture. For example, if such a point is *A*

[24] G. R. Irwin, "Fracture Mechanics," *Proceedings, First Symposium on Naval Structural Mechanics* (Long Island City, NY: Pergamon, 1958), 557. Also see *A Symposium on Fracture Toughness Testing and Its Applications,* American Society for Testing and Materials Special Technical Publication No. 381 (Phila-delphia, PA: American Society for Testing and Materials and Washington, DC: National Aeronautics and Space Administration, 1965).

[25] Adapted from H. Kupfer, and K. Gerstle, "Behavior of Concrete Under Biaxial Stresses," *J. Eng. Mech. Div.* ASCE, 99 (1973): EM4, 863.

[26] C. Bedard, and M. D. Kotsovos, "Application of NLFEA to Concrete Struc-tures," *J. Struct. Div.* ASCE, 111 (ST12) (1985). Z. P. Bažant, ed., *Mechanics of Geomaterials: Rocks, Concrete, Soils* (Chinchester: Wiley, 1985).

[27] As noted earlier, Otto Mohr was also principally responsible for the devel-opment of the stress circle bearing his name.

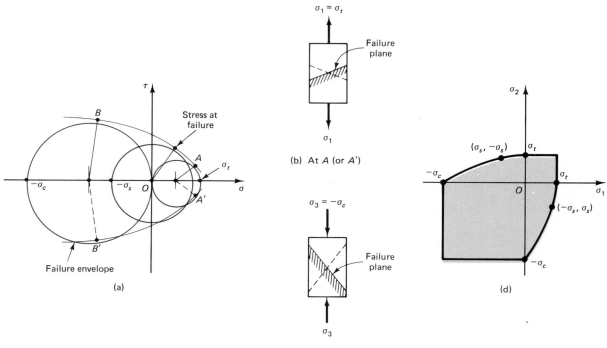

**Fig. 8-38** (a) Mohr envelopes, (b) failure planes at $A$ and $A'$, (c) failure planes at $B$ and $B'$, (d) Mohr envelope solution in principal stress space.

(or $A'$), the stresses and the plane(s) on which they act can be found using the established procedure for Mohr's circle of stress (Section 8-7). The corresponding planes for points $A$ or $A'$ are shown in Fig. 8-38(b), and a material such as duraluminum does fracture in tension at a flat angle as shown. Similarly, by relating the fracture planes to either point $B$ or $B'$, the fracture occurs at a steep angle characteristic of concrete cylinders tested in compression, Fig. 8-38(c). Such agreements with experiments support the assumed approach.

The data from Fig. 8-38(a) can be replotted in the principal stress space as in Fig. 8-38(d). Since in the first quadrant, the *minimum* principal stress $\sigma_3 = 0$, and in the third quadrant, $\sigma_3 = 0$ is the *maximum* principal stress, per Figs. 8-28(a)–(c), in these quadrants, the fracture lines in the principal stress space are similar to those of Fig. 8-29. Moreover, if the material strengths in tension and compression are the same, a hexagon identical to that shown in Fig. 8-29 is obtained. However, whereas the hexagon in Fig. 8-29 gives a *yield* condition for *ductile* materials, in the present context, it defines a *fracture* criterion for *brittle* materials.

Extrapolation of Mohr envelopes beyond the range of test data is not advisable. In many applications, this may mean that parts of the stress circles for tension and compression should be taken as envelope ends.

Interpolation along the failure envelopes between these two partial end stress circles is justified, and a stress circle for other conditions can be placed between them. When more extensive data are lacking conservatively, straight-line envelopes can be used.

The use of straight lines for asymptotes has a rational basis and has been found particularly advantageous in soil mechanics. For a loose granular media such as sand, the straight-line Mohr envelopes correspond to the limiting condition of dry friction, $\mu = \tan\phi$, Fig. 8-39. Any circle tangent to the envelope, as at $B$, gives the state of critical stress. If some cohesion can be developed by the media, the origin $O$ is moved to the right such that at zero stress, the $\tau$ intercept is equal to the cohesion. As soils basically cannot transmit tensile stresses, in specialized literature it is customary to direct the compression axis to the right.

Unlike the maximum distortion-energy theory, the fracture theory based on Mohr envelopes, using the largest principal stress circles, neglects dependence on the intermediate principal stress.

Sometimes the yield and fracture criteria discussed before are inconvenient to apply. In such cases, interaction curves such as in Fig. 7-6 can be used to advantage. Experimentally determined curves of this type, unless complicated by a local or buckling phenomenon, are equivalent to the strength criteria discussed here.

In the design of members in the next chapter, departures will be made from strict adherence to the yield and fracture criteria established here, although, unquestionably, these theories provide the rational basis for design.

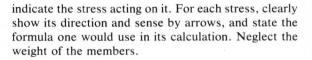

**Fig. 8-39** Mohr envelopes for cohesionless granular media.

# Problems

### Section 8-2

8-1. Infinitesimal elements $A$, $B$, $C$, $D$, and $E$ are shown on the figures for two different members. Draw each element separately, and on the isolated element, indicate the stress acting on it. For each stress, clearly show its direction and sense by arrows, and state the formula one would use in its calculation. Neglect the weight of the members.

(a)

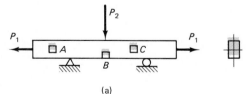

**Fig. P8-1**

(b)

**8-2 through 8-5.** For the infinitesimal elements shown in the figures, find the normal and shear stresses acting on the indicated inclined planes. Use the "wedge" method of analysis discussed in Example 8-1.

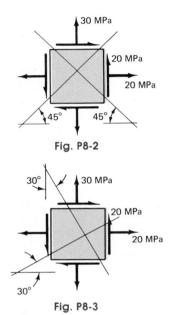

Fig. P8-2

Fig. P8-3

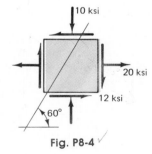

Fig. P8-4

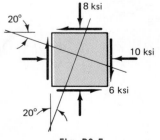

Fig. P8-5

**8-6.** The magnitudes and sense of the stresses at a point are as shown in the figure. Determine the stresses acting on the vertical and horizontal planes.

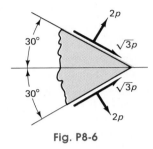

Fig. P8-6

**8-7.** The infinitesimal element shown in the figure is in equilibrium. Determine the normal and shear stresses acting on the vertical plane.

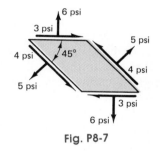

Fig. P8-7

**8-8.** At a particular point in a wooden member, the state of stress is as shown in the figure. The direction of the grain in the wood makes an angle of $+30°$ with the $x$ axis. The allowable shear stress parallel to the grain is 150 psi for this wood. Is this state of stress permissible? Verify your answer by calculations.

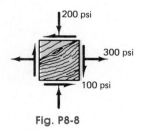

Fig. P8-8

**8-9.** After the erection of a heavy structure, it is estimated that the state of stress in the rock foundation

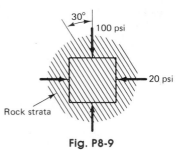

Fig. P8-9

will be essentially two-dimensional and as shown in the figure. If the rock is stratified, the strata making an angle of 30° with the vertical, is the anticipated state of stress permissible? Assume that the static coefficient of friction of rock on rock is 0.50, and along the planes of stratification, cohesion is 85 kN/m$^2$.

## Sections 8-3 through 8-5

**8-10.** Derive Eq. 8-2.

**8-11.** Using Eqs. 8-1 and 8-2, rework Prob. 8-2.

**8-12.** Using Eqs. 8-1 and 8-2, rework Prob. 8-3.

✓ **8-13.** Using Eqs. 8-1 and 8-2, rework Prob. 8-4.

**8-14.** Using Eqs. 8-1 and 8-2, rework Prob. 8-5.

✓ **8-15.** If at a point $\sigma_x = +8$ ksi, $\sigma_y = +2$ ksi, and $\tau = +4$ ksi, what are the principal stresses? Show their magnitude and sense on a properly oriented element.

✓ **8-16.** Determine the maximum (principal) shear stresses and the associated normal stresses for the last problem. Show the results on a properly oriented element.

**8-17 through 8-20.** For the following data, using the stress transformation equations, (a) find the principal stresses and show their sense on properly oriented elements; (b) find the maximum (principal) shear stresses with the associated normal stresses and show the results on properly oriented elements; and (c) check the invariance of the normal stresses for solutions in (a) and (b).

**8-17.** $\sigma_x = -30$ ksi, $\sigma_y = +10$ ksi, and $\tau = -20$ ksi.

**8-18.** $\sigma_x = 0$, $\sigma_y = +20$ ksi, and $\tau = +10$ ksi.

**8-19.** $\sigma_x = -40$ MPa, $\sigma_y = +10$ MPa, and $\tau = +20$ MPa.

**8-20.** $\begin{pmatrix} 20 & -20 \\ -20 & -10 \end{pmatrix}$ MPa

**8.21.** $\begin{pmatrix} 0 & -30 \\ -30 & -40 \end{pmatrix}$ ksi

## Sections 8-6 and 8-7

**8-22 through 8-25.** Draw Mohr's circles for the states of stress shown in the figures. (a) Determine the principal stresses and show their sense on properly oriented isolated elements. (b) Find the maximum (principal) shear stresses with the associated normal stresses and show the results on properly oriented elements. For both cases, check the invariance of the normal stresses.

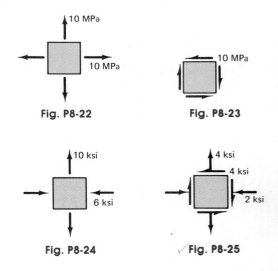

Fig. P8-22

Fig. P8-23

Fig. P8-24

✓ Fig. P8-25

**8-26 through 8-32.** For the following data, using Mohr's circles of stress *and trigonometry*, (a) find the principal stresses and show their sense on properly oriented isolated elements; (b) find the maximum (principal) shear stresses with the associated normal stresses and show the results on properly oriented elements. In each case, check the invariance of the normal stresses.

**8-26.** $\sigma_x = +50$ MPa, $\sigma_y = +30$ MPa, and $\tau = +20$ MPa.

**8-27.** $\sigma_x = +80$ psi, $\sigma_y = +20$ psi, and $\tau = +40$ psi.

✓ **8-28.** $\sigma_x = -30$ ksi, $\sigma_y = +10$ ksi, and $\tau = -20$ ksi.

**8-29.** $\sigma_x = -40$ MPa, $\sigma_y = -30$ MPa, and $\tau = +25$ MPa.

**8-30.** $\sigma_x = -15$ MPa, $\sigma_y = +35$ MPa, and $\tau = +60$ MPa.

**8-31.** $\sigma_x = +20$ ksi, $\sigma_y = 0$, and $\tau = -15$ ksi.

**8-32.** $\sigma_x = 0$, $\sigma_y = -20$ ksi, and $\tau = -10$ ksi.

**8-33 through 8-36.** For the following data, using Mohr's circles of stress, determine the normal and shear stresses acting on the planes defined by the given angle $\theta$. Show the results on isolated elements.

**8-33.** $\sigma_x = \sigma_1 = 0$, $\sigma_y = \sigma_2 = -20$ ksi, for $\theta = +30°$.

**\*8-34.** Rework Prob. 8-4 with $\theta = +30°$.

**\*8-35.** Rework Prob. 8-2 for $\theta = +45°$.

**8-36.** $\sigma_x = \sigma_y = 0$, $\tau = -20$ ksi, for $\theta = 20°$.

**8-37.** For the data shown for Prob. 8-6, using Mohr's circle of stress, find the principal stresses and show the results on a properly oriented element.

**8-38.** For the data shown for Prob. 8-7, using Mohr's circle of stress, find the principal stresses and the orientation of the planes on which these act.

**\*8-39.** Using Mohr's circle, determine the angle between the right-hand face of the element shown in the figure and the plane or planes where the normal stress is zero. Check the result using the "wedge" method. Show the stresses with proper sense on the rotated element(s).

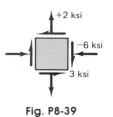

Fig. P8-39

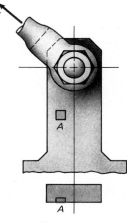

Fig. P8-40

**8-40.** A clevice transmits a force $F$ to a bracket, as shown in the figure. Stress analysis of this bracket gives the following stress components acting on element $A$: 1000 psi due to bending, 1500 psi due to axial force, and 600 psi due to shear. (Note that these are stress magnitudes only; their directions and senses must be determined by inspection.) (a) Indicate the resultant stresses on a drawing of the isolated element $A$. (b) Using Mohr's circle for the state of stress found in (a), determine the principal stresses and the maximum shear stresses with the associated normal stresses. Show the results on properly oriented elements.

**8-41.** At point $A$ on an unloaded edge of an elastic body, oriented as shown in the figure with respect to the $x$–$y$ axes, the maximum shear stress is 3500 kN/m$^2$. (a) Find the principal stresses, and (b) determine the state of stress on an element oriented with its edges parallel to the $x$–$y$ axes. Show the results on a drawing of the element at $A$.

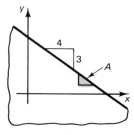

Fig. P8-41

## Sections 8-8 and 8-9

**8-42.** Determine the principal stresses and their directions for the following stress tensor:

$$\begin{pmatrix} 3 & 0 & 0 \\ 0 & 2 & 2 \\ 0 & 2 & 5 \end{pmatrix} \text{ksi}$$

Use the procedure discussed in Section 8-8. The direction cosines should be normalized. (This problem can also be solved using the equations for stress transformation discussed in Sections 8-4 and 8-7.)

**\*8-43.** For the following stress tensor, determine (a) the stress invariants, (b) the principal stresses, and (c) the direction of the largest principal stress. The direction cosines for this principal stress should be normalized.

$$\begin{pmatrix} 10 & 4 & -6 \\ 4 & -6 & 8 \\ -6 & 8 & 14 \end{pmatrix} \text{MPa}$$

**8-44.** For the data in Prob. 8-26, determine the principal stresses and draw the three principal circles of stress.

**8-45.** For the data in Prob. 8-28, determine the principal stresses and draw the three principal circles of stress.

## Sections 8-11 and 8-12

**8-46.** Rederive Eq. 8-28 by assuming that the shear deformation occurs first, then the deformation in the $y$ direction, and finally the deformation in the $x$ direction.

**8-47.** With the aid of Fig. 8-20, show that

$$\beta = -(\varepsilon_x - \varepsilon_y) \sin \theta \cos \theta + \gamma_{xy} \cos^2 \theta$$

✓ **8-48.** If the unit strains are $\varepsilon_x = -120$ μm/m, $\varepsilon_y = +1120$ μm/m, and $\gamma = -200$ μm/m, what are the principal strains and in which directions do they occur? Use Eqs. 8-35 and 8-36.

✓ **8-49.** If the unit strains are $\varepsilon_x = -800$ μm/m, $\varepsilon_y = -200$ μm/m, and $\gamma = +800$ μm/m, what are the principal strains and in which direction do they occur? Use Eqs. 8-35 and 8-36.

**\*\*8-50.** For the following strain tensor, using the method analogous to that described in Section 8-8 for stress transformation, determine (a) the principal strains, and (b) the directions of the maximum and minimum principal strains.

$$\begin{pmatrix} 70 & -10\sqrt{3} & 0 \\ -10\sqrt{3} & 5 & 0 \\ 0 & 0 & -20 \end{pmatrix} \text{μm/m}$$

## Section 8-13

**8-51.** Rework Prob. 8-48 using Mohr's circle of strain.

**8-52.** Rework Prob. 8-49 using Mohr's circle of strain.

## Section 8-14

**8-53.** The measured strains for a rectangular rosette, attached to a stressed steel member, are $\varepsilon_{0°} = -220$ μm/m, $\varepsilon_{45°} = +120$ μm/m, and $\varepsilon_{90°} = +220$ μm/m. What are the principal stresses and in which directions do they act? $E = 30 \times 10^6$ psi and $\nu = 0.3$.

**8-54.** The measured strains for an equiangular rosette, attached to a stressed aluminum alloy member, are $\varepsilon_{0°} = +400$ μm/m, $\varepsilon_{60°} = +400$ μm/m, and $\varepsilon_{120°} = -600$ μm/m. What are the principal stresses and in which directions do they act? $E = 70$ GPa and $\nu = 0.25$.

**8-55.** The data for a strain rosette with four gage lines attached to a stressed aluminum alloy member are $\varepsilon_{0°} = -120$ μm/m, $\varepsilon_{45°} = +400$ μm/m, $\varepsilon_{90°} = +1120$ μm/m, and $\varepsilon_{135°} = +600$ μm/m. Check the consistency of the data. Then determine the principal stresses and the directions in which they act. Use the values of $E$ and $\nu$ given in Prob. 8-54.

**8-56.** At a point in a stressed elastic plate, the following information is known: maximum shear strain $\gamma_{max} = 500$ μm/m, and the sum of the normal stresses on two perpendicular planes passing through the point is 27.5 MPa. The elastic properties of the plate are $E = 200$ GPa, $G = 80$ GPa, and $\nu = 0.25$. Calculate the magnitude of the principal stresses at the point.

## Section 8-17

**8-57.** Recast the stress tensor given in Prob. 8-43 into the spherical and the deviatoric stress tensors.

## Section 8-18

**8-58.** In classical experiments on plasticity, a two-dimensional stress field is often obtained by subjecting a thin-walled tube simultaneously to an axial force and a torque. The results of such experiments are reported on $\sigma_x$–$\tau_{xy}$ plots. If only $\sigma_x$ and $\tau_{xy}$ stresses are studied, how would the theoretical curves based on the Tresca and on the von Mises yield criteria look on such a plot? Derive the two required equations and sketch the results on a diagram.

**8-59.** Ordinarily the Tresca and von Mises yield stresses are made to coincide in simple tension. This gives rise to a discrepancy for pure shear. If, instead, the yield condition is assumed to be the same in shear, what discrepancy will result for simple tension and for $\sigma_1 = \sigma_2$?

**8-60.** A critical element develops the principal stresses, $\sigma_1$, $\sigma_2$, and $\sigma_3$, in the ratio $5:2:-1$, i.e., the stresses are $5p$, $2p$, and $-p$, where $p$ is a parameter. Such loadings are called radial. If this element is subjected to this loading condition, determine the maximum magnitudes the stresses may reach before yielding (a) according to the Tresca yield criterion, and (b) according to the von Mises criterion. Assume that the material yields in tension at 60 ksi.

**8-61.** A metal bar is being compressed along the $x_1$ axis between two rigid walls such that $\varepsilon_3 = 0$ and $\sigma_2 = 0$. This process causes an axial stress $\sigma_1$ and no shear stresses. Determine the apparent yield value of $\sigma_1$ if the material in a conventional compression test exhibits a yield strength $\sigma_{yp}$ and Poisson's ratio $\nu$. Assume that the material is governed by the von Mises yield condition. Find an alternative expression if the Tresca condition is postulated.

## chapter 9

# Elastic Stress Analysis and Design

## 9-1. Introduction

Formulas for determining the state of stress in elastic members traditionally considered in an introductory text on mechanics of solids have been derived in previous chapters. Usually, they give either a normal or a shear stress caused by a single force component acting at a section of a member. For linearly elastic materials, the main formulas are summarized:

1. Normal stresses

    (a) due to an axial force $\qquad \sigma = \dfrac{P}{A}$ $\qquad$ (1-13)

    (b) due to bending • straight members $\sigma = -\dfrac{My}{I}$ $\qquad$ (6-11)

    $\qquad$ • symmetrical curved bars $\qquad \sigma = \dfrac{My}{Ae(R - y)}$ $\qquad$ (6-32)

2. Shear stresses

    (a) due to torque • circular shaft $\qquad \tau = \dfrac{T\rho}{J}$ $\qquad$ (4-4)

    $\qquad$ • rectangular shaft $\tau_{\max} = \dfrac{T}{\alpha bt^2}$ $\qquad$ (4-30)

    $\qquad$ • closed thin-walled tube $\qquad \tau = \dfrac{T}{2\textcircled{A}t}$ $\qquad$ (4-34, 4-35)

    (b) due to shear force in a beam $\qquad \tau = \dfrac{VQ}{It}$ $\qquad$ (7-6)

The superposition of normal stresses caused by axial forces and bending simultaneously using these formulas was discussed in Chapter 6. Likewise, the superposition of shear stresses caused by torque and direct shear acting simultaneously was considered in Chapter 7. In this chapter, the consequences of the simultaneous occurrence of normal *and* shear stresses are examined with the aid of the stress-transformation procedures developed in Chapter 8. This condition commonly occurs in beams and transmission shafts.

In applying the above formulas, particularly in the analysis or design of mechanical equipment, stress-concentration factors must be introduced (see Sections 2-10, 4-7, and 6-6). Because of the problem of fatigue commonly occurring in such cases, reduced stresses are employed. Special consideration must also be given to dynamic loading (see Sections 2-13, 4-11, and 6-7). For such loadings, if occurring in milliseconds, the allowable stresses may be significantly increased.

This chapter is entirely devoted to *elastic* problems, an approach most commonly used at usual working loads. In Part A, the state of stress for some basic cases is discussed from the point of view of stress transformations.

The elastic design of members is considered in Part B. Although the stress-analysis formulas listed before are applicable to both statically determinate and indeterminate problems, discussion will be limited to statically determinate cases. There are at least two reasons for this. First, the more frequently occurring statically indeterminate problems involve beams; these are treated beginning with the next chapter. Second, more significantly, the *design* of statically indeterminate systems in contrast to their stress *analysis* is necessarily complex. As an example, consider an elastic bar of variable cross section, fixed at both ends, and subjected to an axial force $P$, as shown in Fig. 9-1. If the cross-sectional areas of the upper and lower parts of the bar are given, reactions $R_1$ and $R_2$ can be found routinely using the procedures discussed in Chapter 2. After either reaction is known, the problem becomes statically determine, and the bar stresses can be found in the usual manner. However, if this statically indeterminate system were to be *designed*, even this simple problem can become involved. Generally, in a design problem, only the applied force $P$ and the boundary conditions would be known. By varying the two cross-sectional areas of the bar, an infinite number of solutions is possible. Additional constraints in the realm of structural or machine design generally enter the problem. Such problems are, therefore, not considered here.

It is to be emphasized that only the problem of *elastic stresses* is considered in this chapter. Some elastic designs may be governed either by the stiffness or the possible instability of a system. The first requirement commonly arises in deflection control and vibration problems; the second, in lateral instability of members.

The main purpose of this chapter is to provide greater insight into the meaning of stress analysis by solving additional problems. There are extraordinarily many cases where applications of the basic formulas of engineering mechanics of solids listed before lead to useful results. No new

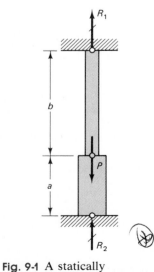

**Fig. 9-1** A statically indeterminate problem.

analytical principles are developed in this chapter. However, some simple design procedures for prismatic beams are given.

It is essential to recognize that in all elastic stress-analysis and design problems, the *material is assumed to be initially stress free.* In many engineering materials, significant residual stresses may be present. These may be caused by the manufacturing processes employed: rolling, welding, forging, temperature or hydration shrinkage, etc. (see Fig. 1-12). *In reality, it is the combination of the residual stresses with those due to the applied forces that cause the initial yield and/or fracture of a member.* In some engineering applications, estimates of residual stresses present a formidable problem.

# Part A    ELASTIC STRESS ANALYSIS

## 9-2. State of Stress for Some Basic Cases

The state of stress for four basic cases in the form of examples is considered in this section. By means of Mohr's principal circles of stress, the states of stress at a point are exhibited graphically. From such representations, the critical stresses can be seen readily and related to the yield or fracture criteria discussed in the preceding chapter. The four cases considered pertain to the uniaxial stress, biaxial stress such as occurs in cylindrical pressure vessels, torsional stresses in circular tubes, and beam stresses caused by bending and shear. Because of the greater complexity of the last problem, some aspects of the solution accuracy are discussed in the next section.

### EXAMPLE 9-1

Consider a state of stress in an axially loaded bar and construct the three principal circles of stress. Relate the critical stresses to yield and fracture criteria.

### Solution

The maximum principal stress $\sigma_1$ in an axially loaded bar can be found using Eq. 1-8. The remaining two principal stresses are each equal to zero, i.e., $\sigma_2 = \sigma_3 = 0$. The basic infinitesimal element for this case, together with its three planar views, is shown in Fig. 9-2(a). The principal circles of stress are drawn in Fig. 9-2(b). Since $\sigma_1$ and $\sigma_2$ are equal, Mohr's circle for these stresses degenerates into a point. For clarity, however, it is shown in the diagram by a small circle of zero diameter.

For this case, the maximum shear stress is equal to $\sigma_1/2$, whereas the maximum normal stress is $\sigma_1$. Therefore, the manner in which a material fails depends on

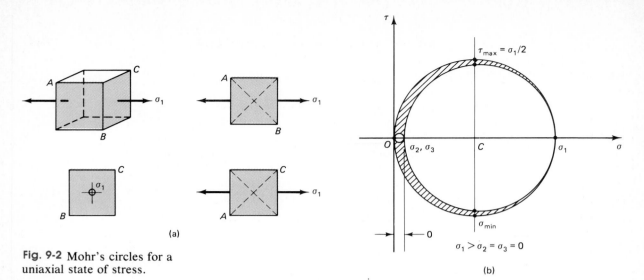

**Fig. 9-2** Mohr's circles for a uniaxial state of stress.

its relative strengths in these two properties. As already pointed out in Section 2-3 and illustrated in Fig. 2-7, a brittle material (cast iron) fails by a cleavage fracture. This is so because it is weaker in tension than in shear. The reverse is true for a ductile material. The cup and cone fractures shown in Fig. 2-7 for steel and aluminum occur approximately along the planes forming a 45° angle with the axis of the specimen. These planes are identified by dashed lines in the elements on the right in Fig. 9-2(a). Greater refinements on the mechanism of fracture are possible by considering the behavior of single crystals within a material.

## EXAMPLE 9-2

Consider a state of stress in a thin-walled cylindrical pressure vessel and construct the three principal circles of stress. Relate the results to a yield criterion.

### Solution

According to Eqs. 3-24 and 3-25, the ratio of the hoop stress $\sigma_1$ to the longitudinal stress $\sigma_2$ is approximately 2. These are the principal stresses as no shear stresses act on the corresponding planes. The third principal stress $\sigma_3$ equals the external or internal pressure $p$, which may be taken as zero since it is small in relation to $\sigma_1$ and $\sigma_2$. A typical infinitesimal element for the vessel and three planar views are shown in Fig. 9-3(a). The principal stress circles are shown in Fig. 9-3(b). The maximum shear stress is found on the *major* stress circle passing through the origin $O$ and $\sigma_1$. Its magnitude is $\sigma_1/2$. The planes on which the maximum shear stresses act are identified by dashed lines in the lower right element. Note that if only the principal stresses $\sigma_1$ and $\sigma_2$ were considered, the maximum shear stress would only be half as large. In design, the yield criterion based on the maximum distortion theory (see Section 8-18) can also be used.

Construction of pressure vessels from brittle materials is generally avoided as such materials provide no accommodation nor warning of failure through yielding before fracture.

It is interesting to note that for a thin-walled spherical pressure vessel, $\sigma_1 = \sigma_2$, and the corresponding principal stress circle degenerates into a point. Nevertheless, the maximum shear stress is $\sigma_1/2$ since the third principal stress is zero.

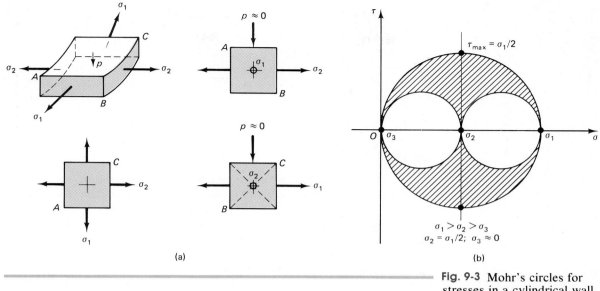

(a)

(b)

**Fig. 9-3** Mohr's circles for stresses in a cylindrical wall of a pressure vessel.

## EXAMPLE 9-3

Examine the state of stress in a circular tube subjected to a torque by constructing the three principal circles of stress. Relate the results to yield and fracture criteria.

### Solution

The shear stresses for this case can be found using Eq. 4-4. A typical infinitesimal element of the tube and three planar views are shown in Fig. 9-4(a). Here the

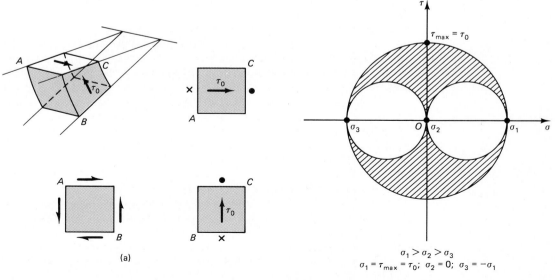

(a)

$$\sigma_1 > \sigma_2 > \sigma_3$$
$$\sigma_1 = \tau_{max} = \tau_0; \quad \sigma_2 = 0; \quad \sigma_3 = -\sigma_1$$

(b)

**Fig. 9-4** Mohr's circles for stresses in a circular tube subjected to torque.

major principal stress circle has a radius equal to the shear stress $\tau_0$ (see Section 8-7 for rules for constructing Mohr's circle). Hence, the two principal stresses are $\sigma_1$ and $\sigma_3$, as shown in Fig. 9-4(b). The middle principal stress $\sigma_2$ is zero, justifying the drawing of the two small circles shown.

For ductile materials, the strength in shear is smaller than in tension, and, as shown earlier in Fig. 4-9, at failure, a square fracture occurs across a member. If, alternatively, the material is stronger in shear than it is in tension, a characteristic fracture along a helix, shown in Fig. 4-10, is observed.

## EXAMPLE 9-4

A $40 \times 300$ mm rectangular elastic beam spans 1000 mm, as shown in mm in Fig. 9-5(a). This beam is braced to prevent lateral buckling. (a) Determine the principal stresses at points $K$, $L$, $M$, $L'$, and $K'$ caused by the application of a concentrated vertical force $P = 80$ kN at midspan to the top of the beam. (b) For the same condition, find the stresses on an inclined plane defined by $\theta = +30°$ for the element $L'$.

## Solution

(a) At section $K$–$K'$, the shear is 40 kN and the bending moment is 10 kN·m acting in the directions shown in Fig. 9-5(c).

No shear stresses act on elements $K$ and $K'$ as they are at the beam boundaries. Therefore, the principal stresses at these points follow directly by applying Eq. 6-21.

$$\sigma_{K \text{ or } K'} = \mp \frac{Mc}{I} = \mp \frac{M}{S} = \mp \frac{6M}{bh^2} = \mp \frac{6 \times 10 \times 10^6}{40 \times 300^2} = \mp 16.67 \text{ MPa}$$

The principal stresses acting in the vertical direction are zero. These results are shown in Figs. 9-5(d) and (h).

The normal stresses acting on elements $L$ and $L'$ follow from the previous results by reducing them by a ratio of the distances from the neutral axis to the elements, i.e., by 140/150. The corresponding shear stresses are obtained using Eq. 7-6 for which the cross-hatched area $A_{fghj}$ and the corresponding $\bar{y}$ are shown in Fig. 9-5(b). Hence,

$$\sigma_{L \text{ or } L'} = \mp \frac{140}{150} \times \sigma_{K'} = \mp 15.56 \text{ MPa}$$

$$\tau_{L \text{ or } L'} = \frac{VA_{fghj}\bar{y}}{It} = \frac{40 \times 10^3 \times 40 \times 10 \times 145}{40 \times 300^3/12 \times 40} = 0.644 \text{ MPa}$$

These results are shown in Figs. 9-5(e) and (g).

Mohr's circle of stress is employed for obtaining the principal stresses at $L$, Fig. 9-5(i), and the results are shown on a rotated element in Fig. 9-5(e). Method 1 of Section 8-7 is used to obtain the results. Note the invariance of the sum of the normal stresses, i.e., $\sigma_x + \sigma_y = \sigma_1 + \sigma_2$ or $-15.56 + 0 = -15.59 + 0.03$. A similar solution for the principal stresses at point $L'$ yields the results shown on the rotated element in Fig. 9-5(g).

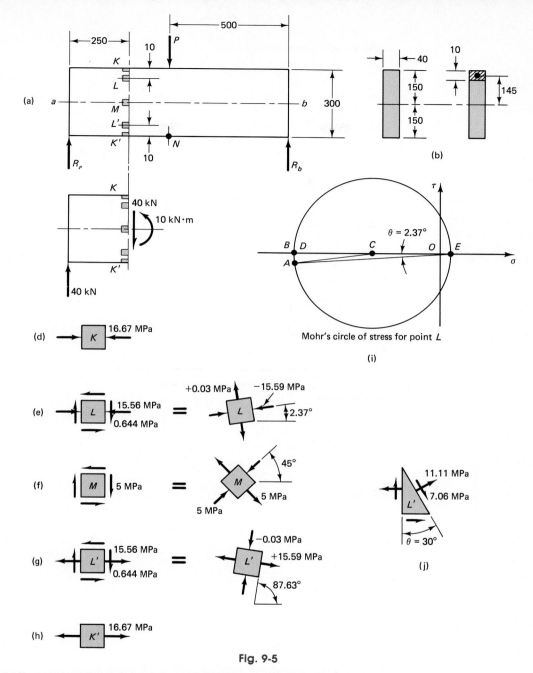

**Fig. 9-5**

Point $M$ lies on the neutral axis of the beam; hence, no flexural stress acts on the corresponding element shown in the first sketch of Fig. 9-5(f). The shear stress on the right face of the element at $M$ acts in the same direction as the internal shear at section $KK'$. Its magnitude can be obtained by applying Eq. 7-6, or directly by using Eq. 7-8a, i.e.,

$$\tau_{max} = \frac{3}{2}\frac{V}{A} = \frac{1.5 \times 40 \times 10^3}{40 \times 300} = 5 \text{ MPa}$$

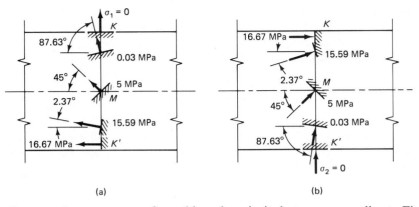

**Fig. 9-6** (a) Behavior of the major principal stress $\sigma_1$. (b) Behavior of the minor principal stress $\sigma_2$.

(a)                                                  (b)

The *pure* shear stress transformed into the principal stresses according to Fig. 8-6 is shown on a rotated element in Fig. 9-5(f).

It is significant to further examine qualitatively the results obtained. For this purpose, the computed principal stresses *acting on the corresponding planes* are shown in Figs. 9-6(a) and (b). In Fig. 9-6(a), the characteristic behavior of the major (tensile) principal stress at a section of a rectangular beam can be seen. This stress progressively diminishes in magnitude from a maximum value at $K'$ to zero at $K$. At the same time, the corresponding directions of $\sigma_1$ gradually turn through 90°. A similar observation can be made regarding the minor (compressive) principal stress $\sigma_2$ shown in Fig. 9-6(b).

(*b*) To find the stresses acting on a plane of $\theta = +30°$ through point $L'$, a direct application of Eqs. 8-1 and 8-2 using the stresses shown on the left element in Fig. 9-5(g) and $2\theta = 60°$ is made.

$$\sigma_\theta = \frac{+15.56}{2} + \frac{+15.56}{2} \cos 60° + (-0.644) \sin 60° = +11.11 \text{ MPa}$$

$$\tau_\theta = \frac{-15.56}{2} \sin 60° + (-0.644) \cos 60° = -7.06 \text{ MPa}$$

These results are shown in Fig. 9-5(j). The sense of the shearing stress $\tau_\theta$ is opposite to that shown in Fig. 8-4(b), since the computed quantity is negative. The "wedge technique" explained in Example 8-1 or the Mohr's circle method in Section 8-7 can be used to obtain the same results.

## 9-3. Comparative Accuracy of Beam Solutions

The solution in the previous example for a beam considering flexure and shear is based on stresses initially obtained using the conventional formulas of engineering mechanics of solids. These formulas are derived essentially assuming that plane sections in a beam remain plane during bending. Since this basic assumption is not entirely true in all cases, these solutions can be referred to as *elementary*. Therefore, it is instructive to compare the obtained results with a more accurate solution. Such a comparison is made here with a finite-element solution, shown in Fig. 9-7. Because of the symmetry of the problem, only one-half of the beam was analyzed using 450 finite elements.[1]

The contour lines for the principal stresses are shown in Fig. 9-7(a). Any point lying on a stress contour has a principal stress of the same magnitude and sign, with tensile stresses being positive. In this diagram, the major principal stresses are shown by black lines, and the minor principal stresses are shown in color. Comparisons between the elementary and finite-element solutions of the normal stress distribution across section $K$–$K'$ are shown in Fig. 9-7(b) and that for the shear stress in Fig. 9-7(c). The agreement is seen to be excellent. However, section $K$–$K'$ is taken midway between the applied concentrated force $P$ and the concentrated reaction $R_a$. At these points, locally large perturbations in stresses occur, resembling those shown earlier in Figs. 2-30 and 2-31. However, according to the Saint-Venant principle, local stresses rapidly diminish and a regular statically equivalent stress pattern sets in. In practice, large stresses at concentrated forces are reduced by applying them over an area to obtain an acceptable bearing pressure. Theoretically, in an elastic body, the stress at a concentrated force is infinite. In reality, some plastic

---

[1] This solution was obtained using the FEAP computer program developed by R. L. Taylor employing isoparametric four-node elements. An automatic mesh-generating technique enables the use of graduated smaller elements at concentrated forces, where the stresses vary more rapidly than elsewhere. Conventional square elements in the FEAP program were used in the previous solutions cited. Since in a two-dimensional plate of finite thickness, the in-plane stresses vary somewhat across the thickness, in the FEAP formulation, the average values of these stresses through the plate thickness are used. Such stresses are called *generalized plane stresses*.

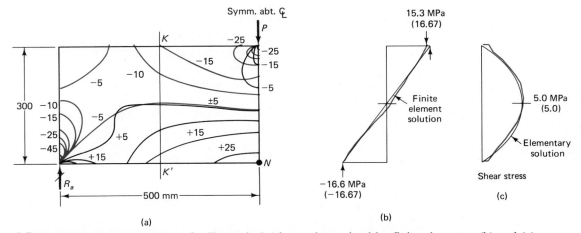

**Fig. 9-7** (a) Principal stress contours for Example 9-4 beam determined by finite elements. (b) and (c) Comparisons between elementary and finite element solutions for normal and shear stresses at section K-K'. (Stresses from elementary solutions in (b) and (c) are given in parentheses).

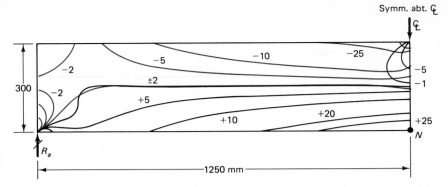

**Fig. 9-8** Principal stress contours for left half of a rectangular simply supported beam loaded in the middle.

yielding, reducing the stress, takes place in the proximity of the applied force.

It is interesting to note from Fig. 9-7(a) that the $\pm 5$ MPa contours coincide for more than half of the span. This condition corresponds to the principal stresses for the middle element $M$ in Fig. 9-5(f), and almost precisely coincides with the neutral axis in the elementary solution. The stress at point $N$ at the bottom of the beam below force $P$ is within 5 percent of the elementary solution.[2]

The beam is relatively short, having a length to depth ratio of 3.33. It is instructive to compare this solution with that for the somewhat longer similar beam having a length to depth ratio of 8.33, which is shown in Fig. 9-8. For this beam of 2500-mm span, the applied concentrated force $P = 32$ kN. One-half of this beam was analyzed using 900 finite elements. According to elementary solutions, the maximum bending stress at point $N$ is the same as in the previous case. However, here the shear stress at the neutral axis is 2 MPa. In the figure, the principal stresses of this magnitude define the neutral axis in the elementary solution. In contrast to the earlier case, it is seen that the neutral plane extends essentially across the entire length of the beam. The stress disturbances caused by the concentrated force as well as reactions are much more localized. Further, the maximum bending stress at point $N$ is within less than 2 percent of the elementary solution. Since point $N$ is a beam depth away from the applied force $P$, this solution again provides an example of Saint-Venant's principle, and the elementary formula is sufficiently accurate.

For beams carrying distributed loads, stress perturbations occur primarily at the supports.

For the previous reasons, the elementary formulas of the technical solid mechanics are generally considered to be sufficiently accurate for the usual design. They are also indispensable for the preliminary design of complex members, where subsequently a member is analyzed by a refined method such as by finite elements.

[2] The Wilson–Stokes analytical and photoelastic solutions developed in the 1890s show that the maximum bending stress caused by a concentrated force in *short* beams is smaller than that given by the elementary flexure theory. The analytical solution shows that it approaches asymptotically the elementary solution with an increasing ratio of the beam length to depth. See M. M. Frocht, *Photoelasticity*, Vol. II (New York: Wiley, 1948), 116.

## **9-4. Experimental Methods of Stress Analysis

In the past, when mathematical procedures became too cumbersome or impossible to apply, the photoelastic method of stress analysis was extensively used to solve practical problems. Many of the stress-concentration factors cited in this text are either drawn or verified by such experimental work. Accurate stresses in an entire specimen can be found using this method.[3] This traditional area of photoelasticity has been largely taken over by modern numerical techniques. An illustration of such an approach using finite elements has been shown a few times in this text, including the two solutions cited in the preceding section. Nevertheless, photoelastic techniques augmented with computers have now advanced and remain useful in special applications. Moreover, several additional experimental procedures became available. Among these, the Moiré, holographic, and laser speckle interferometries are playing an increasingly important role. These methods are discussed in specialized texts.[4] However, some terminology developed primarily in two-dimensional photoelasticity is in general use and is given for reference.

In the preceding section, the principal stresses of the same algebraic magnitude provided a "map" of *stress contours*. Similarly, the points at which the directions of the minor principal stresses form a *constant angle* with the *x* axis can be connected. Moreover, since the principal stresses are mutually perpendicular, the direction of the major principal stresses through the same points also forms a constant angle with the *x* axis. The line so connected is a locus of points along which the principal stresses have *parallel directions*. This line is called an *isoclinic line*. The adjective *isoclinic* is derived from two Greek words, *isos*, meaning equal, and *klino*, meaning slope or incline. Three isoclinic lines can be found by inspection in a rectangular prismatic beam subjected to transverse load acting normal to its axis. The lines corresponding to the upper and lower boundaries of

**Fig. 9-A** Fringe photograph of a rectangular beam. (Photograph by R. W. Clough.)

[3] Figure 9-A shows regularly spaced and perturbed fringes at concentrated load points. These photoelastic fringes provide a map for the *difference* in principal stresses. They do not directly give contours for selected stresses as does the finite-element method.

[4] See A. S. Kobayashi (ed.), *Handbook on Experimental Mechanics* (Englewood Cliffs, NJ: Prentice-Hall, 1987).

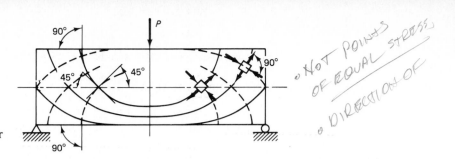

Fig. 9-9 Principal stress trajectories for a rectangular beam.

∘ NOT POINTS OF EQUAL STRESS
∘ DIRECTION OF

a beam form two isoclinic lines as, at the boundary, the flexural stresses are the principal stresses and act parallel to the boundaries. The flexural stress is zero at the neutral axis, where only pure shear stresses exist. These pure shear stresses transform into principal stresses, all of which act at an angle of 45° with the axis of the beam. Hence, another isoclinic line (the 45° isoclinic) is located on the axis of the beam. The other isoclinic lines are curved and are more difficult to determine.

Another set of curves can be drawn for a stressed body for which the magnitude and the sense of the principal stresses are known at a great many points. A curve whose tangent is changing in direction to conform with the direction of the principal stresses is called a *principal stress trajectory* or isostatic line. Like the isoclinic lines, the principal stress trajectories *do not* connect the points of equal stresses, but rather indicate the directions of the principal stresses. Since the principal stresses at any point are mutually perpendicular, the principal stress trajectories for the two principal stresses form a family of orthogonal (mutually perpendicular) curves.[5] An example of idealized stress trajectories for a rectangular beam loaded with a concentrated force at the midspan is shown in Fig. 9-9. The principal stress trajectories corresponding to the tensile stresses are shown in the figure by solid lines; those for the compressive stresses are shown dashed. The trajectory pattern (not shown) is severely disturbed at the supports and at the point of application of load $P$ as can be surmised from Fig. 9-7(a).

## Part B ELASTIC DESIGN OF MEMBERS FOR STRENGTH

### 9-5. Design of Axially Loaded Members

Axially loaded tensile members and short compression members[6] are designed for strength using Eq. 1-16, i.e., $A = P/\sigma_{\text{allow}}$. The *critical* section for an axially loaded member occurs at a section of minimum cross-sec-

---

[5] A somewhat analogous situation is found in fluid mechanics, where in "two-dimensional" fluid flow problems, the *streamlines* and the *equipotential lines* form an orthogonal system of curves, the *flow net*.

[6] Slender compression members are discussed in Chapter 11.

tional area, where the stress is a maximum. This requires the use of *net*, rather than gross, cross-sectional areas. If an abrupt discontinuity in the cross-sectional area is imposed by the design requirements, the use of Eq. 2-19, $\sigma_{max} = KP/A$, is appropriate. The use of the latter formula is necessary in the design of machine parts to account for the local stress concentrations where fatigue failure may occur. In design of static structures, such as buildings, stress concentration factors are seldom considered (see Fig. 2-35).

Besides the normal stresses, given by the previous equations, shear stresses act on inclined planes. Therefore, if a material is weak in shear strength in comparison to its strength in tension or compression, it will fail along planes approximating the planes of the maximum shear stress as discussed in Section 8-20. However, regardless of the type of fracture that may actually take place, the allowable stress for design of axially loaded members is customarily based on the *normal* stress. This design procedure is consistent. The maximum normal stress that a material can withstand at failure is directly related to the *ultimate* strength of the material. Hence, although the actual break may occur on an inclined plane, the maximum normal stress can be considered as the ultimate strength.

If in the design it is necessary to consider the deflection or stiffness of an axially loaded member, the use of Eqs. 2-7 and 2-9 is appropriate. *In some situations, these criteria govern the selection of members.*

## 9-6. Design of Torsion Members

Explicit formulas for elastic design of circular tubular and solid shafts are provided in Section 4-6. Some stress concentration factors essential in design of such members subjected to cyclic loading are given in Section 4-7. Large local shear stresses can develop at changes in the cross-sectional area. Stress-analysis formulas for some noncircular solid and thin-walled tubular members are given in Sections 4-14 and 4-16. In these sections, the corresponding formulas for calculating the stiffnesses of these members are also provided. Except for stress-concentration factors, these formulas are suitable for the design of torsion members for many types of cross sections.

Most torsion members are designed by selecting an *allowable shear stress*. This amounts to a direct use of the maximum shear theory of failure. However, it is well to bear in mind that a state of pure shear stress, which occurs in torsion, can be transformed into the principal stresses, and, in brittle materials, tensile fractures may be caused by the tensile principal stress.

A similar approach is used in the design of shafts for gear trains in mechanical equipment. However, in such cases, the shafts, in addition to carrying a torque, also act as beams. Therefore, this topic is postponed until Section 9-10.

## 9-7. Design Criteria for Prismatic Beams

If a beam is subjected to *pure bending*, its fibers are assumed to be in a state of uniaxial stress. If, further, a beam is prismatic, i.e., of a constant cross-sectional area and shape, the critical section occurs at the section of the greatest bending moment. By assigning an allowable stress, the section modulus of such a beam can be determined using Eq. 6-21, $S = M/\sigma_{\max}$. After the required section modulus is known, a beam of correct proportions can be selected. However, if a beam resists shear in addition to bending, its design becomes more involved.

Consider the prismatic rectangular beam of Example 9-4 at a section 250 mm from the left support, where the beam transmits a bending moment *and* a shear; see Fig. 9-10(a). The principal stresses at points *K, L, M, L', * and *K'* at this section were found before and are reproduced in Fig. 9-10(b). If this section were the critical section, it is seen that the design of this beam, based on the maximum normal stress theory, would be governed by the stresses *at the extreme fibers* as no other stresses exceed these stresses. For a prismatic beam, these stresses depend only on the magnitude of the bending moment and are largest at a section where the maximum bending moment occurs. Therefore, in ordinary design it is *not* necessary to perform the combined stress analysis for interior points. In the example considered, the maximum bending moment is at the middle of the span. The foregoing may be generalized into a basic rule for the design of prismatic beams: *a critical section for a prismatic beam carrying transverse forces acting normal to its axis occurs where the bending moment reaches its absolute maximum.*

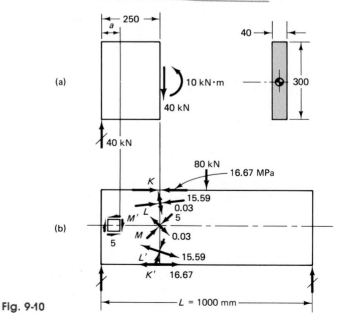

Fig. 9-10

For cross sections without two axes of symmetry, such as T beams, made from material that has different properties in tension than in compression, the *largest* moments of *both senses* (positive or negative) must be examined. Under some circumstances, a smaller bending moment of one sense may cause a more critical stress than a larger moment of another sense. The section at which the extreme fiber stress of either sign in relation to the respective allowable stress is highest is the critical section.

The previous criterion for the design of prismatic beams is incomplete, as attention was specifically directed to the stresses caused by the moment. In some cases, the shear stresses caused by the shear at a section may control the design. In the example considered, Fig. 9-10, the magnitude of the shear remains constant at every section through the beam. At a small distance $a$ from the right support, the maximum shear is still 40 kN, whereas the bending moment, $40a$ kN·m, is small. The maximum shear stress at the neutral axis corresponding to $V = 40$ kN is the same at point $M'$ as it is at point $M$.[7] Therefore, since in a general problem, the bending stresses may be small, they may not control the selection of a beam, and *another critical section for any prismatic beam occurs where the shear is a maximum*. In applying this criterion, it is customary to work directly with the maximum shear stress that may be obtained from Eq. 7-6, $\tau = VQ/It$, and not transform $\tau_{max}$ so found into the principal stresses. For rectangular and I beams, the maximum shear stress given by Eq. 7-6 reduces to Eqs. 7-8a and 7-10, $\tau_{max} = (3/2)V/A$ and $(\tau_{max})_{approx} = V/A_{web}$, respectively.

Whether the section where the bending moment is a maximum or the section where the shear is a maximum governs the selection of a prismatic beam depends on the loading and the material used. Generally the allowable shear stress is less than the allowable bending stress. For example, for steel, the ratio between these allowable stresses is about 1/2, whereas for some woods, it may be as low as 1/15.[8] Regardless of these ratios of stresses, *the bending stresses usually control the selection of a beam*. Only in beams spanning a short distance does shear control the design. For small lengths of beams, the applied forces and reactions have small moment arms, and the required resisting bending moments are small. On the other hand, the shear forces may be large if the applied forces are large.

The two criteria for the design of beams are accurate if the two critical sections are in different locations. However, in some instances the maximum bending moment and the maximum shear occur at the *same* section through the beam. In such situations, sometimes higher combined stresses than $\sigma_{max}$ and $\tau_{max}$, as given by Eqs. 6-21 and 7-6, may exist at the interior points. For example, consider an I beam of negligible weight that carries

[7] At point $M$, the shear stresses are shown transformed into the principal stresses.
[8] Wood is weak in shear strength *parallel* to its grain.

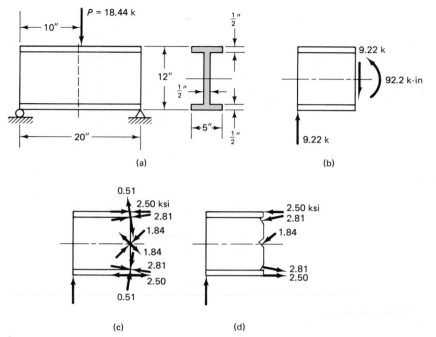

Fig. 9-11

a force $P$ at the middle of the span, Fig. 9-11(a). The maximum bending moment occurs at midspan. Except for sign, the shear is the same on either side of the applied force. At a section just to the right or just to the left of the applied force, the maximum moment *and* the maximum shear occur simultaneously. A section just to the left of $P$, with the corresponding system of forces acting on it, is shown in Fig. 9-11(b). For this section, it can be shown that the stresses at the extreme fibers are 2.50 ksi, whereas the principal stresses at the juncture of the web with the flanges, neglecting stress concentrations, are $\pm 2.81$ ksi and $\pm 0.51$ ksi, acting as shown in Figs. 9-11(c) and (d).

It is customary not to consider directly the effect of the local disturbance on longitudinal stresses in the neighborhood of an applied concentrated force. Instead, as indicated in Section 9-4, the problem of local stresses is resolved by requiring a sufficiently large contact area for the applied force so as to obtain an acceptable bearing stress. For some materials, such as wood, this may require the use of steel bearing plates in order to spread the effect of the concentrated force.

From this example, it is seen that the maximum normal stress does not always occur at the extreme fibers. Nevertheless, only the maximum normal stresses and the maximum shear stress at the neutral axis are investigated in ordinary design. In design codes, the allowable stresses are presumably set sufficiently low so that an adequate factor of safety remains, even though the higher combined stresses are disregarded. By increasing a span for the same applied concentrated force, the flexural stresses increase linearly with the span length, whereas the shear stresses remain constant. Hence, in most cases, the bending stresses rapidly become dominant. Therefore, generally, it is necessary to perform the combined stress analysis only for very short beams or in unusual arrangements.

From the previous discussion, it is seen that, for the design of prismatic beams, the critical sections must be determined in every problem, as the design is entirely based on the stresses developed at these sections. The critical sections are best located with the aid of shear and bending-moment diagrams. The required values of $M_{max}$ and $V_{max}$ can be determined easily from such diagrams. The construction of these diagrams is discussed in Chapter 5.

## 9-8. Design of Prismatic Beams

As noted in the preceding section, the customary approach for design of prismatic beams is controlled by the maximum stresses at the *critical sections*. One such critical section usually occurs where the bending moment is a maximum; the other where the shear is a maximum. These sections are conveniently determined with the aid of shear and moment diagrams.[9] In most cases,[10] the *absolute maximum moment,* i.e., whether positive or negative, is used for selecting a member. Likewise, the above maximum shear is critical for the design. For example, consider a simple beam with a concentrated force, as shown in Fig. 9-12. The shear diagram, neglecting the weight of the beam, is shown in Fig. 9-12(a) as it is ordinarily constructed by assuming the applied force concentrated at a point. In Fig. 9-12(b), an allowance is made for the width of the applied force and reactions, assuming them uniformly distributed along the beam. Note that in either case, the design shears are less than the applied force.

The allowable stresses to be used in design are prescribed by various authorities. In most cases, the designer must follow a code, depending on the location of the installation. In different codes, even for the same material and the same use, the allowable stresses differ.

In elastic design, after the critical values of moment and shear are determined and the allowable stresses are selected, the beam is usually first designed to resist a maximum moment using Eq. 6-13 or 6-21 ($\sigma_{max}$ = $M/S$ or $\sigma_{max} = Mc/I$). Then the beam is *checked* for shear stress. As most beams are governed by flexural stresses, this procedure is convenient. However, in some cases, particularly in timber and concrete design, the shear stress frequently controls the dimensions of the cross section.

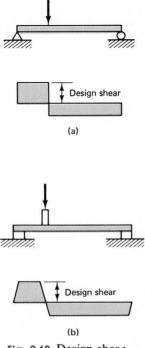

**Fig. 9-12** Design shear.

---

[9] With experience, construction of complete diagrams may be avoided. After reactions are computed and a section where $V = 0$ or a change of sign is determined, the maximum moment corresponding to this section may be found directly by using the method of sections. For simple loadings, various handbooks give formulas for the maximum shear and moment.

[10] This is not always true for materials that have different properties in tension and compression.

The method used in computing the shear stress depends on the type of beam cross section. For rectangular sections, the maximum shear stress is 1.5 times the average stress, Eq. 7-8. For wide flange and I beams, the total allowable vertical shear is taken as the area of the *web* multiplied by an allowable shear stress, Eq. 7-10. For other cases, Eq. 7-6, $\tau = VQ/It$, is used.

*Usually, there are several types or sizes of commercially available members that may be used for a given beam.* Unless specific size limitations are placed on the beam, the lightest member is used for economy. The procedure of selecting a member is a trial-and-error process.

It should also be noted that some beams must be selected on the basis of allowable deflections. This topic is treated in the next chapter.

For beams with statically applied loads, such as occur in buildings, there is an increasing trend to design them on the basis of inelastic (plastic) behavior. This approach is considered in Chapter 13.

### EXAMPLE 9-5

Select a Douglas fir beam of rectangular cross section to carry two concentrated forces, as shown in Fig. 9-13(a). The allowable stress in bending is 8 MPa; in shear, 0.7 MPa; and in bearing perpendicular to the grain of the wood, 1.4 MPa.

### Solution

Shear and moment diagrams for the applied forces are prepared first and are shown, respectively, in Figs. 9-13(b) and (c). From Fig. 9-13(c), it is seen that $M_{max} = 10$ kN·m. From Eq. 6-21,

$$S = \frac{M}{\sigma_{allow}} = \frac{10 \times 10^6}{8} = 1.25 \times 10^6 \text{ mm}^3$$

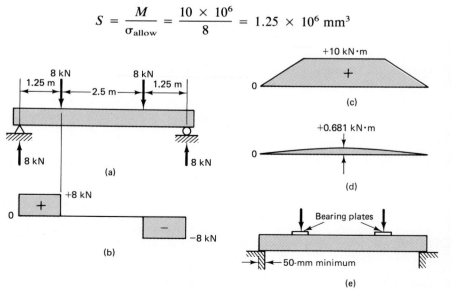

**Fig. 9-13**

By *arbitrarily assuming* that the depth $h$ of the beam is to be two times greater than its width $b$, from Eq. 6-22,

$$S = \frac{bh^2}{6} = \frac{h^3}{12} = 1.25 \times 10^6$$

Hence, the required $h = 247$ mm and $b = 123$ mm.

Let a *surfaced* beam 140 by 240 mm, having a section modulus $S = 1.34 \times 10^6$ mm$^3$, be used to fulfill this requirement. For this beam, from Eq. 7-8a,

$$\tau_{max} = \frac{3V}{2A} = \frac{3 \times 8 \times 10^3}{2 \times 140 \times 240} = 0.357 \text{ MPa}$$

This stress is well within the allowable limit. Hence, the beam is satisfactory.

*Note that other proportions of the beam can be used,* and a more direct method of design is to find a beam of size corresponding to the wanted section modulus directly from a table similar to Table 10, which gives properties of standard dressed sections in the U.S. conventional units.

The analysis was made without regard for the beam's own weight, which initially is unknown. (Experienced designers often make an allowance for the weight of the beam at the outset.) However, this may be accounted for now. Assuming that wood weighs 6.5 kN/m$^3$, the beam selected weighs 0.218 kN per linear meter. This uniformly distributed load causes a parabolic bending-moment diagram, shown in Fig. 9-13(d), where the maximum ordinate is $w_o L^2/8 = 0.218 \times 5^2/8 = 0.681$ kN·m (see Fig. 5-24). This bending-moment diagram should be added to the moment diagram caused by the applied forces. Inspection of these diagrams shows that the maximum bending moment due to both causes is $0.681 + 10 = 10.68$ kN·m. Hence, the required section modulus actually is

$$S = \frac{M}{\sigma_{allow}} = \frac{10.68 \times 10^6}{8} = 1.34 \times 10^6 \text{ mm}^3$$

The surfaced 140 by 240 mm beam already selected provides the required $S$.

In order to avoid the crushing of wood at the supports and at applied concentrated forces, adequate bearing areas for these forces must be provided. Neglecting the weight of the beam, such areas $A$ at the four locations, according to Eq. 1-13, should be

$$A = \frac{P}{\sigma_{allow}} = \frac{8 \times 10^3}{1.4} = 5710 \text{ mm}^2$$

These areas can be provided by conservatively specifying that the beam's ends rest on at least 50 by 140 mm (7000 mm$^2$) supports, whereas at the concentrated forces, 80 by 80 mm (6400 mm$^2$) steel washers be used.

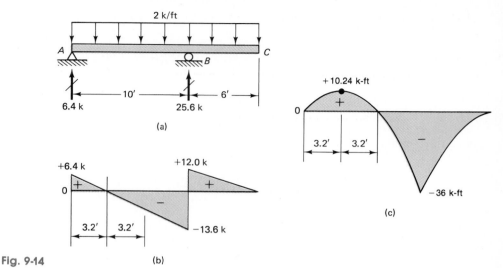

Fig. 9-14

(a)

(b)

(c)

## EXAMPLE 9-6

Select an I beam or a wide-flange steel beam to support the load shown in Fig. 9-14(a). For the beam, $\sigma_{\text{allow}} = 24$ ksi and $\tau_{\text{allow}} = 14.5$ ksi.

Solution

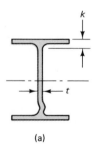

(a)

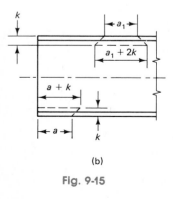

(b)

Fig. 9-15

The shear and the bending-moment diagrams for the loaded beam are shown in Figs. 9-14(b) and (c), respectively. The maximum moment is 36 k-ft. From Eq. 6-21,

$$S = \frac{36 \times 12}{24} = 18.0 \text{ in}^3$$

Examination of Tables 3 and 4 in the Appendix shows that this requirement for a section modulus is met by a 10-in S section weighing 24.7 lb/ft ($S = I/c = 24.7$ in$^3$). However, a lighter 8-in wide-flange W section weighing 24 lb/ft ($S = 20.9$ in$^3$) also is adequate. Therefore, for reasons of economy, the lighter W8 × 24 section will be used. The beam weight is small in comparison with the applied loads and will be neglected in calculations.

From Fig. 9-14(b), $V_{\text{max}} = 13.6$ kips. Hence, from Eq. 7-10,

$$(\tau_{\text{max}})_{\text{approx}} = \frac{V}{A_{\text{web}}} = \frac{13.6}{0.245 \times 7.93} = 7.00 \text{ ksi}$$

This stress is below the allowable value, and the selected beam is satisfactory.

At the supports or concentrated forces, S and wide-flange beams should be checked for crippling of the webs. This phenomenon is illustrated at the bottom of Fig. 9-15(a). Crippling of the webs is more critical for members with thin webs than direct bearing of the flanges, which may be investigated as in the preceding example. To preclude crippling, a design rule is specified by the AISC. It states that the direct stress on area, $(a + k)t$ at the ends and $(a_1 + 2k)t$ at the interior points, must not exceed $0.75\sigma_{yp}$. In these expressions, $a$ and $a_1$ are the respective lengths of bearing of the applied forces at exterior or interior portions of a beam, Fig. 9-15(b), $t$ is the thickness of the web, and $k$ is the distance from the outer face of a flange to the toe of the web fillet. The values of $k$ and $t$ are tabulated in manufacturers' catalogues.

For this example, assuming $\sigma_{yp} = 36$ ksi, the *minimum* widths of the supports, according to the rule, are as follows:

*At support A:*

$$27(a + k)t = 6.4 \quad \text{or} \quad 27(a + 7/8) \times 0.245 = 6.4$$
and $\quad a = 0.09$ in

*At support B:*

$$27(a_1 + 2k)t = 25.6 \quad \text{or} \quad 27(a_1 + 2 \times 7/8) \times 0.245 = 25.6$$
and $\quad a_1 = 2.12$ in

These requirements can easily be met in an actual case.

---

The preceding two examples illustrate the design of beams whose cross sections have two axes of symmetry. In both cases, the bending moments controlled the design, and, since this is usually true, it is significant to note which members are efficient in flexure. A concentration of as much material as possible away from the neutral axis results in the best sections for resisting flexure, Fig. 9-16(a). Material concentrated near the outside fibers works at a high stress. For this reason, I-type sections, which approximate this requirement, are widely used in practice.

The previous statements apply for materials having nearly equal properties in tension and compression. If this is not the case, a deliberate shift of the neutral axis from the midheight position is desirable. This accounts for the wide use of T and channel sections for such materials (see Example 6-5).

Finally, two other items warrant particular attention in the design of beams. In many cases, the loads for which a beam is designed are transient in character. They may be placed on the beam all at once, piecemeal, or in *different locations*. The loads, which are not a part of the "dead weight" of the structure itself, are called *live loads*. They must be so placed as to cause the highest possible stresses in a beam. In many cases, the placement may be determined by inspection. For example, in a simple beam with a single moving load, the placement of the load at midspan causes the largest bending moment, whereas placing the same load very near to a support causes the greatest shear. For most building work, the live load, which supposedly provides for the most severe expected loading condition, is specified in building codes on the basis of a load per unit floor area. Multiplying this live load by the spacing of parallel beams gives the

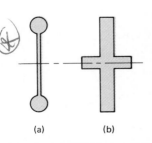

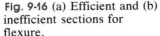
Fig. 9-16 (a) Efficient and (b) inefficient sections for flexure.

*uniformly distributed live load* per unit length of the beam. For design purposes, this load is added to the dead weight of construction. Situations where the applied force is delivered to a beam with a shock or impacts are discussed in Section 10-11.

The second item pertains to *lateral instability* of beams. The beam's flanges, if not held laterally, may be so narrow in relation to the span that a beam may buckle sideways and collapse. Special consideration of this problem is given in Section 11-14.

## 9-9. Design of Nonprismatic Beams

It should be apparent from the preceding discussion that the selection of a prismatic beam is based only on the stresses at the critical sections. At all other sections through the beam, the stresses will be below the allowable level. Therefore, the potential capacity of a given material is not fully utilized. This situation may be improved by designing a beam of variable cross section, i.e., by making the beam nonprismatic. Since flexural stresses control the design of most beams, as has been shown, the cross sections may everywhere be made just strong enough to resist the corresponding moment. Such beams are called *beams of constant strength*. Shear governs the design at sections through these beams where the bending moment is small.

### EXAMPLE 9-7

Design a cantilever of constant strength for resisting a concentrated force applied at the end. Neglect the beam's own weight.

### Solution

A cantilever with a concentrated force applied at the end is shown in Fig. 9-17(a); the corresponding moment diagram is plotted in Fig. 9-17(b). Basing the design

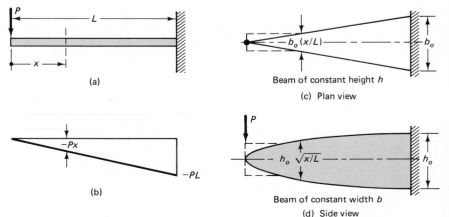

(a)

(b)

Beam of constant height $h$

(c) Plan view

Beam of constant width $b$

(d) Side view

**Fig. 9-17**

on the bending moment, the required section modulus at an arbitrary section is given by Eq. 6-21:

$$S = \frac{M}{\sigma_{\text{allow}}} = \frac{Px}{\sigma_{\text{allow}}}$$

A great many cross-sectional areas satisfy this requirement; so, first, it will be assumed that the beam will be of rectangular cross section and of *constant height h*. The section modulus for this beam is given by Eq. 6-22 as $bh^2/6 = S$; hence,

$$\frac{bh^2}{6} = \frac{Px}{\sigma_{\text{allow}}} \qquad \text{or} \qquad b = \left(\frac{6P}{h^2\sigma_{\text{allow}}}\right) x = \frac{b_o}{L} x \qquad (9\text{-}1)$$

where the expression in parentheses is a constant and is set equal to $b_o/L$, so that when $x = L$, the width is $b_o$. A beam of constant strength with a constant depth in a plan view looks like the wedge[11] shown in Fig. 9-17(c). Near the free end, this wedge must be modified to be of adequate strength to resist the shear force $V = P$.

*If the width or breadth b of the beam is constant,*

$$\frac{bh^2}{6} = \frac{Px}{\sigma_{\text{allow}}} \qquad \text{or} \qquad h = \sqrt{\frac{6Px}{b\sigma_{\text{allow}}}} = h_o\sqrt{\frac{x}{L}} \qquad (9\text{-}2)$$

This expression indicates that a cantilever of constant width loaded at the end is also of constant strength if its height varies parabolically from the free end, Fig. 9-17(d).

Beams of approximately constant strength are used in leaf springs and in many machine parts that are cast or forged. In structural work, an approximation to a beam of constant strength is frequently made. For example, the moment diagram for the beam loaded as shown in Fig. 9-18(a) is given by lines $AB$ and $BC$ in Fig. 9-18(b). By selecting a beam of flexural capacity equal only to $M_1$, the middle portion of the beam is overstressed. However, cover plates can be provided near

[11] Since this beam is not of constant cross-sectional area, the use of the elementary flexure formula is not entirely correct. When the angle included by the sides of the wedge is small, little error is involved. As this angle becomes large, the error may be considerable. An exact solution shows that when the total included angle is 40°, the solution is in error by nearly 10 percent.

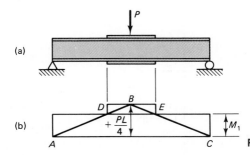

**Fig. 9-18** Coverplated I beam.

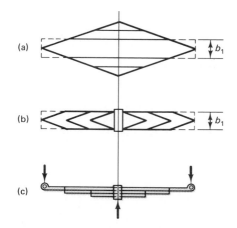

(a)

(b)

(c)

**Fig. 9-19** Leaf spring.

the middle of the beam to boost the flexural capacity of the composite beam to the required value of the maximum moment. For the case shown, the cover plates must extend at least over the length *DE* of the beam, and in practice they are made somewhat longer. A leaf spring, approximating a beam of constant strength as in Fig. 9-17(c), is shown in Fig. 9-19.

## 9-10. Design of Complex Members

In many instances, the design of complex members cannot be carried out in a routine manner as was done in the preceding simple examples. Sometimes the size of a member must be *assumed* and a complete stress analysis performed at sections where the stresses appear critical. Designs of this type may require several revisions. Finite-element analyses with increasing frequency are used in such cases for final design. Alternatively, experimental methods are also resorted to since elementary formulas may not be sufficiently accurate.

As a last example in this chapter, a transmission shaft problem is analyzed. A direct analytical procedure is possible in this problem, which is of great importance in the design of power equipment.

### EXAMPLE 9-8

Select the size of a solid steel shaft to drive the two sprockets shown in Fig. 9-20(a). These sprockets drive $1\frac{3}{4}$-in pitch roller chains,[12] as shown in Figs. 9-20(b) and (c). Pitch diameters of the sprockets shown in the figures are from a manufacturer's catalogue. A 20-hp speed-reducer unit is coupled directly to the shaft and drives it at 63 rpm. At each sprocket, 10 hp is taken off. Assume the maximum shear theory of failure, and let $\tau_{allow} = 6$ ksi.

[12] Similar sprockets and roller chains are commonly used on bicycles.

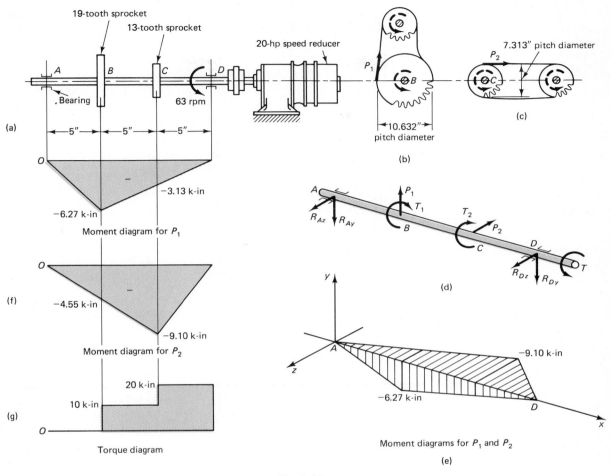

Fig. 9-20

## Solution

According to Eq. 4-11, the torque delivered to shaft segment $CD$ is $T = 63,000(\text{hp}/N) = (63,000)20/63 = 20,000$ lb-in $= 20$ k-in. Hence, torques $T_1$ and $T_2$ delivered to the sprockets are $T/2 = 10$ k-in *each*. Since the chains are arranged as shown in Figs. 9-20(b) and (c), the pull in the chain at sprocket $B$ is $P_1 = T_1/(D_1/2) = 10/(10.632/2) = 1.88$ k-in. Similarly, $P_2 = 10/(7.313/2) = 2.73$ k. The pull $P_1$ on the chain *is equivalent to* a torque $T_1$ and a vertical force at $B$, as shown in Fig. 9-20(d). At $C$, force $P_2$ acts horizontally and exerts a torque $T_2$. A complete free-body diagram for shaft $AD$ is shown in Fig. 9-20(d).

It is seen from the free-body diagram, that this shaft is simultaneously subjected to bending and torque. These effects on the member are best studied with the aid of appropriate diagrams, which are shown in Figs. 9-20(e)–(g). Next, note that although bending takes place in two planes, a *vectorial resultant of the moments* may be used in the flexure formula, since the beam has a circular cross section.

By keeping the last statement in mind, the general Eq. 8-9, giving the maximum shear stress, reduces in this problem of bending and torsion to

$$\tau_{max} = \sqrt{\left(\frac{\sigma_{bending}}{2}\right)^2 + \tau_{torsion}^{\;2}}$$

$$\tau_{max} = \sqrt{\left(\frac{Mc}{2I}\right)^2 + \left(\frac{Tc}{J}\right)^2}$$

However, since for a circular cross section, $J = 2I$, Eq. 6-20, $J = \pi d^4/32$, Eq. 4-2, and $c = d/2$, the last expression reduces to

$$\tau_{max} = \frac{16}{\pi d^3} \sqrt{M^2 + T^2}$$

By assigning the allowable shear stress to $\tau_{max}$, a design formula, based on the maximum shear theory[13] of failure, for a shaft subjected to bending and torsion is obtained as

$$d = \sqrt[3]{\frac{16}{\pi \tau_{allow}} \sqrt{M^2 + T^2}} \tag{9-3}$$

This formula may be used to select the diameter of a shaft simultaneously subjected to bending and torque. In the example investigated, a few trials should convince the reader that the $\sqrt{M^2 + T^2}$ is largest at sprocket $C$; hence, the critical section is at $C$. Thus,

$$\begin{aligned} M^2 + T^2 &= (M_{vert})^2 + (M_{horiz})^2 + T^2 \\ &= (6.27/2)^2 + 9.10^2 + 20^2 = 492 \text{ k}^2\text{-in}^2 \end{aligned}$$

$$d = \sqrt[3]{\frac{16}{6\pi} \sqrt{492}} = 2.66 \text{ in.}$$

A $2\frac{11}{16}$-in diameter shaft, which is a commercial size, should be used.

---

The effect of shock load on the shaft has been neglected in the foregoing analysis. For some equipment, where its operation is jerky, this condition requires special consideration. The initially assumed allowable stress presumably allows for keyways and fatigue of the material.

Although Eq. 9-3 and similar ones based on other failure criteria are

[13] See Prob. 9-50 for the formula based on the maximum stress theory of failure.

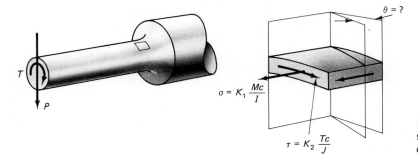

Fig. 9-21 Analysis of a shaft with stress concentrations.

widely used in practice, the reader is cautioned in applying them.[14] In many machines, shaft diameters change abruptly, giving rise to stress concentrations. In stress analysis, this requires the use of stress-concentration factors in bending, which are usually different from those in torsion. Therefore, the problem should be analyzed by considering the actual stresses at the critical section. (See Fig. 9-21.) Then an appropriate procedure, such as Mohr's circle of stress, should be used to determine the significant stress, depending on the selected fracture criteria.

# Problems

## Section 9-2. Miscellaneous Stress Analysis Problems

**9-1.** A concrete cylinder tested in a vertical position failed at a compressive stress of 30 MPa. The failure occurred on a plane of 30° with the vertical. On a clear sketch show the normal and shear stresses that acted on the plane of failure.

**9-2.** In a research investigation on the creep of lead, it was necessary to control the state of stress for the element of a tube. In one such case, a long cylindrical tube with closed ends was pressurized and simultaneously subjected to a torque. The tube was 100 mm in outside diameter with 4-mm walls. What were the principal stresses at the outside surface of the wall of the cylinder if the chamber was pressurized to 1.5 MPa and the externally applied torque was 200 N·m?

**9-3.** A cylindrical thin-walled tank weighing 100 lb/ft is supported as shown in the figure. If, in addition, it is subjected to an internal pressure of 200 psi, what state of stress would develop at points $A$ and $B$? Show the results on isolated elements. The mean radius of the tank is 10 in and its thickness is 0.20 in. Comment on the importance of the dead weight of the tank on the total stresses.

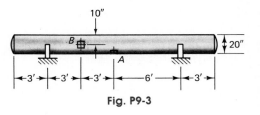

**Fig. P9-3**

[14] For further details on mechanical design, see A. H. Burr, *Mechanical Analysis and Design* (New York: Elsevier, 1982). A. D. Deutschman, W. J. Michaels, and C. E. Wilson, *Machine Design, Theory and Practice* (New York: Macmillan, 1975). J. E. Shigley, *Mechanical Engineering Design*, 3rd ed. (New York: McGraw-Hill, 1977). M. F. Spotts, *Design of Machine Elements*, 6th ed. (Englewood Cliffs, NJ: Prentice Hall, 1985).

For design on steel, aluminum alloys, and wood structures, see the references in the relevant sections of Chapter 11.

References for design in reinforced concrete are given after Example 6-14.

**9-4.** A cylindrical pressure vessel and its contents are lifted by cables, as shown in the figure. The mean diameter of the cylinder is 600 mm and its wall thickness is 6 mm. Determine the state of stress at points $A$ and $B$ and show the calculated results on isolated elements when the vessel is pressurized to 0.50 MPa and the vessel's mass is 102 kg/m.

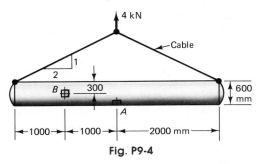

**Fig. P9-4**

**9-5.** A cylindrical pressure vessel of 2500-mm diameter with walls 12 mm thick operates at 1.5 MPa internal pressure. If the plates are butt-welded on a 30° helical spiral (see figure), determine the stresses acting normal and tangential to the weld.

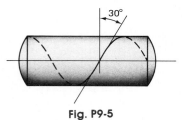

**Fig. P9-5**

**9-6.** A cylindrical thin-walled pressure vessel with mean radius $r = 300$ mm and thickness $t = 6$ mm is hoisted by two cables into the position shown in the figure. If the vessel is pressurized to 0.50 MPa gage pressure and the vessel weighs 102 kg/m, determine the state of stress at point $A$. Show the results on a properly oriented isolated element.

**9-7.** A fractionating column, 45 ft long, is made of a 12-in-inside-diameter standard steel pipe weighing 49.56 lb/ft. (See Table 8 of the Appendix.) This pipe is operating in a vertical position, as indicated on the sketch. If this pipe is internally pressurized to 600 psi and is subjected to a wind load of 40 lb/ft of height,

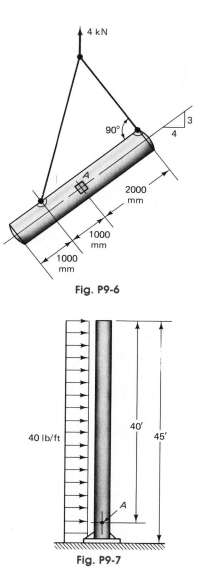

**Fig. P9-6**

**Fig. P9-7**

what is the state of stress at point $A$? Clearly show your calculated stresses on an isolated element; principal stresses need not be found.

**9-8.** A cylindrical thin-walled pressure vessel with a mean diameter of 20 in and thickness of 0.25 in is rigidly attached to a wall, forming a cantilever, as shown in the figure. (a) If an internal pressure of 250 psi is applied and, in addition, an external eccentric force $P = 31.4$ k acts on the assembly, what stresses are caused at point $A$? Show the results on a properly ori-

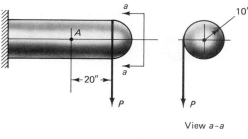

Fig. P9-8

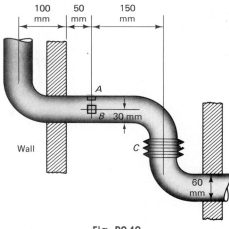

Fig. P9-10

ented element. (b) What maximum shear stress develops in the element? (*Caution:* All three principal Mohr's circles of stress should be examined.)

**9-9.** A cylindrical thin-walled pressure vessel rigidly attached to a wall, as shown in the figure, is subjected to an internal pressure of $p$ and an externally applied torque $T$. Due to these combined causes, the stresses on plane $a$–$a$ are $\sigma_{x'} = 0$, and $\tau_{x'y'} = 10$ MPa. Determine the internal pressure $p$ and the magnitude of the torque $T$. The mean diameter of the vessel is 400 mm and the wall thickness is 6 mm.

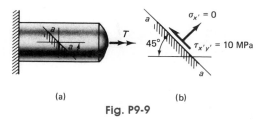

(a)                     (b)

Fig. P9-9

**9-10.** An assembly of seamless stainless steel tubing forming a part of a piping system is arranged as shown in the figure. A flexible expansion joint is inserted at $C$, which is capable of resisting hoop stresses but transmits no longitudinal force. The tubing is 60 mm in outside diameter and is 2 mm thick. If the pipe is pressurized to 2 MPa, determine the state of stress at points $A$ and $B$. Show the results on infinitesimal elements viewed from the outside. No distinction between the inside and outside dimensions of the tube need be made in the calculations. All dimensions shown on the figure are in mm. No stress transformations are required.[15]

[15] Data in the problem are fictitious; however, a major failure occurred at a petrochemical plant due to an oversight of basic behavior of this system of piping. (Courtesy of I. Finnie.)

**9-11.** A simple beam $50 \times 120$ mm spans 1500 mm and supports a uniformly distributed load of 80 kN/m, including its own weight. Determine the principal stresses and their directions at points $A$, $B$, $C$, $D$, and $E$ at the section shown in the figure.

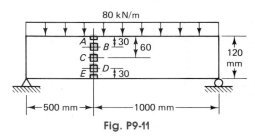

Fig. P9-11

**9-12.** A $100 \times 400$ mm rectangular wooden beam supports a 40-kN load, as shown in the figure. The grain of the wood makes an angle of 20° with the axis of the beam. Find the shear stresses at points $A$ and $B$ along the grain of the wood due to the applied concentrated force.

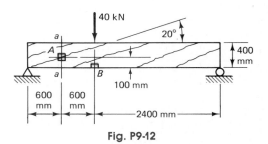

Fig. P9-12

**9-13.** A short I beam cantilever is loaded as shown in the figure. Find the principal stresses and their directions at points $A$, $B$, and $C$. Point $B$ is in the web at the juncture with the flange. Neglect the weight of the beam and ignore the effect of stress concentrations. Use the accurate formula to determine the shear stresses.

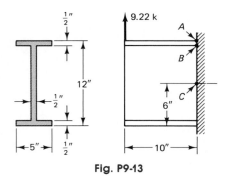

**Fig. P9-13**

**9-14.** The cantilever shown in the figure is loaded by an inclined force $P$ acting in the plane of symmetry of the cross section. (a) What is the magnitude of applied force $P$ if it causes an axial strain of 200 $\mu$m/m at point $A$? $E = 30 \times 10^6$ psi. (b) What is the maximum principal strain at $A$?

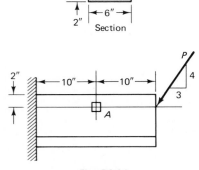

**Fig. P9-14**

**9-15.** The principal shear stress at point $A$ caused by application of force $P$ is 120 psi; see the figure. What is the magnitude of $P$?

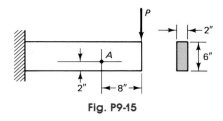

**Fig. P9-15**

**9-16.** At point $A$ on the upstream face of a dam, the water pressure is $-40$ kPa, and a measured tensile stress in the dam parallel to this surface is 20 kPa. Calculate stresses $\sigma_x$, $\sigma_y$, and $\tau_{xy}$ at that point and show them on an isolated element.

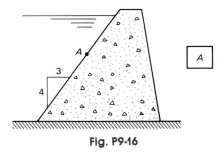

**Fig. P9-16**

**9-17.** A special hoist supports a 15-k load by means of a cable, as shown in the figure. Determine the principal stresses at point $A$ due to this load.

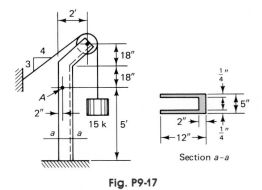

**Fig. P9-17**

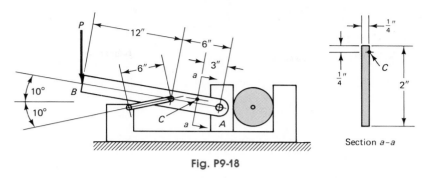

Section *a–a*

Fig. P9-18

**9-18.** By applying a vertical force $P$, the toggle clamp, as shown in the figure, exerts a force of 1000 lb on a cylindrical object. The movable jaw slides in a guide that prevents its upward movement. (a) Determine the magnitude of the applied vertical force $P$ and the downward force component developed at hinge $A$; (b) determine the stresses due to axial thrust, transverse shear, and bending moment acting on an element at point $C$ of section $a–a$; (c) draw an element at point $C$ with sides parallel and perpendicular to the axis of member $BA$ and show the stresses acting on the element; and (d) using Mohr's circle, determine the largest principal stress and the maximum shear stress at $C$.

**9-19.** A 2-in-diameter shaft is simultaneously subjected to a torque and pure bending. At every section of this shaft, the largest principal stress caused by the applied loading is $+24$ ksi and, simultaneously, the largest longitudinal tensile stress is $+18$ ksi. Determine the applied bending moment and torque.

**\*9-20.** Compare the moment-carrying capacity of a 6 $\times$ 6 $\times \frac{1}{2}$ in steel angle in the two different positions shown in the figure. In both cases, the applied vertical load acts through the shear center. (*Hint:* Table 6 gives the least radius of gyration, $r$, for the cross section. Hence, per Eq. 11-19a, $I_{min} = Ar^2_{min}$. Alternatively, $I_{min}$ can be calculated directly by considering the angle to consist of two plates.)

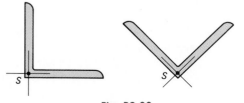

Fig. P9-20

**9-21.** A $\frac{1}{2}$-in-diameter drill bit is inserted into a chuck as shown in the figure. During the drilling operation, an axial force $P = 3.92$ k and a torque $T = 10\pi/128$ k-in act on the bit. If a horizontal force of 35.7 lb is accidentally applied to the plate being drilled, what is the magnitude of the largest principal stress that develops at the top of the drill bit? Determine the critically stressed point on the drill by inspection.

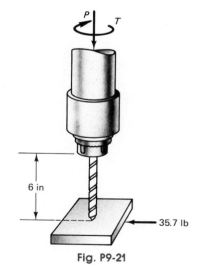

Fig. P9-21

**9-22.** A solid circular shaft is loaded as shown in the figure. At section $ACBD$ the stresses due to the 10-kN force and the weight of the shaft and round drum are found to be as follows: maximum bending stress is 40 MN/m², maximum torsional stress is 30 MN/m², and maximum shear stress due to $V$ is 6 MN/m². (a) Set up elements at points $A$, $B$, $C$, and $D$ and indicate the magnitudes and directions of the stresses acting on them. In each case, state from which direction the ele-

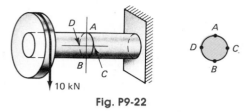

Fig. P9-22

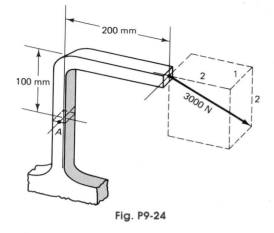

Fig. P9-24

ment is observed. (b) Using Mohr's circle, find directions and magnitudes of the principal stresses and of the maximum shear stress at point $A$.

**9-23.** A circular bar of 2-in diameter with a rectangular block attached at its free end is suspended as shown in the figure. Also a horizontal force is applied eccentrically to the block as shown. Analysis of the stresses at section $ABCD$ gives the following results: maximum bending stress is 1000 psi, maximum torsional stress is 300 psi, maximum shear stress due to $V$ is 400 psi, and direct axial stress is 200 psi. (a) Set up an element at point $A$ and indicate the magnitudes and directions of the stresses acting on it (the top edge of the element to coincide with section $ABCD$). (b) Using Mohr's circle, find the direction and the magnitude of the maximum (principal) shear stresses and the associated normal stresses at point $A$.

(The force $F$ in plan view acts in the direction of the $x$ axis.) Determine the magnitudes and directions of the stresses due to $F$ on the elements $A$ and $B$ at section $a$–$a$. Show the results on elements clearly related to the points on the rod. Principal stresses are not required.

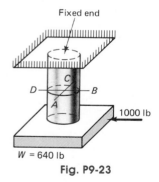

Fig. P9-23

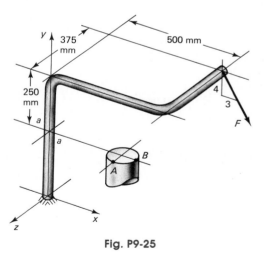

Fig. P9-25

**9-24.** A bent rectangular bar is subjected to an inclined force of 3000 N, as shown in the figure. The cross section of the bar is $12 \times 12$ mm. (a) Determine the state of stress at point $A$ caused by the applied force and show the results on an element. (b) Find the maximum principal stress.

**9-25.** A 50-mm-diameter rod is subjected at its free end to an inclined force $F = 225\pi$ N as in the figure.

**9-26.** A horizontal $12 \times 12$ mm rectangular bar 100 mm long is attached at one end to a rigid support. Two of the bar's sides form an angle of 30° with the vertical, as shown in the figure. By means of an attachment (not shown), a vertical force $F = 4.45$ N is applied acting through a corner of the bar. (See the figure.) Calculate

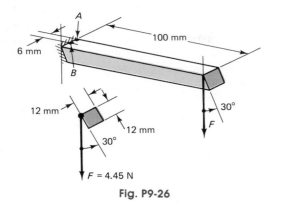

Fig. P9-26

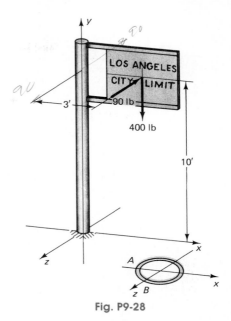

Fig. P9-28

the stress at points $A$ and $B$ caused by the applied force $F$. Neglect stress concentrations. Show the results on the elements viewed from the top. Stress transformations to obtain principal stresses are optional.

**9-27.** A $2 \times 2$ in square bar is attached to a rigid support, as shown in the figure. Determine the principal stresses at point $A$ caused by force $P = 50$ lb applied to the crank.

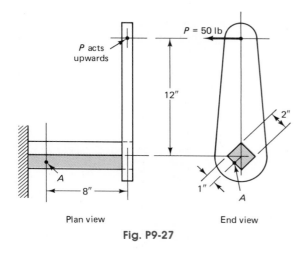

Fig. P9-27

✓ **9-28.** A 400-lb sign is supported by a $2\frac{1}{2}$-in standard-weight steel pipe, as shown in the figure. The maximum horizontal wind force acting on this sign is estimated to be 90 lb. Determine the state of stress caused by this loading at points $A$ and $B$ at the built-in end. Principal stresses are not required. Indicate results on sketches of elements cut out from the pipe

at these points. These elements are to be viewed from outside the pipe.

**\*9-29.** For the circular three-hinged arch rib shown in the figure, determine the principal stresses 75 mm above the centroid of the cross section at section $a$–$a$ due to the applied vertical load on the left half of the structure. Because of the large curvature, the rib at the section investigated can be treated as a straight bar.

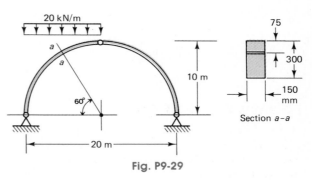

Fig. P9-29

**9-30.** Find the largest bending stress for the beam shown in the figure due to the applied loads. Neglect stress concentrations. All dimensions in the figure are in mm.

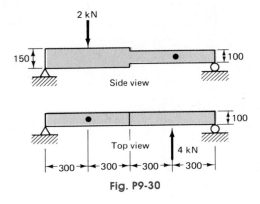

Fig. P9-30

**\*9-31.** In a mechanical device, a horizontal rectangular bar of length $L$ is fixed at the rotating end and is loaded by a strap through a bolt with a vertical force $P$ at the free end, as shown in the figures. Find the angle $\alpha$ for which the normal stress at $A$ is maximum and locate the neutral axis for the beam in this position. Neglect stress concentration, which would have to be considered in an actual problem.

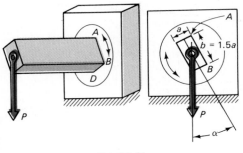

Fig. P9-31

## Sections 9-5 and 9-6.

**\*9-32.** Select the diameter for a solid circular steel shaft to transmit a torque of 6 kN·m and a bending moment of 4 kN·m if the maximum allowable shear stress is 80 MPa.

**\*9-33.** For the loading condition and the allowable shear stress given in the preceding problem, determine the diameter of a hollow circular steel shaft such that the ratio of the inside diameter to the outside diameter is 0.80.

## Section 9-8

**9-34.** A Douglas fir wood beam of rectangular cross section is loaded as shown in the figure. What is the required standard dressed size for the member and what are the minimum sizes of the bearing plates under the concentrated forces and the minimum beam lengths at the supports? In the calculations, consider the weight of the beam. The allowable stress in bending is 1250 psi; in shear, 95 psi; and in compression perpendicular to the grain, 625 psi. Use Table 10 for actual lumber sectional properties.

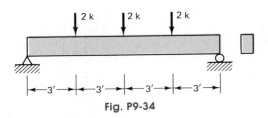

Fig. P9-34

**9-35.** A standard-size wood beam is to be used in the device shown in the figure to transmit a force $P = 250$ lbs. What size member should be used if the allowable stresses are as given in the preceding problem? Neglect the weight of the beam. Select a beam of 3-in nominal width; use Table 10.

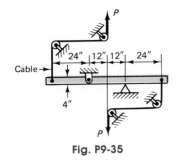

Fig. P9-35

**9-36.** A standard-size wood beam 16 ft long is to carry a uniformly distributed load of 2 k/ft, including its own weight, as shown in the figure. (a) Determine the length $a$ such that the maximum bending moment between the supports is numerically equal to that over the right support. (b) Select the beam size required and calculate the minimum length of the supports. Use the allowable stresses given in Prob. 9-34.

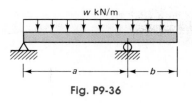

Fig. P9-36

**9-37.** A 4 × 6 in (actual size) wooden beam is to be symmetrically loaded with two equal loads $P$, as shown in the figure. Determine the position of these loads and their magnitudes when a bending stress of 1600 psi and a shearing stress of 100 psi are just reached. Neglect the weight of the beam.

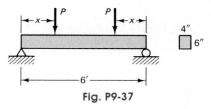

Fig. P9-37

**9-38.** A 12-in deep box beam is fabricated by gluing two pieces of $\frac{3}{4}$-in plywood to two 3 × 8 in nominal-size wood pieces to form the cross section shown in the figure. If the beam is to be used to carry a concentrated force in the middle of a span (see the figure), (a) based on the shear capacity of the section, what may be the magnitude of the applied force $P$, (b) how long may the span be, and (c) what bearing areas should be provided under the concentrated forces? Neglect the weight of the beam. Use Table 10 in the Appendix for dressed sizes of wooden pieces. The allowable stresses are 1200 psi in bending, 120 psi in shear for plywood, 60 psi in shear for glued joints, and 625 psi in compression perpendicular to the grain.

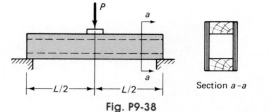

Fig. P9-38

**9-39.** A plastic beam is to be made from two 20 × 60 mm pieces to span 600 mm and to carry an intermittently applied, uniformly distributed load $w$. The pieces can be arranged in two alternative ways, as shown in the figure. The allowable stresses are 4 MPa in flexure, 600 kPa shear in plastic, and 400 kPa shear in glue. Which arrangement of pieces should be used, and what load $w$ can be applied?

Fig. P9-39

**9-40.** Consider two alternative beam designs for spanning 24 ft to support a uniformly distributed load of 1 k/ft. Both beams are simply supported. One of the beams is to be of steel, the other of wood. The allowable stresses for steel are 24 ksi in bending and 14.4 ksi in shear; and those for wood, respectively, are 1250 psi and 95 psi. (a) Find the size required for each beam based on the given strength criteria. (In a comprehensive design, beam deflections are also generally determined.) Consider the beam weights in the calculations; see Tables 4 and 10 in the Appendix. (b) What percentage of the total load is due to the weight of the beam in each case?

**9-41.** Select either an S or a W lightest section for a beam with overhangs for carrying an applied uniformly distributed load of 2 k/ft, as shown in the figure. The specified load includes the beam weight. The allowable stresses are 24 ksi in bending and 14.4 ksi in shear.

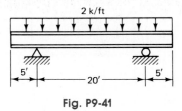

Fig. P9-41

**9-42.** A portion of the floor-framing plan for an office building is shown in the figure. Wooden joists spanning 12 ft are spaced 16 in apart and support a wooden floor above and a plastered ceiling below. Assume that the floor may be loaded by the occupants everywhere by as much as 75 lb per square foot of floor area (live load). Assume further that floor, joists, and ceiling

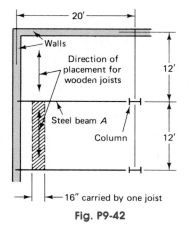

Fig. P9-42

weigh 25 lb per square foot of the floor area (dead load). (a) Determine the depth required for standard commercial joists nominally 2 in wide. For wood, the allowable bending stress is 1200 psi and the shear stress is 100 psi. (b) Select the size required for steel beam *A*. Since the joists delivering the load to this beam are spaced closely, assume that the beam is loaded by a uniformly distributed load. The allowable stresses for steel are 24 ksi and 14.4 ksi for bending and shear, respectively. Use a W or an S beam, whichever is lighter. Neglect the depth of the column.

**9-43.** A bay of an apartment house floor is framed as shown in the figure. Determine the required size of minimum weight for steel beam *A*. Assume that the floor may be loaded everywhere as much as 75 lb/ft² of floor area (live load). Assume further that the weight of the hardwood flooring, structural concrete slab, plastered ceiling below, the weight of the steel beam

being selected, etc., also amounts to approximately 75 lb/ft² of floor area (dead load). Use the allowable stresses given in part (b) of Prob. 9-42.

**9-44.** A four-wheel car running on rails is to be used in light industrial service. When loaded, a force of 10 kN is applied to each bearing. If the bearings are located with respect to the rails as in the figure, what size round axle should be used? Assume the allowable bending stress to be 80 MPa and the allowable shear stress to be 40 MPa.

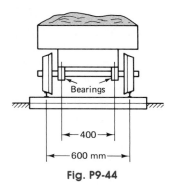

Fig. P9-44

**9-45.** A standard steel beam (S shape) serves as a rail for an overhead traveling crane of 4-ton capacity; see the figure. Determine the required size for the beam and the maximum force on the hanger. Locate the crane so as to cause maximum stresses for each condition. Assume a pinned connection at the wall, and neglect the weight of the beam in calculations. The allowable stress in bending is 16 ksi and that in shear is 9.6 ksi.

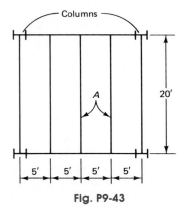

Fig. P9-43

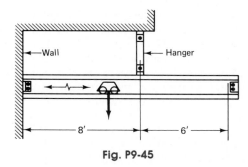

Fig. P9-45

**\*\*9-46.** A glued-laminated wooden beam supports a rail and is loaded by one side of a four-wheel cart, as shown in the figure. The beam is made up from 40 ×

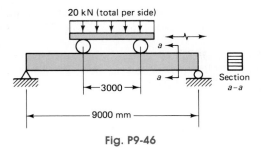

Fig. P9-46

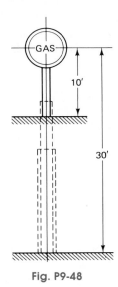

Fig. P9-48

100 mm board laminates. (a) Locate the cart so as to cause the maximum bending moment in the beam. (b) Locate the cart to cause the maximum shear in the beam. (c) Determine the number of board laminates required. The allowable stresses are 14 MPa in bending and 1 MPa in shear. In calculations, neglect the weight of the beam. (*Hint:* Locate the left wheel of the cart a distance $x$ from the left support and write an expression for the bending moment. Setting the derivative of this expression equal to zero determines the position of the cart for the maximum beam moment.)

## Section 9-9

**9-47.** Determine the elevation and plan of a cantilever of uniform flexural stress and circular cross section for resisting a concentrated force $P$ applied at the end. Neglect the weight of the member.

**9-48.** In many engineering design problems, it is very difficult to determine the magnitudes of the loads that will act on a structure or a machine part. Satisfactory performance in an existing installation may provide the basis for extrapolation. With this in mind, suppose that a certain sign, such as shown in the figure, has performed satisfactorily on a 4-in standard steel pipe when its centroid was 10 ft above the ground. What should the size of pipe be if the sign were raised to 30 ft above the ground? Assume that the wind pressure on the sign at the greater height will be 50 percent greater than it was in the original installation. Vary the size of the pipe along the length as required; however, for ease in fabrication, the successive pipe segments must fit into each other. In arranging the pipe segments, also give some thought to aesthetic considerations. For simplicity in calculations, neglect the weight of the pipes and the wind pressure on the pipes themselves.

**9-49.** A W 14 $\times$ 38 beam is coverplated with two 5 $\times$ $\frac{1}{2}$ in plates, as shown in Fig. 9-18(a). If the span is 20 ft, (a) what concentrated force $P$ can be applied to the beam, and where can the coverplates be cut off? Neglect the weight of the beam and assume that the beam section and the coverplates are properly interconnected. Note that coverplates are usually extended a few inches beyond the theoretical cut-off points. (b) Obtain a revised solution if, instead of $P$, a uniformly distributed load were applied. Assume that the allowable bending stress is 24 ksi in both cases.

## Section 9-10

**9-50.** (a) Show that the larger principal stress for a circular shaft simultaneously subjected to a torque and a bending moment is

$$\sigma_1 = (c/J)(M + \sqrt{M^2 + T^2})$$

(b) Show that the design formula for shafts, on the basis of the maximum-stress theory, is

$$d = \sqrt[3]{\frac{16}{\pi \sigma_{\text{allow}}} (M + \sqrt{M^2 + T^2})}$$

**9-51.** At a critical section, a solid circular shaft transmits a torque of 40 kN·m and a bending moment of 10 kN·m. Determine the size of the shaft required so that the maximum shear stress would not exceed 50 MPa.

**9-52.** Rework the preceding problem assuming that $\sigma_{yp} = 100$ MPa and the safety factor is 2 on the von Mises yield criterion given by Eq. 8-54.

**9-53.** The head shaft of an inclined bucket elevator is arranged as shown in the figure. It is driven at A at 11 rpm and requires 60 hp for steady operation. Assuming that one-half of the delivered horsepower is used at each sprocket, determine the size of shaft required so that the maximum shear stress would not exceed 6000 psi. The assigned stress allows for keyways.

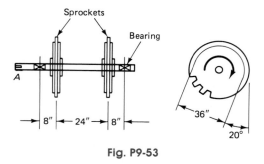

**Fig. P9-53**

**9-54.** A shaft is fitted with pulleys as shown in the figure. The end bearings are self-aligning, i.e., they do not introduce moment into the shaft at the supports. Pulley B is the driving pulley. Pulleys A and C are the driven pulleys and take off 9000 in-lb and 3000 in-lb of torque, respectively. The resultant of the pulls at each pulley is 400 lb acting downward. Determine the size of the shaft required so that the principal shear stress would not exceed 6000 psi.

**9-55.** Two pulleys of $4\pi$ in radius are attached to a 2-in-diameter solid shaft, which is supported by the bearings, as shown in the figure. If the maximum principal shear stress is limited to 6 ksi, what is the largest magnitude that the forces F can assume? The direct shear stress caused by V need not be considered.

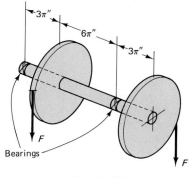

**Fig. P9-55**

**9-56.** A low-speed shaft is acted upon by an eccentrically applied load P caused by a force developed between the gears. Determine the allowable magnitude of force P on the basis of the maximum shear-stress theory if $\tau_{allow} = 6500$ psi. The small diameter of the overhung shaft is 3 in. Consider the critical section to be where the shaft changes diameter, and that $M = 3P$ in-lb and $T = 6P$ in-lb. Note that since the diameter size changes abruptly, the following stress concentration factors must be considered: $K_1 = 1.6$ in bending and $K_2 = 1.2$ in torsion.

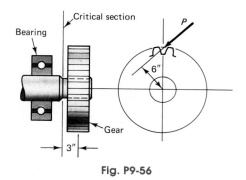

**Fig. P9-56**

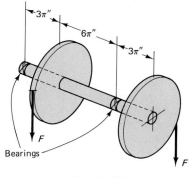

**Fig. P9-54**

**9-57.** A drive shaft for two pulleys is arranged as shown in the figure. The belt tensions are known. Determine the required size of the shaft. Assume that $\tau_{allow} = 6000$ psi for shafts with keyways. Since the shaft will operate under conditions of suddenly applied load, multiply the given loads by a shock factor of $1\frac{1}{2}$.

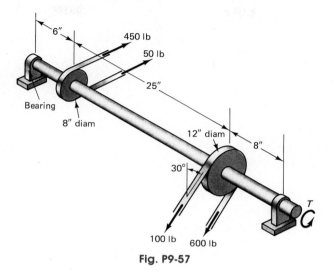

Fig. P9-57

# chapter 10

# Deflection of Beams

## 10-1. Introduction

The axis of a beam deflects from its initial position under action of applied forces. Accurate values for these beam deflections are sought in many practical cases: elements of machines must be sufficiently rigid to prevent misalignment and to maintain dimensional accuracy under load; in buildings, floor beams cannot deflect excessively to avoid the undesirable psychological effect of flexible floors on occupants and to minimize or prevent distress in brittle-finish materials; likewise, information on deformation characteristics of members is essential in the study of vibrations of machines as well as of stationary and flight structures. Deflections are also used in analyses of statically indeterminate problems.

This chapter has two parts. The governing differential equation for the deflection of beams is derived in Part A, and the different types of boundary conditions are identified. Several illustrative examples follow for different kinds of loading and boundary conditions. This includes statically indeterminate beams, presenting no special difficulties using this mathematical approach. A section on the application of singularity functions is provided for symbolic solutions for differential equations having discontinuous loading functions along a span. Methods for solving problems by superposition as well as calculation of deflections for unsymmetrical bending are also presented.

An energy method for calculating beam deflections and the effect of impact loads are briefly introduced. Part A concludes with a discussion of the inelastic deflection of beams. These results are essential for treating the plastic collapse limit states considered in Chapter 13.

Part B is devoted to the discussion of statically determinate and indeterminate beams using the *moment-area method*, also called the *area-moment method*. This specialized procedure is particularly convenient if the deflection of only a few points on a beam or a frame are required. For this reason, it can be used to advantage in the solution of statically indeterminate problems and for deflection check. An excellent insight into the kinematics of deformations is obtained by using this method.

A deflection analysis of slender beams in the presence of axial compressive forces in some instances may cause a profound increase in deflections, causing member instability. This topic is considered in the next chapter.

# Part A    DEFLECTIONS BY INTEGRATION

## 10-2. Moment-Curvature Relation

Beam deflections due to bending are determined from deformations taking place along a span. These are based on the kinematic hypothesis that during bending, plane sections through a beam remain plane. This hypothesis was first introduced in Section 6-2 in deriving the flexure formula for beams having symmetric cross sections, and extended in Section 6-14 to beams of arbitrary cross section for bending about either or both principal axes. For the present, it will be assumed that bending takes place only about one of the principal axes of the cross section. Such a case is illustrated in Fig. 10-1, where it is further assumed that the radius of curvature $\rho$ of the *elastic curve* can change along the span. *Except for a slightly greater generality, the derivation that follows leads to the same*

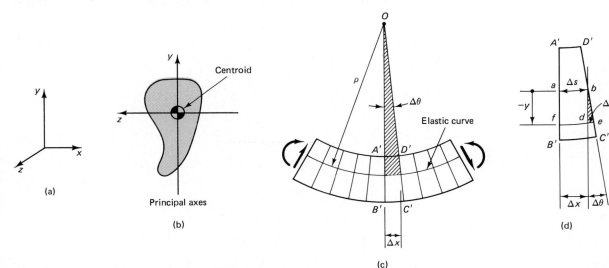

**Fig. 10-1** Deformation of a beam in bending.

*results as found earlier in Section 6-2.* Deflections due to shear are not considered in this development; some consideration of this problem is given in Section 10-11, Example 10-12.

The center of curvature $O$ for the elastic curve for any element can be found by extending to intersection any two adjoining sections such as $A'B'$ and $D'C'$. In the enlarged view of element $A'B'C'D'$ in Fig. 10-1(d), it can be seen that in a bent beam, the included angle between two adjoining sections is $\Delta\theta$. If distance $y$ from the neutral surface to the strained fibers is measured in the usual manner as being positive upwards, the deformation $\Delta u$ of any fiber can be expressed as

$$\Delta u = -y\,\Delta\theta \tag{10-1}$$

For negative $y$'s, this yields elongation, which is consistent with the deformation shown in the figure.

The fibers lying in the curved neutral surface of the deformed beam, characterized in Fig. 10-1(d) by fiber $ab$, are not strained at all. Therefore, arc length $\Delta s$ corresponds to the initial length of all fibers between sections $A'B'$ and $D'C'$. Bearing this in mind, upon dividing Eq. 10-1 by $\Delta s$, one can form the following relations:

$$\lim_{\Delta s \to 0}\frac{\Delta u}{\Delta s} = -y \lim_{\Delta s \to 0}\frac{\Delta\theta}{\Delta s} \quad \text{or} \quad \frac{du}{ds} = -y\frac{d\theta}{ds} \tag{10-2}$$

One can recognize that $du/ds$ is the normal strain in a beam fiber at a distance $y$ from the neutral axis. Hence,

$$\frac{du}{ds} = \varepsilon \tag{10-3}$$

The term $d\theta/ds$ in Eq. 10-2 has a clear geometrical meaning. With the aid of Fig. 10-1(c), it is seen that, since $\Delta s = \rho\,\Delta\theta$,

$$\lim_{\Delta s \to 0}\frac{\Delta\theta}{\Delta s} = \frac{d\theta}{ds} = \frac{1}{\rho} = \kappa \tag{10-4}$$

which is the definition of *curvature*[1] $\kappa$ (kappa) introduced before in Eq. 6-1.

On this basis, upon substituting Eqs. 10-3 and 10-4 into Eq. 10-2, one may express the fundamental relation between curvature of the elastic curve and the normal strain as

$$\boxed{\frac{1}{\rho} = \kappa = -\frac{\varepsilon}{y}} \tag{10-5}$$

[1] Note that both $\theta$ and $s$ must increase in the same direction.

It is important to note that as no material properties were used in deriving Eq. 10-5, *this relation can be used for inelastic as well as for elastic problems*. For the *elastic* case, since $\varepsilon = \varepsilon_x = \sigma_x/E$ and $\sigma_x = -My/I$,

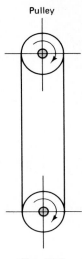

Pulley

$$\frac{1}{\rho} = \frac{M}{EI} \tag{10-6}$$

This equation relates bending moment $M$ at a given section of an elastic beam having a moment of inertia $I$ around the neutral axis to the curvature $1/\rho$ of the elastic curve.

Fig. 10-2

### EXAMPLE 10-1

For cutting metal, a band saw 15 mm wide and 0.60 mm thick runs over two pulleys of 400-mm diameter, as shown in Fig. 10-2. What maximum bending stress is developed in the saw as it goes over a pulley? Let $E = 200$ GPa.

### Solution

In this application, the material must behave elastically. As the thin saw blade goes over the pulley, it conforms to the radius of the pulley; hence, $\rho \approx 200$ mm.

By using Eq. 6-11, $\sigma = -My/I$, together with Eq. 10-6, after some minor simplifications, a generally useful relation follows:

$$\sigma = -\frac{Ey}{\rho} \tag{10-7}$$

With $y = \pm c$, the maximum bending stress in the saw is determined:

$$\sigma_{max} = \frac{Ec}{\rho} = \frac{200 \times 10^3 \times 0.30}{200} = 300 \text{ MPa}$$

The high stress developed in the band saw necessitates superior materials for this application.

## 10-3. Governing Differential Equation

In texts on analytic geometry, it is shown that in Cartesian coordinates, the curvature of a line is defined as

$$\frac{1}{\rho} = \frac{\dfrac{d^2v}{dx^2}}{\left[1 + \left(\dfrac{dv}{dx}\right)^2\right]^{3/2}} = \frac{v''}{[1 + (v')^2]^{3/2}} \tag{10-8}$$

where $x$ and $v$ are the coordinates of a point on a curve. For the problem at hand, distance $x$ locates a point on the elastic curve of a deflected beam, and $v$ gives the deflection of the same point from its initial position.

If Eq. 10-8 were substituted into Eq. 10-6 the exact differential equation for the elastic curve would result. In general, the solution of such an equation is very difficult to achieve. However, since the deflections tolerated in the vast majority of engineering structures are very small, slope $dv/dx$ of the elastic curve is also very small. Therefore, the square of slope $v'$ is a negligible quantity in comparison with unity, and Eq. 10-8 simplifies to

$$\frac{1}{\rho} \approx \frac{d^2v}{dx^2} \tag{10-9}$$

This simplification eliminates the *geometric nonlinearity* from the problem, and the governing differential equation for small deflections of elastic beams[2] using Eq. 10-6 is

$$\boxed{\frac{d^2v}{dx^2} = \frac{M}{EI}} \tag{10-10}$$

where it is understood that $M = M_z$, and $I = I_z$.

Note that in Eq. 10-10, the $xyz$ coordinate system is employed to locate the material points in a beam for calculating the moment of inertia $I$. On the other hand, in the planar problem, it is the $xv$ system of axes that is used to locate points on the elastic curve.

The positive direction of the $v$ axis is taken to have the same sense as

[2] In some texts, the positive direction for deflection $v$ is taken downward with the $x$ axis directed to the right. For such a choice of coordinates, the positive curvature is concave downwards, Fig. 6-51(b). Whereas, if the usual sense for positive moments is retained, Fig. 6-51(a), the corresponding curvature of the bent beam is concave upwards. Therefore, since the curvature induced by positive moments $M$ is opposite to that associated with the positive curvature of the elastic curve, one has

$$\frac{d^2v}{dx^2} = -\frac{M}{EI} \tag{10-10a}$$

Some texts analyze basic beam deflection problems in the $xz$ plane, as shown in Fig. 6-51(b), and define downward deflection $w$ as positive. In this setting, the governing equation also has a negative sign:

$$\frac{d^2w}{dx^2} = -\frac{M}{EI} \tag{10-10b}$$

This notation is particularly favored in the treatment of plates and shells.

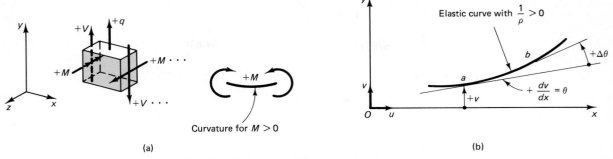

Fig. 10-3 Moment and its relation to curvature.

that of the positive $y$ axis and the positive direction of the applied load $q$, Fig. 10-3. Note, especially, that if the positive slope $dv/dx$ of the elastic curve becomes larger as $x$ increases, curvature $1/\rho \approx d^2v/dx^2$ is positive. This sense of curvature agrees with the induced curvature caused by the applied positive moments $M$. For this reason, the signs are positive on both sides of Eq. 10-10.[3]

Generally, only Eq. 10-10 is used in this text, and if biaxial bending occurs, the deflection directions are determined by inspection.

It is important to note that for the elastic curve, at the level of accuracy of Eq. 10-10, one has $ds = dx$. This follows from the fact that, as before, the square of the slope $dv/dx$ is negligibly small compared with unity, and

$$ds = \sqrt{dx^2 + dv^2} = \sqrt{1 + (v')^2}\, dx \approx dx \qquad (10\text{-}11)$$

Thus, in the small-deflection theory, no difference in length is said to exist between the initial length of the beam axis and the arc of the elastic curve. Stated alternatively, there is no horizontal displacement of the points lying on the neutral surface, i.e., at $y = 0$.

The beam theory discussed here is limited to deflections that are small in relation to span length. However, it is remarkably accurate when compared to exact solutions based on Eq. 10-8. An idea of the accuracy involved may be gained by noting, for example, that there is approximately a 1-percent error from the exact solution if deflections of a simple span are on the order of one-twentieth of its length. By increasing the deflection to one-tenth of the span length, which ordinarily would be considered an intolerably large deflection, the error is raised to approximately 4 percent. As stiff flexural members are required in most engineering applications, this limitation of the theory is not serious. For clarity, however, the deflections of beams will be shown greatly exaggerated on all diagrams.

[3] The equation of the elastic curve was formulated by James Bernoulli, a Swiss mathematician, in 1694. Leonhard Euler (1707–1783) greatly extended its application.

### **$^{**4}$10-4. Alternative Derivation of the Governing Equation

In the classical theories of plates and shells that deal with small deflections, equations analogous to Eq. 10-10 are established. The characteristic approach can be illustrated on the beam problem.

In a deformed condition, point $A'$ on the axis of an unloaded beam, Fig. 10-4, according to Eq. 10-11 is directly above its initial position $A$. The tangent to the elastic curve at the same point rotates through an angle $dv/dx$. A plane section with the centroid at $A'$ also rotates through the same angle $dv/dx$ since during bending sections remain normal to the bent axis of a beam. Therefore, the displacement $u$ of a material point at a distance[5] $y$ from the elastic curve is

$$u = -y \frac{dv}{dx} \tag{10-12}$$

where the negative sign shows that for positive $y$ and $v'$, the displacement $u$ is toward the origin. For $y = 0$, there is no displacement $u$, as required by Eq. 10-11.

Next, recall Eq. 2-6, which states that $\varepsilon_x = du/dx$. Therefore, from Eq. 10-12, $\varepsilon_x = -y\, d^2v/dx^2$ since $v$ is a function of $x$ only.

The same normal strain also can be found from Eqs. 3-14 and 6-11, yielding $\varepsilon_x = -My/EI$. On equating the two alternative expressions for $\varepsilon_x$ and eliminating $y$ from both sides of the equation,

$$\frac{d^2v}{dx^2} = \frac{M}{EI}$$

which is the previously derived Eq. 10-10.

---

[4] Optional section.
[5] Since angle $dv/dx$ is small, its cosine can be taken as unity.

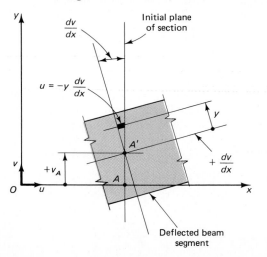

**Fig. 10-4** Longitudinal displacements in a beam due to rotation of a plane section.

## 10-5. Alternative Forms of the Governing Equation

The differential relations among the applied loads, shear, and moment, Eqs. 5-3 and 5-4, can be combined with Eq. 10-10 to yield the following useful sequence of equations:

$$v = \text{deflections of the elastic curve}$$

$$\theta = \frac{dv}{dx} = v' = \text{slope of the elastic curve}$$

$$M = EI\frac{d^2v}{dx^2} = EIv''$$

$$V = \frac{dM}{dx} = \frac{d}{dx}\left(EI\frac{d^2v}{dx^2}\right) = (EIv'')' \qquad (10\text{-}13)$$

$$q = \frac{dV}{dx} = \frac{d^2}{dx^2}\left(EI\frac{d^2v}{dx^2}\right) = (EIv'')''$$

In applying these relations, the sign convention shown in Fig. 10-3 must be adhered to strictly. For beams with *constant* flexural rigidity $EI$, Eq. 10-13 simplifies into three alternative governing equations for determining the deflection of a loaded beam:

$$EI\frac{d^2v}{dx^2} = M(x) \qquad (10\text{-}14a)$$

$$EI\frac{d^3v}{dx^3} = V(x) \qquad (10\text{-}4b)$$

$$EI\frac{d^4v}{dx^4} = q(x) \qquad (10\text{-}14c)$$

The choice of one of these equations for determining $v$ depends on the ease with which an expression for load, shear, or moment can be formulated. Fewer constants of integration are needed in the lower-order equations. Equation 10-14b is seldom used, since it is more convenient to begin a solution either with the load function $q(x)$ or the moment function $M(x)$.

## 10-6. Boundary Conditions

For the solution of beam-deflection problems, in addition to the differential equations, boundary conditions must be prescribed. Several types of homogeneous boundary conditions are as follows:

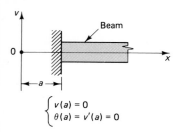

$$\begin{cases} v(a) = 0 \\ \theta(a) = v'(a) = 0 \end{cases}$$

**(a) Clamped support**

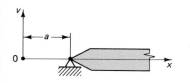

$$\begin{cases} v(a) = 0 \\ M(a) = EIv''(a) = 0 \end{cases}$$

**(b) Simple support**

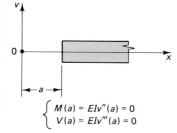

$$\begin{cases} M(a) = EIv''(a) = 0 \\ V(a) = EIv'''(a) = 0 \end{cases}$$

**(c) Free end**

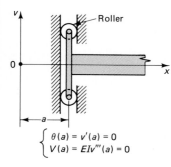

$$\begin{cases} \theta(a) = v'(a) = 0 \\ V(a) = EIv'''(a) = 0 \end{cases}$$

**(d) Guided support**

**Fig. 10-5** Homogeneous boundary conditions for beams with constant *EI*. In (a) both conditions are *kinematic*; in (c) both are *static*; in (b) and (d), conditions are mixed.

1. *Clamped or fixed support:* In this case, the displacement $v$ and the slope $dv/dx$ must vanish. Hence, at the end considered, where $x = a$,

$$v(a) = 0 \qquad v'(a) = 0 \qquad (10\text{-}15a)$$

2. *Roller or pinned support:* At the end considered, no deflection $v$ nor moment $M$ can exist. Hence,

$$v(a) = 0 \qquad M(a) = EIv''(a) = 0 \qquad (10\text{-}15b)$$

Here the physically evident condition for $M$ is related to the derivative of $v$ with respect to $x$ from Eq. 10-14.

3. *Free end:* Such an end is free of moment and shear. Hence,

$$M(a) = EIv''(a) = 0 \qquad V(a) = (EIv'')'_{x=a} = 0 \qquad (10\text{-}15c)$$

4. *Guided support:* In this case, free vertical movement is permitted, but the rotation of the end is prevented. The support is not capable of resisting any shear. Therefore,

$$v'(a) = 0 \qquad V(a) = (EIv'')'_{x=a} = 0 \qquad (10\text{-}15d)$$

The same boundary conditions for beams with *constant EI* are summarized in Fig. 10-5. Note the two basically different types of boundary conditions. Some pertain to the force quantities and are said to be *static boundary conditions*. Others describe geometrical or deformational behavior of an end; these are *kinematic boundary conditions*.

Nonhomogeneous boundary conditions, where a given shear, moment, rotation, or displacement is prescribed at the boundary, also occur in applications. In such cases, the zeros in the appropriate Eqs. 10-15a through 10-15d are replaced by the specified quantity.

These boundary conditions apply both to statically determinate and indeterminate beams. As examples of statically indeterminate single-span beams, consider the three cases shown in Fig. 10-6. The beam in Fig. 10-6(a) is indeterminate to the first degree, as any one of the reactions can be removed and the beam will remain stable. In this example, there are no horizontal forces. The boundary conditions shown in Fig. 10-5(a) apply for end *A*, and those in Fig. 10-5(b), for end *B*.

The vertical reactions for the beam in Fig. 10-6(b) can be found directly from statics. Since the pinned supports cannot move horizontally, there is a tendency for developing horizontal reactions at the supports due to the beam deflection. However, for small beam deflections, according to Eq. 10-11, $ds \approx dx$ and no significant axial strain can develop in trans-

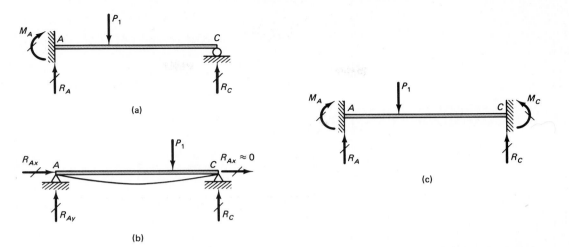

**Fig. 10-6** The beam in (a) is indeterminate to the first degree. If it is assumed that the horizontal reaction component is negligible, the beam in (b) is determinate and in (c) indeterminate to the second degree.

versely loaded beams.[6] Therefore, the horizontal components of the reactions in beams with immovable supports are negligible. On the same basis, no horizontal reactions need be considered for the beam shown in Fig. 10-6(c). Therefore, the beam shown is indeterminate to the second degree. In this case, any two reactive forces can be removed and the beam would remain in equilibrium.

In some problems, discontinuities in the mathematical functions for either load or member stiffness arise along a given span length. Such discontinuities, for example, occur at concentrated forces or moments and at abrupt changes in cross-sectional areas affecting $EI$. In such cases, the boundary conditions must be supplemented by the physical requirements of *continuity of the elastic curve*. This means that at any juncture of the two zones of a beam where a discontinuity occurs, the deflection and the tangent to the elastic curve must be the same regardless of the direction from which the common point is approached. Unacceptable geometry of elastic curves is illustrated in Fig. 10-7.

By using the singularity functions discussed in Section 10-8, the continuity conditions of the elastic curve are identically satisfied.

## 10-7. Direct Integration Solutions

As a general example of calculating beam deflection, consider Eq. 10-14c, $EIv^{iv} = q(x)$. By successively integrating this expression four times, the formal solution for $v$ is obtained. Thus,

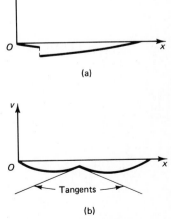

**Fig. 10-7** Unacceptable geometric conditions in a continuous elastic curve.

---

[6] The horizontal force becomes important in thin plates. See S. Timoshenko and S. Woinowsky-Krieger, *Theory of Plates and Shells,* 2nd ed. (New York: McGraw-Hill, 1959), 6.

$$EIv^{\text{iv}} = EI\frac{d^4v}{dx^4} = EI\frac{d}{dx}(v''') = q(x)$$

$$EIv''' = \int_0^x q\,dx + C_1$$

$$EIv'' = \int_0^x dx \int_0^x q\,dx + C_1x + C_2 \qquad\qquad (10\text{-}16)$$

$$EIv' = \int_0^x dx \int_0^x dx \int_0^x q\,dx + C_1x^2/2 + C_2x + C_3$$

$$EIv = \int_0^x dx \int_0^x dx \int_0^x dx \int_0^x q\,dx + C_1x^3/3! + C_2x^2/2! + C_3x + C_4$$

In these equations the constants $C_1$, $C_2$, $C_3$, and $C_4$ have a special physical meaning. Since, per Eq. 10-14b, $EIv''' = V$, by substituting this relation into the second of Eqs. 10-16, and simplifying, Eq. 5-6 is reproduced, i.e.,

$$V = \int_0^x q\,dx + C_1 \qquad\qquad (5\text{-}6)$$

By substituting this relation into Eq. 5-7, and integrating, a different form of Eq. 5-7 is obtained.

$$M = \int_0^x dx \int_0^x q\,dx + C_1x + C_2 \qquad\qquad (10\text{-}17)$$

The right side of this equation is identical to the third of Eqs. 10-16.

These results unequivocally show that the constants $C_1$ and $C_2$ are a part of the equilibrium equations and are the _static boundary_ conditions. At this point, no kinematics nor material properties enter the problem. However, next, by dividing $M$ by $EI$ for substitution into Eq. 10-10, these properties are brought in, limiting the solutions to the _elastic_ behavior of _prismatic_ beams. Thus, rewriting Eq. 10-10, for clarity, in several different forms,

$$\frac{d^2v}{dx^2} = \frac{d}{dx}\left(\frac{dv}{dx}\right) = \frac{dv'}{dx} = \frac{M}{EI} \qquad\qquad (10\text{-}18)$$

Then, using Eq. 10-17, and integrating twice, the last two relations in Eqs. 10-16 are reproduced. These two equations, and the associated new constants of integration $C_3$ and $C_4$, define slope and deflection of the elastic curve, i.e., they describe the kinematics of a laterally loaded beam. These constants are the _kinematic boundary conditions_.

If, instead of Eq. 10-14c, one starts with Eq. 10-14a, $EIv'' = M(x)$, after two integrations the solution is

$$EIv = \int_0^x dx \int_0^x M\,dx + C_3x + C_4 \qquad\qquad (10\text{-}19)$$

In both equations, constants $C_1$, $C_2$, $C_3$, and $C_4$ must be determined from the conditions at the boundaries. In Eq. 10-19, constants $C_1$ and $C_2$ are incorporated into the expression of $M$. Constants $C_1$, $C_2$, $C_3/EI$, and $C_4/EI$, respectively, are usually[7] the initial values of $V$, $M$, $\theta$, and $v$ at the origin.

The first term on the right hand of the last part of Eq. 10-16 and the corresponding one in Eq. 10-19 are the particular solutions of the respective differential equations. The one in Eq. 10-16 is especially interesting as it depends only on the loading condition of the beam. This term remains the same regardless of the prescribed boundary conditions, whereas the *constants* are determined from the boundary conditions.

If the loading, shear, and moment functions are continuous and the flexural rigidity $EI$ is constant, the evaluation of the particular integrals is very direct. When discontinuities occur, solutions can be found for each segment of a beam in which the functions are continuous; the complete solution is then achieved by enforcing continuity conditions at the common boundaries of the beam segments. Alternatively, graphical or numerical procedures,[8] of successive integrations can be used very effectively in the solution of practical problems.

Any one of Eqs. 10-14 or 10-16 can be used for finding beam deflection. The choice depends entirely on the initial data and the amount of work necessary for solving a problem. If one begins with the applied load, all four integrations must be performed. On the other hand, if the bending-moment function is written, the number of required integrations is reduced to two.

## *Procedure Summary*

The same *three basic concepts* of engineering mechanics of solids repeatedly applied before are used in developing the elastic deflection theory of beams. These may be summarized as follows:

1. *Equilibrium conditions* (statics) are used for a beam element to establish the relationships between the applied load and shear, Eq. 5-3, as well as between the shear and bending moment, Eq. 5-4.

2. *Geometry of deformation* (kinematics) is used by assuming that plane sections through a beam element remain plane after deformation. Such planes intersect and define beam strains and the radius of curvature for an element. Although in the above sense the expression for curvature, Eq. 10-4, is exact, the theory is limited to small deflections, since $\sin \theta$ is approximated by $\theta$, Eq. 10-9. No warpage due to shear of sections is accounted for in the formulation.

[7] In certain cases where transcendental functions are used, these constants do not have this meaning. Basically, the whole function, which includes the constants of integration, must satisfy the conditions at the boundary.

[8] Such procedures are useful in complicated problems. For example, see N. M. Newmark, "Numerical Procedure for Computing Deflections, Moments, and Buckling Loads," *Trans. ASCE* 108 (1943): 1161. Finite element solutions of such problems are now widely used in practice.

3. *Properties of materials* (constitutive relations) in the form of Hooke's law, Eq. 2-3, are assumed to apply only to longitudinal normal stresses and strains. Poisson's effect is neglected.

A solution of any one of Eqs. 10-14a, 10-14b, or 10-14c,[9] *subject to the prescribed boundary conditions,* constitutes a solution of a given transversely loaded elastic beam problem. These equations are equally applicable to statically determinate *and* statically indeterminate beam problems.[10] However, the solutions are simpler if the functions $q(x)$ and $I(x)$ are continuous across a span. When discontinuities in either $q(x)$ or $I(x)$ occur, continuity of the elastic curve at such points must be maintained. If $I$ is constant, singularity functions for describing the loads can be effectively used.

It is to be noted that although at load and cross-section discontinuities large local perturbations in strain and stresses develop, beam deflections are less sensitive to these effects. Deflections are determined using integration, a process tending to smooth out the function.

Several illustrative examples of statically determinate and statically indeterminate beam problems follow. The applications of singularity functions for elastic beam deflections is given in the next section.

### EXAMPLE 10-2

A bending moment $M_1$ is applied at the free end of a cantilever of length $L$ and of constant flexural rigidity $EI$, Fig. 10-8(a). Find the equation of the elastic curve.

### Solution

The boundary conditions are recorded near the figure from *inspection* of the conditions at the ends. At $x = L$, $M(L) = +M_1$, a nonhomogeneous condition.

From a free-body diagram of Fig. 10-8(b), it can be observed that the bending moment is $+M_1$ throughout the beam. By applying Eq. 10-14a, integrating successively, and making use of the boundary conditions, one obtains the solution for $v$:

$$EI \frac{d^2v}{dx^2} = M = M_1$$

$$EI \frac{dv}{dx} = M_1 x + C_3$$

But $\theta(0) = 0$; hence, at $x = 0$, one has $EIv'(0) = C_3 = 0$ and

---

[9] The adopted sign convention for applied loads and shear results in all Eqs. 10-13 and 10-14 having positive signs, an advantage in hand calculations.

[10] This is analogous to the axially loaded bar problems discussed in Section 2-19.

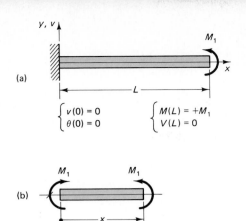

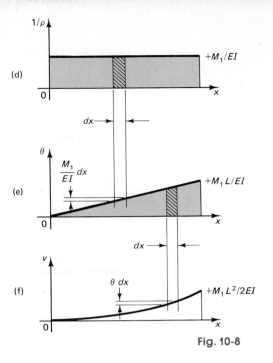

(a)

$$\begin{cases} v(0) = 0 \\ \theta(0) = 0 \end{cases} \qquad \begin{cases} M(L) = +M_1 \\ V(L) = 0 \end{cases}$$

(b)

(c)

(d) $+M_1/EI$

$dx$

(e) $+M_1L/EI$

$\dfrac{M_1}{EI}\,dx$

$dx$

(f) $+M_1L^2/2EI$

$\theta\,dx$

**Fig. 10-8**

$$EI\,\frac{dv}{dx} = M_1 x$$

$$EIv = \frac{1}{2}M_1 x^2 + C_4$$

But $v(0) = 0$; hence, $EIv(0) = C_4 = 0$ and

$$v = \frac{M_1 x^2}{2EI} \qquad\qquad (10\text{-}20)$$

The positive sign of the result indicates that the deflection due to $M_1$ is upward. The largest value of $v$ occurs at $x = L$. The slope of the elastic curve at the free end is $+M_1L/EI$ radians.

Equation 10-20 shows that the elastic curve is a parabola. However, every element of the beam experiences equal moments and deforms alike. Therefore, the elastic curve should be a part of a circle. The inconsistency results from the use of an approximate relation for the curvature $1/\rho$. It can be shown that the error committed is in the ratio of $(\rho - v)^3$ to $\rho^3$. As the deflection $v$ is much smaller than $\rho$, the error is very small.

It is important to associate the above successive integration procedure with a graphical solution or interpretation. This is shown in the sequence of Figs. 10-8(c) through (f). First, the conventional moment diagram is shown. Then using Eqs. 10-9 and 10-10, $1/\rho \approx d^2v/dx^2 = M/EI$, the curvature diagram is plotted in Fig. 10-8(d). For the elastic case, this is simply a plot of $M/EI$. By integrating the curvature diagram, one obtains the $\theta$ diagram. In the next integration, the elastic curve is obtained. In this example, since the beam is fixed at the origin, the conditions $\theta(0) = 0$ and $v(0) = 0$ are used in constructing the diagrams. This graphical approach or its numerical equivalents are very useful in the solution of problems with variable $EI$.

### EXAMPLE 10-3

A simple beam supports a uniformly distributed downward load $w_o$. The flexural rigidity $EI$ is constant. Find the elastic curve by the following three methods: ($a$) Use the second-order differential equation to obtain the deflection of the beam. ($b$) Use the fourth-order equation instead of the one in part ($a$). ($c$) Illustrate a graphical solution of the problem.

### Solution

($a$) A diagram of the beam together with the given boundary conditions is shown in Fig. 10-9(a). The expression for $M$ for use in the second-order differential equation has been found in Example 5-8. From Fig. 5-24,

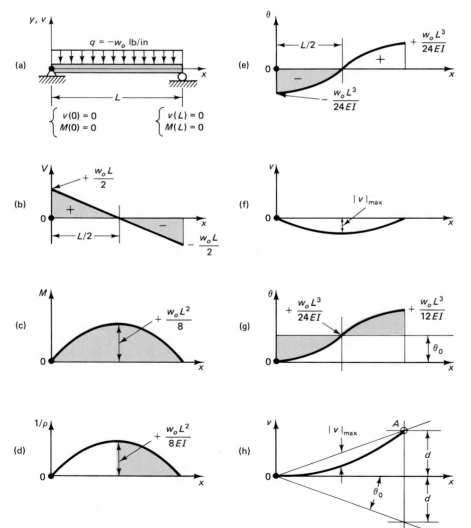

Fig. 10-9

$$M = \frac{w_o L x}{2} - \frac{w_o x^2}{2}$$

Substituting this relation into Eq. 10-14a, integrating it twice, and using the boundary conditions, one finds the equation of the elastic curve:

$$EI \frac{d^2 v}{dx^2} = M = \frac{w_o L x}{2} - \frac{w_o x^2}{2}$$

$$EI \frac{dv}{dx} = \frac{w_o L x^2}{4} - \frac{w_o x^3}{6} + C_3$$

$$EIv = \frac{w_o L x^3}{12} - \frac{w_o x^4}{24} + C_3 x + C_4$$

But $v(0) = 0$; hence, $EIv(0) = 0 = C_4$; and, since $v(L) = 0$,

$$EIv(L) = 0 = \frac{w_o L^4}{24} + C_3 L \quad \text{and} \quad C_3 = -\frac{w_o L^3}{24}$$

$$v = -\frac{w_o x}{24 EI}(L^3 - 2Lx^2 + x^3) \tag{10-21}$$

Because of symmetry, the largest deflection occurs at $x = L/2$. On substituting this value of $x$ into Eq. 10-21, one obtains

$$|v|_{\max} = \frac{5 w_o L^4}{384 EI} \tag{10-22}$$

The condition of symmetry could also have been used to determine constant $C_3$. Since it is known that $v'(L/2) = 0$, one has

$$EIv'(L/2) = \frac{w_o L(L/2)^2}{4} - \frac{w_o (L/2)^3}{6} + C_3 = 0$$

and, as before, $C_3 = -w_o L^3/24$.

(b) Application of Eq. 10-14c to the solution of this problem is direct. The constants are found from the boundary conditions.

$$EI \frac{d^4 v}{dx^4} = q(x) = -w_o$$

$$EI \frac{d^3 v}{dx^3} = -w_o x + C_1$$

$$EI \frac{d^2 v}{dx^2} = -\frac{w_o x^2}{2} + C_1 x + C_2$$

But $M(0) = 0$; hence, $EIv''(0) = 0 = C_2$; and, since $M(L) = 0$,

$$EIv''(L) = 0 = -\frac{w_oL^2}{2} + C_1L \quad \text{or} \quad C_1 = \frac{w_oL}{2}$$

hence,

$$EI\frac{d^2v}{dx^2} = \frac{w_oLx}{2} - \frac{w_ox^2}{2}$$

The remainder of the problem is the same as in part ($a$). In this approach, no preliminary calculation of reactions is required. As will be shown later, this is advantageous in some statically indeterminate problems.

($c$) The steps needed for a graphical solution of the complete problem are in Figs. 10-9(b) through (f). In Figs. 10-9(b) and (c), the conventional shear and moment diagrams are shown. The curvature diagram is obtained by plotting $M/EI$, as in Fig. 10-9(d).

Since, by virtue of symmetry, the slope to the elastic curve at $x = L/2$ is horizontal, $\theta(L/2) = 0$. Therefore, the construction of the $\theta$ diagram can be begun from the center. In this procedure, the right ordinate in Fig. 10-9(e) must equal the shaded area of Fig. 10-4(d), and vice versa. By summing the $\theta$ diagram, one finds the elastic deflection $v$. The shaded area of Fig. 10-9(e) is equal numerically to the maximum deflection. In the above, the condition of symmetry was employed. A generally applicable procedure follows.

After the curvature diagram is established as in Fig. 10-9(d), the $\theta$ diagram can be constructed with an assumed initial value of $\theta$ at the origin. For example, let $\theta(0) = 0$ and sum the curvature diagram to obtain the $\theta$ diagram, Fig. 10-9(g). Note that the shape of the curve so found is identical to that of Fig. 10-9(e). Summing the area of the $\theta$ diagram gives the elastic curve. In Fig. 10-9(h), this curve extends from 0 to $A$. This violates the boundary condition at $A$, where the deflection must be zero. Correct deflections are given, however, by measuring them vertically from a straight line passing through 0 and $A$. This inclined line corrects the deflection ordinates caused by the incorrectly assumed $\theta(0)$. In fact, after constructing Fig. 10-9(h), one knows that $\theta(0) = -d/L = -w_oL^3/24EI$. When this value of $\theta(0)$ is used, the problem reverts to the preceding solution, Figs. 10-9(e) and (f). In Fig. 10-9(h), inclined measurements have no meaning. The procedure described is applicable for beams with overhangs. In such cases, the base line for measuring deflections must pass through the support points.

## EXAMPLE 10-4

A beam fixed at both ends supports a uniformly distributed downward load $w_o$, Fig. 10-10(a). The $EI$ for the beam is constant. ($a$) Find the expression for the elastic curve using the fourth-order governing differential equation. ($b$) Verify the results found using the second-order differential equation.

### Solution

($a$) As discussed in connection with Fig. 10-6(c), this beam is statically indeterminate to the second degree since horizontal reactions are assumed to be zero.

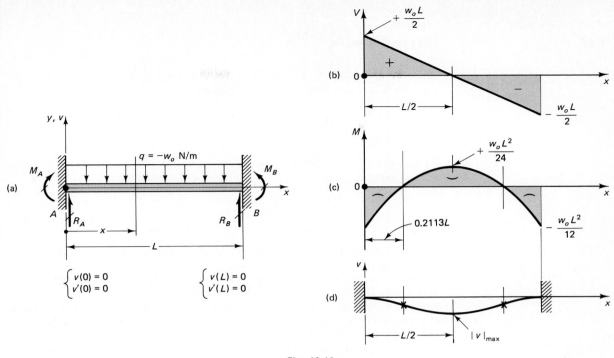

Fig. 10-10

The solution is obtained by four successive integrations of Eq. 10-14c in a manner shown in Eqs. 10-16. Then the constants of integration are found from the boundary conditions.

$$EI \frac{d^4 v}{dx^4} = q(x) = -w_o$$

$$EI \frac{d^3 v}{dx^3} = -w_o x + C_1$$

$$EI \frac{d^2 v}{dx^2} = -\frac{w_o x^2}{2} + C_1 x + C_2$$

$$EI \frac{dv}{dx} = -\frac{w_o x^3}{6} + C_1 \frac{x^2}{2} + C_2 x + C_3$$

$$EIv = -\frac{w_o x^4}{24} + C_1 \frac{x^3}{6} + C_2 \frac{x^2}{2} + C_3 x + C_4$$

Four kinematic boundary conditions are available for determining the constants of integration:

$$EIv(0) = EIv_A = 0 = C_4$$

$$EIv'(0) = EIv'_A = 0 = C_3$$

$$EIv(L) = EIv_B = 0 = -\frac{w_o L^4}{24} + C_1 \frac{L^3}{6} + C_2 \frac{L^2}{2}$$

$$EIv'(L) = EIv'_B = 0 = -\frac{w_o L^3}{6} + C_1 \frac{L^2}{2} + C_2 L$$

Constants $C_3$ and $C_4$ do not enter the last two equations since they are zero. By solving the last two equations simultaneously,

$$C_1 = \frac{w_oL}{2} \quad \text{and} \quad C_2 = -\frac{w_oL^2}{12}$$

By substituting these constants into the equation for the elastic curve, after algebraic simplifications,

$$v = -\frac{w_ox^2}{24EI}(L - x)^2 \tag{10-23}$$

According to Eqs. 10-14a and 10-14b, $EI$ times the second and third derivatives of the deflection $v(x)$ gives, respectively, $M(x)$ and $V(x)$. At $x = 0$, these relations define the reactions at $A$. Hence, $C_1$ is the vertical reaction and $C_2$ is the moment at this support. In this case, because of symmetry, the vertical reactions can be found directly from statics. However, this is not necessary in this typical solution of a *boundary-value problem*. The moment and shear at $B$ can be found from the same expressions at $x = L$.

Shear, moment, and deflection diagrams for this beam are shown in Fig. 10-10. The absolute maximum deflection occurring in the middle of the span is

$$|v|_{\max} = \frac{w_oL^4}{384EI} \tag{10-24}$$

(*b*) This solution is found using Eq. 10-14a, and, although the vertical reaction at $A$ can be determined directly from statics, it will be treated as an unknown. On this basis,

$$EI\frac{d^2v}{dx^2} = M(x) = M_A + R_Ax - \frac{w_ox^2}{2}$$

Integrating twice,

$$EI\frac{dv}{dx} = M_Ax + R_A\frac{x^2}{2} - \frac{w_ox^3}{6} + C_3$$

$$EIv = M_A\frac{x^2}{2} + R_A\frac{x^3}{6} - \frac{w_ox^4}{24} + C_3x + C_4$$

Constants $C_3$ and $C_4$ *as well as* $R_A$ and $M_A$ are found from the four kinematic boundary conditions:

$$EIv(0) = EIv_A = 0 = C_4$$
$$EIv'(0) = EIv'_A = 0 = C_3$$
$$EIv(L) = EIv_B = 0 = M_A\frac{L^2}{2} + R_A\frac{L^3}{6} - \frac{w_oL^4}{24}$$
$$EIv'(L) = EIv'_B = 0 = M_AL + R_A\frac{L^2}{2} - \frac{w_oL^3}{6}$$

Solving the last two equations simultaneously,

$$R_A = \frac{w_o L}{2} \quad \text{and} \quad M_A = -\frac{w_o L^2}{12}$$

Substituting these expressions into the equation for deflection *with* $C_3 = C_4 = 0$, Eq. 10-23 is again obtained.

---

## EXAMPLE 10-5

Determine the equation of the elastic curve for the uniformly loaded continuous beam shown in Fig. 10-11(a). Use the second-order differential equation. *EI* is constant.

### Solution

Because of symmetry, the solution can be confined to determining the deflection for either span. Also, because of symmetry, it can be concluded that at the middle support, not only is the deflection zero, but since the elastic curve cannot rotate in either direction, its slope is also zero. In this manner, the problem can be reduced to the one-degree statically indeterminate problem shown in Fig. 10-11(b) with known boundary conditions.

By using Eq. 10-14a, the solution proceeds in the usual manner. First, an expression for $M(x)$ is formulated and two successive integrations of the differential equation are performed. Boundary conditions provide the necessary information for determining the constants of integration *and* an unknown reaction $R_A$.

Second-order differential-equation solution:

$$EI \frac{d^2 v}{dx^2} = M(x) = R_A x - \frac{w_o x^2}{2}$$

$$EI \frac{dv}{dx} = R_A \frac{x^2}{2} - \frac{w_o x^3}{6} + C_3$$

$$EIv = R_A \frac{x^3}{6} - \frac{w_o x^4}{24} + C_3 x + C_4$$

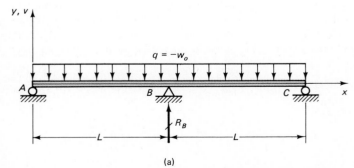

(a)

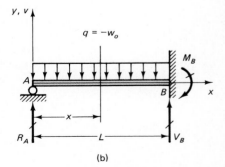

(b)

**Fig. 10-11**

Boundary conditions:

$$EIv(0) = EIv_A = 0 = C_4$$

$$EIv'(L) = EIv'_B = 0 = R_A \frac{L^2}{2} - \frac{w_o L^3}{6} + C_3$$

$$EIv(L) = EIv_B = 0 = R_A \frac{L^3}{6} - \frac{w_o L^4}{24} + C_3 L$$

By solving the last two equations simultaneously,

$$R_A = \frac{3w_o L}{8} \quad \text{and} \quad C_3 = -\frac{w_o L^3}{48}$$

which, upon substitution into the equation for the elastic curve, leads to

$$v = -\frac{w_o x}{48EI}(L^3 - 3Lx^2 + 2x^3) \tag{10-25}$$

From symmetry, the reactions at $A$ and $C$ are equal, and, by using statics, the reaction at $B$ is

$$R_B = \frac{5w_o L}{4} \tag{10-26}$$

This reaction is also numerically equal to $2V_B$.

---

## EXAMPLE 10-6

A simple beam supports a concentrated downward force $P$ at a distance $a$ from the left support, Fig. 10-12(a). The flexural rigidity $EI$ is constant. Find the equation of the elastic curve by successive integration.

### Solution

The solution will be obtained using the second-order differential equation. The reactions and boundary conditions are noted in Fig. 10-12(a). The moment diagram plotted in Fig. 10-12(b) clearly shows a discontinuity in $M(x)$ at $x = a$, requiring two different functions. At first, the solution proceeds independently for each segment of the beam.

| For segment $AD$ | For segment $DB$ |
|---|---|
| $\dfrac{d^2v}{dx^2} = \dfrac{M}{EI} = \dfrac{Pb}{EIL}x$ | $\dfrac{d^2v}{dx^2} = \dfrac{M}{EI} = \dfrac{Pa}{EI} - \dfrac{Pa}{EIL}x$ |
| $\dfrac{dv}{dx} = \dfrac{Pb}{EIL}\dfrac{x^2}{2} + A_1$ | $\dfrac{dv}{dx} = \dfrac{Pa}{EI}x - \dfrac{Pa}{EIL}\dfrac{x^2}{2} + B_1$ |
| $v = \dfrac{Pb}{EIL}\dfrac{x^3}{6} + A_1 x + A_2$ | $v = \dfrac{Pa}{EI}\dfrac{x^2}{2} - \dfrac{Pa}{EIL}\dfrac{x^3}{6} + B_1 x + B_2$ |

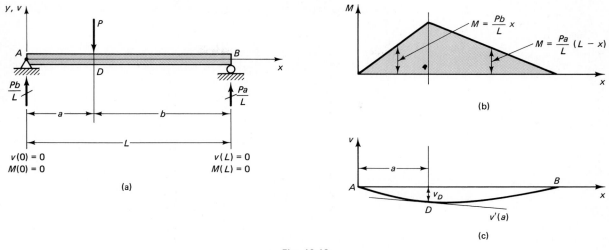

Fig. 10-12

To determine the four constants $A_1$, $A_2$, $B_1$, and $B_2$, two boundary and two continuity conditions must be used.

*For segment AD:*

$$v(0) = 0 = A_2$$

*For segment DB:*

$$v(L) = 0 = \frac{PaL^2}{3EI} + B_1L + B_2$$

Equating deflections for both segments at $x = a$:

$$v_D = v(a) = \frac{Pa^3b}{6EIL} + A_1a = \frac{Pa^3}{2EI} - \frac{Pa^4}{6EIL} + B_1a + B_2$$

Equating slopes for both segments at $x = a$:

$$\theta_D = v(a) = \frac{Pa^2b}{2EIL} + A_1 = \frac{Pa^2}{EI} - \frac{Pa^3}{2EIL} + B_1$$

Upon solving the four equations simultaneously, one finds

$$A_1 = -\frac{Pb}{6EIL}(L^2 - b^2) \qquad A_2 = 0$$

$$B_1 = -\frac{Pa}{6EIL}(2L^2 + a^2) \qquad B_2 = \frac{Pa^3}{6EI}$$

With these constants, for example, the elastic curve for segment $AD$ of the beam, after algebraic simplification, becomes

$$v = -\frac{Pbx}{6EIL}(L^2 - b^2 - x^2) \tag{10-27}$$

Deflection $v_B$ at applied force $P$ is

$$v_B = v(a) = -\frac{Pa^2b^2}{3EIL} \tag{10-28}$$

The largest deflection occurs in the longer segment of the beam. If $a > b$, the point of maximum deflection is at $x = \sqrt{a(a + 2b)/3}$, which follows from setting the expression for the slope equal to zero. The deflection at this point is

$$|\,v\,|_{max} = \frac{Pb(L^2 - b^2)^{3/2}}{9\sqrt{3}\ EIL} \tag{10-29}$$

Usually, the deflection at the center of the span is very nearly equal to the numerically largest deflection. Such a deflection is much simpler to determine, which recommends its use. If force $P$ is applied at the middle of the span, when $a = b = L/2$, by direct substitution into Eq. 10-28 or 10-29,

$$|\,v\,|_{max} = \frac{PL^3}{48EI} \tag{10-30}$$

Here it is well to recall the definition of the *spring constant,* or *stiffness, k* given by Eq. 2-11. In the present context, for a force $P$ placed at an arbitrary distance $a$ from a support,

$$k = \frac{P}{v_B} = \frac{3EIL}{a^2b^2} \tag{10-31}$$

For a particular case, when $a = b = L/2$, this equation reduces to

$$k_o = \frac{48EI}{L^3} \tag{10-32}$$

This expression also follows directly from Eq. 10-30.

The previous equations are useful in static and dynamic analyses and are essential in vibration analysis.

The solution of deflection problems having discontinuous load functions is greatly facilitated with the use of singularity functions discussed in the next section.

## **\*\*EXAMPLE 10-7**

A simply supported beam 5 m long is loaded with a 20-N downward force at a point 4 m from the left support, Fig. 10-13(a). The moment of inertia of the cross section of the beam is $4I_1$ for segment $AB$ and $I_1$ for the remainder of the beam. Determine the elastic curve.

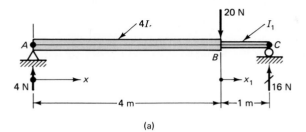

(a)

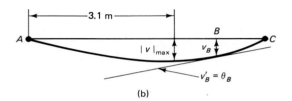

(b)

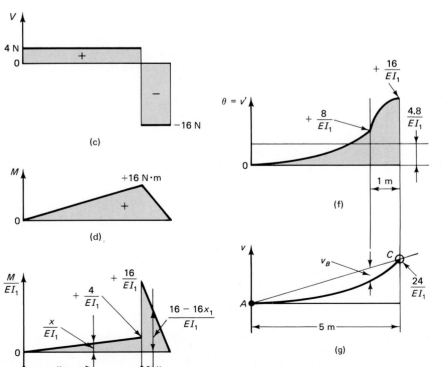

(c)

(d)

(e)

(f)

(g)

**Fig. 10-13**

## Solution

A similar problem was solved in the preceding example. Another useful technique will be illustrated here that is convenient in some complicated problems where different *M/EI* expressions are applicable to several segments of the beam. This method consists of selecting an origin at one end of the beam and carrying out successive integrations until expressions for $\theta$ and $v$ are obtained for the first segment. The values of $\theta$ and $v$ are then determined at the end of the first segment. Due to continuity conditions, these become the initial constants in the integrations carried out for the next segment. This process is repeated until the far end of the beam is reached; then the boundary conditions are imposed to determine the remaining unknown constants. A new origin is used at every juncture of the segments, and all $x$'s are taken to be positive in the same direction.

*For segment AB, $0 < x < 4$:*

$$M = 4x \quad \text{and} \quad EI = 4EI_1$$

$$\frac{d^2v}{dx^2} = \frac{M}{EI} = \frac{x}{EI_1}$$

$$\theta = \frac{dv}{dx} = \frac{x^2}{2EI_1} + A_1$$

$$v = \frac{x^3}{6EI_1} + A_1x + A_2$$

At $x = 0$: $v(0) = v_A = 0$, and $\theta(0) = \theta_A$. Hence, $A_1 = \theta_A$ and $A_2 = 0$.
At the end of segment *AB*:

$$\theta(4) = \theta_B = \frac{8}{EI_1} + \theta_A \quad \text{and} \quad v(4) = v_B = \frac{32}{3EI_1} + 4\theta_A$$

*For segment BC, $0 < x_1 < 1$:*

$$M = 4(4 + x_1) - 20x_1 = 16 - 16x_1 \quad \text{and} \quad EI = EI_1$$

$$\frac{d^2v}{dx_1^2} = \frac{16}{EI_1} - \frac{16x_1}{EI_1}$$

$$\theta = \frac{dv}{dx_1} = \frac{16x_1}{EI_1} - \frac{8x_1^2}{EI_1} + A_3$$

$$v = \frac{8x_1^2}{EI_1} - \frac{8x_1^3}{3EI_1} + A_3x_1 + A_4$$

At $x_1 = 0$: $v(0) = v_B$ and $\theta(0) = \theta_B$. Hence from the solution before, $A_4 = v_B = 32/3EI_1 + 4\theta_A$, and $A_3 = \theta_B = 8/EI_1 + \theta_A$. The expressions for $\theta$ and $v$ in segment *BC* are then obtained as

$$\theta = \frac{16x_1}{EI_1} - \frac{8x_1^2}{EI_1} + \frac{8}{EI_1} + \theta_A$$

$$v = \frac{8x_1^2}{EI_1} - \frac{8x_1^3}{3EI_1} + \frac{8x_1}{EI_1} + \theta_Ax_1 + \frac{32}{3EI_1} + 4\theta_A$$

Finally, the boundary condition at $C$ is applied to determine the value of $\theta_A$. At $x_1 = 1$: $v(1) = v_c = 0$; therefore,

$$0 = \frac{8}{EI_1} - \frac{8}{3EI_1} + \frac{8}{EI_1} + \theta_A + \frac{32}{3EI_1} + 4\theta_A \quad \text{and} \quad \theta_A = -\frac{4.8}{EI_1}$$

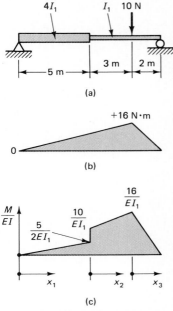

(a)

+16 N·m

(b)

Substituting this value of $\theta_A$ into the respective expressions for $\theta$ and $v$, equations for these quantities can be obtained for either segment. For example, the equation for the slope in segment $AB$ is $\theta = x^2/2EI_1 - 24/5EI_1$. Upon setting this quantity equal to zero, $x$ is found to be 3.1 m. The maximum deflection occurs at this value of $x$, and $|v|_{max} = 9.95/EI_1$. Characteristically, the deflection at the center of the span (at $x = 2.5$ m) is nearly the same, being $9.4/EI_1$.

A self-explanatory graphical procedure is shown in Figs. 10-13(d) through (g). Variations in $I$ cause virtually no complications in the graphical solution, a great advantage in complex problems. Multiple origins can be used as shown in Fig. 10-14 to simplify the numerical work as in the present example.

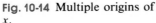

(c)

**Fig. 10-14** Multiple origins of $x$.

## **11 10-8. Singularity Functions for Beams

The possibility for writing symbolic mathematical expressions for discontinuous functions of load, shear, and moment along a beam using singularity functions was introduced in Section 5-16. These functions can be very effectively used for the solution of statically determinate and indeterminate beam deflection problems. However, it is best to limit their applications to *prismatic beams of constant EI*. Otherwise, considerable complexities arise.

Besides the convenience of solving with singularity functions beams of single spans, these functions can also be applied for beams on several supports. In either case, a single symbolic mathematical function for the forces acting on a beam, together with Eq. 10-14c, upon successive integrations, gives the solution for a deflection in a problem.[12]

Two illustrative examples follow.

### **EXAMPLE 10-8

Rework Example 10-6 using singularity functions.

#### Solutions

First, the singularity function for the concerned downward force $P$ is written for the right side of Eq. 10-16. This is followed by successive integrations determining the constants of integrations as convenient.

[11] This section is optional.
[12] Singularity functions can also be used for constructing influence lines for prismatic beams. The required special functions for such problems are given in E. P. Popov, *Introduction to Mechanics of Solids* (Englewood Cliffs, NJ: 1968) 403–405.

$$EI \frac{d^4v}{dx^4} = q(x) = -P\langle x - a \rangle_*^{-1}$$

$$EI \frac{d^3v}{dx_3} = -P\langle x - a \rangle^0 + C_1$$

$$EI \frac{d^2v}{dx^2} = -P\langle x - a \rangle^1 + C_1x + C_2$$

But $M(L) = 0$; hence, $EIv''(0) = 0 = C_2$; and also since $M(L) = 0$,

$$EIv''(L) = -Pb + C_1L = 0 \quad \text{or} \quad C_1 = Pb/L$$

$$EI \frac{dv}{dx} = -\frac{P}{2}\langle x - a \rangle^2 + \frac{Pb}{2L}x^2 + C_3$$

$$EIv = -\frac{P}{6}\langle x - a \rangle^3 + \frac{Pb}{6L}x^3 + C_3x + C_4$$

But $v(0) = 0$; hence, $EIv(0) = 0 = C_4$. Similarly, from $v(L) = 0$,

$$EIv(L) = 0 = -\frac{Pb^3}{6} + \frac{PbL^2}{6} + C_3L \quad \text{or} \quad C_3 = -\frac{Pb}{6L}(L^2 - b^2)$$

$$v = \frac{Pb}{6EIL}\left[ x^3 - (L^2 - b^2)x - \frac{L}{b}\langle x - a \rangle^3 \right]$$

This equation applies to the entire span. For $0 < x < a$, the last term must be omitted. This reduced expression agrees with Eq. 10-27 found earlier.

## **EXAMPLE 10-9

Rework Example 10-5 using a singularity function.

### Solution

In applying Eq. 10-14c using a singularity function, the whole continuous span for this beam is considered. The unknown reaction $R_B$ is treated as a concentrated upward force. Here, besides the four boundary conditions, it should be noted that the deflection at $B$ is zero. This is a general approach as symmetry in the problem is not utilized.

$$EI \frac{d^4v}{dx^4} = q(x) = -w_o + R_B\langle x - L \rangle_*^{-1}$$

$$EI \frac{d^3v}{dx^3} = -w_ox + R_B\langle x - L \rangle^0 + C_1$$

$$EI \frac{d^2v}{dx^2} = -\frac{w_ox^2}{2} + R_B\langle x - L \rangle^1 + C_1x + C_2$$

$$EI \frac{dv}{dx} = -\frac{w_ox^3}{6} + R_B\frac{\langle x - L \rangle^2}{2} + C_1\frac{x^2}{2} + C_2x + C_3$$

$$EIv = -\frac{w_ox^4}{24} + R_B\frac{\langle x - L \rangle^3}{6} + C_1\frac{x^3}{6} + C_2\frac{x^2}{2} + C_3x + C_4$$

Static and kinematic conditions at $A$, $B$, and $C$ provide information for determining the constants of integration.

$$EIv''(0) = 0: \qquad\qquad\qquad\qquad\qquad\qquad C_2 = 0$$
$$EIv(0) = 0: \qquad\qquad\qquad\qquad\qquad\qquad C_4 = 0$$

$$EIv(L) = 0: \qquad -\frac{w_oL^4}{24} + C_1\frac{L^3}{6} + C_3L = 0$$

$$EIv(2L) = 0: \qquad -\frac{2w_oL^4}{3} + R_B\frac{L^3}{6} + C_1\frac{4L^3}{3} + 2C_3L = 0$$

$$EIv''(2L) = 0: \qquad -2w_oL^2 + R_BL + 2C_1L = 0$$

Solving the last three equations simultaneously,

$$C_1 = \frac{3}{8}w_oL \qquad C_3 = -\frac{w_oL^3}{48} \qquad \text{and} \qquad R_B = \frac{5}{4}w_oL$$

Substituting these constants into the equation for beam deflection,

$$v = -\frac{w_o}{48EI}(2x^4 - 3Lx^3 + L^3x - 10L\langle x - L\rangle^3)$$

The first three terms in the parentheses agree with those found in Example 10-5. The last term in this equation applies only for $x > L$ when it becomes $10L(x - L)^3$.

## 10-9. Deflections by Superposition

The integration procedures discussed before for obtaining the elastic deflections of loaded beams are generally applicable. The reader must realize, however, that numerous problems with different loadings have been solved and are readily available.[13] Nearly all the tabulated solutions are made for simple loading conditions. Therefore, in practice, the deflections of beams subjected to several or complicated loading conditions are usually synthesized from the simpler loadings, using the *principle of superposition*. For example, the problem in Fig. 10-15 can be separated into three different cases as shown. The algebraic sum of the three separate solutions gives the total deflection.

[13] See any civil or mechanical engineering handbook.

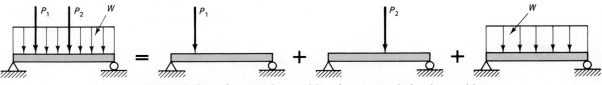

**Fig. 10-15** Resolution of a complex problem into several simpler problems.

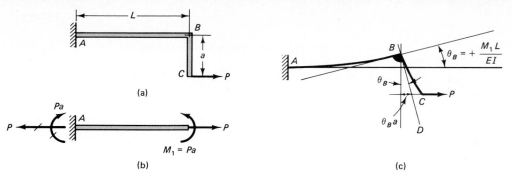

**Fig. 10-16** A method of analyzing deflections of frames.

The superposition procedure for determining elastic deflection of beams can be extended to structural systems consisting of several flexural members. For example, consider the simple frame shown in Fig. 10-16(a), for which the deflection of point $C$ due to applied force $P$ is sought. The deflection of vertical leg $BC$ alone can be found by treating it as a cantilever fixed at $B$. However, due to the applied load, joint $B$ deflects and rotates. This is determined by studying the behavior of member $AB$.

A free-body diagram for member $AB$ is shown in Fig. 10-16(b). This member is seen to resist axial force $P$ and a moment $M_1 = Pa$. Usually, the effect of axial force $P$ on deflections due to bending can be neglected.[14] The axial elongation of a member usually is also very small in comparison with the bending deflections. Therefore, the problem here can be reduced to that of determining the deflection and rotation of $B$ caused by an end moment $M_1$. This solution was obtained in Example 10-2, giving the angle $\theta_B$ shown in Fig. 10-16(c). By multiplying angle $\theta_B$ by length $a$ of the vertical member, the deflection of point $C$ due to rotation of joint $B$ is determined. Then the cantilever deflection of member $BC$ treated alone is increased by $\theta_B a$. The vertical deflection of $C$ is equal to the vertical deflection of point $B$.

In interpreting the shape of deformed structures, such as shown in Fig. 10-16(c), it must be kept clearly in mind that the deformations are greatly exaggerated. In the small deformation theory discussed here, the cosines of all small angles such as $\theta_B$ are taken to equal unity. Both the deflections and the rotations of the elastic curve are small.

Beams with overhangs can also be analyzed conveniently using the concept of superposition in the manner just described. For example, the portion of a beam between the supports, as $AB$ in Fig. 10-17(a), is isolated[15] and rotation of the tangent at $B$ is found. The remainder of the problem is analogous to the case discussed before.

Approximations similar to those just discussed are also made in composite structures. In Fig. 10-18(a), for example, a simple beam rests on a rigid support at one end and on a yielding support with a spring constant $k$ at the other end. If $R_B$ is the reaction at $B$, support $B$ settles $\Delta = R_B/k$, Fig. 10-18(b). A rigid beam would assume the alignment of line $AB'$ making an angle $\theta_1 = \tan^{-1}(\Delta/L) \approx \Delta/L$ radians with the horizontal line.

---

[14] See Section 11-9 on beam-columns.
[15] The effect of the overhang on beam segment $AB$ must be included by introducing bending moment $-Pa$ at support $B$.

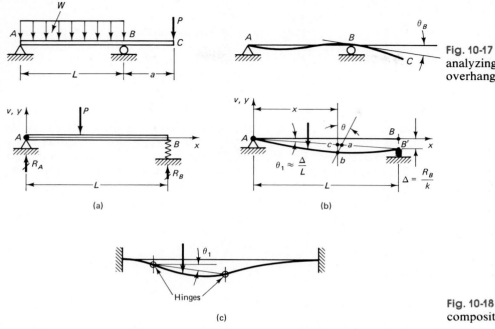

**Fig. 10-17** A method of analyzing deflections of an overhang.

(a)

(b)

(c)

**Fig. 10-18** Deflections in a composite structure.

For an elastic beam, the elastic curve between $A$ and $B'$ may be found in the usual manner. However, since the ordinates, such as $ab$, Fig. 10-18(b), make a very small angle $\theta$ with the vertical, $ab \approx cb$. Hence, the deflection of a point such as $b$ is very nearly $\theta_1 x + cb$. Deflections of beams in situations where hinges are introduced, Fig. 10-18(c), are treated similarly. For these, the tangent to the adjoining elastic curves is *not continuous* across a hinge.

The method of superposition can be effectively used for determining deflections or reactions for statically indeterminate beams. As an illustration, consider the continuous beam analyzed in Example 10-5. By removing support $R_B$, the beam would deflect at the middle, as shown in Fig. 10-19(a). By applying force $R_B$ in an upward direction, the required condition of no displacement at $B$ can be restored. The respective expressions for these deflections are given by Eqs. 10-22 and 10-30. By equating them, $R_B$ is found to be $5w_oL/4$, agreeing with the previous result (see Example 10-5).

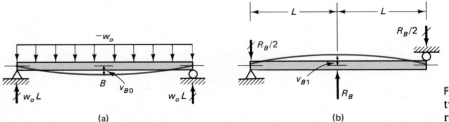

(a)

(b)

**Fig. 10-19** Superposition of two solutions for determining reactions.

### EXAMPLE 10-10

Two cantilever beams $AD$ and $BF$ of equal flexural rigidity $EI = 24 \times 10^{12}$ N·mm², shown in Fig. 10-20(a), are interconnected by a taut steel rod $DC$ ($E = 200$ GPa). Rod $DC$ is 5000 mm long and has a cross section of 300 mm². Find the deflection of cantilever $AD$ at $D$ due to a force $P = 50$ kN applied at $F$.

### Solution

By separating the structure at $D$, the two free-body diagrams in Figs. 10-20(b) and (c) are obtained. In both diagrams, the same unknown force $X$ is shown acting (a condition of statics). The deflection of point $D$ is the same, whether beam $AD$ at $D$ or the top of rod $DC$ is considered. Deflection $\Delta_1$ of point $D$ in Fig. 10-20(b) is caused by $X$. Deflection $\Delta_2$ of point $D$ on the rod is equal to the deflection $v_c$ of beam $BF$ caused by forces $P$ *and* $X$ less the elastic stretch of rod $DC$.

From statics:

$$X_{\text{pull on } AD} = X_{\text{pull on } DC} = X$$

From geometry:

$$\Delta_1 = \Delta_2 \qquad \text{or} \qquad |v_D| = |v_C| - \Delta_{\text{rod}}$$

Beam deflections can be found using the methods discussed earlier in this chapter. Alternatively, from Table 11 of the Appendix, in terms of the notation of this problem, one has

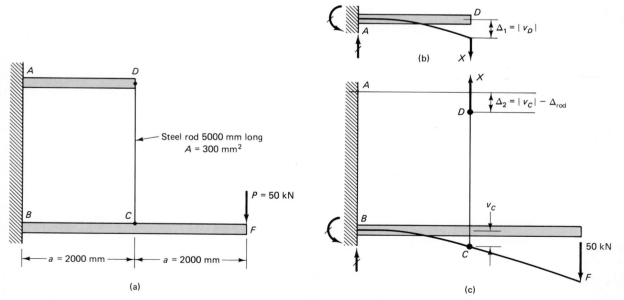

Fig. 10-20

$$v_D = -\frac{Xa^3}{3EI} = -\frac{X \times 2^3 \times 10^9}{3 \times 24 \times 10^{12}} = -1.11 \times 10^{-4}X \text{ mm}$$

$$v_{C \text{ due to } X} = +1.11 \times 10^{-4}X \text{ mm}$$

$$v_{C \text{ due to } P} = -\frac{P}{6EI}[2(2a)^3 - 3(2a)^2a + a^3] = -13.9 \text{ mm}$$

and using Eq. 2-9,

$$\Delta_{\text{rod}} = \frac{XL_{CD}}{A_{CD}E} = \frac{X \times 5000}{300 \times 200 \times 10^3} = 0.833 \times 10^{-4}X \text{ mm}$$

Then, equating deflections and treating the downward deflections as negative,

$$-1.11 \times 10^{-4}X = -13.9 + 1.11 \times 10^{-4}X + 0.833 \times 10^{-4}X$$

Hence, $\qquad\qquad\qquad X = 45.5 \times 10^3 \text{ N}$

and $\qquad v_D = -1.11 \times 10^{-4} \times 45.5 \times 10^3 = -5.05 \text{ mm}$

Note particularly that in these calculations, the deflection of point $C$ is determined by superposing the effects of applied force $P$ at the end of the cantilever *and* the unknown force $X$ at $C$.

---

## *10.10. Deflections in Unsymmetrical Bending

In the preceding discussion, it was assumed that deflections were caused by a beam bending around one of the principal axes. However if unsymmetrical bending takes place, deflections are calculated in each of the principal planes and the deflections so found are *added vectorially*. An example is shown in Fig. 10-21 for a $Z$ section. Here the $y$ and $z$ axes are the principal axes passing through the centroid as well as the shear center of the cross section. A positive deflection $v_1$ is shown for the beam deflection taking place in the $xy$ plane, and, similarly, $w_1$ corresponds to the deflection in the $xz$ plane. Their vectorial sum, $AA'$, is the total beam deflection.

In order to prevent torsion, the applied forces must act through the shear center for the cross section. If not, torsional stresses and deformations, treated in Chapter 4, must also be considered.

Beams having significantly different magnitudes of moments of inertia about the two principal axes of a cross section are very sensitive to load alignment. As is shown in the next example, even a small inclination of the applied force from the vertical causes large lateral displacements (and high stresses).

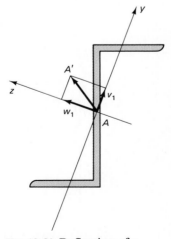

Fig. 10-21 Deflection of a beam subjected to unsymmetrical bending.

## *EXAMPLE 10-11

A C 15 × 33.9 steel channel cantilever 100 in long is subjected to an inclined force $P$ of 2 kips through the shear center, as shown in Figs. 10-22(a) and (b). Determine the tip deflection at the applied force. Let $E = 29 \times 10^6$ psi.

### Solution

The properties for this channel are given in Table 5 in the Appendix: $I_z = 315$ in$^4$ and $I_y = 8.13$ in$^4$. Maximum deflection of a cantilever bent around either principal axis is given in Table 11 of the Appendix: $v_{max} = PL^3/3EI$. Hence, identifying by subscripts $H$ the horizontal, and by $V$ the vertical components of tip deflection $\Delta$ and applied force $P$, one has

$$\Delta_H = \frac{P_H L^3}{3EI_y} = \frac{(2000 \sin 5°) \times 100^3}{3 \times 29 \times 10^6 \times 8.13} = 0.246 \text{ in}$$

$$\Delta_V = \frac{P_V L^3}{3EI_z} = \frac{(2000 \cos 5°) \times 100^3}{3 \times 29 \times 10^6 \times 315} = 0.0727 \text{ in}$$

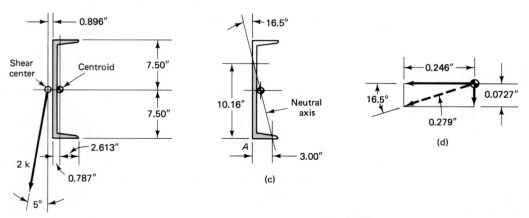

(a)

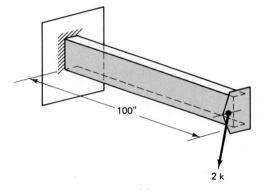

(b)

(c)

(d)

**Fig. 10-22**

These deflections and their vector sum of 0.279 in, making an angle of 16.5° with the horizontal, are shown in Fig. 10-22(d).

It is instructive to note that, as to be expected, the *maximum deflection occurs in the direction normal to the neutral axis*. This axis may be located by performing a stress analysis and finding the points of zero stress. One such point is 10.16 in above $A$ and the other is 3.00 in to the right of $A$, as shown in Fig. 10-22(c). Alternatively, the neutral axis can be located using Eq. 6-43. Using this approach,

$$\tan \beta = \frac{I_z}{I_y} \tan \alpha = \frac{315}{8.13} \tan 5° = 3.39 \quad \text{and} \quad \beta = 73.6°$$

Hence, $90° - \beta = 16.4°$ compares well with the angle shown in Fig. 10-22(d). (The small discrepancy can be attributed to roundoff error.)

## *10-11. Energy Method for Deflections and Impact

A comprehensive treatment of the energy method for finding beam deflections is given in Chapter 12. Without establishing the necessary theorem, it is possible to solve only a very limited class of problems. Unless special conditions such as symmetry are at hand, direct solutions based on the principle of conservation of energy must be limited to the action of a single force or moment. This limited approach has been found useful in the axial force, torsion, and pure bending problems in Chapters 2, 4, and 6. In beams, one can go a step further and include, if needed, both the bending and the shear strain energies. The procedure based on equating the internal strain energy $U$ to the external work $W_e$ remains the same.

This method permits an assessment of deflections caused by bending in relation to that caused by shear. The following example is concerned with such a problem, where in the solution, it is assumed that the force is *gradually applied*. By contrast, in the second example, an impact on a beam caused by a falling mass is considered; in this example, the effect of shear deformation is neglected.

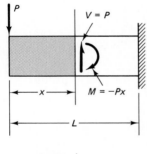
(a)

### *EXAMPLE 10-12

Find the maximum deflection due to force $P$ applied at the end of a cantilever having a rectangular cross section, Fig. 10-23. Consider the effect of the flexural and shear deformations.

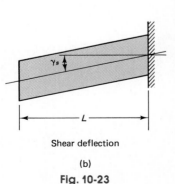
Shear deflection

(b)

**Fig. 10-23**

#### Solution

If force $P$ is gradually applied to the beam, the external work $W_e = \frac{1}{2}P\Delta$, where $\Delta$ is the total deflection of the end of the beam. The internal strain energy consists of two parts. One part is due to the bending stresses; the other is caused by the shear stresses. These strain energies may be directly superposed.

The strain energy in pure bending is obtained from Eq. 6-24, $U_{bending} = \int M^2 \, dx/2EI$, by noting that $M = -Px$. The strain energy in shear is found from Eq. 3-5, $U_{shear} = \int (\tau^2/2G) \, dV$. In this particular case, the shear at every section is equal to applied force $P$, and the shear stress $\tau$, according to Eq. 7-7, is distributed parabolically as

$$\tau = \frac{P}{2I}\left[\left(\frac{h}{2}\right)^2 - y^2\right]$$

At any one level $y$, this shear stress does not vary across breadth $b$ and length $L$ of the beam. Therefore, the infinitesimal volume $dV$ in the shear energy expression is taken as $Lb \, dy$. By equating the sum of these two internal strain energies to the external work, the total deflection is obtained:

$$U_{bending} = \int_0^L \frac{M^2 \, dx}{2EI} = \int_0^L \frac{(-Px)^2 \, dx}{2EI} = \frac{P^2L^3}{6EI}$$

$$U_{shear} = \int_{vol} \frac{\tau^2}{2G} \, dV = \frac{1}{2G} \int_{-h/2}^{+h/2} \left\{\frac{P}{2I}\left[\left(\frac{h}{2}\right)^2 - y^2\right]\right\}^2 Lb \, dy$$

$$= \frac{P^2Lb}{8GI^2}\frac{h^5}{30} = \frac{P^2Lbh^5}{240G}\left(\frac{12}{bh^3}\right)^2 = \frac{3P^2L}{5AG} \qquad (10\text{-}34)$$

where $A = bh$ is the cross section of the beam. Then

$$W_e = U = U_{bending} + U_{shear}$$

$$\frac{P\Delta}{2} = \frac{P^2L^3}{6EI} + \frac{3P^2L}{5AG} \qquad \text{or} \qquad \Delta = \frac{PL^3}{3EI} + \frac{6PL}{5AG}$$

The first term in this answer, $PL^3/3EI$, is the deflection of the beam due to flexure. The second term is the deflection due to shear, assuming no warping restraint at the built-in end. The factor $\alpha = 6/5$ varies for different shapes of the cross section, since it depends on the nature of the shear-stress distribution.

It is instructive to recast the expression for the total deflection $\Delta$ as

$$\boxed{\Delta = \frac{PL^3}{3EI}\left(1 + \frac{3E}{10G}\frac{h^2}{L^2}\right)} \qquad (10\text{-}35)$$

where, as before, the last term gives the deflection due to shear.

To gain further insight into this problem, in the last expression, replace the ratio $E/G$ by 2.5, a typical value for steels. Then

$$\Delta = (1 + 0.75h^2/L^2)\Delta_{bending} \qquad (10\text{-}35a)$$

From this equation, it can be seen that for a short beam (for example, one with $L = h$), the total deflection is 1.75 times that due to bending alone. Hence, shear deflection is very important in comparable cases. On the other hand, if $L = 10h$, the deflection due to shear is less than 1 percent. Small deflections due to shear are typical for ordinary slender beams. This fact can be noted further from the original equation for $\Delta$. There, the deflection due to bending increases as the cube of the span length, whereas the deflection due to shear increases directly. Hence, as beam length increases, the bending deflection quickly becomes dominant. For this reason, it is usually possible to neglect the deflection due to shear.

## *EXAMPLE 10-13

Find the instantaneous maximum deflections and bending stresses for the 50 × 50 mm steel beam shown in Fig. 10-24 when struck by a 15.3-kg mass falling from a height 75 mm above the top of the beam, if (a) the beam is on rigid supports, and (b) the beam is supported at each end on springs. Constant $k$ for each spring is 300 N/mm. Let $E = 200$ GPa.

### Solution

The deflection of the system due to a statically applied force of $15.3\ g = 15.3 \times 9.81 = 150$ N is computed first. In the first case, this deflection is that of the beam only; see Table 11 of the Appendix. In the second case, the static deflection of the beam is augmented by the deflection of the springs subjected to a 75-N force each. The impact factors are then computed from Eq. 2-27 or 2-28. Static deflections and stresses are multiplied by the impact factors to obtain the answers.

(a)
$$\Delta_{st} = \frac{PL^3}{48EI} = \frac{150 \times 1000^3}{48 \times 200 \times 10^3 \times 50^4/12} = 0.030 \text{ mm}$$

$$\text{impact factor} = 1 + \sqrt{1 + \frac{2h}{\Delta_{st}}}$$

$$= 1 + \sqrt{1 + \frac{2 \times 75}{0.030}} = 71.7$$

(b)
$$\Delta_{st} = \Delta_{beam} + \Delta_{spr} = 0.030 + \frac{75}{300} = 0.280 \text{ mm}$$

$$\text{impact factor} = 1 + \sqrt{1 + \frac{2 \times 75}{0.280}} = 24.2$$

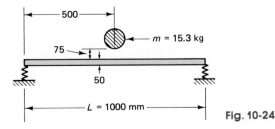

Fig. 10-24

For either case, the maximum bending stress in the beam due to a static application of $P$ is

$$(\sigma_{\max})_{\text{st}} = \frac{M}{S} = \frac{PL}{4S} = \frac{150 \times 1000}{4 \times 50^3/6} = 1.800 \text{ MPa}$$

Multiplying the static deflections and stress by the respective impact factors gives the required results.

| | Static | | Dynamic | |
|---|---|---|---|---|
| | With Springs | No Springs | With Springs | No Springs |
| $\Delta_{\max}$, mm | 0.280 | 0.030 | 6.78 | 2.15 |
| $\sigma_{\max}$, MPa | 1.80 | 1.80 | 43.6 | 129 |

It is apparent from this table that large deflections and stresses are caused by a dynamically applied load. The stress for the condition with no springs is particularly large; however, owing to the flexibility of the beam, it is not excessive.

## *[16]10-12. Inelastic Deflection of Beams

All the preceding solutions for beam deflections apply only if the material behaves elastically. This limitation is the result of introducing Hooke's law into the strain-curvature relation, Eq. 10-5, to yield the moment-curvature equation, Eq. 10-6. The subsequent procedures for approximating the curvature as $d^2v/dx^2$ and the integration schemes do not depend on the material properties.

Superposition does not apply to inelastic problems, since deflections are not linearly related to the applied forces. As a consequence, in some cases piecewise linear solutions for small load or displacement increments are made until the desired level of load or displacement is reached. Such stepwise linear calculations are made with the aid of a computer. Alternatively, time-consuming trial-and-error solutions are used to calculate deflections in indeterminate beams. However, it is possible to develop simple solutions for *ultimate strengths* of statically determinate and indeterminate beams and frames *assuming ideal plastic behavior of material*. For such a method, a relationship between the bending moment and curvature at a section of a beam must be developed. An illustration defining such a relationship is given in the next example. Essentially, it is this approach that is relied upon in Chapter 13 for plastic limit state analyses of statically determinate and indeterminate beams and simple frames.

The second example that follows discusses the deflection analysis of a

[16] This section is optional.

statically determinate elastic-plastic beam. The solution demonstrates that as long as at least a part of a beam's cross section remains elastic, the deflections remain bounded, i.e., finite, and can be calculated.

## EXAMPLE 10-14

Determine and plot the moment-curvature relationship for an elastic-ideally plastic rectangular beam.

### Solution

In a rectangular elastic-plastic beam at $y_o$, where the juncture of the elastic and plastic zones occurs, the linear strain $\varepsilon_x = \pm \varepsilon_{yp}$; see Fig. 6-30. Therefore, according to Eq. 10-5, with the curvature $1/\rho = \kappa$,

$$\frac{1}{\rho} = \kappa = -\frac{\varepsilon_{yp}}{y_o} \quad \text{and} \quad \kappa_{yp} = -\frac{\varepsilon_{yp}}{h/2}$$

where the last expression gives the curvature of the member at impending yielding when $y_o = h/2$. From these relations,

$$\frac{y_o}{h/2} = \frac{\kappa_{yp}}{\kappa}$$

By substituting this expression into Eq. 6-40, one obtains the required moment-curvature relationship:

$$M = M_p \left[ 1 - \frac{1}{3} \left( \frac{y_o}{h/2} \right)^2 \right] = \frac{3}{2} M_{yp} \left[ 1 - \frac{1}{3} \left( \frac{\kappa_{yp}}{\kappa} \right)^2 \right] \quad (10\text{-}36)$$

This function is plotted in Fig. 10-25. Note how rapidly it approaches the asymptote. At curvature just double that of the impending yielding, eleven-twelfths, or

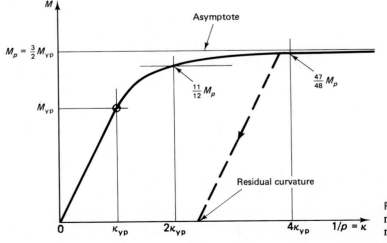

Fig. 10-25 Moment-curvature relation for an elastic-plastic rectangular beam.

91.6 percent, of the ultimate plastic moment $M_p$ is already reached. At this point, the middle half of the beam remains elastic.

On releasing an applied moment, the beam rebounds elastically, as shown in the figure. On this basis, residual curvature can be determined.

The reader should recall that the ratio of $M_p$ to $M_{\text{yp}}$ varies for different cross sections. For example, for a typical steel wide-flange beam, $M_p/M_{\text{yp}}$ is about 1.14. Establishing the asymptotes for plastic moments gives a practical basis for finding the ultimate plastic limit state for beams and frames discussed in Chapter 13.

### EXAMPLE 10-15

A 3-in wide mild-steel cantilever beam has the dimensions shown in Fig. 10-26(a). Determine the tip deflection caused by applying two loads of 5 kips each. Assume $E = 30 \times 10^3$ ksi and $\sigma_{\text{yp}} = \pm 40$ ksi.

### Solution

The moment diagram is shown in Fig. 10-26(b). From $\sigma_{\max} = Mc/I$, it is found that the largest stress in beam segment $ab$ is 24.4 ksi, which indicates elastic behavior. An analogous calculation for the shallow section of the beam gives a stress of 55 ksi, which is not possible as the material yields at 40 ksi.

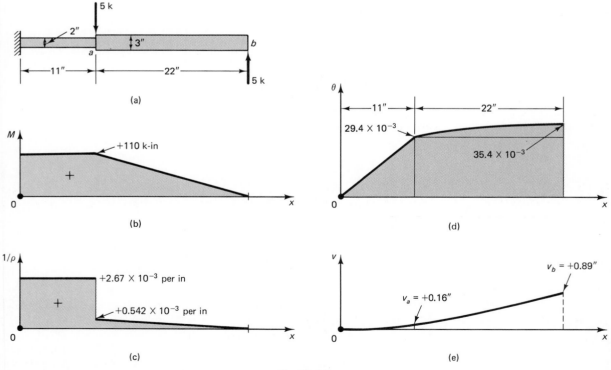

**Fig. 10-26**

A check of the ultimate capacity for the 2-in deep section based on Eq. 6-38 gives

$$M_p = M_{ult} = \sigma_{yp} \frac{bh^2}{4} = \frac{40 \times 3 \times 2^2}{4} = 120 \text{ k-in}$$

This calculation shows that although the beam yields partially, it can carry the applied moment. The applied moment is $\frac{11}{12}M_p$. According to the results found in the preceding example, this means that the curvature in the 2-in deep section of the beam is twice that at the beginning of yielding. Therefore, the curvature in the 11-in segment of the beam adjoining the support is

$$\frac{1}{\rho} = 2\kappa_{yp} = 2\frac{\varepsilon_{yp}}{h/2} = 2\frac{\sigma_{yp}}{Eh/2} = \frac{2 \times 40}{30 \times 10^3 \times 1} = 2.67 \times 10^{-3} \text{ per in}$$

The maximum curvature for segment $ab$ is

$$\frac{1}{\rho} = \frac{M_{max}}{EI} = \frac{\sigma_{max}}{Ec} = \frac{24.4}{3 \times 10^3 \times 1.5} = 0.542 \times 10^{-3} \text{ per in}$$

These data on curvatures are plotted in Fig. 10-26(c). On integrating this twice with $\theta(0) = 0$ and $v(0) = 0$, the deflected curve, Fig. 10-26(e), is obtained. The tip deflection is 0.89 in upward.

If the applied loads were released, the beam would rebound elastically. As can be verified by elastic analysis, this would cause a tip deflection of 0.64 in. Hence a residual tip deflection of 0.25 in would remain. The residual curvature would be confined to the 2-in deep segment of beam.

If the end load were applied alone, the 165 k-in moment at the left end would exceed the plastic moment capacity of 120 k-in and the beam would collapse. Superposition cannot be used to solve this problem.

---

## **\*\*17Part B**     **DEFLECTIONS BY THE MOMENT-AREA METHOD**

## **\*\*10-13. Introduction to the Moment-Area Method**

In numerous engineering applications where deflections of beams must be determined, the loading is complex, and the cross-sectional areas of the beam vary. This is the usual situation in machine shafts, where gradual or stepwise variations in the shaft diameter are made to accommodate rotors, bearings, collars, retainers, etc. Likewise, haunched or tapered

[17] Part B is optional.

beams are frequently employed in aircraft as well as in bridge construction. By interpreting semigraphically the mathematical operations of solving the governing differential equation, an effective procedure for obtaining deflections in complicated situations has been developed. Using this alternative procedure, one finds that problems with load discontinuities and arbitrary variations of inertia of the cross-sectional area of a beam cause no complications and require only a little more arithmetical work for this solution. The solution of such problems is the objective in the following sections on the moment-area method.[18]

The method to be developed is generally used to obtain only the displacement and rotation at a single point on a beam. It may be used to determine the equation of the elastic curve, but no advantage is gained in comparison with the direct solution of the differential equation. Often, however, it is the deflection and/or the angular rotation of the elastic curve, or both, at a particular point of a beam that are of greatest interest in the solution of practical problems.

The method of moment areas is just an alternative method for solving the deflection problem. It possesses the same approximations and limitations discussed earlier in connection with the solution of the differential equation of the elastic curve. By applying it, one determines only the deflection due to the flexure of the beam; deflection due to shear is neglected. Application of the method will be developed for statically determinate and indeterminate beams.

## **10-14. Moment-Area Theorems

The necessary theorems are based on the geometry of the elastic curve and the associated $M/EI$ diagram. Boundary conditions do not enter into the derivation of the theorems since the theorems are based only on the interpretation of definite integrals. As will be shown later, further geometrical considerations are necessary to solve a complete problem.

For deriving the theorems, Eq. 10-10, $d^2v/dx^2 = M/EI$, can be rewritten in the following alternative forms:

$$\frac{d^2v}{dx^2} = \frac{d}{dx}\left(\frac{dv}{dx}\right) = \frac{d\theta}{dx} = \frac{M}{EI} \quad \text{or} \quad d\theta = \frac{M}{EI}\,dx \quad (10\text{-}37)$$

From Fig. 10-27(a), quantity $(M/EI)\,dx$ corresponds to an infinitesimal area of the $M/EI$ diagram. According to Eq. 10-37, this area is equal to the change in angle between two adjoining tangents. The contribution of an angle change in one element to the deformation of the elastic curve is shown in Fig. 10-27(b).

[18] The development of the moment-area method for finding deflections of beams is due to Charles E. Greene, of the University of Michigan, who taught it to his classes in 1873. Somewhat earlier, in 1868, Otto Mohr, of Dresden, Germany, developed a similar method that appears to have been unknown to Professor Greene.

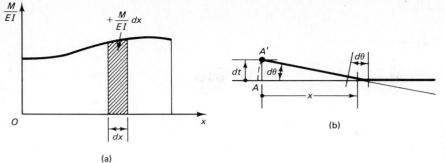

Fig. 10-27 Interpretation of a small angle change in an element.

If the small angle change $d\theta$ for an element is multiplied by a distance $x$ from an arbitrary origin to the same element, a vertical distance $dt$ is obtained; see Fig. 10-27(b). As only small deflections are considered, no distinction between arc $AA'$ and vertical distance $dt$ need be made. Based on this geometrical reasoning, one has

$$dt = x\, d\theta = \frac{M}{EI}x\, dx \qquad (10\text{-}38)$$

Formally integrating Eqs. 10-37 and 10-38 between any two points such as $A$ and $B$ on a beam (see Fig. 10-28), yields the two moment-area theorems. The **first moment-area theroem** is

$$\int_A^B d\theta = \theta_B - \theta_A = \Delta\theta_{B/A} = \int_A^B \frac{M}{EI}\, dx \qquad (10\text{-}39)$$

where $\Delta\theta_{B/A}$ is the *angle change between B and A*. This change in angle measured in radians between any two tangents at points $A$ and $B$ on the

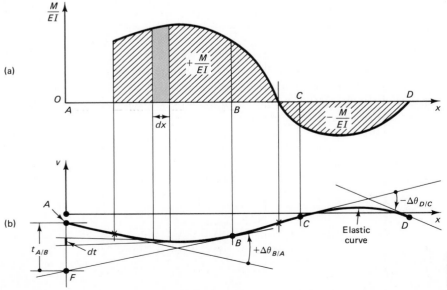

Fig. 10-28 Relationship between the $M/EI$ diagram and the elastic curve.

**539**

elastic curve is equal to the $M/EI$ area bounded by the ordinates through $A$ and $B$. Further, if slope $\theta_A$ of the elastic curve at $A$ is known, slope $\theta_B$ at $B$ is given as

$$\theta_B = \theta_A + \Delta\theta_{B/A} \qquad (10\text{-}40)$$

The first theorem shows that a numerical evaluation of the $M/EI$ area bounded between the ordinates through any two points on the elastic curve gives the angular rotation between the corresponding tangents. In performing this summation, areas corresponding to the positive bending moments are taken positive and those corresponding to the negative moments are taken negative. If the sum of the areas between any two points such as $A$ and $B$ is positive, the tangent on the right rotates in the counterclockwise direction; if negative, the tangent on the right rotates in a clockwise direction; see Fig. 10-28(b). If the net area is zero, the two tangents are parallel.

The quantity $dt$ in Fig. 10-28(b) is due to the effect of curvature of an element. By summing this effect for all elements from $A$ to $B$, vertical distance $AF$ is obtained. Geometrically, this distance represents the displacement or deviation of a point $A$ from a tangent to the elastic curve at $B$. Henceforth, it will be termed the *tangential deviation* of a point $A$ from a tangent at $B$ and will be designated $t_{A/B}$. The foregoing, in mathematical form, gives the **second moment-area theorem**:

$$t_{A/B} = \int_A^B d\theta \; x = \int_A^B \frac{M}{EI} x \; dx \qquad (10\text{-}41)$$

This states that the tangential deviation of a point $A$ on the elastic curve from a tangent through another point $B$ also on the elastic curve is equal to the statical (or first) moment of the bounded section of the $M/EI$ diagram around a vertical line through $A$. In most cases, the tangential deviation is not in itself the desired deflection of a beam.

Using the definition of the center of gravity of an area, one may for convenience restate Eq. 10-41 for numerical applications in a simpler form as

$$t_{A/B} = \Phi\bar{x} \qquad (10\text{-}42)$$

where $\Phi$ is the total area of the $M/EI$ diagram between the two points considered and $\bar{x}$ is the horizontal distance to the centroid of this area *from $A$.*

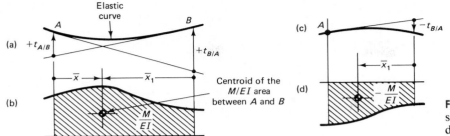

Fig. 10-29 Interpretation of signs for tangential deviation.

By analogous reasoning, the *deviation of a point B from a tangent at A* is

$$t_{B/A} = \Phi \bar{x}_1 \qquad (10\text{-}43)$$

where the same *M/EI* area is used, but $\bar{x}_1$ is measured from the vertical line through point $B$; see Fig. 10-29. Note carefully the order of the subscript letters for $t$ in these two equations. The point whose deviation is being determined is written first.

In the previous equations, distances $\bar{x}$ and $\bar{x}_1$ are always taken positive, and as $E$ and $I$ intrinsically are also positive quantities, therefore the sign of the tangential deviation depends on the sign of the bending moments. A positive value for the tangential deviation indicates that a given point lies above a tangent to the elastic curve drawn through the other point, and vice versa; see Fig. 10-29.

The previous two theorems are applicable between any two points on a *continuous* elastic curve of any beam for any loading. They apply between and beyond the reactions for overhanging and continuous beams. However, it must be emphasized that only relative rotation of the tangents and only tangential deviations are obtained directly. A further consideration of the geometry of the elastic curve at the supports to include the boundary conditions is necessary in every case to determine deflections. This will be illustrated in the examples that follow.

In applying the moment-area method, a carefully prepared sketch of the elastic curve is essential. Since no deflection is possible at a pinned or a roller support, the elastic curve is drawn passing through such supports. At a fixed support, neither displacement nor rotation of the tangent to the elastic curve is permitted, so the elastic curve must be drawn tangent to the direction of the unloaded axis of the beam. In preparing a sketch of the elastic curve in the above manner, it is customary to exaggerate the anticipated deflections. On such a sketch the deflection of a point on a beam is usually referred to as being above or below its initial position, without emphasis on the signs. To aid in the application of the method, useful properties of areas enclosed by curves and centroids are assembled in Table 2 of the Appendix.

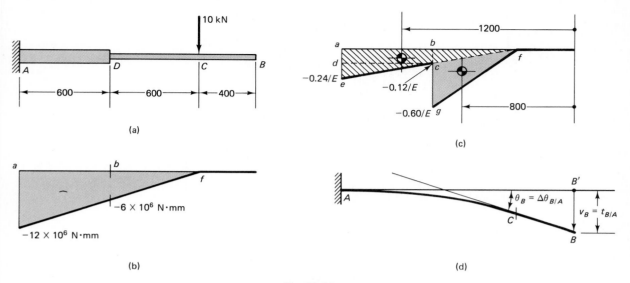

Fig. 10-30

### EXAMPLE 10-16

Consider an aluminum cantilever beam 1600 mm long with a 10-kN force applied 400 mm from the free end, as shown in Fig. 10-30(a). For a distance of 600 mm from the fixed end, the beam is of greater depth than it is beyond, having $I_1 = 50 \times 10^6$ mm$^4$. For the remaining 1000 mm of the beam, $I_2 = 10 \times 10^6$ mm$^4$. Find the deflection and the angular rotation of the free end. Neglect the weight of the beam, and assume $E$ for aluminum at 70 GPa.

### Solution

The bending-moment diagram is in Fig. 10-30(b). By dividing all ordinates of the $M$ diagram by $EI$, the $M/EI$ diagram in Fig. 10-30(c) is obtained. Two ordinates appear at point $D$. One, $-0.12/E$, is applicable just to the left of $D$; the other, $-0.60/E$, applies just to the right of $D$. Since the bending moment is negative from $A$ to $C$, the elastic curve throughout this distance is concave down; see Fig. 10-30(d). At fixed support $A$, the elastic curve must start out tangent to the initial direction $AB'$ of the unloaded beam. The unloaded straight segment $CB$ of the beam is tangent to the elastic curve at $C$.

After the foregoing preparatory steps, from the geometry of the sketch of the elastic curve, it may be seen that distance $BB'$ represents the desired deflection of the free end. However, $BB'$ is *also* the tangential deviation of point $B$ from the tangent at $A$. Therefore, the second moment-area theorem may be used to obtain $t_{B/A}$, which in this special case represents the deflection of the free end. Also, from the geometry of the elastic curve, it is seen that the angle included between lines $BC$ and $AB'$ is the angular rotation of segment $CB$. This angle is the same as the one included between the tangents to the elastic curve at points $A$ and $B$, and the first moment-area theorem may be used to compute this quantity.

It is convenient to extend line $ec$ in Fig. 10-30(c) to point $f$ for computing the area of the $M/EI$ diagram. This gives two triangles, the areas of which are easily calculated.

The area of triangle $afe$: $\quad \Phi_1 = -\dfrac{1200 \times 0.24}{2E} = -\dfrac{144}{E}$

The area of triangle $feg$: $\quad \Phi_2 = -\dfrac{600 \times 0.48}{2E} = -\dfrac{144}{E}$

$$\theta_B = \Delta\theta_{B/A} = \int_A^B \frac{M}{EI}\,dx = \Phi_1 + \Phi_2 = -\frac{288}{70 \times 10^3} = -4.11 \times 10^{-3}\ \text{rad}$$

$$v_B = t_{B/A} = \Phi_1\bar{x}_1 + \Phi_2\bar{x}_2$$

$$= \left(-\frac{144}{E}\right) \times 1200 + \left(-\frac{144}{E}\right) \times 800 = -4.11\ \text{mm}$$

The negative sign of $\Delta\theta$ indicates clockwise rotation of the tangent at $B$ in relation to the tangent at $A$. The negative sign of $t_{B/A}$ means that point $B$ is below a tangent through $A$.

## EXAMPLE 10-17

Find the deflection due to the concentrated force $P$ applied as shown in Fig. 10-31(a) at the center of a simply supported beam. The flexural rigidity $EI$ is constant.

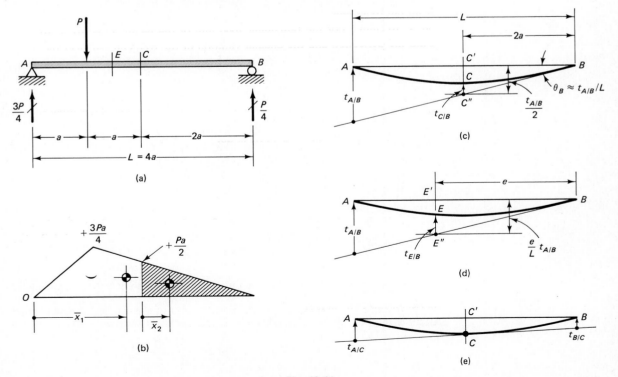

Fig. 10-31

## Solution

The bending-moment diagram is in Fig. 10-31(b). Since $EI$ is constant, the $M/EI$ diagram need not be made, as the areas of the bending-moment diagram divided by $EI$ give the necessary quantities for use in the moment-area theorems. The elastic curve is in Fig. 10-31(c). It is concave upward throughout its length as the bending moments are positive. This curve must pass through the points of the support at $A$ and $B$.

It is apparent from the diagram of the elastic curve that the desired quantity is represented by distance $CC'$. Moreover, from purely geometrical or kinematic considerations, $CC' = C'C'' - C''C$, where distance $C''C$ is measured from a tangent to the elastic curve passing through the point of support $B$. However, since the deviation of a support point from a tangent to the elastic curve at the other support may always be computed by the second moment-area theorem, a distance such as $C'C''$ may be found by proportion from the geometry of the figure. In this case, $t_{A/B}$ follows by taking the whole $M/EI$ area between $A$ and $B$ and multiplying it[19] by its $\bar{x}$ measured from a vertical through $A$; hence, $C'C'' = \frac{1}{2}t_{A/B}$. By another application of the second theorem, $t_{C/B}$, which is equal to $C''C$, is determined. For this case, the $M/EI$ area is hatched in Fig. 10-31(b), and, for it, $\bar{x}$ is measured from $C$. Since the right reaction is $P/4$ and the distance $CB = 2a$, the maximum ordinate for the shaded triangle is $+Pa/2$.

$$v_c = C'C'' - C''C = t_{A/B}/2 - t_{C/B}$$

$$t_{A/B} = \Phi_1 \bar{x}_1 = \frac{1}{EI}\left(\frac{4a}{2}\frac{3Pa}{4}\right)\frac{a + 4a}{3} = +\frac{5Pa^3}{2EI}$$

$$t_{C/B} = \Phi_2 \bar{x}_2 = \frac{1}{EI}\left(\frac{2a}{2}\frac{Pa}{2}\right)\frac{2a}{3} = +\frac{Pa^3}{3EI}$$

$$v_C = \frac{t_{A/B}}{2} - t_{C/B} = \frac{5Pa^3}{4EI} - \frac{Pa^3}{3EI} = \frac{11Pa^3}{12EI}$$

The positive signs of $t_{A/B}$ and $t_{C/B}$ indicate that points $A$ and $C$ lie above the tangent through $B$. As may be seen from Fig. 10-31(c), the deflection at the center of the beam is in a downward direction.

The slope of the elastic curve at $C$ can be found from the slope of one of the ends and from Eq. 10-40. For point $B$ on the right,

$$\theta_B = \theta_C + \Delta\theta_{B/C} \quad \text{or} \quad \theta_C = \theta_B - \Delta\theta_{B/C}$$

$$\theta_C = \frac{t_{A/B}}{L} - \Phi_2 = \frac{5Pa^2}{8EI} - \frac{Pa^2}{2EI} = \frac{Pa^2}{8EI} \quad \text{(counterclockwise)}$$

The previous procedure for finding the deflection of a point on the elastic curve

[19] See Table 2 of the Appendix for the centroid of the whole triangular area. Alternatively, by treating the whole $M/EI$ area as two triangles,

$$t_{A/B} = \frac{1}{EI}\left(\frac{a}{2}\frac{3Pa}{4}\right)\frac{2a}{3} + \frac{1}{EI}\left(\frac{3a}{2}\frac{3Pa}{4}\right)\left(a + \frac{3a}{3}\right) = +\frac{5Pa^3}{2EI}$$

is generally applicable. For example, if the deflection of point $E$, Fig. 10-31(d), at a distance $e$ from $B$ is wanted, the solution may be formulated as

$$v_E = E'E'' - E''E = (e/L)t_{A/B} - t_{E/B}$$

By locating point $E$ at a variable distance $x$ from one of the supports, the equation of the elastic curve can be obtained.

To simplify the arithmetical work, some care in selecting the tangent at a support must be exercised. Thus, although $v_C = t_{B/A}/2 - t_{C/A}$ (not shown in the diagram), this solution would involve the use of the unshaded portion of the bending-moment diagram to obtain $t_{C/A}$, which is more tedious.

## Alternative Solution

The solution of the foregoing problem may be based on a different geometrical concept. This is illustrated in Fig. 10-31(e), where a tangent to the elastic curve is drawn at $C$. Then, since distances $AC$ and $CB$ are equal,

$$v_C = CC' = (t_{A/C} + t_{B/C})/2$$

i.e., distance $CC'$ is an average of $t_{A/C}$ and $t_{B/C}$. The tangential deviation $t_{A/C}$ is obtained by taking the first moment of the unshaded $M/EI$ area in Fig. 10-31(b) about $A$, and $t_{B/C}$ is given by the first moment of the shaded $M/EI$ area about $B$. The numerical details of this solution are left for completion by the reader. This procedure is usually longer than the first.

Note particularly that if the elastic curve is not symmetrical, the tangent at the center of the beam is *not horizontal*.

## EXAMPLE 10-18

For a prismatic beam loaded as in the preceding example, find the maximum deflection caused by applied force $P$; see Fig. 10-32(a).

## Solution

The bending-moment diagram and the elastic curve are shown in Figs. 10-32(b) and (c), respectively. The elastic curve is concave up throughout its length, and the maximum deflection occurs where the tangent to the elastic curve is horizontal. This point of tangency is designated in the figure by $D$ and is located by the unknown horizontal distance $d$ measured from the right support $B$. Then, by drawing a tangent to the elastic curve through point $B$ at the support, one sees that $\Delta \theta_{B/D} = \theta_B$ since the line passing through the supports is horizontal. However, the slope $\theta_B$ of the elastic curve at $B$ may be determined by obtaining $t_{A/B}$ and dividing it by the length of the span. On the other hand, by using the first moment-area theorem, $\Delta \theta_{B/D}$ may be expressed in terms of the shaded area in Fig. 10-32(b). Equating $\Delta \theta_{B/D}$ to $\theta_B$ and solving for $d$ locates the horizontal tangent at $D$. Then, again from geometrical considerations, it is seen that the maximum deflection represented by $DD'$ is equal to the tangential deviation of $B$ from a horizontal tangent through $D$, i.e., $t_{B/D}$.

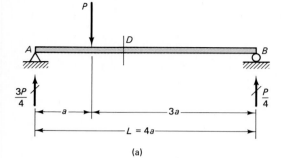

(a)

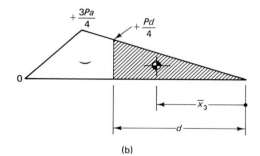

(b)

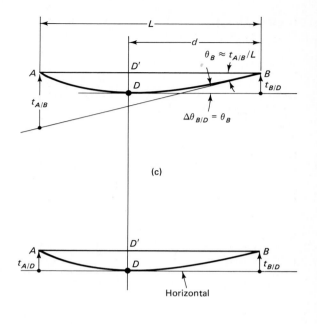

(c)

(d)

**Fig. 10-32**

$$t_{A/B} = \Phi_1 \bar{x}_1 = +\frac{5Pa^3}{2EI} \qquad \text{(see Example 10-17)}$$

$$\theta_B = \frac{t_{A/B}}{L} = \frac{t_{A/B}}{4a} = \frac{5Pa^2}{8EI}$$

$$\Delta\theta_{B/D} = \frac{1}{EI}\left(\frac{d}{2}\frac{Pd}{4}\right) = \frac{Pd^2}{8EI} \qquad \text{(area between } D \text{ and } B)$$

Since $\theta_B = \theta_D + \Delta\theta_{B/D}$ and it is required that $\theta_D = 0$,

$$\Delta\theta_{B/D} = \theta_B \qquad \frac{Pd^2}{8EI} = \frac{5Pa^2}{8EI} \qquad \text{hence, } d = \sqrt{5}a$$

$$v_{\max} = v_D = DD' = t_{B/D} = \Phi_3 \bar{x}_3$$
$$= \frac{1}{EI}\left(\frac{d}{2}\frac{Pd}{4}\right)\frac{2d}{3} = \frac{5\sqrt{5}\,Pa^3}{12EI} = \frac{11.2Pa^3}{12EI}$$

After distance $d$ is found, the maximum deflection may also be obtained as $v_{\max} = t_{A/D}$, or $v_{\max} = (d/L)t_{A/B} - t_{D/B}$ (not shown). Also note that using the condition $t_{A/D} = t_{B/D}$, Fig. 10-32(d), an equation may be set up for $d$.

It should be apparent from this solution that it is easier to calculate the deflection at the center of the beam, which was illustrated in Example 10-17, than to determine the maximum deflections. Yet, by examining the end results, one sees that, numerically, the two deflections differ little: $v_{\text{center}} = 11Pa^3/12EI$ as opposed to $v_{max} = 11.2Pa^3/12EI$. For this reason, in many practical problems of simply supported beams, where all the applied forces act in the same direction, it is often sufficiently accurate to calcuate the deflection at the center instead of attempting to obtain the true maximum.

## EXAMPLE 10-19

In a simply supported beam, find the maximum deflection and rotation of the elastic curve at the ends caused by the application of a uniformly distributed load of $w_o$ lb/ft; see Fig. 10-33(a). Flexural rigidity $EI$ is constant.

### Solution

The bending-moment diagram is in Fig. 10-33(b). As established in Example 5-8, it is a second-degree parabola with a maximum value at the vertex of $w_oL^2/8$. The elastic curve passing through the points of supports $A$ and $B$ is shown in Fig. 10-33(c).

In this case, the $M/EI$ diagram is symmetrical about a vertical line passing through the center. Therefore, the elastic curve must be symmetrical, and the tangent to this curve at the center of the beam is horizontal. From the figure, it is seen that $\Delta\theta_{B/C}$ is equal to $\theta_B$, and the rotation of $B$ is equal to one-half the area[20] of the whole $M/EI$ diagram. Distance $CC'$ is the desired deflection, and from the geometry of the figure, it is seen to be equal to $t_{B/C}$ (or $t_{A/C}$, not shown).

$$\Phi = \frac{1}{EI}\left(\frac{2}{3}\frac{L}{2}\frac{w_oL^2}{8}\right) = \frac{w_oL^3}{24EI}$$

$$\theta_B = \Delta\theta_{B/C} = \Phi = +\frac{w_oL^3}{24EI}$$

$$v_C = v_{max} = t_{B/C} = \Phi\bar{x} = \frac{w_oL^3}{24EI}\frac{5L}{16} = \frac{5w_oL^4}{384EI}$$

[20] See Table 2 of the Appendix for a formula giving an area enclosed by a parabola as well as for $\bar{x}$.

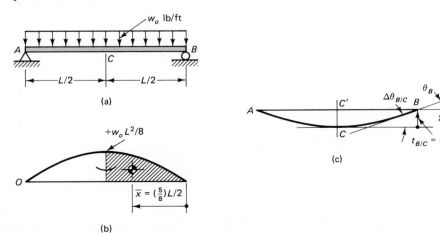

(a)

$+w_oL^2/8$

$\bar{x} = \left(\frac{5}{8}\right)L/2$

(b)

(c)

Fig. 10-33

The value of the deflection agrees with Eq. 10-22, which expresses the same quantity derived by the integration method. Since point $B$ is above the tangent through $C$, the sign of $v_C$ is positive.

### EXAMPLE 10-20

Find the deflection of the free end $A$ of the beam shown in Fig. 10-34(a) caused by the applied forces. $EI$ is constant.

### Solution

The bending-moment diagram for the applied forces is shown in Fig. 10-34(b). The bending moment changes sign at $a/2$ from the left support. At this point, an inflection in the elastic curve occurs. Corresponding to the positive moment, the curve is concave up, and vice versa. The elastic curve is so drawn and passes over the supports at $B$ and $C$, Fig. 10-34(c). To begin, the inclination of the tangent to the elastic curve at support $B$ is determined by finding $t_{C/B}$ as the statical moment of the areas with the proper signs of the $M/EI$ diagram between the verticals through $C$ and $B$ about $C$.

$$t_{C/B} = \Phi_1 \bar{x}_1 + \Phi_2 \bar{x}_2 + \Phi_3 \bar{x}_3$$
$$= \frac{1}{EI}\left[\frac{a}{2}(+Pa)\frac{2a}{3} + \frac{1}{2}\frac{a}{2}(+Pa)\left(a + \frac{1}{3}\frac{a}{2}\right)\right.$$
$$\left. + \frac{1}{2}\frac{a}{2}(-Pa)\left(\frac{3a}{2} + \frac{2}{3}\frac{a}{2}\right)\right]$$
$$= +\frac{Pa^3}{6EI}$$

The positive sign of $t_{C/B}$ indicates that point $C$ is *above the tangent at B*. Hence, a corrected diagram of the elastic curve is made, Fig. 10-34(d), where it is seen that the deflection sought is given by distance $AA'$ and is equal to $AA'' - A'A''$. Further, since triangles $A'A''B$ and $CC'B$ are similar, distance $A'A'' = t_{C/B}/2$. On the other hand, distance $AA''$ is the deviation of point $A$ from the tangent to the elastic curve at support $B$. Hence,

$$v_A = AA' = AA'' - A'A'' = t_{C/B}/2$$
$$t_{A/B} = \frac{1}{EI}(\Phi_4 \bar{x}_4) = \frac{1}{EI}\left[\frac{a}{2}(-Pa)\frac{2a}{3}\right] = -\frac{Pa^3}{3EI}$$

where the negative sign means that point $A$ is below the tangent through $B$. This sign is not used henceforth, as the geometry of the elastic curve indicates the direction of the actual displacements. Thus, the deflection of point $A$ *below the line passing through the supports* is

$$v_A = \frac{Pa^3}{3EI} - \frac{1}{2}\frac{Pa^3}{6EI} = \frac{Pa^3}{4EI}$$

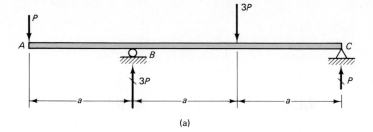

(a)

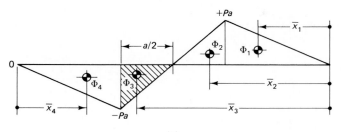

(b)

(c)

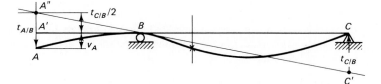

(d)

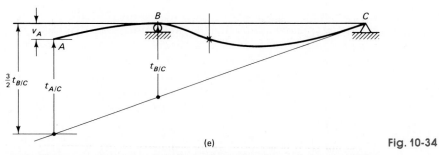

(e)

Fig. 10-34

This example illustrates the necessity of watching the signs of the quantities computed in the applications of the moment-area method, although usually less difficulty is encountered than in this example. For instance, if the deflection of end $A$ is established by first finding the rotation of the elastic curve at $C$, no ambiguity in the direction of tangents occurs. This scheme of analysis is shown in Fig. 10-34(e), where $v_A = \frac{3}{2}t_{B/C} - t_{A/C}$.

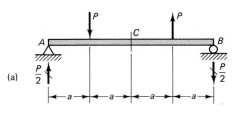

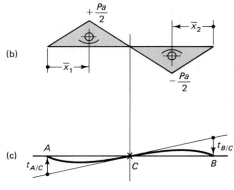

(b)

(c)

**Fig. 10-35**

### EXAMPLE 10-21

A simple beam supports two equal and opposite forces $P$ at the quarter points; see Fig. 10-35(a). Find the deflection of the beam at the middle of the span. $EI$ is constant.

Solution

The bending-moment diagram and elastic curve with a tangent at $C$ are shown in Figs. 10-35(b) and (c), respectively. Then, since the statical moments of the positive and negative areas of the bending-moment diagram around $A$ and $B$, respectively, are numerically equal, i.e., $|t_{A/C}| = |t_{B/C}|$, the deflection of the beam at the center of the span is *zero*. The elastic curve in this case is *anti*symmetrical. Noting this, much work may be avoided in obtaining the deflections at the *center of the span*. The deflection of any other point on the elastic curve can be found in the usual manner.

---

The foregoing examples illustrate the manner in which the moment-area method can be used to obtain the deflection of any statically determinate beam. No matter how complex the $M/EI$ diagrams may become, the previous procedures are applicable. In practice, any $M/EI$ diagram whatsoever may be approximated by a number of rectangles and triangles. It is also possible to introduce concentrated angle changes at hinges to account for discontinuities in the directions of the tangents at such points. The magnitudes of the concentrations can be found from kinematic requirements.[21]

For complicated loading conditions, deflections of elastic beams determined by the moment-area method are often best found by superposition. In this manner, the areas of the separate $M/EI$ diagrams may be-

[21] For a systematic treatment of more complex problems see, for example, A. C. Scordelis and C. M. Smith, "An Analytical Procedure for Calculating Truss Displacements," *Proc. ASCE* 732 (July 1955): 732-1 to 732-17.

come simple geometrical shapes. In the next section, superposition is used in solving statically indeterminate problems.

The method described here can be used very effectively in determining the inelastic deflection of beams, provided the *M/EI* diagrams are replaced by the curvature diagrams such as in Fig. 10-26(c).

## **\*\*10-15.** Statically Indeterminate Beams

Statically indeterminate beams can readily be solved for unknown reactions using the moment-area method by employing superposition. After the redundant reactions are determined, the beam deflections and rotations can be found in the usual manner, again, often employing superposition. Two different procedures for finding the redundant reactions are considered in this section. In the more widely used procedure, it is recognized that restrained[22] and continuous beams differ from simply supported beams mainly by the presence of redundant moments at the supports. Therefore, bending-moment diagrams for these beams may be considered to consist of two independent parts—one part for the moment caused by all of the applied loading on a beam assumed to be simply supported, the other part for the redundant end moments. Thus, the effect of redundant end moments is superposed on a beam assumed to be simply supported. Physically, this notion can be clarified by imagining an indeterminate beam cut through at the supports while the vertical reactions are maintained. The continuity of the elastic curve of the beam is preserved by the redundant moments.

Although the critical ordinates of the bending-moment diagrams caused by the redundant moments are not known, their shape is known. Application of a redundant moment at an end of a simple beam results in a triangular-shaped moment diagram, with a maximum at the applied moment, and a zero ordinate at the other end. Likewise, when end moments are present at both ends of a simple beam, two triangular moment diagrams superpose into a trapezoidal-shaped diagram.

The known and the unknown parts of the bending-moment diagram together give a complete bending-moment diagram. This whole diagram can then be used in applying the moment-area theorems to the continuous elastic curve of a beam. The geometrical conditions of a problem, such as the continuity of the elastic curve at the support or the tangents at built-in ends that cannot rotate, permit a rapid formulation of equations for the unknown values of the redundant moments at the supports.

An alternative method for determining the redundant reactions employs a procedure of plotting the bending-moment diagrams by parts. In applying this method, only one of the existing fixed supports is left in place, creating a cantilever. Then separate bending-moment diagrams for *each one* of the applied forces as well as for the unknown reactions at the

---

[22] Indeterminate beams with one or more fixed ends are called *restrained beams*.

unsupported beam end are drawn. The sum of *all* of these moment diagrams for the *cantilever* make up the *complete* bending-moment diagram then used in the usual manner.

In either method, for beams of variable flexural rigidity, the moment diagrams must be divided by the corresponding $EI$'s.

Both methods of solving for the redundant reactions are illustrated in the following examples.

### EXAMPLE 10-22

Find the maximum downward defllection of the small aluminum beam shown in Fig. 10-36(a) due to an applied force $P = 100$ N. The beam's constant flexural rigidity $EI = 60$ N·m².

### Solution

The solution of this problem consists of two parts. First, a redundant reaction must be determined to establish the numerical values for the bending-moment diagram; then the usual moment-area procedure is applied to find the deflection.

By assuming the beam is released from the redundant end moment, a simple beam-moment diagram is constructed above the base line in Fig. 10-36(b). The moment diagram of known shape due to the unknown redundant moment $M_A$ is shown on the same diagram below the base line. One assumes $M_A$ to be positive, since in this manner, its correct sign is obtained automatically according to the beam sign convention. The composite diagram represents a *complete* bending-moment diagram.

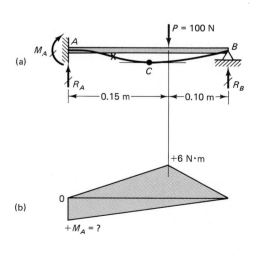

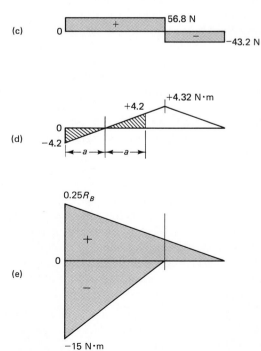

**Fig. 10-36**

The tangent at the built-in end remains horizontal after the application of force $P$. Hence, the geometrical condition is $t_{B/A} = 0$. An equation formulated on this basis yields a solution for $M_A$.[23] The equations of static equilibrium are used to compute the reactions. The final bending-moment diagram, Fig. 10-36(d), is obtained in the usual manner after the reactions are known. Thus, since $t_{B/A} = 0$,

$$\frac{1}{EI}\left[\frac{1}{2}(0.25)(6)\frac{1}{3}(0.25 + 0.10) + \frac{1}{2}(0.25)M_A\frac{2}{3}(0.25)\right] = 0$$

Hence, $M_A = -4.2$ N·m. Since, initially, $M_A$ was assumed to be positive, and is so shown in Figs. 10-36(a) and (b), this result indicates that actually $M_A$ has an *opposite* sense. The correct sense for $M_A$ must be used in the equations of statics that follow and is reflected in the shear and moment diagrams constructed in Figs. 10-36(c) and (d), respectively.

$$\sum M_A = 0 \;\circlearrowright + \qquad 100(0.15) - R_B(0.25) - 4.2 = 0 \qquad R_B = 43.2 \text{ N}$$
$$\sum M_B = 0 \;\circlearrowleft + \qquad 100(0.10) + 4.2 - R_A(0.25) = 0 \qquad R_A = 56.8 \text{ N}$$
$$Check: \quad \sum F_y = 0 \;\uparrow + \qquad 43.2 + 56.8 - 100 = 0$$

The maximum deflection occurs where the tangent to the elastic curve is horizontal, point $C$ in Fig. 10-36(a). Hence, by noting that the tangent at $A$ is also horizontal and using the first moment-area theorem, point $C$ is located. This occurs when the hatched areas in Fig. 10-36(d) having opposite signs are equal, i.e., at a distance $2a = 2(4.2/56.8) = 0.148$ m from $A$. The tangential deviation $t_{A/C}$ (or $t_{C/A}$) gives the deflection of point $C$.

$$v_{\max} = v_C = t_{A/C}$$
$$= \frac{1}{EI}\left[\frac{1}{2} \times 0.074(+4.2)\left(0.074 + \frac{2}{3} \times 0.074\right)\right.$$
$$\left. + \frac{1}{2} \times 0.074(-4.2)\frac{1}{3} \times 0.074\right]$$
$$= (15.36)10^{-3}/EI = 0.256 \text{ mm} \qquad \text{(down)}$$

## Alternative Solution

A rapid solution can also be obtained by plotting the moment diagram by cantilever parts. This is shown in Fig. 10-36(e). Note that one of the ordinates is in terms of the redundant reaction $R_B$. Again, using the geometrical condition $t_{B/A} = 0$, one obtains an equation yielding $R_B$. Other reactions follow by statics. From $t_{B/A} = 0$,

$$\frac{1}{EI}\left[\frac{1}{2}(0.25)(+0.25R_B)\frac{2}{3}(0.25) + \frac{1}{2}(0.15)(-15)\left(0.1 + \frac{2}{3} \times 0.15\right)\right] = 0$$

Hence, $R_B = 43.2$ N, acting up as assumed.

[23] See Table 2 of the Appendix for the centroidal distance of a whole triangle.

$$\sum M_A = 0 \circlearrowleft + \qquad M_A + 43.2(0.25) - 100(0.15) = 0 \qquad M_A = 4.2\,\text{N·m}$$

Here $M_A$, within the equation of statics for the summation of moments, is considered positive since it is assumed to act in a *counterclockwise* direction. However, in the *beam sign convention*, such an end moment at $A$ is negative.

After the combined moment diagram is constructed, Fig. 10-36(d), the remainder of the work is the same as in the preceding solution.

### EXAMPLE 10-23

Find the moments at the supports for a fixed-end beam loaded with a uniformly distributed load of $w_o$ N/m; see Fig. 10-37(a).

### Solution

The moments at the supports are called fixed-end moments, and their determination is of great importance in structural theory. Due to symmetry in this problem, the fixed-end moments are equal, as are the vertical reactions, which are $w_o L/2$ each. The moment diagram for this beam, considered to be simply supported, is a parabola, as shown in Fig. 10-37(b), while the assumed positive fixed-end moments give the rectangular diagram shown in the same figure.

Although this beam is statically indeterminate to the second degree, because of symmetry, a single equation based on a geometrical condition is sufficient to yield the redundant moments. From the geometry of the elastic curve, any one of the following conditions may be used: $\Delta \theta_{A/B} = 0$,[24] $t_{B/A} = 0$, or $t_{A/B} = 0$. From the first condition, $\Delta \theta_{A/B} = 0$,

$$\frac{1}{EI}\left[\frac{2}{3}L\left(+\frac{w_o L^2}{8}\right) + L(+M_A)\right] = 0$$

[24] Also since the tangent at the center of the span is horizontal, $\Delta \theta_{A/C} = 0$ and $\Delta \theta_{C/B} = 0$.

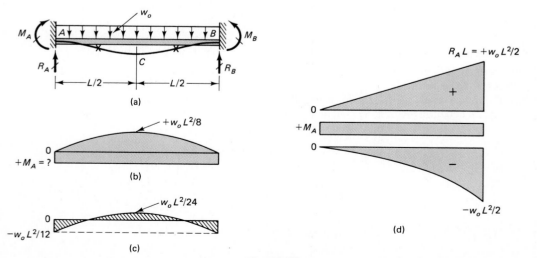

Fig. 10-37

then
$$M_A = M_B = -\frac{w_o L^2}{12} \qquad (10\text{-}44)$$

The negative sign for these moments indicates that their sense is opposite from that assumed in Figs. 10-37(a) and (b).

The composite moment diagram is shown in Fig. 10-37(c). In comparison with the maximum bending moment of a simple beam, a considerable reduction in the magnitude of the critical moments occurs.

### Alternative Solution

The moment diagram by cantilever parts is shown in Fig. 10-37(d). Noting that $R_A = R_B = w_o L/2$, and using the same geometrical condition as above, $\Delta\theta_{A/B} = 0$, one can verify the former solution as follows:

$$\frac{1}{EI}\left[\frac{1}{2}L\left(+\frac{w_o L^2}{2}\right) + L(+M_A) + \frac{1}{3}L\left(-\frac{w_o L^2}{2}\right)\right] = 0$$

and
$$M_A = -\frac{w_o L^2}{12}$$

---

## EXAMPLE 10-24

A beam fixed at both ends carries a concentrated force $P$, as shown in Fig. 10-38. Find the fixed-end moments. $EI$ is constant.

### Solution

By treating beam $AB$ as a simple beam, the moment diagram due to $P$ is shown above the base line in Fig. 10-38(b). The assumed positive fixed-end moments are *not equal* and result in the trapezoidal diagram. Three geometrical conditions for the elastic curve are available to solve this problem, which is indeterminate to the second degree:

(a) $\Delta\theta_{A/B} = 0$, since the change in angle between the tangents at $A$ and $B$ is zero.

(b) $t_{B/A} = 0$, since support $B$ does not deviate from a fixed tangent at $A$.

(c) Similarly, $t_{A/B} = 0$.

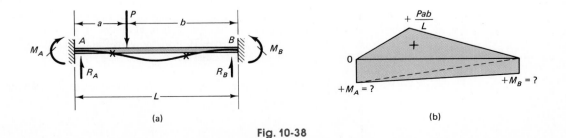

(a)                                                   (b)

**Fig. 10-38**

*Any two* of these conditions may be used; arithmetical simplicity of the resulting equations governs the choice. Thus, by using condition (a), which is always the simplest, and condition (b), the two equations are[25]

$$\Delta\theta_{A/B} = \frac{1}{EI}\left(\frac{1}{2}L\frac{Pab}{L} + \frac{1}{2}LM_A + \frac{1}{2}LM_B\right) = 0$$

or

$$M_A + M_B = -\frac{Pab}{L}$$

$$t_{B/A} = \frac{1}{EI}\left[\frac{1}{2}L\frac{Pab}{L}\frac{1}{3}(L+b) + \frac{1}{2}LM_A\frac{2}{3}L + \frac{1}{2}LM_B\frac{1}{3}L\right] = 0$$

or

$$2M_A + M_B = -\frac{Pab}{L^2}(L+b)$$

Solving the two reduced equations simultaneously gives

$$M_A = -\frac{Pab^2}{L^2} \quad \text{and} \quad M_B = -\frac{Pa^2b}{L^2} \tag{10-45}$$

These negative moments have an opposite sense from that initially assumed and shown in Figs. 10-38(a) and (b).

---

## EXAMPLE 10-25

Plot moment and shear diagrams for a continuous beam loaded as shown in Fig. 10-39(a). *EI* is constant for the whole beam.

### Solution

This beam is statically indeterminate to the second degree. By treating each span as a simple beam with the redundant moments assumed positive, the moment diagram of Fig. 10-39(c) is obtained. For each span, these diagrams are similar to the ones shown earlier in Figs. 10-36(b) and 10-38(b). No end moments exist at *A* as this end is on a roller. The clue to the solution is contained in two geometrical conditions for the elastic curve for the whole beam, Fig. 10-39(d):

(a) $\theta_B = \theta_B'$. Since the beam is physically continuous, there is a line at support *B* that is tangent to the elastic curve in *either* span.

(b) $t_{B/C} = 0$, since support *B* does not deviate from a fixed tangent at *C*.

To apply condition (a), $t_{A/B}$ and $t_{C/B}$ are determined, and, by *dividing these quantities by the respective span lengths,* the two angles $\theta_B$ and $\theta_B'$ are obtained. These angles are equal. However, although $t_{C/B}$ is algebraically expressed as a positive quantity, the tangent through point *B* is *above* point *C*. Therefore, this deviation must be considered negative. Hence, by using condition (a), one equation with the redundant moments is obtained.

---

[25] See Table 2 of the Appendix for the centroidal distance of a whole triangle.

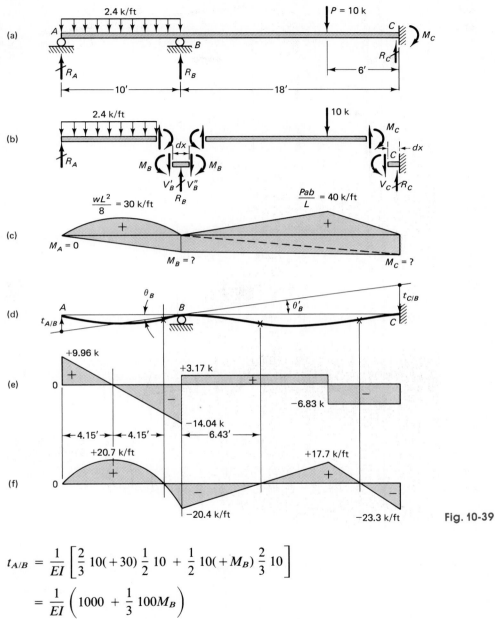

Fig. 10-39

$$t_{A/B} = \frac{1}{EI} \left[ \frac{2}{3} 10(+30) \frac{1}{2} 10 + \frac{1}{2} 10(+M_B) \frac{2}{3} 10 \right]$$

$$= \frac{1}{EI} \left( 1000 + \frac{1}{3} 100M_B \right)$$

$$t_{C/B} = \frac{1}{EI} \left[ \frac{1}{2} 18(+40) \frac{1}{3} (18 + 6) + \frac{1}{2} 18(+M_B) \frac{2}{3} 18 + \frac{1}{2} 18(+M_C) \frac{1}{3} 18 \right]$$

$$= \frac{1}{EI} (2880 + 108M_B + 54M_C)$$

Since $\theta_B = \theta_B'$ or $\dfrac{t_{A/B}}{L_{A/B}} = -\dfrac{t_{C/B}}{L_{C/B}}$

$$\frac{1}{EI}\left(\frac{1000 + 100M_B/3}{10}\right) = -\frac{1}{EI}\left(\frac{2880 + 108M_B + 54M_C}{18}\right)$$

or $28M_B/3 + 3M_C = -260$

Using condition (b) for span $BC$ provides another equation, $t_{B/C} = 0$, or

$$\frac{1}{EI}\left[\frac{1}{2}18(+40)\frac{1}{3}(18 + 12) + \frac{1}{2}18(+M_B)\frac{1}{3}18 + \frac{1}{2}18(+M_C)\frac{2}{3}18\right] = 0$$

or $3M_B + 6M_C = -200$

Solving the two reduced equations simultaneously,

$$M_B = -20.4 \text{ ft-lb} \quad \text{and} \quad M_C = -23.3 \text{ ft-lb}$$

where the signs agree with the convention of signs used for beams. These moments with their proper sense are shown in Fig. 10-39(b).

After the redundant moments $M_A$ and $M_B$ are found, no new techniques are necessary to construct the moment and shear diagrams. *However, particular care must be exercised to include the moments at the supports while computing shears and reactions.* Usually, isolated beams, as shown in Fig. 10-39(b), are the most convenient free-bodies for determining shears. Reactions follow by adding the shears on the adjoining beams.

*For free body AB:*

$$\sum M_B = 0 \circlearrowright + \quad 2.4(10)5 - 20.4 - 10R_A = 0 \quad R_A = 9.96 \text{ k} \uparrow$$
$$\sum M_A = 0 \circlearrowleft + \quad 2.4(10)5 + 20.4 - 10V_B' = 0 \quad V_B' = 14.04 \text{ k} \uparrow$$

*For free body BC:*

$$\sum M_C = 0 \circlearrowright + \quad 10(6) + 20.4 - 23.3 - 18V_B'' = 0$$
$$V_B'' = 3.17 \text{ k} \uparrow$$
$$\sum M_B = 0 \circlearrowleft + \quad 10(12) - 20.4 + 23.3 - 18V_C = 0$$
$$V_C = R_C = 6.83 \text{ k} \uparrow$$

*Check:* $R_A + V_B' = 24 \text{ k} \uparrow$ and $V_B'' + R_C = 10 \text{ k} \uparrow$

From above, $R_B = V_B' + V_B'' = 17.21$ kips $\uparrow$.

The complete shear and moment diagrams are shown in Figs. 10-39(e) and (f), respectively.

Generalizing the procedure used in the preceding example, a recurrence formula, i.e., an equation which may be repeatedly applied for every two adjoining spans, may be derived for continuous beams. For any $n$ number of spans, $n - 1$ such equations may be written. This gives enough simultaneous equations for the solution of redundant moments over the supports. This recurrence formula is called the *three-moment equation* because three unknown moments appear in it.[26]

# Problems

## Section 10-2

**10-1.** A $2 \times 6$ mm steel strip 3142 mm long is clamped at one end as shown in the figure. What is the required end moment to force the strip to touch the wall? What would be the maximum stress when the strip is in the bent condition? $E = 200$ GPa.

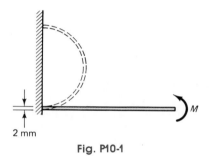

2 mm

**Fig. P10-1**

**10-2.** A round aluminum bar of 6 mm diameter is bent into a circular ring having a mean diameter of 3 m. What is the maximum stress in the bar? $E = 70$ GPa.

**10-3.** What will be the radius of curvature of a W 8 × 17 beam bent around the $X$–$X$ axis if the stress in the extreme fibers is 36 ksi? $E = 29 \times 10^6$ psi.

**\*10-4.** Assume that a straight rectangular bar after severe cold working has a residual stress distribution such as was found in Example 6-12; see Fig. 6-29. (a) If one-sixth of the thickness of this bar is machined off on the top and on the bottom, reducing the bar to two-thirds of its original thickness, what will be the curvature ρ of the machined bar? Assign the necessary parameters to solve this problem in general terms. (b) For the previous conditions, if the bar is 1 in² and 40 in long, what will be the deflection of the bar at the center from the chord through the end? Let $\sigma_{yp} = 54$ ksi and $E = 27 \times 10^6$ psi. Note that for small deflec-

tions, the maximum deflection from a chord $L$ long of a curve bent into a circle of radius $R$ is approximately[27] $L^2/(8R)$. (*Hint:* The machining operation removes the internal microresidual stresses.)

## Section 10-7

**10-5.** If the equation of the elastic curve for a simply supported beam of length $L$ having a constant $EI$ is $v = (k/360EI)(-3x^5 + 10x^3L^2 - 7xL^4)$, how is the beam loaded?

**10-6.** An elastic beam of constant $EI$ and of length $L$ has the deflected shape $EIv(x) = M_o(x^3 - x^2L)/4L$. (a) Determine the loading and support conditions. (b) Plot the shear and moment diagrams for the beam and sketch the deflected shape.

**10-7.** Rework Example 10-2 by taking the origin of the coordinate system at the free end.

**\*10-8.** Using the exact differential equation, Eq. 10-8, show that the equation of the elastic curve in Example 10-2 is $x^2 + (v - \rho)^2 = \rho^2$, where ρ is a constant. (*Hint:* Let $dv/dx = \tan \theta$ and integrate.)

**10-9 through 10-29.** (a) Determine the equations of the elastic curves for the beams shown in the figures due to the applied loading for the given boundary conditions. Unless directed otherwise, use Eq. 10-14a or 10-14c, whichever is simpler to apply. For all cases, $EI$ is constant, except that in Prob. 10-20, $EI$ varies. Wherever applicable, take advantage of symmetry or antisymmetry. (b) For statically indeterminate cases only, plot shear and moment diagrams, giving all critical ordinates.

---

[27] This follows by retaining the first term of the expansion of $R(1 - \cos \theta)$, where θ is one-half the included angle.

---

[26] For discussion of this procedure, for example, see E. P. Popov, *Mechanics of Materials*, 2nd ed. (Englewood Cliffs, NJ: 1976) 435–440.

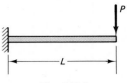

Fig. P10-9

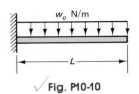

✓ Fig. P10-10

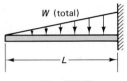

Fig. P10-11

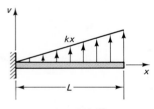

Fig. P10-12

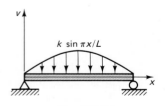

Fig. P10-13

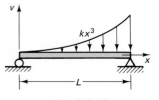

Fig. P10-14

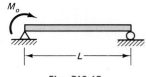

Fig. P10-15

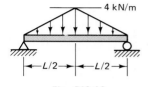

Fig. P10-16

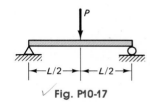

✓ Fig. P10-17

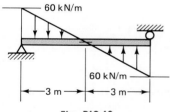

Fig. P10-18

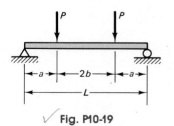

✓ Fig. P10-19

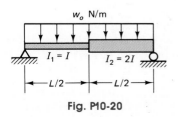

Fig. P10-20

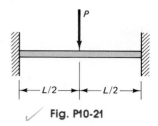

Fig. P10-21

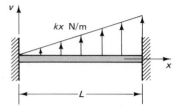

Fig. P10-22

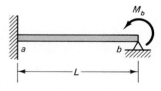

Fig. P10-23

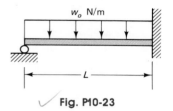

Fig. P10-24

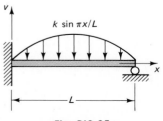

Fig. P10-25

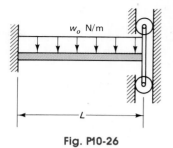

Fig. P10-26

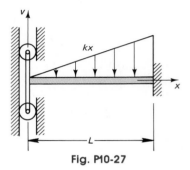

Fig. P10-27

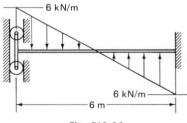

Fig. P10-28

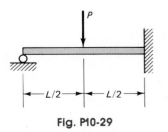

Fig. P10-29

**10-30 through 10-32.** (a) Determine equations for the elastic curves due to an imposed small vertical displacement Δ of the end for the beams of length $L$ and of constant $EI$ shown in the figures. (b) Plot shear and moment diagrams.

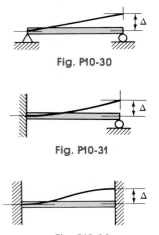

Fig. P10-30

Fig. P10-31

Fig. P10-32

**10-33.** If in Prob. 10-17, the cross-sectional area of the beam is constant, and the left half of the span is made of steel ($E = 30 \times 10^6$ psi) and the right half is made of aluminum ($E = 10 \times 10^6$ psi), determine the equation of the elastic curve.

**10-34.** What is the equation of the elastic curve for the cantilever of constant width and flexural strength loaded at the end by a concentrated force $P$? See Figs. 9-17(a) and (d). Neglect the effect of the required increase in beam depth at the end for shear.

**10-35.** An overhanging beam of constant flexural rigidity $EI$ is loaded as shown in the figure. For portion $AB$ of the beam, (a) find the equation of the elastic curve due to the applied load of $2w_0$ N/m, and (b) determine the maximum deflection between the supports and the deflection midway between the supports.

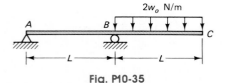

Fig. P10-35

**10-36.** A beam with an overhang of constant flexural rigidity $EI$ is loaded as shown in the figure. (a) Determine length $a$ of the overhang such that the elastic curve would be horizontal over support $B$. (B) Determine the maximum deflection between the supports.

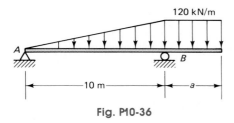

Fig. P10-36

**10-37.** Using a semigraphical procedure, such as shown in Figs. 10-9 and 10-13, find the deflection of the beam at the point of the applied load; see the figure. Let $I_1 = 400$ in$^4$, $I_2 = 300$ in$^4$, and $E = 30 \times 10^6$ psi.

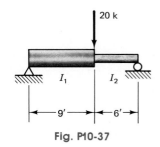

Fig. P10-37

**10-38.** Using a semigraphical procedure, such as shown in Figs. 10-9 and 10-13, find the deflection at the center of the span for the beam loaded as shown in the figure. Neglect the effect of the axial force on deflection. $EI$ for the beam is constant.

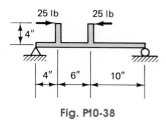

Fig. P10-38

**10-39.** A steel beam is to span 30 ft and support a 1.2 k/ft uniformly distributed load, including its own weight. Select the required W section of minimum weight, using the abridged Table 4 in the Appendix, for bending around its strong axis. The allowable bending stress is 24 ksi and that for shear is 14.4 ksi. It is also required that the maximum deflection does not exceed 1 in. This requirement corresponds to 1/360-th of the span length and is often used to limit deflection due to the applied load in building design. $E = 29 \times 10^3$ ksi.

**10-40.** A wooden beam is to span 24 ft and to support a 1 k/ft uniformly distributed load, including its own weight. Select the size required from Table 10 in the Appendix. The allowable bending stress is 2000 psi and that in shear is 100 psi. The deflection is limited to 1/360-th of the span length.

**10-41.** The maximum deflection for a simple beam spanning 24 ft and carrying a uniformly distributed load of 40 k total, including its own weight, is limited to 0.5 in. (a) Specify the required steel I beam. Let $E = 30 \times 10^6$ psi. (b) What size aluminum-alloy beam would be needed for the same requirements? Let $E = 10 \times 10^6$ psi, and use Table 3 in the Appendix for section properties. (c) Determine the maximum stresses in both cases.

**10-42.** A uniformly loaded $6 \times 12$ in (nominal size) wooden beam spans 10 ft and is considered to have satisfactory deflection characteristics. Select an aluminum-alloy I beam, a steel I beam, and a polyester-plastic I beam having the same deflection characteristics. In making the beam selections, neglect the differences in their own weights. Let $E = 1.5 \times 10^6$ psi for wood and polyester plastic, $E = 10 \times 10^6$ psi for aluminum, and $E = 30 \times 10^6$ psi for steel. For section properties of all I beams, use Table 4 in the Appendix.

## Section 10-8

**10-43.** Using singularity functions, rework Prob. 10-19.

**10-44.** Using singularity functions, rework Prob. 10-29.

**10-45 through 10-50.** Using singularity functions, obtain equations for the elastic curves for the beams loaded as shown in the figures. $EI$ is constant for all beams.

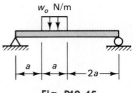

**Fig. P10-45**

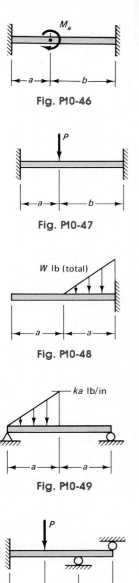

**Fig. P10-46**

**Fig. P10-47**

**Fig. P10-48**

**Fig. P10-49**

**Fig. P10-50**

## Section 10-9. Use the deflection equations in Examples 10-2 through 10-6 and Table 11 in the Appendix.

**10-51.** (a) From the solution given in Table 11 in the Appendix for a cantilever loaded by a concentrated force $P$ at the end, show that the free-end deflection at $A$ for the cantilever shown in the figure is

$$v_A = \frac{Pb^2}{6EI}(3L - b)$$

(b) Show that the deflection at $A$ due to force $P$ at $B$ is equal to the deflection at $B$ due to force $P$ at $A$. (See Section 13-4 on Maxwell's theorem of reciprocal deflections.)

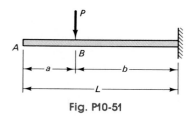

**Fig. P10-51**

**10-52.** The data for a beam loaded as shown in Fig. 10-17 are $w_0 = 30$ kN/m, $P = 25$ kN, $L = 3$ m, and $a = 1.2$ m. If the beam is made from a W 8 × 24 section ($I = 29 \times 10^6$ mm$^4$), what is the deflection of the free end $C$ caused by the applied loads? $E = 200$ GPa.

**10-53.** A W 8 × 40 steel beam is loaded as shown in the figure. Calculate the deflection at the center of the span. $E = 29 \times 10^3$ ksi.

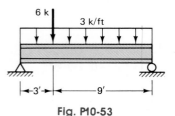

**Fig. P10-53**

**10-54.** Using the results found in Example 10-6 for deflection of a beam due to a concentrated force $P$,

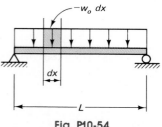

**Fig. P10-54**

determine the deflection at the center of the beam caused by a uniformly distributed downward load $w_o$. (Treat $w_o\, dx$ as an infinitesimal concentrated force, and integrate.) This method of *influence coefficients* (so named by Maxwell) can be effectively used for many distributed-load problems.

**10-55.** Using the method outlined in the preceding problem, determine the deflection at the center of the beam for the loading given in Prob. 10-49.

**10-56.** An elastic prismatic beam with an overhang is loaded with a concentrated end moment $M_o$. Determine the deflection of the free end. The spring constant $k = 48EI/L^3$.

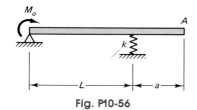

**Fig. P10-56**

**10-57.** An L-shaped member, such as shown in Fig. 10-16(a), is made up from a bar of constant cross section. Determine the horizontal and vertical deflections of point $C$ caused by applied force $P$. Let $a = L/4$. Neglect the effect on deflection of the axial force in the horizontal member and of the shear in the vertical member. Express the results in terms of $P$, $L$, $E$, and $I$.

**10-58.** A vertical rod with a concentrated mass at its free end is attached to a rotating plane, as shown in the figure. (a) At what angular velocity $\omega$ will the maximum bending stress in the rod just reach yield? (b) For the condition in (a), determine the deflection of the mass. The rod is 5 mm in diameter and its mass is $150 \times 10^{-6}$ kg/mm. The mass at the top of the rod is $60 \times 10^{-3}$ kg and can be considered to be concentrated at a point. $\sigma_{yp} = 1$ GPa and $E = 200$ GPa.

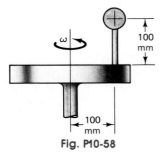

**Fig. P10-58**

**10-59.** Two identical, horizontal, simply supported beams span 3.6 m each. The beams cross each other at right angles at their respective midspans. When erected, there is a 6-mm gap between the two beams. If a concentrated downward force of 50 kN is applied at midspan to the upper beam, how much will the lower beam carry? $EI$ for each beam is 6000 kN·m².

**10-60.** The midpoint of a cantilever beam 6 m long rests on the midspan of a simply supported beam 8 m long. Determine the deflection of point $A$, where the beams meet, which results from the application of a 40-kN force at the end of the cantilever beam. State the answer in terms of $EI$, which is the same and is constant for both beams.

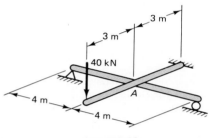

**Fig. P10-60**

**10-61.** A 30-in cantilever of constant flexural rigidity, $EI = 10^7$ lb-in², initially has a gap of 0.02 in between its end and the spring. The spring constant $k = 10$ k/in. If a force of 100 lb is applied to the cantilever, as shown in the figure, how much of this force will be carried by the spring? (*Hint:* See Prob. 10-51.)

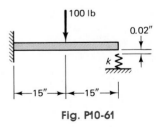

**Fig. P10-61**

**10-62.** A steel wire 5 m in length with a cross-sectional area equal to 160 mm² is stretched tightly between the midpoint of the simple beam and the free end of the cantilever, as shown in the figure. Determine the deflection of the end of the cantilever as a result of a temperature drop of 50°C. For steel wire: $E = 200$

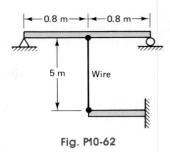

**Fig. P10-62**

GPa, $\alpha = 12 \times 10^{-6}$ per °C. For both beams: $I = 10 \times 10^6$ mm⁴ and $E = 10$ GPa.

**10-63.** One end of a W 18 × 50 beam is cast into concrete. It was intended to support the other end with a 1-in² steel rod 12 ft long, as shown in the figure. During the installation, however, the nut on the rod was poorly tightened and in the unloaded condition there is a ½-in gap between the top of the nut and the bottom of the beam. What tensile force will develop in the rod because a force of 15 kips is applied at the middle of the beam? $E = 30 \times 10^6$ psi. (*Hint:* See Prob. 10-51.)

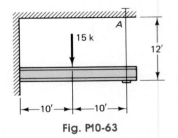

**Fig. P10-63**

**10-64.** A steel piano wire 30 in long is stretched from the middle of an aluminum beam $AB$ to a rigid support at $C$, as shown in the figure. What is the increase in stress in the wire if the temperature drops 100°F? The cross-sectional area of the wire is 0.0001 in² and $E = 30 \times 10^6$ psi. For the aluminum beam, $EI = 1040$ lb-in². Let $\alpha_{St} = 6.5 \times 10^{-6}$ per °F, $\alpha_{Al} = 12.9 \times 10^{-6}$ per °F.

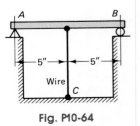

**Fig. P10-64**

**10-65.** A flexible steel bar is suspended by three steel rods, as shown in the figure, with the dimensions given in mm. If, initially, the rods are taut, what additional forces will develop in the rods due to the application of the force $F = 1500$ N and a drop in temperature of 50°C in the right rod? The cross-sectional area of each rod is 10 mm², and $\alpha = 12 \times 10^{-6}$ per °C. For the bar, $I = 2 \times 10^4$ mm⁴, and for steel, $E = 200 \times 10^3$ N/mm².

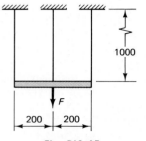

**Fig. P10-65**

**10-66.** An L-shaped steel rod of 2.125 in diameter is built-in at one end to a rigid wall and is simply supported at the other end, as shown in the figure. In plan the bend is 90°. What bending moment will be developed at the built-in end due to the application of a 2000-lb force at the corner of the rod? Assume $E = 30 \times 10^6$ psi, $G = 12 \times 10^6$ psi, and, for simplicity, let $I = 1.00$ in⁴ and $J = 2.00$ in⁴.

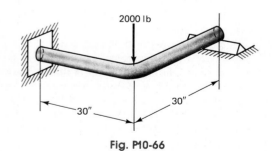

**Fig. P10-66**

**10-67.** Two parallel circular steel shafts of the same length are fixed at one end and are interconnected at the other end by means of a taut vertical wire, as shown in the figure. The shafts are 40 mm in diameter; the radius of the rigid pulley keyed to the upper shaft is 100 mm. The cross-sectional area of the interconnecting wire is 5 mm². If a vertical pull $P$ of 100 N is applied to the lower shaft, how much of the applied

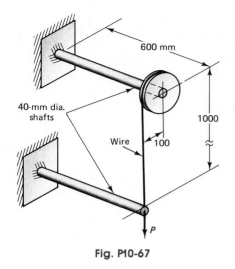

**Fig. P10-67**

force will be carried by the upper shaft? $E = 200$ GPa and $G = 80$ GPa.

**10-68.** A horizontal L-shaped rod is connected by a taut wire to a cantilever, as shown in the figure. If a drop in temperature of 100°C takes place and a downward force $P = 250$ N is applied at the end of the cantilever, what maximum bending stress will this cause at the cantilever support? Assume elastic behavior and neglect stress concentrations. All dimensions shown in the figure are in mm. The diameter of

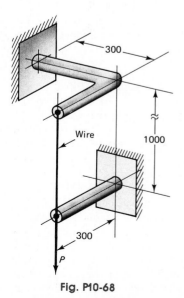

**Fig. P10-68**

the bent rod, as well as that of the cantilever, is 20 mm. The cross-sectional area of the wire is 0.40 mm². The assembly is made from steel having $E = 200$ GPa, $\nu = 0.25$, $G = 80$ GPa, and $\alpha = 11.7 \times 10^{-6}$ per °C.

**10-69.** The temperature in a furnace is measured by means of a stainless-steel wire placed in it. The wire is fastened to the end of a cantilever beam outside the furnace. The strain measured by the strain gage glued to the outside of the beam is a measure of the temperature. Assuming that the full length of the wire is heated to the furnace temperature, what is the change in furnace temperature if the gage records a change in strain of $-100 \times 10^{-6}$ in/in? Assume that the wire has sufficient initial tension to perform as intended. The mechanical properties of the materials are as follows: $\alpha_{SS} = 9.5 \times 10^{-6}$ per °F, $\alpha_{Al} = 12 \times 10^{-6}$ per °F, $E_{SS} = 30 \times 10^6$ psi, $E_{Al} = 10 \times 10^6$ psi, $A_{wire} = 5 \times 10^{-4}$ in², $I_{beam} = 6.5 \times 10^{-4}$ in⁴. The depth of the small beam is 0.25 in.

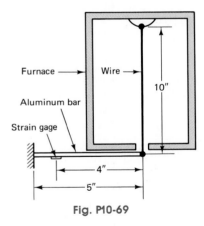

**Fig. P10-69**

**10-70.** With the aid of the first two solutions given in Table 11 of the Appendix, (a) find the reaction at $A$, and (b) plot the shear and moment diagrams and show the deflected shape of the beam.

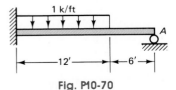

**Fig. P10-70**

## Section 10-10

**10-71.** A 5-ft-long cantilever is loaded at the end with a force $P = 1000$ lb forming an angle $\alpha$ with the vertical. The member is an S 8 × 18.4 steel beam. Determine the total tip deflection for $\alpha = 0°$, $10°$, $45°$, and $90°$ caused by the applied force. $E = 29 \times 10^6$ psi. (b) Verify that deflections are normal to the neutral axes.

**Fig. P10-71**

**10-72.** Determine the maximum deflection for the wooden beam in Example 6-15, Fig. 6-36. $E = 12$ GPa for the wood. Verify that the maximum deflection is normal to the neutral axis.

**10-73.** Consider an aluminum-alloy Z section having the dimensions given in Prob. 7-41. If a 100-in horizontal cantilever employing this section is fixed at one end and is subjected to a vertical downward 20-lb force at the centroid of the other end, what is the maximum deflection? How does the direction of this force relate to the neutral axis? $E = 10 \times 10^3$ ksi.

## Section 10-11

**10-74.** Consider a W 18 × 35 short steel cantilever fixed at one end and loaded at the free end, as shown in Fig. 10-23. Determine the length of this cantilever such that the deflection due to flexure is the same as that due to shear. The steel yields at 36 ksi in tension or compression and at 21 ksi in shear. Note that unlike a rectangular beam, it can be assumed that only the web yields uniformly in shear. (Although this solution is not exact, the results are representative of actual conditions.)

**10-75.** A heavy object weighing 4000 lb is dropped in the middle of a 20-ft simple span through a distance of 1 in. If the supporting beam is a W 10 × 33 steel beam, what is the impact factor? Assume elastic behavior. $E = 29 \times 10^6$ psi.

## Section 10-12

**10-76.** A 1-in square bar of a linearly elastic-plastic material is to be wrapped around a round mandrel, as shown in the figure. (a) What mandrel diameter $D$ is required so that the outer thirds of the cross sections become plastic, i.e., the elastic core is $\frac{1}{3}$ in deep by 1 in wide? Assume the material to be initially stress-free with $\sigma_{yp} = 40$ ksi. Let $E = 30 \times 10^6$ psi. The pitch of the helix angle is so small that only the bending of the bar in a plane need be considered. (b) What will be the diameter of the coil after the release of the forces used in forming it? Stated alternatively, determine the coil diameter after the elastic springback.

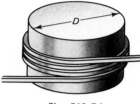

**Fig. P10-76**

**10-77.** A rectangular, weightless, simple beam of linearly elastic-plastic material is loaded in the middle by force $P$, as shown in the figure. (a) Determine the magnitude of force $P$ that would cause the plastic zone to penetrate one-fourth of the beam depth from each side. (b) For the previous loading condition, sketch the moment-curvature diagram, clearly showing it for the plastic zone.

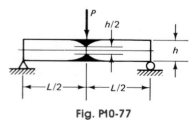

**Fig. P10-77**

## Section 10-14. Beam deflections for specified points in many of the problems for sections 10-7 and 10-8 can be assigned for solution by the moment-area method.

**10-78 through 10-89.** Using the moment-area method, determine the deflection and the slope of the elastic curves at points $A$ due to the applied loads for the beams, as shown in the figures. Specify the direction of deflection and of rotation for the calculated quantities. If neither the size of a beam nor its moment of inertia are given, $EI$ is constant. Wherever needed, let $E = 29 \times 10^3$ ksi or 200 GPa. In all cases, a well-prepared sketch of the elastic curve, showing the inflection points, should be made.

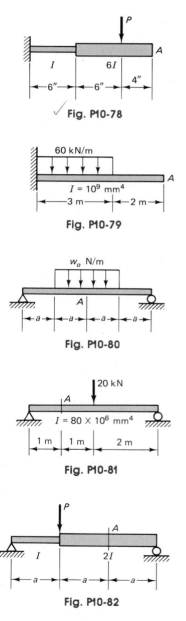

**Fig. P10-78**

**Fig. P10-79**

**Fig. P10-80**

**Fig. P10-81**

**Fig. P10-82**

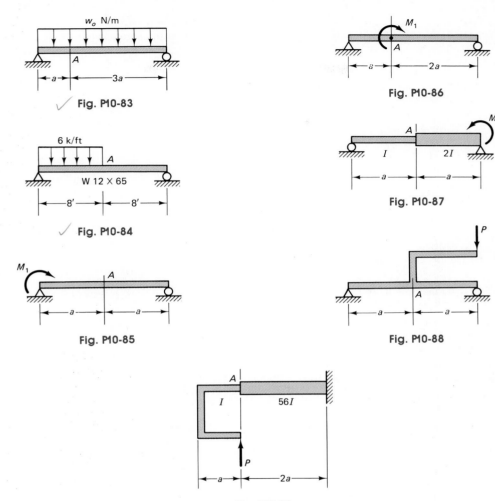

Fig. P10-83

Fig. P10-86

Fig. P10-84

Fig. P10-87

Fig. P10-85

Fig. P10-88

Fig. P10-89

**10-90.** Determine the deflection at the midspan of a simple beam, loaded as shown in the figure, by solving the two separate problems indicated and superimposing the results. Use the moment-area method. $EI$ is constant. (*Note:* Solutions of complex problems by subdividing them into a symmetrical part and an unsymmetrical part is often very advantageous because it reduces the numerical work.)

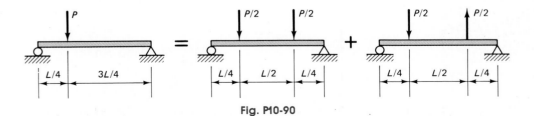

Fig. P10-90

**10-91.** Determine the elastic deflection at the center of the span for the beam loaded as shown in the figure if $I_1 = 10 \times 10^6$ mm$^4$, $I_2 = 20 \times 10^6$ mm$^4$, and $E = 70$ GPa. All given dimensions are in meters.

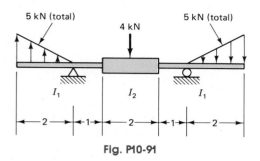

Fig. P10-91

**10-92.** Using the moment-area method, establish the equation of the elastic curve for the beam in Prob. 10-9.

**\*10-93.** Using the moment-area method, establish the equation of the elastic curve for the beam in Prob. 10-83.

**10-94.** Using the moment-area method, determine the maximum deflection for the beam in Prob. 10-85.

**10-95.** Using the moment-area method, determine the maximum deflection for the beam in Prob. 10-86.

**10-96.** Using the moment-area method, determine the maximum deflection for the beam in Prob. 10-82.

**10-97.** Using the moment-area method, determine the maximum deflection for the beam in Prob. 10-87.

**10-98.** Using the moment-area method, rework Prob. 10-38, and, in addition, determine the maximum deflection.

**10-99.** For the beam loaded as shown in the figure, determine (a) the deflection at the center of the span, (b) the deflection at the point of inflection of the elastic curve, and (c) the maximum deflection. $EI = 1,800$ lb-in$^2$.

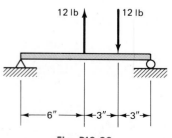

Fig. P10-99

**10-100 and 10-101.** Using the moment-area method, determine the deflection and slope of the overhang at point $A$ for the beams loaded as shown in the figures. $EI$ in the second problem is constant.

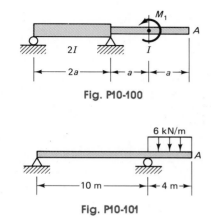

Fig. P10-100

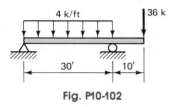

Fig. P10-101

**10-102.** Determine the maximum *upward* deflection for the overhang of a beam loaded as shown in the figure. $E$ and $I$ are constant.

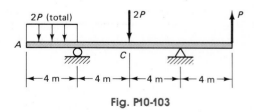

Fig. P10-102

**10-103.** For the elastic beam of constant flexural rigidity $EI$, loaded as shown in the figure, find the deflection and the slope at points $A$ and $C$.

Fig. P10-103

**10-104.** A structure is formed by joining a simple beam to a cantilever with a hinge, as shown in the figure. If a 10-kN force is applied at the center of the simple

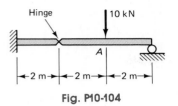

**Fig. P10-104**

span, determine the deflection at $A$ caused by this force. $EI$ is constant over the entire structure.

**10-105.** A hinged beam system is loaded as shown in the figure. Determine the deflection and slope of the elastic curve at point $A$.

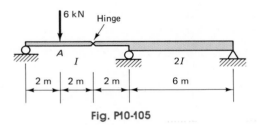

**Fig. P10-105**

**10-106.** Beam $AB$ is subjected to an end moment at $A$ and an unknown concentrated moment $M_C$, as shown in the figure. Using the moment-area method, determine the magnitude of the bending moment $M_C$ so that the deflection at point $B$ will be equal to zero. $EI$ is constant.

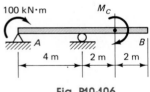

**Fig. P10-106**

**10-107.** The beam shown in the figure has a constant $EI = 3600 \times 10^6$ lb-in$^2$. Determine the distance $a$ such that the deflection at $A$ would be 0.25 in if the end were subjected to a concentrated moment $M_A = 15$ k-ft.

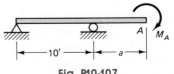

**Fig. P10-107**

**10-108.** A light pointer is attached only at $A$ to a $6 \times 6$ in (actual) wooden beam, as shown in the figure. Determine the position of the end of the pointer after a concentrated force of 1200 lb is applied. $E = 1.2 \times 10^6$ psi.

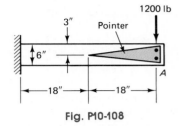

**Fig. P10-108**

**10-109.** Beam $ABCD$ is initially horizontal. Load $P$ is then applied at $C$, as shown in the figure. It is desired to place a vertical force at $B$ to bring the position of the beam at $B$ back to the original level $ABCD$. What force is required at $B$?

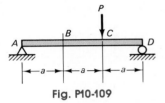

**Fig. P10-109**

## Section 10-15

**10-110 and 10-111.** For the beams loaded as shown in the figures, using the moment-area method, determine the redundant reactions and plot shear and moment diagrams. In both problems, $EI$ is constant.

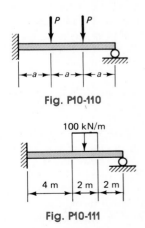

**Fig. P10-110**

**Fig. P10-111**

**10-112.** For the beam loaded as shown in the figure, (a) determine the ratio of the moment at the fixed end to the applied moment $M_A$; (b) determine the rotation of the end $A$. $EI$ is constant.

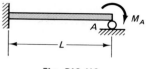

Fig. P10-112

**10-113.** (a) Using the moment-area method, determine the redundant moment at the fixed end for the beam shown in the figure, and plot the shear and moment diagrams. Neglect the weight of the beam. (b) Select a W beam using an allowable bending stress of 18,000 psi and a shearing stress of 12,000 psi. (c) Determine the maximum deflection of the beam between the supports and the maximum deflection of the overhang. $E = 29 \times 10^6$ psi.

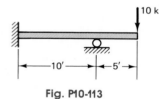

Fig. P10-113

**10-114.** (a) Using the moment-area method, determine the redundant moment at the fixed end for the beam shown in the figure and plot the shear and moment diagrams. Neglect the weight of the beam. (b) Select the depth for a 200-mm-wide wooden beam using an allowable bending stress of 8000 kN/m² and a shear stress of 1000 kN/m².

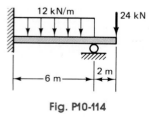

Fig. P10-114

**10-115.** For the beam of constant flexural rigidity $EI$ and loaded as shown in the figure, (a) determine the reaction and the deflection at point $A$. The spring con-

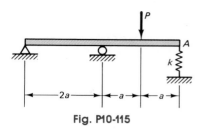

Fig. P10-115

stant $k = 3EI/a^3$. (b) Plot the shear and moment diagrams. Show the deflected shape of the beam.

**10-116.** For the beam loaded as shown in the figure, (a) plot the shear and bending moment diagrams, (b) sketch the shape of the elastic curve showing the point of inflection, and (c) determine the rotation of end $A$.

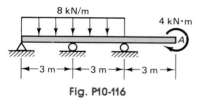

Fig. P10-116

**10-117.** For the beam loaded as shown in the figure, (a) determine the ratio of the moment at the fixed end to the applied moment $M_A$; (b) determine the rotation of end $A$.

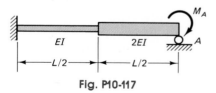

Fig. P10-117

**10-118.** Using the moment-area method, show that the maximum deflection of a beam fixed at both ends and carrying a uniformly distributed load is one-fifth the maximum deflection of the same beam simply supported. $EI$ is constant.

**10-119 and 10-120.** For the beams of constant $EI$ shown in the figures, using the moment-area method, (a) determine the fixed-end moments due to the applied loads and plot shear and bending-moment diagrams. (b) Find the maximum deflections.

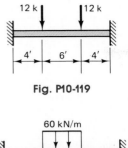

Fig. P10-119

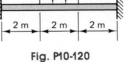

Fig. P10-120

**10-121 through 10-124.** For the beams of constant flexural rigidity shown in the figures, plot the shear and bending-moment diagrams. Locate points of inflection and sketch the elastic curves.

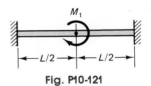

Fig. P10-121

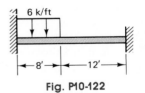

Fig. P10-122

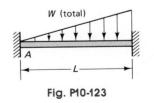

Fig. P10-123

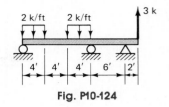

Fig. P10-124

**10-125.** A beam of constant flexural rigidity $EI$ is fixed at both its ends, a distance $L$ apart. If one of the supports settles vertically downward an amount $\Delta$ relative to the other support (without causing any rotation), what moments will be induced at the ends?

**10-126 and 10-127.** Plot the shear and bending-moment diagrams for the beams of variable flexural rigidities shown in the figures. Locate points of inflection and sketch the elastic curves.

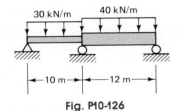

Fig. P10-126

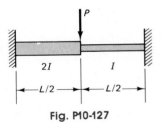

Fig. P10-127

**10-128.** Rework Example 10-25 after assuming that right support $C$ is pinned.

**10-129.** After assuming that the left support is fixed, rework Prob. 10-124.

**10-130.** Assuming that both supports $A$ and $C$ are fixed, rework Example 10-25.

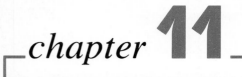

# Stability of Equilibrium: Columns

## 11-1. Introduction

The selection of structural and machine elements is based on three characteristics: **strength, stiffness,** and **stability.** The procedures of stress and deformation analyses in a state of *stable* equilibrium were discussed in some detail in the preceding chapters. But not all structural systems are necessarily stable. For example, consider a square-ended metal rod of say 10 mm in diameter. If such a rod were made 20 mm long to act as an axially compressed member, no question of instability would enter, and a considerable force could be applied. On the other hand, if another rod of the same material were made 1000 mm long to act in compression, then, at a much smaller load than the short piece could carry, the long rod would buckle laterally and could collapse. A slender measuring stick, if subjected to an axial compression, could fail in the same manner. The consideration of material strength alone is not sufficient to predict the behavior of such members. Stability considerations are primary in some structural systems.

The phenomenon of structural instability occurs in numerous situations where compressive stresses are present. Thin sheets, although fully capable of sustaining tensile loadings, are very poor in transmitting compression. Narrow beams, unbraced laterally, can turn sidewise and collapse under an applied load. Vacuum tanks, as well as submarine hulls, unless properly designed, can severely distort under external pressure and can assume shapes that differ drastically from their original geometry. A thin-walled tube can wrinkle like tissue paper when subjected either to axial compression or a torque; see Fig. 11-1.[1] During some stages of firing, the

[1] Figures are adapted from L. A. Harris, H. W. Suer, and W. T. Skene, "Model Investigations of Unstiffened and Stiffened Circular Shells," *Experimental Mechanics* (July 1961): 3 and 5.

(a)  (b)

Fig. 11-1 Typical buckling patterns for thin-walled cylinders (a) in compression and (b) in torsion for a pressurized cylinder. (Courtesy L. A. Harris of North American Aviation, Inc.)

thin casings of rockets are critically loaded in compression. These are crucially important problems for engineering design. Moreover, often the buckling or wrinkling phenomena observed in loaded members occur rather suddenly. For this reason, many structural instability failures are spectacular and very dangerous.

A vast number of the structural instability problems suggested by the preceding listing of problems are beyond the scope of this text.[2] Essentially, only the column problem will be considered here.

For convenience, this chapter is divided into two parts. Part A is devoted to the theory of column buckling, and Part B deals with design applications. First, however, examples of possible instabilities that may occur in straight prismatic members with different cross sections will be discussed. This will be followed by establishing the stability criteria for static equilibrium. The purpose of the next two introductory sections is

[2] F. Bleich, *Buckling Strength of Metal Structures* (New York: McGraw-Hill, 1952). D. O. Brush, and B. O. Almroth, *Buckling of Bars, Plates, and Shells* (New York: McGraw-Hill, 1975). A. Chajes, *Principles of Structural Stability* (Englewood Cliffs, NJ: Prentice-Hall, 1974). G. Gerard et al., *Handbook of Structural Stability*, Parts I–VI, NACA TN, 3781-3786, (Washington, D.C.: NASA 1957–1958). B. G. Johnston (ed.), *Design Criteria for Metal Compression Members*, 4th ed. (New York: Wiley, 1988). S. P. Timoshenko, and J. M. Gere, *Theory of Elastic Stability*, 2nd ed. (New York: McGraw-Hill, 1961). A. S. Volmir, *Flexible Plates and Shells*, Air Force Flight Dynamics Laboratory (trans.), Technical Report No. 66-216, Wright-Patterson Air Force Base, 1967.

to clarify for the reader the aspects of column instability considered in the remainder of the chapter.

## *11-2. Examples of Instability

Analysis of the general instability problem of even straight *prismatic* columns discussed in this chapter is rather complex, and it is important to be aware, at least in a *qualitative* way, of the complexities involved to understand the limitations of the subsequently derived equations. Buckling of straight columns is strongly influenced by the type of cross section and some considerations of this problem follow.

In numerous engineering applications, compression members have tubular cross sections. If the wall thickness is thin, the plate-like elements of such members can buckle locally. An example of this behavior is illustrated in Fig. 11-2(a) for a square thin-walled tube. At a sufficiently large axial load, the side walls tend to subdivide into a sequence of alternating inward and outward buckles. As a consequence, the plates carry a smaller axial stress in the regions of large amount of buckling displacement away from corners; see Fig. 11-2(b). For such cases, it is customary to approximate the complex stress distribution by a constant allowable stress acting over an *effective width w* next to the corners or stiffeners.[3] In this text, except for the design of aluminum-alloy columns, it will be assumed that the thicknesses of a column plate element are sufficiently large to exclude the need for considering this local buckling phenomenon.

Some aspects requiring attention in a general column instability problem are illustrated in Fig. 11-3. Here the emphasis is placed on the kind of buckling that is possible in prismatic members. A plank of limited flexural but adequate torsional stiffness subjected to an axial compressive force is shown to buckle in a bending mode; see Fig. 11-3(a). If the same plank is subjected to end moments, Fig. 11-3(b), in addition to a flexural buckling mode, the cross sections also have a tendency to twist. This is a torsion-bending mode of buckling, and the same kind of buckling may occur for the eccentric force $P$, as shown in Fig. 11-3(c). Lastly, a pure torsional buckling mode is illustrated in Fig. 11-3(d). This occurs when the torsional stiffness of a member is small. As can be recalled from Section 4-14, thin-walled *open sections* are generally poor in torsional stiffness. In contrast, thin-walled *tubular* members are excellent for resisting torques and are torsionally stiff. Therefore, a tubular member, such as shown in Fig. 11-2, generally, will not exhibit torsional buckling. A number of the open thin-walled sections in Fig. 11-4 are next examined for their susceptibility to torsional buckling.

Two sections having biaxial symmetry, where centroids $C$ and shear centers $S$ coincide, are shown in Fig. 11-4(a). Compression members hav-

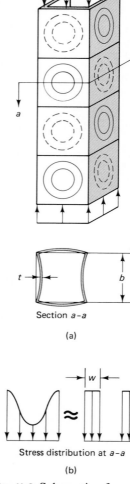

Section *a–a*

(a)

Stress distribution at *a–a*

(b)

**Fig. 11-2** Schematic of buckled thin-walled square tube.

[3] T. von Karman, E. E. Sechler, and L. H. Donnell, "The Strength of Thin Plates in Compression," *Trans. ASME* 54, APM-54-5 (1932): 53 to 57. See also references given in the last footnote.

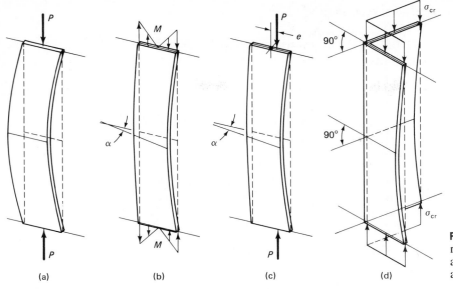

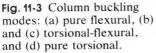

Fig. 11-3 Column buckling modes: (a) pure flexural, (b) and (c) torsional-flexural, and (d) pure torsional.

(a)  (b)  (c)  (d)

ing such cross sections buckle either in pure flexure, Fig. 11-3(a), or twist around $S$, Fig. 11-3(d). For *thin-walled* members, when the torsional stiffness (Section 4-14) is smaller than the flexural stiffness, a column may twist before exhibiting flexural buckling. Generally, this is more likely to occur in columns with cruciform cross sections than in I-shaped sections. However, the torsional mode of buckling generally does not control the design, since the usual rolled or extruded metal cross sections are relatively thick.

The cross sections shown in Figs. 11-4(b) and (c) have their centroids $C$ and shear centers $S$ in different locations. Flexural buckling would occur for the sections in Fig. 11-4(b) if the smallest flexural stiffness around the major principal axis is less than the torsional stiffness. Otherwise, simultaneous flexural and torsional buckling would develop, with the member twisting around $S$. For the sections in Fig. 11-4(c), buckling always occurs in the latter mode. In the subsequent derivations, it will be assumed that the wall thicknesses of members are sufficiently large to exclude the possibility of torsional or torsional-flexural buckling. Compression members having cross sections of the type shown in Fig. 11-4(c) are not considered.

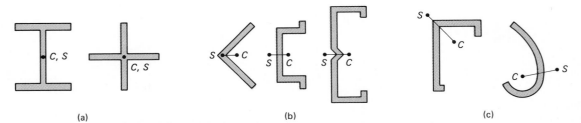

(a)  (b)  (c)

Fig. 11-4 Column sections exhibiting different buckling modes.

**577**

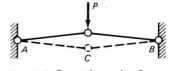

**Fig. 11-5** Snap-through of compression bars.

**Fig. 11-6** Spiral spatial twist-buckling of a slender shaft.

The following interesting cases of possible buckling of straight members are also excluded from consideration in this text. One of these is shown in Fig. 11-5, where two bars with pinned joints at the ends form a very small angle with the horizontal. In this case, it is possible that applied force $P$ can reach a magnitude such that the deformed compressed bars become horizontal. Then, on a slightly further increase in $P$, the bars *snap-through* to a new equilibrium position. This kind of instability is of great importance in shallow thin-walled shells and curved plates. Another possible buckling problem is shown in Fig. 11-6, where a slender circular bar is subjected to torque $T$. When applied torque $T$ reaches a critical value, the bar snaps into a helical spatial curve.[4] This problem is of importance in the design of long slender transmission shafts.

## 11-3. Criteria for Stability of Equilibrium

In order to clarify the stability criteria for static equilibrium,[5] consider a rigid vertical bar with a torsional spring of stiffness $k$ at the base, as shown in Fig. 11-7(a). The behavior of such a bar subjected to vertical force $P$ and horizontal force $F$ is shown in Fig. 11-7(b) for a large and a small $F$. The question then arises: How will this system behave if $F = 0$?

To answer this question analytically, the system must be *deliberately displaced* a small (infinitesimal) amount consistent with the boundary con-

[4] A. G. Greenhill, "On the Strength of Shafting when Exposed Both to Torsion and to End Thrust. Appendix: Theoretical Investigation of the Stability of Shafting under Given Forces," *Proceedings*, (London: Institution of Mechanical Engineers, 1883), 190–209.

[5] Some readers may find it advantageous to study Sections 11-9 and 11-10 first, where columns subjected to axial and transverse loads acting simultaneously are considered. Column buckling is the limiting (degenerate) case in such problems.

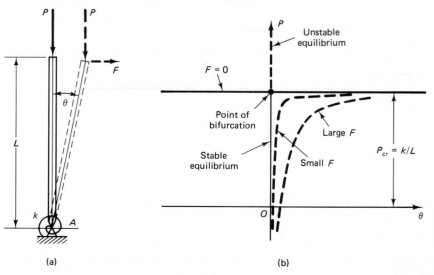

**Fig. 11-7** Buckling behavior of a rigid bar.

(a)

(b)

ditions. Then, if the restoring forces are greater than the forces tending to upset the system, the system is stable, and vice versa.

The rigid bar shown in Fig. 11-7(a) can only rotate. Therefore, it has only one degree of freedom. For an assumed small rotation angle $\theta$, the *restoring moment* is $k\theta$, and, with $F = 0$, the *upsetting moment* is $PL \sin \theta \approx PL\theta$. Therefore, if

$$k\theta > PL\theta \qquad \text{the system is } \textit{stable} \qquad (11\text{-}1)$$

and if

$$k\theta < PL\theta \qquad \text{the system is } \textit{unstable} \qquad (11\text{-}2)$$

Right at the transition point, $k\theta = PL\theta$, and the equilibrium is neither stable nor unstable, but is *neutral*. The force associated with this condition is the *critical*, or *buckling*, *load*, which will be designated $P_{cr}$. For the bar system considered,

$$P_{cr} = k/L \qquad (11\text{-}3)$$

In the presence of horizontal force $F$, the $P - \theta$ curves are as shown by the dashed lines in Fig. 11-7(b) becoming asymptotic to the horizontal line at $P_{cr}$. Similar curves would result by placing the vertical force $P$ eccentrically with respect to the axis of the bar. In either case, even for unstable systems, $\theta$ cannot become infinitely large, as there is always a point of equilibrium at $\theta$ somewhat less than $\pi$. The apparent discrepancy in the graph is caused by assuming in Eqs. 11-1 and 11-2 that $\theta$ is small and that $\sin \theta \approx \theta$, and $\cos \theta \approx 1$. The condition found for neutral equilibrium when $F = 0$ can be further elaborated upon by making reference to Fig. 11-8.

It is convenient to relate the process for determining the kind of stability to a ball resting on differently shaped frictionless surfaces; see Fig. 11-8. In this figure, in all three cases, the balls in position 1 are in equilibrium. In order to determine the kind of equilibrium, it is necessary to displace the balls an infinitesimal distance $\delta\theta$ to either side. In the first case, Fig. 11-8(a), the ball would roll back to its initial position, and the equilibrium is *stable*. In the second case, Fig. 11-8(b), the ball once displaced will not return to its initial position, and the equilibrium is *unstable*. In the last case, Fig. 11-8(c), the ball can remain in its displaced position, where it is again in equilibrium. Such an equilibrium is *neutral*. Therefore, by analogy, a structural system is in a state of neutral equilibrium when it has at least *two neighboring equilibrium positions* an infinitesimal distance

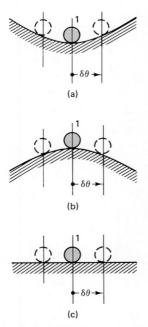

Fig. 11-8 (a) Stable, (b) unstable, and (c) neutral equilibrium.

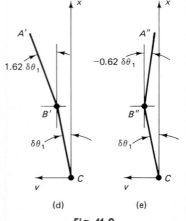

apart. This criterion for neutral equilibrium is applicable only for *infinitesimal* displacements, as at large displacements, different conditions may prevail (Fig. 11-5).

Based on the previous reasoning, the horizontal line for $F = 0$ shown in Fig. 11-7(b) is purely schematic for defining $P_{cr}$. Theoretically it has meaning only within an infinitesimal distance from the vertical axis.

To demonstate this again, consider the rigid vertical bar shown in Fig. 11-7(a) and set $F = 0$. Then, in order to determine *neutral* equilibrium, *displace* the bar in either direction through an angle $\delta\theta$ (*not* through the angle $\theta$ shown in the figure) and formulate the equation of equilibrium:

$$PL\,\delta\theta - k\,\delta\theta = 0 \qquad \text{or} \qquad \boxed{(PL - k)\,\delta\theta = 0} \qquad (11\text{-}4)$$

This equation has *two* distinct solutions: first, when $\delta\theta = 0$ and $P$ is arbitrary, Fig. 11-7(b), and, second, when the expression in parentheses vanishes. This second solution yields $P_{cr} = k/L$. For this value of the axial force, $\delta\theta$ is arbitrary. Therefore, there are *two* equilibrium positions at $P_{cr}$. One of these is for a straight bar, and the other for a bar inclined at an angle $\delta\theta$. Since at $P_{cr}$, there are these two branches of the solution, such a point is called the *bifurcation (branch) point*.[6]

In the previous illustration, the rigid bar has only one degree of freedom, since for an arbitrary infinitesimal displacement, the system is completely described by angle $\delta\theta$. A problem with two degrees of freedom is analyzed in the following example.

(a)          (b)

(c)

## **EXAMPLE 11-1

Two rigid bars, each of length $L$, forming a straight vertical member as shown in Fig. 11-9(a), have torsional springs of stiffness $k$ at ideal pinned joints $B$ and $C$. Determine the critical vertical force $P_{cr}$ and the shape of the buckled member.

### Solution

In order to determine the critical buckling force $P_{cr}$, the system must be given a displacement compatible with the boundary conditions. Such a displacement with positive sense is shown as $A'B'C$ in Fig. 11-9(a). Bar $BC$ rotates through an angle $\delta\theta_1$, and bar $AB$ independently rotates through an angle $\delta\theta_2$. Therefore, this system has *two* degrees of freedom. Free-body diagrams for members $AB$ and $BC$ in deflected positions are drawn in Figs. 11-9(b) and (c). Then, assuming that the

(d)          (e)

**Fig. 11-9**

[6] The static criterion for neutral equilibrium is not applicable to nonconservative systems where bifurcation does not occur and dynamic criteria must be used. Such cases arise, for example, when applied force $P$ remains tangent to the axis of the deflected bar at the point of application. This problem was extensively studied by H. Ziegler. See his book, *Principles of Structural Stability* (Waltham, Mass.: Blaisdell, 1968).

member rotations are infinitesimal, equations of equilibrium are written for each member. In writing these equations, it should be noted that $M_1 = k(\delta\theta_2 - \delta\theta_1)$, where the terms in parentheses constitute the infinitesimal rotation angle *between* the two bars. On this basis,

$$\sum M_{B'} = 0: \qquad PL\,\delta\theta_2 - k(\delta\theta_2 - \delta\theta_1) = 0$$

and

$$\sum M_C = 0: \qquad PL\,\delta\theta_1 + k(\delta\theta_2 - \delta\theta_1) - k\,\delta\theta_1 = 0$$

Rearranging,

$$k\,\delta\theta_1 - (k - PL)\,\delta\theta_2 = 0$$
$$-(2k - PL)\,\delta\theta_1 + k\,\delta\theta_2 = 0$$

These two homogeneous linear equations possess a trivial solution, $\delta\theta_1 = \delta\theta_2 = 0$, as well as a nonzero solution[7] if the determinant of the coefficients is zero, i.e.,

$$\begin{vmatrix} k & -(k - PL) \\ -(2k - PL) & k \end{vmatrix} = 0$$

On expanding this determinant, one obtains the *characteristic equation*

$$P^2 - 3\frac{k}{L}P + \frac{k^2}{L^2} = 0$$

The roots of such an equation are called *eigenvalues*, the *smallest* of which is the critical buckling load. In this case, there are two roots:

$$P_1 = \frac{3 - \sqrt{5}}{2}\frac{k}{L} \qquad \text{and} \qquad P_2 = \frac{3 + \sqrt{5}}{2}\frac{k}{L}$$

and $P_{cr} = P_1$.

Substituting the roots into either one of the simultaneous equations determines the ratios between the rotations of the bars. Thus, for $P_1$, $\delta\theta_2/\delta\theta_1 = 1.62$, and for $P_2$, $\delta\theta_2/\delta\theta_1 = -0.62$. The corresponding deflected modes are shown in Figs. 11-9(d) and (e). The one in Fig. 11-9(d) corresponds to $P_{cr}$. These mode shapes are called *eigenvectors* and are often written in matrix form as

$$\begin{Bmatrix} \delta\theta_1 \\ \delta\theta_2 \end{Bmatrix} = \begin{Bmatrix} 1 \\ 1.62 \end{Bmatrix}\delta\theta_1 \qquad \text{and} \qquad \begin{Bmatrix} \delta\theta_1 \\ \delta\theta_2 \end{Bmatrix} = \begin{Bmatrix} 1 \\ -0.62 \end{Bmatrix}\delta\theta_1$$

where $\delta\theta_1$ is an arbitrary constant.

[7] Heuristically, this can be demonstrated in the following manner. Let two homogeneous linear equations be
$$Ax + By = 0 \qquad \text{and} \qquad Cx + Dy = 0$$
The first one of these equations requires that $y/x = -A/B$, whereas the second that $y/x = -C/D$. For the two equations to be consistent, $A/B = C/D$, or $AD - CB = 0$, which is the value of the expanded determinant for the coefficients in the simultaneous equations. There also is a trivial solution for $x = y = 0$.

As can be readily surmised, by increasing the number of hinged bars with springs to represent a column, the degrees of freedom increase. In the limit, a continuous elastic column has an infinite number of degrees of freedom. However, unlike vibration problems, in buckling analysis, only the smallest root is important. The buckling loads for elastic columns with different boundary conditions will be derived in the sections that follow.

Before proceeding with the derivation for critical column loads based on the concept of neutral equilibrium, it is significant to examine the meaning of such analyses. Critical loads do not describe the postbuckling process. However, by using the exact (nonlinear) differential equations for curvature, it can be shown[8] that for *elastic* columns, one can find equilibrium positions *above* $P_{cr}$. The results of such an analysis are illustrated in Fig. 11-10. Note, especially, that increasing $P_{cr}$ by a mere 1.5 percent causes a maximum sideways deflection of 22 percent of the column length.[9] For practical reasons, such enormous deflections can seldom be tolerated. Moreover, the material usually cannot resist the induced bending stresses. Therefore, failure of real columns would be inelastic. Generally there is little additional post-buckling strength for real columns, and the use of $P_{cr}$ for column capacity is acceptable. This contrasts with the behavior of plates and shells where significant post-buckling strength may develop.

Another illustration of the meaning of $P_{cr}$ in relation to the behavior of elastic[10] and elastic-plastic[11] columns based on nonlinear analyses is

[8] J. L. Lagrange, "On the Shapes of Columns," *Oevres de Lagrange*, Vol. 1 (Paris, 1867).

[9] The fact that an elastic column continues to carry a load beyond the buckling stage can be demonstrated by applying a force in excess of the buckling load to a flexible bar or plate such as a carpenter's saw.

[10] Discussion of elastic deflection of columns, referred to as *Lagrange Elastica*, may be found in S. P. Timoshenko, and J. M. Gere, *Theory of Elastic Stability*, 2nd ed. (New York: McGraw-Hill, 1961).

[11] T. von Karman, "Untersuchungen Ueber Knickfestigkeit," *Collected Works of Theodore von Karman*, Volume I, 1902–1913, (London: Butterworth Scientific Publications, 1956, 90–140). See also the previous footnote.

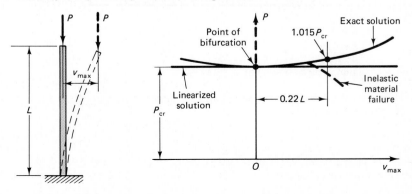

**Fig. 11-10** Behavior of an ideal elastic column.

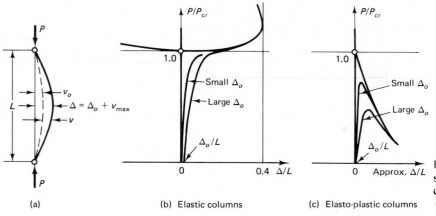

**Fig. 11-11** Behavior of straight and initially curved columns where $(v_0)_{max} = \Delta_0$.

(a)  (b) Elastic columns  (c) Elasto-plastic columns

shown in Fig. 11-11. In these plots, columns that are initially bowed into sinusoidal shapes with a maximum center deflection of $\Delta_o$ are considered. The paths of equilibrium for these cases vary, depending on the extent of the initial curvature. However, regardless of the magnitude of $\Delta_o$, critical load $P_{cr}$ serves as an asymptote for columns with a small amount of curvature, which are commonly encountered in engineering problems; see Fig. 11-11(b). It is to be noted that a *perfectly elastic* initially straight long column with pinned ends, upon buckling into approximately a complete "circle," attains the intolerable deflection of 0.4 of the column length. Behavior of elastic-plastic columns is entirely different; see Fig. 11-11(c). Only a perfectly straight column can reach $P_{cr}$ and thereafter drop precipitously in its carrying capacity. Column imperfections such as crookedness drastically reduce the carrying capacity. Nevertheless, in either case, $P_{cr}$ provides the essential parameter for determining column capacity. With appropriate safeguards, design procedures can be devised employing this key parameter.

# Part A    BUCKLING THEORY FOR COLUMNS

## 11-4. Euler Load for Columns with Pinned Ends

At the critical load, a column that is circular or tubular in its cross-sectional area may buckle sideways in any direction. In the more general case, a compression member does not possess equal flexural rigidity in all directions. The moment of inertia is a maximum around one centroidal axis and of the cross-sectional area a minimum around the other; see Fig. 11-12. The significant flexural rigidity $EI$ of a column depends on the *minimum I*, and at the critical load a column buckles either to one side

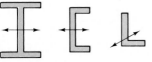

**Fig. 11-12** Flexural column buckling occurs in plane of major axis.

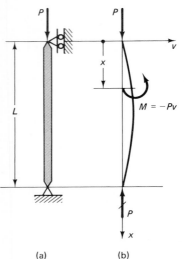

P

P

$v$

$x$

$M = -Pv$

$L$

$P$

$x$

(a)                    (b)

**Fig. 11-13** Column pinned at both ends.

or the other in the plane of the major axis. The use of a minimum $I$ in the derivation that follows is understood.

Consider the ideal perfectly straight column with pinned supports at both ends; see Fig. 11-13(a). The *least* force at which a buckled mode is possible is the *critical* or *Euler buckling load*.

In order to determine the critical load for this column, the compressed column is *displaced* as shown in Fig. 11-13(b). In this position, the bending moment according to the beam sign convention[12] is $-Pv$. By substituting this value of moment into Eq. 10-10, the differential equation for the elastic curve for the initially straight column becomes

$$\frac{d^2v}{dx^2} = \frac{M}{EI} = -\frac{P}{EI}v \qquad (11\text{-}5)$$

by letting $\lambda^2 = P/EI$, and transposing, gives

$$\frac{d^2v}{dx^2} + \lambda^2 v = 0 \qquad (11\text{-}6)$$

This is an equation of the same form as the one for simple harmonic motion, and its solution is

$$v = A \sin \lambda x + B \cos \lambda x \qquad (11\text{-}7)$$

where $A$ and $B$ are arbitrary constants that must be determined from the boundary conditions. These conditions are

$$v(0) = 0 \qquad \text{and} \qquad v(L) = 0$$

Hence,        $v(0) = 0 = A \sin 0 + B \cos 0$     or     $B = 0$

and                              $v(L) = 0 = A \sin \lambda L \qquad (11\text{-}8)$

This equation can be satisfied by taking $A = 0$. However, with $A$ and $B$ each equal to zero, as can be seen from Eq. 11-7, this is a solution for a straight column, and is usually referred to as a trivial solution. An alternative solution is obtained by requiring the sine term in Eq. 11-8 to vanish. This occurs when $\lambda n$ equals $n\pi$, where $n$ is an integer. Therefore, since $\lambda$ was defined as $\sqrt{P/EI}$, the $n$th critical force $P_n$ that makes the deflected shape of the column possible follows from solving $\sqrt{P/EI} \, L = n\pi$. Hence,

[12] For the positive direction of the deflection $v$ shown, the bending moment is negative. If the column were deflected in the opposite direction, the moment would be positive. However, $v$ would be negative. Hence, to make $Pv$ positive, it must likewise be treated as a negative quantity.

$$P_n = \frac{n^2\pi^2 EI}{L^2} \qquad (11\text{-}9)$$

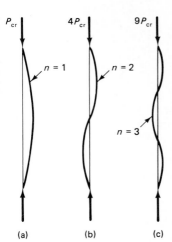

These $P_n$'s are the *eigenvalues* for this problem. However, since in stability problems only the *least value* of $P_n$ is of importance, $n$ must be taken as unity, and the *critical* or *Euler load*[13] $P_{cr}$ for an initially *perfectly straight elastic* column with pinned ends becomes

$$\boxed{P_{cr} = \frac{\pi^2 EI}{L^2}} \qquad (11\text{-}10)$$

where $E$ is the elastic modulus of the material, $I$ is the *least* moment of inertia of the constant cross-sectional area of a column, and $L$ is its length. This case of a column pinned at both ends is often referred to as the *fundamental case*.

**Fig. 11-14** First three buckling modes for a column pinned at both ends.

According to Eq. 11-7, at the critical load, since $B = 0$, the equation of the buckled elastic curve is

$$v = A \sin \lambda x \qquad (11\text{-}11)$$

This is the characteristic, or *eigenfunction*, of this problem, and, since $\lambda = n\pi/L$, $n$ can assume any integer value. There is an infinite number of such functions. In this linearized solution, amplitude $A$ of the buckling mode remains indeterminate. For the fundamental case $n = 1$, the elastic curve is a half-wave sine curve. This shape and the modes corresponding to $n = 2$ and 3 are shown in Fig. 11-14. The higher modes have no physical significance in buckling problems, since the least critical buckling load occurs at $n = 1$.

## 11-5. Euler Loads for Columns with Different End Restraints

The same procedure as that discussed before can be used to determine the critical axial loads for columns with different boundary conditions. The solutions of these problems are very sensitive to the end restraints. Consider, for example, a column with one end fixed and the other pinned, as shown in Fig. 11-15, where the buckled column is drawn in a *deflected* position. Here the effect of unknown end moment $M_o$ and the reactions must be considered in setting up the differential equation for the elastic curve at the critical load:

$$\frac{d^2v}{dx^2} = \frac{M}{EI} = \frac{-Pv + M_o(1 - x/L)}{EI} \qquad (11\text{-}12)$$

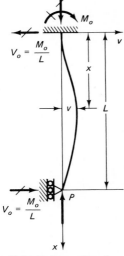

[13] This formula was derived by the great mathematician Leonhard Euler in 1757.

**Fig. 11-15** Column fixed at one end and pinned at the other.

Letting $\lambda^2 = P/EI$ as before, and transposing, gives

$$\frac{d^2v}{dx^2} + \lambda^2 v = \frac{\lambda^2 M_o}{P}\left(1 - \frac{x}{L}\right) \tag{11-13}$$

The *homogeneous solution* of this differential equation, i.e., when the right side is zero, is the same as that given by Eq. 11-7. The *particular solution*, due to the nonzero right side, is given by dividing the term on that side by $\lambda^2$. The complete solution then becomes

$$v = A \sin \lambda x + B \cos \lambda x + (M_o/P)(1 - x/L) \tag{11-14}$$

where $A$ and $B$ are arbitrary constants, and $M_o$ is the unknown moment at the fixed end. The three kinematic boundary conditions are

$$v(0) = 0 \qquad v(L) = 0 \qquad \text{and} \qquad v'(0) = 0$$

Hence,     $v(0) = 0 = B + M_o/P$
$$v(L) = 0 = A \sin \lambda L + B \cos \lambda L$$

and     $v'(0) = 0 = A\lambda - M_o/PL$

Solving these equations simultaneously, one obtains the following transcendental equation

$$\lambda L = \tan \lambda L \tag{11-15}$$

which must be satisfied for a nontrivial equilibrium shape of the column at the critical load. The smallest root of Eq. 11-15 is

$$\lambda L = 4.493$$

from which the corresponding least eigenvalue or critical load for a *column fixed at one end and pinned at the other* is

$$P_{\text{cr}} = \frac{20.19EI}{L^2} = \frac{2.05\pi^2 EI}{L^2} \tag{11-16}$$

It can be shown that in the case of a *column fixed at both ends*, Fig. 11-16(d), the critical load is

$$P_{\text{cr}} = \frac{4\pi^2 EI}{L^2} \tag{11-17}$$

The last two equations show that by restraining the ends the critical loads are substantially larger than those in the fundamental case, Eq. 11-10. On

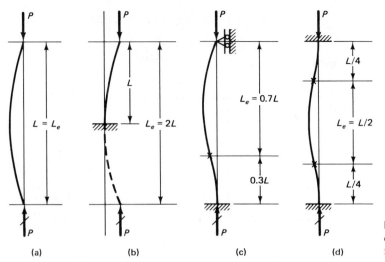

(a)   (b)   (c)   (d)

Fig. 11-16 Effective lengths of columns with different restraints.

the other hand, the critical load for a *free-standing column*,[14] Fig. 11-16(b), with a load at the top is

$$P_{cr} = \frac{\pi^2 EI}{4L^2} \qquad (11\text{-}18)$$

In this extreme case, the critical load is only one-fourth of that for the fundamental case.

All the previous formulas can be made to resemble the fundamental case, provided that the *effective column lengths* are used instead of the actual column length. This length turns out to be the distance between the inflection points on the elastic curves. The effective column length $L_e$ for the fundamental case is $L$, but for the cases discussed it is $0.7L$, $0.5L$, and $2L$, respectively. For a general case, $L_e = KL$, where $K$ is the effective length factor, which depends on the end restraints.

In contrast to the classical cases shown in Fig. 11-16, actual compression members are seldom truly pinned or completely fixed against rotation at the ends. Because of the uncertainty regarding the fixity of the ends, 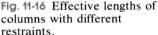 columns are often assumed to be pin-ended. With the exception of the case shown in Fig. 11-16(b), where it cannot be used, this procedure is conservative.

## Procedure Summary

Column buckling loads in this and the preceding section are found using the same curvature-moment relation that was derived for the deflection of beams, Eqs. 10-10. However, the bending moments are written for

---

[14] A telephone pole having no external braces and with a heavy transformer at the top is an example.

axially loaded columns in slightly *deflected* positions. Mathematically this results in an entirely different kind of second order differential equation than that for beam flexure. The solution of this equation shows that, for the *same load*, two neighboring equilibrium configurations are possible for a column. One of these configurations corresponds to a straight column, the other to a slightly bent column. The axial force associated simultaneously with the bent and the straight shape of the column is the critical buckling load. This occurs at the bifurcation (branching) point of the solution.

In the developed formulation, the columns are assumed to be linearly *elastic*, and to have the same cross section throughout the column length. Only the flexural deformations of a column are considered.

For the second order differential equations considered in this treatment the same *kinematic* boundary conditions are applicable as for beams in flexure, Fig. 10-5.

Elastic buckling load formulas are truly remarkable. Although they *do not depend on the strength of a material*, they determine the carrying capacity of columns. The only material property involved is the elastic modulus $E$, which physically represents the stiffness characteristic of a material.

The previous equations *do not apply* if the axial column stress exceeds the proportional limit of the material. This problem is discussed in the next section.

## 11-6. Limitations of the Euler Formulas

The elastic modulus $E$ was used in the derivation of the Euler formulas for columns; therefore, all the reasoning presented earlier is applicable *while the material behavior remains linearly elastic*. To bring out this significant limitation, Eq. 11-10 is rewritten in a different form. By definition, $I = Ar^2$, where $A$ is the cross-sectional area, and $r$ is its *radius of gyration*. Substitution of this relation into Eq. 11-10 gives

$$P_{\text{cr}} = \frac{\pi^2 EI}{L^2} = \frac{\pi^2 EAr^2}{L^2}$$

or

$$\sigma_{\text{cr}} = \frac{P_{\text{cr}}}{A} = \frac{\pi^2 E}{(L/r)^2} \tag{11-19}$$

where the *critical stress* $\sigma_{\text{cr}}$ for a column is defined as $P_{\text{cr}}/A$, i.e., as an *average* stress over the cross-sectional area $A$ of a column at the critical load $P_{\text{cr}}$. The length of the column is $L$, and $r$ is the *least* radius of gyration of the cross-sectional area, since the original Euler formula is in terms of the minimum $I$. By using the effective length $L_e$, the expression becomes

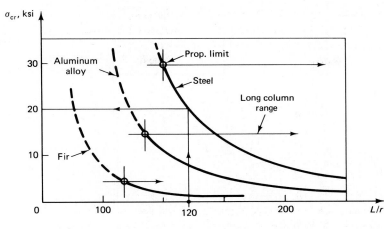

Fig. 11-17 Variation of critical column stress with slenderness ratio for three different materials.

general. The ratio $L/r$ of the column length to the *least* radius of gyration is called the column *slenderness ratio*. *No factor of safety is included in the last equation*.

A graphical interpretation of Eq. 11-19 is shown in Fig. 11-17, where the critical column stress is plotted versus the slenderness ratio for three different materials. For each material, $E$ is constant, and the resulting curve is a hyperbola. However, since Eq. 11-19 is based on the elastic behavior of a material, $\sigma_{cr}$ determined by this equation cannot exceed the proportional limit of a material. Therefore, the hyperbolas shown in Fig. 11-17 are drawn dashed beyond the individual material's proportional limit, and these portions of the curves *cannot be used*. The necessary modifications of Eq. 11-19 to include inelastic material response will be discussed in the next section.

The useful portions of the hyperbolas do not represent the behavior of one column, but rather the behavior of an infinite number of ideal columns. For example, a particular steel column, say, with an $L/r = 120$, may at the most carry a load of $\sigma_1 A$. Note that $\sigma_{cr}$ *always decreases with increasing ratios of* $L/r$. Moreover, note that a precise definition of a long column is now possible with the aid of these diagrams. Thus, a column is said to be long if the elastic Euler formula applies. The beginning of the long-column range is shown for three materials in Fig. 11-17.

## EXAMPLE 11-2

Find the shortest length $L$ for a steel column with pinned ends having a cross-sectional area of 60 by 100 mm, for which the elastic Euler formula applies. Let $E = 200$ GPa and assume the proportional limit to be 250 MPa.

## Solution

The minimum moment of inertia of the cross-sectional area $I_{min} = 100 \times 60^3/12 = 1.8 \times 10^6$ mm⁴. Hence,

$$r = r_{\min} = \sqrt{\frac{I_{\min}}{A}} \tag{11-19a}$$

and
$$r_{\min} = \sqrt{\frac{1.8 \times 10^6}{60 \times 100}} = \sqrt{3} \times 10 \text{ mm}$$

Then, using Eq. 11-19, $\sigma_{cr} = \pi^2 E/(L/r)^2$. Solving for the $L/r$ ratio at the proportional limit,

$$\left(\frac{L}{r}\right)^2 = \frac{\pi^2 E}{\sigma_{cr}} = \frac{\pi^2 \times 200 \times 10^3}{250} = 800\pi^2$$

or
$$\frac{L}{r} = 88.9 \quad \text{and} \quad L = 88.9\sqrt{3} \times 10 = 1540 \text{ mm}$$

Therefore, if this column is 1.54 m or more in length, it will buckle elastically as, for such dimensions of the column, the critical stress at buckling will not exceed the proportional limit for the material.

## 11-7. Generalized Euler Buckling-Load Formulas

A typical compression stress-strain diagram for a specimen that is prevented from buckling is shown in Fig. 11-18(a). In the stress range from $O$ to $A$, the material behaves elastically. If the stress in a column at buckling does not exceed this range, the column buckles elastically. The hyperbola expressed by Eq. 11-19, $\sigma_{cr} = \pi^2 E/(L/r^2)$, is applicable in such a case. This portion of the curve is shown as $ST$ in Fig. 11-18(b). It is important to recall that this curve does not represent the behavior of one column, but rather the behavior of an infinite number of ideal columns of different lengths. The hyperbola beyond the useful range is shown in the figure by dashed lines.

A column with an $L/r$ ratio corresponding to point $S$ in Fig. 11-18(b) is the shortest column of a given material and size that will buckle elastically. A shorter column, having a still smaller $L/r$ ratio, will not buckle at the proportional limit of the material. On the compression stress-strain diagram, Fig. 11-18(a), this means that the stress level in the column has passed point $A$ and has reached some point $B$ perhaps. At this higher stress level, it may be said that a column of different material has been created, since the stiffness of the material is no longer represented by the elastic modulus. At this point, the material stiffness is given instantaneously by the tangent to the stress-strain curve, i.e., by the *tangent modulus* $E_t$; see Fig. 11-18(a). The column remains stable if its new flexural rigidity $E_t I$ at $B$ is sufficiently large, and it can carry a higher load. As the load is increased, the stress level rises, whereas the tangent modulus decreases. A column of ever "less stiff material" is acting under an increasing load. Substitution of the tangent modulus $E_t$ for the elastic modulus $E$ is then the only modification necessary to make the elastic buckling formulas applicable in the inelastic range. Hence, the *generalized Euler buckling-load formula*, or the *tangent modulus formula*,[15] becomes

$$\sigma_{cr} = \frac{\pi^2 E_t}{(L/r)^2} \qquad (11\text{-}20)$$

Since stresses corresponding to the tangent moduli can be obtained from the compression stress-strain diagram, the $L/r$ ratio at which a column will buckle with these values can be obtained from Eq. 11-20. A plot representing this behavior for low and intermediate ratios of $L/r$ is shown

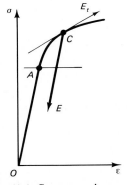

**Fig. 11-A** Stress-strain behavior in buckled column.

---

[15] The tangent modulus formula gives the carrying capacity of a column at the *instant it tends to buckle*. As a column deforms further, the stiffness of the fibers on the concave side continue to approximately exhibit the tangent modulus $E_t$. The fibers on the convex side, however, on being relieved of some stress, rebound with the original elastic modulus $E$, as shown in Fig. 11-A at point $C$. Inasmuch as two moduli, $E_t$ and $E$, are used in developing this theory (see F. Bleich, *Buckling Strength of Metal Structures* [New York: McGraw-Hill, 1952]), it is referred to either as the *double-modulus* or the *reduced-modulus theory* of column buckling. For the same column slenderness ratio, this theory always gives a slightly higher column buckling capacity than the tangent-modulus theory. The discrepancy between the two solutions is not very large. The reason for this discrepancy was explained by F. R. Shanley (see his paper, "Inelastic Column Theory," *J. Aero. Sci.* 14/5 (May 1947):261–267). According to his concept, buckling proceeds *simultaneously* with the increasing axial load. The applied load given by the tangent-modulus theory increases asymptotically to that given by the double-modulus theory; see Fig. 11-B. However, prior to reaching the load given by the double-modulus theory, one can anticipate a material yield or failure, making the tangent modulus an attractive choice. It is convenient that in the tangent-modulus theory the mechanical properties for the whole cross section are the same, wheras they vary differently for different cross-sections in the double-modulus theory.

The maximum load lying between the tangent-modulus load and the double-modulus load for any time-independent elastic-plastic material and cross section was accurately determined by T. H. Lin (see his paper on "Inelastic Column Buckling," *J. Aeron. Sci.*, Vol. 17, No. 3, 1950, 159–172). J. E. Duberg and T. W. Wilder (see their paper on "Column Behavior in the Plastic Stress Range," *J. Aeron. Sci.*, Vol. 17, No. 6, 1950, 323–327) have further concluded that for materials whose stress-strain curves change gradually in the inelastic range, the maximum column load can be appreciably above the tangent-modulus load. If, however, the material in the inelastic range tends to rapidly exhibit plastic behavior, the maximum load is only slightly higher than the tangent-modulus load.

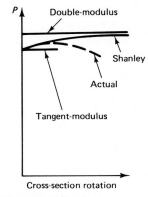

**Fig. 11-B** Inelastic buckling loads by different theories.

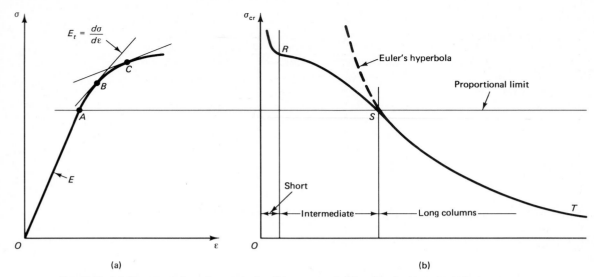

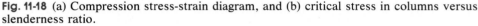

**Fig. 11-18** (a) Compression stress-strain diagram, and (b) critical stress in columns versus slenderness ratio.

in Fig. 11-18(b) by the curve from $R$ to $S$. Tests on individual columns verify this curve with remarkable accuracy.

As mentioned earlier, columns that buckle elastically are generally referred to as *long columns*. Columns having small $L/r$ ratios exhibiting no buckling phenomena are called *short columns*. The remaining columns are of *intermediate length*. At small $L/r$ ratios, ductile materials "squash out" and can carry very large loads.

If length $L$ in Eq. 11-20 is treated as the effective length of a column, different end conditions can be analyzed. Following this procedure for comparative purposes, plots of critical stress $\sigma_{cr}$ versus the slenderness ratio $L/r$ for fixed-ended columns and pin-ended ones are shown in Fig. 11-19. It is important to note that the carrying capacity for these two cases per Eqs. 11-10 and 11-17 is in a ratio of 4 to 1 only for columns having the slenderness ratio $(L/r)_1$ or greater. For smaller $L/r$ ratios, progressively less benefit is derived from restraining the ends. At small $L/r$ ratios, the curves merge. It makes little difference whether a "short block" is pinned or fixed at the ends, as strength rather than buckling determines the behavior.

## *11-8. Eccentric Loads and the Secant Formula

A different method of analysis may be used to determine the capacity of a column than was discussed before. Since no column is perfectly straight nor are the applied forces perfectly concentric, the behavior of real columns may be studied with some statistically determined imperfections or possible misalignments of the applied loads. Then, for the design of an

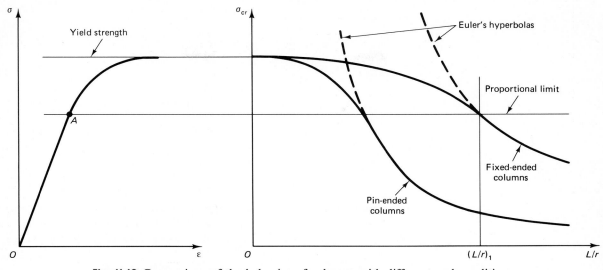

**Fig. 11-19** Comparison of the behavior of columns with different end conditions.

actual column, which is termed "straight," a probable crookedness or an effective load eccentricity may be assigned. Also, there are many columns where an eccentric load is deliberately applied. Thus, an eccentrically loaded column can be studied and its capacity determined on the basis of an allowable elastic stress. *This does not determine the ultimate capacity of a column*.

To analyze the behavior of an eccentrically loaded column, consider the column shown in Fig. 11-20. If the origin of the coordinate axes is taken at the upper force $P$, the bending moment at any section is $-Pv$, and the differential equation for the elastic curve is the same as for a concentrically loaded column, i.e.,

$$\frac{d^2v}{dx^2} = \frac{M}{EI} = -\frac{P}{EI}v \qquad (11\text{-}5)$$

where, by again letting $\lambda = \sqrt{P/EI}$, the general solution is as before:

$$v = A \sin \lambda x + B \cos \lambda x \qquad (11\text{-}7)$$

However, the remainder of the problem is not the same, since *the boundary conditions are now different*. At the upper end, $v$ is equal to the eccentricity of the applied load, i.e., $v(0) = e$. Hence, $B = e$, and

$$v = A \sin \lambda x + e \cos \lambda x \qquad (11\text{-}21)$$

Next, because of symmetry, the elastic curve has a vertical tangent at the midheight of the column, i.e.,

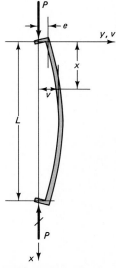

**Fig. 11-20** Eccentrically loaded column.

$$v'(L/2) = 0$$

Therefore, by setting the derivative of Eq. 11-21 equal to zero at $x = L/2$, it is found that

$$A = e\,\frac{\sin \lambda L/2}{\cos \lambda L/2}$$

Hence, the equation for the elastic curve is

$$v = e\left(\frac{\sin \lambda L/2}{\cos \lambda L/2}\,\sin \lambda x + \cos \lambda x\right) \qquad (11\text{-}22)$$

No indeterminacy of any constants appears in this equation, and the maximum deflection $v_{max}$ can be found from it. This maximum deflection occurs at $L/2$, since at this point, the derivative of Eq. 11-22 is equal to zero. Hence,

$$v(L/2) = v_{max} = e\left(\frac{\sin^2 \lambda L/2}{\cos \lambda L/2} + \cos\frac{\lambda L}{2}\right) = e\,\sec\frac{\lambda L}{2} \qquad (11\text{-}23)$$

For the column shown in Fig. 11-20, the largest bending moment $M$ is developed at the point of maximum deflection and numerically is equal to $Pv_{max}$. Therefore, since the direct force and the largest bending moment are now known, the *maximum* compressive stress occurring in the column (contrast this with the *average stress* $P/A$ acting on the column) can be computed by the usual formula as

$$\sigma_{max} = \frac{P}{A} + \frac{Mc}{I} = \frac{P}{A} + \frac{Pv_{max}c}{Ar^2} = \frac{P}{A}\left(1 + \frac{ec}{r^2}\,\sec\frac{\lambda L}{2}\right)$$

But $\lambda = \sqrt{P/EI} = \sqrt{P/EAr^2}$, hence,

$$\boxed{\;\sigma_{max} = \frac{P}{A}\left(1 + \frac{ec}{r^2}\,\sec\frac{L}{r}\sqrt{\frac{P}{4EA}}\right)\;} \qquad (11\text{-}24)$$

This equation, because of the secant term, is known as *the secant formula for columns*, and it applies to columns of any length, provided the maximum stress does not exceed the elastic limit. A condition of equal eccentricities of the applied forces in the same direction causes the largest deflection.

Note that in Eq. 11-24, the radius of gyration *r may not be minimum*, since it is obtained from the value of $I$ associated with the axis around which bending occurs. In some cases, a more critical condition for buckling can exist in the direction of no definite eccentricity. Also note that in Eq. 11-24, *the relation between* $\sigma_{\max}$ *and P is not linear*; $\sigma_{\max}$ *increases faster than P. Therefore, the solutions for maximum stresses in columns caused by different axial forces cannot be superposed*; instead, the forces must be superposed first, and then the stresses can be calculated.

For an allowable force $P_a$ on a column, where $n$ is the factor of safety, $nP_a$, must be substituted for $P$ in Eq. 11-24, and $\sigma_{\max}$ must be set at the yield point of a material, i.e.,

$$\sigma_{\max} = \sigma_{yp} = \frac{nP_a}{A}\left(1 + \frac{ec}{r^2}\sec\frac{L}{r}\sqrt{\frac{nP_a}{4EA}}\right) \qquad (11\text{-}25)$$

This procedure assures a correct factor of safety for the applied force, since such a force can be increased $n$ times before a critical stress is reached. Note the term $nP_a$ appearing under the radical.

Application of Eqs. 11-24 and 11-25 is cumbersome, requiring a trial-and-error procedure. Alternatively, they can be studied graphically, as shown in Fig. 11-21.[16] From this plot, note the large effect that load eccentricity has on short columns and the negligible one on very slender columns. Graphs of this kind form a suitable aid in practical design. The secant equation covers the whole range of column lengths. The greatest

[16] This figure is adapted from D. H. Young, "Rational Design of Steel Columns," *Trans. ASCE* 101 (1936):431.

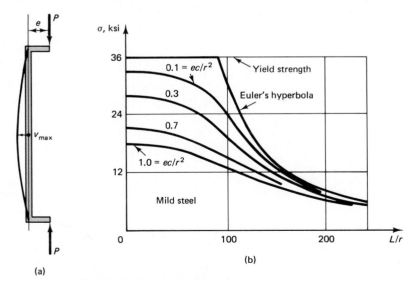

**Fig. 11-21** Results of analyses for different columns by the secant formula.

handicap in using this formula is that some eccentricity $e$ must be assumed even for supposedly straight columns, and this is a difficult task.[17]

The secant formula for *short* columns reverts to a familiar expression when $L/r$ approaches zero. For this case, the value of the secant approaches unity; hence, in the limit, Eq. 11-24 becomes

$$\sigma_{max} = \frac{P}{A} + \frac{Pec}{Ar^2} = \frac{P}{A} + \frac{Mc}{I} \qquad (11\text{-}26)$$

a relation normally used for short blocks.

## **11-9. Beam-Columns

In the preceding section, the problem of an axially loaded column subjected to equal end moments was considered. This is a special case of a member acted upon simultaneously by an axial force and transverse forces or moments causing bending. Such members are referred to as *beam-colums*. The behavior of beam-columns and the linearized solutions that are generally employed for their analysis can be clarified by the simple example of the rigid bar shown in Fig. 11-22(a). This bar of length $L$ is initially held in a vertical position by a spring at $A$ having a torsional spring constant $k$. When vertical force $P$ and horizontal force $F$ are applied to the top of the bar, it rotates and the equilibrium equation must be wrriten for the *deformed state*, a form similar to that used in stability analysis. Bearing in mind that $k\theta$ is the resisting moment developed by the spring at $A$, one obtains

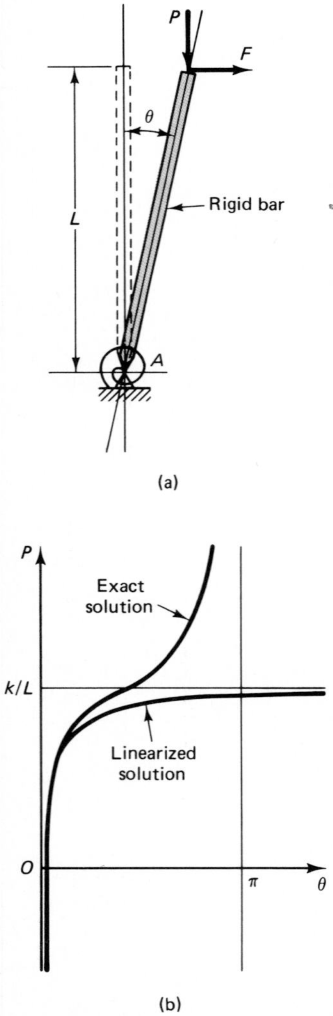

$$\sum M_A = 0 \circlearrowleft + \qquad PL \sin\theta + FL \cos\theta - k\theta = 0$$

or

$$P = \frac{k\theta - FL \cos\theta}{L \sin\theta} \qquad (11\text{-}27)$$

The qualitative features of this result are shown in Fig. 11-22(b), and the corresponding curve is labeled as the *exact solution*. It is interesting to note that as $\theta \to \pi$, provided the spring continues to function, a very large force $P$ can be supported by the system. For a force $P$ applied in an upward direction, plotted downward in the figure, angle $\theta$ decreases as $P$ increases.

The solution expressed by Eq. 11-27 is for arbitrarily large deformations. In complex problems, it is difficult to achieve solutions of such

(a)

(b)

**Fig. 11-22** Rigid bar with one degree of freedom.

[17] Moreover, there is some question as to the philosophical correctness of the secant formula. The fact that the stress reaches a certain value does not mean that the column buckles, i.e., stress is not a measure of buckling load in every case. It can be shown that an additional axial load can be resisted beyond the point where the maximum stress at the critical section is reached. See F. Bleich, *Buckling Strength of Metal Structures* (New York: McGraw-Hill, 1952), Chapter 1.

generality. Moreover, in the majority of applications, large deformations cannot be tolerated. Therefore, it is usually possible to limit the investigation of the behavior of systems to small and moderately large deformations. In this problem, this can be done by setting $\sin \theta \approx \theta$, and $\cos \theta \approx 1$. In this manner, Eq. 11-27 simplifies to

$$P = \frac{k\theta - FL}{L\theta} \qquad \text{or} \qquad \theta = \frac{FL}{k - PL} \qquad (11\text{-}28)$$

For small finite values of $\theta$, this solution is quite acceptable. On the other hand, as $\theta$ increases, the discrepancy between this linearized solution and the exact one becomes very large and loses its physical meaning.

Analogous to this, for the analysis of elastic beam-columns, where the deflections are small to moderate, it is generally sufficiently accurate to employ the usual linear differential equation for elastic deflection of beams. However, in applying this equation, the bending moments caused by the transverse loads as well as the axial forces *must be written for a deflected member*. Such a procedure is illustrated in the next example.

## EXAMPLE 11-3

A beam-column is subjected to an axial force $P$ and an upward transverse force $F$ at its midspan; see Fig. 11-23(a). Determine the equation of the elastic curve, and the critical axial force $P_{cr}$. $EI$ is constant.

## Solution

The free-body diagram for the *deflected* beam-column is shown in Fig. 11-23(b). This diagram assists with formulation of the total bending moment $M$, which includes the effect of the axial force $P$ multiplied by the deflection $v$. Thus, using the relation $M = EIv''$, Eq. 10-10, and noting that for the left side of the span, $M = -(F/2)x - Pv$, one has

$$EIv'' = M = -Pv - (F/2)x \qquad 0 \leq x \leq L/2$$

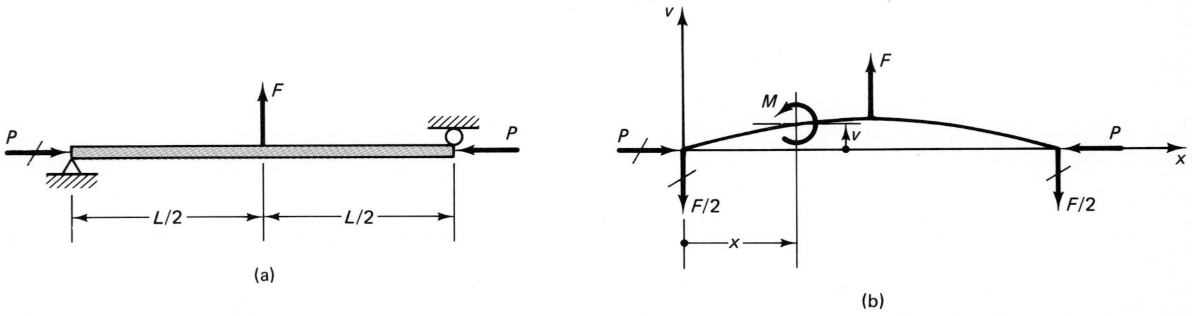

(a)

(b)

Fig. 11-23

or
$$EIv'' + Pv = -(F/2)x$$

By dividing through by $EI$ and letting $\lambda^2 = P/EI$, after some simplification, the governing differential equation becomes

$$\frac{d^2v}{dx^2} + \lambda^2 v = -\frac{\lambda^2 F}{2P}x \qquad 0 \le x \le L/2 \qquad (11\text{-}29)$$

The homogeneous solution of this differential equation is the same as that of Eq. 11-6, and the particular solution equals the right-hand term divided by $\lambda^2$. Therefore, the complete solution is

$$v = C_1 \sin \lambda x + C_2 \cos \lambda x - (F/2P)x \qquad (11\text{-}30)$$

Constants $C_1$ and $C_2$ follow from the boundary condition, $v(0) = 0$, and from a condition of symmetry, $v'(L/2) = 0$. The first condition gives

$$v(0) = C_2 = 0$$

Since

$$v' = C_1\lambda \cos \lambda x - C_2\lambda \sin \lambda x - F/2P$$

with $C_2$ already known to be zero, the second condition gives

$$v'(L/2) = C_1\lambda \cos \lambda L/2 - F/2P = 0$$

or
$$C_1 = F/[2P\lambda \cos (\lambda L/2)]$$

On substituting this constant into Eq. 11-30,

$$v = \frac{F}{2P\lambda}\frac{\sin \lambda x}{\cos \lambda L/2} - \frac{F}{2P}x = \frac{F}{2P\lambda}\left(\frac{\sin \lambda x}{\cos \overline{u}} - \lambda x\right) \qquad (11\text{-}31)$$

where the last relationship is obtained by setting

$$\lambda L/2 = \overline{u} \qquad (11\text{-}32)$$

Since the maximum deflection occurs at $x = L/2$, after some simplifications,

$$v_{\max} = \frac{F}{2P\lambda}(\tan \overline{u} - \overline{u}) \qquad (11\text{-}33)$$

and the absolute maximum bending moment occurring at midspan is

$$M_{\max} = \left| -\frac{FL}{4} - Pv_{\max} \right| = \frac{F}{2\lambda}\tan \overline{u} \qquad (11\text{-}34)$$

Equations 11-31, 11-33, and 11-34 become infinite when $\bar{u}$ is a multiple of $\pi/2$, since then $\cos \bar{u}$ is equal to zero and $\tan \bar{u}$ is infinite. In conformity with this requirement, for an $n$th mode, where $n$ is an integer,

$$\bar{u} = \frac{\lambda L}{2} = \sqrt{\frac{P_n}{EI}} \frac{L}{2} = \frac{n\pi}{2} \tag{11-35}$$

Solving the last two expressions for $P_n$, and setting $n = 1$, the critical buckling load is obtained.

$$P_n = \frac{n^2 \pi^2 EI}{L^2} \quad \Rightarrow \quad P_{cr} = \frac{\pi^2 EI}{L^2} \tag{11-36}$$

This procedure shows that a solution of the *linearized* differential equation yields the *Euler buckling load* causing infinite deflections and moments. For tensile forces, on the other hand, the deflections are reduced. These trends are similar to those shown in Fig. 11-22(b).

Next it is of considerable practical importance to obtain an approximate solution to this problem that can then be generalized for a great many beam-column problems for finding deflections and maximum moments. For this purpose, expand $\tan \bar{u}$ into the Maclaurin (Taylor) series and subtitute the result into Eq. 11-33, making note of Eq. 11-35.

$$\tan \bar{u} = \bar{u} + \frac{1}{3}\bar{u}^3 + \frac{2}{5}\bar{u}^5 + \frac{17}{315}\bar{u}^7 + \frac{62}{2835}\bar{u}^9 + \cdots \tag{11-37}$$

$$v_{max} = \frac{F}{2P\lambda} \frac{1}{3} \left(\frac{\lambda L}{2}\right)^3 \left(1 + \frac{2}{5}\bar{u}^2 + \frac{17}{105}\bar{u}^4 + \frac{62}{945}\bar{u}^6 + \cdots\right) \tag{11-38}$$

However, in view of Eqs. 11-35 and 11-36,

$$\bar{u}^2 = \frac{\lambda^2 L^2}{4} = \frac{PL^2}{4EI} = 2.4674 \frac{P}{P_{cr}} \tag{11-39}$$

By substituting the last equation into Eq. 11-38 and simplifying,

$$v_{max} = \frac{FL^3}{48EI} \left[0.9870\left(\frac{P}{P_{cr}}\right) + 0.9857\left(\frac{P}{P_{cr}}\right)^2 \right. $$
$$\left. + 0.9855\left(\frac{P}{P_{cr}}\right)^3 + \cdots\right] \tag{11-40}$$

By approximating the coefficients[18] in the bracketed expression by unity and

---

[18] A. Chajes, *Principles of Structural Stability* (Englewood Cliffs, NJ: Prentice-Hall, 1974). For discussion of elastic-plastic beam columns, see K. Jezek, "Die Tragfaehigkeit axial gedrueckter und auf Biegung beanspruchter Stahlstaebe," *Der Stahlbau* 9 (1936):12, 22, and 39.

recalling that the sum of the resulting power series[19] can be written in a compact form, one has

$$v_{max} \approx \frac{FL^3}{48EI} \left( \frac{1}{1 - P/P_{cr}} \right) \tag{11-41}$$

In this expression, it can be recognized (see Table 11 in the Appendix) that the coefficient in front of the bracket is the beam center deflection *without* the axial force. The bracketed expression gives the *deflection magnification factor* caused by the applied axial force $P$. When this force reaches $P_{cr}$, the deflection becomes infinite. This magnification factor can be used with virtually any kind of transverse loadings as long as they are applied in the same direction, and the results are remarkably accurate for small and moderate deflections.

After the approximate maximum deflection is obtained using Eq. 11-41, the maximum bending moment follows from statics as

$$M_{max} = \left| -\frac{FL}{4} - Pv_{max} \right| \tag{11-42}$$

where the first term is due to transverse loading, and the second to the axial force in a deflected member. For stocky beam-columns, the last term becomes unimportant.

It is important to note that the differential equations, such as Eq. 11-29, for beam-columns are of different kind than those used for beams loaded transversely only. For this reason, the singularity functions previously presented cannot be applied in these problems.

## **11-10. Alternative Differential Equations for Beam-Columns

For some solutions of beam-column problems, it is convenient to recast the governing differential equations into different forms from that discussed in the previous section. In order to derive such equations, consider the beam-column element shown in Fig. 11-24, and make the following small-deflection approximations:

$$dv/dx = \tan \theta \approx \sin \theta \approx \theta \qquad \cos \theta \approx 1 \qquad \text{and} \qquad ds \approx dx$$

On this basis, the two equilibrium equations are

$$\sum F_y = 0 \uparrow + \qquad q\, dx + V - (V + dV) = 0$$
$$\sum M_A = 0 \circlearrowleft + \qquad M - P\, dv + V\, dx + q\, dx\, dx/2 - (M + dM) = 0$$

[19] This can be verified by dividing unity by the denominator.

The first one of these equations yields

$$\frac{dV}{dx} = q \tag{11-43}$$

which is identical to Eq. 5-3. The second, on neglecting the infinitesimals of higher order, gives

$$V = \frac{dM}{dx} + P\frac{dv}{dx} \tag{11-44}$$

Therefore, for beam-columns, shear $V$, in addition to depending on the rate of change in moment $M$ as in beams, now also depends on the *magnitude of the axial force and the slope of the elastic curve*. The latter term is the component of $P$ along the inclined sections shown in Fig. 11-24.

On substituting Eq. 11-44 into Eq. 11-43 and using the usual beam curvature-moment relation $d^2v/dx^2 = M/EI$, one obtains the two alternative governing differential equations for beam-columns:

$$\boxed{\frac{d^2M}{dx^2} + \lambda^2 M = q} \tag{11-45}$$

or

$$\boxed{\frac{d^4v}{dx^4} + \lambda^2\frac{d^2v}{dx^2} = \frac{q}{EI}} \tag{11-46}$$

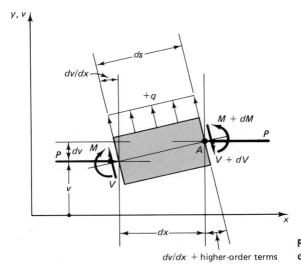

**Fig. 11-24** Beam-column element.

where, for simplicity, $EI$ is assumed to be constant, and, as before, $\lambda^2 = P/EI$. The boundary conditions for these equations are the same as for beams in flexure (see Fig. 10-5), *except for shear* where Eq. 11-44 applies. By again making use of the beam curvature-moment relation, Eq. 11-44 in more appropriate alternative form can be written as

$$V = EI\frac{d^3v}{dx^3} + P\frac{dv}{dx} \qquad (11\text{-}44a)$$

If $P = 0$, Eqs. 11-44a, 11-45, and 11-46 revert, respectively, to Eqs. 10-14b, 5-5, and 10-14c for transversely loaded beams.

For future reference, the homogeneous solution of Eq. 11-46 and several of its derivatives are

$$v = C_1 \sin \lambda x + C_2 \cos \lambda x + C_3 x + C_4 \qquad (11\text{-}47a)$$
$$v' = C_1\lambda \cos \lambda x - C_2\lambda \sin \lambda x + C_3 \qquad (11\text{-}47b)$$
$$v'' = -C_1\lambda^2 \sin \lambda x - C_2\lambda^2 \cos \lambda x \qquad (11\text{-}47c)$$
$$v''' = -C_1\lambda^3 \cos \lambda x + C_2\lambda^3 \sin \lambda x \qquad (11\text{-}47d)$$

These relations are useful for expressing the boundary conditions in evaluating constants $C_1$, $C_2$, $C_3$, and $C_4$. The use of Eq. 11-44a rather than Eq. 10-14b is essential when shear at a boundary must be considered.

Solutions of homogeneous Eqs. 11-45 or 11-46 for particular boundary conditions lead to critical buckling loads for elastic *prismatic* columns. These solutions have the same meaning as discussed earlier in connection with the equivalent solutions of the second order differential equations in Sections 11-4 and 11-5.

### EXAMPLE 11-4

A slender bar of constant $EI$ is simultaneously subjected to end moments $M_o$ and axial force $P$, as shown in Fig. 11-25(a). Determine the maximum deflection and the largest bending moment.

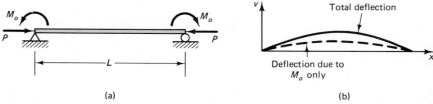

(a)

(b)

**Fig. 11-25**

## Solution

Within the span, there is no transverse load. Therefore, the right-hand term of Eq. 11-46 is zero, and the homogeneous solution of this equation given by Eq. 11-47a is the complete solution. The boundary conditions are

$$v(0) = 0 \qquad v(L) = 0 \qquad M(0) = -M_o \qquad \text{and} \qquad M(L) = -M_o$$

Since $M = EIv''$, with the aid of Eqs. 11-47a and 11-47c, these conditions yield:

$$
\begin{aligned}
v(0) &= &+ C_2 &&+ C_4 &= 0 \\
v(L) &= +C_1 \sin \lambda L &+ C_2 \cos \lambda L + C_3 L &&+ C_4 &= 0 \\
M(0) &= &- C_2 EI\lambda^2 &&&= -M_o \\
M(L) &= -C_1 EI\lambda^2 \sin \lambda L &- C_2 EI\lambda^2 \cos \lambda L &&&= -M_o
\end{aligned}
$$

Solving these four equations simultaneously,

$$C_1 = \frac{M_o}{P} \frac{1 - \cos \lambda L}{\sin \lambda L} \qquad C_2 = -C_4 = \frac{M_o}{P} \qquad \text{and} \qquad C_3 = 0$$

Therefore, the equation of the elastic curve is

$$v = \frac{M_o}{P} \left( \frac{1 - \cos \lambda L}{\sin \lambda L} \sin \lambda x + \cos \lambda x - 1 \right) \qquad (11\text{-}48)$$

The maximum deflection occurs at $x = L/2$. After some simplifications, it is found to be

$$v_{max} = \frac{M_o}{P} \left( \frac{\sin^2 \lambda L/2}{\cos \lambda L/2} + \cos \frac{\lambda L}{2} - 1 \right) = \frac{M_o}{P} \left( \sec \frac{\lambda L}{2} - 1 \right) \qquad (11\text{-}49)$$

The largest bending moment also occurs at $x = L/2$. Its absolute maximum is

$$M_{max} = | -M_o - Pv_{max} | = M_o \sec \lambda L/2 \qquad (11\text{-}50)$$

This solution is directly comparable to that given in Section 11-8 for an eccentrically loaded column. Two differences in the details of the solutions, however, should be noted. The end moments $M_o = Pe$ of the earlier solution and the $x$ axis of the eccentrically loaded column is at a distance $e$ away from the column axis. Then, with the use of some trigonometric identities, it can be shown that Eqs. 11-22 and 11-48 lead to the same results.

The results again show that in slender members, bending moments can be substantially increased in the presence of axial compressive forces. Similar to the condition encountered in Example 11-3, when $\lambda L/2 = \pi/2$, axial force $P = P_{cr}$, and $v_{max}$ and $M_{max}$ become infinite.

If the applied axial forces are tensile instead of compressive, the sign of $P$ changes and so does the character of Eqs. 11-45 and 11-46. For such cases, the deflections are reduced with increasing axial force $P$.

Next Eq. 11-49 is recast into an approximate form in the same manner as has been done in Example 11-3. For this purpose, sec $\lambda L/2 = \text{sec } \bar{u}$ is expanded into the Maclaurin (Taylor) series, and, after substituting into Eq. 11-49, is simplified using Eq. 11-39. Thus,

$$\text{sec } \bar{u} = 1 + \frac{1}{2}\bar{u}^2 + \frac{5}{24}\bar{u}^4 + \frac{61}{720}\bar{u}^6 + \cdots \tag{11-51}$$

and

$$v_{\max} = \frac{M_o}{P}\frac{1}{2}\left(\frac{\lambda L}{2}\right)^2\left[1 + 1.028\left(\frac{P}{P_{cr}}\right)\right. $$
$$\left. + 1.032\left(\frac{P}{P_{cr}}\right)^2 + \cdots \right\} \tag{11-52}$$

Again, all the coefficients in the bracketed expression can be approximated by unity, and the power series summed, giving

$$v_{\max} \approx \frac{M_o L^2}{8EI}\left(\frac{1}{1 - P/P_{cr}}\right) \tag{11-53}$$

The coefficient in front of the bracketed expression is the deflection at the middle of the span due to the end moments $M_o$ (see Table 11 in the Appendix). The deflection magnification factor due to the axial force $P$ in the brackets is identical to that found earlier in Example 11-3. When force $P$ reaches the Euler buckling load, the deflection becomes infinite according to this *linear small deflection theory*.

The maximum bending moment at the center of the beam follows from statics:

$$M_{\max} = \left| -M_o - Pv_{\max} \right| \tag{11-54}$$

## EXAMPLE 11-5

By using Eq. 11-46 in homogeneous form, determine the Euler buckling load for a column with pinned ends.

## Solution

For this purpose, Eq. 11-46 can be written as

$$\frac{d^4v}{dx^4} + \lambda^2\frac{d^2v}{dx^2} = 0 \tag{11-55}$$

The solution of this equation and several of its derivatives are given by Eqs. 11-47. For a pin-ended column, the boundary conditions are

$$v(0) = 0 \qquad v(L) = 0 \qquad M(0) = EIv''(0) = 0$$

and
$$M(L) = EIv''(L) = 0$$

Using these conditions with Eqs. 11-47a and 11-47c, one obtains

$$
\begin{array}{cccc}
 & C_2 & + C_4 = 0 \\
C_1 \sin \lambda L & + C_2 \cos \lambda L + C_3 L + C_4 = 0 \\
 & - C_2 \lambda^2 EI & = 0 \\
- C_1 \lambda^2 EI \sin \lambda L - C_2 \lambda^2 EI \cos \lambda L & = 0
\end{array}
$$

To obtain a nontrivial solution requires that the determinant of the coefficients for this set of homogeneous algebraic equations be equal to zero (see Example 11-1). Therefore, with $\lambda^2 EI = P$,

$$
\begin{vmatrix}
0 & 1 & 0 & 1 \\
\sin \lambda L & \cos \lambda L & L & 1 \\
0 & -P & 0 & 0 \\
-P \sin \lambda L & -P \cos \lambda L & 0 & 0
\end{vmatrix} = 0
$$

The evaluation of this determinant leads to $\sin \lambda L = 0$, which is precisely the same condition as given by Eq. 11-8.

This approach is advantageous in problems with different boundary conditions, where the axial force and $EI$ remain constant throughout the length of the column. The method cannot be applied directly if the axial force extends over only a part of a member.

## Part B    DESIGN OF COLUMNS

## [20]11-11. General Considerations

For other than short columns and blocks, the buckling theory for columns shows that their cross-sectional areas should have the largest possible least radius of gyration $r$. Such a provision for columns assures the smallest possible slenderness ratio, $L_e/r$, permitting the use of higher stresses. However, as discussed in Section 11-2, limitations must be placed on the minimum thickness of the material to prevent local plate buckling. Since conventional rolled shapes generally have wall-thickness ratios sufficiently large to prevent such buckling, only a brief treatment of this problem as it applies to aluminum alloy compression members will be given

[20] The remainder of this chapter is optional.

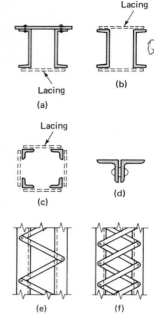

Lacing

(a)

Lacing

Lacing

(b)

Lacing

(c)

(d)

(e)     (f)

**Fig. 11-26** Typical built-up column cross sections.

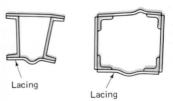

Lacing

Lacing

**Fig. 11-27** Lattice instability.

here. Torsional buckling modes that may control the capacity of columns made from thin plate elements and open unsymmetrical cross sections are excluded from consideration (see Section 11-2).

Since tubular members have a large radius of gyration in relation to the amount of material in a cross section, they are excellent for use as columns. Wide flange sections (sometimes referred to as H sections) are also very suitable for use as columns, and are superior to I sections, which have narrow flanges, resulting in larger ratios of $L_e/r$. In order to obtain a large radius of gyration, columns are often built up from rolled or extruded shapes, and the individual pieces are spread out to obtain the desired effect. Cross sections for typical bridge compression members are shown in Figs. 11-26(a) and (b), for a derrick boom or a radio tower in Fig. 11-26(c), and for an ordinary truss in Fig. 11-26(d). The angles in the latter case are separated by spacers. The main longitudinal shapes in the other members are separated by plates, or are laced (latticed) together by light bars as shown in Figs. 11-26(e) and (f). Local instability must be carefully guarded against to prevent failures in lacing bars, as shown in Fig. 11-27. Such topics are beyond the scope of this text.[21]

Unavoidable imperfections must be recognized in the practical design of columns. Therefore, specifications usually stipulate not only the quality of material, but also fabrication tolerances for permissible out-of-straightness. The residual stresses caused by the manufacturing process must also be considered. For example, steel wide-flange sections, because of uneven cooling during a hot rolling operation, develop residual stress patterns of the type shown in Fig. 11-28. The maximum residual compressive stresses may be on the order of $0.3\sigma_{yp}$ in such members. Welds in aluminum alloy members reduce the mechanical properties of the material in the heat affected zone. For these reasons, experimental results on column buckling have a large scatter.

After initially accepting the Euler buckling-load formula beyond its range of applicability, a chaotic situation existed for many years with

[21] B. G. Johnson (ed.), *Stability Design Criteria for Metal Structures,* 4th ed. (New York: Wiley, 1988).

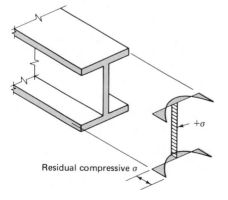

$+\sigma$

Residual compressive $\sigma$

**Fig. 11-28** Schematic residual stress pattern.

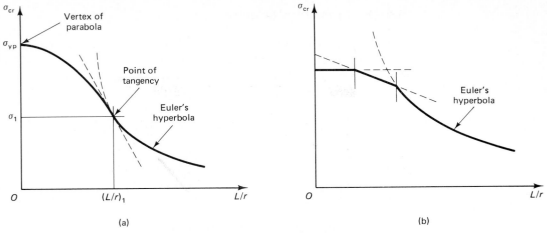

**Fig. 11-29** Typical column-buckling curves for design.

regard to column-design formulas. Now that the column-buckling phenomenon is more clearly understood, only a few column formula types are in common use. For steel, it is now customary to specify two formulas. One of these is for use for short and intermediate-length columns; the other, for slender columns, Fig. 11-29(a). For the lower range of column length, usually a parabola, and, in a few instances, an inclined straight line, is specified. In this manner, the basic compressive strength of the material, residual stresses, and fabrication tolerances are accounted for. For slender (long) columns, the Euler elastic buckling load provides the basis for the critical stress. In this range of column lengths, the residual stresses play a relatively minor role. The dominant parameters are the material stiffness, $E$, and geometric fabrication imperfections. Often the two specified complementary equations have a common tangent at a selected value of $L/r$. Such a condition cannot be fulfilled if a straight line is used instead of a parabola. In a few specifications, the more conservative approach of using the elastic formula and an allowable stress is made by assuming an accidental eccentricity based on manufacturing tolerances.

For some materials, a sequence of three different equations is specified for the design of columns, Fig. 11-29(b). One of these equations for short columns defines the basic compressive strength of a material. Another equation, specifically applicable for the long column range, is based on the Euler buckling load. An empirical relation, such as an inclined straight line shown in the figure, or a parabola, is specified for columns of intermediate lengths. Such a type of formula is generally given for aluminum alloys and wood.

In applying the design formulas, it is important to observe the following items:

1. The material and fabrication tolerances for which the formula is written.

2. Whether the formula gives the working load (or stress) or whether

it estimates the ultimate carrying capacity of a member. If the formula is of the latter type, a safety factor must be introduced.

3. The range of the applicability of the formula. Some empirical formulas can lead to unsafe design if used beyond the specified range.

## *11-12. Concentrically Loaded Columns

As examples of column-design formulas for nominally concentric loading, representative formulas for structtural steel, an aluminum alloy, and wood follow. Formulas for eccentrically loaded columns are considered in the next section.

### Column Formulas for Structural Steel

The American Institute of Steel Construction (AISC)[22] provides two sets of column formulas with two formulas in each set. One of these sets is for use in the _allowable stress design_ (ASD) and the other for the _load and resistance factor design_ (LRFD). In the second approach, an implicit probabilistic determination of the reliability of column capacity based on load and resistance factors is made (see Section 1-12). These two sets of formulas follow. Since steels of several different yield strengths are manufactured, the formulas are stated in terms of $\sigma_{yp}$, which varies for different steels. The elastic modulus $E$ for all steels is approximately the same, and is taken to be $29 \times 10^3$ ksi (200 MPa).

    **AISC ASD Formulas for Columns.** The AISC formula for _allowable_ stress, $\sigma_{\text{allow}}$, for _slender_ columns is based on the Euler elastic buckling load with a safety factor of $23/12 = 1.92$. Slender columns are defined as having the slenderness ratio $(L_e/r)_1 = C_c = \sqrt{2\pi^2 E/\sigma_{yp}}$ or greater. Constant $C_c$ corresponds to the critical stress $\sigma_{cr}$ at the Euler load equal to one-half the steel yield stress $\sigma_{yp}$.

    The formula for long columns when $(L_e/r) > C_c$ is

$$\sigma_{\text{allow}} = \frac{12\pi^2 E}{23(L_e/r)^2} \tag{11-56}$$

where $L_e$ is the effective column length, and $r$ is the least radius of gyration for the cross-sectional area. No columns are permitted to exceed an $L_e/r$ of 200.

    For an $L_e/r$ ratio less than $C_c$, AISC specifies a parabolic formula:

$$\sigma_{\text{allow}} = \frac{[1 - (L_e/r)^2/2C_c^2]\sigma_{yp}}{\text{F.S.}} \tag{11-57}$$

[22] For ASD formulas, see _AISC Manual of Steel Construction,_ 9th ed. (Chicago: AISC, 1989). For LRFD formulas, see _AISC LRFD Manual of Steel Construction_ (Chicago: AISC, 1986). See also B. J. Johnston, F. J. Lin, and T. V. Galambos, _Basic Steel Design,_ 3rd ed. (Englewood Cliffs, NJ: Prentice-Hall, 1986). C. G. Salmon, and J. E. Johnson, _Steel Structures,_ 2nd ed. (New York: Harper and Row, 1980). W-W. Yu, _Cold-Formed Steel Design_ (New York: Wiley, 1985).

where F.S., the factor of safety, is defined as

$$\text{F.S.} = \frac{5}{3} + \frac{3(L_e/r)}{8C_c} - \frac{(L_e/r)^3}{8C_c^3}$$

Note that F.S. varies, being more conservative for the larger ratios of $L_e/r$. The equation chosen for F.S. approximates a quarter sine curve, with the value of 1.67 at zero $L_e/r$ and 1.92 at $C_c$. An allowable stress versus slenderness ratio for axially loaded columns of several kinds of structural steels is shown in Fig. 11-30.

Since, in practical applications, the ideal restraint of the column ends, assumed in Section 11-5, cannot always be relied upon, conservatively, AISC specifies modification of the effective lengths, for example, as follows:

*For columns built in at both ends:* $L_e = 0.65L$

*For columns built in at one end and pinned at the other:* $L_e = 0.80L$

No modification need be made for columns pinned at both ends, where $L_e = L$. Modifications for other end restraints may be found in the AISC Specifications.

*AISC LRFD Formulas for Columns.* Here, again, there are two equations governing column strength, one for elastic and the other for inelastic buckling. The boundary between the inelastic and elastic instability is at $\lambda_c = 1.5$, where the *column slenderness parameter* $\lambda_c$ is defined as

$$\lambda_c = \frac{L_e}{r\pi} \sqrt{\frac{\sigma_{yp}}{E}} \tag{11-58}$$

This expression results from normalizing the slenderness ratio $L_e/r$ with respect to the slenderness ratio for the Euler elastic critical stress, assuming $\sigma_{cr} = \sigma_{yp}$.

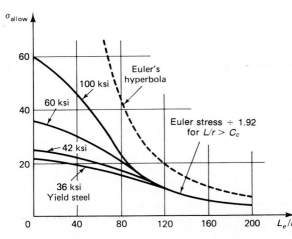

**Fig. 11-30** Allowable stress for concentrically loaded columns per AISC specifications.

For $\lambda_c > 1.5$, the *critical* buckling stress $\sigma_{cr}$ is based on the Euler load and is given as

$$\sigma_{cr} = \left[ \frac{0.877}{\lambda_c^2} \right] \sigma_{yp} \qquad (11\text{-}59)$$

where the factor 0.877 is introduced to account for the initial out-of-straightness of the column, see Fig. 11-11(c), and the effects of residual stresses.

For $\lambda_c = 1.5$, an empirical relationship based on extensive experimental and probabilistic studies is given as

$$\sigma_{cr} = (0.658^{\lambda_c^2})\sigma_{yp} \qquad (11\text{-}60)$$

This equation includes the effects of residual stresses and initial out-of-straightness.

Both of the previous formulas give the nominal axial strength (capacity) of columns and must be used in conjunction with factored loads and a resistance factor $\phi_c$ of 0.85 (see Example 1-7). The effective slenderness ratios $L_e/r$ are determined as for the ASD.

### Column Formulas for Aluminum Alloys

A large number of aluminum alloys are available for engineering applications. The yield and the ultimate strengths of such alloys vary over a wide range. The elastic modulus for the alloys, however, is reasonably constant. The Aluminum Association (AA)[23] provides a large number of column design formulas for different aluminum alloys. In all of these formulas, the allowable stress varies with the column slenderness ratio, as shown in Fig. 11-29(b). A representative set of three equations is given here for 6061-T6 alloy. As identified by the first number, the major alloying elements in this aluminum alloy are magnesium and silicon. T6 designates that this alloy has been thermally treated to produce stable temper. This alloy finds its greatest use for heavy-duty structures requiring good corrosion resistance as in trucks, pipelines, buildings, etc. Alloys such as 2024 and 7075 in their various tempers are used in aircraft, where similar formulas are employed.

The three basic column formulas for 6061-T6 alloy are

$$\sigma_{allow} = 19 \text{ ksi} \qquad\qquad 0 \le L/r \le 9.5 \qquad (11\text{-}61\text{a})$$
$$\sigma_{allow} = 20.2 - 0.126L/r \text{ ksi} \qquad 9.5 \le L/r \le 66 \qquad (11\text{-}61\text{b})$$
$$\sigma_{allow} = \frac{51{,}000}{(L/r)^2} \text{ ksi} \qquad\qquad 66 \le L/r \qquad (11\text{-}61\text{c})$$

[23] *Aluminum Construction Manual*, Section 1, "Specifications for Aluminum Structures," 5th ed., April 1982; Section 2, "Illustrative Examples of Design," April 1978; and "Engineering Data for Aluminum Structures," 5th ed., November 1981 (Washington, DC: The Aluminum Association, Inc.).

For aluminum alloy compression members, the effective lengths are approximated in the same manner as recommended by the AISC.

These stresses in Eqs. 11-61a and 11-61b are reduced to 12 ksi within 1 in of a weld.

In designing aluminum alloy columns, it is also recommended to check local buckling of the column components. Therefore, formulas are also given by the Aluminum Association for the allowable stresses for outstanding flanges or legs and column webs, i.e., flat plates with supported legs. These formulas, in groups of three, are similar to Eqs. 11-61, except that in place of the slenderness ratios $L/r$, the ratios $b/t$ are used, where $b$ is the width of a plate, and $t$ is its thickness. The allowable stresses given by such formulas may govern the design if such stresses are smaller than those required in Eqs. 11-61. Two basic groups of formulas for determining local buckling for 6061-T6 alloy are

*For outstanding legs or flanges:*

$$\sigma_{\text{allow}} = 19 \text{ ksi} \qquad\qquad 0 \leq b/t \leq 5.2 \qquad (11\text{-}62a)$$
$$\sigma_{\text{allow}} = 23.1 - 0.79b/t \text{ ksi} \qquad 5.2 \leq b/t \leq 12 \qquad (11\text{-}62b)$$
$$\sigma_{\text{allow}} = 1970/(b/t)^2 \text{ ksi} \qquad 12 \leq b/t \qquad\qquad (11\text{-}62c)$$

*For edge-supported plates[24]:*

$$\sigma_{\text{allow}} = 19 \text{ ksi} \qquad\qquad 0 \leq b/t \leq 16 \qquad (11\text{-}63a)$$
$$\sigma_{\text{allow}} = 23.1 - 0.25b/t \text{ ksi} \qquad 16 \leq b/t \leq 33 \qquad (11\text{-}63b)$$
$$\sigma_{\text{allow}} = 490/(b/t) \text{ ksi} \qquad 33 \leq b/t \qquad\qquad (11\text{-}63c)$$

Since all three groups of these formulas are given for the *allowable* stresses on gross sections, they include factors of safety for the intended usage.

### Column Formulas for Wood

The National Forest Products Association (NFPA)[25] provides the necessary information for the design of wood columns. Here attention will be limited to solid *rectangular* columns. In treating such columns, it is convenient to recast the design formulas in terms of the slenderness ratio $L_e/d$, where $L_e$ is the effective column length, and $d$ is the *least* dimension of the cross section; see Fig. 11-31. On this basis,

$$r_{\text{min}} = \sqrt{\frac{I_{\text{min}}}{A}} = \sqrt{\frac{bd^3}{12}\frac{1}{bd}} = \frac{d}{\sqrt{12}} \qquad (11\text{-}64)$$

---

[24] Such as column webs.

[25] See *National Design Specifications for Wood Construction* and *Design Values for Wood Construction, NDS Supplement, National Forest Products Association* DC. Table compiled by National Forest Products Association. See also D. E. Breyer, *Design of Wood Structures,* 2nd ed (New York: McGraw-Hill, 1988).

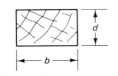

**Fig. 11-31** Cross section of a wooden column.

Substituting this relation into Eq. 11-19 and dividing the critical stress by the recommended factor of safety of 2.74,

$$\sigma_{\text{allow}} = \frac{\pi^2 E}{2.74 (L_e/r)^2} = \frac{\pi^2 E}{2.74 \times 12(L_e/d)^2} = \frac{0.30E}{(L_e/d)^2} \quad (11\text{-}65)$$

Since this stress is deduced from the *elastic* Euler formula, a limitation on its use must be placed for the smaller values of $L_e/d$. In the NFPA *design specifications,* this is achieved by requiring that at a slenderness ratio $L_e/d$, designated as $K$, the allowable stress does not exceed two-thirds of the design stress $F_c$ for a short wood block in compression parallel to the grain. In the form of an equation, using Eq. (11-65), this means that

$$\frac{2}{3} F_c = \frac{0.3E}{(L_e/d)^2_{\min}} = \frac{0.3E}{K^2}$$

Hence,
$$K = \sqrt{\frac{0.45E}{F_c}} = 0.671 \sqrt{\frac{E}{F_c}} \quad (11\text{-}66)$$

The value of $K$ provides the boundary for the least slender column for use in Eq. 11-65. Note that since $E$ and $F_c$ for different woods vary, $K$ assumes different values.

A qualitative graphical representation for the allowable stresses for columns over the permissible range of column slenderness ratios $L_e/d$ is shown in Fig. 11-32. Note that for short columns, a constant stress is specified; for the intermediate and long slenderness ratios, a curve with an inflection point at $K$ is shown. There is a small discontinuity at $L_e/d = 11$.

The allowable stresses in axially loaded wooden rectangular columns are

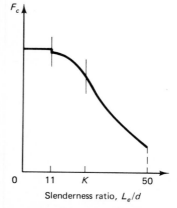

$F_c$

Slenderness ratio, $L_e/d$

**Fig. 11-32** Typical allowable stress for concentrically loaded wood columns per NFPA specifications.

$$\sigma_{\text{allow}} = F_c \qquad\qquad 0 \le L_e/d \le 11 \quad (11\text{-}67\text{a})$$

$$\sigma_{\text{allow}} = F_c \left[ 1 - \frac{1}{3} \left( \frac{L_e/d}{K} \right)^4 \right] \qquad 11 < L_e/d \le K \quad (11\text{-}67\text{b})$$

$$\sigma_{\text{allow}} = \frac{0.30E}{(L_e/d)^2} \qquad\qquad K \le L_e/d \le 50 \quad (11\text{-}67\text{c})$$

where $F_c$ is the allowable design stress for a short block in compression parallel to grain, $E$ is the modulus of elasticity, and $K$ is defined by Eq. 11-66. Note that the maximum allowable slenderness ratio $L_e/d$ is 50.

It must be recognized that $F_c$ and $E$ for wood are highly variable quantities, depending on species, grading rules, moisture, service conditions, temperature, duration of load, etc. Therefore, in actual applications, the reader should consult texts dealing with such problems in more detail.

The effective lengths are approximated in the same manner as recommended by the AISC.

The following examples illustrate some applications of the design formulas for axially loaded columns.

## EXAMPLE 11-6

(*a*) Determine the allowable axial loads for two 15-ft W 14 × 159 steel columns using AISC ASD formulas when one of the columns has pinned ends and the other has one end fixed and the other pinned. (*b*) Repeat the solution for two 40-ft W 14 × 159 columns. For the given section, $A = 46.7$ in$^2$ and $r_{min} = 4.00$ in. Assume A36 steel having $\sigma_{yp} = 36$ ksi.

### Solution

For both cases, it is necessary to calculate $C_c$ to determine whether Eq. 11-56 or 11-57 is applicable.

$$C_c = \sqrt{2\pi^2 E/\sigma_{yp}} = \sqrt{2\pi^2 \times 29 \times 10^3/36} = 126.1$$

(*a*) For the W 14 × 159 shape, the *minimum* $r = 4.00$ in. Hence, for the 15-ft column with pinned ends, $L_e/r = 15 \times 12/4 = 45 < C_c$, and Eq. 11-57 applies. Hence,

$$\sigma_{allow} = \frac{[1 - 45^2/(2 \times 126.1^2)]36}{5/3 + 3 \times 45/(8 \times 126.1) - 45^3/(8 \times 126.1^3)} = 18.78 \text{ ksi}$$

and

$$P_{allow} = A\sigma_{allow} = 46.7 \times 18.78 = 877 \text{ kips}$$

For the column with one end fixed and the other pinned, according to the AISC, the effective length $L_e = 0.8L = 12$ ft. Hence, $L_e/r = 12 \times 12/4 = 36$, and again applying Eq. 11-57, $\sigma_{allow} = 19.50$ ksi, and $P_{allow} = A\sigma_{allow} = 46.7 \times 19.50 = 911$ kips.

Here the allowable axial force is increased by 3.9 percent by fixing one of the column ends.

(*b*) For a 40-ft column with pinned ends, $L_e/r = 40 \times 12/4 = 120 < C_c$. Hence, on using Eq. 11-57 again, it can be determined that $\sigma_{allow} = 10.28$ ksi and $P_{allow} = A\sigma_{allow} = 46.7 \times 10.28 = 480$ kips. Similarly, since for a column fixed at one end and pinned at the other, $L_e/r = 0.8 \times 120 = 96$, Eq. 11-57 gives $\sigma_{allow} = 13.48$ ksi and $P_{allow} = A\sigma_{allow} = 46.7 \times 13.48 = 630$ kips.

For this case, the allowable axial force is increased 31.2 percent by fixing one of the column ends. This contrasts with the 3.9 percent found earlier for the shorter columns. This finding is in complete agreement with the generalized Euler theory for columns, Section 11-7. As can be noted from Fig. 11-19, by restraining the ends of *long* columns, a large increase in their strength is obtained at large values of $L_e/r$. Restraining the ends of short columns results only in a modest increase in their strength.

## EXAMPLE 11-7

Using the AISC ASD column formulas, select a 15-ft long pin-ended column to carry a concentric load of 200 kips. The structural steel is to be A572, having $\sigma_{yp}$ = 50 ksi.

### Solution

The required size of the column can be found directly from the tables in the *AISC Steel Construction Manual*. However, this example provides an opportunity to demonstrate the trial-and-error procedure that is so often necessary in design, and the solution presented follows from using this method.

*First try:* Let $L/r = 0$ (a poor assumption for a column 15 ft long). Then, from Eq. 11-57, since F.S. = 5/3, $\sigma_{allow}$ = 50/F.S. = 30 ksi and $A = P/\sigma_{allow}$ = 200/30 = 6.67 in². From Table 4 in the Appendix, this requires a W 8 × 24 section, whose $r_{min}$ = 1.61 in. Hence, $L/r$ = 15(12)/1.61 = 112. With this $L/r$, the allowable stress is found using Eq. 11-56 or 11-57, whichever is applicable depending on $C_c$:

$$C_c = \sqrt{2\pi^2 E/\sigma_{yp}} = \sqrt{2\pi^2 \times 29 \times 10^3/50} = 107 < L/r = 112$$

Hence, using Eq. 11-56, $\sigma_{allow} = \dfrac{12\pi^2 \times 29 \times 10^3}{23 \times 112^2} = 11.9$ ksi

This is much smaller than the initially assumed stress of 30 ksi, and another section must be selected.

*Second try:* Let $\sigma_{allow}$ = 11.9 ksi as found before. Then $A$ = 200/11.9 = 16.8 in², requiring a W 8 × 58 section having $r_{min}$ = 2.10 in. Now $L/r$ = 15(12)/2.10 = 85.7, which is less than $C_c$ found before. Therefore, Eq. 11-57 applies, and

$$\text{F.S.} = 5/3 + 3(85.7)/(8 \times 107) - (85.7)^3/(8 \times 107^3) = 1.90$$

and $\qquad \sigma_{allow} = [1 - (85.7)^2/(2 \times 107^2)]50/1.90 = 17.9$ ksi

This stress requires $A$ = 200/17.9 = 11.2 in², which is met by a W 8 × 40 section with $r_{min}$ = 2.04 in. A calculation of the capacity for this section shows that the allowable axial load for it is 204 kips, which meets the requirements of the problem.

---

## *EXAMPLE 11-8

Determine the design compressive strength $P_u$ for a 15-ft W 14 × 159 steel column pinned at both ends based on the AISC LRFD provisions. For this section, $A$ = 46.7 in² and $r_{min}$ = 4.00 in. Assume A36 steel having $\sigma_{allow}$ = 36 ksi.

## Solution

The column slenderness parameter as defined by Eq. 11-58 is

$$\lambda_c = \frac{15 \times 12}{4\pi} \sqrt{\frac{36}{29 \times 10^3}} = 0.5047$$

Since $\lambda_c$ is less than 1.5, Eq. 11-60 applies for determining the critical stress and

$$\sigma_{cr} = (0.658^{0.5047^2})36 = 32.36 \text{ ksi}$$

Hence, for this column, the *nominal* compressive strength

$$P_n = A\sigma_{cr} = 46.7 \times 32.36 = 1510 \text{ kips}$$

and since the resistance factor $\phi_c = 0.85$, the column-design compressive strength

$$P_u = \phi_c P_n = 0.85 \times 1510 = 1289 \text{ kips.}$$

By dividing $P_u$ by $P_{\text{allow}} = 877$ kips for a comparable column analyzed in the preceding Example 11-6, one obtains 1.46. This load factor gives an indication of the relationship between the ASD and the LRFD for this case.

---

## *EXAMPLE 11-9

Determine the allowable axial loads for two compression members made from 6061-T6 aluminum alloy having 5 in × 5 in × 5.366 lb/ft wide-flange section with the dimensions shown in Fig. 11-33. One of the members is 20 in long and the other is 60 in. Assume each strut to be pinned at both ends. For the given section, $A = 4.563$ in$^2$, and the minimum radius of gyration $r_{\text{min}} = 1.188$ in.

## Solution

Regardless of the column length, for aluminum alloys, it is necessary to investigate *local buckling*. For the given section, two calculations must be made, one for the outstanding legs of the flanges, and the other for the web. In both instances, $b/t$ values determine the allowable compressive stress. For the flanges,

$$\frac{b_f}{t} = \frac{5 - 0.312}{2 \times 0.312} = 7.51$$

This ratio requires the use of Eq. 11-62b; hence,

$$(\sigma_{\text{allow}})_{\text{flanges}} = 23.1 - 0.79 \times 7.51 = 17.2 \text{ ksi}$$

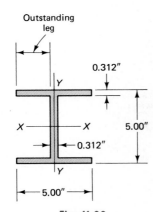

**Fig. 11-33**

The web width-thickness ratio is

$$\frac{b_w}{t} = \frac{5 - 2 \times 0.312}{0.312} = 14.0$$

Since this ratio is less than 16, according to Eq. 11-63a,

$$(\sigma_{allow})_{web} = 19 \text{ ksi}$$

Overall buckling is investigated using Eqs. 11-47, which depend on $L/r$, and for a 20-in strut is

$$\frac{L}{r} = \frac{20}{1.188} = 16.8$$

Hence, using Eq. 11-61b,

$$\sigma_{allow} = 20.2 - 0.126 \times 16.8 = 18.1 \text{ ksi}$$

For this case of a well-balanced design, the allowable stress for local flange buckling controls. Therefore,

$$P_{allow} = 4.563 \times 17.2 = 78.5 \text{ kips}$$

The slenderness ratio for the 60-in strut is

$$\frac{L}{r} = \frac{60}{1.188} = 50.5$$

Hence, again using Eq. 11-61b,

$$\sigma_{allow} = 20.2 - 0.126 \times 50.5 = 13.8 \text{ ksi}$$

This stress is smaller than those for local buckling; hence,

$$P_{allow} = 4.563 \times 13.8 = 63.0 \text{ kips}$$

---

## *11-13. Eccentrically Loaded Columns

In the past, the secant-type formulas discussed in Section 11-8 were used as a rational method for the design of eccentrically loaded columns. Two other methods that have found a wide use follow.

### *Allowable Stress Method*

A procedure for designing eccentrically loaded columns is obtained by adapting the elastic solution for short blocks subjected to bending with axial loads, Eq. 6-45, and setting the maximum compressive stress equal to or less than for an axially loaded column. For a planar case, this becomes

$$\sigma_x = \frac{P}{A} + \frac{Mc}{I} \leq \sigma_{\text{allow}} \qquad (11\text{-}68)$$

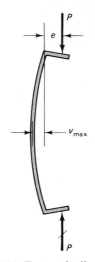

The compressive stresses in the last equation are treated as positive quantities. If only an eccentric force $P$ is applied, the bending moment $M = Pe$, where $e$ is the load eccentricity; see Fig. 11-34. The allowable stress $\sigma_{\text{allow}}$ is determined from an appropriate formula, such as given in the preceding section for axially loaded columns of different materials. Usually, the solution of Eq. 11-68 requires a trial-and-error procedure.

For short and intermediate length columns, the previous procedure is usually conservative, since $\sigma_{\text{allow}}$ for compressive stresses is generally less than the allowable bending stress. On the other hand, this procedure may become unconservative for slender columns, where the deflections are magnified due to the axial force. For such cases, it is appropriate to determine the extent of bending moment magnification caused by column deflection using the approximate deflection magnification factor derived in Example 11-3 or 11-4. *p. 597*

**Fig. 11-34** Eccentrically loaded column.

### Interaction Method

In an eccentrically loaded column, much of the total stress may result from the applied moment. However, *the allowable stress in flexure is usually higher than the allowable axial stress*. Hence, for a particular column, it is desirable to accomplish some balance between the two stresses, depending on the relative magnitudes of the bending moment and the axial force. Thus, since in bending, $\sigma = Mc/I = Mc/Ar_1^2$, where $r_1$ is the radius of gyration *in the plane of bending*, in effect, area $A_b$ required by bending moment $M$ is

$$A_b = \frac{Mc}{\sigma_{ab}r_1^2}$$

where $\sigma_{ab}$ is the allowable *maximum stress in bending*. (See also Section 11-14.) Similarly, area $A_a$ required for axial force $P$ is

$$A_a = \frac{P}{\sigma_{aa}}$$

where $\sigma_{aa}$ is the *allowable axial stress for the member acting as a column*, and which depends on the $L/r$ ratio. Therefore, the *total* area $A$ required for a column subjected to an axial force and a bending moment is

$$A = A_a + A_b = \frac{P}{\sigma_{aa}} + \frac{Mc}{\sigma_{ab}r_1^2} \qquad (11\text{-}69)$$

By dividing by $A$,

$$\frac{P/A}{\sigma_{aa}} + \frac{Mc/Ar_1^2}{\sigma_{ab}} = 1 \quad \text{or} \quad \frac{\sigma_a}{\sigma_{aa}} + \frac{\sigma_b}{\sigma_{ab}} = 1 \qquad (11\text{-}70)$$

where $\sigma_a$ is the axial stress caused by the applied vertical loads, and $\sigma_b$ is the bending stress caused by the applied moment. If a column is carrying only an axial load and the applied moment is zero, the formula indicates that the column is designed for the stress $\sigma_{aa}$. On the other hand, the allowable stress becomes the flexural stress $\sigma_{ab}$ if there is no direct compressive force acting on the column. Between these two extreme cases, Eq. 11-70 measures the relative importance of the two kinds of action and specifies the nature of their interaction. Hence, it is often referred to as an *interaction formula* and serves as the basis for the specifications in the AISC ASD manual, where it is stated that the sum of these two stress ratios must not exceed unity. The same philosophy has found favor in applications other than those pertaining to structural steel. The Aluminum Association suggests a similar relation. The National Forest Products Association developed a series of formulas to serve the same purpose.

In terms of the notations used by the AISC, Eq. 11-70 is rewritten as

$$\frac{f_a}{F_a} + \frac{f_b}{F_b} \leq 1.0 \qquad (11\text{-}71)$$

In practice, the eccentricity of the load on a column may be such as to cause bending moments about both axes of the cross section. Equation 11-71 is then modified to

$$\frac{f_a}{F_a} + \frac{f_{bx}}{F_{bx}} + \frac{f_{by}}{F_{by}} \leq 1.0 \qquad (11\text{-}72)$$

Subscripts $x$ and $y$ combined with subscript $b$ indicate the axis of bending about which a particular stress applies, and

$F_a$ = allowable axial stress if the axial force alone existed
$F_b$ = allowable compressive bending stress if the bending moment alone existed
$f_a$ = computed axial stress
$f_b$ = computed bending stress

At points that are braced in the plane of bending, $F_a$ is equal to 60 percent of $F_y$, the yield stress of the material, and

$$\frac{f_a}{0.6F_y} + \frac{f_{bx}}{F_{bx}} + \frac{f_{by}}{F_{by}} \leq 1.0 \qquad (11\text{-}73)$$

At intermediate points in the length of a compression member, the secondary bending moments due to deflection (see Fig. 11-34) can contribute significantly to the combined stress. Following the AISC specifications, this contribution is neglected in cases where $f_a/F_a$ is less than 0.15, i.e., the axial stress is small in relation to the allowable axial stress, and Eq. 11-73 can still be used. When $f_a/F_a$ is greater than 0.15, the effect of the additional secondary bending moments may be approximated by multiplying both $f_{bx}$ and $f_{by}$ by an *amplification factor*, $C_m/(1 - f_a/F'_e)$, which takes into account the slenderness ratio in the plane of bending and also the nature of the end moments. The term in the denominator of the amplification factor brings in the effect of the slenderness ratio through the use of $F'_e$, the Euler buckling stress (using $L_e/r$ in the plane of bending) divided by 23/12, or 1.92, which is the AISC factor of safety for a very long column with $L_e/r$ greater than $C_c$. (See Section 11-12 for a definition of $C_c$.) It can be noted that the amplification factor increases as $f_a$ increases and *blows up* as $f_a$ approaches $F'_e$. The term $C_m$ in the numerator is a correction factor that takes into account the ratio of the end moments as well as their relative sense of direction. The term $C_m$ is larger if the end moments are such that they cause a single curvature of the member, and smaller if they cause a reverse curvature. The formula for $f_a/F_a >$ 0.15 then becomes

$$\frac{f_a}{F_a} + \frac{C_{mx}f_{bx}}{(1 - f_a/F'_{ex})F_{bx}} + \frac{C_{my}f_{by}}{(1 - f_a/F'_{ey})F_{by}} \leq 1.0 \qquad (11\text{-}74)$$

According to the AISC specifications,[26] the value of $C_m$ shall be taken as follows:

1. For compression members in frames subject to joint translation (sidesway), $C_m = 0.85$.

2. For restrained compression members in frames braced against joint translation and not subject to transverse loading between their supports in the plane of bending,

$$C_m = 0.6 - 0.4M_1/M_2$$

(but not less than 0.4), where $M_1/M_2$ is the ratio of the smaller to larger moments at the ends of that portion of the member unbraced in the plane of bending under consideration. $M_1/M_2$ is positive when the member is bent in reverse curvature and negative when it is bent in single curvature.

3. For compression members in frames braced against joint translation in the plane of loading and subjected to transverse loading between their supports, the value of $C_m$ can be determined by rational anal-

[26] *AISC Steel Construction Manual*, 9th ed. (Chicago: AISC, 1989), 5–27.

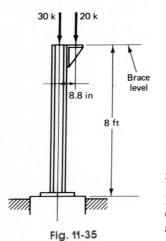

8.8 in

8 ft

Brace level

**Fig. 11-35**

ysis. However, in lieu of such analysis, the following values may be used: (a) for members whose ends are restrained, $C_m = 0.85$; (b) for members whose ends are unrestrained, $C_m = 1.0$.

## *EXAMPLE 11-10

Select a 6061-T6 aluminum alloy column for the loading shown in Fig. 11-35 using the allowable stress method. Assume the column to be pinned and laterally supported at both ends.

### Solution

In this problem, Eq. 11-68 must be satisfied with $\sigma_{\text{allow}}$ given by one of Eqs. 11-61. By assuming that the column length is in the intermediate range, Eq. 11-61b applies, and the following relation can be written:

$$\frac{30 + 20}{A} + \frac{20 \times 8.8}{S} = \frac{50}{A} + \frac{176}{S} \leq 20.2 - 0.126 \frac{L}{r}$$

where $A$, $S$, and $r$ depend on the selected column cross section. Note that $S$ applicable to the plane of bending must be used. A trial-and-error procedure is used to solve the problem.

*First Try:* It is convenient to recast the last equation into the following form:

$$\frac{50}{A} + \frac{176}{A}\left(\frac{A}{S}\right) \leq 20.2 - 0.126\frac{L}{r}$$

where $A/S = B$ defines a *bending factor*.[27] These factors are reasonably constant for a whole class of cross sections. Therefore, the solution can begin by selecting a plausible size for a member, which then provides data for $A/S$, and the above equation can be solved for a trial value of $A$. Following this procedure, *assume* here an 8 in × 8.5 in × 8.32 lb/ft aluminum alloy wide-flange section. The Aluminum Association Construction Manual gives the following data for this section: $A = 7.08$ in$^2$, $S_x = 21.04$ in$^3$, and $r_{\text{min}} = 1.61$ in. (Geometrically, this cross section is very similar to the W 8 × 24 steel section given in Table 4 of the Appendix. The corresponding values given there are $A = 7.08$ in$^2$, $S_x = 20.9$ in$^3$, and $r_{\text{min}} = 1.61$ in.) Based on this data, $B = A/S_x = 7.08/21.04 = 0.337$. Hence, the basic design equation becomes

$$\frac{50}{A} + \frac{176}{A} \times 0.337 = \frac{109.3}{A} = 20.2 - 0.126 \times \frac{8 \times 12}{1.61} = 12.69$$

The solution of this equation gives $A = 8.61$ in$^2$, which is larger than that provided by the assumed section, and requires another trial.

---

[27] Bending factors are tabulated for many cross sections in the *AISC Manual of Steel Construction* or may be calculated for an assumed section when $A$ and $S$ are known.

*Second Try:* Select 8 in × 8 in × 10.72 lb/ft, the next larger available section, with $A = 9.12$ in$^2$, $S_x = 27.41$ in$^3$, and $r_{min} = 2.01$ in. (A W 8 × 31 steel section has the same properties.) Substituting these quantities into the first equation formulated before shows that

$$\frac{50}{9.12} + \frac{176}{27.41} = 11.9 \leq 20.2 - 0.126 \times \frac{8 \times 12}{2.01} = 14.2 \text{ ksi}$$

Therefore, this section is satisfactory. For a complete solution of this problem, local buckling of flanges and webs should also be checked, as was done in Example 11-9. Such a solution, not given here, shows that the local buckling stresses are larger than the allowable axial stress and do not control the design.

## *EXAMPLE 11-11

Select a steel column for the loading shown in Fig. 11-36 using the AISC ASD interaction method. Assume the column to be pinned and laterally braced at both ends. Let $F_y = 50$ ksi and $F_b = 30$ ksi.

### Solution

In this problem, the interaction formulas, Eq. 11-72 or Eq. 11-73, must be satisfied, depending upon whether $f_a/F_a$ is less than or greater than 0.15. The solution is obtained by trial-and-error process as is outlined.

*First Try:* Let $L_e/r = 0$, although it is a poor assumption for a 15-ft column. Corresponding to this value of the slenderness ratio, $F_a$ can be calculated, using Eq. 11-57, as $F_a = 50/(5/3) = 30$ ksi. The required area of the section can then be computed using Eq. 11-72:

$$1.0 \geq \frac{f_a}{F_a} + \frac{f_b}{F_b} \quad \text{or} \quad A \geq \frac{Af_a}{F_a} + \frac{Af_b}{F_b}$$

Since

$$f_b = \frac{M}{S_x} = \frac{M}{A}\frac{A}{S_x} = \frac{M}{A}B_x \quad \text{and} \quad f_a = \frac{P}{A}$$

$$A \geq \frac{P}{F_a} + \frac{M}{F_b}B_x \qquad B_x = \frac{A}{S_x}$$

For any one depth of section, the bending factor $B_x$ does not vary a good deal. Therefore, if a W 10 section is to be chosen, a typical value of $B_x$ is about 0.264 (check a few values of $A/S_x$ in Table 4 of the Appendix). Then

$$A = \frac{200}{30} + \frac{800 \times 0.264}{30} = 13.7 \text{ in}^2$$

Select a W 10 × 49 section with $A = 14.4$ in$^2$, $r_{min} = 2.54$ in, $r_x = 4.35$ in, and $B_x = 0.264$, and carry out the necessary calculations to determine whether the interaction Eq. 11-72 or 11-74 governs.

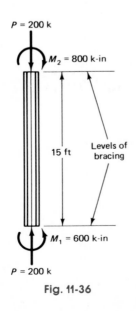

*P* = 200 k

$M_2$ = 800 k-in

15 ft

Levels of bracing

$M_1$ = 600 k-in

*P* = 200 k

**Fig. 11-36**

$$f_a = \frac{P}{A} = \frac{200}{14.4} = 13.9 \text{ ksi} \qquad f_b = \frac{MB_x}{A} = \frac{800 \times 0.264}{14.4} = 14.7 \text{ ksi}$$

$$\frac{L_e}{r_{\min}} = \frac{15 \times 12}{2.54} = 70.9 < C_c \qquad C_c = \sqrt{2\pi^2 E/F_y} = 107$$

Using Eq. 11-57, $F_a = 19.3$ ksi, $f_a/F_a = 13.9/19.3 = 0.72 > 0.15$; hence, the interaction formula of Eq. 11-74 must be checked. For this purpose, using $L_e/r_x$ in the plane of bending, one determines

$$F'_e = \frac{12\pi^2 E}{23(L_e/r_x)^2} = \frac{149 \times 10^3}{(15 \times 12/4.35)^2} = \frac{149 \times 10^3}{(41.4)^2} = 86.9 \text{ ksi}$$

Then, since the end moments subject the column to a single curvature, $M_1/M_2 = -600/800 = -0.75$, and

$$C_m = 0.6 - 0.4 M_1/M_2 = 0.6 - (0.4)(-0.75) = 0.9$$

With bending taking place in one plane only, Eq. 11-74 reduces to

$$\frac{f_a}{F_a} + \frac{C_m f_b}{(1 - f_a/F'_e)F_b} \leq 1.0$$

On substituting the appropriate quantities into this relation,

$$\frac{13.9}{19.3} + \frac{0.9 \times 14.7}{(1 - 13.9/86.9)30} = 0.72 + 0.52 = 1.24 > 1.0$$

Since Eq. 11-74 is violated, a larger section must be used.

*Second Try:* As an aid in choosing a larger section, assume $F_a = 19.3$ ksi, which is the value computed for the section in the previous trial. Also, using $B_x = 0.264$ for W 10 sections,

$$A \geq \frac{P}{F_a} + \frac{MB_x}{F_b} = \frac{200}{19.3} + \frac{800 \times 0.264}{30} = 17.4 \text{ in}^2$$

Now select a W 10 × 60 section with $A = 17.6$ in$^2$, $r_{\min} = 2.57$ in, $r_x = 4.39$ in, and $B_x = 0.264$, and proceed as in the first trial to check the interaction formula.

$$f_a = \frac{P}{A} = \frac{200}{17.6} = 11.4 \text{ ksi} \qquad f_b = \frac{MB_x}{A} = \frac{800 \times 0.264}{17.6} = 12.0 \text{ ksi}$$

$$\frac{L_e}{r_{\min}} = \frac{15 \times 12}{2.57} = 70.0 < C_c$$

Using Eq. 11-57, $F_a = 19.4$ ksi, $f_a/F_a = 11.4/19.4 = 0.59 > 0.15$; hence, Eq. 11-74 must be checked.

$$F'_e = \frac{149 \times 10^3}{(L_e/r_x)^2} = \frac{149 \times 10^3}{(15 \times 12/4.39)^2} = 88.6 \text{ ksi}$$

Again, using Eq. 11-74 for bending in one plane and substituting into it the appropriate quantities, one has

$$\frac{11.4}{19.4} + \frac{0.9 \times 12.0}{(1 - 11.4/88.6)30} = 0.59 + 0.41 = 1.00$$

Since this relation satisfies Eq. 11-74, the W 10 × 60 section is satisfactory.

## *11-14. Lateral Stability of Beams

The strength and deflection theory of beams developed in this text applies only if such beams are in *stable equilibrium*. Narrow or slender beams that do not have occasional lateral supports may buckle sideways and thus become unstable; see Fig. 11-37. Theoretical and experimental studies of this problem show that, within limits, reduced bending stresses can be used to maintain the stability of such beams. The nature of the reduced stresses resembles the curves displayed for columns in Figs. 11-29 and 11-32. The key parameter for stress reduction depends on the material properties, geometry of the cross section, and moment gradient. Several of the references cited in this chapter discuss this topic. In this section, only a simple criterion for avoiding the problem of lateral torsional buckling for compact steel beams is given.

**Fig. 11-37** Lateral-torsional buckling of a narrow beam.

According to the AISC ASD specifications, in order for a compact beam to qualify for the maximum allowable bending stress, intermittent lateral supports shall be provided at intervals not exceeding the value

$$\frac{76b_f}{\sqrt{F_y}} \quad \text{nor} \quad \frac{20{,}000}{(d/A_f)F_y} \tag{11-75}$$

where $A_f$ is the area of a compression flange, $b_f$ is the flange width, $d$ is the depth of a beam, and $F_y$ is the yield stress for the material.

## Problems

### Section 11-3

**11-1.** A rigid bar hinged at the base is held in a vertical position by two springs: one has a stiffness $k$ N/mm and the other, $2k$ N/mm, as shown in the figure. Determine the critical force $P_{cr}$ for this system.

**11-2 through 11-4.** Rigid-bar segments of equal lengths $a$ are connected at the joints and at the bottoms by frictionless hinges and are maintained in straight po-

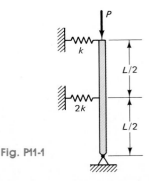

**Fig. P11-1**

sitions by torsional springs of the stiffnesses shown in the figures. Determine the eigenvalues for these systems and show the eigenfunctions on separate diagrams. Identify the critical loads.

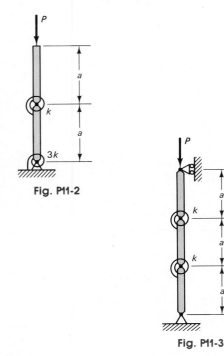

**Fig. P11-2**

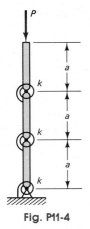

**Fig. P11-4**

**Fig. P11-3**

**\*11-5.** A weightless prismatic elastic column can be approximated by a series of rigid bars each of length $a$, with an appropriate torsional spring constant $k$ at each joint, as shown in the figure. Set up the determinental equation for finding the critical load for a system having $n$ degrees of freedom.

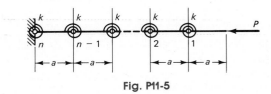

**Fig. P11-5**

**11-6.** An ideal column is pinned at the base and guyed at the top by four wires, as shown in the figure. The 3000-mm-long column has a solid circular cross section

of 80 mm in diameter. For the column and the wires, $E = 200$ GPa and $\sigma_{yp} = 400$ MPa. What should be the diameter of the wires such that a perfectly concentric buckling load $P_{cr}$ could be reached simultaneously with lateral displacement at the the top? Assume that

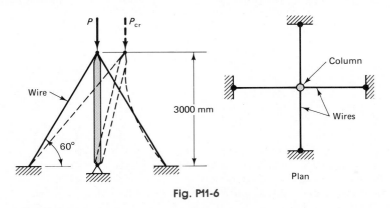

**Fig. P11-6**

the lateral displacement of the top is prevented by one wire only with the diametrically opposite wire becoming slack, as shown by the dashed curve. Consider the column to be perfectly rigid during lateral displacement of the top. (*Note:* Load eccentricity and column crookedness should be considered in actual applications.)

**11-7.** A 1-in round steel bar 4 ft long acts as a spreader bar in the arrangement shown in the figure. If cables and connections are properly designed, what pull $F$ can be applied to the assembly? Use Euler's formula and assume a factor of safety of 3. $E = 29 \times 10^6$ psi.

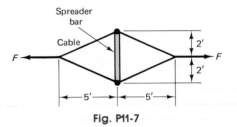

**Fig. P11-7**

**11-8.** A boom is made from an aluminum pipe of 60 mm outside diameter and having a 4-mm wall thickness, and is part of an arrangement for lifting weights, as shown in the figure. Determine the magnitude of the force $F$ that could be applied to this planar system as controlled by the capacity of the boom. Assume a factor of safety of 3 for the Euler buckling load. $E = 75$ GPa. All dimensions are shown in mm.

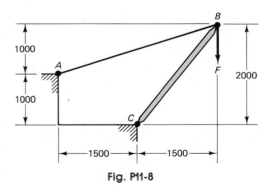

**Fig. P11-8**

**11-9.** The mast of a derrick is made of a standard rectangular $4 \times 2$ in steel tubing weighing 6.86 lb/ft. ($A = 2.02$ in$^2$, $I_x = 1.29$ in$^4$, $\bar{I}_z = 3.87$ in$^4$.) If this derrick is assembled as indicated in the figure, what vertical force $F$, governed by the size of the mast, can be applied at $A$? Assume that all joints are pin-connected and that the connection details are so made that the mast is loaded concentrically. The top of the mast is braced to prevent sidewise displacement. Use Euler's formula with a factor of safety of 3.3. $E = 29 \times 10^6$ psi.

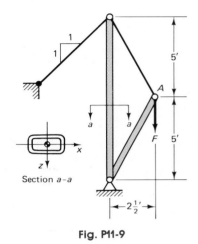

**Fig. P11-9**

**11-10.** What force $F$ can be applied to the system shown in the figure, governed by the $25 \times 16$ mm aluminum-alloy bar $AB$? The factor of safety on the Euler buckling load is to be 2.5. Assume the ends are pinned. $E = 70$ GPa.

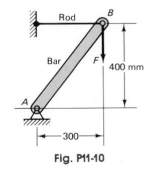

**Fig. P11-10**

**11-11.** Governed by the steel T section, what force $F$ can be applied to the system shown in the figure? The factor of safety on the buckling load must be 2. Assume that the ends are pinned and that the applied force is concentrically applied. $E = 200$ GPa. Neglect the possibility of torsional buckling.

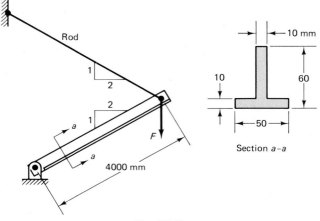

**Fig. P11-11**

**11-12.** A thin bar of stainless steel is axially precompressed 100 N between two plates that are fixed at a constant distance of 150 mm apart; see the figure. This assembly is made at 20°C. How high can the temperature of the bar rise, so as to have a factor of safety of 2 with respect to buckling? Assume $E = 200$ GPa and $\alpha = 15 \times 10^{-6}$ per °C.

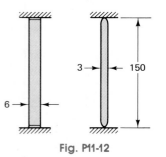

**Fig. P11-12**

**11-13.** What size standard steel pipe should be used for the horizontal member of the jib crane shown in the figure for supporting the maximum force of 4 k, which includes an impact factor? Use the Euler buckling for-

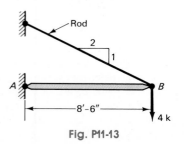

**Fig. P11-13**

mula for columns with pinned ends and a factor of safety of 2.5. Neglect the weight of construction. $E = 29 \times 10^6$ psi.

**11-14.** Select a W steel section for member $AB$ for the system shown in the figure to resist a vertical force of 150 k. The system is laterally braced at $B$ and $C$. Neglect the weight of the members. Assume pinned ends and a factor of safety of 2. $E = 29 \times 10^3$ ksi.

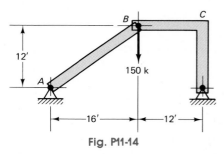

**Fig. P11-14**

**11-15.** Select standard steel pipes for the tripod shown in the figure to support a vertical load $F = 4.75$ k with

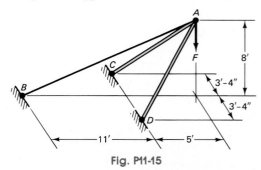

**Fig. P11-15**

a factor of safety of 3 on the Euler buckling load. Neglect the weight of the members. $E = 29 \times 10^3$ ksi.

**11-16.** A tripod is to be made up from $3 \times 3$ in steel angles, each 10 ft long, to support a vertical load $F = 8$ k at the center, as shown in the figure. Using the Euler buckling formula with a factor of safety of 3 to account for impact, determine the required thickness of the angles. Neglect the weight of the angles, assume that they are loaded concentrically, and that the ends are pinned. $E = 30 \times 10^6$ psi.

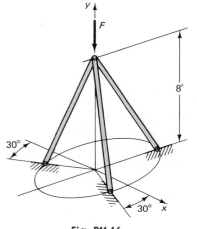

**Fig. P11-16**

**\*11-17.** A simple beam of flexural rigidity $EI_b$ is propped up at the middle by a slender rod of flexural rigidity $EI_c$. Estimate the deflection of the beam at the center if a force $F$ double the Euler load for the column is applied to this system.

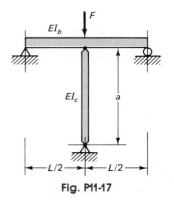

**Fig. P11-17**

## Section 11-5

**11-18.** Derive Eq. 11-17 using Eq. 11-5 in the form $EIv'' + Pv = M_o$, where $M_o$ is the moment at the end.

**11-19.** Derive Eq. 11-18 using Eq. 11-5 in the form $EIv'' = P(\delta - v)$, where $\delta$ is the end deflection.

**\*11-20.** Determine the critical buckling load for the column shown in the figure. (*Hint:* See the preceding problem; enforce continuity conditions at a change in $EI$.)

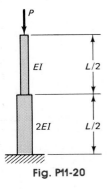

**Fig. P11-20**

**\*11-21.** Determine the transcendental equation for finding the critical buckling load for bar $AB$ of constant $EI$ due to the application of axial force $P$ through rigid link $BC$. (*Hint:* In a deflected position, note the presence of a shear force at $B$.)

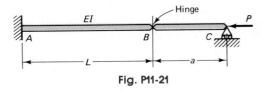

**Fig. P11-21**

**11-22.** An allowable axial load for a 4-m-long pin-ended column of a certain linearly elastic material is 20 kN. Five different columns made of the same material and having the same cross section have the supporting conditions shown in the figure. Using the column capacity for the 4-m column as the criterion, what are the allowable loads for the five columns shown?

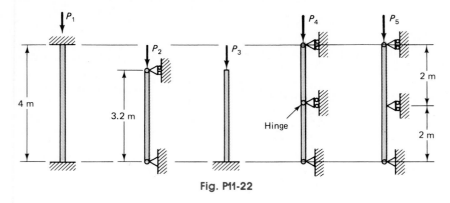

**Fig. P11-22**

**11-23.** A machine bracket of steel alloy is to be made as shown in the figure. The compression member $AB$ is so arranged that it can buckle as a pin-ended column in the plane $ABC$, but as a fixed-ended column in the direction perpendicular to this plane. (a) If the thickness of the member is $\frac{1}{2}$ in, what should be its height $h$ to have equal probability of buckling in the two mutually perpendicular directions? (b) If $E = 28 \times 10^6$, and the factor of safety on instability is 2, what force $F$ can be applied to the bracket? Assume that the bar designed in (a) controls the capacity of the assembly.

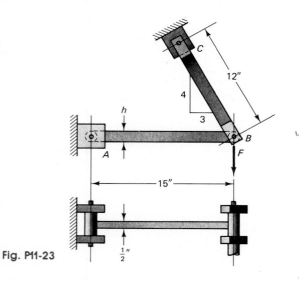

**Fig. P11-23**

**11-24.** A piece of mechanical equipment is to be supported at the top of a 5-in nominal-diameter standard steel pipe, as shown in the figure. The equipment and its supporting platform weigh 5500 lb. The base of the pipe will be anchored in a concrete pad and the top

**Fig. P11-24**

end will be unsupported. If the factor of safety required against buckling is 2.5, what is the maximum height of the column on which the equipment can be supported? $E = 30 \times 10^6$ psi. (*Note:* Solution becomes inaccurate if the height of the rigid mass is significant in relation to the height of the column.)

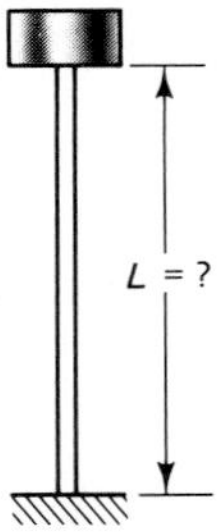

## Sections 11-6 through 11-8

**11-25.** Find the shortest lengths for columns with pinned ends such that the Euler elastic buckling formula would apply. Consider three different cases: (a) a $2 \times 4$ in wooden strut of nominal size (see Table 10 of the Appendix) if $E = 1.8 \times 10^6$ psi and the maximum compression stress is 1500 psi, (b) a solid aluminum-alloy shaft 50 mm in diameter if $E = 70$ GPa and $\sigma_{yp} = 360$ MPa, and (c) a W $14 \times 193$ steel section (see Table 4 of the Appendix) if $E = 29 \times 10^3$ ksi and $\sigma_{yp} = 36$ ksi.

**11-26.** Two grades of steel are in common use for columns in buildings: A36 steel with $\sigma_{yp} = 36$ ksi, and A572 with $\sigma_{yp} = 50$ ksi. For each steel determine the smallest slenderness ratios for which the Euler elastic buckling formula applies when the column is pinned at both ends and when it is fixed at both ends.

***11-27.*** The stress-strain curve in simple tension for an aluminum alloy is shown in the figure, where, for convenience, $\varepsilon \times 10^3 = e$. The alloy is linearly elastic for stresses up to 40 ksi; the ultimate stress is 50 ksi. (a) Idealize the stress-strain relation by fitting a parabola to the curve so that $\sigma$ and $d\sigma/de = E_t$ is continuous at the proportional limit and so that the $\sigma = 50$-ksi line is tangent to the parabola. (b) Plot $E_t(\sigma)/E$ against $\sigma/\sigma_{ult}$, where $E$ is the elastic modulus, $\sigma_{ult}$ the ultimate stress, and $E_t$ the tangent modulus at stress $\sigma$. (c) Plot in one graph $\sigma_{cr}$ against $L/r$ for fixed-fixed and pinned-pinned columns, where $\sigma_{cr}$ is based on $E_t$.

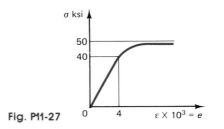

**Fig. P11-27**

***11-28.*** For some materials, the stress-strain relationship in normalized form can be expressed[28] as $\sigma = 1 - \exp(-c\varepsilon)$, where $c$ is an arbitrary constant. Setting $c = 500$, plot the stress-strain diagram for $\varepsilon$ from 0 to 0.01, and the normalized-stress-vs-column slenderness ratio $L_e/r$ from 0 to 200. (*Note:* $\exp(x) = e^x$.)

***11-29.*** Using Eq. 11-24, obtain the average stress $P/A$ for $L/r = 0$ and 75. Assume $ec/r^2 = 0.05$.

## Section 11-9

***11-30.*** A high-strength thin-walled steel tube 1250 mm long is loaded as shown in Fig. 11-23. The axial force $P = 25$ kN and the transverse $F = 500$ N. The outside diameter of the tube is 37 mm, and its cross-sectional area is 223 mm². For this tube, $I = 34.2 \times 10^3$ mm⁴ and $E = 200$ GPa. (a) Determine the maximum deflection and bending moment using Eqs. 11-33 and 11-34. (b) Compare the results in (a) with the results using the approximate Eqs. 11-41 and 11-42. (c) Calculate the combined stresses due to the axial force and the maximum bending moment. Neglect local stress concentrations.

***11-31.*** Show that for a beam-column loaded by an end moment $M_B$, as shown in the figure, the deflection is

[28] Courtesy of F. C. Filippou.

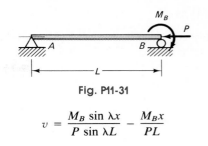

**Fig. P11-31**

$$v = \frac{M_B \sin \lambda x}{P \sin \lambda L} - \frac{M_B x}{PL}$$

and the bending moment is

$$M = -\frac{M_B \sin \lambda x}{\sin \lambda L}$$

***11-32.*** Consider the thin-walled tube having the mechanical properties given in Prob. 11-30 subjected to an end moment $M_o = 250$ N·m and an axial force $P = 30$ kN, as shown in the figure. (a) Determine the maximum deflection and then the maximum bending moment using an approximate method. Use Table 11 of the Appendix for beam deflection due to an end moment. (b) Compare the results in (a) with those using the accurate expressions found in the preceding example. Note that the maximum moment occurs at $dM/dx = 0$. (c) Calculate the maximum in-span stresses due to the axial force and bending.

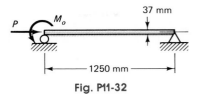

**Fig. P11-32**

***11-33.*** If an elastic bar is initially curved as shown in the figure, show that the total deflection

$$v = v_o + v_1 = \left(\frac{1}{1 - P/P_{cr}}\right) a \sin \frac{\pi x}{L}$$

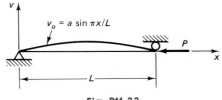

**Fig. P11-33**

## Section 11-10

**11-34.** Show that since the character of Eqs. 11-45 and 11-46 changes if instead of a compression axial force, a tensile force is applied, the homogeneous solution of the differential equation for deflection is

$$v = C_1 \sinh \lambda x + C_2 \cosh \lambda x + C_3 x + C_4$$

where constants $C_1$, $C_2$, $C_3$, and $C_4$ are determined from the boundary conditions.

**11-35.** Show that if in Example 11-3, axial force $P$ were tensile, the deflection

$$v = \frac{F}{2P\lambda} \operatorname{sech} \frac{\lambda L}{2} \sinh \lambda x - \frac{Fx}{2P}$$

**11-36.** Verify Eq. 11-48 by superposing the deflections due to the moments applied at each end using the expression for the deflection found in Prob. 11-31. This special case demonstrates that the solutions for beam-column deflections can be found by superposition for identical members subjected to the *same* axial force.

**11-37.** Show that the equation of the elastic curve for an elastic beam-column of constant $EI$ subjected to a sinusoidal load as shown in the figure is

$$v = \frac{1}{1 - P/P_{cr}} \frac{q_o L^4}{\pi^4 EI} \sin \frac{\pi x}{L}$$

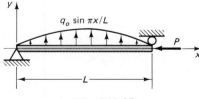

**Fig. P11-37**

**11-38.** Using Eq. 11-45, show that the equation for the bending moment for an elastic beam column subjected to a uniformly varying increasing load to the right is given as

$$M = -\frac{q_o \sin \lambda x}{\lambda^2 \sin \lambda L}$$

where $q = q_o x/L$.

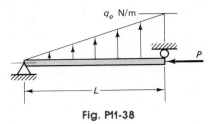

**Fig. P11-38**

**11-39.** (a) Using Eq. 11-45, show that the equation for the bending moment for a uniformly loaded elastic beam-column is given as

$$M = -\frac{q_o}{\lambda^2} \left( \frac{\cos \lambda L - 1}{\sin \lambda L} \sin \lambda x - \cos \lambda x + 1 \right)$$

(b) How can the equation of the elastic curve be easily found from the preceding result? (*Hint:* See Eq. 11-50.)

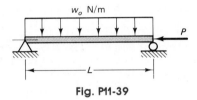

**Fig. P11-39**

**11-40.** Rework Example 11-4 using Eq. 11-45, and show that for $P = 0$, Eq. 11-49 reduces to $v_{max} = M_o L^2/8EI$.

**11-41.** Using Eq. 11-55, rederive Eq. 11-16.

**11-42.** Using homogeneous Eq. 11-45, determine the critical buckling load for the column of variable stiffness shown in the figure. (*Hint:* Enforce continuity conditions at the change in $EI$.)

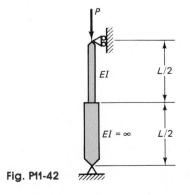

**Fig. P11-42**

**11-43.** A pin-ended bar of constant $EI$ is supported along its length by an elastic foundation, as shown in the figure. The foundation modulus is $k$ lb/in² and is such that when the bar deflects by an amount $v$, a restoring force $kv$ lb/in is exerted by the foundation normal to the bar. First, satisfy yourself that the governing homogeneous differential equation for this problem is

$$EIv^{iv} + Pv'' + kv = 0$$

Then, show that the required eigenvalue of the differential equation is

$$P_{cr} = \frac{\pi^2 EI}{L^2}\left[n^2 + \frac{1}{n^2}\left(\frac{kL^4}{\pi^4 EI}\right)\right]$$

Note that if $k = 0$, the minimum value of $P_{cr}$ becomes the classical Euler buckling load.

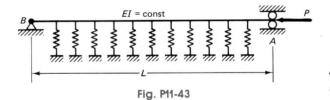

Fig. P11-43

## Sections 11-11 and 11-12

**11-44.** (a) If a pin-ended solid circular shaft is 1.5 m long and its diameter is 50 mm, what is the shaft's slenderness ratio? (b) If the same amount of material as in (a) is reshaped into a square bar of the same length, what is the slenderness ratio of the bar?

**11-45.** The cross section of a compression member for a small bridge is made as shown in Fig. 11-26(a). The top cover plate is $\frac{1}{2} \times 18$ in and the two C 12 × 20.7 channels are placed 10 in from back to back. If this member is 20 ft long, what is its slenderness ratio? (Check $L/r$ in two directions.)

**11-46.** Consider two axially loaded columns made from W 10 × 112 sections of A36 steel, where $\sigma_{yp} = 36$ ksi. One of the columns is 12 ft long and the other is 40 ft long. Both columns are braced at the pin ends. (a) Using the AISC ASD, determine the allowable loads for these columns. (b) What would be the allowable loads if A572 grade steel having $\sigma_{yp} = 50$ ksi is used instead? (*Note:* This illustrates the reason why

A572 grade steel is often used in building construction where the columns are stocky.)

**11-47.** A 14 × 193 column of A36 ($\sigma_{yp} = 36$ ksi) steel is laterally braced 12 ft apart in the weak direction of buckling and 24 ft apart in the strong direction, as shown in the figure. (a) Determine the allowable axial load for this column per AISC ASD. (b) Is this a well-balanced design?

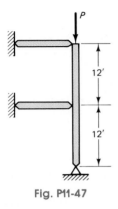

Fig. P11-47

**11-48.** A standard 12-in-nominal-diameter steel pipe (see Table 8 of the Appendix) supports a water tank, as shown in the figure. Assuming that the effective length of the free-standing pipe-column is 30 ft, what weight of water can be supported per AISC ASD? Let $\sigma_{yp} = 36$ ksi. (*Note:* In a complete design, wind load should also be considered.)

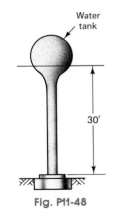

Fig. P11-48

**11-49.** For A36 steel, $\sigma_{yp} = 36$ ksi and $E = 29 \times 10^3$ ksi. (a) Determine the ratio $L_e/r$ for the transition point between Eqs. 11-56 and 11-57 for AISC ASD formulas.

(These formulas are constructed using the concept shown in Fig. 11-29(a).) (b) Show that the AISC LRFD Eq. 11-58 reduces to $L_e/r = 89.2\lambda_c$, and then determine the ratio $L_e/r$ for the transition point between Eqs. 11-59 and 11-60.

**11-50.** (a) Using the AISC LRFD method, determine the nominal axial column strengths (factored loads) $P_n = \phi_c A\sigma_{cr}$, where $A$ is the cross section for the two columns in Prob. 11-46. (b) Determine the ratios between the factored and the allowable axial loads for the corresponding columns in Prob. 11-46. Such allowable axial loads are 593 k and 153 k, respectively, for the short and the long columns. Interpret the results with the aid of Eq. 1-28.

**11-51.** Using the AISC LRFD formulas, rework Prob. 11-47 and form the ratio between the factored and the allowable axial loads. Interpret the result with the aid of Eq. 1-28. (See the preceding problem.)

**11-52.** Using AISC LRFD formulas, rework Prob. 11-48.

**11-53.** Two A36 steel C 10 × 15.3 channels form a 24-ft-long square compression member; the channel flanges are turned in, and are adequately laced together. Using the AISC ASD formulas, what is the allowable axial force on this member? $\sigma_{yp} = 36$ ksi and $E = 29 \times 10^3$ ksi.

**11-54.** A compression member is made up from two A572 steel C 8 × 11.5 channels arranged as shown in Fig. 11-26(b). (a) Determine the distance back to back of the channels so that the moments of inertia for the section about the two principal axes are equal. (b) If the member is 32 ft long, what is the nominal axial compressive strength of the member according to AISC LRFD provisions? $\sigma_{yp} = 50$ ksi and $E = 29 \times 10^3$ ksi.

**11-55.** A boom for an excavating machine is made up from four $2\frac{1}{2} \times 2\frac{1}{2} \times \frac{1}{2}$ in A36 steel angles, as shown in Fig. 11-26(c). Out-to-out dimensions of the square column, excluding lacing bars, is 14 in. According to AISC ASD formulas, what axial load can be applied to this member if it is 52 ft long? $\sigma_{yp} = 36$ ksi and $E = 29 \times 10^3$ ksi.

**11-56.** A compression chord of a small truss consists of two 4 × 4 × $\frac{3}{8}$ in steel angles arranged as shown in Fig. 11-26(d). The vertical legs of the angles are separated by spacers $\frac{1}{2}$ in apart. If the length of this member between braced points is 8 ft, what axial load may be applied according to the AISC ASD code? $\sigma_{yp} = 36$ ksi and $E = 29 \times 10^3$ ksi.

**11-57.** Using Aluminum Association formulas, determine the allowable axial loads for two 8 in × 8 in × 10.72 lb/ft 6061-T6 aluminum-alloy pin-ended columns that are 10 and 30 ft long. For cross-sectional properties of the columns, use Table 4 of the Appendix for W 8 × 31 steel section.

**11-58.** Using the NFPA formulas, determine the allowable axial loads for three 6 × 6 in Douglas Fir columns of different lengths: 5, 12, and 20 ft. Each column is braced at both ends, and $F_c = 1000$ psi and $E = 1.6 \times 10^6$ psi.

## Sections 11-13 and 11-14

**11-59.** An observation platform 6 ft in diameter is attached to the top of a standard 6-in pipe 20 ft long supported by a footing. Governed by the strength of the pipe, what weight, including a person or persons, can be placed on the platform? Locate the live load 3 ft from the pipe centerline. Neglect the weight of construction. Use Eq. 11-68 with the allowable stress given by the Euler formula with FS = 3. $E = 29 \times 10^6$ psi and $\sigma_{yp} = 36$ ksi.

**11-60.** A W 12 × 85 column 20 ft long is subjected to an eccentric load of 180 k located as shown in the figure. Using the AISC ASD interaction formula, determine whether this column is adequate. Use the same allowable stresses as in Example 11-11.

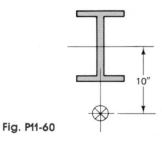

**Fig. P11-60**

**11-61.** A W 14 × 68 column made of A36 grade steel ($\sigma_{yp} = 36$ ksi) is 20 ft long and is loaded eccentrically as shown in the figure. Determine the allowable load

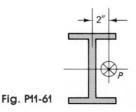

**Fig. P11-61**

*P* using the AISC ASD formulas. Assume pin-ended conditions. Let $F_b = 27$ ksi.

**11-62.** A W 12 × 40 column has an effective length of 20 ft. Using the AISC ASD formulas, determine the magnitude of an eccentric load that can be applied to this column at *A*, as shown in the figure, in addition to a concentric load of 20 k. The column is braced at top and bottom. The allowable bending stress $F_b = 17$ ksi.

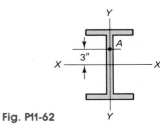

**Fig. P11-62**

**11-63.** What is the magnitude of the maximum beam reaction that can be carried by a W 10 × 49 column having an effective length of 14 ft, according to the AISC ASD interaction formula? Assume that the beam delivers the reaction at the outside flange of the column, as shown in the figure, and is concentric with respect to the minor axis. The top and bottom of the column are held laterally. Assume $F_y = 36$ ksi and $F_b = 22$ ksi.

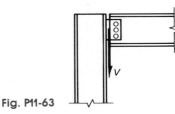

**Fig. P11-63**

**11-64.** Using the AISC ASD code, select a W shape column to carry a concentric load of 60 k and an ec-

centric load of 25 k applied on the *Y–Y* axis at a distance of 6 in from the *X–X* axis. The column is braced top and bottom and is 14 ft long. The allowable bending stress is 22 ksi and $\sigma_{yp} = 36$ ksi.

***11-65.** A narrow rectangular beam, such as shown in the figure, can collapse when loaded through lateral instability by twisting and displacing sidewise. It can be shown[29] that for this case, the critical force that may be applied at the end is

$$P_{cr} = 4.013\sqrt{B_1 C} / L^2$$

where $B_1 = hb^3 E/12$ is the flexural stiffness of the beam around the vertical axis, and $C = \beta hb^3 G$ is the torsional stiffness. (For rectangular sections, coefficient $\beta$ is given in a table in Section 4-14.)

A 5 × $\frac{1}{2}$ in narrow rectangular cantilever is made from steel ($\sigma_{yp} = 36$ ksi and $E = 30 \times 10^3$ ksi) and is loaded as shown in the figure. (a) Determine the critical load $P_{cr}$ and the critical length $L_{cr}$, where both the strength and the stability criteria are equally applicable. (b) Plot *P* vs *L* in the neighborhood of $P_{cr}$ and $L_{cr}$ for the two criteria. (Note that the smaller of the *P* values governs the design.)

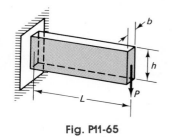

**Fig. P11-65**

**11-66.** Using Eq. 11-75, determine the maximum distance between intermittent lateral supports for the compression flange of a laterally unsupported W 24 × 76 beam spanning 24 ft.

[29] See Timoshenko and Gere, *Theory of Elastic Stability*, p. 260.

# chapter 12

# Energy and Virtual Work Methods

## 12-1. Introduction

In a few instances in the preceding chapters, the deflection of members was obtained by invoking the law of conservation of energy and equating the internal strain energy to the external work. This Lagrangian approach of employing *scalar functions* can be greatly extended, resulting in some of the most effective procedures for the analysis of deformable bodies. In Part A, the previously encountered concept of elastic strain energy is discussed from a somewhat more general point of view. This is followed by a specialized statement of the law of conservation of energy for deformable bodies, and the reason for the need to develop additional methods based on work and energy concepts to solve deflection problems.

Part B serves as an introduction to the two virtual work methods for deformable bodies. One of these, the method of *virtual forces*, is very useful for determining deflections caused by any kind of deformation and is not limited at all to elastic behavior. This method is one of the best available for calculating deflections of members. The conjugate method of *virtual displacements*, of great importance in the matrix analysis of structures and in finite elements, is also discussed. The duality of these two methods is illustrated by considering discrete structural systems.

In Part C of this chapter, the classical energy methods for treating similar problems to those susceptible to analysis by virtual work methods are discussed. These methods are based on considering the internal strain energy or the complementary strain energy and the corresponding external work. The derived equations are specialized for *linearly elastic* systems and are known as Castigliano's theorems. Illustrative examples are given for both statically determinate and indeterminate cases. A brief discussion on an application of the elastic energy concepts for determining column buckling loads concludes the chapter.

# Part A     ELASTIC STRAIN ENERGY AND EXTERNAL WORK

## 12-2. Elastic Strain Energy

The elastic strain-energy density $U_o$, i.e., the strain energy per unit volume, for a three-dimensional body was given by Eq. 8-49 in terms of principal stresses and strains. For a cartesian element in a general state of stress, the main energy density is

$$U_o = \frac{1}{2}(\sigma_x \varepsilon_x + \sigma_y \varepsilon_y + \sigma_z \varepsilon_z + \tau_{xy}\gamma_{xy} + \tau_{yz}\gamma_{yz} + \tau_{zx}\gamma_{zx}) \quad (12\text{-}1)$$

Therefore, the general expression for the total *internal* strain energy in a *linearly elastic* body is

$$U = \frac{1}{2} \iiint_V (\sigma_x \varepsilon_x + \sigma_y \varepsilon_y + \sigma_z \varepsilon_z$$
$$+ \tau_{xy}\gamma_{xy} + \tau_{yz}\gamma_{yz} + \tau_{zx}\gamma_{zx})\, dx\, dy\, dz \quad (12\text{-}2)$$

Integration extends over the volume of a body. Such a general expression is used in elasticity. In engineering mechanics of solid, a less general class of problems is considered and Eq. 12-2 simplifies. An expression

$$U = \frac{1}{2} \iiint_V (\sigma_x \varepsilon_x + \tau_{xy}\gamma_{xy})\, dx\, dy\, dz \quad (12\text{-}3)$$

is sufficient for determining the strain energy in axially loaded bars as well as in bent and sheared beams. Moreover, the last term of Eq. 12-3 written in the appropriate coordinates is all that is needed in the torsion problem of a circular shaft and for thin-walled tubes. These cases include the major types of problems treated in this text.

For linearly elastic material, for uniaxial stress, $\varepsilon_x = \sigma_x/E$, and for pure shear, $\gamma_{xy} = \tau_{xy}/G$. Thus, Eq. 12-3 can be recast in the following form:

$$U = \underbrace{\iiint_V \frac{\sigma_x^2}{2E}\, dx\, dy\, dz}_{\substack{\text{for axial loading and} \\ \text{bending of beams}}} + \underbrace{\iiint_V \frac{\tau_{xy}^2}{2G}\, dx\, dy\, dz}_{\text{for shear in beams}} \quad (12\text{-}4)$$

or $\qquad U = \iiint_V \frac{E\varepsilon_x^2}{2}\, dx\, dy\, dz + \iiint_V \frac{G\gamma_{xy}^2}{2}\, dx\, dy\, dz \quad (12\text{-}5)$

These equations can be specialized for the solutions encountered in engineering mechanics of solids, where it is generally customary to work with stress resultants $P$, $V$, $M$, and $T$. In this manner, the triple integrals are reduced to single integrals. Assuming that $E$ and $G$ are constant, some special cases of the last two equations follow.

### Strain Energy for Axially Loaded Bars

In this problem, $\sigma_x = P/A$ and $A = \iint dy\, dz$. Therefore, since axial force $P$ and cross-sectional area $A$ can only be functions of $x$,

$$U = \iiint_V \frac{\sigma_x^2}{2E}\, dV = \int_L \frac{P^2}{2AE}\, dx \tag{12-6}$$

where an integration along bar length $L$ gives the required quantity.

If $P$, $A$, and $E$ are constant, and, since for such cases, per Eq. 2-9, bar elongation $\Delta = PL/AE$, alternatively,

$$U = \frac{P^2 L}{2AE} = \frac{AE\Delta^2}{2L} \tag{12-7}$$

### Strain Energy for Beams in Bending

According to Eq. 6-24, the elastic strain energy in pure bending of a beam around one of its principal axes reduces to an integral along the beam length $L$, i.e.,

$$U = \iiint_V \frac{\sigma_x^2}{2E}\, dV = \int_L \frac{M^2}{2EI}\, dx \tag{12-8}$$

where $M$ is the bending moment, and $I$ is the moment of inertia for the cross section.

### Strain Energy for Beams in Shear

The expression given by Eq. 10-34 for a rectangular beam subjected to a constant shear can be generalized using the last term in Eq. 12-4 to read

$$U = \iiint_V \frac{\tau^2}{2G}\, dx\, dy\, dz = \alpha \int_L \frac{V^2}{2GA}\, dx \tag{12-9a}$$

where factor $\alpha$ depends on the cross-sectional area of a beam, and was shown to be 6/5 in Example 10-12 for a rectangular beam.[1] In this equation, both shear $V$ and area $A$ can vary along the span of length $L$.

### Strain Energy for Circular Tubes in Torsion

For this case, the basic expression for the shearing strain energy is analogous to the last term of Eq. 12-4. Such an expression has been used previously in Example 4-11. By substituting $\tau = T\rho/J$, Eq. 4-4, after some simplifications, becomes

$$U = \iiint_V \frac{\tau^2}{2G}\, dV = \int_L \frac{T^2}{2GJ}\, dx \qquad (12\text{-}9\text{b})$$

## 12-3. Displacements by Conservation of Energy

The law of conservation of energy, which states that energy can be neither created nor destroyed, can be adopted for determining the displacements of elastic systems due to the applied forces. The first law of thermodynamics expresses this principle as

$$\text{work done } = \text{ change in energy} \qquad (12\text{-}10)$$

For an adiabatic process[2] and when no heat is generated in the system, with the forces applied in a quasistatic manner,[3] the special form of this law for conservative systems[4] reduces to

$$W_e = U \qquad (12\text{-}11)$$

where $W_e$ is the total work done by the externally applied forces during the loading process, and $U$ is the total strain energy stored in the system.

It is significant to note that the total work $W$ must be zero, and

$$W = W_e + W_i = 0 \qquad (12\text{-}12)$$

---

[1] For a circular cross section, $\alpha = 10/9$, and for I beams and box sections, $\alpha = 1$, provided only web area $A_{\text{web}}$ is used in Eq. 12-9a.

[2] No heat is added or subtracted from the system.

[3] These forces are applied to the body so slowly that the kinetic energy can be neglected.

[4] In a conservative system, there are no dissipative forces such as those due to friction. More generally, in a conservative system no work is done in moving the system around any closed path.

where $W_e$ is the external work, and $W_i$ is the internal work. Therefore, from Eqs. 12-11 and 12-12, one has

$$U = -W_i \qquad\qquad (12\text{-}13)$$

where $W_i$ has a negative sign because the deformations are opposed by the internal forces. (See the discussion in connection with Fig. 12-2.)

Some formulations for determining the internal elastic strain energy $U$ were given in the preceding section. For linearly elastic systems, when a force, or a couple, is gradually applied, the external work $W_e$ is equal to one-half the total force multiplied by the displacement in the direction of its action. The possibility of formulating both $W_e$ and $U$ provides the basis for applying Eq. 12-11 for determining displacements.

This procedure was used in Example 2-10 for finding the deflection of an axially loaded bar, and again in Example 4-11 for determining the twist of a circular shaft. A general relation, Eq. 4-37, was derived using this procedure for twist of a thin-walled hollow member subjected to a torque. Lastly, this method was applied in Example 10-12 for finding the deflection caused by bending and shear in a cantilever loaded by a concentrated force at the end. In all of these cases, the procedure was limited to the determination of elastic deflections caused by a single concentrated force at the point of its application. Otherwise, intractable equations are obtained. For example, for two forces $P_1$ and $P_2$, $P_1 \Delta_1/2 + P_2 \Delta_2/2 = U$, where $\Delta_1$ and $\Delta_2$ are, respectively, the unknown deflections of the two forces. An additional relationship between $\Delta_1$ and $\Delta_2$, except in cases of symmetry, is not available. This requires development of the more general methods discussed in the remainder of the chapter.

## Part B          VIRTUAL WORK METHODS

### *12-4. Virtual Work Principle

Direct use of external work and internal strain energy for determining deflections breaks down if several deflections and/or rotations are sought at different points in a deformed body subjected to one or more forces. It is possible, however, to devise extraordinarily effective means for solving such problems by replacing true or real work and strain energy by external and internal virtual (imaginary) work. Two different procedures for applying the virtual work principle are described, resulting in the virtual displacement method and the virtual force method.

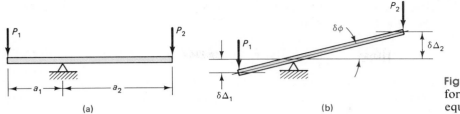

**Fig. 12-1** Alternative means for determining static equilibrium.

### Virtual Displacement Method

The conventional solution of static equilibrium problems usually follows the concepts introduced by Archimedes in his studies of levers. On this basis, the forces shown in Fig. 12-1(a) are related as

$$P_1 a_1 = P_2 a_2 \tag{12-14}$$

An alternative method[5] consists of rotating the lever through an *imaginary* or *virtual* angle $\delta\phi$. Here, as elsewhere, for emphasis, all such virtual quantities are expressed as $\delta\phi$ rather than the $d\phi$ employed in usual differential notation. The rotation shown causes *virtual displacements* $\delta\Delta_1$ and $\delta\Delta_2$ at the points of load application. Then, assuming that the system is conservative, the *virtual work* $\delta W$ done by real forces moving through virtual displacements in the direction of the applied forces is zero. Such work is positive when the direction of forces and displacements coincide. Applying this principle to the rigid bar shown in Fig. 12-2(b),

$$\delta W = P_1\,\delta\Delta_1 - P_2\,\delta\Delta_2 = 0 \tag{12-15}$$

However, since $\delta\Delta_1 = a_1\,\delta\phi$ and $\delta\Delta_2 = a_2\,\delta\phi$, and $P_1$ and $P_2$ *do not change* during the application of $\delta\phi$,

$$(P_1 a_1 - P_2 a_2)\,\delta\phi = 0 \tag{12-16}$$

Inasmuch as $\delta\phi$ is *perfectly arbitrary*, bearing no relation to the applied forces, the expression in parentheses must be zero, reverting to Eq. 12-14. Stated differently, for a system in equilibrium, the virtual displacement equation simply leads to an *equation of statics* multiplied by an arbitrary function $\delta\phi$.

For deformable bodies, the virtual-displacement equation must be generalized. For such systems, the total virtual work $\delta W$, consisting of the

---

[5] This approach apparently was considered by Leonardo da Vinci (1452–1519), Stevinus (1548–1620), Galileo (1564–1642), and Johann Bernoulli, who in 1717 introduced the notion of virtual displacements (velocities) in his letter to Varignon.

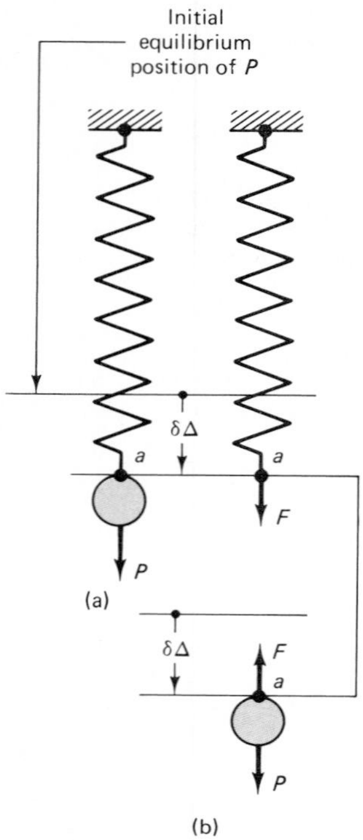

Initial equilibrium position of P

$\delta\Delta$

a

a

$F$

$P$

(a)

$\delta\Delta$

$F$

a

$P$

(b)

**Fig. 12-2** Virtual displacement mass-spring model.

external virtual work $\delta W_e$ and the internal virtual work $\delta W_i$, is zero. In the form of an equation,

$$\delta W = \delta W_e + \delta W_i = 0 \qquad (12\text{-}17)$$

This equation can be interpreted by making reference to Fig. 12-2, where a weightless spring supports a rigid mass. This mass applies a force $P$ to the spring, and *the system is in equilibrium.* Then a *virtual displacement* $\delta\Delta$ is imposed on this system, as shown in Fig. 12-2(a). During this displacement, force $P$ and internal forces $F$, shown on isolated parts in Fig. 12-2(b), *remain constant.*

As can be seen from the isolated mass in Fig. 12-2(b), the *external* virtual work $\delta W_e$ done by force $P$ is $P\,\delta\Delta$. On the other hand, the *internal* virtual work $\delta W_i$ done by $F$ is $-F\,\delta\Delta$. Therefore, this internal virtual work is negative.

However, it can be noted that the work done by $F$ acting on the spring, shown in the upper diagram of Fig. 12-2(b), has an opposite sign. Therefore, by calling this *internal* virtual work caused by the *external* force as $\delta W_{ie}$, it follows that

$$\delta W_i = -\delta W_{ie} \qquad (12\text{-}18)$$

By substituting this relation into Eq. 12-17,

$$\boxed{\delta W_e = \delta W_{ie}} \qquad (12\text{-}19)$$

Applying Eq. 12-19 to the simple system in Fig. 12-2,

$$P\,\delta\Delta = F\,\delta\Delta \qquad \text{or} \qquad (P - F)\,\delta\Delta = 0$$

This relation is analogous to Eq. 12-14. Here $\delta\Delta$ is arbitrary, so $P - F = 0$, an equation of equilibrium.

The *virtual displacement method* for deformable systems expressed by Eq. 12-19 establishes the *equations for static equilibrium.*

It is essential to note that during a virtual displacement, the magnitudes and the directions of applied forces do not change. It is to be emphasized that constitutive relations do not enter into the deviation of the virtual work equations.

### Virtual Force Method

For deformable bodies, virtual work can be formulated in two alternative ways. In the previous discussion, virtual work was determined by multiplying real forces by virtual displacements. Here the virtual work is obtained as a product of the virtual forces and real displacements. This approach leads to the *virtual force method*. In this method, again, no restrictions are placed on constitutive relations, and problems with thermal deformations, as well as settlement of supports and lack of member fit, can be analyzed.

In the virtual force method, the total virtual force $\delta W^*$, consisting of the *external* virtual work $\delta W_e^*$ and the *internal* work $\delta W_i^*$, is zero. In order to differentiate between the virtual work in this method with that in the virtual displacement method, the work quantities are identified by asterisks. For this case, analogous to Eq. 12-17,

$$\delta W^* = \delta W_e^* + \delta W_{ie}^* = 0 \qquad (12\text{-}20)$$

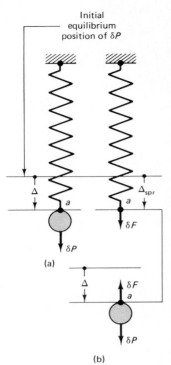

Initial equilibrium position of $\delta P$

(a)

(b)

**Fig. 12-3** Virtual force mass-spring model.

This equation can be clarified with the aid of Fig. 12-3, where a weightless spring supports a rigid mass. However, unlike the previous case, virtual force $\delta P$ is placed on the system first and is in equilibrium with internal forces $\delta F = \delta P$, as shown on isolated parts in Fig. 12-3(b). The deformation of the system is permitted to take place *after* force $\delta P$ is applied. Thereafter, $\delta P$ remains *constant*.

In the next step, *real displacement* $\Delta$ is allowed to occur when force $\delta P$ does the *external* virtual work:

$$\delta W_e^* = \delta P \, \Delta$$

During the same process, as can be seen from the isolated mass in Fig. 12-3(b), the *internal* virtual work $\delta W_i^*$ done by $\delta F$ is $-\delta F \, \Delta$. Therefore, here, again, the internal virtual work is negative. However, this sign can be reversed by considering the *internal* virtual work $\delta W_{ie}^*$ caused by the *external* force $\delta P = \delta F$, i.e.,

$$\delta W_i^* = -\delta W_{ie}^* \qquad (12\text{-}21)$$

Therefore, since from Eq. 12-20,

$$\delta W_e^* = -\delta W_i^*$$

the basic virtual work equation for the *virtual force method* is

$$\boxed{\delta W_e^* = \delta W_{ie}^*} \qquad (12\text{-}22)$$

where $\delta W_e^*$ is the *external* virtual work, and $\delta W_{ie}^*$ is the *internal* virtual work calculated in the sense described before.

In applying Eq. 12-22 to the simple system in Fig. 12-3, it is known *a priori* that $\delta F = \delta P$. Then, since the virtual work equation is $\delta P \, \Delta = \delta F \, \Delta_{\text{spr}}$, $\Delta = \Delta_{\text{spr}}$, an *equation of compatibility*.

It is important to recognize that only one deformable element is considered in each of the simple systems in Figs. 12-2 and 12-3. Typically, there are several such elements and calculations for the internal virtual work must extend over all of them.

To summarize: in the *virtual displacement* method, the use of kinematically admissible (plausible) displacements assures compatibility, and solutions lead to equations for static equilibrium. By contrast, in the *virtual force* method, the requirements of statics are fulfilled by assuming the virtual force system in equilibrium, and solutions lead to conditions of compatibility for the systems.

In applying Eqs. 12-19 and 12-22, the terms $\delta W_{ie}$ and $\delta W_{ie}^*$ will be simply referred to as the *internal* virtual work. It is to be understood, however, that these terms are calculated in accordance with the definitions given in connection with Eqs. 12-18 and 12-21.

In the next five sections, self-contained development of the two virtual work methods at an introductory level is given. For more advanced treatment of this important subject, the reader is referred to other texts.[6] In this text, the applications are limited to small deformations.

## 12-5. Virtual Forces for Deflections

The virtual work principle can be simply stated in words as

$$\text{external virtual work = internal virtual work} \qquad (12\text{-}23)$$

For the virtual force method, virtual work is obtained by multiplying *virtual forces* by *real displacements*. An algebraic implementation of this equation enables one to calculate deflection (or rotation) of any point (or element) on a deformed body. The deformations may be due to any cause, such as a temperature change, misfit of parts, or external forces deforming a body. *The method is not limited to the solution of elastic problems.* For this reason, this method has an exceptionally broad range of applications. By confining the discussion to typical problems of engineering mechanics of solids involving stress resultants, the basic virtual work equation, corresponding to general Eq. 12-23, can be readily derived. For this purpose, consider, for example, a body such as shown in Fig. 12-4 for which the deflection of some point $A$ in direction $AB$ caused by deformation or distortion of the body is sought. For this, the virtual work equation can be formulated by employing the following sequence of reasoning.

---

[6] For rigorous mathematical treatment of virtual work for three-dimensional elastic problems requiring the use of the divergence (Green's) theorem, see J. T. Oden, and E. A. Ripperger, *Mechanics of Elastic Structures*, 2nd ed. (New York: McGraw-Hill, 1981). For an extensive exposition of virtual work, see G. A. O. Davies, *Virtual Work in Structural Analysis* (Chichester: Wiley, 1982).

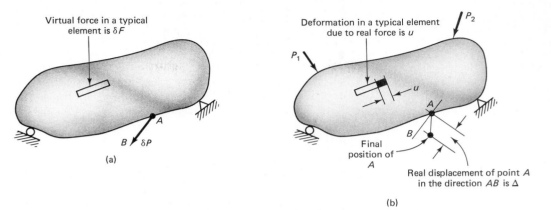

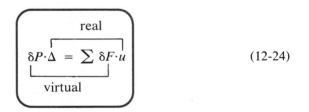

**Fig. 12-4** Virtual forces and real displacements.

First, apply an imaginary or virtual force $\delta P$ at $A$ acting in the desired direction $AB$ to the unloaded body. This force causes internal forces throughout the body. These internal forces, designated as $\delta F$, Fig. 12-4(a), can be found in any statically determinate systems.

Next, with the virtual force remaining on the body, apply the actual or real forces, Fig. 12-4(b), or introduce the specified deformations, such as those due to a change in temperature. This causes real internal deformations $u$, which can be computed. Owing to these deformations, the virtual force system does external and internal work.

Therefore, since the external work done by virtual force $\delta P$ moving a real amount $\Delta$ in the direction of this force is equal to the total work done on the internal elements by the virtual forces $\delta F$'s moving their respective real amounts $u$, the special form of the virtual-work equation becomes

$$
\underbrace{\delta P \cdot \overbrace{\Delta}^{\text{real}} = \sum \delta F \cdot u}_{\text{virtual}}
\qquad (12\text{-}24)
$$

Since all virtual forces attain their full values before real deformations are imposed, virtual work is a product of these quantities. Summation, or, in general, integration, of the right side of the equation indicates that *all internal* virtual work must be included.

Note that in Eq. 12-24, the ratios between $\delta F$'s and $\delta P$ remain constant regardless of the value of $\delta P$; hence, these virtual quantities need not be infinitesimal. Therefore, it is particularly convenient in applications to choose the applied virtual force $\delta P$ equal to unity, and to restate Eq. 12-24 as

$$\boxed{\overset{\text{real}}{\overline{1}\cdot\Delta} \;=\; \sum \underset{\text{virtual}}{\overline{p}\cdot u}} \qquad (12\text{-}25)$$

where   $\overline{1}$ = virtual unit force

$\Delta$ = real displacement of a point in the direction of the applied virtual unit force

$\overline{p}$ = virtual internal forces in equilibrium with the virtual unit force

$u$ = real internal displacements of a body

For simplicity, the symbols designating virtual quantities are redefined and are barred as shown instead of being identified by $\delta$'s. The real deformations can be due to any cause, with the elastic ones being a special case. Tensile forces and elongations of members are taken positive. A positive result indicates that the deflection occurs in the same direction as the applied virtual force.

In determining the angular rotations of a member, a unit couple is used instead of a unit force. In practice, the procedure of using a virtual unit force or a virtual unit couple in conjunction with virtual work is referred to as the *unit-dummy-load method*.

## 12-6. Virtual Force Equations for Elastic Systems

Equation 12-25 can be specialized for linearly elastic systems to facilitate the solution of problems. This is done here for axially loaded and for flexural members. Application examples follow.

### Trusses

A virtual unit force must be applied at a point in the direction of the deflection to be determined.

For linearly elastic bars of constant cross section $A$ subjected to real axial forces $F$, according to Eq. 2-9, the real axial bar deformations $u = FL/AE$. Therefore, Eq. 12-25 becomes

$$\boxed{\overline{1} \times \Delta = \sum_{i=1}^{n} \frac{\overline{p}_i F_i L_i}{A_i E_i}} \qquad (12\text{-}26)$$

where $\overline{p}_i$ is the axial force in a member due to the virtual unit force, and

$F_i$ is the force in the same member due to the real loads. The summation extends over all members of a truss.

## Beams

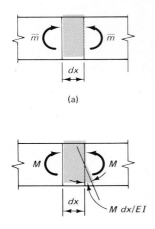

Fig. 12-5 Beam elements. (a) Virtual bending moments $m$. (b) Real bending moments $M$ and the rotation of sections they cause.

If the deflection of a point on an elastic beam is wanted by the virtual work method, a virtual unit force must be applied first in the direction in which the deflection is sought. This virtual force will set up internal bending moments at various sections of the beam designated by $\overline{m}$, as is shown in Fig. 12-5(a). Next, as the real forces are applied to the beam, bending moments $M$ rotate the "plane sections" of the beam $M\,dx/EI$ radians, Eq. 10-37. Hence, the virtual work done on an element of a beam by the virtual moments $\overline{m}$ is $\overline{m}M\,dx/EI$. Integrating this over the length of the beam gives the internal work on the elements. Hence, the special form of Eq. 12-25 for beams becomes

$$\overline{1} \times \Delta = \int_0^L \frac{\overline{m}M\,dx}{EI} \qquad (12\text{-}27)$$

An analogous expression may be used to find the angular rotation of a particular section in a beam. For this case, instead of applying a virtual unit force, a virtual unit couple is applied to the beam at the section being investigated. This virtual couple sets up internal moments $\overline{m}$ along the beam. Then, as the real forces are applied, they cause rotations $M\,dx/EI$ of the cross sections. Hence, the same integral expression as in Eq. 12-27 applies here. The external work by the virtual unit couple is obtained by multiplying it by the real rotation $\theta$ of the beam at this couple. Hence,

$$\overline{1} \times \theta = \int_0^L \frac{\overline{m}M\,dx}{EI} \qquad (12\text{-}28)$$

In Eqs. 12-27 and 12-28, $\overline{m}$ is the bending moment due to the virtual loading, and $M$ is the bending moment due to the real loads. Since both $\overline{m}$ and $M$ usually vary along the length of the beam, both must be expressed by appropriate functions.

## EXAMPLE 12-1

Find the vertical deflection of point $B$ in the pin-jointed steel truss shown in Fig. 12-6(a) due to the following causes: (*a*) the elastic deformation of the members, (*b*) a shortening by 0.125 in of member $AB$ by means of a turnbuckle, and (*c*) a

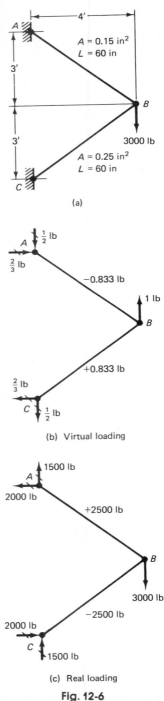

(a)

(b) Virtual loading

(c) Real loading

**Fig. 12-6**

drop in temperature of 120 °F occurring in member $BC$. The coefficient of thermal expansion of steel is $6.5 \times 10^{-6}$ inch per inch per degree Fahrenheit. Neglect the possibility of lateral buckling of the compression member. Let $E = 30 \times 10^6$ psi.

### Solution

(*a*) A virtual unit force is applied in the upward vertical direction, as shown in Fig. 12-6(b), and the resulting forces $\bar{p}$ are determined and recorded on the same diagram. Then the forces in each member due to the real force are also determined and recorded, Fig. 12-5(c). The solution follows by means of Eq. 12-26. The work is carried out in the table.

| Member | $\bar{p}$, lb | $F$, lb | $L$, in | $A$, in² | $\bar{p}FL/A$ |
|--------|------|------|------|------|--------|
| AB | −0.833 | +2500 | 60 | 0.15 | −833,000 |
| BC | +0.833 | −2500 | 60 | 0.25 | −500,000 |

From this table, $\sum \bar{p}FL/A = -1,333,000$. Hence,

$$\bar{1} \times \Delta = \sum \frac{\bar{p}FL}{AE} = \frac{-1,333,000}{30 \times 10^6} = -0.0444 \text{ lb-in}$$

and

$$\Delta = -0.0444 \text{ in}$$

The negative sign means that point $B$ deflects down. In this case, "negative work" is done by the virtual force acting upward when it is displaced in a downward direction. Note particularly the units and the signs of all quantities. Tensile forces in members are taken positive, and vice versa.

(*b*) Equation 12-25 is used to find the vertical deflection of point $B$ due to the shortening of member $AB$ by 0.125 in. The forces set up in the bars by the virtual force acting in the direction of the deflection sought are shown in Fig. 12-6(b). Then, since $u$ is $-0.125$ in (shortening) for member $AB$ and is zero for member $BC$,

$$\bar{1} \times \Delta = (-0.833)(-0.125) + (+0.833)(0) = +0.1042 \text{ lb-in}$$

and

$$\Delta = +0.1042 \text{ in up}$$

(*c*) Again, using Eq. 12-25, and noting that due to the drop in temperature, Eq. 2-18, $\Delta_T = -6.5 \times 10^{-6} \times 120 \times 60 = -0.0468$ in in member $BC$,

$$\bar{1} \times \Delta = (+0.833)(-0.0468) = -0.0390 \text{ lb-in}$$

and

$$\Delta = -0.0390 \text{ in down}$$

By superposition, the net deflection of point $B$ due to all three causes is $-0.0444$

$+ 0.1042 - 0.0390 = +0.0208$ in up. To find this quantity, all three effects could have been considered simultaneously in the virtual work equation.

## EXAMPLE 12-2

Find the deflection and rotation at the middle of the cantilever beam loaded as shown in Fig. 12-7. $EI$ for the beam is constant.

### Solution

The downward virtual force is applied at point $A$, whose deflection is sought, Fig. 12-7(b). The $\overline{m}$ diagram and the $M$ diagram are shown in Figs. 12-7(c) and (d), respectively. For these functions, the same origin of $x$ is taken at the free end of the cantilever. After these moments are determined, Eq. 12-27 is applied to find the deflection.

$$M = -\frac{x}{2}\frac{w_o x}{L}\frac{x}{3} = -\frac{w_o x^3}{6L} \qquad 0 \le x \le L$$

$$\overline{m} = 0 \qquad\qquad 0 \le x \le L/2$$

$$\overline{m} = -1(x - L/2) \qquad L/2 \le x \le L$$

$$\overline{1} \times \Delta = \int_0^L \frac{\overline{m}M\,dx}{EI} = \frac{1}{EI}\int_0^{L/2} (0)\left(-\frac{w_o x^3}{6L}\right) dx$$

$$+ \frac{1}{EI}\int_{L/2}^L \left(-x + \frac{L}{2}\right)\left(-\frac{w_o x^3}{6L}\right) dx$$

$$= \frac{49 w_o L^4}{3840 EI}\ \text{N}\cdot\text{m}$$

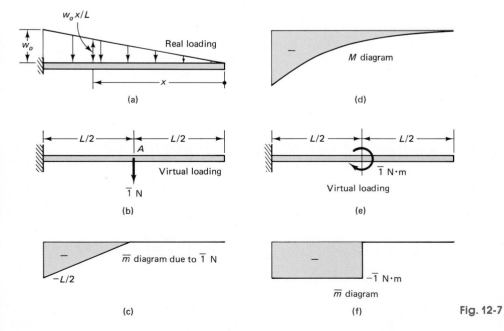

$w_o x/L$

$w_o$

Real loading

$x$

(a)

$M$ diagram

(d)

— $L/2$ — $L/2$ —

$A$

Virtual loading

$\overline{1}$ N

(b)

— $L/2$ — $L/2$ —

$\overline{1}$ N·m

Virtual loading

(e)

$-L/2$

$\overline{m}$ diagram due to $\overline{1}$ N

(c)

$-\overline{1}$ N·m

$\overline{m}$ diagram

(f)

**Fig. 12-7**

The deflection of point $A$ is numerically equal to this quantity. The deflection due to shear has been neglected.

To find the beam rotation at the middle of the beam, a virtual unit couple is applied at $A$, Fig. 12-7(e). The corresponding $\overline{m}$ diagram is shown in Fig. 12-7(f). The real bending moment $M$ is the same as in the previous part of the problem. The virtual moment $\overline{m} = 0$ for $x$ between 0 and $L/2$, and $\overline{m} = -1$ for the remainder of the beam. Using these moments, and applying Eq. 12-28, determines the rotation of the beam at $A$.

$$M = -\frac{w_o x^3}{6L} \quad \text{and} \quad \overline{m} = -1 \quad L/2 \leq x \leq L$$

$$1 \times \theta = \int_0^L \frac{\overline{m}M \, dx}{EI} = \frac{1}{EI} \int_{L/2}^L (-1)\left(-\frac{w_o x^3}{6L}\right) dx = \frac{15 w_o L^3}{384 EI} \text{ N·m}$$

The rotation of the beam at $A$ is numerically equal to this result.

## EXAMPLE 12-3

An aluminum beam is supported by a pin at one end and an inclined aluminum bar at a third point, as shown in Fig. 12-8(a). Find the deflection at $C$ caused by the application of the downward force of 2 kN at that point. The cross section

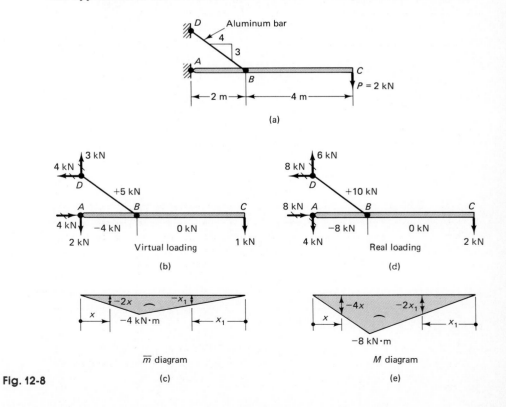

**Fig. 12-8**

of the beam is 5000 mm$^2$ (50 × 10$^{-4}$ m$^2$), and that of the bar, 500 mm$^2$ (5 × 10$^{-4}$ m$^2$). The moment of inertia for the beam around the horizontal axis is 60 × 10$^6$ mm$^4$ (60 × 10$^{-6}$ m$^4$). Neglect deflection caused by shear. Let $E$ = 70 GPa.

## Solution

A unit virtual force of 1 kN is applied vertically downward at $C$. This force causes an axial force in member $DB$ and in part $AB$ of the beam, Fig. 12-8(b). Owing to this force, bending moments are also caused in beam $AC$, Fig. 12-8(c). Similar computations are made and are shown in Figs. 12-8(d) and (e) for the applied real force. The deflection of point $C$ depends on the deformations caused by the axial forces, as well as flexure; hence, the virtual work equation is

$$\bar{1} \times \Delta = \sum \frac{\bar{p}FL}{AE} + \int_0^L \frac{\bar{m}M\,dx}{EI}$$

The first term on the right side of this equation is computed in the table. Then the integral for the internal virtual work due to bending is found. For the different parts of the beam, two origins of $x$'s are used in writing the expressions for $\bar{m}$ and $M$; see Figs. 12-8(c) and (e), respectively.

| Member | $\bar{p}$, kN | $F$, kN | $L$, m | $A$, m$^2$ | $\bar{p}FL/A$ |
|--------|------|------|------|-----------|-----------|
| DB | +5 | +10 | 2.5 | 5 × 10$^{-4}$ | +250,000 |
| AB | −4 | −8 | 2.0 | 50 × 10$^{-4}$ | +12,800 |

From the table, $\sum \bar{p}FL/A$ = +262,800

or
$$\sum \bar{p}FL/AE = 3.75 \times 10^{-3} \text{ kN·m}$$

$$\int_0^L \frac{\bar{m}M\,dx}{EI} = \int_0^2 \frac{(-2x)(-4x)\,dx}{EI} + \int_0^4 \frac{(-x_1)(-2x_1)\,dx_1}{EI}$$
$$= + 15.25 \times 10^{-3} \text{ kN·m}$$

Therefore, $\bar{1} \times \Delta = (3.75 + 15.25)10^{-3} = 19 \times 10^{-3}$ kN·m and point $C$ deflects $19 \times 10^{-3}$ m = 19 mm down.

Note that the work due to the two types of action was superposed. Also note that the origins for the coordinate system for moments may be chosen as convenient; however, the same origin must be used for the corresponding $\bar{m}$ and $M$.

## EXAMPLE 12-4

Find the horizontal deflection, caused by concentrated force $P$, of the end of the curved bar shown in Fig. 12-9(a). The flexural rigidity $EI$ of the bar is constant. Neglect the effect of axial force and shear on the deflection.

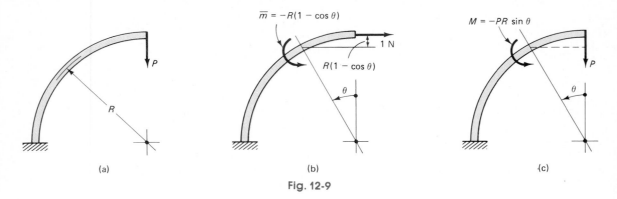

$$\overline{m} = -R(1 - \cos\theta)$$

1 N

$$R(1 - \cos\theta)$$

θ

(b)

$$M = -PR\sin\theta$$

P

θ

(c)

**Fig. 12-9**

## Solution

If the radius of curvature of a bar is large in comparison with the cross-sectional dimensions (Section 6-9), ordinary beam deflection formulas may be used, replacing $dx$ by $ds$. In this case, $ds = R\,d\theta$.

Applying a horizontal virtual force at the end in the direction of the deflection wanted, Fig. 12-9(b), it is seen that $\overline{m} = -R(1 - \cos\theta)$. Similarly, for the real load, from Fig. 12-9(c), $M = -PR\sin\theta$. Therefore,

$$
\begin{aligned}
\overline{1} \times \Delta &= \int_0^L \frac{\overline{m}M\,ds}{EI} \\
&= \int_0^{\pi/2} \frac{-R(1 - \cos\theta)(-PR\sin\theta)R\,d\theta}{EI} = +\frac{PR^3}{2EI}\,\text{N·m}
\end{aligned}
$$

The deflection of the end to the right is numerically equal to this expression.

## 12-7. Virtual Forces for Indeterminate Problems

The unit-dummy-load method derived using the virtual force concept can be used to advantage for the solution of statically indeterminate problems. Here the procedure is illustrated on a problem statically indeterminate to the first degree. The basic procedure is essentially the same as that already described in Section 2-15 on the *force* method of analysis for statically indeterminate axially loaded bar systems (see Fig. 2-43). Applications of the *force* (or *flexibility*) method to problems of higher degree of statical indeterminancy is discussed in Sections 13-2 and 13-3. In general, this method is best suited for linearly elastic problems, where superposition is valid.

### EXAMPLE 12-5

(*a*) Find the forces in the bars of the pin-jointed steel structure shown in Fig. 12-10. (*b*) Determine the deflection of joint (nodal point) *B*. Let $E = 30 \times 10^6$ psi.

## Solution

(*a*) The structure can be rendered statically determinate by cutting bar *DB* at *D*. Then the forces in the members are as shown in Fig. 12-10(b). In this determinate structure, the deflection of point *D* must be found. This can be done by applying a vertical virtual force at *D*, Fig. 12-10(c), and using the virtual force method. However, since the $\bar{p}FL/AE$ term for member *BD* is zero, the vertical deflection of point *D* is the same as that of *B*. In Example 12-1, the latter quantity was found to be $44.4 \times 10^{-3}$ in down and is so shown in Fig. 12-10(b).

The deflection of point *D*, shown in Fig. 12-10(b), violates a boundary condition of the problem, and a vertical force must be applied at *D* to restore it. If $f_{DD}$ is the deflection of point *D* due to a unit (real) force at *D*, it defines the flexibility of this system. It is necessary to multiply $f_{DD}$ by a factor $X_D$ to chose the gap $\Delta_{DP} = 44.4 \times 10^{-3}$ in at *D* caused by the force *P* in the determinate system. This simply means that the deflection $\Delta_D$ at *D* becomes zero. Stated algebraically,

$$\Delta_D = f_{DD}X_D + \Delta_{DP} = 0$$

Hence, the problem resolves into finding $f_{DD}$. This can be done by applying a $\bar{1}$-lb virtual force at *D*, then applying a 1-lb real force at the same point, and then using Eq. 12-26. The forces set up in the determinate structure by the virtual and the real forces are numerically the same, Fig. 12-10(c). To differentiate between the two, forces in members caused by a virtual force are designated by $\bar{p}$ and the real force by $p$. The solution is carried out in the following table.

| Member | $\bar{p}$, lb | $p$, lb | $L$, in | $A$, in² | $\bar{p}pL/A$ |
|--------|--------|--------|--------|--------|--------|
| AB | −0.833 | −0.833 | 60 | 0.15 | +278 |
| BC | +0.833 | +0.833 | 60 | 0.25 | +167 |
| BD | +1.000 | +1.000 | 40 | 0.10 | +400 |

From the table, $\sum \bar{p}pL/A = +845$. Therefore, since

$$\bar{1} \times \Delta = \sum \frac{\bar{p}pL}{AE} = \frac{+845}{30 \times 10^6} = 28.1 \times 10^{-6} \text{ lb-in}$$

$$f_{DD} = 28.1 \times 10^{-6} \text{ in} \quad \text{and} \quad 28.1 \times 10^{-6} X_D - 44.4 \times 10^{-3} = 0$$

To close the gap of $44.4 \times 10^{-3}$ in, the 1-lb real force at *D* must be increased $0.0444/0.0000281 = 1580$ times. Therefore the actual force in the member *DB* is 1580 lb. The forces in the other two members may now be determined from statics or by superposition of the forces shown in Fig. 12-10(b) with $X_D$ times the $p$ forces shown in Fig. 12-10(c). By either method, the force in *AB* is found to be $+1180$ lb (tension), and in *BC*, $-1180$ lb (compression).

(*b*) Three different virtual systems are employed to determine the deflection of nodal point *B* caused by the applied force. The simplest of the three consists of recognizing that since the force in vertical member *BD* is known to be 1580 lb, the deflection at *B*

$$\Delta_B = \frac{FL}{AE} = \frac{1580 \times 40}{0.10 \times 30 \times 10^6} = 21 \times 10^{-3} \text{ in}$$

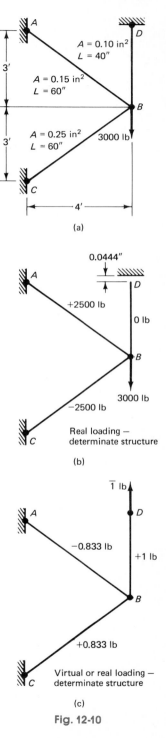

$A = 0.10$ in²
$L = 40''$

$A = 0.15$ in²
$L = 60''$

$A = 0.25$ in²
$L = 60''$

3000 lb

(a)

0.0444"

+2500 lb

0 lb

3000 lb

−2500 lb

Real loading —
determinate structure

(b)

$\bar{1}$ lb

−0.833 lb

+1 lb

+0.833 lb

Virtual or real loading —
determinate structure

(c)

**Fig. 12-10**

This solution from the point of view of a virtual force system means that the virtual force in member *DB* is unity and is zero in the other members.

Alternatively, the virtual force system may consist of active members *AB* and *BC*, with a zero virtual force in member *BD*. Then, by assuming that the virtual unit force acts upwards at *B*, the virtual forces in members *AB* and *BC* are as shown in Fig. 12-10(c) (or Fig. 12-6(b)). (Remember that the force in member *BD* is assumed to be zero.) From the solution for (*a*), the real forces in members *AB*, *BC*, and *BD* are known to be, respectively, +1180 lb, −1180 lb, and +1580 lb. The solution of Eq. 12-26 to obtain the deflection at *B* is carried out in the following table.

| Member | $\bar{p}$, lb | $F$, lb | $L$, in | $A$, in$^2$ | $\bar{p}FL/A$ |
|--------|------|------|------|------|------|
| *AB* | −0.833 | +1180 | 60 | 0.15 | −393,000 |
| *BC* | +0.833 | −1180 | 60 | 0.25 | −236,000 |
| *BD* | 0 | +1580 | 40 | 0.10 | 0 |

Hence, 
$$\Delta_B = \frac{\sum \bar{p}FL/A}{E} = -\frac{629,000}{30 \times 10^6} = -21 \times 10^{-3} \text{ in}$$

The negative sign shows that the deflection is downward.

Lastly, let the virtual force system consist of all three bars. The virtual forces in the bars due to a unit downward force can be found by dividing the bar forces due to the applied forces by 3000; for example, for member *AB*, such a virtual force is 1180/3000 = 0.393 lb. Again, the solution is carried out in tabular form.

| Member | $\bar{p}$, lb | $F$, lb | $L$, in | $A$, in$^2$ | $\bar{p}FL/A$ |
|--------|------|------|------|------|------|
| *AB* | +0.393 | +1180 | 60 | 0.15 | +185,000 |
| *BC* | −0.393 | −1180 | 60 | 0.25 | +111,000 |
| *BD* | +0.527 | +1580 | 40 | 0.10 | +333,000 |

Hence, 
$$\Delta_B = \frac{\sum \bar{p}FL/A}{E} = \frac{629,000}{30 \times 10^6} = 21 \times 10^{-3} \text{ in}$$

The results are the same by three entirely different virtual force systems that are in static equilibrium. This is true in general. *Any* self-equilibrating virtual system can be used provided its displacements go through the prescribed real displacements.

In any given case, to make certain that the elastic analysis is applicable, maximum stresses must be determined. For the solution to be correct, these must be in the linearly elastic range for the material used.

## **\*\*12-8.** **Virtual Displacements for Equilibrium**

The virtual work principle can be adapted for developing the virtual displacement method of structural analysis. The derivation of this method can begin by restating the virtual work principle in words:

$$\text{external virtual work} = \text{internal virtual work} \qquad (12\text{-}23)$$

Virtual work for the virtual displacement method is determined by multiplying the virtual displacements by real forces. This is to be contrasted with the virtual force method, where virtual work is found by multiplying the virtual forces by real displacements. Because of this, a number of differences arise in the virtual work equations.

In the virtual displacement method, both the virtual and the real displacements must be compatible with the special requirements of a problem. This means that the member displacements must conform to the boundary conditions and the displacements of the load points. Since the boundary conditions are simplest to satisfy at pin-ended axially loaded bars, only such problems are considered here. Moreover, the discussion is limited to bar assemblies meeting at a single pinned joint where an external force is applied. Such a joint is referred to as a *nodal point*. Although the discussion is limited to the simplest class of problems, the described procedure provides an introduction to the most widely used method in the matrix analysis of structures and finite elements, where it is indispensable.

In the virtual displacement method, besides an accurate definition of the real and the virtual displacements, member forces must be defined as functions of the nodal displacements. This is achieved with the aid of constitutive relationships. Here such relationships are *strictly limited to linearly elastic behavior*.

As with Archimedes' lever, the virtual work equations provide the equations of equilibrium. In this manner, the three basic requirements of equilibrium, compatibility, and constitutive relations are satisfied.

In order to construct the basic virtual work equation, by analogy to Eq. 12-25, one can formulate it by going directly to the *unit-dummy-displacement method*. Here the real external force at its full value moves through a virtual unit displacement in the direction of the force. Simultaneously, the real internal forces at their full values move through the virtual displacements caused by the unit virtual displacement. This yields the following virtual work equation:

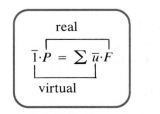

$$\overline{1} \cdot P = \sum \overline{u} \cdot F \qquad (12\text{-}29)$$

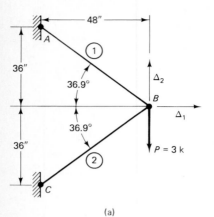

(a)

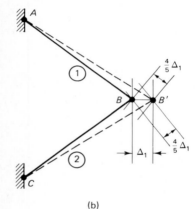

(b)

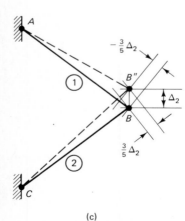

(c)

**Fig. 12-11**

where    $\bar{1}$ = virtual unit displacement at a nodal point in the direction of $P$

$P$ = real external (nodal) force

$\bar{u}$ = virtual internal displacements compatible with the virtual unit displacement

$F$ = real internal forces in equilibrium with $P$

Deformations of flexural members as well as those of finite elements for a continuum generally require more than one nodal point for their definitions. Therefore, the examples that follow consider only axially loaded pin-ended bars.

## EXAMPLE 12-6

Using the virtual displacement method, determine the bar forces in the pin-jointed linearly elastic steel truss of Example 12-1; see Fig. 12-11. $E = 30 \times 10^3$ ksi.

### Solution

In this truss, bar $AB$ has the cross-sectional area $A_1 = 0.15$ in$^2$, and bar $BC$ has an area $A_2 = 0.25$ in$^2$. Because of this lack of bar symmetry, during a loading process, joint $B$ can move both horizontally and vertically, Figs. 12-11(b) and (c). Hence, this system has two degrees of freedom, or two degrees of kinematic indeterminancy (see Section 2-17). These displacement components for nodal point $B$ are designated, respectively, as $\Delta_1$ and $\Delta_2$, with their positive sense shown in Fig. 12-11(a).

The compatibility requirements for the problem are complied with by permitting nodal point $B$ to move, as shown in Figs. 12-11(b) or (c). A linear combination of these displacements is appropriate.

The constitutive requirements for the problem are defined by the bar stiffnesses, Eq. 2-12, which for bars 1 and 2, respectively, are

$$k_1 = \frac{A_1 E}{L_1} = \frac{0.15 \times 30 \times 10^3}{60} = 75 \text{ k/in}$$

$$k_2 = \frac{A_2 E}{L_2} = \frac{0.25 \times 30 \times 10^3}{60} = 125 \text{ k/in}$$

On this basis, the internal bar forces $F_1$ and $F_2$ can be determined as functions of the joint displacements $\Delta_1$ and $\Delta_2$. The bar deformations $u_1$ and $u_2$ corresponding to these displacements are

$$u_1 = 0.8\Delta_1 - 0.6\Delta_2 \quad \text{and} \quad u_2 = 0.8\Delta_1 + 0.6\Delta_2$$

Hence,    $F_1 = k_1 u_1 = 75(0.8\Delta_1 - 0.6\Delta_2) = 60\Delta_1 - 45\Delta_2$
$F_2 = k_2 u_2 = 125(0.8\Delta_1 + 0.6\Delta_2) = 100\Delta_1 + 75\Delta_2$

As this problem is kinematically indeterminate to the second degree, the virtual displacement Eq. 12-29 must be applied twice. The equilibrium equation for forces acting in the horizontal direction follows by taking $\Delta_1 = \bar{1}$, causing virtual bar displacements $\bar{u}_1 = \bar{u}_2 = 0.8$. Noting that no horizontal force is applied at $B$,

$$\bar{1} \times 0 = \bar{u}_1 F_1 + \bar{u}_2 F_2$$
$$0.8(60\Delta_1 - 45\Delta_2) + 0.8(100\Delta_1 + 75\Delta_2) = 0$$

or

$$128\Delta_1 + 24\Delta_2 = 0$$

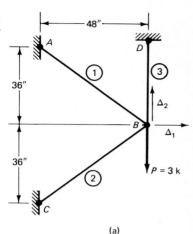

(a)

Similarly, the equilibrium equation for forces acting in the vertical direction is obtained by setting $\Delta_2 = \bar{1}$, resulting in virtual bar displacements $\bar{u}_1 = -0.6$ and $\bar{u}_2 = 0.6$. Again, applying Eq. 12-29,

$$\bar{1} \times (-3) = \bar{u}_1 F_1 + \bar{u}_2 F_2$$
$$-0.6(60\Delta_1 - 45\Delta_2) + 0.6(100\Delta_1 + 75\Delta_2) = -3$$

or

$$24\Delta_1 + 72\Delta_2 = -3$$

Solving the two reduced equations simultaneously,

$$\Delta_1 = 8.33 \times 10^{-3} \text{ in} \quad \text{and} \quad \Delta_2 = -44.4 \times 10^{-3} \text{ in}$$

Hence,

$$F_1 = 60\Delta_1 - 45\Delta_2 = 2.5 \text{ k}$$
$$F_2 = 100\Delta_1 + 75\Delta_2 = -2.5 \text{ k}$$

These results are in complete agreement with those given in Example 12-1 for bar forces as well as for the vertical deflection. The advantages of this method are more apparent in the next example.

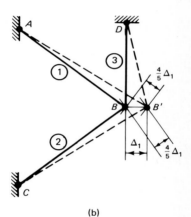

(b)

## EXAMPLE 12-7

Using the virtual displacement method, determine the bar forces in the statically indeterminate pin-jointed linearly elastic steel truss of Example 12-5; see Fig. 12-12. Let $E = 30 \times 10^3$ ksi.

### Solution

Although this problem is statically indeterminate to the first degree, as in the previous example of a statically determinate case, the system remains kinematically indeterminate to the second degree. In both cases, nodal point $B$ has two deflection components, a horizontal and a vertical, and both systems have two degrees of freedom. Therefore, the solution becomes only slightly more complicated than that in the previous example. Hence, proceeding as before, and noting that the cross-sectional area for the third bar, $DB$, is 0.10 in², 

$$k_3 = \frac{A_3 E}{L_3} = \frac{0.10 \times 30 \times 10^3}{40} = 75 \text{ k/in}$$

From the previous example, $k_1 = 75$ k/in and $k_2 = 125$ k/in.

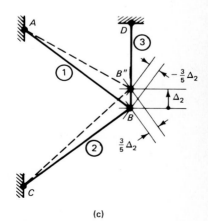

(c)

**Fig. 12-12**

The displacements of the system in the horizontal and vertical directions are shown in Figs. 12-12(b) and (c). It is to be noted that the length of bar *DB* is considered not to change because $\Delta_1$ is very small (see Fig 2-24). Bar deformations $u_1$ and $u_2$ due to $\Delta_1$ and $\Delta_2$, respectively, remain the same as in the previous example, whereas $u_3$ is equal to $\Delta_2$. Summarizing these results,

$$u_1 = 0.8\Delta_1 - 0.6\Delta_2 \qquad u_2 = 0.8\Delta_1 + 0.6\Delta_2 \qquad \text{and} \qquad u_3 = \Delta_2$$

Hence, recalling the earlier results, and adding only a new term for bar *DB*,

$$F_1 = 60\Delta_1 - 45\Delta_2 \qquad F_2 = 100\Delta_1 + 75\Delta_2 \qquad F_3 = k_3 u_3 = 75\Delta_2$$

By noting that for the virtual displacement $\Delta_1 = \bar{1}$, $\bar{u}_1 = \bar{u}_2 = 0.8$ and $\bar{u}_3 = 0$, the equilibrium equation for the forces acting in the horizontal direction, based on Eq. 12-29, becomes

$$\bar{1} \times 0 = \bar{u}_1 F_1 + \bar{u}_2 F_2 + \bar{u}_3 F_3$$

However, since $u_3 = 0$, this equation reduces to the one given before, reading

$$128\Delta_1 + 24\Delta_2 = 0$$

Since for $\Delta_2 = \bar{1}$, $\bar{u}_1 = -0.6$, $\bar{u}_2 = 0.6$, and $\bar{u}_3 = 1$, the equilibrium equation for the forces acting in the vertical direction, using Eq. 12-29, becomes

$$\bar{1} \times (-3) = \bar{u}_1 F_1 + \bar{u}_2 F_2 + \bar{u}_3 F_3$$
$$0.6(60\Delta_1 - 45\Delta_2) + 0.6(100\Delta_1 + 75\Delta_2) + 75\Delta_2 = -3$$

or
$$24\Delta_1 + 147\Delta_2 = -3$$

Solving the reduced equations simultaneously,

$$\Delta_1 = 3.947 \times 10^{-3} \text{ in} \qquad \text{and} \qquad \Delta_2 = -21.05 \times 10^{-3} \text{ in}$$

Hence,
$$F_1 = 60\Delta_1 - 45\Delta_2 = 1.18 \text{ k}$$
$$F_2 = 100\Delta_1 + 75\Delta_2 = 1.18 \text{ k}$$
$$F_3 = 75\Delta_2 = 1.58 \text{ k}$$

These results are in complete agreement with those found earlier in Example 12-5.

The kinematic indeterminacy of this problem would remain the same regardless of the number of bars meeting at joint *B*.

Problems of higher degree of statical indeterminacy are discussed in Section 13-6 in connection with the *displacement* method of analysis.

## **12-9. Virtual Work for Discrete Systems

The virtual displacement and virtual force methods both stem from the same virtual work principle. It is instructive, therefore, to show the interrelationship between the two methods using discrete structural systems, i.e., on structural systems with a finite number of applied forces, members, and nodal displacements. This is done here by employing matrix notation. The required matrix definitions and operations required for this purpose follow.

A *matrix* is an ordered array of numbers, such as encountered earlier in Eqs. 1-1a, 1-1b, and 3-12. The special matrix in Eq. 1-1a is commonly referred to as a *column vector*. The matrices in Eqs. 1-1b and 3-12 are known as *square matrices*. Here matrices are identified by braces for column vectors and by brackets for square matrices, or are shown in boldface type.

A *matrix product* of a 2 × 2 square matrix by a 2 × 1 column vector results in a 2 × 1 column vector:

$$\begin{bmatrix} a_{11} & a_{12} \\ a_{21} & a_{22} \end{bmatrix} \begin{Bmatrix} b_1 \\ b_2 \end{Bmatrix} = \begin{bmatrix} a_{11}b_1 + a_{12}b_2 \\ a_{21}b_1 + a_{22}b_2 \end{bmatrix}$$

A product of a 4 × 4 symmetric square matrix by a 4 × 1 column vector is displayed in Eq. 2-42.

The *transpose* of a column vector such as $\mathbf{P}$ is denoted by $\mathbf{P}^T$ and is obtained by interchanging the rows and columns of $\mathbf{P}$. Therefore, if

$$\mathbf{P} = \begin{Bmatrix} P_1 \\ P_2 \\ \vdots \\ P_n \end{Bmatrix} \quad \text{and} \quad \mathbf{\Delta} = \begin{Bmatrix} \Delta_1 \\ \Delta_2 \\ \vdots \\ \Delta_n \end{Bmatrix}$$

then

$$\mathbf{P}^T = [P_1 \ P_2 \ P_3 \ \cdots \ P_n] \quad \text{and} \quad \mathbf{\Delta}^T = [\Delta_1 \ \Delta_2 \ \Delta_3 \ \cdots \ \Delta_n]$$

The following products of these two functions lead to the same *scalar function*:

$$\mathbf{P}^T\mathbf{\Delta} = P_1\Delta_1 + P_2\Delta_2 + P_3\Delta_3 + \cdots + P_n\Delta_n$$
$$\mathbf{\Delta}^T\mathbf{P} = \Delta_1 P_1 + \Delta_2 P_2 + \Delta_3 P_3 + \cdots + \Delta_n P_n$$

These scalar functions are associated with the work term in the discussion that follows.

The *transpose of the product* of the two matrices needed in the subsequent development is defined[7] as the product of the transposed matrices taken in the reverse order, i.e., if $\mathbf{F} = \mathbf{bP}$,

---

[7] For proof, see any text on linear algebra such as B. Noble, *Applied Linear Algebra* (Englewood Cliffs, NJ: Prentice-Hall, 1969).

$$\mathbf{F}^T = \mathbf{P}^T\mathbf{b}^T$$

The *duality* of the virtual force and virtual displacement methods can be readily shown with the aid of this matrix notation. By recalling, first, Eq. 12-23 in words, a parallel development employing virtual work is given. For the virtual force method, the forces are virtual and the displacements are real, whereas for the virtual displacement method, the forces are real and the displacements are virtual. Except for forming the virtual work in a different manner, the matrix operations for the two methods are identical. The following is an outline for the two methods.

$$\textbf{external virtual work = internal virtual work} \qquad (12\text{-}23)$$

| **Virtual Force Method** | **Virtual Displacement Method** |
|---|---|
| $\delta W_e^* = \delta W_{ie}^* \qquad (12\text{-}22)$ | $\delta W_e = \delta W_{ie} \qquad (12\text{-}19)$ |
| *Statics*: | *Kinematics*: |
| $\overline{\mathbf{F}} = \mathbf{b}\overline{\mathbf{P}} \qquad (12\text{-}30\text{a})$ | $\overline{\mathbf{u}} = \mathbf{a}\overline{\boldsymbol{\Delta}} \qquad (12\text{-}30\text{b})$ |
| $\{\mathbf{F}\}$ = internal member forces <br> $[\mathbf{b}]$ = force transformation matrix <br><br> $\{\mathbf{P}\}$ = external forces at nodes | $\{\mathbf{u}\}$ = member distortions <br> $[\mathbf{a}]$ = displacement transformation matrix <br> $\{\boldsymbol{\Delta}\}$ = nodal displacements |
| $\delta W_e^* = \overline{\mathbf{P}}^T\boldsymbol{\Delta}$ and $\delta W_{ie}^* = \overline{\mathbf{F}}^T\mathbf{u}$ | $\delta W_e = \overline{\boldsymbol{\Delta}}^T\mathbf{P}$ and $\delta W_{ie} = \overline{\mathbf{u}}^T\mathbf{F}$ |
| *Equating*: | *Equating*: |
| $\overline{\mathbf{P}}^T\boldsymbol{\Delta} = \overline{\mathbf{F}}^T\mathbf{u}$ | $\overline{\boldsymbol{\Delta}}^T\mathbf{P} = \overline{\mathbf{u}}^T\mathbf{F}$ |
| *From Eq. 12-30a*: | *From Eq. 12-30b*: |
| $\overline{\mathbf{F}}^T = \overline{\mathbf{P}}^T\mathbf{b}^T$ | $\overline{\mathbf{u}}^T = \overline{\boldsymbol{\Delta}}^T\mathbf{a}^T$ |
| *Hence*, | *Hence*, |
| $\overline{\mathbf{P}}^T\boldsymbol{\Delta} = \overline{\mathbf{P}}^T\mathbf{b}^T\mathbf{u}$ or $\overline{\mathbf{P}}^T(\boldsymbol{\Delta} - \mathbf{b}^T\mathbf{u}) = 0$ | $\overline{\boldsymbol{\Delta}}^T\mathbf{P} = \overline{\boldsymbol{\Delta}}^T\mathbf{a}^T\mathbf{F}$ or $\overline{\boldsymbol{\Delta}}^T(\mathbf{P} - \mathbf{a}^T\mathbf{F}) = 0$ |
| and | and |
| $\boldsymbol{\Delta} = \mathbf{b}^T\mathbf{u} \qquad (12\text{-}31\text{a})$ | $\mathbf{P} = \mathbf{a}^T\mathbf{F} \qquad (12\text{-}31\text{b})$ |

Since $\overline{\mathbf{P}}^T$ and $\overline{\boldsymbol{\Delta}}^T$ are perfectly arbitrary, bearing no relation to the applied forces, the preceding expressions in parentheses must vanish, and Eqs. 12-31a and 12-31b are in *real variables* only.

In the outline, the internal forces {**F**} are determined by conventional statics and are related to the externally applied nodal forces {**P**} by the matrix [**b**]. Similarly, member distortions (deformations) {**u**} are related by kinematics to the nodal displacements {**Δ**} through a displacement transformation matrix [**a**]. Symbols designating virtual quantities are barred.

The parallel development in the two methods is striking, but whereas the virtual force method leads to *equations of compatibility*, the virtual displacement method determines the *equations of equilibrium*.

An example illustrating the application of these procedures follows. It is confined to a statically determinate problem, as additional matrix operations are required for statically indeterminate problems; such procedures are discussed in texts on finite element analysis.[8]

### EXAMPLE 12-8

(*a*) Using the virtual force method in matrix notation, determine the displacement components for nodal point *B* for the pin-ended elastic truss system shown in Fig. 12-13. (*b*) Using the virtual displacement method in matrix notation, find nodal forces $P_1$ and $P_2$ for static equilibrium of the same system if the *elastic elongation* of bar *AB* is $5L/AE$, and that of bar *BC* is $25L/AE$. For both bars, lengths *L* and cross-sectional areas *A* are the same, and *E* is constant.

### Solution

(*a*) From statics,

$$F_{AB} = F_1 = \frac{5}{8}P_1 - \frac{5}{6}P_2 \quad \text{and} \quad F_{BC} = F_2 = \frac{5}{8}P_1 + \frac{5}{6}P_2$$

[8] See, for example, J. L. Meek, *Matrix Structural Analysis* (New York: McGraw-Hill, 1971).

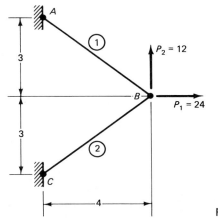

Fig. 12-13

Hence,
$$\mathbf{F} = \begin{Bmatrix} F_1 \\ F_2 \end{Bmatrix} = \mathbf{bP} = \begin{bmatrix} \dfrac{5}{8} & -\dfrac{5}{6} \\ \dfrac{5}{8} & \dfrac{5}{6} \end{bmatrix} \begin{Bmatrix} P_1 \\ P_2 \end{Bmatrix}$$

By using Eq. 2-9, the constitutive relations for the bars are

$$u_1 = \frac{F_1 L}{AE} = \left( \frac{5}{8} \times 24 - \frac{5}{6} \times 12 \right) \frac{L}{AE} = \frac{5L}{AE}$$

$$u_2 = \frac{F_2 L}{AE} = \left( \frac{5}{8} \times 24 + \frac{5}{6} \times 12 \right) \frac{L}{AE} = \frac{25L}{AE}$$

Hence,
$$\mathbf{u} = \begin{Bmatrix} u_1 \\ u_2 \end{Bmatrix} = \begin{Bmatrix} 5 \\ 25 \end{Bmatrix} \frac{L}{AE}$$

By using Eq. 12-31a, for the compatibility of the system,

$$\mathbf{\Delta} = \begin{Bmatrix} \Delta_1 \\ \Delta_2 \end{Bmatrix} = \mathbf{b}^T \mathbf{u} = \begin{bmatrix} \dfrac{5}{8} & \dfrac{5}{8} \\ -\dfrac{5}{6} & \dfrac{5}{6} \end{bmatrix} \begin{Bmatrix} 5 \\ 25 \end{Bmatrix} \frac{L}{AE} = \begin{Bmatrix} 18.75 \\ 16.67 \end{Bmatrix} \frac{L}{AE}$$

(*b*) Compatible bar deformations as functions of nodal displacements, with the aid of Figs. 12-12(b) and (c), are determined to be

$$u_1 = \frac{4}{5} \Delta_1 - \frac{3}{5} \Delta_2 \qquad \text{and} \qquad u_2 = \frac{4}{5} \Delta_1 + \frac{3}{5} \Delta_2$$

Hence,
$$\mathbf{u} = \begin{Bmatrix} u_1 \\ u_2 \end{Bmatrix} = \mathbf{a\Delta} = \begin{bmatrix} \dfrac{4}{5} & -\dfrac{3}{5} \\ \dfrac{4}{5} & \dfrac{3}{5} \end{bmatrix} \begin{Bmatrix} \Delta_1 \\ \Delta_2 \end{Bmatrix}$$

From constitutive relations for *linearly elastic* bars, since $u_1 = 5L/AE$ and $u_2 = 25L/AE$,

$$F_1 = \frac{AE}{L} u_1 = 5 \qquad \text{and} \qquad F_2 = \frac{AE}{L} u_2 = 25$$

Hence, using Eq. 12-31b, for static equilibrium, the joint forces are

$$\mathbf{P} = \begin{Bmatrix} P_1 \\ P_2 \end{Bmatrix} = \mathbf{a}^T \mathbf{F} = \begin{bmatrix} \dfrac{4}{5} & \dfrac{4}{5} \\ -\dfrac{3}{5} & \dfrac{3}{5} \end{bmatrix} \begin{Bmatrix} 5 \\ 25 \end{Bmatrix} = \begin{Bmatrix} 24 \\ 12 \end{Bmatrix}$$

This result corresponds to the joint forces given for part (a). This means that the two processes are completely reversible for *elastic* systems.

In the matrix analyses of structures and especially in finite element applications, equations similar to the previous ones contain a large number of unknowns. The use of computers is essential for the solution of such systems of equations.

## Part C    ELASTIC ENERGY METHODS

### *[9]12-10. General Remarks

Elastic strain-energy equations for applications in engineering mechanics of solids are summarized in Section 12-2. The direct use of these equations in determining deflections in conjunction with the law of conservation of energy in equating the total elastic energy of the system to the total work done by the externally applied forces is very limited. This was pointed out in Section 12-3. An effective approach to enlarging the scope of possible applications was discussed in Part B. Here an alternative classical approach based on elastic strain energy and complementary strain energy is considered. This requires the derivation of appropriate theorems for solving problems similar to those treated in Part B describing the virtual work methods. The derived equations will be found useful in the next chapter in considering problems with a high degree of indeterminancy. An example of the use of the potential energy approach for determining buckling loads is given at the end of this part.

### *12-11. Strain Energy and Complementary Strain Energy Theorems

In Section 2-5, it was indicated that some materials during loading and unloading respond in a nonlinear manner along the same stress-strain curve, Fig. 2-11(b). Such materials are elastic although they do not obey Hooke's law. It is advantageous to consider such *nonlinearly elastic* materials in deriving the two theorems based on strain energy concepts. In this manner, the distinction between elastic strain energy and complementary strain energy is clearly evident. The derivation of the two basic theorems for nonlinear and linear elastic systems is essentially the same. These theorems are specialized in the next section for the solution of the linearly elastic problems considered in this text.

As a rudimentary example for deriving the theorems, consider the axially loaded bar shown in Fig. 12-14(a). The nonlinear *elastic* stress-strain diagram for the material of this bar is shown in Fig. 12-14(b). By multiplying the normal stress $\sigma$ in the bar by the cross-sectional area $A$ of the

---

[9] The remainder of this chapter is optional.

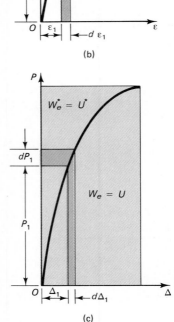

bar, one obtains axial force $P$. Similarly, the product of the axial strain $\varepsilon$ by bar length $L$ gives bar elongation $\Delta$. The $P$-$\Delta$ diagram in Fig. 12-14(c), except for scale, corresponds in detail to the $\sigma$-$\varepsilon$ diagram. Since in engineering mechanics of solids it is customary to work with *stress resultants*, only the $P$-$\Delta$ diagram is considered in the following discussion.

According to the diagram in Fig. 12-14(c), when force $P_1$ is increased by $dP_1$, the bar elongates $d\Delta_1$. Therefore, neglecting infinitesimals of higher order, the increment in the external work $dW_e = P_1\,d\Delta_1$. According to Eq. 12-11, this increment in the external work must equal the increase in the strain energy of the system, since $dW_e = dU$. The colored vertical strip in Fig. 12-14(c) corresponds to $P_1\,d\Delta_1$, and the area under the curve is the total strain energy $U$, which is equal to $W_e$. Stated mathematically,

$$W_e = U = \int_O^\Delta P_1\,d\Delta_1 \qquad (12\text{-}32)$$

A derivative of this relation with respect to the upper limit gives

$$\frac{dU}{d\Delta} = P \qquad (12\text{-}33)$$

A generalization of this procedure, which follows, establishes the first basic theorem.

An analogous expression can be formulated by increasing $\Delta_1$ by $d\Delta_1$, causing a force increment $dP_1$, Fig. 12-14(c). Then, by *defining* an increment in the *complementary* external work $dW_e^* = \Delta_1\,dP_1$, it can be noted that this quantity is represented by the horizontal colored strip in the figure. By analogy to $U$, the integral of this infinitesimal area above the curve *defines* the *complementary strain energy* $U^*$, and it follows that $U^* = W_e^*$. On this basis, one can write

$$U^* = W_e^* = \int_O^P \Delta_1\,dP_1 \qquad (12\text{-}34)$$

A derivative of this relation with respect to the upper limit gives

$$\frac{dU^*}{dP} = \Delta \qquad (12\text{-}35)$$

This is the prototype of the second basic theorem.[10]

In order to generalize these results for problems where several forces (and/or moments) are applied simultaneously, consider the externally statically determinate body shown in Fig. 12-15. The stress resultants in such a member, or group of members, in any given problem must be related

[10] The concept of complementary energy and derivation of this equation is generally attributed to F. Engesser's 1889 paper. See S. P. Timoshenko, *History of Strength of Materials* (New York: McGraw-Hill, 1953), 292.

**Fig. 12-14** Work and complementary work, and strain energy and complementary strain energy.

to displacement by the *same nonlinear (or linear) function*. On this basis, a general theorem corresponding to Eq. 12-35 is derived.

The complementary strain energy $U^*$ for a statically determinate body, such as shown in Fig. 12-16(a), is defined to be a function of the externally applied forces $P_1, P_2, \ldots, P_k, \ldots, P_n; M_1, M_2, \ldots, M_p$, i.e.,

$$U^* = U^*(P_1, P_2, \ldots, P_k, \ldots, P_n; M_1, M_2, \ldots, M_j, \ldots, M_p)$$
$$(12\text{-}36)$$

An infinitesimal increase in this function $\delta U^*$ is given by the total differential as

$$\delta U^* = \frac{\partial U^*}{\partial P_1} \delta P_1 + \frac{\partial U^*}{\partial P_2} \delta P_2 + \cdots + \frac{\partial U^*}{\partial P_k} \delta P_k$$
$$+ \cdots + \frac{\partial U^*}{\partial M_j} \delta M_j + \cdots \quad (12\text{-}37)$$

In this expression, $\delta P$'s, and $\delta M$'s are used instead of ordinary differentials to emphasize the *linear independence* of these quantities. From

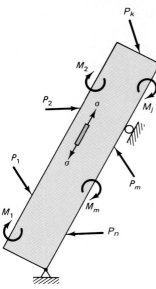

**Fig. 12-15** Statically determinate member.

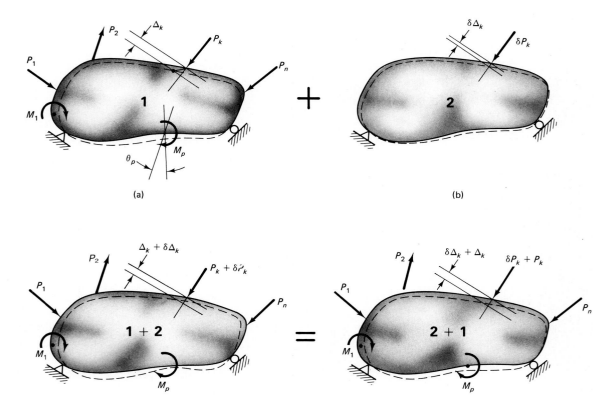

**Fig. 12-16** Alternative loading sequences for an elastic system.

**663**

this point of view, if only force $P_k$ were increased by $\delta P_k$, the complementary strain energy increment would be

$$\delta U^* = \frac{\partial U^*}{\partial P_k} \, \delta P_k \qquad (12\text{-}38)$$

The addition of the incremental force $\delta P_k$ is illustrated in Fig. 12-16(c). If the order of load application were reversed, Fig. 12-16(d), infinitesimal force $\delta P_k$ would be applied to the system first, Fig. 12-16(b). Then, being already applied to the system, it would do work by moving through a deflection $\Delta_k$ caused by the application of the loads shown in Fig. 12-16(a). This work may be likened to the horizontal strip in Fig. 12-14(c), and, by definition, is an increment in the *complementary work* $\delta W_e^*$. Hence,

$$\delta W_e^* = \Delta_k \, \delta P_k \qquad (12\text{-}39)$$

However, since $W_e^* = U_e^*$, setting Eqs. 12-38 and 12-39 equal, and cancelling $\delta P_k$,

$$\Delta_k = \frac{\partial U^*}{\partial P_k} \qquad (12\text{-}40)$$

which is the generalization of Eq. 12-35, and gives deflection $\Delta_k$ in the direction of force $P_k$.

By retaining a derivative with respect to $M_j$ in Eq. 12-37 and proceeding as before,

$$\theta_j = \frac{\partial U^*}{\partial M_j} \qquad (12\text{-}41)$$

where $\theta_j$ is the rotation in the direction of moment $M_j$.

In an analogous manner to the previous derivation, strain energy $U$ can be defined as a function of displacements $\Delta_k$ and/or rotations $\theta_j$, as well as known members' constitutive relations, see Eq. 12-7. On this basis,

$$U = U(\Delta_1, \Delta_2, \ldots, \Delta_k, \ldots, \Delta_n; \theta_1, \theta_2, \ldots, \theta_j, \ldots, \theta_m) \qquad (12\text{-}42)$$

The total differential for this case is

$$\delta U = \frac{\partial U}{\partial \Delta_1} \delta \Delta_1 + \frac{\partial U}{\partial \Delta_2} \delta \Delta_2 + \cdots + \frac{\partial U}{\partial \Delta_k} \delta \Delta_k + \cdots + \frac{\partial U}{\partial \theta_j} \delta \theta_j + \cdots \qquad (12\text{-}43)$$

If only one displacement were allowed to occur with the other remaining fixed, the last relation reduces to

$$\delta U = \frac{\partial U}{\partial \Delta_k} \delta \Delta_k \qquad (12\text{-}44)$$

For this case, external work $\delta W_e = P_k \, \delta \Delta_k$ and corresponds to the vertical colored strip in Fig. 12-13(c). Therefore, since $\delta W_e = \delta U$, after substitution of the previous quantities into this relation and simplifications

$$P_k = \frac{\partial U}{\partial \Delta_k} \qquad (12\text{-}45)$$

This relation is a generalization of Eq. 12-33, and gives the external force acting at point $k$ if $U$ is expressed as a continuous function of displacements and rotations. A similar expression can be written for an external moment acting at a point by taking a derivative of $U$ with respect to a rotation angle such as $\theta_j$.

In the next section, the general expressions are specialized for linearly elastic materials.

## *12-12. Castigliano's Theorems

Castigliano's theorems[11] apply to linearly elastic systems for small deformations. The mathematical statements of these theorems are the same as those derived in the previous section for nonlinear elastic materials. However, as shown in Fig. 12-17, for linearly elastic material, the elastic strain energy $U$ is equal to the complementary strain energy $U^*$, i.e.,

$$U = U^* \qquad (12\text{-}46)$$

The external work $W_e$ is also equal to the complementary external work $W_e^*$. Therefore, using Eqs. 12-40 and 12-41, because of Eq. 12-46, one can express the *second* Castigliano's theorem for linearly elastic material as

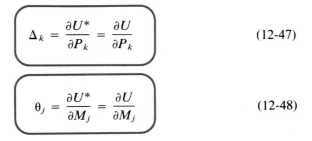

$$\Delta_k = \frac{\partial U^*}{\partial P_k} = \frac{\partial U}{\partial P_k} \qquad (12\text{-}47)$$

$$\theta_j = \frac{\partial U^*}{\partial M_j} = \frac{\partial U}{\partial M_j} \qquad (12\text{-}48)$$

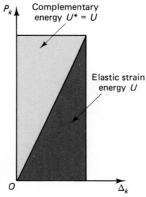

Fig. 12-17 Energies in linearly elastic materials.

[11] These theorems were first derived by Italian engineer C. A. Castigliano in 1879. Extension of the theorems to the nonlinear elastic cases developed in the preceding section, as noted earlier, is generally attributed to F. Engesser of Karlsruhe. Further developments of this approach are due to H. M. Westergaard and J. H. Argyris. See J. H. Argyris, "Energy Theorems and Structural Analysis," *Aircraft Engineering* 26 (1954) and 27 (1955). These articles, combining joint papers with S. Kelsey, were republished in book form by Butterworth & Co. in 1960.

In both equations, if $U$ is expressed as a function of externally applied forces, $\Delta_k$ (or $\theta_j$) is the deflection (or rotation) in the direction of the force (or moment) $P_k$ (or $M_j$).

The expression for the *first* Castigliano's theorem remains the same as before for nonlinear elastic materials, and Eq. (12-45) is repeated here for reference.

$$P_k = \frac{\partial U}{\partial \Delta_k} \qquad\qquad (12\text{-}45)$$

where, if $U$ is expressed as a function of displacements, $P_k$ is the force (or moment) in the direction of the deflection (or rotation) $\Delta_k$ (or $\theta_k$).

It should be further noted from Eqs. 12-5–12-9 that the strain energy $U$ for linearly elastic materials is of quadratic form. Therefore, in applying Castigliano's theorems, it is advantageous to form derivatives of $U$ before carrying out a complete solution of the problem.

It is also important to note that if a deflection (rotation) is required where no force (moment) is acting, a *fictitious* force (moment) must be applied at the point in question. Then, after applying Eq. 12-47 or 12-48, the fictitious force is set equal to zero in order to obtain the desired results.

Several examples follow illustrating the application of Castigliano's second theorem to statically determinate linearly elastic problems. An application of Castigliano's first theorem for a statically indeterminate case is given in Example 12-17, where the use of the theorem is more appropriate.

### EXAMPLE 12-9

By applying Castigliano's second theorem, verify the results of Examples 2-10, 4-11, and 10-12.

Solution

In all these examples, the expressions for the internal strain energy $U$ have been formulated. Therefore, a direct application of Eq. 12-47 or 12-48 is all that is necessary to obtain the required results. In all cases, the material obeys Hooke's law.

Deflection of an axially loaded bar ($P = $ constant):

$$U = \frac{P^2 L}{2AE} \qquad \text{hence,} \qquad \Delta = \frac{\partial U}{\partial P} = \frac{PL}{AE}$$

Angular rotation of a circular shaft ($T = $ constant):

$$U = \frac{T^2 L}{2JG} \qquad \text{hence,} \qquad \varphi \equiv \theta = \frac{\partial U}{\partial T} = \frac{TL}{JG}$$

Deflection of a rectangular cantilever due to end load $P$:

$$U = \frac{P^2L^3}{6EI} + \frac{3P^2L}{5AG} \qquad \text{hence,} \qquad \Delta = \frac{\partial U}{\partial P} = \frac{PL^3}{3EI} + \frac{6PL}{5AG}$$

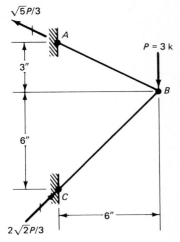

Fig. 12-18

## EXAMPLE 12-10

The bracket of Example 1-3 is shown schematically in Fig. 12-18. Verify the deflection of point $B$ caused by applied force $P = 3$ kips using Castigliano's second theorem with the result found in Example 2-2. Assume that each bar is of constant cross-sectional area, with $A_{AB} = A_1 = 0.125$ in$^2$, and $A_{BC} = A_2 = 0.219$ in$^2$. As before, let $E = 10.6 \times 10^3$ ksi.

## Solution

From Eq. 12-7, the elastic strain energy is

$$U = U^* = \sum_{k=1}^{2} \frac{P_k^2 L_k}{2A_k E_k} = \frac{P_1^2 L_1}{2A_1 E} + \frac{P_2^2 L_2}{2A_2 E} \qquad (12\text{-}49a)$$

By differentiating with respect to $P$, an expression for the vertical deflection $\Delta$ at $B$ is determined.

$$\Delta = \frac{\partial U^*}{\partial P} = \frac{P_1 L_1}{A_1 E} \frac{\partial P_1}{\partial P} + \frac{P_2 L_1}{A_2 E} \frac{\partial P_2}{\partial P} \qquad (12\text{-}49b)$$

By statics, the forces in bars as *functions* of applied force $P$ are

$$P_{AB} = P_1 = \frac{\sqrt{5}}{3} P \qquad \text{and} \qquad P_{BC} = P_2 = -\frac{2\sqrt{2}}{3} P$$

Here $A_1 = 0.125$ in$^2$, $A_2 = 0.219$ in$^2$, $L_1 = 3\sqrt{5}$ in, and $L_2 = 6\sqrt{2}$ in.

Subsituting the above quantities into Eq. 12-49b and carrying out the necessary operations, the deflection $\Delta$ for $P = 3$ kips is found.

$$\Delta = \frac{(\sqrt{5}\, P/3) \times (3\sqrt{5})}{0.125 \times 10.6 \times 10^3} \left(\frac{\sqrt{5}}{3}\right) + \frac{(-2\sqrt{2}\, P/3) \times (6\sqrt{2})}{0.219 \times 10.6 \times 10^3} \left(-\frac{2\sqrt{2}}{3}\right)$$

$$= 0.002813P + 0.003249P = 0.006062P = 18.2 \times 10^{-3} \text{ in}$$

This more easily obtained result, except for a small discrepancy because of roundoff errors, is in agreement with that found by an entirely different method in Example 2-2.[12]

---

[12] A solution of this problem by the virtual force method requires the use of Eq. 12-26. For a downward virtual unit force applied at $B$, $\bar{p}_1 = \sqrt{5}/3$ and $\bar{p}_2 = -2\sqrt{2}/3$, resulting in an identical expression for $\Delta$.

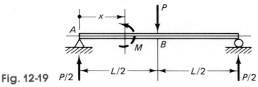

Fig. 12-19

## EXAMPLE 12-11

A linearly elastic prismatic beam is loaded as shown in Fig. 12-19. Using Castigliano's second theorem, find the deflection due to bending caused by applied force $P$ at the center.

### Solution

The expression for the internal strain energy in bending is given by Eq. 12-8. Since, according to Castigliano's theorem, the required deflection is a derivative of this function, it is advantageous to differentiate the expression for $U$ before integrating. In problems where $M$ is a complex function, this scheme is particularly useful. For this purpose, the following relation becomes applicable:

$$\Delta = \frac{\partial U}{\partial P} = \int_0^L \frac{M}{EI} \frac{\partial M}{\partial P} \, dx \qquad (12\text{-}50)$$

Proceeding on this basis, one has, from $A$ to $B$:

$$M = + \frac{P}{2} x \qquad \text{and} \qquad \frac{\partial M}{\partial P} = \frac{x}{2}$$

On substituting these relations[13] into Eq. 12-50 and observing the symmetry of the problem,

$$\Delta = 2 \int_0^{L/2} \frac{P x^2}{4EI} \, dx = + \frac{P L^3}{48 EI}$$

The positive sign indicates that the deflection takes place in the direction of applied force $P$.

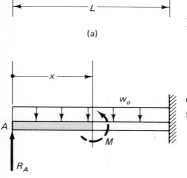

(a)

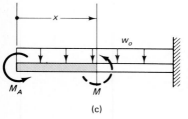

(b)

(c)

Fig. 12-20

## EXAMPLE 12-12

Using Castigliano's second theorem, determine the deflection and the angular rotation of the end of a uniformly loaded cantilever, Fig. 12-20(a). $EI$ is constant.

---

[13] Note, again, that for a downward virtual unit force at the middle of the span, $\overline{m} = x/2$, corresponding to $\partial M/\partial P$.

## Solution

No forces are applied at the end of the cantilever where the displacements are to be found. Therefore, in order to be able to apply Castigliano's theorem, a fictitious force[14] must be added corresponding to the displacement sought. Thus, as shown in Fig. 12-20(b), in addition to the specified loading, force $R_A$ has been introduced. This permits determining $\partial U/\partial R_A$, which with $R_A = 0$, gives the vertical deflection of point $A$. Applying Eq. 12-50 in this manner, one has

$$M = -\frac{w_o x^2}{2} + R_A x \qquad \text{and} \qquad \frac{\partial M}{\partial R_A} = +x$$

$$\Delta_A = \frac{\partial U}{\partial R_A} = \frac{1}{EI} \int_0^L \left( -\frac{w_o x^2}{2} + R_A x^0 \right)(+x)\, dx = -\frac{w_o L^4}{8EI}$$

where the negative sign shows that the deflection is in the opposite direction to that assumed for force $R_A$. If $R_A$ in the integration were not set equal to zero, the end deflection due to $w_o$ and $R_A$ would be found.

The angular rotation of the beam at $A$ can be found in an analogous manner. A fictitious moment $M_A$ is applied at the end, Fig. 12-20(c), and the calculations are made in the same manner as before:

$$M = -\frac{w_o x^2}{2} - M_A \qquad \text{and} \qquad \frac{\partial M}{\partial M_A} = -1$$

$$\Delta_A = \frac{\partial U}{\partial M_A} = \frac{1}{EI} \int_0^L \left( -\frac{w_o x^2}{2} - M_A^0 \right)(-1)\, dx = +\frac{w_o L^3}{6EI}$$

where the sign indicates that the sense of the rotation of the end coincides with the assumed sense of the fictitious moment $M_A$.

## EXAMPLE 12-13

Using Castigliano's second theorem, determine the horizontal deflection for the elastic frame shown in Fig. 12-21(a). Consider only the deflection caused by bending. The flexural rigidity $EI$ of both members is equal and constant.

[14] Application of a fictitious force or a fictitious couple at $A$ precisely corresponds, respectively, to the application of a virtual unit force or a virtual unit couple at $A$ in the virtual force method.

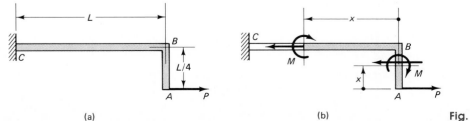

(a)  (b)  **Fig. 12-21**

Solution

The strain energy function is a scalar. Therefore, the separate strain energies for the different elements of an elastic system can be added algebraically. After the total strain energy is determined, its partial derivative with respect to a force gives the displacement of that force. For the problems at hand, Eq. 12-50 is appropriate.

From $A$ to $B$:

$$M = +Px \qquad \text{and} \qquad \partial M/\partial P = +x$$

From $B$ to $C$[15]:

$$M = +\frac{PL}{4} \qquad \text{and} \qquad \frac{\partial M}{\partial P} = +\frac{L}{4}$$

$$\Delta_A = \frac{\partial U}{\partial P} = \frac{1}{EI} \int_0^{L/4} (+Px)(+x) \, dx$$
$$+ \frac{1}{EI} \int_0^L \left( +\frac{PL}{4} \right)\left( +\frac{L}{4} \right) dx = +\frac{13PL^3}{192EI}$$

Note the free choice in location of the $x$-coordinate axes and the sign convention for bending moments. If the elastic strain energy included the energy due to the axial force in member $BC$ and the shear energy in member $AB$, the deflection caused by these effects would also be found. However, deflection $\Delta_A$ due to bending is generally dominant.

If the vertical deflection of point $A$ were required, a fictitious vertical force $F$ at $A$ would have to be applied. Then, as in the preceding example, $\partial U/\partial F$, with $F = 0$, would give the desired result. In a similar manner, the rotation of any normal section for this beam may be obtained.

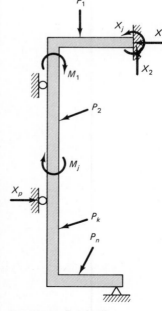

**Fig. 12-22** Statically indeterminate system.

## *12-13. Statically Indeterminate Systems

Castigliano's second theorem can be generalized for statically indeterminate linear elastic systems. The necessary modifications consist of expressing the strain energy not only to be a function of $n$ externally applied forces (and/or moments) $P_1$, $P_2$, $P_3$, . . . , $P_n$, but also of $p$ statically indeterminate *redundant forces* (and/or moments) $X_1$, $X_2$, . . . , $X_p$. A possible system of such forces is shown in Fig. 12-22. The necessary number of selected supports for maintaining static equilibrium of the primary section (Section 2-15) are excluded from the enumeration of the redundant forces. On this basis, strain energy $U$ for an indeterminate system can be defined as

$$U^* = U = U(P_1, P_2, P_3, \ldots, P_n; X_1, X_2, \ldots, X_j, \ldots, X_p)$$

$$(12\text{-}51)$$

[15] The reader should check the correspondence of the $\partial M/\partial P$ terms with those caused by a horizontal virtual unit force applied at $A$.

Using this function and Castigliano's second theorem, $p$ displacements (and/or rotations) at the points of application of redundant forces (and/or couples) $X_j$ in the direction of these forces can be found. If these displacements are zero,

$$\frac{\partial U}{\partial X_j} = 0 \qquad (j = 1, 2, 3, \ldots, p) \qquad (12\text{-}52)$$

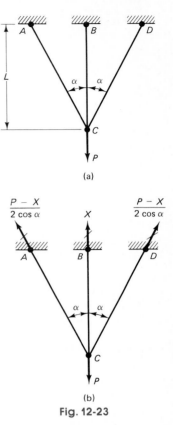

These $p$ equations are equal to the degree of statical indeterminancy of the system.[16] By solving these equations simultaneously, the magnitudes of the redundants are obtained.

Castigliano's first theorem can be used directly for the solution of statically indeterminate problems.

Several examples follow illustrating these procedures.

## EXAMPLE 12-14

Using Castigliano's second theorem, verify the bar forces found in Example 2-14 caused by applied force $P$. The planar system of three elastic bars is repeated in Fig. 12-23(a). The cross-sectional area $A$ of each bar is the same, and their elastic modulus is $E$.

### Solution

It is convenient to visualize the system to be cut at $B$ and to designate the unknown force in bar $BC$ by $X$. From statics, the forces in the inclined bars then have the magnitudes shown in Fig. 12-23(b). Hence, using Eq. 12-49a, with an appropriate change, the complementary strain energy is

$$U^* = U = \frac{X^2 L}{2AE} + 2\left[\frac{(P - X)^2 L}{2 \times 2^2 AE \cos^3 \alpha}\right]$$

Since the deflection of the system at point $B$ is zero, by applying Eq. 12-52,

$$\frac{\partial U}{\partial X} = \frac{XL}{AE} \times 1 + \frac{(P - X)L}{2AE \cos^3 \alpha} \times (-1) = 0$$

$$X = \frac{P}{1 + 2\cos^3 \alpha}$$

This expression is identical to that given for $F_1$ in Eq. 2-34. Here the procedure for obtaining the result is more direct.

---

[16] Italian mathematician L. F. Menabrea (1809–1896) proved that the total work for a problem solved in this manner is a minimum. His theorem is known as the *principle of least work*. Castigliano employed this principle in the solution of statically indeterminate problems.

### EXAMPLE 12-15

Consider an elastic uniformly loaded beam clamped at one end and simply supported at the other, as represented in Fig. 12-20(b). Determine the reaction at $A$. Use Eq. 12-52.

### Solution

The solution is analogous to that of Example 12-12 except that $R_A$ must be treated as the unknown and not permitted to vanish. The key kinematic condition per Eq. 12-52 is

$$\Delta_A = \partial U / \partial R_A = 0$$

which states that no deflection occurs at $A$ due to the applied load $w_o$ and $R_A$.

$$M = -\frac{w_o x^2}{2} + R_A x \quad \text{and} \quad \frac{\partial M}{\partial R_A} = +x$$

$$\Delta_A = \frac{\partial U}{\partial R_A} = \frac{1}{EI} \int_0^L \left( -\frac{p_o x^2}{2} + R_A x \right)(+x)\, dx = -\frac{w_o L^4}{8EI} + \frac{R_A L^3}{3EI} = 0$$

Therefore, $R_A = +3w_o L/8$, the result found in Example 10-5.

### EXAMPLE 12-16

Consider an elastic beam fixed at both ends and subjected to a uniformly increasing load to one end, as shown in Fig. 12-24. Determine the reactions at end $A$ using Eq. 12-52. $EI$ for the beam is constant.

### Solution

This problem is statically indeterminate to the second degree. It is convenient to take reactions $R_A$ and $M_A$ as the redundant forces and to express $M$ as a function of these forces as well as of the applied load. The kinematic conditions require that the vertical displacement and the rotation at $A$ be zero. These two conditions can be fulfilled by applying Eq. 12-52 twice and setting deflection $\Delta_A$ and rotation $\theta_A$ at $A$ equal to zero. This provides two simultaneous equations for determining $R_A$ and $M_A$. Proceeding in this manner,

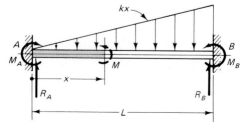

**Fig. 12-24**

$$M = M_A + R_A x - kx^3/6$$

Hence,
$$\frac{\partial M}{\partial M_A} = 1 \quad \text{and} \quad \frac{\partial M}{\partial R_A} = x$$

$$\Delta_A = \frac{\partial U}{\partial R_A} = \int_0^L \frac{M}{EI} \frac{\partial M}{\partial R_A} dx = \frac{1}{EI} \int_0^L \left( M_A + R_A x - \frac{1}{6}kx^3 \right) x \, dx = 0$$

$$\theta_A = \frac{\partial U}{\partial M_A} = \int_0^L \frac{M}{EI} \frac{\partial M}{\partial M_A} dx = \frac{1}{EI} \int_0^L \left( M_A + R_A x - \frac{1}{6}kx^3 \right)(1) \, dx = 0$$

Carrying out the indicated operations and simplifying,

$$M_A/2 + R_A L/3 = kL^3/30$$
$$M_A + R_A L/2 = kL^3/24$$

Solving these two equations simultaneously,

$$R_A = 3kL^2/20 \quad \text{and} \quad M_A = -kL^3/30 \tag{12-53}$$

where the negative sign of $M_A$ shows that this end moment has a counterclockwise sense.

## EXAMPLE 12-17

Rework Example 12-14 using Castigliano's first theorem. See Fig. 12-25.

## Solution

In applying Castigliano's first theorem, the elastic strain energy in all three bars must be expressed in terms of the vertical elongation $\Delta_1$ of the center bar. For small deflections,

$$\Delta_2 = \Delta_1 \cos \alpha$$

and using Eq. 12-7 expressed as a function of displacement $\Delta_1$, and noting that $L_{AC} = L_{DC} = L/\cos \alpha$,

$$U = \sum_{k=1}^{3} \frac{A_k E_k \Delta_k^2}{2L_k} = \frac{AE \Delta_1^2}{2L} + 2 \frac{AE \Delta_1^2 \cos^3 \alpha}{2L} \tag{12-54}$$

and
$$\frac{\partial U}{\partial \Delta_1} = \frac{AE}{L} \Delta_1 + 2 \frac{AE}{L} \cos^3 \alpha \, \Delta_1 = P$$

Hence,
$$\Delta_1 = \frac{PL}{AE} \frac{1}{1 + 2 \cos^3 \alpha}$$

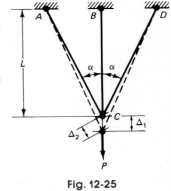

Fig. 12-25

Then the force in the vertical bar using Eq. 2-10 is

$$X = k\,\Delta_1 = \frac{AE}{L}\,\Delta_1 = \frac{P}{1 + 2\cos^3\alpha}$$

This result is in agreement with that found in Example 12-14.

This type of solution is easily extended to any number of symmetrically inclined bars with $\Delta_1$ remaining as the only unknown, regardless of the degree of indeterminacy.

If the inclined bars lack symmetry or the applied force forms an angle with the vertical, the problem is more complex. In such cases, the elongation in each bar is determined from two separate displacements at the load point. This procedure is analogous to that shown in Example 12-7 and illustrated in Fig. 12-12. Further discussion of this approach can be found in Section 13-5, where the *displacement* method of analysis is considered.

## **12-14. Elastic Energy for Buckling Loads

Stability problems can be treated in a very general manner using the energy or the virtual work methods. As an introduction to such methods, the basic criteria for determining the stability of equilibrium are derived in this article for conservative linearly elastic systems using an energy method.

To establish the stability criteria, a function $\Pi$, called the *total potential* of the system, must be formulated. This function is expressed as the sum of the internal energy $U$ (strain energy) and the potential energy $\Omega$ (omega) of the external forces that act on a system, i.e.,

$$\Pi = U + \Omega \tag{12-55}$$

Disregarding a possible additive constant, $\Omega = -W_e$, i.e., the loss of potential energy during the application of the forces is equal to the work done on the system by the external forces. Hence, Eq. 12-55 can be rewritten as

$$\Pi = U - W_e \tag{12-56}$$

As is known from classical mechanics, for equilibrium, total potential $\Pi$ must be stationary;[17] therefore, its variation $\delta\Pi$ must equal zero, i.e.,

$$\boxed{\delta\Pi = \delta U - \delta W_e = 0} \tag{12-57}$$

---

[17] In terms of the ordinary functions, this simply means that a condition exists where the derivative of a function with respect to an independent variable is zero and the function itself has a maximum, a minimum, a minimax, or a constant value.

For conservative, elastic systems, this relation is in agreement with Eq. 12-11. This condition can be used to determine the position of equilibrium. However, Eq. 12-57 cannot discern the type of equilibrium and thereby establish the condition for the stability of equilibrium. Only by examining the higher order terms in the expression for the change $\Delta\Pi$ in the total potential $\Pi$ can this be determined. Therefore, the more complete expression for the increment in $\Pi$ as given by Taylor's expansion must be examined. Such an expression is

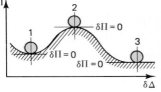

**Fig. 12-26** Different equilibrium conditions.

$$\Delta\Pi = \delta\Pi + \frac{1}{2!}\delta^2\Pi + \frac{1}{3!}\delta^3\Pi + \cdots \qquad (12\text{-}58)$$

Since for any type of equilibrium, $\delta\Pi = 0$, it is the first nonvanishing term of this expansion that determines the type of equilibrium. For linear elastic systems, the second term suffices. Thus, from Eq. 12-58, the stability criteria are

$$
\begin{array}{l|l}
\delta^2\Pi > 0 & \text{for stable equilibrium} \\
\delta^2\Pi < 0 & \text{for unstable equilibrium} \qquad\qquad (12\text{-}59) \\
\delta^2\Pi = 0 & \text{for neutral equilibrium associated with} \\
& \quad \text{the critical load}
\end{array}
$$

The meaning of these expressions may be clarified by making reference to Fig. 12-26, where the curve represents the potential function $\Pi$. The origin of this function is shown below the curve, since the absolute value of $\Pi$ is arbitrary. Three different possible positions of equilibrium for the ball are shown in this figure.[18] The first derivative of $\Pi$ at points of equilibrium is zero for all three cases; it is the second derivative that determines the type of equilibrium.

For simple functions of $\Pi$, the procedures for forming the derivatives, differentials, and variations are alike. If, however, the function of $\Pi$ is expressed by integrals, the problem becomes mathematically much more complicated, requiring the use of the calculus of variations or finite elements. The treatment of such problems is beyond the scope of this text.[19]

## EXAMPLE 12-18

Using the energy method, verify the critical load found before for a rigid bar with a torsional spring at the base, Fig. 11-7(a).

[18] A point on a curve resulting from a combination, for example, of the curve to the left of position 1 with that of the curve to the right of position 2 defines a *minimax*. In stability analysis, such a point corresponds to the condition of *unstable* equilibrium.

[19] H. L. Langhaar, *Energy Methods in Applied Mechanics* (New York: Wiley, 1962). K. Washizu, *Variational Methods in Elasticity and Plasticity,* 2nd ed. (New York: Pergamon, 1975). J. S. Przemieniecki, "Discrete-Element Methods for Stability Analysis of Complex Structures," *Aeron. J.* 72(1968):1077.

### Solution

For a displaced position of the bar, the strain energy in the spring is $k\theta^2/2$. For the same displacement, force $P$ lowers an amount $L - L \cos \theta = L(1 - \cos \theta)$. Therefore,

$$\Pi = U - W_e = \frac{1}{2} k\theta^2 - PL(1 - \cos \theta)$$

If the study of the problem is confined to small (infinitesimal) displacements and $\cos \theta = 1 - \theta^2/2! + \theta^4/4! + \cdots$, the total potential $\Pi$ to a consistent order of accuracy simplifies to

$$\Pi = \frac{k\theta^2}{2} - \frac{PL\theta^2}{2}$$

Note especially that the $\frac{1}{2}$ in the last term is due to the *expansion of the cosine into the series*. Full external force $P$ acts on the bar as $\theta$ is permitted to change.

Having the expression for the total potential, one must solve two distinctly different problems. In the first problem, a position of equilibrium is found. For this purpose, Eq. 12-57 is applied:

$$\delta\Pi = \frac{d\Pi}{d\theta} \delta\theta = (k\theta - PL\theta) \delta\theta = 0 \qquad \text{or} \qquad (k - PL)\theta \, \delta\theta = 0$$

At this point of the solution, $k$, $P$, and $L$ must be considered constant, and $\delta\theta$ cannot be zero. Therefore, an equilibrium position occurs at $\theta = 0$.

In the second, distinctly different, phase of the solution, according to the last part of Eq. 12-59, for neutral equilibrium,

$$\delta^2\Pi = \frac{d^2\Pi}{d\theta^2} \delta\theta^2 + \frac{d\Pi}{d\theta} \delta^2\theta = 0$$
$$(k - PL)(\delta\theta)^2 + (k - PL)\theta \, \delta^2\theta = 0$$

For equilibrium at $\theta = 0$, the second term on the left side vanishes; whereas, since $\delta^2\theta$ cannot be zero, the first term yields $P = k/L$, which is the critical buckling load.

---

## Problems

### Sections 12-2 and 12-3

**12-1.** A solid circular bar bent 90° at two points is built in at one end as shown in the figure. Application of force $P$ at the free end causes an axial force, direct shear, bending, and torsion in the three bar segments.

(a) Using Eq. 12-11, obtain the expression for the deflection of the free end. Constants $A$, $I$, $J$, $E$, and $G$ are given for the bar. (*Hint:* See Examples 2-10 and 10-12.) (b) If $L = 100$ mm and the diameter $d = 40$ mm, in percentage, what amount of deflection is due to each of the four causes enumerated earlier? Assume

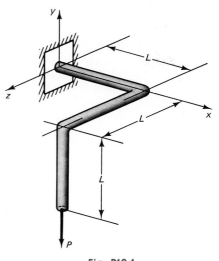

Fig. P12-1

$E/G = 2.5$. (c) Repeat part (b) for $L = 500$ mm and $d = 40$ mm. Neglect the effect of local stress concentrations on deflection.

**12-2.** Using Eq. 12-11, determine the vertical deflection, in mm, of the free end of the cantilever shown in the figure due to the application of force $P = 500$ N. Consider only flexural deformation. $E = 200$ GPa.

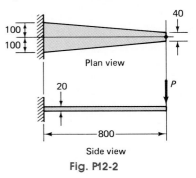

Fig. P12-2

**12-3.** Using Eq. 12-11 and taking advantage of symmetry, determine the flexural deflections at the load points due to the application of both forces $P$ for the elastic beam shown in the figure.

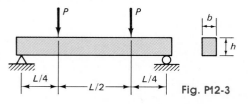

Fig. P12-3

**12-4.** Using Eq. 12-11 and taking advantage of symmetry, determine the flexural deflections at the load points due to the application of both forces $P$ for the elastic beam shown in the figure. The moment of inertia of the cross section in the middle half of the beam is $I_o$.

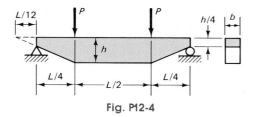

Fig. P12-4

**12-5.** A man weighing 180 lb jumps onto a diving board, as shown in the figure, from a height of 2 ft. What maximum bending stress will this cause in the board? The diving board is $2 \times 12$ in cross section, and its $E = 1.6 \times 10^6$ psi. Use Eq. 12-11 to determine the deflection characteristics of the board.

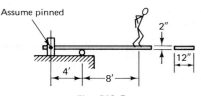

Fig. P12-5

## Sections 12-5 and 12-6

*All problems for these two sections should be solved using the virtual-force method. For planar problems, the following notation applies: $\Delta_V$ and $\Delta_H$ are, respectively, the vertical and the horizontal deflections, and $\theta$ is the rotation of an element at a specified point.* In each case, clearly indicate the *direction* and *sense* of the computed quantity.

**Trusses.** *Consider axial deformations only.*

**12-6.** In Example 12-1, determine $\Delta_H$ for point $B$ due to the enumerated three causes.

**12-7.** For the planar mast and boom arrangement shown in the figure, (a) determine $\Delta_V$ of load $W$ caused by lengthening rod $AB$ a distance of 0.5 in. (b) By how much must rod $BC$ be shortened to bring weight $W$ to its original position?

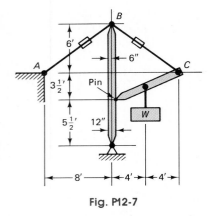

**Fig. P12-7**

**12-8.** A pin-joined system of three bars, each having the same cross section $A$, is loaded as shown in the figure. (a) Determine $\Delta_V$ and $\Delta_H$ of joint $B$ due to applied force $P$. (b) If by means of a turnbuckle the length of member $AC$ is shortened by 0.5 in, what $\Delta_V$ and $\Delta_H$ take place at point $B$?

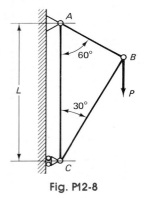

**Fig. P12-8**

**12-9.** For the planar truss shown in the figure, deter-

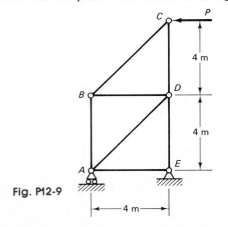

**Fig. P12-9**

mine $\Delta_V$ and $\Delta_H$ of joint $C$ due to applied force $P = 10$ kN. For simplicity, assume $AE = 1$ for all members.

**12-10.** For the truss in Problem 12-9, determine the relative deflection between joints $B$ and $E$ caused by applied force $P = 10$ kN. (*Hint:* Place equal and opposite unit forces, one at joint $B$ and the other at joint $E$, along a line joining them.)

**12-11.** For the truss shown in the figure, determine $\Delta_V$ of joint $B$ due to applied vertical force $P = 9$ k at $B$. For simplicity, assume $L/AE$ is unity for all members.

**Beams.** *Consider flexural deformations only.*

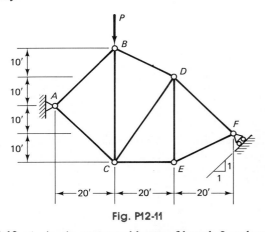

**Fig. P12-11**

**12-12.** A simply supported beam of length $L$ and constant $EI$ supports a downward uniformly distributed load $w_o$. Find the maximum $\Delta_V$ due to $w_o$.

**\*12-13.** For the beam in Problem 12-3, find the maximum $\Delta_V$ due to the two applied forces.

**12-14.** For the beam in Problem 12-3, find $\Delta_V$ and/or $\theta$, as assigned, at the left force $P$ due to both applied forces.

**12-15.** A simply suppported beam of length $L$ and constant $EI$ supports a downward uniformly distributed load $w_o$. Determine $\Delta_V$ and/or $\theta$, as assigned, due to $w_o$ at a distance $L/3$ from the left support.

**12-16 and 12-17.** Determine $\Delta_V$ and/or $\theta$, as assigned, at the center of the span due to the applied loads shown in the figures. $EI$ is constant.

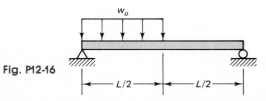

**Fig. P12-16**

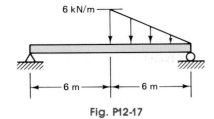

6 kN/m

6 m | 6 m

**Fig. P12-17**

**12-18.** Find $\Delta_V$ and/or $\theta$, as assigned, at the point of application of force $P$ for the beam of variable cross section shown in the figure.

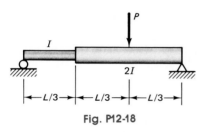

*P*

*I*

*2I*

L/3 | L/3 | L/3

**Fig. P12-18**

**12-19.** For the cantilever shown in the figure, determine (a) $\Delta_V$ at the applied force, and (b) $\Delta_V$ at the tip. $EI$ is constant.

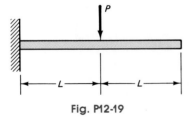

*P*

L | L

**Fig. P12-19**

**12-20.** Find the deflection at the point of application of force $P$. $EI$ is constant.

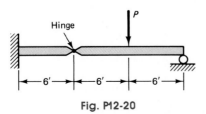

Hinge

*P*

6' | 6' | 6'

**Fig. P12-20**

**12-21.** For the overhanging beam shown in the figure, find $\Delta_V$ and/or $\theta$, as assigned, at the point of application of couple $M_o$. $EI$ is constant.

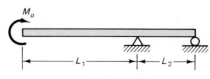

$M_o$

$L_1$ | $L_2$

**Fig. P12-21**

**Frames.** *Consider flexural deformations only.*

**12-22.** A planar bent bar of constant $EI$ has the dimensions shown in the figure. Determine $\Delta_V$, $\Delta_H$, or $\theta$, as assigned, at the tip due to the application of force $P$. Comment on the virtual-force method in comparison to the geometric approach based on the differential equations and superposition discussed in Chapter 10.

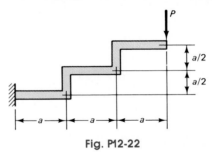

*P*

a/2 | a/2

a | a | a

**Fig. P12-22**

**12-23 through 12-30.** For the planar frames shown in the figures, determine $\Delta_V$, $\Delta_H$, or $\theta$ for point $A$, as assigned, due to the applied loading. For all cases, as-

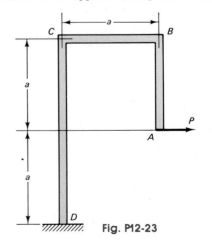

C | a | B

a

a

*P*

A

D

**Fig. P12-23**

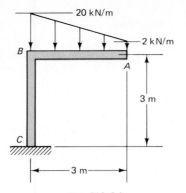

Fig. P12-24

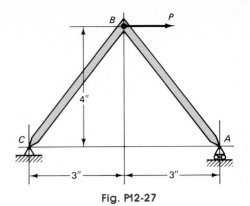

Fig. P12-27

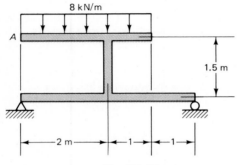

Fig. P12-25

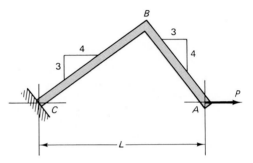

Fig. P12-28

Fig. P12-26

$M_1 = 1200$ k-ft

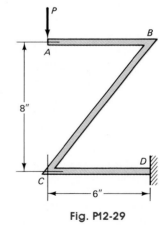

Fig. P12-29

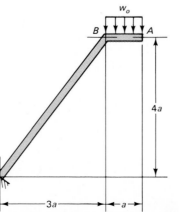

Fig. P12-30

sume $EI$ constant. (*Hint:* For ease of solution, for each frame segment, locate the origin of $x$ to obtain the simplest expressions for $m(x)$ and $M(x)$.)

**12-31 and 12-32.** For the planar frames shown in the figures, determine $\Delta_V$, $\Delta_H$, or $\theta$ for points $A$ and $B$, as assigned, due to the applied loading. For both cases, $EI$ is constant.

**Frames.** *Consider axial and flexural deformations.*

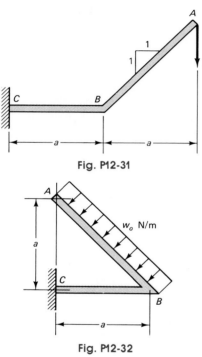

**Fig. P12-31**

**Fig. P12-32**

**12-33.** For the aluminum alloy, planar structural system shown in the figure, determine the vertical deflection of $D$ due to the applied force of 12 k. For the rod, $A = 0.5$ in$^2$; for the beam, $A = 4$ in$^2$ and $I = 15$ in$^4$. Let $E = 10 \times 10^3$ ksi.

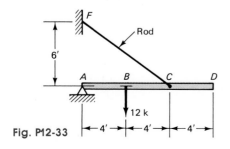

**Fig. P12-33**

**12-34.** In the preceding problem, it was determined that due to the applied force of 12 k, end $D$ moves 1.57 in up. If, without removing this force, it is necessary to return point $D$ to its initial position to make a connection, what is the required change in the length of rod $CF$? This change in length can be accomplished by means of a turnbuckle.

**12-35.** An inclined steel bar 2 m long, having a cross section of 4000 mm$^2$ and an $I$ of $8.53 \times 10^6$ mm$^4$, is supported as shown in the figure. The inclined steel hanger $DB$ has a cross section of 600 mm$^2$. Determine the downward deflection of point $C$ due to the application of the vertical force of $2\sqrt{2}$ kN. Let $E = 200$ GPa.

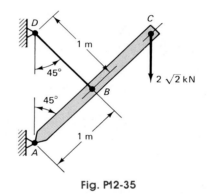

**Fig. P12-35**

**12-36.** A planar system consists of an inclined cantilever and rods $BC$ and $CD$, as shown in the figure. Determine $\Delta_V$ and/or $\Delta_H$, as assigned, of joint $C$ due to the application of force $P = 300$ N. The cross-sectional area of each rod is 10 mm$^2$ and that of the cantilever 400 mm$^2$. For the cantilever, $I = 10^4$ mm$^4$. For each member, $E = 200$ GPa.

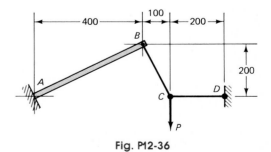

**Fig. P12-36**

**12-37.** A planar structure consists of a moment-resisting frame $ABC$ and a truss $CDE$ with pinned joints. Determine $\Delta_V$ and/or $\Delta_H$ at $C$, as assigned, due to the horizontal force 8 k at $D$. For all members, $EI = 800$ k-ft² and $EA = 500$ k. Work the problem using the units of k and ft.

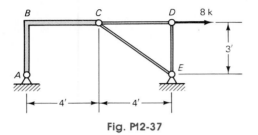

Fig. P12-37

**12-38.** For the data given in Problem 12-37, determine the rotation of member $CD$ due to the applied force at $D$. (*Hint:* Apply equal and opposite forces at $C$ and $D$, generating a unit couple.)

## Curved Members. *Neglect deformations due to direct shear.*

**12-39.** A U-shaped member of constant $EI$ has the dimensions shown in the figure. Determine the deflection of the applied forces away from each other due to flexure.

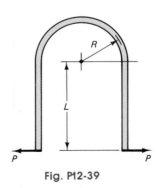

Fig. P12-39

**12-40.** In order to install a split ring used as a retainer on a machine shaft, it is necessary to open a gap of $\Delta$ by applying forces $P$, as shown in the figure. If $EI$ of the cross section of the ring is constant, determine the required magnitude of forces $P$. Consider only flexural effects.

**12-41.** A bar having a circular cross section is bent into a semicircle and is built in at one end as shown in the

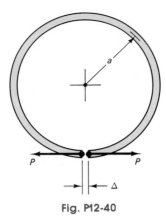

Fig. P12-40

figure. Determine the deflection of the free end caused by the application of force $P$ acting normal to the plane of the semicircle. Neglect the contribution of shear deformation.

## Deformations in Three-dimensions. *Neglect deformations due to direct shear.*

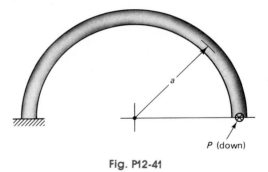

Fig. P12-41

**12-42.** A solid bar of circular cross section is bent into the shape of a right angle and is built in at one end, as shown in the figure. Determine the three transla-

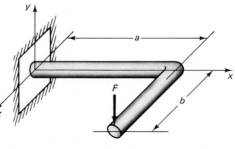

Fig. P12-42

tions, $\Delta_x$, $\Delta_y$, and $\Delta_z$, and the three rotations, $\theta_{xy}$, $\theta_{yz}$, and $\theta_{zx}$, of the free end due to applied force $F$. Constants $E$, $G$, $I$, and $J$ for the bar are given.

**\*12-43.** A solid circular bar is bent into the shape shown in the figure for Problem 12-1 and is built in at one end. Determine the three translations, $\Delta_x$, $\Delta_y$, and $\Delta_z$, and the three rotations, $\theta_{xy}$, $\theta_{yz}$, and $\theta_{zx}$, of the free end due to applied force $P$. Constants $A$, $I$, $J$, $E$, and $G$ for the bar are given.

## Section 12-7

**12-44.** A system of steel rods, each having a cross-sectional area of 0.20 in², is arranged as shown in the figure. At 50°F, joint $D$ is 0.10 in away from its support. (a) At what temperature can the connection be made without stressing any of the members? Let $E = 30 \times 10^6$ psi, and $\alpha = 6.5 \times 10^{-6}/°F$. (b) What stresses will develop in the members if after making the connections at $D$, the temperature drops to $-10°F$?

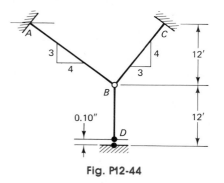

Fig. P12-44

**12-45.** Find the force in bar $AC$ for the planar truss shown in the figure due to the 30-kN horizontal force at $C$. For member $AC$, let the relative $L/AE = 0.50$, and for all other members, unity.

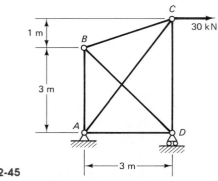

Fig. P12-45

**12-46.** (a) For the planar truss shown in the figure, determine the axial forces in all members due to the 18-kN vertical force at $B$. (b) By using at least two different virtual systems, find the vertical deflection at $B$ caused by the applied force at $B$. Let the relative values of $L/A$ be as follows: 1 for $AB$, 2 for $DB$, and 3 for $CB$. Consider member $BC$ to be redundant.

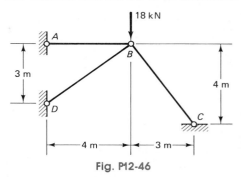

Fig. P12-46

**12-47.** For the planar truss shown in the figure, determine the reaction at $A$, treating it as redundant, due to the applied vertical force at $B$. For all members, $L/A$ is unity.

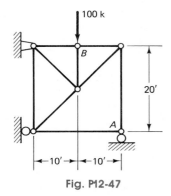

Fig. P12-47

**12-48.** For the beam shown in the figure, (a) determine the reaction at $A$, treating it as redundant. (b) Determine the moment at $B$, treating it as redundant. (*Hint:* Use the solution given to Problem 10-51.)

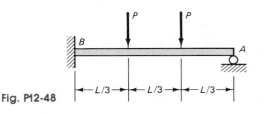

Fig. P12-48

**12-49.** A uniformly loaded beam fixed at both ends has the reactions shown in the figure. By using a simply supported beam with a unit load in the middle as a virtual system, determine the maximum deflection for the real beam.

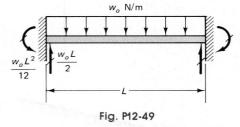

Fig. P12-49

**12-50.** A small pipe expansion joint in a plane can be idealized as shown in the figure. Hinge support points $A$ and $B$ are immovable. Derive an expression for the horizontal abutment reactions $R$ caused by the change in temperature $\delta T$ in the pipe. The coefficient of thermal expansion for the pipe is $\alpha$ and its flexural rigidity is $EI$. Consider flexural deformations only.

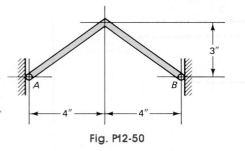

Fig. P12-50

**12-51.** A circular ring of a linearly elastic material is loaded by two equal and opposite forces $P$, as shown in the figure. For this ring, both $A$ and $I$ are constant. (a) Determine the largest bending moment caused by

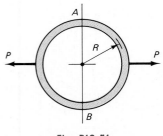

Fig. P12-51

the applied forces, and plot the entire moment diagram. (b) Find the decrease in diameter $AB$ caused by the applied forces. Consider only flexural deformations. (*Hint:* Take advantage of symmetry and consider the moment at $A$ as redundant.)

## Section 12-8

**12-52.** Using the virtual displacement method, (a) determine the forces in members $BD$ and $BC$ in Problem 12-46 assuming that member $BA$ is removed from the system, and (b) find the forces in all three members in the complete framing.

**12-53.** For the elastic truss shown in the figure and using the virtual displacement method, (a) determine the forces in members $AC$, $AD$, and $AE$ assuming that $AB$ is inactive due to the applied force at $A$, and (b) find the forces in all four members in the complete framing. The relative values $L/A$ are as follows: 0.40 for $AB$ and $AD$, 0.20 for $AC$, and 0.80 for $AE$.

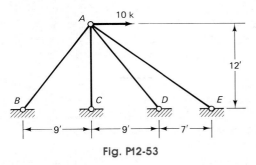

Fig. P12-53

## Section 12-12

**12-54.** Using Castigliano's second theorem, in Example 12-10, determine the horizontal deflection of $B$ due to the applied vertical force $P = 3$ k.

**The following problems are for solution by Castigliano's second theorem:**

**12-55.** Rework Problem 12-8.

**12-56.** Rework Problem 12-19.

**12-57.** Rework Problem 12-16.

**12-58.** Rework Problem 12-17.

**12-59.** Rework Problem 12-21.

**12-60.** Rework Problem 12-24.

**12-61.** Rework Problem 12-28.

**12-62.** Rework Problem 12-29.

**12-63.** Rework Problem 12-30.

**12-64.** Rework Problem 12-31.

**12-65.** Rework Problem 12-32.

**12-66.** Rework Problem 12-33.

**12-67.** Rework Problem 12-34.

## Section 12-13

**12-68.** Using Eq. 12-52, determine the forces in the elastic bar system shown in the figure due to applied force $P$. $L/AE$ is the same for each bar.

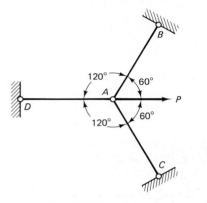

Fig. P12-68

**12-69.** A two-span continuous beam is loaded with a uniformly distributed downward load $w_0$ N/m. If the left span is $L$ and the right one is $2L$, what is the reaction at the middle support? Use Eq. 12-52 to obtain the solution. Draw shear and moment diagrams for this beam.

**12-70.** Without taking advantage of symmetry and using Eq. 12-52, determine the reaction components on the left for the beam in Problem 12-49 due to applied load $w_0$.

**12-71.** Assuming that in Example 12-16 end $B$ is simply supported, determine the reactions with the aid of Eq. 12-52.

**12-72.** Rework Example 12-17 after assuming that the cross section of bar $BC$ is twice as large as that of bar $AC$ or $DC$.

## Section 12-14

**12-73.** Using Eqs. 12-57 and 12-59 for neutral equilibrium, determine the critical buckling load in Problem 11-1.

**\*\*12-74.** Using Eq. 12-59 for neutral equilibrium, determine the critical Euler buckling load $P_{cr}$ for an elastic column of constant $EI$ with pinned ends, as shown in Fig. 11-13. Assume that the deflected shape for a slightly bent column in a neighboring equilibrium position is $v = A \sin \pi x/L$, where $A$ is an arbitrary constant. (*Hint:* Axial shortening of a column due to lateral deflection $v$ is given by

$$\Delta_v = \frac{1}{2} \int_0^L \left(\frac{dv}{dx}\right)^2 dx$$

and heuristically[20] the external work $\delta W_e = P \Delta_v$. By noting from Eq. 11-5 that $M = -Pv$, the expression for $\delta U$ then follows from Eq. 12-8.)

**\*\*12-75.** Find an approximate solution to the preceding problem by assuming the deflected shape of the column to be a parabola, $v = A(x^2 - xL)$, *satisfying the kinematic boundary conditions.* (*Note:* Energy solutions are not very sensitive to the assumed deflected shape provided one takes $M = -Pv$ and **not** as $M = EIv''$, where the second derivative of an assumed function is used. Numerous approximate solutions of column-buckling problems can be found in this manner.)

[20] For further study of this problem, see K. Marguerre, *Neuere Festigkeitsprobleme des Ingenieurs* (Berlin/Göttingen/Heidelberg: Springer, 1980), 189–229.

# chapter 13

# Statically Indeterminate Problems

## *[1]13-1. Introduction

The force and displacement methods for solving linearly elastic statically indeterminate problems previously encountered in this text are extended in Part A of this chapter to more complex cases. These two methods are particularly important in the matrix analysis of structures and in finite element formulations. They are directly applicable in satisfying elastic design criteria often based on the maximum allowable stress. Such a criterion is referred to as the *limit state* for the maximum stress. In other cases, the limit state may be the maximum allowable deflection, or the effect of system stiffness on vibrations. Such criteria are *serviceability limit states.*

If the *strength* of a member or members for emergency overloads is the only controlling parameter, the elastic maximum stress limit state may lead to an unduly conservative design. For *ductile* materials, where the fatigue problem does not arise, merely reaching the maximum stress at a point or a few points of a member does not necessarily exhaust the strength capacity of a system. The ultimate strength of such systems can be reasonably well approximated by considering the material rigid-ideally plastic. A few such cases were encountered earlier. This approach is discussed in Part B of this chapter for beams and frames.[2] The plastic limit state is of considerable importance in understanding the ultimate behavior of ductile structures, especially as it applies to seismic design and other emergency overload situations.

[1] This chapter is optional.
[2] For further discussion of beams, frames, as well as plates and shells, see P. G. Hodge, Jr., *Plastic Analysis of Structures* (New York: McGraw-Hill, 1959).

# Part A   ELASTIC METHODS OF ANALYSIS

## *13-2. Two Basic Methods for Elastic Analysis

Structural systems that experience only small deformations and are composed of linearly elastic materials are linear structural systems. The principle of superposition is applicable for such structures and forms the basis for two of the most effective methods for the analysis of indeterminate systems.

In the first of these methods, a statically indeterminate system is reduced initially to one that is determinate by removing redundant (superfluous) reactions or internal forces for maintaining static equilibrium; see Fig. 2-42. Then these redundant forces are considered as externally applied, and their magnitudes are so adjusted as to satisfy the prescribed deformation conditions at their points of application. Once the redundant reactions are determined, the system is statically determinate and can be analyzed for strength or stiffness characteristics by the methods introduced earlier. This widely used method is commonly referred to as the *force method*, or the flexibility method; see Section 2-15.

In the second method, referred to as the *displacement method*, or the stiffness method, the joint displacements of a structure are treated as the unknowns; see Sections 2-16 and 2-17. The system is first reduced to a series of members whose *joints* are imagined to be completely restrained from any movement. The joints are then released to an extent sufficient to satisfy the force equilibrium conditions at each joint. This method is extremely well-suited for computer coding and, hence, is even more widely used in practice than the force method, especially for the analysis of large-scale structures.

While some of the older classical methods continue to have some utility, the force and displacement methods are the two modern approaches to the solution of indeterminate structural systems.

## *13-3. Force Method

The first step in the analysis of structural systems using the force method is the determination of the degree of statical indeterminacy, which is the same as the number of redundant reactions, as discussed in Sections 1-9 and 2-15. The redundant reactions[3] are temporarily removed to obtain a statically determinate structure, which is referred to as the *released* or

---

[3] In the analysis of beams and frames, the bending moments at the supports are often treated as redundants. In such cases, rotations of tangents at the supports are considered instead of deflections.

*primary* structure. Then, since this structure is artificially reduced to statical determinacy, it is possible to find any desired displacement by the methods previously discussed. For example, the beam shown in Fig. 13-1(a) is indeterminate to the first degree. For this beam to remain in stable static equilibrium, only one of the vertical reactions can be removed. Removing the vertical reactions at $b$, Fig. 13-1(b), deflection $\Delta_{bP}$ at $b$ caused by applied forces $P$ can be calculated. By reapplying the removed redundant reaction $R_b$ to the *unloaded* member, Fig. 13-1(c), deflection $\Delta_{bb}$ at $b$ due to $R_b$ can be found. Since the deflection at $b$ of the given beam must be zero, by *superposing* the deflections and requiring that $\Delta_{bP} + \Delta_{bb} = 0$, the magnitude of $R_b = X_b$ can be determined.

This procedure can be generalized to any number of redundant reactions. However, it is essential in such cases to recognize that *the displacement of every point on the primary structure is affected by each reapplied redundant force*. This also holds true for the rotation of elements. As an example, consider the beam in Fig. 13-2(a).

By removing any two of the redundant reactions such as $R_b$ and $R_c$, the beam becomes determinate and the deflections at $b$ and $c$ can be computed, Fig. 13-2(b). These deflections are designated $\Delta_{bP}$ and $\Delta_{cP}$, respectively, where the first letter of the subscript indicates the point where the deflection occurs, and the second, the cause of the deflection. By reapplying $R_b$ to the same beam, the deflections at $b$ and $c$ due to $R_b$ at $b$ can be found, Fig. 13-2(c). These deflections are designated $\Delta_{bb}$ and $\Delta_{cb}$, respectively. Similarly, $\Delta_{bc}$ and $\Delta_{cc}$, due to $R_c$, can be established, Fig. 13-2(d). Superposing the deflections at each support and setting the sum equal to zero, since points $b$ and $c$ actually do not deflect, one obtains two equations:

$$\Delta_b = \Delta_{bP} + \Delta_{bb} + \Delta_{bc} = 0 \qquad (13\text{-}1)$$
$$\Delta_c = \Delta_{cP} + \Delta_{cb} + \Delta_{cc} = 0$$

These can be rewritten in a more meaningful form using *flexibility coefficients* $f_{bb}$, $f_{bc}$, $f_{cb}$, and $f_{cc}$, which are defined as the deflections shown in Figs. 13-1(e) and (f) due to unit forces applied in the direction of the

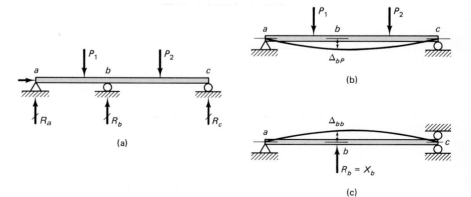

(b)

(a)

(c)

Fig. 13-1 Superposition for the force method.

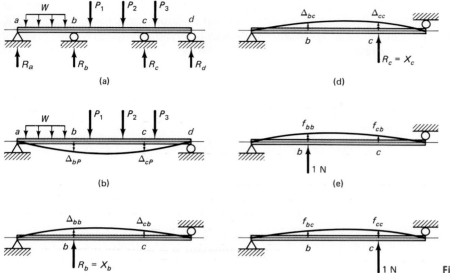

Fig. 13-2 Superposition for a continuous beam.

redundants. Then, since a linear structural system is being considered, the deflection at point $b$ due to the redundants can be expressed as

$$\Delta_{bb} = f_{bb}X_b \quad \text{and} \quad \Delta_{bc} = f_{bc}X_c \qquad (13\text{-}2)$$

and, similarly, at point $c$ as

$$\Delta_{cb} = f_{cb}X_b \quad \text{and} \quad \Delta_{cc} = f_{cc}X_c \qquad (13\text{-}3)$$

where $X_b$ and $X_c$ are the redundant quantities. Using this notation, Eq. 13-1 becomes

$$\Delta_b = f_{bb}X_b + f_{bc}X_c + \Delta_{bP} = 0 \qquad (13\text{-}4)$$
$$\Delta_c = f_{cb}X_b + f_{cc}X_c + \Delta_{cP} = 0$$

where the only unknown quantities are $X_b$ and $X_c$; simultaneous solution of these equations constitutes the solution of the problem.

Generalizing the above results for determining the unknown forces for systems with $n$ redundants, using superposition, the following *compatibility equations*[4] can be formed:

$$\Delta_a = f_{aa}X_a + f_{ab}X_b + \cdots + f_{an}X_n + \Delta_{aP}$$
$$\Delta_b = f_{ba}X_a + f_{bb}X_b + \cdots + f_{bn}X_n + \Delta_{bP} \qquad (13\text{-}5)$$
$$\vdots$$
$$\Delta_n = f_{na}X_b + f_{nb}X_b + \cdots + f_{nn}X_n + \Delta_{nP}$$

[4] Sometimes these expressions are referred to as the Maxwell-Mohr equations.

When the redundant supports are immovable, the left column of the equation is zero. Alternatively, such deflections can be prescribed.[5] The terms for deflections in the right column can be calculated for the primary structure. The flexibility coefficients[6] $f_{ij}$ are for the whole primary system. All quantities in these equations represent either deflections or angular rotations, depending on whether they are associated with forces or couples.

For the force method, it is customary to express Eq. 13-5 in matrix form as

$$
\begin{bmatrix}
f_{aa} & f_{ab} & \cdots & f_{an} \\
f_{ba} & f_{bb} & \cdots & f_{bn} \\
\cdots & \cdots & \cdots & \cdots \\
f_{na} & f_{nb} & \cdots & f_{nn}
\end{bmatrix}
\begin{Bmatrix}
X_a \\
X_b \\
\cdots \\
X_n
\end{Bmatrix}
=
\begin{Bmatrix}
\Delta_a - \Delta_{aP} \\
\Delta_b - \Delta_{bP} \\
\cdots \\
\Delta_n - \Delta_{nP}
\end{Bmatrix}
\tag{13-6}
$$

Because the square matrix is made up of the flexibility coefficients, this method is often called the *flexibility method* of structural analysis.

It should be clearly understood that the previous equations are applicable only to linearly elastic systems that undergo small displacements. It should be noted further that the matrix exhibited by Eq. 13-6 is a *system* or *global* flexibility matrix. Such matrices can be readily constructed directly only for the simpler problems. For treatment of more complex problems, the reader is referred to previously cited texts on finite elements or structural matrix analysis in Sections 2-10 and 12-9.

Before proceeding with examples, it is shown next that the matrix of the flexibility coefficients $f_{ij}$ is *symmetric,* i.e., $f_{ij} = f_{ji}$.

## *13-4. Flexibility Coefficients Reciprocity

According to the definition for flexibility coefficients, for linearly elastic systems, the displacement $\Delta_i$ at $i$ due to forces $P_i$ at $i$ and $P_j$ at $j$ patterned after Eq. 13-5 can be expressed as

$$
\Delta_i = f_{ii}P_i + f_{ij}P_j
\tag{13-5a}
$$

Similarly, the deflection at $j$ is

$$
\Delta_j = f_{ji}P_i + f_{jj}P_j
\tag{13-5b}
$$

where $f_{ii}$, $f_{ij}$, $f_{ji}$, and $f_{jj}$ are the flexibility coefficients of a given system.

If the strain energy of the system due to the application of these forces is $U$, according to Castigliano's second theorem, Eq. 12-47, the same qualities are also given as

---

[5] If an elastic support is provided at an $i$th point, the flexibility coefficient at the support is increased by adding the flexibility of such a support.

[6] The flexibility coefficients are also called the *deflection influence coefficients*.

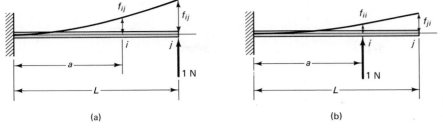

Fig. 13-3 Reciprocal deflections $f_{ij} = f_{ji}$.

(a)                                      (b)

$$\Delta_i = \frac{\partial U}{\partial P_i} \quad \text{and} \quad \Delta_j = \frac{\partial U}{\partial P_j}$$

By taking partial derivatives of $\Delta_i$ with respect to $P_j$ in Eq. 13-5a and the preceding equation, the following equality is obtained:

$$\frac{\partial \Delta_i}{\partial P_j} = f_{ij} = \frac{\partial^2 U}{\partial P_j \, \partial P_i}$$

and, similarly,

$$\frac{\partial \Delta_j}{\partial P_i} = f_{ji} = \frac{\partial^2 U}{\partial P_i \, \partial P_j}$$

However, since the order of differentiation is immaterial,

$$f_{ij} = f_{ji} \tag{13-7}$$

As illustrated in Fig. 13-3, this relation states that the displacement at any point $i$ due to a unit force at any point $j$ is equal to the displacement of $j$ due to a unit force at $i$, provided the directions of the forces and deflections in each of the two cases coincide. It can be noted that this relationship holds true for several cases considered earlier in applications of virtual force equations in Section 12-6. For example, in calculating flexibility coefficients, using Eq. 12-26, by setting $F_i$ equal to unit force $p_i$, its role with $\bar{p}_j$ is interchangable. This is also true in the use of Eqs. 12-27 and 12-28.

The derived relationship is often called *Maxwell's theorem of reciprocal displacements.*[7]

## EXAMPLE 13-1

For the simply supported elastic beam shown in Fig. 13-4, show that the rotation of the tangent to the elastic curve at the support $i$, caused by applying a unit force at $j$, is equal to the deflection at $j$ caused by applying a unit couple at $i$.

---

[7] This relationship was discovered by James Clerk Maxwell in 1864. The more general case was demonstrated by E. Betti in 1872.

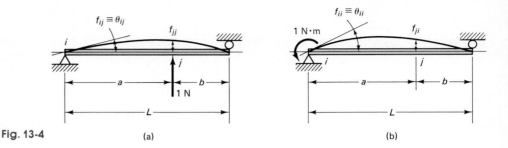

**Fig. 13-4**                    (a)                                    (b)

## Solution

The deflection of the beam due to a concentrated force $P$ at $j$ is given by Eq. 10-27, applicable for $0 \le x \le a$. The derivative of this equation with respect to $x$ gives the slope for the elastic curve. The slope of this function at $x = 0$ gives the rotation $\theta_{ij}$, defining $f_{ij}$, when the applied force $P = -1$.

A-13

$$v = -\frac{Pb}{6EIL}(L^2x - b^2x - x^3) \quad \text{and} \quad v' = -\frac{Pb}{6EIL}(L^2 - b^2 - 3x^2)$$

$$v'(0) = \theta(0) = -\frac{Pb}{6EIL}(L^2 - b^2) = -\frac{Pab}{6EIL}(a + 2b)$$

and

$$\theta_{ij} \equiv f_{ij} = \frac{ab}{6EIL}(a + 2b) \tag{13-8}$$

The equation for an elastic curve for a beam subjected to an end moment is derived next. By proceeding as before, the deflection at $j$ is found. Assuming that a counterclockwise moment $M_o$ is applied at $j$,

$$M = -M_o + M_ox/L \quad \text{and} \quad EIv'' = M = -M_o + M_ox/L$$
$$EIv' = -M_ox + M_ox^2/2L + C_3$$

Hence,          $EIv = -M_ox^2/2 + M_ox^3/6L + C_3x + C_4$

From $v(0) = 0$, $C_4 = 0$, and from $v(L) = 0$, $C_3 = M_oL/3$, and

$$EIv = \frac{M_ox}{6EIL}(-3Lx + x^2 + 2L^2) \tag{13-9}$$

Therefore, for $v(a)$ and $M_o = 1$,

$$f_{ji} = \frac{a}{6EIL}(-3La + a^2 + 2L^2) = \frac{ab}{6EIL}(a + 2b) \tag{13-10}$$

This result is identical to that given by Eq. 13-8.

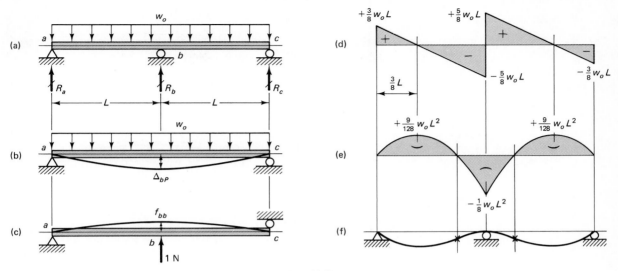

**Fig. 13-5**

## EXAMPLE 13-2

A two-span continuous elastic beam on simple supports carries a uniformly distributed load, as shown in Fig. 13-5. Determine the reactions and plot shear and moment diagrams. $EI$ for the beam is constant.

## Solution

Reaction $R_b$ at $b$ is removed to make the beam statically determinate. The deflection at $b$ for the primary structure using Eq. 10-22 is

$$\Delta_{bP} = -\frac{5w_o(2L)^2}{384EI} = -\frac{5w_oL^4}{24EI}$$

The deflection at $b$ due to a concentrated force is given by Eq. 10-30. Therefore, by setting $P = 1$, the flexibility coefficient is

$$f_{bb} = \frac{1 \times (2L)^3}{48EI} = \frac{L^3}{6EI}$$

By using Eq. (13-5) and assuming that the supports are immovable,

$$\Delta_b = f_{bb}X_b + \Delta_{bP} = 0 \quad \text{and} \quad X_b \equiv R_b = -\Delta_{bP}/f_{bb} = 5w_oL/4$$

From statics, $R_a = R_c = 3w_oL/8$ and the shear and moment diagrams are as in Figs. 13-5(d) and (e), respectively. The elastic curve is shown in Fig. 13-5(f).

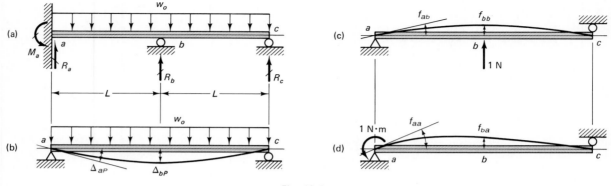

Fig. 13-6

## EXAMPLE 13-3

A two-span continuous beam is clamped at one end and simply supported at two other points; see Fig. 13-6(a). Determine the reactions caused by the application of a uniformly distributed load $w_o$. $EI$ for the beam is constant.

### Solution

This beam is statically indeterminate to the second degree. Therefore, two redundants must be removed to proceed. A convenient choice is to remove $M_a$ and $R_b$, resulting in a simply supported beam, Fig. 13-6(b). Using the results found in Example 10-3, and summarized in Table 11 of the Appendix,

$$\Delta_{aP} \equiv \theta_{aP} = -\frac{w_o(2L)^3}{24EI} = -\frac{w_oL^3}{3EI}$$

$$\Delta_{bP} = -\frac{5w_o(2L)^3}{384EI} = -\frac{5w_oL^4}{24EI}$$

One set of flexibility coefficients is determined by applying a unit force at $b$, Fig. 13-6(c), and determining the rotation at $a$ and the deflection at $b$ using the equations in Table 11. This process is repeated by applying a unit moment at $a$, Fig. 13-6(d), and finding the rotation at $a$ and the deflection at $b$. Thus,

$$f_{ab} = \frac{(2L)^2}{16EI} = \frac{L^2}{4EI} \quad \text{and} \quad f_{bb} = \frac{(2L)^3}{48EI} = \frac{L^3}{6EI}$$

$$f_{aa} = \frac{(2L)}{3EI} = \frac{2L}{3EI} \quad \text{and} \quad f_{ba} = \frac{L[(2L)^2 - L^2]}{6EI(2L)} = \frac{L^2}{4EI}$$

Note that as to be expected, $f_{ab} = f_{ba}$.

Forming two equations for compatibility of displacements at $a$ and $b$ using Eq. 13-5,

$$\Delta_a \equiv \theta_a = \Delta_{aP} + f_{aa}X_a + f_{ab}X_b = 0$$
$$\Delta_b = \Delta_{bP} + f_{ba}X_a + f_{bb}X_b = 0$$

A-13

Substituting the relevant quantities from before, the required equations are

$$\frac{2L}{3EI} X_a + \frac{L^2}{4EI} X_b = \frac{w_o L^3}{3EI}$$

$$\frac{L^2}{4EI} X_a + \frac{L^3}{6EI} X_b = \frac{5w_o L^4}{24EI}$$

Solving these two equations simultaneously,

$$X_a = M_a = 0.0714 w_o L^2 \quad \text{and} \quad X_b = R_b = 1.143 w_o L$$

The positive signs of these quantities indicate agreement with the assumed direction of unit forces.

## EXAMPLE 13-4

Consider the planar elastic pin-ended bar system shown in Fig. 13-7(a). Determine the bar forces caused by the application of inclined force $P = 10\sqrt{5}$ kN at joint $e$. All bars can resist either tensile or compressive forces. For simplicity in calculations, let $L/EA$ for each member be unity.

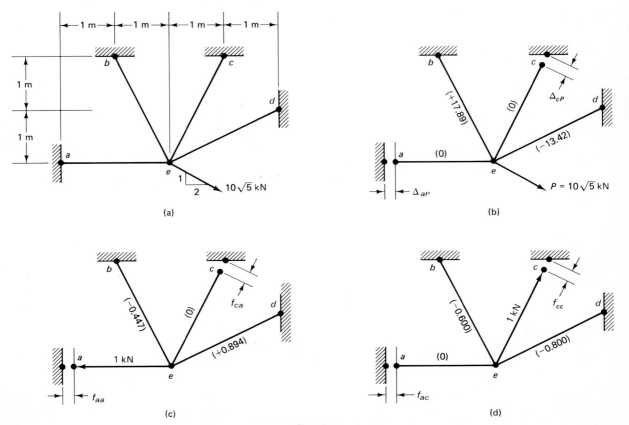

Fig. 13-7

Solution

This problem is statically indeterminate to the second degree, and, in this solution, bars *ae* and *ce* are assumed to be redundant. Therefore, the bar system with the bars cut at *a* and *c*, shown in Fig. 13-7(b), is the *primary* system. The bar forces for this condition are shown on the diagram in parentheses. In this primary system, the possible displacements that may develop at *a* and *c* are noted and must be removed to restore the required compatibility conditions. Therefore, the behavior of the *unloaded* primary system due to the application of unit axial forces is studied first, as shown in Figs. 13-7(c) and (d). Again, the axial bar forces for each case are shown directly on the figures. Note that when the force in bar *ae* is unity, the force in bar *ce* is zero, Fig. 13-7(c); conversely, when the force in the bar *ce* is unity, the force in bar *ae* is zero.

    Calculations for the required deflections and flexibility coefficients are carried out in tabular form using the virtual force method and following the solution pattern of Example 12-5.

| Bar | $F$, kN | $\bar{p}_a$ or $p_a$ | $\bar{p}_c$ or $p_c$ | $\bar{p}_a F$ | $\bar{p}_b F$ | $\bar{p}_a p_a$ | $\bar{p}_c p_c$ | $\bar{p}_a p_c$ or $\bar{p}_c p_a$ |
|-----|---------|----------------------|----------------------|---------------|---------------|-----------------|-----------------|-------------------------------------|
| *ae* | 0 | +1 | 0 | 0 | 0 | +1 | 0 | 0 |
| *be* | +17.89 | −0.447 | −0.60 | −8.00 | −10.73 | +0.20 | +0.36 | +0.268 |
| *ce* | 0 | 0 | +1 | 0 | 0 | 0 | +1 | 0 |
| *de* | −13.42 | +0.894 | −0.80 | −12.00 | +10.73 | +0.80 | +0.64 | −0.716 |
| | | | Sum: | −20.00 | 0 | +2.00 | +2.00 | −0.448 |

    Since for each bar, $L/AE = 1$, the relative deflections and flexibility coefficients according to Eq. 12-26 are

$$\Delta_{aP} = -20 \qquad \Delta_{cP} = 0 \qquad f_{aa} = f_{cc} = 2$$

and
$$f_{ac} = f_{ca} = -0.448$$

    Therefore, for bar forces $F_{ac} = X_a$ and $F_{cc} = X_c$, the required conditions of compatibility at *a* and *c*, using Eq. 13-5 gives

$$\Delta_a = f_{aa}X_a + f_{ac}X_c + \Delta_{aP} = 2X_a - 0.448X_c - 20 = 0$$

and

$$\Delta_c = f_{ca}X_a + f_{cc}X_c + \Delta_{cP} = -0.448X_a + 2X_c + 0 = 0$$

By solving these two equations simultaneously,

$$F_{ae} = X_a = +10.52 \text{ kN} \qquad \text{and} \qquad F_{ce} = X_b = +2.36 \text{ kN}$$

Using superposition, the forces in the other two bars are

$$F_{be} = 17.89 + 10.52 \times (-0.447) + 2.36 \times (-0.600) = +11.77 \text{ kN}$$

and

$$F_{de} = -13.42 + 10.52 \times 0.894 + 2.36 \times (-0.800) = -5.90 \text{ kN}$$

Computer solutions are commonly used for problems with a high degree of indeterminacy.

## *13-5. Introduction to the Displacement Method

In the force method discussed in Section 13-3, the redundant forces were assumed to be the unknowns. In the displacement method, on the other hand, the displacement—both linear and/or angular—of the joints or nodal points are taken as the unknowns. The first step in applying this method is to prevent these joint displacements, which are called *kinematic indeterminants* or *degrees of freedom*. The suppression of these degrees of freedom results in a modified system that is composed of a series of members each of whose end points are restrained from translations and rotations. Calculation of reactions at these artificially restrained ends due to externally applied loads can be carried out using any of the previously described methods. The results of such calculations are usually available for a large variety of loading conditions and a few are given in Table 12 of the Appendix. In beam analysis by this method, counterclockwise moments and upward reactions acting *on either end of a member* are taken as positive. *This beam sign convention differs from that used previously in this text,* and is necessary for a consistent formulation of the super-position equations.

Sometimes this sign convention is referred to as "analyst's" to distinguish it from "designer's" used previously throughout. The designer's sign convention conveniently differentiates between tensile and compressive regions in flexural members.

The procedure for applying a displacement method for a beam with one degree of kinematic indeterminancy is illustrated in Fig. 13-8(a). First,

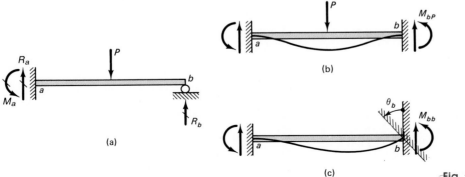

Fig. 13-8

the support at $b$ is restrained, Fig. 13-8(b), reducing the problem to that of a fixed-end beam. Both the vertical reactions and end moments in such a beam can be found by methods discussed previously. Moment $M_{bP}$ is an example of such a reactive force at an end, where the first letter of the subscript designates the location and the second identifies the cause. Such end moments and reactions are referred to as *fixed-end actions* at beam ends. For general use, fixed-end actions are identified here by a letter $A$ with two subscripts. For the above case, $A_{bP} \equiv M_{bP}$. Subscript $P$ refers to *any kind* of applied lateral load.

Next, moment $M_{bb}$ at $b$, Fig. 13-8(c), is determined as a function of the applied rotation $\theta_b$. In this notation, the first letter of the subscript identifies the location of the fixed-end force (moment) and the second identifies the location of the applied displacement. Two basic cases for end moments and reactions caused either by applied end rotation or displacement are given in Table 12 in the Appendix.

Finally, an equation for static equilibrium is written. In this case, since the beam is simply supported at $b$, total moment $M_b$ must be zero. For general use, such force quantities are identified as $P_b$, i.e., $M_b \equiv P_b$. Therefore, assuming that the system is linearly elastic and undergoes small dislacements, for equilibrium at joint $b$,

$$M_b = M_{bP} + M_{bb} = 0 \qquad (13\text{-}11a)$$

or in generalized notation,

$$P_b = A_{bP} + A_{bb} = 0 \qquad (13\text{-}11b)$$

## EXAMPLE 13-5

Using a displacement method, determine the reactions for a uniformly loaded beam fixed at one end and simply supported at the other, Fig. 13-9(a). $EI$ is constant.

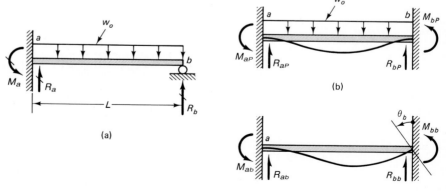

(a)

(b)

(c)

Fig. 13-9

## Solution

Since joint rotation at $b$ is the only kinematic unknown at the supports, Fig. 13-9(a), this beam is kinematically indeterminate to the first degree. Using Table 12 in the Appendix, the fixed-end actions due to the applied load, Fig. 13-9(b), and the end moments and reactions due to $\theta_b$, Fig. 13-9(c), are

$$M_{aP} \equiv A_{aP} = w_oL^2/12 \quad \text{and} \quad M_{bP} \equiv A_{bP} = -w_oL^2/12$$
$$M_{ab} \equiv A_{ab} = 2EI\theta_b/L \quad \text{and} \quad M_{bb} \equiv A_{bb} = 4EI\theta_b/L$$
$$R_{ab} = 6EI\theta_b/L^2 \quad \text{and} \quad R_{bb} = -6EI\theta_b/L^2$$

For moment equilibrium at the end $b$, using Eq. 13-11a,

$$M_b \equiv P_b = M_{bP} + M_{bb} = -\frac{w_oL^2}{12} + \frac{4EI}{L}\theta_b = 0$$

Hence,
$$\theta_b = w_oL^2/48EI$$

Using this $\theta_b$ in the superposition equations,

$$M_a = M_{aP} + M_{ab} = \frac{w_oL^2}{12} + \frac{2EI}{L}\theta_b = \frac{w_oL^2}{8}$$

$$R_a = R_{aP} + R_{ab} = \frac{w_oL}{2} + \frac{6EI}{L}\theta_b = \frac{5w_oL}{8}$$

$$R_b = R_{bP} + R_{bb} = \frac{w_oL}{2} - \frac{6EI}{L}\theta_b = \frac{3w_oL}{8}$$

The sign of $M_a$ is opposite from that of the designer's beam sign convention.

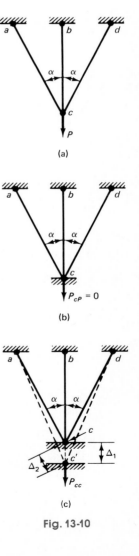

## EXAMPLE 13-6

Three elastic pin-jointed bars are symmetrically arranged in a plane to form the system shown in Fig. 13-10(a). The cross-sectional area $A$ of each bar is the same, and the elastic modulus is $E$. Verify the bar forces found in Examples 2-14, 12-23, and 12-17 caused by applied force $P$.

## Solution

Because of symmetry, this system has only one degree of kinematic freedom and joint $c$ can only displace in the vertical direction. In this solution, first, the joint is restrained from displacement, Fig. 13-10(b). Here all of the fixed-end actions are zero, and $A_{cP} \equiv P_{cP} = 0.$[8]

Force $P_{cc}$ for the system is determined next as a function of deflection $\Delta_1$ for bar $bc$. As in Example 12-17, for geometric compatibility at joint $c$, $\Delta_2 =$

---

[8] If an axial force were applied somewhere between $b$ and $c$, $P_{cP}$ would not be zero. For example, if a downward force $P_1$ were applied at a distance $L_{bc}/4$ above $c$, per Eq. 2-39, the fixed-end downward force at $c$ would be $R_{cP} = 3P_1/4$.

**Fig. 13-10**

Fig. 13-11

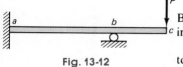

Fig. 13-12

$\Delta_1 \cos \alpha$. Hence, using Eq. 2-12, the bar forces $F_1$ and $F_2$ in members $bc$ and $ac$ (or $dc$), respectively, are

$$F_1 = k_1 \Delta_1 = \frac{AE}{L} \Delta_1 \qquad \text{and} \qquad F_2 = k_2 \Delta_1 \cos \alpha = \frac{AE}{L/\cos \alpha} \Delta_1 \cos \alpha$$

Then for vertical force equilibrium at joint $c$,

$$P_c = P_{cP} + F_1 + 2F_2 \cos \alpha = P$$

Substituting the values of $P_{cP}$, $F_1$, and $F_2$ into the above equation, and solving for $\Delta_1$, one finds

$$\Delta_1 = \frac{PL}{AE(1 + 2 \cos^3 \alpha)}$$

By substituting this value of $\Delta_1$ into the relations for the bar forces, the results in Example 2-23 are verified.

If in this example there were no symmetry about the vertical axis (either due to lack of symmetry in the structure itself or due to application of force $P$ at an angle), a horizontal displacement would also have developed at the joint. Two force equilibrium equations, one in the horizontal direction and the other in the vertical direction, must then be set up and solved simultaneously for the horizontal and vertical displacements. Such cases are illustrated in Figs. 12-11 and 12-12, and are also considered in the next section.

It should be noted that by adding additional bars to the system, as shown in Fig. 13-11, does not increase the kinematic indeterminacy, and it remains at two. In the force method, on the other hand, the number of statical redundants increases, as does the number of simultaneous equations for determining redundants. However, this does not imply that the displacement method always involves the solution of fewer equations compared to the force method. Consider, for example, the case of a propped cantilever with an overhang; see Fig. 13-12. This beam is statically indeterminate only to the first degree, but kinematically indeterminate to the third degree (rotations at $b$ and $c$, and a vertical deflection at $c$); hence, only one equation is needed for solution by the force method, but three simultaneous equations are required using the displacement method.

## *13-6. Further Remarks on the Displacement Method

The displacement method is extended to problems with several degrees of kinematic indeterminancy in this section. For this purpose, consider the beam shown in Fig. 13-13(a), where the guided support at $c$ allows for vertical displacement but no rotation of the beam. The other degree of freedom of this beam is the rotation of its tangent at support $b$. This beam is thus kinematically indeterminate to the second degree. Upon restraining these two degrees of freedom, one obtains a system consisting

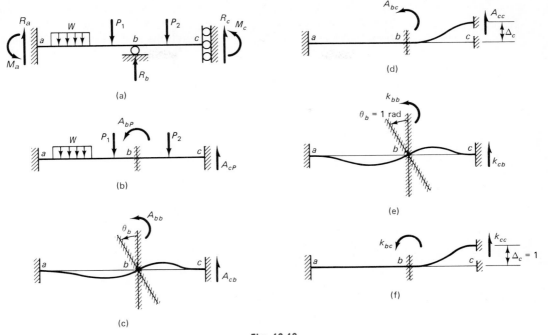

Fig. 13-13

of two fixed-end beams, $ab$ and $bc$, Fig. 12-13(b). The effect of the externally applied loads on these two fixed-end beams is to produce a set of reactive forces at the supports. The fixed-end action (moment) $A_{bP}$ at $b$ is the *sum* of the fixed-end moments in beams $ba$ and $bc$ at $b$ caused by the applied loads. Similarly, the fixed-end action $A_{cP}$ is the vertical reaction at $c$ restraining vertical displacement. Since the support is capable of developing a moment at $c$, it does not enter the problem at this level.

Next the support at $b$ is rotated through an angle $\theta_b$, giving rise to the fixed-end actions (moments) $A_{bb}$ and $A_{cb}$ at points $b$ and $c$, respectively, as shown in Fig. 13-13(c). Similarly, $A_{bc}$ and $A_{cc}$ are caused by the vertical displacement $\Delta_c$ at $c$, Fig. 13-13(d).

Since no external moment $M_b$ is applied at $b$, nor a vertical reaction $P_c$ at $c$, the resultant forces $M_b \equiv P_b$ and $P_c$ at these points are equal to zero. These two forces may be found by superposing three separate analyses, shown in Figs. 13-13(b)–(d), leading to two simultaneous equilibrium equations:

$$P_b = A_{bP} + A_{bb} + A_{bc} = 0 \qquad (13\text{-}12)$$
$$P_c = A_{cP} + A_{cb} + A_{cc} = 0$$

These equations can be rewritten in more meaningful form using *stiffness coefficients* $k_{bb}$, $k_{bc}$, and $k_{cc}$, defined as the fixed-end actions shown in

Figs. 13-13(e) and (f) due to the unit displacements (linear or angular) corresponding to the kinematic indeterminants. Then for a linear system, the moments at $b$ and the vertical reactions at $c$, caused by displacements $\Delta_b = \theta_b$ and $\Delta_c$, are

$$A_{bb} = k_{bb}\,\Delta_b \qquad \text{and} \qquad A_{bc} = k_{bc}\,\Delta_c \qquad (13\text{-}13)$$

$$A_{cb} = k_{cb}\,\Delta_b \qquad \text{and} \qquad A_{cc} = k_{cc}\,\Delta_c \qquad (13\text{-}14)$$

By substituting these relations into Eq. 13-12,

$$P_b = k_{bb}\,\Delta_b + k_{bc}\,\Delta_c + A_{bP} = 0 \qquad (13\text{-}15)$$

$$P_c = k_{cb}\,\Delta_b + k_{cc}\,\Delta_c + A_{cP} = 0$$

These equations can be solved simultaneously for unknowns $\Delta_b$ and $\Delta_c$.

By extending this approach to systems having $n$ degrees of kinematic indeterminancy, the force equilibrium equations for determining the unknown nodal displacements $\Delta_i$ are

$$P_a = k_{aa}\,\Delta_a + k_{ab}\,\Delta_b + \cdots + k_{an}\,\Delta_n + A_{aP}$$

$$P_b = k_{ba}\,\Delta_a + k_{bb}\,\Delta_b + \cdots + k_{bn}\,\Delta_n + A_{bP} \qquad (13\text{-}16)$$

$$\vdots$$

$$P_n = k_{na}\,\Delta_a + k_{nb}\,\Delta_b + \cdots + k_{nn}\,\Delta_n + A_{nP}$$

where terms $P_a$, $P_b$, . . . , $P_n$ correspond to the *external* forces applied *at the nodal points*. In the absence of such forces, these terms are zero. The stiffness coefficients $k_{ij}$ are associated either with a displacement or a rotation. The fixed-end actions $A_{aP}$, $A_{bP}$, . . . , $A_{nP}$ are caused by the externally applied loads.

In matrix form, Eq. 13-16 for the displacement method can be written as

$$\begin{bmatrix} k_{aa} & k_{ab} & \cdots & k_{an} \\ k_{ba} & k_{bb} & \cdots & k_{bn} \\ \cdots & \cdots & \cdots & \cdots \\ k_{na} & k_{nb} & \cdots & k_{nn} \end{bmatrix} \begin{Bmatrix} \Delta_1 \\ \Delta_2 \\ \cdots \\ \Delta_n \end{Bmatrix} = \begin{Bmatrix} P_a - A_{aP} \\ P_b - A_{bP} \\ \cdots \\ P_n - A_{nP} \end{Bmatrix} \qquad (13\text{-}17)$$

Because the square matrix consists entirely of stiffness coefficients, the displacement method is often referred to as the *stiffness method*. For general procedures for constructing the stiffness matrix, the reader is referred to previously cited texts on finite elements or structural matrix analysis in Sections 2-10 and 12-9. In this text, only the simpler problems are considered.

Before proceeding with examples, it will be shown that the stiffness matrix is *symmetric*, i.e., $k_{ij} = k_{ji}$, and that it is related to the flexibility matrix.

## *13-7. Stiffness Coefficients Reciprocity

For linearly elastic systems, an analogous relationship for stiffness coefficients can be obtained similar to that found in Section 13-4 for flexibility coefficients. Thus, if the system's elastic energy is $U$, according to Castigliano's first theorem, Eq. 12-45, the displacement of forces $P_i$ and $P_j$ in the respective directions of $\Delta$'s are

$$P_i = \frac{\partial U}{\partial \Delta_i} \quad \text{and} \quad P_j = \frac{\partial U}{\partial \Delta_j} \tag{13-18}$$

Alternatively, it can be seen from Eq. 13-16 that a partial derivative of $P_i$ with respect to $\Delta_j$ is $k_{ij}$. Similarly, a partial derivative of $P_j$ with respect to $\Delta_i$ is $k_{ji}$. Carrying out these operations with Eqs. 13-18 establishes the following equalities:

$$\frac{\partial P_i}{\partial \Delta_j} = \frac{\partial^2 U}{\partial \Delta_j \, \partial \Delta_i} = k_{ij}$$

and

$$\frac{\partial P_j}{\partial \Delta_i} = \frac{\partial^2 U}{\partial \Delta_i \, \partial \Delta_j} = k_{ji} \tag{13-19}$$

Since the order of differentiation for the mixed derivatives is immaterial,

$$\boxed{k_{ij} = k_{ji}} \tag{13-20}$$

This relation proves that the matrix of stiffness coefficients is *symmetric*, a very important property for analysis of structural systems.

The relationship between the stiffness and flexibility coefficients is illustrated in the next example. It is more complex than that for systems with one degree of kinematic and static indeterminancy.

### EXAMPLE 13-7[9]

Show the relationship between the flexibility and the stiffness matrices for the two-spring system shown in Fig. 13-14. The externally applied forces are $P_1$ and $P_2$, and the linearly elastic flexibilities and stiffnesses for each spring are shown in the figure.

### Solution

The displacement of nodal points $b$ and $c$ for the loaded system can be written using spring flexibilities as

[9] Adapted from M. F. Rubinstein, *Matrix Computer Analysis of Structures*, (Englewood Cliffs, N. J.: Prentice-Hall, 1966), 60–63.

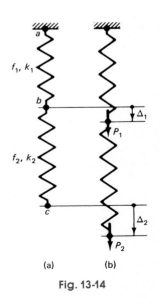

Fig. 13-14

$$\Delta_1 = (P_1 + P_2)f_1 \quad \text{and} \quad \Delta_2 = (P_1 + P_2)f_1 + P_2 f_2$$

where force $P_1 + P_2$ acts on spring $ab$.

Similarly, the equilibrium equations for each nodal point $b$ and $c$, using spring stiffnesses, are

$$P_1 = k_1 \Delta_1 - k_2(\Delta_2 - \Delta_1) \quad \text{and} \quad P_2 = k_2(\Delta_2 - \Delta_1)$$

where the stretch of spring $bc$ is $\Delta_2 - \Delta_1$.

Recasting these equations into matrix form gives

$$\begin{Bmatrix} \Delta_1 \\ \Delta_2 \end{Bmatrix} = \begin{bmatrix} f_1 & f_1 \\ f_1 & f_1 + f_2 \end{bmatrix} \begin{Bmatrix} P_1 \\ P_2 \end{Bmatrix}$$

and

$$\begin{Bmatrix} P_1 \\ P_2 \end{Bmatrix} = \begin{bmatrix} k_1 + k_2 & -k_2 \\ -k_2 & k_2 \end{bmatrix} \begin{Bmatrix} \Delta_1 \\ \Delta_2 \end{Bmatrix}$$

Next it can be noted that the individual spring flexibilities can be replaced by the reciprocals of the spring constants. Then the flexibility matrix, expressed in terms of spring constants, is multiplied by the stiffness matrix using the rules of matrix multiplication, giving

$$\begin{bmatrix} \dfrac{1}{k_1} & \dfrac{1}{k_1} \\ \dfrac{1}{k_1} & \dfrac{1}{k_1} + \dfrac{1}{k_2} \end{bmatrix} \begin{bmatrix} k_1 + k_2 & -k_2 \\ -k_2 & k_2 \end{bmatrix} = \begin{bmatrix} 1 & 0 \\ 0 & 1 \end{bmatrix}$$

This shows that a product of a flexibility matrix by a stiffness matrix leads to an *identity* matrix. All diagonal elements of this *unit* matrix are unity, and all others are zero.

This result means that a flexibility matrix is an *inverse* of a stiffness matrix or vice versa. For these symmetric matrices, this can be symbolically written as

$$[\mathbf{f}] = [\mathbf{k}]^{-1} \quad \text{or} \quad [\mathbf{k}] = [\mathbf{f}]^{-1} \tag{13-21}$$

For problems with single degrees of static and kinematic indeterminacy, these expressions degenerate into simple reciprocals of these quantities.

## EXAMPLE 13-8

Using the displacement method, calculate the rotations at $b$ and $c$ for the continuous beam of constant $EI$ loaded as shown in Fig. 13-15(a), and determine the moments at $a$ and $b$.

## Solution

At supports $b$ and $c$, the beam is free to rotate, making the system kinematically indeterminate to the second degree. By temporarily restraining these supports

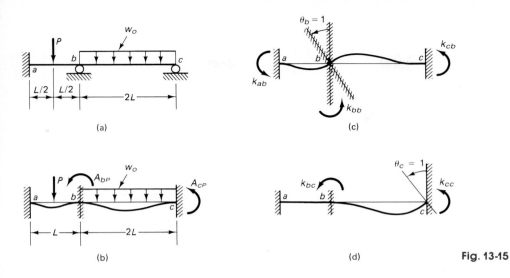

**Fig. 13-15**

against rotations, a system of two fixed-end beams is obtained, Fig. 13-15(b). The fixed-end actions for these beams can be obtained with the aid of Table 12 in the Appendix. In the following, the first letter of the subscript outside the brackets identifying a beam designates the end where the fixed-end action applies.

*For beam ab:*

$$[A_{aP}]_{ab} = + PL/8 \quad \text{and} \quad [A_{bP}]_{ba} = - PL/8$$

*For beam bc:*

$$[A_{bP}]_{bc} = + w_o(2L)^2/12 = + w_oL^2/3 \quad \text{and} \quad [A_{cP}]_{cb} = - w_oL^3/3$$

*For joint b:*

$$A_{bP} = [A_{bP}]_{ba} + [A_{bP}]_{bc} = - PL/8 + w_oL^2/3$$

The *stiffness coefficients* can be calculated by subjecting the temporarily fixed ends *b* and *c*, to unit rotations *one at a time*, Figs. 13-15(c) and (d). Again, using formulas in Table 12 of the Appendix and by noting that the two adjoining spans contribute to the stiffness of the joint at *b*, one has

$$k_{bb} = \left[\frac{4EI}{L}\right]_{ba} + \left[\frac{4EI}{2L}\right]_{bc} = \frac{6EI}{L} \quad k_{bc} = \left[\frac{2EI}{2L}\right]_{bc} = \frac{EI}{L}$$

$$k_{cb} = \left[\frac{2EI}{2L}\right]_{cb} = \frac{EI}{L} \quad k_{cc} = \left[\frac{4EI}{2L}\right]_{cc} = \frac{2EI}{L}$$

Similarly, for the member *ab*, due to a unit rotation at *b*,

$$k_{ba} = \left[\frac{4EI}{L}\right]_{ba} = \frac{4EI}{L} \quad \text{and} \quad k_{ab} = \left[\frac{2EI}{L}\right]_{ab} = \frac{2EI}{L}$$

Since there are no externally applied forces (moments) at $b$ and $c$, for equilibrium at these joints, Eqs. 13-15 become

$$P_b = \frac{6EI}{L} \Delta_b + \frac{EI}{L} \Delta_c - \frac{PL}{8} + \frac{w_o L^2}{3} = 0$$

$$P_c = \frac{EI}{L} \Delta_b + \frac{2EI}{L} \Delta_c - \frac{w_o L^2}{3} = 0$$

By solving these two equations simultaneously,

$$\Delta_b \equiv \theta_b = \frac{L^2}{11EI}\left(\frac{P}{4} - w_o L\right) \quad \text{and} \quad \Delta_c \equiv \theta_c = \frac{L^2}{11EI}\left(-\frac{P}{8} + \frac{7}{3} w_o L\right)$$

By substituting these displacement values into the member superposition equations, the end moments in all members are found.

$$M_{ab} = [A_{aP}]_{ab} + k_{ab}\theta_b = \frac{15}{88} PL - \frac{2}{11} w_o L^2$$

$$M_{ba} = [A_{bP}]_{ba} + k_{ba}\theta_b = -\frac{3}{88} PL - \frac{4}{11} w_o L^2$$

$$M_{bc} = [A_{bP}]_{bc} + k_{bc}\theta_b + k_{cb}\theta_c = +\frac{3}{88} PL + \frac{4}{11} w_o L^2$$

$$M_{cb} = [A_{cP}]_{cb} + k_{cb}\theta_b + k_{cc}\theta_c = 0$$

Note that with the analyst's beam sign convention employed in this solution, $M_{ba} + M_{bc} = 0$, since they are of opposite sign.

## EXAMPLE 13-9

Rework Example 13-4 using the displacement method of analysis; see Fig. 13-16(a).

### Solution

In this problem, since nodal point $e$ can move horizontally and vertically, the system has two degrees of freedom. As noted in Example 13-4, this is also statically indeterminate to the second degree. Each additional bar emanating from $e$ would increase the statical indeterminancy by one, however, the kinematic degree of indeterminacy would remain at two.

The horizontal and vertical positive displacements, $\Delta_1$ and $\Delta_2$, shown in Figs. 13-16(b) and (c), respectively, are the unknowns. For displacement $\Delta_1$, if end $e$ of bar $ie$ is *constrained* to move only horizontally, as in the upper diagram of Fig. 13-16(d), the bar elongates by $\Delta_1 \sin \alpha_i$. This would develop a bar axial force $P_1^i = k^i \Delta_1 \sin \alpha_i$, where the bar spring constant $k^i = A_i E_i / L_i$. This bar force, $P_1^i$, can be resolved, respectively, into horizontal and vertical components $P_{11}^i = P_1^i \sin \alpha_i$ and $P_{12}^i = P_1^i \cos \alpha_i$. Therefore,

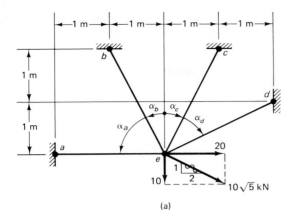

(a)

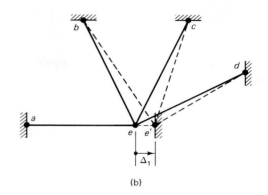

(b)

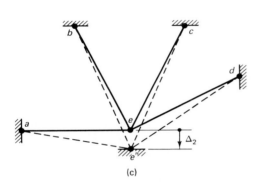

(c)

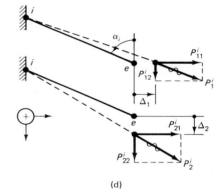

(d)

**Fig. 13-16**

$$P_{11}^i = \left(\frac{A_i E_i}{L_i} \sin^2 \alpha_i\right) \Delta_1 = k_{11}^i \Delta_1$$

$$P_{12}^i = \left(\frac{A_i E_i}{L_i} \sin \alpha_i \cos \alpha_i\right) \Delta_1 = k_{12}^i \Delta_1$$

(13-22a)

where $k_{11}^i$ and $k_{12}^i$ are bar stiffnesses for a horizontal displacement $\Delta_1$.

By the same reasoning, if end $e$ of bar $ie$ is *constrained* to move $\Delta_2$ in the vertical direction only, as shown in the lower diagram of Fig. 13-16(d), the respective horizontal and vertical force components for the bar are

$$P_{22}^i = \left(\frac{A_i E_i}{L_i} \cos^2 \alpha_i\right) \Delta_2 = k_{22}^i \Delta_2$$

$$P_{21}^i = \left(\frac{A_i E_i}{L_i} \cos \alpha_i \sin \alpha_i\right) \Delta_2 = k_{21}^i \Delta_2$$

(13-22b)

where $k_{22}^i$ and $k_{22}^i$ are bar vertical and horizontal stiffnesses, respectively, for vertical displacement $\Delta_2$.

To solve this problem, these equations must be applied to each of the four bars and summed to obtain the horizontal and vertical stiffnesses of the system. This is carried out in the table.

| Bar | $\alpha_i$, degrees | $\sin \alpha_i$ | $\cos \alpha_i$ | $\sin^2 \alpha_i$ | $\cos^2 \alpha_i$ | $\sin \alpha_i \cos \alpha_i$ |
|-----|------|------|------|------|------|------|
| ae | 90. | 1. | 0. | 1. | 0. | 0. |
| be | 26.565 | 0.4472 | 0.8944 | 0.200 | 0.800 | 0.400 |
| ce | −26.565 | −0.4472 | 0.8944 | 0.200 | 0.800 | −0.400 |
| de | −63.435 | −0.8944 | 0.4472 | 0.800 | 0.200 | −0.400 |
| | | | Sum: | 2.200 | 1.800 | −0.400 |

The relative bar stiffness $A_i E_i / L_i$ for each bar is unity. Therefore, from the table, the system's horizontal stiffness $k_{11} = \sum k_{11}^i = \sum \sin^2 \alpha_i = 2.2$, and, similarly, $k_{22} = \sum \cos^2 \alpha_i = 1.8$, and $k_{12} = k_{21} = \sum \sin \alpha_i \cos \alpha_i = -0.4$.

Writing these results in matrix form,

$$\begin{bmatrix} k_{11} & k_{12} \\ k_{21} & k_{22} \end{bmatrix} \begin{Bmatrix} \Delta_1 \\ \Delta_2 \end{Bmatrix} = \begin{Bmatrix} P_1 \\ P_2 \end{Bmatrix} \quad \text{or} \quad \begin{bmatrix} 2.2 & -0.4 \\ -0.4 & 1.8 \end{bmatrix} \begin{Bmatrix} \Delta_1 \\ \Delta_2 \end{Bmatrix} = \begin{Bmatrix} 20 \\ 10 \end{Bmatrix}$$

The solution for this matrix equation gives $\Delta_1 = 10.536$ and $\Delta_2 = 7.895$. Therefore, again, since for each bar, $A_i E_i / L_i = 1$,

$$F_i = \Delta_1 \sin \alpha_i + \Delta_2 \sin \alpha_2$$

Using this equation, $F_{ae} = +10.53$ kN, $F_{be} = +11.77$ kN, $F_{ce} = +2.35$ kN, and $F_{de} = -5.88$ kN. These results agree with those found in Example 13-4 by the force method.

Application of the displacement method to a similar problem with more bars would only be slightly more complex.

# Part B      PLASTIC LIMIT ANALYSIS

## *13-8. Plastic Limit Analysis of Beams

Procedures for determining ultimate loads for axially loaded bar systems of elastic-ideally plastic material are given in Examples 2-18 and 2-23. These ultimate loads are the plastic limit states. In the process of obtaining these loads, the entire range of elastic-plastic system behavior under an increasing load is considered. As can be seen from Fig. 2-54(e) or 2-59(c), there are three distinct regions of response. At first, these systems respond in a linearly elastic manner. Then a part of the structural system yields as the remainder continues to deform elastically. This is the range of contained plastic flow. Finally, a structure continues to yield at no further increase in applied load. At this stage of behavior of ideally plastic structures, the deformations become unbounded. This condition is the plastic limit state. In this analytical idealization, the effects of strain hardening and changes in structure geometry are neglected.

As is shown in the previous examples, a direct calculation of the plastic limit state for ideally plastic materials is both possible and rather simple. From the practical point of view, such calculations provide an insight into the collapse mode of ductile structures. However, such direct solutions for plastic limit load do not provide complete information on inelastic behavior. If at a service or working load some prior yielding had occurred, the deflections and distribution of forces remain unknown. Only step-by-step computer solutions, or solutions for simple cases, as in Examples 2-18 and 2-23, can provide complete history of force and deflection distributions.

The same general behavior is exhibited by elastic-ideally plastic beams and frames, and here the objective is to develop simplified procedures for determining directly the plastic limit states for such members. By bypassing the elastic, and the elastic-plastic stages of loading, and determining the plastic limit loads, the procedure becomes relatively simple. Some previously established results are reexamined for background.

Typical moment-curvature relationships, normalized with respect to $M_{yp}$, for elastic-perfectly plastic beams are shown in Fig. 13-17 for three different cross sections. Basic results for a rectangular beam were established in Example 10-14 (see Fig. 10-25). Results for the other two cases can be found using the same procedure. Curves normalized with respect

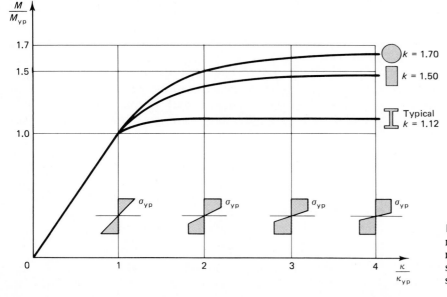

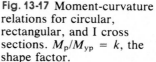

Fig. 13-17 Moment-curvature relations for circular, rectangular, and I cross sections. $M_p/M_{yp} = k$, the shape factor.

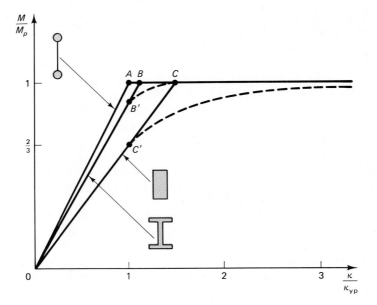

**Fig. 13-18** Moment-curvature idealizations for plastic analyses of beams of different cross sections.

to $M_p$ are shown in Fig. 13-18. The behavior of an idealized cross section with large flanges and a negligibly thin web is added in this diagram.

In both diagrams, as the cross sections plastify, a rapid ascent of the curves toward their respective asymptotes occurs. This means that shortly after reaching the elastic capacity of a beam, a rather constant moment, very near to $M_p$, is both achieved and maintained. This is particularly true for the important case of an I beam. As can be noted from Fig. 13-18, for this cross section, the elastic-plastic behavior is essentially confined to the range between $B'$ and $C$; for the remainder, the moment is essentially $M_p$. The influence of the elastic core next to the beam neutral axis is more pronounced for members with rectangular or round cross sections, whose shape factors, $k$, are larger than those for an I beam, Fig. 13-17. Nevertheless, in the plastic limit analysis of members subjected to bending, it is generally assumed that *an abrupt transition from elastic to ideally plastic behavior occurs at $M_p$.* Therefore, member behavior between $M_{yp}$ and $M_p$ is considered to be elastic. It is further assumed that when $M_p$ is reached, a *plastic hinge* is formed in the member. In contrast to a frictionless hinge permitting free rotation, the plastic hinge allows large rotations to occur at a *constant plastic moment $M_p$.*

In a plastic limit analysis of beams, the elastic displacements in relation to the plastic ones are small and can be neglected. Detailed analyses have shown[10] that it is sufficiently accurate to consider beams rigid-plastic, with plasticity confined to plastic hinges at *points.* In reality, plastic hinges extend along short lengths of beams and depend on loading conditions.

[10] See, for example, L. S. Beedle, *Plastic Design of Steel Frames* (New York: Wiley, 1966) or S. J. Moy, *Plastic Methods for Steel and Concrete Structures* (New York: Wiley, 1981).

The approximate theory discussed here is applicable to beams as well as columns subjected to moderate axial forces. When a cross section lacks biaxial symmetry, the positive and negative moments differ in their magnitudes and should be accounted for in the analysis. A method for determining the reduced plastic capacity of members in the presence of axial forces is discussed in Section 6-13.

By inserting a plastic hinge at a plastic limit load into a statically determinate beam, a kinematic mechanism permitting an unbounded displacement of the system can be formed. This is commonly referred to as the *collapse mechanism*.[11] For each degree of static indeterminancy of a beam, an additional plastic hinge must be added to form a collapse mechanism. The insertion of plastic hinges must be such as to obtain a kinematically *admissible* (plausible) collapse mechanism. The use of kinematically admissible collapse mechanisms is illustrated in the examples to follow.

In plastic limit design, it is necessary to multiply working loads by a load factor larger than unity to obtain design *factored loads*. This is analogous to the use of a factor of safety in elastic analyses. This issue is discussed in Section 1-11.

There are two common methods of plastic limit analysis. One is based on conventional statics and the other on virtual work. In either method the bending moments anywhere along a member cannot exceed the plastic moment $M_p$, and the conditions of equilibrium must always be satisfied. The procedure for forming kinematically admissible mechanisms, somewhat similar to continuity conditions in elastic analysis, is illustrated in the following examples.

## EXAMPLE 13-10

A concentrated force $P$ is applied at the middle of a simply supported prismatic beam, as shown in Fig. 13-19(a). If the beam is of a ductile material, what is the plastic limit load $P_{ult}$? Obtain the solution using ($a$) the equilibrium method and ($b$) the virtual work method. Consider only flexural behavior, i.e., neglect the effect of shear forces. Neglect beam weight.

## Solution

($a$) The shape of the moment diagram is the same regardless of the load magnitude. For any value of $P$, the maximum moment $M = PL/4$, and if $M \leq M_{yp}$, the beam behaves elastically. When the moment is at $M_{yp}$, the force at yield

$$P_{yp} = 4M_{yp}/L$$

When $M_{yp}$ is exceeded, contained yielding of the beam commences and continues until the plastic moment $M_p$ is reached, Fig. 13-19(c).

---

[11] In seismic analyses, the plastic hinges dissipate energy. Therefore, it is preferable to call such mechanisms *energy dissipating mechanisms*.

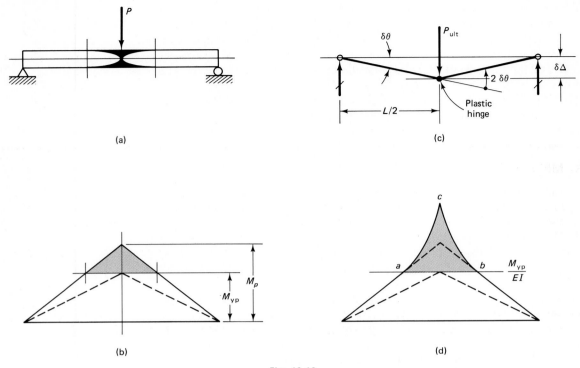

**(a)**          **(c)**

**(b)**          **(d)**

**Fig. 13-19**

The curvature diagram prior to reaching $M_p$ at the middle of the beam resembles that shown in Fig. 13-19(d). Since the elastic curvature can at most be $M_{\text{yp}}/EI$, it is exceeded as shown above line $ab$. At $M_p$, the fully plastic part of the beam near the middle is shown in black in Fig. 13-19(a). This region is considerably narrower for I beams than for the rectangular cross section implied in this figure because most of the bending moment is carried in the flanges. The curvature at the middle of the beam becomes very large as it rapidly approaches $M_p$ and continues to grow without bound (see Fig. 13-18). By setting the plastic moment $M_p$ equal to $PL/4$ with $P = P_{\text{ult}}$, one obtains the result sought:

$$P_{\text{ult}} = 4M_p/L$$

Note that consideration of the actual plastic region indicated in Fig. 13-19(a) is unnecessary in this calculation. A comparison of this result with $P_{\text{yp}}$ shows that

$$P_{\text{ult}} = \frac{M_p}{M_{\text{yp}}} P_{\text{yp}} = k\, P_{\text{yp}}$$

where the difference between the two forces depends only on the shape factor $k$.

(*b*) An admissible virtual kinematic mechanism assuming a rigid-plastic beam is shown in Fig. 13-19(c). The external virtual work is $P_{ult} \, \delta\Delta$, where from geometry $\delta\Delta = L \, \delta\theta/2$. The internal virtual work is caused by rotating $M_p$ through an angle of $2 \, \delta\theta$. Hence, per Eq. 12-19, equating the previous expressions for work,

$$P_{ult} \, \delta\Delta = P_{ult} \, L \, \delta\theta/2 = M_p(2 \, \delta\theta)$$

On solving the last two expressions for $P_{ult}$, as before,

$$P_{ult} = 4M_p/L$$

## EXAMPLE 13-11

A prismatic beam of ductile material, fixed at one end and simply supported at the other, carries a concentrated force in the middle, as shown in Fig. 13-20(a). Determine the plastic limit load $P_{ult}$ using (a) the equilibrium method and (b) the virtual displacement method. Compare the result with that of an elastic solution. Neglect beam weight.

## Solution

(a) The results of an elastic analysis are shown in Fig. 13-20(b). The same results are replotted in Fig. 13-20(c) from horizontal baseline $AB$. In both diagrams, the colored portions of the diagrams represent the net result. Note that the auxiliary

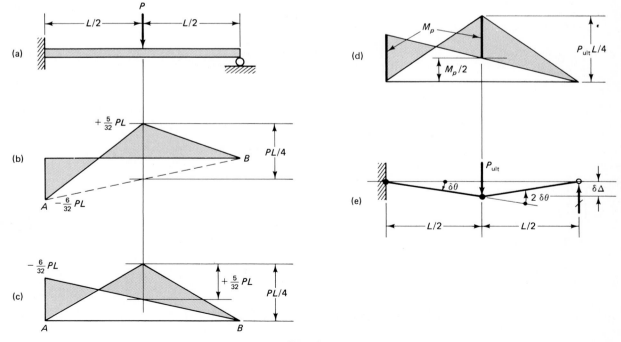

**Fig. 13-20**

ordinates $PL/4$ have precisely the value of the maximum moment in a simple beam with a concentrated force in the middle.

By setting the maximum elastic moment equal to $M_{yp}$, one obtains force $P_{yp}$ at impending yield:

$$P_{yp} = \frac{16M_{yp}}{3L}$$

When the load is increased above $P_{yp}$, the moment at the built-in end increases and can reach but cannot exceed $M_p$. This is also true of the moment at the middle of the span. These limiting conditions are shown in Fig. 13-20(d). At the plastic limit load, it is necessary to have a kinematically admissible mechanism. With the two plastic hinges and a roller on the right, this condition is satisfied, Fig. 13-20(e).

From the geometric construction in Fig. 13-20(d), in the middle of the span, $M_p + M_p/2 = P_{ult}L/4$. Hence,

$$P_{ult} = 6M_p/L$$

Comparing this result with $P_{yp}$, one has

$$P_{ult} = \frac{9M_p}{8M_{yp}} P_{yp} = \frac{9}{8} kP_{yp}$$

The increase in $P_{ult}$ over $P_{yp}$ is due to two causes; the shape factor $k$ and the *equalization of the maximum moments*. (Compare the moment diagrams in Figs. 13-20(c) and (d)).

(*b*) For the virtual displacement shown in Fig. 13-20(e), the external virtual work at plastic limit load is $P_{ult}\,\delta\Delta$. The internal virtual work takes place in the plastic hinges at the left support and in the middle of the span. Equating these expressions of work per Eq. 12-19,

$$P_{ult}\,\delta\Delta = P_{ult}L\,\delta\theta/2 = M_p\,\delta\theta + M_p(2\,\delta\theta)$$

giving, as before,

$$P_{ult} = 6M_p/L$$

## EXAMPLE 13-12

A prismatic beam of ductile material is loaded as shown in Fig. 13-21(a). Using the virtual displacement method, determine the plastic limit loads. Neglect the weight of the beam.

### Solution

In this case, several kinematic displacement mechanisms are possible, and the solution is found by a trial-and-error process. The correct mechanism is one where the assumed virtual displacement generates a compatible moment diagram.

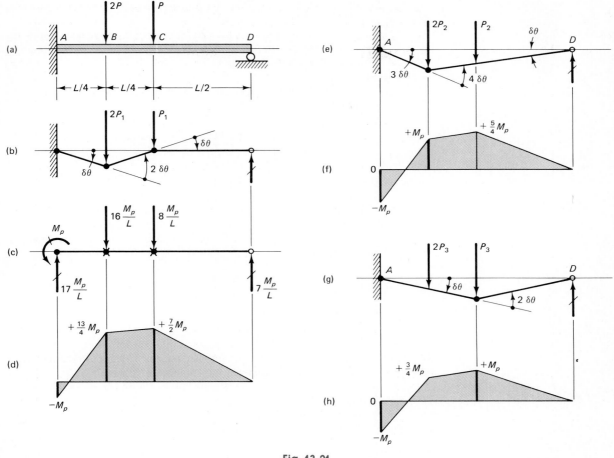

**Fig. 13-21**

An admissible mechanism is shown in Fig. 13-21(b). By equating the external and the internal virtual work per Eq. 12-19, and identifying the plastic limit loads for this case as $P_1$ and $2P_1$, one has

$$(2P_1)(L\,\delta\theta/4) = M_p\,\delta\theta + M_p 2\,\delta\theta + M_p\,\delta\theta$$

where the three terms on the right apply, respectively, to the plastic hinges at $A$, $B$, and $C$. The solution of this equation gives

$$P_1 = 8M_p/L \qquad \text{and} \qquad 2P_1 = 16M_p/L$$

By applying the forces to the beam and assuming hinges $B$ and $C$ rigid, Fig. 13-21(c), the resulting bending moment diagram is as shown in Fig. 13-21(d). This diagram shows that with $P_1$ and $2P_1$, the moments at $B$ and $C$ are greater than $M_p$. This is an *upper bound* solution that asserts that a load found on the basis of an assumed admissible kinematic mechanism is always greater than or at best equal to the plastic limit load.[12]

By reducing $P_1$ and $2P_1$ by a ratio of 2/7, the conditions for the plastic moment capacity of the member and that of equilibrium *are satisfied*. Since such a solution occurs prior to the full development of a kinematic mechanism, it gives the *lower bound*[13] for the applied loads.

Based on this reasoning, the obtained results with the assumed mechanism have the following lower and upper bounds:

$$\frac{2}{7} \times 8 \frac{M_p}{L} = \frac{16}{7} \frac{M_p}{L} < P_1 < 8 \frac{M_p}{L}$$

A similar relation applies for $2P_i$. These bounds are rather far apart, and alternative mechanisms, shown in Figs. 13-21(e) and (g), are tried.

By following the earlier procedure, the results for the mechanism in Fig. 13-21(e) give $P_2 = 3.5M_p/L$ and $2P_2 = 7M_p/L$. The moment diagram corresponding to these forces is shown in Fig. 13-21(f). These results establish better bounds for the solution, which are

$$\frac{4}{5} \times 3.5 \frac{M_p}{L} = 2.8 \frac{M_p}{L} < P_2 < 3.5 \frac{M_p}{L}$$

By carrying out a solution for the mechanism in Fig. 13-2(g), it can be shown that $P_3 = 3M_p/L$ and $2P_3 = 6M_p/L$. The moment diagram for these forces in Fig. 13-21(h) confirms the correct choice of the mechanism, since the moments at $A$ and $C$ are each equal to $M_p$. Therefore, the solution is "exact."

The mechanism in Fig. 13-21(b) is not a good choice for this problem. However, even this solution, as can be seen from Fig. 13-21(c), indicates that the plastic hinges within the span should be at $C$. By taking advantage of such observations, the exact result could have been obtained more quickly.

This problem can be easily solved by the equilibrium method. For such a solution, assuming the beam simply supported, the moment diagram is prepared first. Then an inclined line, as shown in Fig. 13-20(d) is drawn such that equal moments $M_p$ develop at $A$ and $C$.

---

### EXAMPLE 13-13

A prismatic beam of ductile material, fixed at one end and simply supported at the other, carries a uniformly distributed load, as shown in Fig. 13-22(a). Find the plastic limit load $w_{ult}$ using (*a*) the equilibrium method and (*b*) the virtual force method.

---

[12] For proof, see any of the cited references on plastic analysis, and H. J. Greenberg and W. Prager, "Limit Design of Beams and Frames," *Trans. ASCE* 117 (1952):447–458.

[13] Proof and formal statement of the lower bound theorem can be found in the previously cited references.

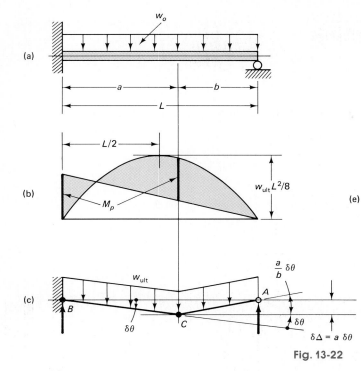

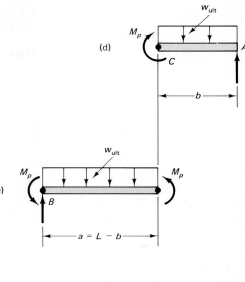

Fig. 13-22

## Solution

(*a*) In this problem, two plastic hinges are required to create a collapse mechanism. One of these hinges is at the built-in end. The location of the hinge associated with the maximum positive moment is not known, since the moment varies gradually and there is no distinct peak. However, one can assume an admissible mechanism, such as shown in Fig. 13-22(c), which is compatible with the moment diagram of Fig. 13-22(b).

For purposes of analysis, the beam with the assumed plastic hinges is separated into two parts, as shown in Figs. 13-22(d) and (e). Then, by noting that no shear is possible at $C$, since it is the point of maximum moment for a continuous function, one can write two equations of static equilibrium:

$$\sum M_A = 0 \; \circlearrowright + \qquad M_p - w_{\text{ult}} b^2/2 = 0$$
$$\sum M_B = 0 \; \circlearrowleft + \qquad 2M_p - w_{\text{ult}}(L - b)^2/2 = 0$$

Simultaneous solution of these equations locates the plastic hinge $C$ at $b = (\sqrt{2} - 1)L$. Either one of these equations yields the limit load

$$w_{\text{ult}} = \frac{2M_p}{b^2} = \frac{2M_p}{[(\sqrt{2} - 1)L]^2}$$

(*b*) On the *average*, the uniformly distributed plastic limit load $w_{ult}$ goes through a virtual displacement of $\delta\Delta/2$, Fig. 13-22(c). Hence, for use in Eq. 12-19,

$$\delta W_e = w_{ult}\,L\,\frac{\delta\Delta}{2} = w_{ult}L\,\frac{a\,\delta\theta}{2}$$

The internal virtual work is done by plastic moments $M_p$ at plastic hinges $B$ and $C$, going through their respective rotations, Fig. 13-22(c). Hence,

$$\delta W_{ie} = M_p\,\delta\theta + M_p\left(1 + \frac{a}{b}\right)\delta\theta$$

By equating the previous two relations and solving for $w_{ult}$, after some simplifications,

$$w_{ult} = \frac{2M_p}{L}\left(\frac{2L - a}{La - a^2}\right)$$

The unknown distance $a$ can be found by taking a derivative of $w_{ult}$ with respect to $a$ and setting it equal to zero. Thus,

$$\frac{dw_{ult}}{da} = 0$$

After carrying out the differentiation and simplifications,

$$-a^2 + 4aL - 2L^2 = 0$$

By solving this quadratic equation and retaining the root falling within the span,

$$a = (2 - \sqrt{2})L \quad \text{and} \quad b = L - a = (\sqrt{2} - 1)L$$

as before. Hence $w_{ult}$ found previously applies to this solution as well.

The virtual work solutions for distributed loads, such as just shown, are somewhat complex for routine applications. Two alternative procedures, however, are possible. In one, the distributed load can be approximated by a series of concentrated forces, where possible plastic hinge locations are more easily identified. Alternatively, the location of a plastic hinge can be estimated, leading to a simple solution. The accuracy of such a solution can be judged by calculating the upper and the lower bounds, as has been illustrated in Example 13-12.

### EXAMPLE 13-14

Rework the previous example by assuming that a plastic hinge for a positive bending moment occurs in the middle of the span; see Fig. 13-23. Determine the bounds on this approximate solution.

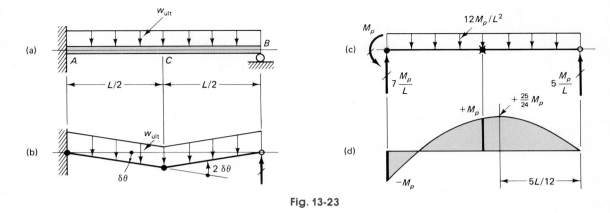

Fig. 13-23

## Solution

By applying Eq. 12-19, i.e., equating the external and the internal virtual work,

$$w_{\text{ult}}L\left(\frac{L\,\delta\theta}{2}\frac{1}{2}\right) = M_p(\delta\theta + 2\,\delta\theta)$$

where $L\,\delta\theta/2$ is the maximum deflection at $C$, and the factor of $\frac{1}{2}$ reduces this to an average deflection for the distributed load. Solution of the last equation gives an upper bound for the plastic limit load $w_{\text{ult}}$ for the assumed mechanism, and

$$w_{\text{ult}} = 12M_p/L^2$$

By assuming the plastic hinge $C$ rigid and applying the above load to the beam, Fig. 13-23(c), the resulting moment diagram is as shown in Fig. 13-23(d). Since, in this diagram, the maximum positive bending moment exceeds $M_p$, the applied load in Fig. 13-23(c) must be reduced by a factor of 24/25 to obtain the lower-bound solution. Therefore, the lower-bound solution for the plastic limit load is $(24/25)12M_p/L^2 = 11.52M_p/L^2$.

Summarizing, the bounds for this solution are

$$11.52\,\frac{M_p}{L^2} < w_{\text{ult}} < 12\,\frac{M_p}{L^2}$$

By taking the plastic hinge at the point of the maximum positive moment in Fig. 13-23(d) and repeating the calculations, nearly an exact plastic limit load is found.

## EXAMPLE 13-15

A prismatic uniformly loaded beam is fixed at both ends, as shown in Fig. 13-24(a). (*a*) Determine the plastic limit load using the equilibrium method, and compare the results with elastic analysis. (*b*) Verify the plastic limit load using the virtual work method.

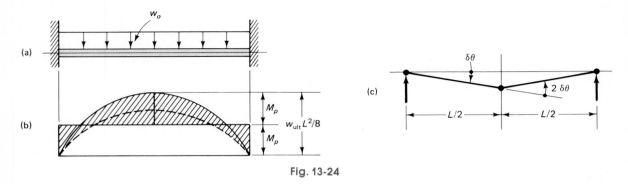

Fig. 13-24

### Solution

According to the analysis in Example 10-23, Fig. 10-37(c), the maximum bending moments occur at the built-in ends and are equal to $w_o L^2/12$. The maximum positive moment develops at the middle of the span and is $w_o L^2/24$. Therefore, at yield, based on the maximum moment,

$$M_{\mathrm{yp}} = w_{\mathrm{yp}} L^2/12 \qquad \text{or} \qquad w_{\mathrm{yp}} = 12 M_{\mathrm{yp}}/L^2$$

By increasing the load, plastic hinges develop at the supports. The collapse mechanism is not formed, however, until a plastic hinge also develops in the middle of the span, Fig. 13-24(c).

The maximum moment for a simply supported uniformly loaded beam is $w_o L^2/8$. Therefore, as can be seen from Fig. 13-24(b), to obtain the limit load in a clamped beam, this quantity must be equated to $2M_p$, with $w_o = w_{\mathrm{ult}}$. Hence,

$$w_{\mathrm{ult}} L^2/8 = 2M_p \qquad \text{or} \qquad w_{\mathrm{ult}} = 16 M_p/L^2$$

Comparing this result with $w_{\mathrm{yp}}$, one has

$$w_{\mathrm{ult}} = \frac{4M_p}{3M_{\mathrm{yp}}}\, w_{\mathrm{yp}} = \frac{4}{3}\, k\, w_{\mathrm{yp}}$$

As in Example 13-11, the increase of $w_{\mathrm{ult}}$ over $w_{\mathrm{yp}}$ depends on shape factor $k$ and the *equalization* of the maximum moments.

(*b*) Because of symmetry, the precise location of the plastic hinges is as shown in Fig. 13-24(c). By writing a virtual work equation, Eq. 12-19, one has

$$w_{\mathrm{ult}} L \left( \frac{L\, \delta\theta}{2}\, \frac{1}{2} \right) = M_p (\delta\theta + 2\, \delta\theta + \delta\theta)$$

and $w_{\mathrm{ult}} = 16 M_p/L^2$ as before.

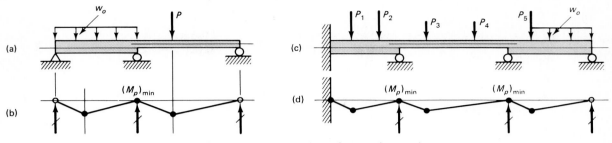

Fig. 13-25 Collapse mechanisms for continuous beams.

## *13-9. Continuous Beams and Frames

The procedures discussed in the preceding section, and illustrated by examples, can be extended to the simpler cases for plastic limit analysis of continuous beams and frames. Usually, the kinematic mechanisms in continuous beams, associated with a collapse mode, occur locally in only one beam. For the two-span continuous beam shown in Fig. 13-25(a), the plastic moment at the middle support is limited to $(M_p)_{min}$ of the smaller beam. Then, whether the kinematic mechanism would develop in the right or the left span depends on the relative beam sizes as well as the magnitudes of the applied loads. The solution in either case follows the procedure discussed in Example 13-11, 13-12, or 13-13.

The beams, restrained at both ends, usually develop the kinematic mechanisms shown in Fig. 13-25(d) for the two left spans in Fig. 13-25(c). The solution of such problems resembles that of Example 13-15, except that the end moments for each span are not necessarily equal. For example, for the left span of Fig. 13-25(c), the plastic moment on the left is determined by the large beam, whereas that on the right depends on the plastic moment of the center span beam.

Plastic limit analysis of frames may become rather complex as the number of members, joints, and different loading conditions increases. For analysis of such problems, the reader is referred to the previously cited texts. As a reasonably simple illustration of the plastic limit state frame analysis, an example follows.

### EXAMPLE 13-16

Consider a rigid jointed planar frame of ductile material fixed at $A$ and pinned at $E$, and loaded as shown in Fig. 13-26(a). All members are of the same size and can develop full $M_p$, i.e., the effect of the axial forces on $M_p$ can be neglected. Determine the plastic limit loads.

### Solution

The solution to this problem is obtained by assuming different kinematically admissible mechanisms and searching for the one that satisfies both equilibrium and plastic member capacity.

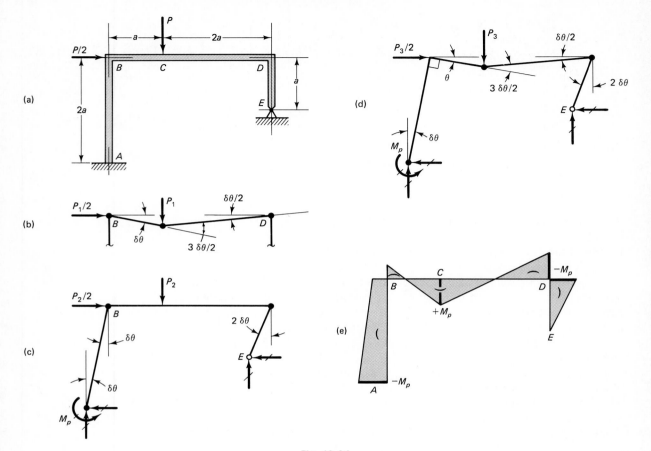

**Fig. 13-26**

A virtual work solution for an assumed mechanism provides an upper bound for the plastic limit loads. With these loads, a static analysis is then performed on members or parts of the frame separated at plastic hinges. Since the moment at each plastic hinge is $M_p$, a complete moment diagram can be constructed for the frame. The lower bound solution is obtained by reducing the upper-bound loads by a factor such that nowhere is $M_p$ exceeded.

The virtual work equation for the admissible kinematic (beam) mechanism in Fig. 13-26(b) is

$$P_1 a\ \delta\theta = M_p(\delta\theta + \delta\theta/2) \qquad \text{and} \qquad P_1 = 3M_p/2a$$

Applying $P_1$ and $P_2/2$ to the frame and separating it at joints $B$ and $D$, where the moments are $M_p$, the moment diagram for the frame is found (not shown). This solution, using the designer's sign convention, such as shown in Fig. 5-27(b), gives the following moments at critical points $A$, $B$, $C$, and $D$:

$$M_A = -2M_p \qquad M_B = -M_p \qquad M_C = M_p \qquad \text{and} \qquad M_D = -M_p$$

Since $M_A$ is twice as large as $M_p$, the upper bound solution found before must be reduced by a factor of 2 in order to obtain the lower bound solution. On this basis, for the assumed mechanism, the bounds for the solution are

$$\frac{3M_p}{2a} < P_1 < \frac{3M_p}{a}$$

By proceeding in the same manner using the (sway) mechanism for the frame shown in Fig. 13-26(c),

$$\frac{P_2}{2} \, 2a \, \delta\theta = M_p(\delta\theta + \delta\theta + 2\,\delta\theta) \qquad \text{and} \qquad P_2 = \frac{4M_p}{a}$$

For this upper bound solution, the moments at the critical points are

$$M_A = -M_p \qquad M_B = M_p \qquad M_C = 3M_p \qquad \text{and} \qquad M_D = -M_p$$

Since $M_C$ is three times greater than $M_p$, the upper bound solution for the assumed mechanism must be reduced by a factor of 3 to obtain the lower bound solution. Hence, for this case,

$$\frac{4M_p}{3a} < P_2 < \frac{4M_p}{a}$$

This solution is no better than the first. However, it is possible to *combine* the previous two mechanisms, such as to eliminate the plastic hinge at $B$, leading to better results. For a proper combination of these mechanisms, the internal plastic work in hinges can be reduced. Such a mechanism is shown in Fig. 13-26(d). The virtual work equation for this case is

$$\frac{P_3}{2} \, 2a \, \delta\theta + P_3 a \, \delta\theta = M_p \left( \delta\theta + \frac{3\,\delta\theta}{2} + \frac{\delta\theta}{2} + 2\,\delta\theta \right) \qquad \text{and} \qquad P_3 = \frac{5M_p}{2a}$$

The moment diagram corresponding to $P_3$ is shown in Fig. 13-26(e), where $M_A = -M_p$, $M_C = M_p$, and $M_D = -M_p$.

The last solution satisfies the three basic conditions of plastic limit analysis, consisting of the requirements of an admissible mechanism, equilibrium, and all moments being at most $M_p$. Therefore, this is an exact solution.

# PROBLEMS

## Sections 13-3 and 13-4

**13-1.** Show that for a linearly elastic simply supported beam, the angle of rotation $\theta_{ji}$ of the elastic curve at $j$ due to a couple acting at $i$, see the figure, is equal to the angle of rotation $\theta_{ij}$ at $i$ due to the same couple at $j$. (*Hint:* Use the results in Example 13-1, and determine $\theta_{ij}$ by the moment area or singularity functions.)

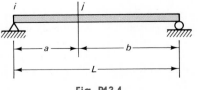

Fig. P13-1

**13-2.** For the planar elastic structure shown in the figure, (a) determine the reactions, and (b) draw the final moment diagram. Both members have the same constant $EI$. (*Hint:* Use the virtual-force method for finding deflections.)

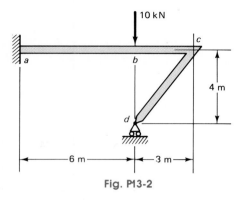

Fig. P13-2

**13-3.** A 10 × 12 in (actual size) rectangular wooden beam is braced by a 1-in round steel rod and an 8 × 8 in (actual size) wooden post, as shown in the figure. Determine the force that would develop in the post by applying a concentrated force $P = 10$ k at the center of the span. For wood, $E_w = 1500$ ksi and steel, $E_{st} = 30 \times 10^3$ ksi. For purposes of calculation, consider post $bd$ to be 5 ft long.

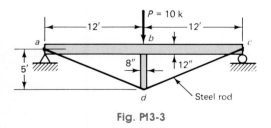

Fig. P13-3

**13-4.** Using the force method, rework Example 13-3. Consider the reactions at $b$ and $c$ as redundants. (*Hint:* Use the solution given in Problem 10-51.)

**13-5.** For a planar structure consisting of rod $ab$ and frame $bcde$, as shown in the figure, (a) determine the reactions, and (b) plot the bending moment diagram for the frame. All members are of the same material. $EI$ for the frame is constant and $AE$ for the rod is $IE/5$. Assume the force in the bar as redundant. Work in k-ft units.

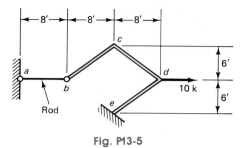

Fig. P13-5

**13-6.** Assuming elastic behavior and using the force method, rework Problem 2-64.

**13-7.** Using the force method, rework Problem 12-53. Consider the forces in $AB$ and $AD$ as redundants.

**13-8.** For the planar system of six elastic bars shown in the figure, determine the forces in the vertical bars due to applied force $P = 30$ kN. The bars are pinned at the ends. The cross-sectional area $A$ of each bar is 100 mm² and $E = 200$ GPa. (*Hint:* Take advantage of symmetry and use Eq. 2-34.)

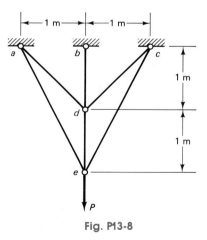

Fig. P13-8

**13-9.** The planar pin-ended bar system of Example 13-4 is augmented by adding member $fe$, as shown in the figure. (a) Assuming members $ae$, $ce$, and $fe$ as redundants, determine the numerical values for a 3 × 3 square matrix of the flexibility coefficients and set-up the corresponding column vectors as in Eq. 13-6. (b) If assigned, find the forces in all bar members.

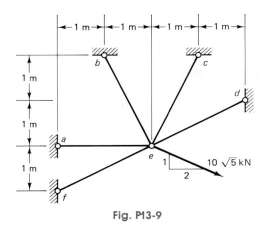

Fig. P13-9

**13-10.** Rework Example 13-4 after assuming that $L/AE$ = 1 for members $ae$ and $be$, and $L/AE$ = 2 for members $ce$ and $de$.

**13-11.** For the planar structure shown in the figure, determine the reactions at the support, and, if assigned, plot the moment diagram. Neglect axial and shear deformations, and assume $EI$ for the members is constant. (*Hint:* Use the virtual-force method for finding deflections.)

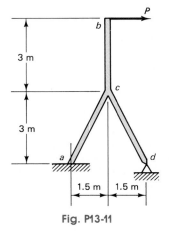

Fig. P13-11

## Sections 13-5 and 13-6

**13-12.** Rework Example 13-8 assuming that end $c$ is fixed.

**13-13.** Using the displacement method, determine the rotation of the elastic curve at $b$ and the moments at $a$, $b$, and $c$ for the continuous beam shown in the figure. $EI$ is constant.

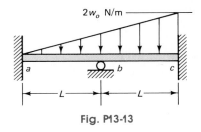

Fig. P13-13

**13-14.** Rework Problem 13-13 assuming that end $c$ is simply supported.

**13-15.** Determine the deflection and rotation at the end of the cantilever shown in the figure due to applied force $P$. Use the displacement method. (*Hint:* Both equations of equilibrium pertain to end $b$; one requires that $M_b$ = 0 and the other that $R_b$ = $-P$.)

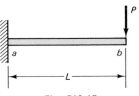

Fig. P13-15

**13-16.** Let the following conditions apply for the continuous beam $ac$ shown in Fig. 13-13(a): (a) Both spans are of equal length $L$, (b) span $ab$ is loaded with a uniformly distributed downward load $w_o$, (c) span $bc$ is loaded with a concentrated downward force $P$ in the middle of the span, and (d) $EI$ is constant. The boundary conditions remain as shown. Determine the rotation at $b$ and the vertical displacement at $c$ due to the applied loads. Calculate the moments at $a$, $b$, and $c$.

**13-17.** A propped cantilever of constant $EI$ is loaded with a concentrated force $P$ = 100 N, as shown in the figure. (a) Using the force method, determine the reaction at $b$. Then calculate the rotation of the elastic curve at $b$ and $c$ and the deflection at $c$. (*Hint:* For rotations and deflection, use the moment-area method.) (b) Using the displacement method, determine the rotations at $b$ and $c$, and the deflection at $c$.

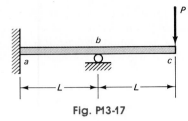

**Fig. P13-17**

Then calculate the moments at $a$ and $b$, and the reaction at $b$. (*Hint:* The three external forces applied at $b$ and $c$ are $M_b = 0$, $M_c = 0$, and $R_c = -100$ N.)

**13-18.** Rework Example 13-9 assuming that $L/AE = 1$ for members $ae$ and $be$ and $L/AE = 2$ for members $ce$ and $de$.

**13-19.** Using the displacement method, find the bar forces for the pin-ended bar system given in Problem 13-9.

## Section 13-8

**13-20.** Rework Example 13-12 after removing the concentrated force $P$ at $C$.

**13-21.** A ductile prismatic beam is simply supported at one end and fixed at the other, as shown in the figure. (a) Determine the position $x$ where the smallest concentrated force $P$ would cause a collapse mechanism. (b) Find the ultimate moments for the critical position of applied force $P$.

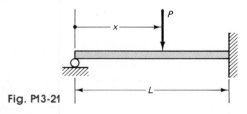

**Fig. P13-21**

**13-22.** A ductile prismatic beam is fixed at both ends. For a concentrated force $P$ placed at the third point of the span, as shown in the figure, determine the plastic limit load $P_{ult}$. Demonstrate that the result satisfies both the upper and lower bound criteria.

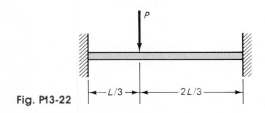

**Fig. P13-22**

**13-23.** A T beam fixed at both ends is loaded by a uniformly distributed load $w$ including its own weight. (a) What load $w_1$ can this beam carry when the stress in the middle just reaches yield and plastic moment point hinges develop at the built in ends? The yield strength of the material is 50 ksi. (b) What is the mid-span deflection due to $w_1$? Let $E = 30 \times 10^3$ ksi. (c) What is the plastic limit load $w_{ult}$?

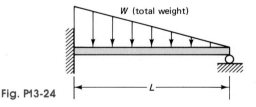

**Fig. P13-23**

**13-24.** A prismatic beam of ductile material, fixed at one end and simply supported at the other, carries a uniformly increasing load, as shown in the figure. Determine the plastic limit load $W_{ult}$ using the virtual-force method and assuming that one of the plastic hinges forms in the middle of the span. Check the result using the equilibrium method.

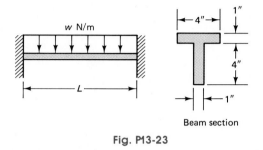

**Fig. P13-24**

**13-25.** A prismatic beam of ductile material is partially loaded, as shown in the figure. (a) Determine the upper and lower bound solutions by assuming a plastic hinge in the middle of the span. Let $M_p = 1000$ in-lb. (b) If assigned, refine the solution by placing the plastic hinge at the point of maximum positive moment found for the lower bound solution in part (a).

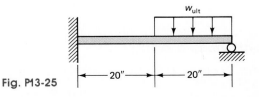

**Fig. P13-25**

## Section 13-9

**13-26.** Using limit analysis, calculate the value of $P$ that would cause flexural collapse of the two-span beam shown. The beam has a rectangular cross section 120 mm wide and 300 mm deep. The yield stress is 15 MPa. Neglect the weight of the beam.

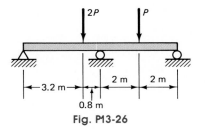

**Fig. P13-26**

**13-27.** Using limit analysis, select a steel W section for the loading condition shown in the figure. Let $\sigma_{yp}$ = 40 ksi, the shape factor be 1.10, and the load factor be 2. The beam size is the same throughout.

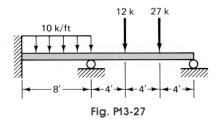

**Fig. P13-27**

**13-28.** Determine the ultimate plastic moment for the governing factored load for the prismatic continuous beam of ductile material shown in the figure.

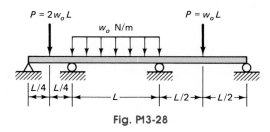

**Fig. P13-28**

**13-29.** For the structure shown in the figure, assume that at collapse, plastic hinges form at $A$, $B$, and $C$. Based on this assumption, establish the upper and lower bound on load $w_{ult}$. The plastic moment for beam $AC$ is 150 and that for column $DE$ is 50. Assume that all quantities are given in a consistent system of units.

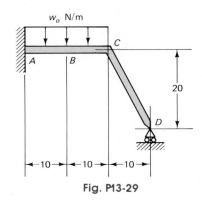

**Fig. P13-29**

**13-30.** Rework Example 13-16 assuming that the horizontal force at $B$ is $P$ and column $DE$ is fixed at $E$. The vertical force $P$ remains at $C$.

**13-31.** A portal frame pinned at $A$ and $F$ carries three concentrated forces $P$, each as shown in the figure. If $M_p$ of all members is the same throughout, obtain the collapse value of $P$. Substantiate your results by applying both the upper and lower bound theorems.

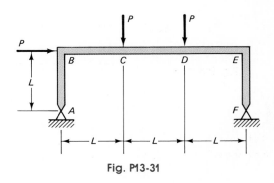

**Fig. P13-31**

# Appendix

**Tables**

*Acknowledgement:* Data for Tables 3 through 9 are taken from the *AISC Manual of Steel Construction* and are reproduced by permission of the American Institute of Steel Construction, Inc. The tables are compiled for use with this text. Original sources should be consulted for actual design.

**Table 1A. Typical Physical Properties of and Allowable Stresses for Some Common Materials[a]**
(In U.S. Customary System of Units)

| Material | Unit Weight, lb/in³ | Ultimate Strength, ksi Tens. | Comp.[c] | Shear | Yield Strength,[a] ksi Tens.[h] | Shear | Allow Stresses,[i] psi Tens. or Comp. | Shear | Elastic Moduli × 10⁶ psi Tens. or Comp. | Shear | Coef. of Thermal Expans. × 10⁻⁶ per °F |
|---|---|---|---|---|---|---|---|---|---|---|---|
| Aluminum alloy (extruded) 2024-T4 | 0.100 | 60 | — | 32 | 44 | 25 | | | 10.6 | 4.00 | 12.9 |
| 6061-T6 | | 38 | — | 24 | 35 | 20 | | | 10.0 | 3.75 | 13.0 |
| Cast iron Gray | 0.276 | 30 | 120 | —[e] | — | — | | | 13 | 6 | 5.8 |
| Malleable | | 54 | — | 48 | 36 | 24 | | | 25 | 12 | 6.7 |
| Concrete[b] 8 gal/sack | 0.087 | — | 3 | —[e] | — | — | −1,350[f] | 66 | 3 | — | 6.0 |
| 6 gal/sack | | — | 5 | — | — | — | −2,250[f] | 86 | 5 | — | |
| Magnesium alloy, AM100A | 0.065 | 40 | — | 21 | 22 | — | ±24,000 | 14,500 | 6.5 | 2.4 | 14.0 |
| Steel 0.2% Carbon (hot-rolled) | 0.283 | 65 | — | 48 | 36 | 24 | | | 30[k] | 12 | 6.5 |
| 0.6% Carbon (hot-rolled) | | 100 | — | 80 | 60 | 36 | | | | | |
| 0.6% Carbon (quenched) | | 120 | — | 100 | 75 | 45 | | | | | |
| 3½% Ni, 0.4% C | | 200 | — | 150 | 150 | 90 | | | | | |
| Wood Douglas fir (coast) | 0.018 | — | 7.4[d] | 1.1[f] | — | — | ±1,900[j] | 120[f] | 1.76 | — | — |
| Southern Pine (longleaf) | 0.021 | — | 8.4[d] | 1.5[f] | — | — | ±2,250[j] | 135[f] | 1.76 | — | — |

[a]Mechanical properties of metals depend not only on composition but also on heat treatment, previous cold working, etc. Data for wood are for clear 2 × 2 in specimens at 12-percent moisture content. True values vary.

[b]8 gal/sack means 8 gallons of water per 94-lb sack of Portland cement. Values are for 28-day-old concrete.

[c]For short blocks only. For ductile materials, the ultimate strength in compression is indefinite; may be assumed to be the same as that in tension.

[d]Compression parallel to grain on short blocks. Compression perpendicular to the grain at proportional limit 950 psi, 1190 psi, respectively. Values from *Wood Handbook*, U.S. Dept. of Agriculture. [e]Fails in diagonal tension. [f]Parallel to grain. [g]For most materials, at 0.2 percent set.

[h]For ductile materials, compressive yield strength may be assumed the same.

[i]Much lower stresses required in machine design because of fatigue properties and dynamic loadings. [j]For static loads only. No tensile stress is allowed in concrete. Timber stresses are for select or dense grade. [k]AISC recommends the value of 29 × 10⁶ psi.

**Table 1B. Typical Physical Properties of and Allowable Stresses for Some Common Materials[a]**
(In SI System of Units)

| Material | Unit Mass ×10³ kg/m³ | Ultimate Strength, MPa | | | Yield Strength,[g] MPa | | Allow Stresses,[i] MPa | | Elastic Moduli, GPa | | Coef. of Thermal Expans. ×10⁻⁶ per °C |
|---|---|---|---|---|---|---|---|---|---|---|---|
| | | Tens. | Comp.[c] | Shear | Tens.[h] | Shear | Tens. or Comp. | Shear | Tens. or Comp. | Shear | |
| Aluminum alloy (extruded) 2014-T6 | 2.77 | 414 | — | 220 | 300 | 170 | | | 73 | 27.6 | 23.2 |
| 6061-T6 | | 262 | — | 165 | 241 | 138 | | | 70 | 25.9 | 23.4 |
| Cast iron Gray | 7.64 | 210 | 825 | —[e] | — | — | | | 90 | 41 | 10.4 |
| Malleable | | 370 | — | 330 | 250 | 165 | | | 170 | 83 | 12.1 |
| Concrete[b] 0.70 water-cement ratio | 2.41 | — | 20 | —[e] | — | — | −9.31[j] | 0.455 | 20 | — | 10.8 |
| 0.53 water-cement ratio | | — | 35 | — | — | — | −15.5[j] | 0.592 | 35 | — | |
| Magnesium alloy, AM100A | 1.80 | 275 | — | 145 | 150 | — | | | 45 | 17 | 25.2 |
| Steel 0.2% Carbon (hot-rolled) | 7.83 | 450 | — | 330 | 250 | 165 | ±165 | | 200 | 83 | 11.7 |
| 0.6% Carbon (hot-rolled) | | 690 | — | 550 | 415 | 250 | | 100 | | | |
| 0.6% Carbon (quenched) | | 825 | — | 690 | 515 | 310 | | | | | |
| 3½% Ni; 0.4% C | | 1380 | — | 1035 | 1035 | 620 | | | | | |
| Wood Douglas Fir (coast) | 0.50 | — | 51[d] | 7[f] | — | — | ±13.1[k] | 0.825[f] | 12.1 | — | — |
| Southern Pine (longleaf) | 0.58 | — | 58[d] | 10[f] | — | — | ±15.5[k] | 0.930[f] | 12.1 | — | — |

[a]Mechanical properties of metals depend not only on composition but also on heat treatment, previous cold working, etc. Data for wood are for clear 50 × 50 mm specimens at 12-percent moisture content. True values vary. Where SI values are not yet available, a soft conversion of values currently accepted in industry was used in constructing this table.
[b]Water-cement ratio by weight for concrete with a 75 to 100 mm slump. Values are for 28-day-old concrete.
[c]For short blocks only. For ductile materials, the ultimate strength in compression is indefinite; may be assumed to be the same as that in tension.
[d]Compression parallel to grain on short blocks. Compression perpendicular to the grain at proportional limit 6.56 MPa, 8.20 MPa, respectively. Soft conversion of values from *Wood Handbook*, U.S. Dept. of Agriculture.   [e]Fails in diagonal tension.   [f]Parallel to grain   [g]For most materials, at 0.2 percent offset.   [h]For ductile materials, compressive yield strength may be assumed the same.
[i]For static loads only. Much lower stresses required in machine design because of fatigue properties and dynamic loadings.
[j]No tensile stress is allowed in concrete.   [k]In bending only. Timber stresses are for select and dense grade.

### Table 2. Useful Properties of Areas

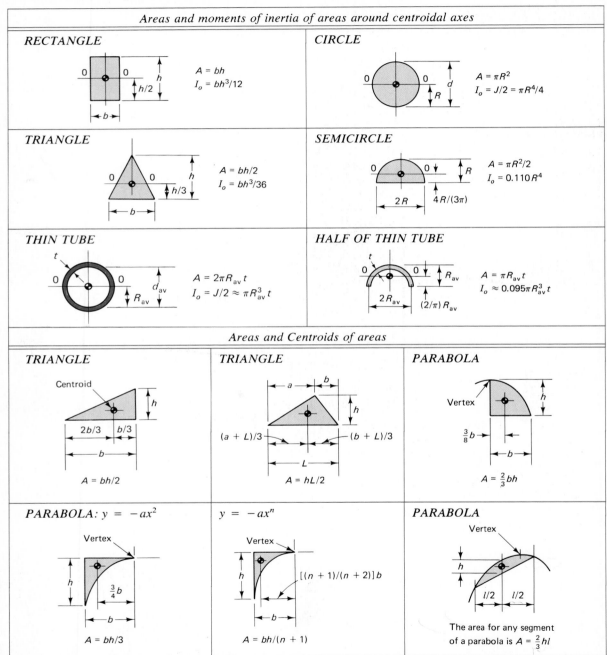

**Areas and moments of inertia of areas around centroidal axes**

**RECTANGLE**

$A = bh$
$I_o = bh^3/12$

**CIRCLE**

$A = \pi R^2$
$I_o = J/2 = \pi R^4/4$

**TRIANGLE**

$A = bh/2$
$I_o = bh^3/36$

**SEMICIRCLE**

$A = \pi R^2/2$
$I_o = 0.110 R^4$

$4R/(3\pi)$

**THIN TUBE**

$A = 2\pi R_{av} t$
$I_o = J/2 \approx \pi R_{av}^3 t$

**HALF OF THIN TUBE**

$A = \pi R_{av} t$
$I_o \approx 0.095\pi R_{av}^3 t$

$(2/\pi) R_{av}$

**Areas and Centroids of areas**

**TRIANGLE**

Centroid

$2b/3$ $b/3$

$A = bh/2$

**TRIANGLE**

$(a + L)/3$ $(b + L)/3$

$A = hL/2$

**PARABOLA**

Vertex

$\frac{3}{8}b$

$A = \frac{2}{3}bh$

**PARABOLA:** $y = -ax^2$

Vertex

$\frac{3}{4}b$

$A = bh/3$

$y = -ax^n$

Vertex

$[(n + 1)/(n + 2)]b$

$A = bh/(n + 1)$

**PARABOLA**

Vertex

$l/2$ $l/2$

The area for any segment
of a parabola is $A = \frac{2}{3}hl$

# Table 3. American Standard Steel Beams, S Shapes, Properties for Designing

| Designation* | Area | Depth | Flange Width | Flange Thickness | Web Thickness | Axis X-X I | Axis X-X S=I/c | Axis X-X r | Axis Y-Y I | Axis Y-Y S=I/c | Axis Y-Y r |
|---|---|---|---|---|---|---|---|---|---|---|---|
| | in² | in | in | in | in | in⁴ | in³ | in | in⁴ | in³ | in |
| S24 × 121 | 35.6 | 24.50 | 8.050 | 1.090 | 0.800 | 3160 | 258 | 9.43 | 83.3 | 20.7 | 1.53 |
| × 106 | 31.2 | 24.50 | 7.870 | 1.090 | 0.620 | 2840 | 240 | 9.71 | 77.7 | 19.6 | 1.57 |
| S24 × 100 | 29.3 | 24.00 | 7.245 | 0.870 | 0.745 | 2390 | 199 | 9.02 | 47.7 | 13.2 | 1.27 |
| × 90 | 26.5 | 24.00 | 7.125 | 0.870 | 0.625 | 2250 | 187 | 9.21 | 44.9 | 12.6 | 1.30 |
| × 80 | 23.5 | 24.00 | 7.000 | 0.870 | 0.500 | 2100 | 175 | 9.47 | 42.2 | 12.1 | 1.34 |
| S20 × 96 | 28.2 | 20.30 | 7.200 | 0.920 | 0.800 | 1670 | 165 | 7.71 | 50.2 | 13.9 | 1.33 |
| × 86 | 25.3 | 20.30 | 7.060 | 0.920 | 0.660 | 1580 | 155 | 7.89 | 46.8 | 13.3 | 1.36 |
| S20 × 75 | 22.0 | 20.00 | 6.385 | 0.795 | 0.635 | 1280 | 128 | 7.62 | 29.8 | 9.32 | 1.16 |
| × 66 | 19.4 | 20.00 | 6.255 | 0.795 | 0.505 | 1190 | 119 | 7.83 | 27.7 | 8.85 | 1.19 |
| S18 × 70 | 20.6 | 18.00 | 6.251 | 0.691 | 0.711 | 926 | 103 | 6.71 | 24.1 | 7.72 | 1.08 |
| × 54.7 | 16.1 | 18.00 | 6.001 | 0.691 | 0.461 | 804 | 89.4 | 7.07 | 20.8 | 6.94 | 1.14 |
| S15 × 50 | 14.7 | 15.00 | 5.640 | 0.622 | 0.550 | 486 | 64.8 | 5.75 | 15.7 | 5.57 | 1.03 |
| × 42.9 | 12.6 | 15.00 | 5.501 | 0.622 | 0.411 | 447 | 59.6 | 5.95 | 14.4 | 5.23 | 1.07 |
| S12 × 50 | 14.7 | 12.00 | 5.477 | 0.659 | 0.687 | 305 | 50.8 | 4.55 | 15.7 | 5.74 | 1.03 |
| × 40.8 | 12.0 | 12.00 | 5.252 | 0.659 | 0.462 | 272 | 45.4 | 4.77 | 13.6 | 5.16 | 1.06 |
| S12 × 35 | 10.3 | 12.00 | 5.078 | 0.544 | 0.428 | 229 | 38.2 | 4.72 | 9.87 | 3.89 | 0.980 |
| × 31.8 | 9.35 | 12.00 | 5.000 | 0.544 | 0.350 | 218 | 36.4 | 4.83 | 9.36 | 3.74 | 1.00 |
| S10 × 35 | 10.3 | 10.00 | 4.944 | 0.491 | 0.594 | 147 | 29.4 | 3.78 | 8.36 | 3.38 | 0.901 |
| × 25.4 | 7.46 | 10.00 | 4.661 | 0.491 | 0.311 | 124 | 24.7 | 4.07 | 6.79 | 2.91 | 0.954 |
| S 8 × 23 | 6.77 | 8.00 | 4.171 | 0.426 | 0.441 | 64.9 | 16.2 | 3.10 | 4.31 | 2.07 | 0.798 |
| × 18.4 | 5.41 | 8.00 | 4.001 | 0.426 | 0.271 | 57.6 | 14.4 | 3.26 | 3.73 | 1.86 | 0.831 |
| S 7 × 20 | 5.88 | 7.00 | 3.860 | 0.392 | 0.450 | 42.4 | 12.1 | 2.69 | 3.17 | 1.64 | 0.734 |
| × 15.3 | 4.50 | 7.00 | 3.662 | 0.392 | 0.252 | 36.7 | 10.5 | 2.86 | 2.64 | 1.44 | 0.766 |
| S 6 × 17.25 | 5.07 | 6.00 | 3.565 | 0.359 | 0.465 | 26.3 | 8.77 | 2.28 | 2.31 | 1.30 | 0.675 |
| × 12.5 | 3.67 | 6.00 | 3.332 | 0.359 | 0.232 | 22.1 | 7.37 | 2.45 | 1.82 | 1.09 | 0.705 |
| S 5 × 14.75 | 4.34 | 5.00 | 3.284 | 0.326 | 0.494 | 15.2 | 6.09 | 1.87 | 1.67 | 1.01 | 0.620 |
| × 10 | 2.94 | 5.00 | 3.004 | 0.326 | 0.214 | 12.3 | 4.92 | 2.05 | 1.22 | 0.809 | 0.643 |
| S 4 × 9.5 | 2.79 | 4.00 | 2.796 | 0.293 | 0.326 | 6.79 | 3.39 | 1.56 | 0.903 | 0.646 | 0.569 |
| × 7.7 | 2.26 | 4.00 | 2.663 | 0.293 | 0.193 | 6.08 | 3.04 | 1.64 | 0.764 | 0.574 | 0.581 |
| S 3 × 7.5 | 2.21 | 3.00 | 2.509 | 0.260 | 0.349 | 2.93 | 1.95 | 1.15 | 0.586 | 0.468 | 0.516 |
| × 5.7 | 1.67 | 3.00 | 2.330 | 0.260 | 0.170 | 2.52 | 1.68 | 1.23 | 0.455 | 0.390 | 0.522 |

*American Standard I-shaped beams are referred to as S shapes, and are designed by the letter S followed by their depth in inches, with their weight in pounds per linear foot given last. For example, S 24 × 100 means that this S shape is 24 in deep and weighs 100 lb/ft.

**A-5**

# Table 4. American Wide-Flange Steel Beams, W Shapes, Properties for Designing

| Designation* | Area | Depth | Flange Width | Flange Thickness | Web Thickness | Axis X-X $I$ | Axis X-X $S = I/c$ | Axis X-X $r$ | Axis Y-Y $I$ | Axis Y-Y $S = I/c$ | Axis Y-Y $r$ |
|---|---|---|---|---|---|---|---|---|---|---|---|
| | $in^2$ | $in$ | $in$ | $in$ | $in$ | $in^4$ | $in^3$ | $in$ | $in^4$ | $in^3$ | $in$ |
| W36 × 230 | 67.6 | 35.90 | 16.470 | 1.260 | 0.760 | 15000 | 837 | 14.9 | 940 | 114 | 3.73 |
| × 150 | 44.2 | 35.85 | 11.975 | 0.940 | 0.625 | 9040 | 504 | 14.3 | 270 | 45.1 | 2.47 |
| W33 × 201 | 59.1 | 33.68 | 15.745 | 1.150 | 0.715 | 11500 | 684 | 14.0 | 749 | 95.2 | 3.56 |
| × 130 | 38.3 | 33.10 | 11.510 | 0.855 | 0.580 | 6710 | 406 | 13.2 | 218 | 37.9 | 2.38 |
| W30 × 173 | 50.8 | 30.44 | 14.985 | 1.065 | 0.655 | 8200 | 539 | 12.7 | 598 | 79.8 | 3.43 |
| × 108 | 31.8 | 29.82 | 10.484 | 0.760 | 0.548 | 4470 | 300 | 11.9 | 146 | 27.9 | 2.15 |
| W27 × 146 | 42.9 | 27.78 | 13.965 | 0.975 | 0.605 | 5630 | 411 | 11.4 | 443 | 63.5 | 3.21 |
| × 94 | 27.7 | 26.92 | 9.990 | 0.745 | 0.490 | 3270 | 243 | 10.9 | 124 | 24.8 | 2.12 |
| W24 × 131 | 38.5 | 24.48 | 12.855 | 0.960 | 0.605 | 4020 | 329 | 10.2 | 340 | 53.0 | 2.97 |
| × 76 | 22.4 | 23.92 | 8.990 | 0.680 | 0.440 | 2100 | 176 | 9.69 | 82.5 | 18.4 | 1.92 |
| W21 × 111 | 32.7 | 21.51 | 12.340 | 0.875 | 0.550 | 2670 | 249 | 9.05 | 274 | 44.5 | 2.90 |
| × 62 | 18.3 | 20.99 | 8.240 | 0.615 | 0.400 | 1330 | 127 | 8.54 | 57.5 | 13.9 | 1.77 |
| W18 × 97 | 28.5 | 18.59 | 11.145 | 0.870 | 0.535 | 750 | 188 | 7.82 | 201 | 36.1 | 2.65 |
| × 50 | 14.7 | 17.99 | 7.495 | 0.570 | 0.355 | 800 | 88.9 | 7.38 | 40.1 | 10.7 | 1.65 |
| × 35 | 10.3 | 17.70 | 6.000 | 0.425 | 0.300 | 510 | 57.6 | 7.04 | 15.3 | 5.12 | 1.22 |
| W16 × 100 | 29.4 | 16.97 | 11.425 | 0.985 | 0.585 | 1490 | 175 | 7.10 | 186 | 35.7 | 2.51 |
| × 50 | 14.7 | 16.25 | 7.070 | 0.630 | 0.380 | 659 | 81.0 | 6.68 | 37.2 | 10.5 | 1.59 |
| × 36 | 10.6 | 15.86 | 6.985 | 0.430 | 0.295 | 448 | 56.5 | 6.51 | 24.5 | 7.00 | 1.52 |
| × 26 | 7.68 | 15.69 | 5.500 | 0.345 | 0.250 | 301 | 38.4 | 6.26 | 9.59 | 3.49 | 1.12 |
| W14 × 730 | 215.0 | 22.42 | 17.890 | 4.910 | 3.070 | 14300 | 1280 | 8.17 | 4720 | 527 | 4.69 |
| × 455 | 134.0 | 19.02 | 16.835 | 3.210 | 2.015 | 7190 | 756 | 7.33 | 2560 | 304 | 4.38 |
| × 311 | 91.4 | 17.12 | 16.230 | 2.260 | 1.410 | 4330 | 506 | 6.88 | 1610 | 199 | 4.20 |
| × 193 | 56.8 | 15.48 | 15.710 | 1.440 | 0.890 | 2400 | 310 | 6.50 | 931 | 119 | 4.05 |
| × 159 | 46.7 | 14.98 | 15.565 | 1.190 | 0.745 | 1900 | 254 | 6.38 | 748 | 96.2 | 4.00 |
| × 90 | 26.5 | 14.02 | 14.520 | 0.710 | 0.440 | 999 | 143 | 6.14 | 362 | 49.0 | 3.70 |

| Designation | | | | | | | | | | | |
|---|---|---|---|---|---|---|---|---|---|---|---|
| W14 × 74 | 21.8 | 14.17 | 10.070 | 0.785 | 0.450 | 796 | 112 | 6.04 | 134 | 26.6 | 2.48 |
| × 68 | 20.0 | 14.04 | 10.035 | 0.720 | 0.415 | 723 | 103 | 6.01 | 121 | 24.2 | 2.46 |
| × 61 | 17.9 | 13.89 | 9.995 | 0.645 | 0.375 | 640 | 92.2 | 5.98 | 107 | 21.5 | 2.45 |
| × 53 | 15.6 | 13.92 | 8.060 | 0.658 | 0.370 | 542 | 77.8 | 5.90 | 57.5 | 14.3 | 1.92 |
| × 43 | 12.6 | 13.66 | 7.995 | 0.530 | 0.305 | 428 | 62.7 | 5.82 | 45.2 | 11.3 | 1.89 |
| W14 × 38 | 11.2 | 14.10 | 6.770 | 0.515 | 0.310 | 385 | 54.6 | 5.87 | 26.7 | 7.86 | 1.54 |
| × 34 | 10.0 | 13.98 | 6.745 | 0.455 | 0.285 | 340 | 48.6 | 5.83 | 23.3 | 6.91 | 1.53 |
| × 30 | 8.85 | 13.84 | 6.730 | 0.385 | 0.270 | 291 | 42.0 | 5.73 | 19.6 | 5.82 | 1.49 |
| W12 × 87 | 25.6 | 12.53 | 12.125 | 0.810 | 0.515 | 740 | 118 | 5.38 | 241 | 39.7 | 3.07 |
| × 65 | 19.1 | 12.12 | 12.000 | 0.605 | 0.390 | 533 | 87.9 | 5.28 | 174 | 29.1 | 3.02 |
| × 53 | 15.6 | 12.06 | 9.995 | 0.575 | 0.345 | 425 | 70.6 | 5.23 | 95.8 | 19.2 | 2.48 |
| × 40 | 11.8 | 11.94 | 8.005 | 0.515 | 0.295 | 310 | 51.9 | 5.13 | 44.1 | 11.0 | 1.93 |
| W12 × 35 | 10.3 | 12.50 | 6.560 | 0.520 | 0.300 | 285 | 45.6 | 5.25 | 24.5 | 7.47 | 1.54 |
| × 30 | 8.79 | 12.34 | 6.520 | 0.440 | 0.260 | 238 | 38.6 | 5.21 | 20.3 | 6.24 | 1.52 |
| × 26 | 7.65 | 12.22 | 6.490 | 0.380 | 0.230 | 204 | 33.4 | 5.17 | 17.3 | 5.34 | 1.51 |
| W10 × 112 | 32.9 | 11.36 | 10.415 | 1.250 | 0.755 | 716 | 126 | 4.66 | 238 | 45.3 | 2.68 |
| × 100 | 29.4 | 11.10 | 10.340 | 1.120 | 0.680 | 623 | 112 | 4.60 | 207 | 40.0 | 2.65 |
| × 88 | 25.9 | 10.84 | 10.265 | 0.990 | 0.605 | 534 | 98.5 | 4.54 | 179 | 34.8 | 2.63 |
| × 77 | 22.6 | 10.60 | 10.190 | 0.870 | 0.530 | 455 | 85.9 | 4.49 | 154 | 30.1 | 2.60 |
| × 60 | 17.6 | 10.22 | 10.080 | 0.680 | 0.370 | 341 | 66.7 | 4.39 | 116 | 23.0 | 2.57 |
| × 49 | 14.4 | 9.98 | 10.000 | 0.560 | 0.340 | 272 | 54.6 | 4.35 | 93.4 | 18.7 | 2.54 |
| W10 × 45 | 13.3 | 10.10 | 8.020 | 0.620 | 0.350 | 248 | 49.1 | 4.32 | 53.4 | 13.3 | 2.01 |
| × 39 | 11.5 | 9.92 | 7.985 | 0.530 | 0.315 | 209 | 42.1 | 4.27 | 45.0 | 11.3 | 1.98 |
| × 33 | 9.71 | 9.73 | 7.960 | 0.435 | 0.290 | 170 | 35.0 | 4.19 | 36.6 | 9.20 | 1.94 |
| W10 × 30 | 8.84 | 10.47 | 5.810 | 0.510 | 0.300 | 170 | 32.4 | 4.38 | 16.7 | 5.75 | 1.37 |
| × 22 | 6.49 | 10.17 | 5.750 | 0.360 | 0.240 | 118 | 23.2 | 4.27 | 11.4 | 3.97 | 1.33 |
| W 8 × 67 | 19.7 | 9.00 | 8.280 | 0.935 | 0.570 | 272 | 60.4 | 3.72 | 88.6 | 21.2 | 2.12 |
| × 58 | 17.1 | 8.75 | 8.220 | 0.810 | 0.510 | 228 | 52.0 | 3.65 | 75.1 | 18.3 | 2.10 |
| × 48 | 14.1 | 8.50 | 8.110 | 0.685 | 0.400 | 184 | 43.3 | 3.61 | 60.9 | 15.0 | 2.08 |
| × 40 | 11.7 | 8.25 | 8.070 | 0.560 | 0.360 | 146 | 35.5 | 3.53 | 49.1 | 12.2 | 2.04 |
| × 35 | 10.3 | 8.12 | 8.020 | 0.495 | 0.310 | 127 | 31.2 | 3.51 | 42.6 | 10.6 | 2.03 |
| × 31 | 9.13 | 8.00 | 8.000 | 0.435 | 0.285 | 110 | 27.5 | 3.47 | 37.1 | 9.27 | 2.02 |
| W 8 × 28 | 8.25 | 8.06 | 6.535 | 0.465 | 0.285 | 98.0 | 24.3 | 3.45 | 21.7 | 6.63 | 1.62 |
| × 24 | 7.08 | 7.93 | 6.405 | 0.400 | 0.245 | 82.8 | 20.9 | 3.42 | 18.3 | 5.63 | 1.61 |
| W 8 × 21 | 6.16 | 8.28 | 5.270 | 0.400 | 0.250 | 75.3 | 18.2 | 3.49 | 9.77 | 3.71 | 1.26 |
| × 18 | 5.26 | 8.14 | 5.250 | 0.330 | 0.230 | 61.9 | 15.2 | 3.43 | 7.97 | 3.04 | 1.23 |

*American wide-flange I- or H-shaped steel beams are referred to as W shapes, and are designed by the letter W followed by their *nominal* depth in inches, with their weight in pounds per linear foot given last. For example, W21 × 62 means that this W shape is 21 in deep and weighs 62 lb/ft. This list is abridged.

# Table 5. American Standard Steel Channels, Properties for Designing

| Designation* | Area | Depth | Flange Width | Flange Average Thickness | Web Thickness | Axis X-X $I$ | Axis X-X $S = I/c$ | Axis X-X $r$ | Axis Y-Y $I$ | Axis Y-Y $S = I/c$ | Axis Y-Y $r$ | Axis Y-Y $x$ |
|---|---|---|---|---|---|---|---|---|---|---|---|---|
| | $in^2$ | $in$ | $in$ | $in$ | $in$ | $in^4$ | $in^3$ | $in$ | $in^4$ | $in^3$ | $in$ | $in$ |
| C15 × 50 | 14.7 | 15.00 | 3.716 | 0.650 | 0.716 | 404 | 53.8 | 5.24 | 11.0 | 3.78 | 0.867 | 0.799 |
| × 40 | 11.8 | 15.00 | 3.520 | 0.650 | 0.520 | 349 | 46.5 | 5.44 | 9.23 | 3.36 | 0.886 | 0.778 |
| × 33.9 | 9.96 | 15.00 | 3.400 | 0.650 | 0.400 | 315 | 42.0 | 5.62 | 8.13 | 3.11 | 0.904 | 0.787 |
| C12 × 30 | 8.82 | 12.00 | 3.170 | 0.501 | 0.510 | 162 | 27.0 | 4.29 | 5.14 | 2.06 | 0.763 | 0.674 |
| × 25 | 7.35 | 12.00 | 3.047 | 0.501 | 0.387 | 144 | 24.1 | 4.43 | 4.47 | 1.88 | 0.780 | 0.674 |
| × 20.7 | 6.09 | 12.00 | 2.942 | 0.501 | 0.282 | 129 | 21.5 | 4.61 | 3.88 | 1.73 | 0.799 | 0.698 |
| C10 × 30 | 8.82 | 10.00 | 3.033 | 0.436 | 0.673 | 103 | 20.7 | 3.42 | 3.94 | 1.65 | 0.669 | 0.649 |
| × 25 | 7.35 | 10.00 | 2.886 | 0.436 | 0.526 | 91.2 | 18.2 | 3.52 | 3.36 | 1.48 | 0.676 | 0.617 |
| × 20 | 5.88 | 10.00 | 2.739 | 0.436 | 0.379 | 78.9 | 15.8 | 3.66 | 2.81 | 1.32 | 0.691 | 0.606 |
| × 15.3 | 4.49 | 10.00 | 2.600 | 0.436 | 0.240 | 67.4 | 13.5 | 3.87 | 2.28 | 1.16 | 0.713 | 0.634 |
| C 9 × 20 | 5.88 | 9.00 | 2.648 | 0.413 | 0.448 | 60.9 | 13.5 | 3.22 | 2.42 | 1.17 | 0.642 | 0.583 |
| × 15 | 4.41 | 9.00 | 2.485 | 0.413 | 0.285 | 51.0 | 11.3 | 3.40 | 1.93 | 1.01 | 0.661 | 0.586 |
| × 13.4 | 3.94 | 9.00 | 2.433 | 0.413 | 0.233 | 47.9 | 10.6 | 3.48 | 1.76 | 0.962 | 0.668 | 0.601 |
| C 8 × 18.75 | 5.51 | 8.00 | 2.527 | 0.390 | 0.487 | 44.0 | 11.0 | 2.82 | 1.98 | 1.01 | 0.599 | 0.565 |
| × 13.75 | 4.04 | 8.00 | 2.343 | 0.390 | 0.303 | 36.1 | 9.03 | 2.99 | 1.53 | 0.853 | 0.615 | 0.553 |
| × 11.5 | 3.38 | 8.00 | 2.260 | 0.390 | 0.220 | 32.6 | 8.14 | 3.11 | 1.32 | 0.781 | 0.625 | 0.571 |
| C 7 × 14.75 | 4.33 | 7.00 | 2.299 | 0.366 | 0.419 | 27.2 | 7.78 | 2.51 | 1.38 | 0.779 | 0.564 | 0.532 |
| × 12.25 | 3.60 | 7.00 | 2.194 | 0.366 | 0.314 | 24.2 | 6.93 | 2.60 | 1.17 | 0.702 | 0.571 | 0.525 |
| × 9.8 | 2.87 | 7.00 | 2.090 | 0.366 | 0.210 | 21.3 | 6.08 | 2.72 | 0.968 | 0.625 | 0.581 | 0.541 |
| C 6 × 13 | 3.83 | 6.00 | 2.157 | 0.343 | 0.437 | 17.4 | 5.80 | 2.13 | 1.05 | 0.642 | 0.525 | 0.514 |
| × 10.5 | 3.09 | 6.00 | 2.034 | 0.343 | 0.314 | 15.2 | 5.06 | 2.22 | 0.865 | 0.564 | 0.529 | 0.500 |
| × 8.2 | 2.40 | 6.00 | 1.920 | 0.343 | 0.200 | 13.1 | 4.38 | 2.34 | 0.692 | 0.492 | 0.537 | 0.512 |
| C 5 × 9 | 2.64 | 5.00 | 1.885 | 0.320 | 0.325 | 8.90 | 3.56 | 1.83 | 0.632 | 0.449 | 0.489 | 0.478 |
| × 6.7 | 1.97 | 5.00 | 1.750 | 0.320 | 0.190 | 7.49 | 3.00 | 1.95 | 0.478 | 0.378 | 0.493 | 0.484 |
| C 4 × 7.25 | 2.13 | 4.00 | 1.721 | 0.296 | 0.321 | 4.59 | 2.29 | 1.47 | 0.432 | 0.343 | 0.450 | 0.459 |
| × 5.4 | 1.59 | 4.00 | 1.584 | 0.296 | 0.184 | 3.85 | 1.93 | 1.56 | 0.319 | 0.283 | 0.449 | 0.458 |
| C 3 × 6 | 1.76 | 3.00 | 1.596 | 0.273 | 0.356 | 2.07 | 1.38 | 1.08 | 0.305 | 0.268 | 0.416 | 0.455 |
| × 5 | 1.47 | 3.00 | 1.498 | 0.273 | 0.258 | 1.85 | 1.24 | 1.12 | 0.247 | 0.233 | 0.410 | 0.438 |
| × 4.1 | 1.21 | 3.00 | 1.410 | 0.273 | 0.170 | 1.66 | 1.10 | 1.17 | 0.197 | 0.202 | 0.404 | 0.437 |

*American Standard Steel Channels are designated by the letter C followed by their depth in inches, with their weight per linear foot given last. For example, C10 × 15.3 means that this channel is 10 in deep and weighs 15.3 lb/ft.

## Table 6. Steel Angles with Equal Legs, Properties for Designing

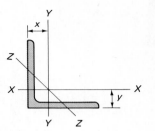

| Size and Thickness | Weight per Foot | Area | Axis X-X and Axis Y-Y | | | | Axis Z-Z |
|---|---|---|---|---|---|---|---|
| | | | $I$ | $S = I/c$ | $r$ | x or y | $r$ |
| in | lb | in² | in⁴ | in³ | in | in | in |
| L 8 × 8 × 1⅛ | 56.9 | 16.7 | 98.0 | 17.5 | 2.42 | 2.41 | 1.56 |
| 1 | 51.0 | 15.0 | 89.0 | 15.8 | 2.44 | 2.37 | 1.56 |
| ⅞ | 45.0 | 13.2 | 79.6 | 14.0 | 2.45 | 2.32 | 1.57 |
| ¾ | 38.9 | 11.4 | 69.7 | 12.2 | 2.47 | 2.28 | 1.58 |
| ⅝ | 32.7 | 9.61 | 59.4 | 10.3 | 2.49 | 2.23 | 1.58 |
| ½ | 26.4 | 7.75 | 48.6 | 8.36 | 2.50 | 2.19 | 1.59 |
| L 6 × 6 × 1 | 37.4 | 11.0 | 35.5 | 8.57 | 1.80 | 1.86 | 1.17 |
| ⅞ | 33.1 | 9.73 | 31.9 | 7.63 | 1.81 | 1.82 | 1.17 |
| ¾ | 28.7 | 8.44 | 28.2 | 6.66 | 1.83 | 1.78 | 1.17 |
| ⅝ | 24.2 | 7.11 | 24.2 | 5.66 | 1.84 | 1.73 | 1.18 |
| ½ | 19.6 | 5.75 | 19.9 | 4.61 | 1.86 | 1.68 | 1.18 |
| ⅜ | 14.9 | 4.36 | 15.4 | 3.53 | 1.88 | 1.64 | 1.19 |
| L 5 × 5 × ⅞ | 27.2 | 7.98 | 17.8 | 5.17 | 1.49 | 1.57 | 0.973 |
| ¾ | 23.6 | 6.94 | 15.7 | 4.53 | 1.51 | 1.52 | 0.975 |
| ½ | 16.2 | 4.75 | 11.3 | 3.16 | 1.54 | 1.43 | 0.983 |
| ⅜ | 12.3 | 3.61 | 8.74 | 2.42 | 1.56 | 1.39 | 0.990 |
| ⁵⁄₁₆ | 10.3 | 3.03 | 7.42 | 2.04 | 1.57 | 1.37 | 0.994 |
| L 4 × 4 × ¾ | 18.5 | 5.44 | 7.67 | 2.81 | 1.19 | 1.27 | 0.778 |
| ⅝ | 15.7 | 4.61 | 6.66 | 2.40 | 1.20 | 1.23 | 0.779 |
| ½ | 12.8 | 3.75 | 5.56 | 1.97 | 1.22 | 1.18 | 0.782 |
| ⅜ | 9.8 | 2.86 | 4.36 | 1.52 | 1.23 | 1.14 | 0.788 |
| ⁵⁄₁₆ | 8.2 | 2.40 | 3.71 | 1.29 | 1.24 | 1.12 | 0.791 |
| ¼ | 6.6 | 1.94 | 3.04 | 1.05 | 1.25 | 1.09 | 0.795 |
| L 3½ × 3½ × ½ | 11.1 | 3.25 | 3.64 | 1.49 | 1.06 | 1.06 | 0.683 |
| ⅜ | 8.5 | 2.48 | 2.87 | 1.15 | 1.07 | 1.01 | 0.687 |
| ⁵⁄₁₆ | 7.2 | 2.09 | 2.45 | 0.976 | 1.08 | 0.990 | 0.690 |
| ¼ | 5.8 | 1.69 | 2.01 | 0.794 | 1.09 | 0.968 | 0.694 |
| L 3 × 3 × ½ | 9.4 | 2.75 | 2.22 | 1.07 | 0.898 | 0.932 | 0.584 |
| ⁷⁄₁₆ | 8.3 | 2.43 | 1.99 | 0.954 | 0.905 | 0.910 | 0.585 |
| ⅜ | 7.2 | 2.11 | 1.76 | 0.833 | 0.913 | 0.888 | 0.587 |
| ⁵⁄₁₆ | 6.1 | 1.78 | 1.51 | 0.707 | 0.922 | 0.869 | 0.589 |
| ¼ | 4.9 | 1.44 | 1.24 | 0.577 | 0.930 | 0.842 | 0.592 |
| ³⁄₁₆ | 3.71 | 1.09 | 0.962 | 0.441 | 0.939 | 0.820 | 0.596 |
| L 2½ × 2½ × ⅜ | 5.9 | 1.73 | 0.984 | 0.566 | 0.753 | 0.762 | 0.487 |
| ⁵⁄₁₆ | 5.0 | 1.46 | 0.849 | 0.482 | 0.761 | 0.740 | 0.489 |
| ¼ | 4.1 | 1.91 | 0.703 | 0.394 | 0.769 | 0.717 | 0.491 |
| ³⁄₁₆ | 3.07 | 0.92 | 0.547 | 0.303 | 0.778 | 0.694 | 0.495 |

Table 7. Steel Angles with Unequal Legs, Properties for Designing

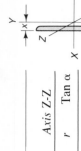

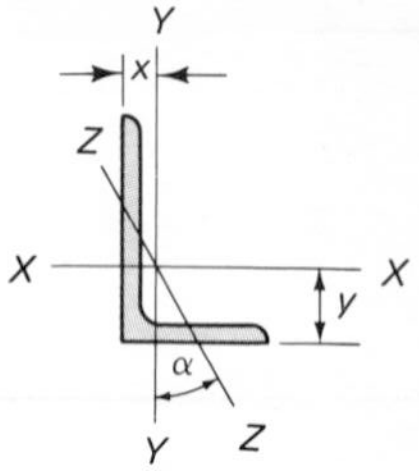

| Size and Thickness* | Weight per Foot | Area | Axis X-X | | | | Axis Y-Y | | | | Axis Z-Z | |
|---|---|---|---|---|---|---|---|---|---|---|---|---|
| | | | $I$ | $S=I/c$ | $r$ | $y$ | $I$ | $S=I/c$ | $r$ | $x$ | $r$ | Tan $\alpha$ |
| in | lb | in² | in⁴ | in³ | in | in | in⁴ | in³ | in | in | in | in |
| L 8 × 6 × 1 | 44.2 | 13.0 | 80.8 | 15.1 | 2.49 | 2.65 | 38.8 | 8.92 | 1.73 | 1.65 | 1.28 | 0.543 |
| ¾ | 33.8 | 9.94 | 63.4 | 11.7 | 2.53 | 2.56 | 30.7 | 6.92 | 1.76 | 1.56 | 1.29 | 0.551 |
| ½ | 23.0 | 6.75 | 44.3 | 8.02 | 2.56 | 2.47 | 21.7 | 4.79 | 1.79 | 1.47 | 1.30 | 0.558 |
| L 8 × 4 × 1 | 37.4 | 11.0 | 69.6 | 14.1 | 2.52 | 3.05 | 11.6 | 3.94 | 1.03 | 1.05 | 0.846 | 0.247 |
| ¾ | 28.7 | 8.44 | 54.9 | 10.9 | 2.55 | 2.95 | 9.36 | 3.07 | 1.05 | 0.953 | 0.852 | 0.258 |
| ½ | 19.6 | 5.75 | 38.5 | 7.49 | 2.59 | 2.86 | 6.74 | 2.15 | 1.08 | 0.859 | 0.865 | 0.267 |
| L 6 × 4 × ¾ | 23.6 | 6.94 | 24.5 | 6.25 | 1.88 | 2.08 | 8.68 | 2.97 | 1.12 | 1.08 | 0.860 | 0.428 |
| ½ | 16.2 | 4.75 | 17.4 | 4.33 | 1.91 | 1.99 | 6.27 | 2.08 | 1.15 | 0.987 | 0.870 | 0.440 |
| L 5 × 3 × ½ | 12.8 | 3.75 | 9.45 | 2.91 | 1.59 | 1.75 | 2.58 | 1.15 | 0.829 | 0.750 | 0.648 | 0.357 |
| ⅜ | 9.8 | 2.86 | 7.37 | 2.24 | 1.61 | 1.70 | 2.04 | 0.888 | 0.845 | 0.704 | 0.654 | 0.364 |
| ¼ | 6.6 | 1.94 | 5.11 | 1.53 | 1.62 | 1.66 | 1.44 | 0.614 | 0.861 | 0.657 | 0.663 | 0.371 |
| L 4 × 3½ × ½ | 11.9 | 3.50 | 5.32 | 1.94 | 1.23 | 1.25 | 3.79 | 1.52 | 1.04 | 1.00 | 0.722 | 0.750 |
| ⅜ | 9.1 | 2.67 | 4.18 | 1.49 | 1.25 | 1.21 | 2.95 | 1.17 | 1.06 | 0.955 | 0.727 | 0.755 |
| ¼ | 6.2 | 1.81 | 2.91 | 1.03 | 1.27 | 1.16 | 2.09 | 0.808 | 1.07 | 0.909 | 0.734 | 0.759 |
| L 4 × 3 × ½ | 11.1 | 3.25 | 5.05 | 1.89 | 1.25 | 1.33 | 2.42 | 1.12 | 0.827 | 0.864 | 0.639 | 0.543 |
| ⅜ | 8.5 | 2.48 | 3.96 | 1.46 | 1.26 | 1.28 | 1.92 | 0.866 | 0.879 | 0.782 | 0.644 | 0.551 |
| ¼ | 5.8 | 1.69 | 2.77 | 1.00 | 1.28 | 1.24 | 1.36 | 0.599 | 0.896 | 0.736 | 0.651 | 0.558 |
| L 3½ × 2½ × ½ | 9.4 | 2.75 | 3.24 | 1.41 | 1.09 | 1.20 | 1.36 | 0.760 | 0.704 | 0.705 | 0.534 | 0.486 |
| ⁷⁄₁₆ | 8.3 | 2.43 | 2.91 | 1.26 | 1.09 | 1.18 | 1.23 | 0.677 | 0.711 | 0.682 | 0.535 | 0.491 |
| ⅜ | 7.2 | 2.11 | 2.56 | 1.09 | 1.10 | 1.16 | 1.09 | 0.592 | 0.719 | 0.660 | 0.537 | 0.496 |
| ⁵⁄₁₆ | 6.1 | 1.78 | 2.19 | 0.927 | 1.11 | 1.14 | 0.939 | 0.504 | 0.727 | 0.637 | 0.540 | 0.501 |
| ¼ | 4.9 | 1.44 | 1.80 | 0.755 | 1.12 | 1.11 | 0.777 | 0.412 | 0.735 | 0.614 | 0.544 | 0.506 |
| L 3 × 2½ × ⅜ | 6.6 | 1.92 | 1.66 | 0.810 | 0.928 | 0.956 | 1.04 | 0.581 | 0.736 | 0.706 | 0.522 | 0.676 |
| ⁵⁄₁₆ | 5.6 | 1.62 | 1.42 | 0.688 | 0.937 | 0.933 | 0.898 | 0.494 | 0.744 | 0.683 | 0.525 | 0.680 |
| ¼ | 4.5 | 1.31 | 1.17 | 0.561 | 0.945 | 0.911 | 0.743 | 0.404 | 0.753 | 0.661 | 0.528 | 0.684 |
| L 3 × 2 × ⅜ | 5.9 | 1.73 | 1.53 | 0.781 | 0.940 | 1.04 | 0.543 | 0.371 | 0.559 | 0.539 | 0.430 | 0.428 |
| ⁵⁄₁₆ | 5.0 | 1.46 | 1.32 | 0.664 | 0.948 | 1.02 | 0.470 | 0.317 | 0.567 | 0.516 | 0.432 | 0.435 |
| ¼ | 4.1 | 1.19 | 1.09 | 0.542 | 0.957 | 0.993 | 0.392 | 0.260 | 0.574 | 0.493 | 0.435 | 0.440 |
| ³⁄₁₆ | 3.07 | 0.902 | 0.842 | 0.415 | 0.966 | 0.970 | 0.307 | 0.200 | 0.583 | 0.470 | 0.439 | 0.446 |
| L 2½ × 2 × ⅜ | 5.3 | 1.55 | 0.912 | 0.547 | 0.768 | 0.831 | 0.514 | 0.363 | 0.577 | 0.581 | 0.420 | 0.614 |
| ⁵⁄₁₆ | 4.5 | 1.31 | 0.788 | 0.466 | 0.776 | 0.809 | 0.446 | 0.310 | 0.584 | 0.559 | 0.422 | 0.620 |
| ¼ | 3.62 | 1.06 | 0.654 | 0.381 | 0.784 | 0.787 | 0.372 | 0.254 | 0.592 | 0.537 | 0.424 | 0.626 |
| ³⁄₁₆ | 2.75 | 0.809 | 0.509 | 0.293 | 0.793 | 0.764 | 0.291 | 0.196 | 0.600 | 0.514 | 0.427 | 0.631 |

*This list is abridged.

## Table 8. Standard Steel Pipe

| | Dimensions | | | | Properties | | |
|---|---|---|---|---|---|---|---|
| Nom. Diam. | Outside Diam. | Inside Diam. | Thick-ness | Weight per Foot | $I$ | $A$ | $r$ |
| in | in | in | in | lb | $in^4$ | $in^2$ | in |
| $\frac{1}{8}$ | 0.405 | 0.269 | 0.068 | 0.24 | 0.001 | 0.072 | 0.12 |
| $\frac{1}{4}$ | 0.540 | 0.364 | 0.088 | 0.42 | 0.003 | 0.125 | 0.16 |
| $\frac{3}{8}$ | 0.675 | 0.493 | 0.091 | 0.57 | 0.007 | 0.167 | 0.21 |
| $\frac{1}{2}$ | 0.840 | 0.622 | 0.109 | 0.85 | 0.017 | 0.250 | 0.26 |
| $\frac{3}{4}$ | 1.050 | 0.824 | 0.113 | 1.13 | 0.037 | 0.333 | 0.33 |
| 1 | 1.315 | 1.049 | 0.133 | 1.68 | 0.087 | 0.494 | 0.42 |
| $1\frac{1}{4}$ | 1.660 | 1.380 | 0.140 | 2.27 | 0.195 | 0.669 | 0.54 |
| $1\frac{1}{2}$ | 1.900 | 1.610 | 0.145 | 2.72 | 0.310 | 0.799 | 0.62 |
| 2 | 2.375 | 2.067 | 0.154 | 3.65 | 0.666 | 1.07 | 0.79 |
| $2\frac{1}{2}$ | 2.875 | 2.469 | 0.203 | 5.79 | 1.53 | 1.70 | 0.95 |
| 3 | 3.500 | 3.068 | 0.216 | 7.58 | 3.02 | 2.23 | 1.16 |
| $3\frac{1}{2}$ | 4.000 | 3.548 | 0.226 | 9.11 | 4.79 | 2.68 | 1.34 |
| 4 | 4.500 | 4.026 | 0.237 | 10.79 | 7.23 | 3.17 | 1.51 |
| 5 | 5.563 | 5.047 | 0.258 | 14.62 | 15.2 | 4.30 | 1.88 |
| 6 | 6.625 | 6.065 | 0.280 | 18.97 | 28.1 | 5.58 | 2.25 |
| 8 | 8.625 | 7.981 | 0.322 | 28.55 | 72.5 | 8.40 | 2.94 |
| 10 | 10.750 | 10.020 | 0.365 | 40.48 | 161. | 11.9 | 3.67 |
| 12 | 12.750 | 12.000 | 0.375 | 49.56 | 279. | 14.6 | 4.38 |

## Table 9. Plastic Section Moduli Around the $X$-$X$ Axis; $\tau_{yp} = 36$ ksi

| Shape | Plastic Modulus Z, $in^3$ | Shape | Plastic Modulus Z, $in^3$ |
|---|---|---|---|
| W 36 × 230 | 943 | W 24 × 84 | 224 |
| W 33 × 221 | 855 | W 24 × 76 | 200 |
| W 36 × 194 | 767 | W 24 × 68 | 177 |
| W 36 × 182 | 718 | W 21 × 68 | 160 |
| W 36 × 170 | 668 | W 24 × 62 | 153 |
| W 36 × 160 | 624 | W 24 × 55 | 134 |
| W 36 × 150 | 581 | W 21 × 57 | 129 |
| W 33 × 141 | 514 | W 18 × 55 | 112 |
| W 36 × 135 | 509 | W 21 × 44 | 95.4 |
| W 33 × 130 | 467 | W 18 × 40 | 78.4 |
| W 33 × 118 | 415 | W 16 × 40 | 72.9 |
| W 30 × 116 | 378 | W 18 × 35 | 66.5 |
| W 30 × 108 | 346 | W 16 × 31 | 54.0 |
| W 30 × 99 | 312 | W 14 × 26 | 40.2 |
| W 27 × 94 | 278 | W 14 × 22 | 33.2 |
| W 24 × 94 | 254 | W 8 × 24 | 23.2 |
| W 27 × 84 | 244 | W 8 × 18 | 17.0 |

## Table 10. Properties of Structural Lumber (Abridged List). Sectional Properties of American Standard Dressed (S4S)* Sizes.

| Nominal Size | Standard Dressed Size | Area of Section | Moment of Inertia | Section Modulus | Weight per Foot | Nominal Size | Standard Dressed Size | Area of Section | Moment of Inertia | Section Modulus | Weight per Foot |
|---|---|---|---|---|---|---|---|---|---|---|---|
| in | in × in | in² | in⁴ | in³ | lb | in | in × in | in² | in⁴ | in³ | lb |
| 2 × 4 | 1½ × 3½ | 5.25 | 5.36 | 3.06 | 1.46 | 10 × 10 | 9½ × 9½ | 90.3 | 679 | 143 | 25.1 |
| 6 | 5½ | 8.25 | 20.8 | 7.56 | 2.29 | 12 | 11½ | 109 | 1204 | 209 | 30.3 |
| 8 | 7¼ | 10.9 | 47.6 | 13.1 | 3.02 | 14 | 13½ | 128 | 1948 | 289 | 35.6 |
| 10 | 9¼ | 13.9 | 98.9 | 21.4 | 3.85 | 16 | 15½ | 147 | 2948 | 380 | 40.9 |
| 12 | 11¼ | 16.9 | 178 | 31.6 | 4.69 | 18 | 17½ | 166 | 4243 | 485 | 46.1 |
| 14 | 13¼ | 19.9 | 291 | 43.9 | 5.52 | 20 | 19½ | 185 | 5870 | 602 | 51.4 |
| | | | | | | 22 | 21½ | 204 | 7868 | 732 | 56.7 |
| 3 × 4 | 2½ × 3½ | 8.75 | 8.93 | 5.10 | 2.43 | 24 | 23½ | 223 | 10274 | 874 | 62.0 |
| 6 | 5½ | 13.8 | 34.7 | 12.6 | 3.82 | | | | | | |
| 8 | 7¼ | 18.1 | 79.4 | 21.9 | 5.04 | 12 × 12 | 11½ × 11½ | 132 | 1458 | 253 | 36.7 |
| 10 | 9¼ | 23.1 | 165 | 35.7 | 6.42 | 14 | 13½ | 155 | 2358 | 349 | 43.1 |
| 12 | 11¼ | 28.1 | 297 | 52.7 | 7.81 | 16 | 15½ | 178 | 3569 | 460 | 49.5 |
| 14 | 13¼ | 33.1 | 485 | 73.2 | 9.20 | 18 | 17½ | 201 | 5136 | 587 | 55.9 |
| 16 | 15¼ | 38.1 | 739 | 96.9 | 10.6 | 20 | 19½ | 224 | 7106 | 729 | 62.3 |
| | | | | | | 22 | 21½ | 247 | 9524 | 886 | 68.7 |
| 4 × 4 | 3½ × 3½ | 12.3 | 12.5 | 7.15 | 3.40 | 24 | 23½ | 270 | 12437 | 1058 | 75.0 |
| 6 | 5½ | 19.3 | 48.5 | 17.6 | 5.35 | | | | | | |
| 8 | 7¼ | 25.4 | 111 | 30.7 | 7.05 | 14 × 14 | 13½ × 13½ | 182 | 2768 | 410 | 50.6 |
| 10 | 9¼ | 32.4 | 231 | 49.9 | 8.94 | 16 | 15½ | 209 | 4189 | 541 | 58.1 |
| 12 | 11¼ | 39.4 | 415 | 73.8 | 10.9 | 18 | 17½ | 236 | 6029 | 689 | 65.6 |
| 14 | 13¼ | 46.4 | 678 | 102 | 12.9 | 20 | 19½ | 263 | 8342 | 856 | 73.1 |
| 16 | 15¼ | 53.4 | 1034 | 136 | 14.8 | 22 | 21½ | 290 | 11181 | 1040 | 80.6 |
| | | | | | | 24 | 23½ | 317 | 14600 | 1243 | 88.1 |
| 6 × 6 | 5½ × 5½ | 30.3 | 76.3 | 27.7 | 8.40 | | | | | | |
| 8 | 7½ | 41.3 | 193 | 51.6 | 11.4 | 16 × 16 | 15½ × 15½ | 240 | 4810 | 621 | 66.7 |
| 10 | 9½ | 52.3 | 393 | 82.7 | 14.5 | 18 | 17½ | 271 | 6923 | 791 | 75.3 |
| 12 | 11½ | 63.3 | 697 | 121 | 17.5 | 20 | 19½ | 302 | 9578 | 982 | 83.9 |
| 14 | 13½ | 74.3 | 1128 | 167 | 20.6 | 22 | 21½ | 333 | 12837 | 1194 | 92.5 |
| 16 | 15½ | 85.3 | 1707 | 220 | 23.6 | 24 | 23½ | 364 | 16763 | 1427 | 101 |
| 18 | 17½ | 96.3 | 2456 | 281 | 26.7 | | | | | | |
| | | | | | | 18 × 18 | 17½ × 17½ | 306 | 7816 | 893 | 85.0 |
| 8 × 8 | 7½ × 7½ | 56.3 | 264 | 70.3 | 15.6 | 20 | 19½ | 341 | 10813 | 1109 | 94.8 |
| 10 | 9½ | 71.3 | 536 | 113 | 19.8 | 22 | 21½ | 376 | 14493 | 1348 | 105 |
| 12 | 11½ | 86.3 | 951 | 165 | 23.9 | 24 | 23½ | 411 | 18926 | 1611 | 114 |
| 14 | 13½ | 101.3 | 1538 | 228 | 28.0 | | | | | | |
| 16 | 15½ | 116.3 | 2327 | 300 | 32.0 | 20 × 20 | 19½ × 19½ | 380 | 12049 | 1236 | 106 |
| 18 | 17½ | 131.3 | 3350 | 383 | 36.4 | 22 | 21½ | 419 | 16150 | 1502 | 116 |
| 20 | 19½ | 146.3 | 4634 | 475 | 40.6 | 24 | 23½ | 458 | 21089 | 1795 | 127 |
| | | | | | | 24 × 24 | 23½ × 23½ | 552 | 25415 | 2163 | 153 |

* Surfaced four sides. All properties and weights given are for dressed sizes only. The weights given are based on an assumed average weight of 40 lb per cubic foot. Based on a table compiled by the National Forest Products Association.

## Table 11. Deflections and Slopes of Elastic Curves for Variously Loaded Beams

| Loading | Equation of Elastic Curve | |
|---|---|---|
| | Maximum Deflection | Slope at End |
| | $v = \dfrac{P}{6EI}(2L^3 - 3L^2x + x^3)$ <br><br> $v_{max} = v(0) = -\dfrac{PL^3}{3EI}$ | $\theta(0) = -\dfrac{PL^2}{2EI}$ |
| | $v = \dfrac{q_o}{24EI}(x^4 - 4L^3x + 3L^4)$ <br><br> $v_{max} = v(0) = \dfrac{q_oL^4}{8EI}$ | $\theta(0) = -\dfrac{q_oL^3}{6EI}$ |
| | $v = \dfrac{q_o x}{24EI}(L^3 - 2Lx^2 + x^3)$ <br><br> $v_{max} = v(L/2) = \dfrac{5q_oL^4}{384EI}$ | See Example 10-3. <br><br> $\theta(0) = -\theta(L) = \dfrac{q_oL^3}{24EI}$ |
| | When $0 \leq x \leq a$, then <br> $v = \dfrac{Pbx}{6EIL}(L^2 - b^2 - x^2)$ <br> When $a = b = \dfrac{L}{2}$, then <br> $v = \dfrac{Px}{48EI}(3L^2 - 4x^2)$ <br><br> $v_{max} = v(L/2) = \dfrac{PL^3}{48EI}$ | See Example 10-6. <br><br> $\left(0 \leq x \leq \dfrac{L}{2}\right)$ <br><br> $\theta(0) = -\theta(L) = \dfrac{PL^2}{16EI}$ |
| | $v = -\dfrac{M_o x}{6EIL}(L^2 - x^2)$ <br><br> $v_{max} = v(L/\sqrt{3}) = -\dfrac{M_oL^2}{9\sqrt{3}\,EI}$ | See Example 13-1. <br><br> $\theta(0) = -\dfrac{\theta(L)}{2} = -\dfrac{M_oL}{6EI}$ |
| | $v_a = v(a) = \dfrac{Pa^2}{6EI}(3L - 4a)$ <br><br> $v_{max} = v(L/2) = \dfrac{Pa}{24EI}(3L^2 - 4a^2)$ | $\theta(0) = \dfrac{Pa}{2EI}(L - a)$ |

## Table 12. Fixed-End Actions of Prismatic Beams*

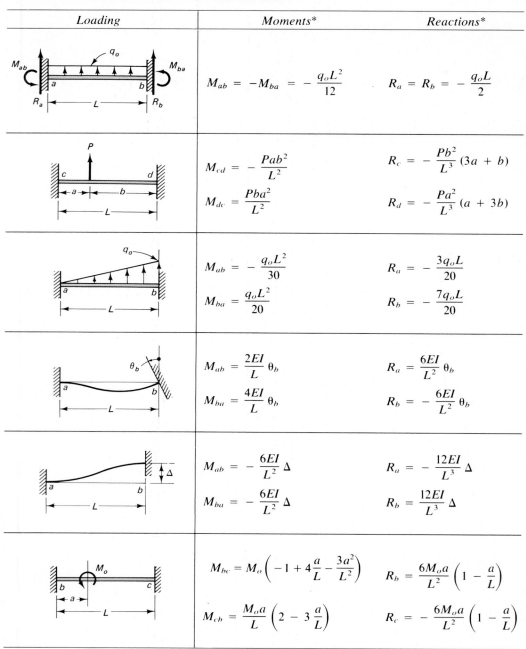

| Loading | Moments* | Reactions* |
|---|---|---|
| | $M_{ab} = -M_{ba} = -\dfrac{q_o L^2}{12}$ | $R_a = R_b = -\dfrac{q_o L}{2}$ |
| | $M_{cd} = -\dfrac{Pab^2}{L^2}$ <br><br> $M_{dc} = \dfrac{Pba^2}{L^2}$ | $R_c = -\dfrac{Pb^2}{L^3}(3a+b)$ <br><br> $R_d = -\dfrac{Pa^2}{L^3}(a+3b)$ |
| | $M_{ab} = -\dfrac{q_o L^2}{30}$ <br><br> $M_{ba} = \dfrac{q_o L^2}{20}$ | $R_a = -\dfrac{3q_o L}{20}$ <br><br> $R_b = -\dfrac{7q_o L}{20}$ |
| | $M_{ab} = \dfrac{2EI}{L}\theta_b$ <br><br> $M_{ba} = \dfrac{4EI}{L}\theta_b$ | $R_a = \dfrac{6EI}{L^2}\theta_b$ <br><br> $R_b = -\dfrac{6EI}{L^2}\theta_b$ |
| | $M_{ab} = -\dfrac{6EI}{L^2}\Delta$ <br><br> $M_{ba} = -\dfrac{6EI}{L^2}\Delta$ | $R_a = -\dfrac{12EI}{L^3}\Delta$ <br><br> $R_b = \dfrac{12EI}{L^3}\Delta$ |
| | $M_{bc} = M_o\left(-1 + 4\dfrac{a}{L} - \dfrac{3a^2}{L^2}\right)$ <br><br> $M_{cb} = \dfrac{M_o a}{L}\left(2 - 3\dfrac{a}{L}\right)$ | $R_b = \dfrac{6M_o a}{L^2}\left(1 - \dfrac{a}{L}\right)$ <br><br> $R_c = -\dfrac{6M_o a}{L^2}\left(1 - \dfrac{a}{L}\right)$ |

* For all the cases tabulated, the positive senses of the end moments and reactions are the same as those shown in the first diagram for uniformly distributed loading. *The special sign convention used here is that adopted for the displacement method in Section 13-5.*

# Answers to Odd-numbered Problems

## Chapter 1

**1-3.** $\sigma_\theta/\tau_\theta$ at 0° and 180°: 1/0; at 45°: 0.5/0.5; at 90°: 0/0.

**1-5.** $\pm\sigma_1$ at 0° and 180°; $\pm\sigma_1/2$ at 45° and 135°.

**1-7.** 17.7 ksi, 2.34 ksi, 6.43 ksi.

**1-9.** 12.1 ksi, 18.2 ksi.

**1-11.** 21.3 MPa; 30 mm, 35 mm.

**1-13.** $\sigma_{max} = 25$ MPa, $\sigma_{min} = 10$ MPa.

**1-15.** 2.11 kPa.

**1-17.** - 109 psi, - 198 psi, - 119 psi.

**1-19.** (a) 30 MPa, (b) 40.9 MPa, (c) 71.6 MPa, (d) 113 MPa.

**1-21.** 6.5 in.

**1-23.** 2.83 ksi.

**1-25.** 4.81 ksi, 5.97 ksi.

**1-27.** $\sigma_{AB} \approx \sigma_{BC} = 100.3$ MPa

**1-29.** 73.6 psi, 111 psi, 902.1 psi, 184 psi.

**1-31.** 10 k, 4.08 ksi.

**1-33.** (a) 2.22 MPa, (b) 0.707 m.

**1-35.** 10.7 in.

**1-37.** 16.4 mm.

**1-39.** 13.2 kg.

**1-41.** $A_{AB} = 5300$ mm², $A_{BC} = 7950$ mm², $A_{BE} = 3640$ mm².

**1-43.** 0.123 in.

**1-45.** 27 mm.

**1-47.** $\sigma_{max} = 14.5$ MPa.

**1-49.** 0.909 in² vs 0.864 in².

**1-51.** 38.9 k.

**1-53.** 8.57 in, 18.2 in.

## Chapter 2

**2-1.** $29.3 \times 10^6$ psi.

**2-3.** (a) 10 mm, (b) 1.67 kN/mm.

**2-5.** 50 kN/mm.

**2-7.** 18.75 kN/mm.

**2-9.** (a) 0.0289 in, (b) 8.57 in from left.

**2-11.** 9.20 mm.

**2-13.** 0.0363 in.

**2-15.** $\Delta_D = 3.7$ mm.

**2-17.** 597 ft.

**2-19.** (a) 0.122 in, (b) 0.028 in.

**2-21.** $7.64 \times 10^3 \, \gamma g/E$.

**2-23.** $10^5 \gamma/E$ mm.

**2-25.** $- 4Pa/9AE$.

**2-27.** 0.00375 in, 0.100 in. Try a graphic solotion as in Fig. 2-24(c). Use large scale.

**2-29.** (a) 4 kN/mm, (b) 4 mm.

**2-31.** 2 mm

**2-33.** 0.165 in, 0.0714 in. Try a graphic solution as in Fig. 2-29(b).

**2-35.** 29.3 kN.

**2-37.** 9.9 kN.

**2-39.** (a) 28.7 ksi, (b) 0.0367 in, (c) 0.565 in, (d) 0.448 in.

**2-41.** 0.04 in.

**2-43.** 4 mm.

**2-45.** 194 MPa, 129 MPa, 228 MPa (lower rod).

**2-47.** $R_L = 1.2P$.

**2-49.** (a) - 120 kN, -70 kN.

**2-51.** (a) 5 k, -25 k.

**2-53.** (a) 1250 N (middle wire)

**2-55.** $0.2274P$, $0.6062P$.

**2-59.** 1/3 k.

**2-61.** $R_L = 0.65P$.

**2-63.** 4.8 kN.

**2-65.** $R_L = - 4P$.

**2-67.** $P/[(a/b)^n + 1]$.

**2-69.** (a) 480 k, (b) 1100 k.

**2-71.** 83.3 N, 333 N, 583 N.

**2-73.** 37.8° C.

**2-75.** $P_{yp} = 376$ kN, $P_{ult} = 518$ kN at 5 mm.

**2-77.** $2\gamma w^2 L^3/3gE$.

## Chapter 3

**3-1.** 51.2 N/mm.

**3-3.** (a) - 3.50 μm, (b) 147 kN.

**3-5.** 0.325, 0.25.

**3-7.** 444 psi.

**3-9.** (a) 0.20 in, use 0.25 plate, (b) 0.212 in.

**3-11.** 14.

**3-13.** 0.080 in, 0.025 in.

**3-15.** 70.7 MPa.

**3-17.** 25.9 MPa.

**3-23.** (a) $\sigma_r = 20(1 - 0.05/r^2)$, (b) 180 MPa, (c) 91.7 μm, 41.0 μm.

**3-25.** (a) $|\sigma_r|_{max} = -160$ MPa, $|\sigma_t|_{max} = -60$ MPa, (b) 90 MPA, (c) 30.3 $\mu$m, $-26.3$ $\mu$m.
**3-27.** (a) 12 in, (b) 7.80 in.
**3-29.** Add term $(1 - \nu^2)\gamma\omega^2 r/gE$ to Eq. 3-38.

## Chapter 4

**4-1.** 17.9 MPa.
**4-3.** 1:0.375.
**4-5.** 2 in.
**4-7.** 3.71 in, 5.35 in, 3.72°.
**4-9.** 0.310 in.
**4-11.** (a) 0, 40.4 MPa, 8.1 MPa, 16.2 MPa, (b) 8.6°.
**4-13.** (a) $r_i = 21$ mm, (b) strength.
**4-15.** (a) 0.75°.
**4-17.** (a) 0.264°, (b) $17 \times 10^{-6}$ rad/lb-in.
**4-19.** $kL^3/6JG$.
**4-21.** $720/JG$.
**4-23.** $0.837T$.
**4-25.** 29.4 k-in, 10.6 k-in, 5.88 $\times 10^{-3}$ rad.
**4-27.** (a) 62.4 lb-in, (b) $\phi_{max} = 0.302$ rad.
**4-29.** (a) $t_0L/3$, (b) $|\phi|_{max} = 2t_0L^2/9\sqrt{3}JG$.
**4-33.** $3t_0L/8$ and $-t_0L/8$.
**4-35.** (a) 186 k-in, (b) 736 hp.
**4-37.** 573 MPa.
**4-39.** 1°, 26 MPa.
**4-41.** (a) 85 MPa, (b) $89 \times 10^{-6}$ rad.
**4-43.** $3R/t$, $t^2/3R^2$.
**4-45.** $470 \times 10^3$ mm$^4$.
**4-47.** 11.1 psi, $0.691/G$ rad/in.
**4-49.** (a) 33 and 67%, (b) 5.3 $\times 10^{-6}$ rad/mm.

## Chapter 5

Only the largest values of $P$, $V$, and $M$ are given in many problems.

**5-1.** $R_{Ay} = 5.11$ k.
**5-3.** 6 k, 24 k, 6 k.
**5-5.** 0, 0, 18 k-ft; 0, - 4.5 k, 9 k-ft.
**5-7.** 4 k, - 2.8 k, - 15.2 k-ft; 4k, - 0.8 k, - 15.2 k-ft;

4 k, - 0.8 k, - 17.6 k-ft; 0, - 0.8 k, 2.4 k-ft.
**5-9.** - 416 lb, 139 lb, 832 lb-ft.
**5-11.** $-P/2$, $-P/2$, $-PR$; 0, $\sqrt{2}$ $P/2$, $-PR/2$.
**5-13.** 34 k, -12 k, 30 k-ft.
**5-15.** - 7.2 k, 9.6 k - 24 k-ft; - 3.15 k, - 12.6 k, 30 k-ft.
**5-17.** - 3 k, - 3 k, 6 k-ft.
**5-19.** 6.75 kN, - 3 kN, 3 kN·m.
**5-21.** $3P$, $-2PL$.
**5-23.** $- w_0L$, $- w_0L^2/2$.
**5-25.** $\pm P/2$, $|M|_{max} = Pa/2$.
**5-27.** $2x$, $2x - (x - 3)^3/12$.
**5-29.** $M_A + w_0Lx/2 - w_0x^2/2$.
**5-31.** $M_A + R_Ax - kLx^2/2 + kx^3/3$.
**5-33.** $M(\theta) = - PR(\sin \theta + 1 - \cos \theta)$.
**5-35.** (a) $F$, $- F(L - x)$, $Fa$; $F$, $- F(a - z)$, 0, (b) $F + w(a + L)$, $- (F + wa + wL/2)L$, $(F + wa/2)a$.
**5-37.** $M = - kx^4/12 + kL^3x/3 - kL^4/4$.
**5-39.** $M = - (kL^2/4\pi^2) \sin 2\pi x/L$.
**5-41.** $- M_1/L$, $- M_1x/L$.
**5-43.** $2P$, $- P(2a + b)$.
**5-45.** $|V|_{max} = 10$ k, 48 k-ft.
**5-47.** $V_{max} = 2P/3$, $\pm Pa/3$.
**5-49.** $|V|_{max} = 600$ kN, - 2.8 MN·m.
**5-51.** $|V|_{max} = 1.98$ k, 10.5 k-ft.
**5-53.** $|V|max = 3$ kN, $\pm 2$ kN·m.
**5-55.** $\pm 4$ k, 10 k-ft.
**5-57.** $\pm 6$ k, - 24 k-ft.
**5-59.** $\pm 2q_0a$, $2.5q_0a^2$.
**5-61.** $1.5q_0a$, $0.625q_0a^2$.
**5-63.** - 700 lb, 625 lb-ft.
**5-65.** 20 kN, 40 kN·m.
**5-67.** - 10 k, 50 k-ft.
**5-69.** $R_A = 35$ k, $R_B = 20$ k.
**5-71.** 50 N, - 10 N·m.
**5-73.** - 68 kN, $\pm 24$ kN, 60 kN·m.
**5-75.** $|V|_{max} = 2P/3$, $4Pa/3$.
**5-77.** 100 kN, 400 kN·m.
**5-79.** 2 k, 70 k- in.
**5-81.** $|V|_{max} = 160$ N, 3.2 N·m.
**5-83.** $M = 900x - 4 \langle x - 10\rangle^3/3$.

**5-85.** $M = 2q_1ax/9 - q_1x^3/24a + 16q_1a \langle x - 3a\rangle^1/9$.
**5-87.** $M = -2x + 8 \langle x - 4\rangle^1 - \langle x - 4\rangle^2/2 + \langle x - 12\rangle^2/2$.

## Chapter 6

**6-1.** 367 kN·m.
**6-3.** 1010 k-ft.
**6-5.** All OK.
**6-7.** 10.6 k/ft.
**6-9.** 231 k-in.
**6-11.** (a) 142 kN, (b) 8.90 kN.
**6-13.** 704 kN, 71.0 mm below NA.
**6-15.** 59.4 kN, 29 mm below NA.
**6-17.** $\sqrt{2}$.
**6-19.** Negative, - 22.5 ksi.
**6-21.** 26.9 k.
**6-23.** 119 MPa, - 96.8 MPa.
**6-25.** 15 mm.
**6-27.** $U = \dfrac{\sigma_{max}^2}{2E} \left(\dfrac{\text{Vol}}{9}\right)$.
**6-29.** NA at 40 mm, 469 MPa.
**6-31.** 40.3 MPa.
**6-33.** 633 k-in.
**6-35.** 122 k-ft.
**6-39.** 1.7.
**6-41.** 1.12.
**6-43.** 1.80.
**6-45.** (a) 73.0 kN·m, (c) $\pm 35.7$ MPa, $\pm 33.2$ MPa.
**6-47.** 102 kN.
**6-49.** (a) 54 kN·m, (b) 108 kN·m.
**6-51.** $\pm 12$ MPa, NA through $C$ and $D$.
**6-53.** 126 MPa.
**6-55.** $\pm 24.6$ MPa, $\pm 9.57$ MPa, 28 mm.
**6-57.** 282 MPa.
**6-59.** 15.6 mm.
**6-61.** - 18 ksi.
**6-63.** - 174 psi.
**6-65.** 150 kN, 5 kN.
**6-67.** 420 N.
**6-69.** 16.0 ksi, - 5.34 ksi.
**6-71.** From - 50 to 100 mm.
**6-73.** $R/4$.
**6-75.** 27.6 ft.
**6-77.** (a) 40.5 kN, $e = 25$ mm, (b) 0.667 $\mu$m/m, 1.33 $\mu$m/m.

**6-79.** 282 MPa.

**6-81.** $b^2h^2/24$, $-b^2h^2/72$.

**6-83.** $I_z = 560 \cdot 10^3$ mm$^4$, $I_y = 290 \cdot 10^3$ mm$^4$, $I_{yz} = 300 \cdot 10^3$ mm$^4$, $I_{z'} = 753.9757 \cdot 10^3$ mm$^4$, $I_{y'} = 96.0243 \cdot 10^3$ mm$^4$, $\theta = 32.8862°$.

## Chapter 7

**7-1.** 35.2 lb/in, 105 lb.

**7-3.** Use (a), 2.44 in.

**7-5.** (a) 1.6 in, (b) 8 in.

**7-7.** 3.4 in, 9.08 in.

**7-9.** 10.2 kN.

**7-13.** 0, 7.06 MPa, 11.3 or 33.9 MPa, 35.3 MPa, 27.3 MPa, 0.

**7-15.** 132 MPa, 139 MPa.

**7-17.** 0, 78.2 kPa, 125 kPa, 140 kPa, 125 kPa.

**7-19.** 51.1 k, 197 k-in.

**7-21.** 144 kN.

**7-23.** 637 kPa.

**7-25.** 4.44 MPa, 2.22 MPa.

**7-27.** 10 kN ←, 240 kN ←, 250 kN →ea, 20 kN ↑, 20 kN ↓.

**7-29.** (a) 254 psi, (b) 31 psi.

**7-31.** (a) $1.82 \times 10^6$ mm$^4$, (b) 5.49 MPa, 7.69 MPa, 18.7 MPa.

**7-33.** Angles 2.5 MPa, plate 48.2 MPa.

**7-35.** (a) 83.0 kN/m, 194 kN/m, 415 kN/m.

**7-37.** 5.30 mm.

**7-39.** $e = 2R(\sin \alpha - \alpha \cos \alpha)/(\alpha - \sin \alpha \cos \alpha)$.

**7-43.** $e = \alpha R/\sin \alpha$.

**7-45.** 0, -154 kPa; -700 kPa, -50 kPa.

**7-47.** 0, -195 kPa; -1500 kPa, -120 kPa.

**7-49.** 151 N/mm, 1.24 kN.

## Chapter 8

**8-1.** $P_1/A$; $P_1/A + My/I$; $P_1/A$, $VQ/It$; $My/I$; $Tr/J$; $Tr/J$.

**8-3.** 39.8 MPa, 14.3 MPa; 10.2 MPa, -14.3 MPa.

**8-5.** -12.1 ksi, 5.24 ksi; -5.91 ksi, -5.24 ksi.

**8-7.** -2 psi, 3 psi.

**8-9.** 225 kPa, not permissible.

**8-11.** 45 MPa, 5 MPa; 5 MPa, -5 MPa.

**8-13.** 22.9 ksi, 6.99 ksi.

**8-15.** 10 ksi on $\theta = 26.6°$, 0.

**8-17.** (a) 18.3 ksi, -38.3 ksi on $\theta = 22.5°$, (b) ±28.3 ksi, -10 ksi.

**8-19.** (a) 17 MPa, -47 MPa on $\theta = -19.3°$, (b) ±32 MPa, -15 MPa.

**8-21.** (a) 16.1 ksi on $\theta = -28.2°$, -56.1 ksi, (b) ±36.1 ksi, -20 ksi.

**8-23.** (a) ±10 MPa at 45°, (b) 10 MPa, 0.

**8-25.** (a) 6 ksi on $\theta = 116.6°$, -4 ksi, (b) 5 ksi, 1 ksi.

**8-27.** (a) 100 psi on $\theta = 26.6°$, 0, (b) ±50 psi, 50 psi.

**8-29.** (a) -9.5 MPa on $\theta = 50.6°$, -60.5 MPa, (b) ±25.5 MPa, -35 MPa.

**8-31.** (a) 28 ksi on 28.2°, -8 ksi, (b) × 18 ksi, 10 ksi.

**8-33.** -15 ksi, 8.66 ksi.

**8-35.** 45 MPa, 5 MPa.

**8-37.** $5p$, $p$.

**8-39.** 51.6° or 14.8°.

**8-41.** (a) 7 MPa, 0, (b) 4.48 MPa, 2.52 MPa, -3.36 MPa.

**8-43.** (a) 18, -120, -1872, (b) 19.1646 MPa, 9.3181 MPa, -10.4827 MPa, (c) (-0.47, 0.20, 0.86).

**8-45.** 18.3 ksi, 0, -38.3 ksi.

**8-49.** 0, -1000 μm/m; -26.6°.

**8-51.** 1128 μm/m, -128 μm/m; 4.58°; (0, 0, 1).

**8-53.** ±5.76 ksi; 75.7°.

**8-55.** 81.9 MPa, 11.4 MPa; 4.57°.

**8-57.** $\begin{pmatrix} 6 & 0 & 0 \\ 0 & 6 & 0 \\ 0 & 0 & 6 \end{pmatrix} + \begin{pmatrix} 4 & 4 & -6 \\ 4 & -12 & 8 \\ -6 & 8 & 8 \end{pmatrix}$

**8-59.** $2\tau_0$, $\sqrt{3}\,\tau_0$.

**8-61.** $\sigma_{yp}/(1 - \nu + \nu^2)^{1/2}$, $\sigma_{yp}/(1 - \nu)$.

## Chapter 9

**9-1.** -7.5 MPa, 13.0 MPa.

**9-3.** 10.0 ksi, 5.06 ksi; 10 ksi, 5 ksi, 47.7 psi.

**9-5.** 97.7 MPa, 33.8 MPa.

**9-7.** 9.90 ksi, 4.81 ksi, 0.22 ksi.

**9-9.** 1.2 MPa, 45.2 kN·m.

**9-11.** -1.6 MPa, -167 MPa; 0.168 MPa, -83.5 MPa; ±5 MPa.

**9-13.** -2.5 ksi, 0; -2.81 ksi, 0.51 ksi; ±1.83 ksi.

**9-15.** 764 lb.

**9-17.** 0.272 ksi, -4.93 ksi.

**9-19.** 14.1 k-in, 18.8 k-in.

**9-21.** -40 ksi.

**9-23.** (b) 241 psi, -41 psi; 141 psi, 100 psi.

**9-25.** (a) 17 MPa, -6.2 MPa, (b) -32 MPa, -6.5 MPa.

**9-27.** -212 psi, -167 psi.

**9-29.** -16.4 MPa, 0.01 MPa.

**9-31.** 56.3°.

**9-33.** 88.9 mm outside diameter.

**9-35.** 3 × 4 in.

**9-37.** 1600 lb, 2 ft.

**9-39.** Rectangular section, 2.13 N/mm.

**9-41.** W12 × 30.

**9-43.** S10 × 25.4.

**9-45.** S18 × 54.7.

**9-47.** $d = d_0(x/L)^{1/3}$.

**9-49.** (a) 44 + 6 = 50 in, (b) 73 + 6 = 79 in from center.

**9-51.** 161 mm.

**9-53.** 6.8 in.

**9-55.** 600 lb.

**9-57.** 2 in.

## Chapter 10

**10-1.** 800 mm, 200 MPa.

**10-3.** 483 ft.

**10-5.** $-kx$.

**10-7.** $EIv = M_1(x^2 - 2Lx + L^2)/2$.

**10-9.** $EIv = -P(x^3 - 3L^2x + 2L^3)/6$ for origin at $P$.

**10-11.** $EIv = -W(x^5 - 5L^4x + 4L^5)/60L^2$.

**10-13.** $EIv = -k(L/\pi)^4 \times \sin \pi x/L$.

**10-15.** $EIv = M_0(x^3/L - Lx)/6$ for origin on the right.

**10-17.** $EIv = P(4x^3 - 3L^2x)/48$ for $0 \le x \le L/2$.

**10-19.** $EIv = P(x^3/6 + a^2x/2 - aLx/2)$ for $0 \le x \le a$; $EIv = Pa(x^2/2 - Lx/2 + a^2/6)$ for $a \le x \le L-a$.

**10-21.** $EIv = P(x^3/12 - Lx^2/16)$ for $0 \le x \le L/2$.

**10-23.** $EIv = -w_0(x^4/24 - 3Lx^3/48 + L^3x/48)$.

**10-25.** $EIv = -(kL/\pi^3)[(L^3/\pi)\sin \pi x/L - x^3/2 + 3Lx^2/2 - L^2x]$.

**10-27.** $EIv = -kx^5/120 + kL^3x^2/48 - kL^5/80$.

**10-29.** $EIv = P(5x^3/3 - L^2x)/32$ for $0 \le x \le L/2$; $EIv = P[5x^3/3 - 16(x - L/2)^3/3 - L^2x]/32$ for $L/2 \le x \le L$.

**10-31.** (a) $v = \delta(3x^2/2L^2 - x^3/2L^3)$.

**10-33.** $E_sIv = P(x^3 - 5L^2x/4)/12$ for $0 \le x \le L/2$; $E_AIv = P[(L-x)^3 - 7L^2(L - x)/12]/12$ for $L/2 \le x \le L$.

**10-35.** (a) $EIv = w_0(L^3x-Lx^3)/6$; (b) $v_{max} = w_0L^4/9\sqrt{3}$, $v(L/2) = w_0L^4/16$.

**10-39.** W18 × 50.

**10-41.** (a) S18 × 70, (b) S24, (c) 14.2 ksi, 6.14 ksi.

**10-45.** $EIv = w_0(10ax^3 - 95a^3x - 4\langle x - a\rangle^4)/96$.

**10-47.** $EIv = \dfrac{Pb^2(b - 3a)x^3}{6(a + b)^3} - \dfrac{Pab^2x^2}{2(a + b)^2} - \dfrac{P}{6}\langle x - a\rangle^3$

**10-49.** $EIv = -kx^5/120 + ka^2x^3/18 - 41ka^4x/360 + ka\langle x-a\rangle^4/24 + k\langle x-a\rangle^5/120$.

**10-53.** - 0.391 in.

**10-55.** $-ka^5/15EI$.

**10-57.** $13PL^3/192EI$.

**10-59.** 6.48 kN.

**10-61.** 8.13 lb.

**10-63.** 2.03 k.

**10-65.** 531 N in middle bar.

**10-67.** 47.1 N.

**10-69.** 96.8° F.

**10-71.** 0.0437 in, 0.121 in, 0.0462 in, 0.653 in; $\beta = 0°, 69.2°, 86.2°, 90°$.

**10-73.** 1.28 in, $\beta = 78.8°$.

**10-75.** 4.1.

**10-77.** (a) $P = 11bh^2\sigma_{yp}/12L$.

**10-79.** 5.74 mm, 0.00137 rad.

**10-81.** 1.15 mm, 0.00094 rad.

**10-83.** $|v_A| = 19w_0a^4/8EI$, $|\theta_A| = 11w_0a^3/6EI$.

**10-85.** $|v_A| = M_1a^2/4EI$, $|\theta_A| = M_1a/12EI$.

**10-87.** $|v_A| = M_1a^2/6EI$, 0.

**10-89.** $|v_A| = Pa^3/12EI$, $|\theta_A| = 13Pa^2/28EI$.

**10-91.** 3 mm.

**10-93.** $|v| = w_0x(x^3 - 8ax^2 + 64a^3)/24\ EI$.

**10-95.** $|v|_{max} = 2\sqrt{2}\ M_1a^2/9EI$.

**10-97.** $|v|_{max} = M_1a^2/6EI$.

**10-99.** $|v|_{max} = 0.078$ in.

**10-101.** $|v_A| = 832/EI$, $|\theta_A| = 224/EI$.

**10-103.** $|v_C| = 64/3EI$, $|\theta_C| = 8/3EI$.

**10-105.** $|v_A| = 18/EI$, $|\theta_A| = 5/EI$.

**10-107.** 5.64 ft.

**10-109.** $7P/8$.

**10-111.** $R_L = 93.36$ kN.

**10-113.** (a) 25 k-ft, (b) W16 × 26, (c) 0.036 in, 0.207 in.

**10-115.** (a) $R_A = 0.382P$.

**10-117.** $M_B/M_A = 2/5$, $|\theta_A| = 11M_AL/80EI$.

**10-119.** (a) - 240/7 k-ft, (b) $208/EI$.

**10-121.** End moments: $\pm M_1/4$.

**10-123.** $M_A = -WL/15$, $R_A = 3W/10$.

**10-125.** $\pm 6EI\Delta/L^2$.

**10-127.** $M_A = -5PL/33$, $M_B = -7PL/66$.

**10-129.** $M_A = -14.33$ k-ft, $M_B = -8.67$ k-ft.

## Chapter 11

**11-1.** $3k/2$.

**11-3.** $k/a$ and $3k/a$.

**11-5.** $\begin{bmatrix} Pa - k & k & & \\ Pa & Pa - k & & \\ \vdots & & \vdots & \\ Pa & & Pa & \\ 0 & 0 & . & 0 \\ k & 0 & . & 0 \\ \vdots & \vdots & \vdots & \vdots \\ Pa & Pa & . & Pa - k \end{bmatrix}$

**11-7.** 5.08 k.

**11-9.** 10.4 k.

**11-13.** 2.5 in.

**11-15.** 2 in.

**11-17.** For rigid bar buckling, $\Delta \approx \pi^2 I_c L^3/48I_ba^2$.

**11-21.** $\tan \lambda L = \lambda(a + L)$.

**11-23.** 53.3 k.

**11-25.** (a) 9.16 ft, (b) 548 mm, (c) 30.1 ft.

**11-27.** (a) $|\sigma| = -2.5e^2 + 30e - 40$.

**11-29.** 34.3 ksi, 30.4 ksi.

**11-45.** 51.5.

**11-47.** 1109 k vs 1070 k.

**11-49.** (a) 126, (b) 134.

**11-51.** $P_n/P_{allow} = 1.47$.

**11-53.** 144 k.

**11-55.** 117 k.

**11-57.** 116 k, 14.7 k.

**11-59.** 742 lb.

**11-61.** 146 k.

**11-63.** 121 k.

**11-65.** (a) $P_{cr} = 733$ lb at $L_c/r = 102.3$, (b) strength governs below $L_c/r = 102.3$.

## Chapter 12

**12-1.** (a) $\Delta = PL/AE + 20PL/9GA + 2PL^3/3EI + PL^3/GJ$; (b) 0.5, 2.8, 33.6, 63.3; (c) 0.02, 0.1, 34.7, 65.1.

**12-3.** $PL^3/48EI$.

**12-5.** 8.54 ksi.

**12-7.** - 0.348 in.

**12-9.** 120, $160(2 + \sqrt{2})$.

**12-11.** 11.2.

**12-13.** $11PL^3/384EI$.

**12-15.** $\theta = 13\ w_0L^3/648EI$.

**12-17.** $518/EI$.

**12-19.** (a) $PL^3/3EI$, (b) $5PL^3/6EI$.

**12-21.** $\Delta_v = M_0L_1(3L_1 + 2L_2)/6EI$.

**12-23.** $\Delta_H = 2Pa^3/EI$.

**12-25.** $\Delta_V = 92/EI$.

**12-27.** $\Delta_H = 80P/3EI$. $\theta = 5P/EI$.

**12-29.** $\theta = 66P/EI$.

**12-31.** $\Delta_{AH} = 1.97Pa^3/EI$, $\theta_A = Pa^2(\sqrt{2} + 3)/2EI$, $\theta_B = 3Pa^2/2EI$.

**12-33.** 1.57 in.

**12-35.** 0.604 mm.

**12-37.** 0.0587 ft.

**12-39.** $EI\Delta = P[2L^3/3 + R(\pi L^2 + \pi R^2/2 + 4LR)]$.

**12-41.** $\Delta = \pi Pa^3(1/EI + 3/GJ)/2$.

**12-43.** $\Delta_x = -PL^3/2EI$, $\Delta_y = 2PL^3/3EI + PL^3/GJ$, $\Delta_z = 0$.

**12-45.** 33.6 k.

**12-47.** 18.75 k.

**12-49.** $w_0L^4/384EI$.

**12-51.** (a) $PR/\pi$, (b) $\Delta = PR^3(2/\pi - 1/2)/EI$.

**12-53.** (a) $F_{AC} = 10.3$ k, $F_{AD} = -7.98$ k, $F_{AE} = -6.52$ k; (b) $F_{AB} = 6.41$ k, $F_{AC} = 1.36$ k, $F_{AD} = -5.32$ k, $F_{AE} = -3.71$ k.

**12-55.** $\Delta_V = 1.34PL/AE$, $\Delta_H = 0.482PL/AE$.

**12-57.** $\theta = w_0L^3/384EI$.

**12-59.** See answer for **12-21**.

**12-61.** $EI\theta = 0.336PL^2/EI$.

**12-63.** $\Delta_H = 25w_0a^4/EI$.

**12-65.** $EI\theta_A = w_0a^3/(\sqrt{2}/3 + 1/2)$.

**12-67.** 0.628 in.

**12-69.** Center reaction $R = 33w_0L/16$.

**12-71.** $R_A = 9kL^2/40$, $M_A = -13kL^3/120$.

**12-73.** $3kL/2$.

**12-75.** $P_{cr} = 10EI/L^2$.

## Chapter 13

**13-1.** $(2L^2 + 3a^2 - 6La)/6L$.

**13-3.** 9.01 k.

**13-5.** $R_a = 3$ k.

**13-7.** See answer for **12-53**(b).

**13-9.** $\begin{bmatrix} 2 & 0.89 & -0.80 \\ 0.89 & 2 & -0.45 \\ -0.80 & -0.45 & 2 \end{bmatrix}$

**13-11.** $R_{dy} = 2.29P$, $R_{dx} = -0.714P$.

**13-13.** $M_{ab} = w_0L^2/60$, $M_{bc} = w_0L^2/12$, $M_{cb} = 3w_0L^2/20$.

**13-15.** $\theta_b = -PL^2/2EI$, $\Delta_b = -PL^3/3EI$.

**13-17.** $R_b = 250$ N, $EI\theta_b = -25L^2$, $EI\theta_c = -75L^2$, $EI\Delta_c = -58.33L^3$.

**13-19.** $F_{ae} = 8.96$ kN, $F_{be} = 11.68$ kN, $F_{ce} = 3.67$ kN, $F_{de} = -4.17$ kN, $F_{fe} = 4.17$ kN.

**13-21.** (a) $0.41L$, (b) $0.172PL$.

**13-23.** (a) $w_1 = 3.90$ k/in, (b) 0.055 in.

**13-25.** $11.25 < w_{ult} < 15$ lb/in.

**13-27.** W 14 × 26.

**13-29.** $4.92 < P_{ult} < 5$.

**13-31.** Plastic hinges at $C$ and $E$.

# Index

# SI Units
# Système International d'Unités

## BASE SI UNITS

| Quantity | Unit (Symbol) |
|----------|---------------|
| length   | meter (m)     |
| mass     | kilogram (kg) |
| force*   | newton (N)*   |
| time     | second (s)    |

\* Derived unit ($kg \cdot m/s^2$)

## RECOMMENDED MULTIPLE AND SUBMULTIPLE UNITS

| Multiplication Factor | Prefix | SI Symbol |
|-----------------------|--------|-----------|
| 1 000 000 000 | giga | G |
| 1 000 000 | mega | M |
| 1 000 | kilo | k |
| 0.001 | milli | m |
| 0.000 001 | micro | μ |
| 0.000 000 001 | nano | n |

## SOME RULES FOR SI STYLE AND USAGE*

A dot is to be used to separate units that are multiplied together. Thus, for example, a newton-meter is written N·m and must not be confused with mN which stands for millinewtons.

Use of prefixes is to be avoided in the denominator of compound units, except for kg since kg is a base SI unit.

For numbers having four or more digits, the digits should be placed in groups of three, separated by spaces instead of commas, counting both to the left and to the right of the decimal point. Thus, for example, write 37 638.246 15 instead of 37,638.24615 as written in the U.S. Customary system of units.

\* For further details see *Standard for Metric Practice*, E380-86, ASTM, Philadelphia, PA, and *Recommended Practice for the Use of Metric (SI) Units in Building Design and Construction*, NBS Technical Note 938, U.S. Department of Commerce, Washington, D.C.